Harnessing NanoOmics and Nanozymes for Sustainable Agriculture

Vishnu D. Rajput
Academy of Biology and Biotechnology, Southern Federal University, Rostov-on-Don, Russia

Abhishek Singh
Faculty of Biology, Yerevan State University, Armenia

Karen Ghazaryan
Faculty of Biology, Yerevan State University, Armenia

Athanasios Alexiou
Department of Science and Engineering, Novel Global Community Educational Foundation, Hebersham, Australia

Abdel Rahman Mohammad Said Al-Tawaha
Department of Biological Sciences, Al Hussein bin Talal University, Jordan

IGI Global
PUBLISHER of TIMELY KNOWLEDGE

A volume in the Advances in Environmental Engineering and Green Technologies (AEEGT) Book Series

Published in the United States of America by
IGI Global
Engineering Science Reference (an imprint of IGI Global)
701 E. Chocolate Avenue
Hershey PA, USA 17033
Tel: 717-533-8845
Fax: 717-533-8661
E-mail: cust@igi-global.com
Web site: http://www.igi-global.com

Library of Congress Cataloging-in-Publication Data

CIP Data Pending

Harnessing NanoOmics and Nanozymes for Sustainable Agriculture
Vishnu Rajput, Abhishek Singh, Karen Ghazaryan, Athanasios Alexiou, Abdel Rahman Al Tawaha
2024 Engineering Science Reference

ISBN: 979-8-3693-1890-4
eISBN: 979-8-3693-1891-1

This book is published in the IGI Global book series Advances in Environmental Engineering and Green Technologies (AEEGT) (ISSN: 2326-9162; eISSN: 2326-9170)

British Cataloguing in Publication Data
A Cataloguing in Publication record for this book is available from the British Library.

Advances in Environmental Engineering and Green Technologies (AEEGT) Book Series

Sang-Bing (Jason) Tsai
Zhongshan Institute, University of Electronic Science and Technology of China, China & Wuyi University, China
Ming-Lang Tseng
Lunghwa University of Science and Technology, Taiwan
Yuchi Wang
University of Electronic Science and Technology of China Zhongshan Institute, China

ISSN:2326-9162
EISSN:2326-9170

MISSION

Growing awareness and an increased focus on environmental issues such as climate change, energy use, and loss of non-renewable resources have brought about a greater need for research that provides potential solutions to these problems. Research in environmental science and engineering continues to play a vital role in uncovering new opportunities for a "green" future.

The **Advances in Environmental Engineering and Green Technologies (AEEGT)** book series is a mouthpiece for research in all aspects of environmental science, earth science, and green initiatives. This series supports the ongoing research in this field through publishing books that discuss topics within environmental engineering or that deal with the interdisciplinary field of green technologies.

COVERAGE

- Air Quality
- Policies Involving Green Technologies and Environmental Engineering
- Contaminated Site Remediation
- Pollution Management
- Electric Vehicles
- Water Supply and Treatment
- Radioactive Waste Treatment
- Alternative Power Sources
- Industrial Waste Management and Minimization
- Renewable Energy

IGI Global is currently accepting manuscripts for publication within this series. To submit a proposal for a volume in this series, please contact our Acquisition Editors at Acquisitions@igi-global.com or visit: http://www.igi-global.com/publish/.

The Advances in Environmental Engineering and Green Technologies (AEEGT) Book Series (ISSN 2326-9162) is published by IGI Global, 701 E. Chocolate Avenue, Hershey, PA 17033-1240, USA, www.igi-global.com. This series is composed of titles available for purchase individually; each title is edited to be contextually exclusive from any other title within the series. For pricing and ordering information please visit http://www.igi-global.com/book-series/advances-environmental-engineering-green-technologies/73679. Postmaster: Send all address changes to above address.

Titles in this Series

For a list of additional titles in this series, please visit: http://www.igi-global.com/book-series/advances-environmental-engineering-green-technologies/73679

Convergence Strategies for Green Computing and Sustainable Development
Vishal Jain (Sharda University, India) Murali Raman (Asia Pacific University of Technology, Malaysia) Akshat Agrawal (Amity University, Guragon, India) Meenu Hans (K.R. Mangalam University, India) and Swati Gupta (Center of Excellence, K.R. Mangalam University, India)
Engineering Science Reference • copyright 2024 • 332pp • H/C (ISBN: 9798369303382) • US $270.00 (our price)

Biosorption Processes for Heavy Metal Removal
Pinki Saini (University of Allahabad, India)
Engineering Science Reference • copyright 2024 • 371pp • H/C (ISBN: 9798369316184) • US $275.00 (our price)

Agriculture and Aquaculture Applications of Biosensors and Bioelectronics
Alex Khang (Global Research Institute of Technology and Engineering, USA)
Engineering Science Reference • copyright 2024 • 576pp • H/C (ISBN: 9798369320693) • US $265.00 (our price)

Innovations in Engineering and Food Science
Shilpa Mehta (Auckland University of Technology, New Zealand) Fakhar Islam (NUR International University, Pakistan) and Ali Imran (Government College University, Pakistan)
Engineering Science Reference • copyright 2024 • 424pp • H/C (ISBN: 9798369308196) • US $265.00 (our price)

Modeling and Monitoring Extreme Hydrometeorological Events
Carmen Maftei (Transilvania University of Brasov, Romania) Radu Muntean (Transilvania University of Braşov, Romania) and Ashok Vaseashta (International Clean Water Institute, USA)
Engineering Science Reference • copyright 2024 • 338pp • H/C (ISBN: 9781668487716) • US $250.00 (our price)

Novel AI Applications for Advancing Earth Sciences
Sudesh Yadav (Department of Higher Education, Government PG College, Ateli, India) Satya Prakash Yadav (GL Bajaj Institute of Technology and Management, India) Pethuru Raj (Edge AI Division, Reliance Jio Platforms Ltd., Bangalore, India) Prabhakar Tiwari (Madan Mohan Malaviya University of Technology, India) and Victor Hugo C. de Albuquerque (Department of Teleinformatics Engineering (DETI), Federal University of Ceará, Brazil)
Engineering Science Reference • copyright 2024 • 405pp • H/C (ISBN: 9798369318508) • US $265.00 (our price)

Exploring Ethical Dimensions of Environmental Sustainability and Use of AI

IGI Global
PUBLISHER of TIMELY KNOWLEDGE

701 East Chocolate Avenue, Hershey, PA 17033, USA
Tel: 717-533-8845 x100 • Fax: 717-533-8661
E-Mail: cust@igi-global.com • www.igi-global.com

Table of Contents

Nanotechnology for Soil and Plant Conservation: Nano-Omics and Non-Enzymes for Human Welfare

Detailed Table of Contents

Shivangi Singh, Sardar Vallabhbhai Patel University of Agriculture and Technology, India
Omkar Singh, Sardar Vallabhbhai Patel University of Agriculture and Technology, India
Sakshi Singh, Tilak Dhari P.G. College, India
Abhishek Singh, Yerevan State University, Armenia
Vishnu D. Rajput, Southern Federal University, Russia
Karen Ghazaryan, Yerevan State University, Armenia
Vaishali Singh, Sardar Vallabhbhai Patel University of Agriculture and Technology, India
Athanasios Alexiou, Funogen, Athens, Greece
Abdel Rahman Mohammad Said Al-Tawaha, Al Hussein bin Talal University, Jordan
Aleksandr Yesayan, Yerevan State University, Armenia
Armine David Chakhmakhchyan, Research Institute of Biology, Yerevan State University, Armenia
Hassan El-Ramady, Soil and Water Dept., Faculty of Agriculture, Kafrelsheikh University, Egypt

Nano-enzymes and nanomics have emerged as promising tools for sustainable agriculture, offering novel solutions to enhance crop productivity, protect the environment, and promote food security. This chapter discusses the applications of nano-enzymes and nanomics in agriculture, including crop protection, improving soil fertility, precision farming, and post-harvest management. It also gives an overview of their current state and future prospects. The chapter also discusses the drawbacks and restrictions of various technologies, such as toxicity and safety risks, cost-effectiveness and scaling issues, and regulatory obstacles. Strong instruments for sustainable agriculture could be developed by integrating nanotechnology with other cutting-edge fields like biotechnology and the internet of things (IoT).

The global population's exponential growth poses a challenge, demanding innovative solutions for sustainable crop yield and nutritional enhancement. Nano-omics, rooted in nanotechnology, revolutionizes agriculture. Nanoparticles (NPs) penetrate plant tissues for precise nutrient delivery, catalyzing peroxidase and catalase synthesis. This scavenges reactive oxygen species, reducing lipid peroxidation, preserving cell membrane integrity, enhancing water and nutrient uptake, and stimulating carbohydrate partitioning. Omics techniques (genomics, transcriptomics, proteomics, and metabolomics) explore molecular interactions. Genomic analysis reveals altered gene expression related to growth, yield, and stress response. Transcriptomic studies show modulation of transcription factors and pathways by nano-zymes. Proteomic investigations focus on protein profile changes induced by nano-zymes, deepening understanding of plant responses. It unveils the interplay between nano-zymes and plants, providing insights into mechanisms for augmented crop growth and nutritional enhancement.

Traditional fertilizers often lead to nutrient runoff, leaching, and soil degradation, exacerbating environmental issues such as water pollution and greenhouse gas emissions. In this chapter's introduction, the authors highlight the pressing challenges facing global agriculture, including rapid population growth, climate change, and resource constraints, necessitating innovative solutions for sustainable food production. Nanotechnology emerges as a promising avenue to address these challenges, offering tailored approaches to nutrient management and soil conditioning. Nano-fertilizers, nanocarriers, and nanostructures present novel solutions by leveraging the unique properties of nanoparticles. These nanoparticles provide controlled release mechanisms, improved solubility, and targeted delivery, optimizing nutrient uptake efficiency and minimizing environmental impact. The chapter delves into the formulation and composition of nano-fertilizers, exploring advanced techniques such as nanoencapsulation and stimuli-responsive materials. Mechanisms of nutrient release and uptake by plants are elucidated, showcasing how nanoparticles interact with plant roots, modify rhizosphere chemistry, and enhance mycorrhizal associations. Moreover, the role of nanoparticles in promoting nutrient uptake, particularly micronutrients, underscores their potential to improve crop growth and development, even in adverse conditions like saline soils. Furthermore, nano-enhanced soil amendments offer a transformative approach to soil management, contributing to soil health, structure, and fertility. By integrating nanotechnology into agricultural practices, researchers

crop yield. Nanotechnology's focus extends to the development of innovative nanomaterials, enhancing soil nutrient availability, seed germination, and seedling growth while concurrently addressing biotic and abiotic stress. Furthermore, the application of diverse nano-based fertilizers and pesticides directly onto crops at reduced dosages emerges as a key strategy, mitigating risks to both human health and the environment. The exploration of advanced nanodiagnostic systems and nanobiosensors plays a pivotal role in monitoring plant diseases, soil quality parameters, and stress, thereby offering a substantial advantage in precision agriculture. Nanosensors are composed of nanoscale particles such as nanoscale wires, carbon nanotubes, thin films, and nanoparticles.

Ayesha Shafqat, School of Botany, Minhaj University, Lahore, Pakistan
Arshad Rasool, School of Botany, Minhaj University, Lahore, Pakistan
Rabia Fatima, Clarkson University, USA
Muhammad Majeed, University of Gujrat, Pakistan
Khursheed Ur Rahman, Hazara University, Pakistan
Altaf Hussain Narejo, University of Karachi, Pakistan
Muhammad Ishtiaq, University of Bhimber, Pakistan
Shazia Kousar, University of Baltistan, Pakistan
Afshan Farid, PMAS Arid Agriculture University, Pakistan

The escalating challenges in global agriculture necessitate the exploration of environmentally sustainable alternatives for pest and disease management. Nanomaterials, characterized by their unique properties at the nanoscale, offer a promising avenue for addressing these challenges in a sustainable manner. This chapter reviews recent developments in the application of nanomaterials for sustainable pest and disease management in agriculture. The authors examine the diverse range of nanomaterials, including nanoparticles, nanocomposites, and nanostructured coatings, and their efficacy against various agricultural pests and pathogens. Nanomaterials demonstrate targeted delivery mechanisms, increased stability, and prolonged effectiveness, minimizing the ecological impact associated with conventional agrochemicals. Consideration is given to the interactions between nanomaterials and plants, as well as their effects on soil microbiota, to assess their overall impact on ecosystem health.

Mohamed Nouri, Sultan Moulay Slimane University, Morocco

The soil's composition, structure, and nutrients have altered throughout time. These changes are natural and anthropogenic. Economic, environmental, and social benefits, including ecosystem services, come from restoring degraded land and exploiting marginal land. Most physiochemical remedies for damaged land are difficult, costly, and time-consuming. Nano-remediation and nano-restoration are novel, efficient, cost-effective, eco-friendly, and dependable for toxin remediation and risk reduction. The high surface area/volume ratio, increased reactivity, customizable physical qualities, and adaptability of nanoscale entities make them attractive for soil remediation. Different nanomaterials (NMs) and nanotools can clean up pollutants. Both foreign chemicals and polluted location affect these methods. Decontaminating soil contaminants with nanoscale entities reduces their detrimental effects on humans, plants, and animals. It also discusses nanoparticles (NPs) and ex- and in-situ cleanup. The authors discuss nanoscale item uses in soil remediation and land restoration in this chapter.

Role of Nanotechnological Aspects, Nano-Omics, and Nano-Enzymes for Crop Production and Protection

lowering environmental impacts. Additionally, nanoparticles can be functionalized with biomolecules like antibodies or aptamers to ensure target selectivity and specificity. Despite these benefits, there remain certain challenges associated with the use of nano devices for plant protection. Foremost among these is the insufficient research on the potential toxicity of certain nanomaterials, such as nano silver and nano gold, to plants, animals, and ecosystems. Accumulation of nanomaterials in plant and animal tissues could potentially enter the food chain, necessitating rigorous safety assessments and consumer education efforts. Nonetheless, the adoption of non-toxic materials, such as starch, chitin, or nano clays, as alternatives to metals, can mitigate such risks.

Nanotechnology for Soil and Plant Conservation: Nano-Omics and Non-Enzymes for Human Welfare

The soil degradation and loss of fertile land pose severe threats to global agriculture and environmental sustainability. Soil quality degradation and loss of arable land have become pressing environmental concerns, call for innovative, sustainable, and transformative approaches to soil management, remediation, and restoration. Nanotechnology can be a transformative approach for soil remediation and land restoration. Nanomaterials remarkable properties offers a paradigm shift in soil science and environmental restoration. With their extraordinarily high surface area, tailored reactivity, and adjustable physicochemical characteristics, nanomaterials present a profound opportunity to revolutionize our approach to soil remediation and land rejuvenation. It delves into their intricate synthesis processes, advanced characterization techniques, and the underlying mechanisms governing their actions of nanomaterials and nanocomposites within the soil matrix. Within the realm of soil remediation, nanotechnology emerges as a versatile and potent tool. It excels in adsorbing, immobilizing, and degrading contaminants, including stubborn heavy metals, persistent organic pollutants, and recalcitrant pesticides. Additionally, the controlled-release capabilities of nano fertilizers come to the fore, facilitating precise nutrient delivery to plants, thereby enhancing fertility of soil and promoting plant growth. Apart from contamination remediation, nano-materials have immense potential to improve soil structure by enhancing water retention, promoting aeration and facilitating nutrient diffusion. This multifaceted approach extends to the realm of bioremediation, where nanotechnology amplifies microbial-mediated processes, elevating enzymatic activities and accelerating contaminant degradation.

Globally, soil contamination threatens the ecosystem and human health. Ecological, financial, and efficiency issues plague excavation, chemical, and bioremediation methods. Nanotechnology is a promise for soil remediation, offering advantages over existing approaches. Nanotechnology technologies for soil pollution treatment and environmental conservation are covered in this chapter. It examines heavy metals, organic pollutants, pesticides, and radioactive contamination in soil and the limitations of current remediation methods. The chapter next discusses nanotechnology in soil remediation, including nanoparticles, nanomaterials for adsorption and degradation, nano-bioremediation, and nanosensors for pollution monitoring. Case studies and research show nanotechnology's success in soil remediation. Future nanotechnology for soil remediation views and difficulties highlight the need for more research and development to address scalability, cost-effectiveness, and environmental safety.

Phytomedicine refers to the use of plants to treat, diagnose and prevent diseases for thousands of years. Medicinal plants, often known as the traditional healers, are potentially rich in phytochemicals with promising pharmacological properties and therefore have been incorporated in traditional medicines. Various medicinal plants and trees are well known for their phytomedicinal benefits and are used in various drugs with lesser side effects. Active components like alkaloids, carotenoids, flavonoids, phenolic acids, tannins, proanthocyanins, sesquiterpenes, glycosides, saponins, volatile oils, minerals, etc. are present in medicinal plants. Some common medicinal plants are used widely such as Azadirachta indica (Neem), Murraya koinigii (Curry leaves), Emblica officinalis (Amla), Myrica esculenta (Kaphal, Bayberry), Terminalia bellerica (Bahera), Terminalia chebula (Haritaki), Mimosa pudica (Humble plant), and Phyllanthus niruri. In this chapter, some phytomedicinal plants and their medicinal properties as well as efficacy are reviewed and discussed.

Nano-Omics-Based Abiotic Stress Management

Drought, salt, temperature extremes, nutritional inadequacies, and heavy metal toxicity hinder worldwide agricultural output and food security. Using nanostructures' unique properties, nanotechnology offers various creative solutions to these issues. This chapter details how nanotechnology can reduce drought, salt, temperature, nutrient deficits, and heavy metal toxicity in agriculture. The chapter covers nanotechnology applications including water retention materials, sensors, fertilisers, plant growth regulators, seed coatings, adsorbents, and chelating agents. Case studies and research demonstrate how these methods boost crop stress tolerance and output. The chapter examines risks, toxicity, regulatory challenges, and commercialization limits of creating and exploiting nanotechnology in agriculture. Interdisciplinary collaboration and stakeholder engagement are crucial to nanotechnology's agricultural potential, according to the chapter. Further research and development to solve knowledge gaps and promote responsible innovation are also supported. This chapter provides a comprehensive and current overview of nanotechnology applications for abiotic stress management in agriculture to motivate academics, practitioners, and policymakers. To solve global issues, it develops sustainable and resilient agriculture systems.

In the face of escalating pressures from climate change, environmental degradation, and escalating pest and disease burdens, traditional agricultural practices are proving insufficient to ensure food security and sustainable crop production. Nanoomics, a pioneering interdisciplinary field, offers a promising avenue

Future Role of Nanopriming-Based Approaches for Food Security and Sustainable Agriculture

Foreword

I am writing to express my sincere appreciation and admiration for the edited book titled ***"Harnessing NanoOmics and Nanozymes for Sustainable Agriculture,"*** edited by **Vishnu D. Rajput (Russia), Abhishek Singh (Armenia), Karen Ghazaryan (Armenia), Athanasios Alexiou (Greece), and Abdel Rahman Mohammad Said Al-Tawaha (Jordan)**. This exceptional publication not only brings together the collective wisdom of esteemed researchers and experts but also offers invaluable insights into the innovative realm of nano-enabled solutions in agriculture.

The book provides a comprehensive exploration of NanoOmics and Nanozymes and their application in sustainable agriculture. It covers a diverse range of topics, including the underlying mechanisms, methodology, optimization strategies, practical applications, and future directions, ensuring a holistic and well-rounded understanding of these cutting-edge approaches. Each chapter, skillfully authored by established scientists who have made significant contributions to the field, is a testament to their expertise and commitment to advancing sustainable agriculture. The meticulous editing process has ensured that the chapters are not only well-structured and coherent but also provide readers with valuable insights supported by thorough research, data analysis, and evidence-based discussions.

"Harnessing NanoOmics and Nanozymes for Sustainable Agriculture" showcases the latest advancements in nano-enabled solutions and sheds light on their potential implications for addressing pressing challenges faced by the agriculture industry, such as climate change, dwindling natural resources, and increasing food demand. The book contribution to the existing literature in the field cannot be overstated, as it offers a comprehensive overview of nano-enabled solutions, their benefits, and their vast potential applications.

I would like to extend my heartfelt appreciation to the editors and the entire editorial team for their dedication and meticulous efforts in curating and refining this outstanding collection of chapters. Their commitment to excellence and attention to detail has resulted in a book that significantly advances the knowledge and understanding of sustainable agriculture.

I am confident that ***"Harnessing NanoOmics and Nanozymes for Sustainable Agriculture"*** will serve as an indispensable resource for researchers, soil scientists, agriculture biotechnologists, ecologists, nanotechnologists, agronomists, policymakers, and professionals who are passionate about sustainable agricultural practices. It is with great pleasure that I recommend this book for publication and commend the editors for their exceptional work in well-renowned publisher IGI Global USA.

Rafayel Barkhudaryan
Yerevan State University, Armenia

Preface

In this dynamic landscape of agriculture, innovative solutions are imperative to tackle the multifaceted challenges of climate change, dwindling natural resources, and a burgeoning population. This book, *Harnessing NanoOmics and Nanozymes for Sustainable Agriculture,* edited by Vishnu D. Rajput, Abhishek Singh, Karen Ghazaryan, Athanasios Alexiou, and Abdel Rahman Mohammad Said Al-Tawaha represents a collaborative effort by leading experts in the field to elucidate the transformative potential of nano-enabled solutions in agriculture. NanoOmics offers a holistic approach to deciphering the intricate molecular mechanisms underpinning plant-microbe interactions, stress responses, and crop performance. By integrating advanced analytical techniques with high-throughput omics methodologies, NanoOmics empowers researchers to unravel the complexities of biological systems with unprecedented resolution, providing invaluable insights for crop improvement and sustainable agriculture. Furthermore, Nanozymes offer versatile platforms for various agricultural applications, including crop protection, nutrient management, and environmental remediation. Leveraging the exceptional catalytic efficiency, substrate specificity, and operational stability of Nanozymes, researchers can develop innovative solutions to enhance crop productivity, mitigate environmental pollution, and promote sustainable agricultural practices. This book features cutting-edge research, insightful perspectives, and practical applications aimed at addressing the multifaceted challenges confronting modern agriculture. By fostering interdisciplinary collaboration and knowledge exchange, this book aims to catalyse innovation, foster scientific dialogue, and accelerate the translation of nanotechnology-enabled solutions from the laboratory to the field. Ultimately, this book is a valuable resource for researchers, educators, policymakers, and stakeholders interested in harnessing the power of NanoOmics and Nanozymes to address the grand challenges facing global agriculture. By embracing innovation, collaboration, and sustainability, we can pave the way towards a resilient, productive, and environmentally conscious agricultural future.

Vishnu D. Rajput
Academy of Biology and Biotechnology, Southern Federal University, Rostov-on-Don, Russia

Abhishek Singh
Faculty of Biology, Yerevan State University, Armenia

Karen Ghazaryan
Faculty of Biology, Yerevan State University, Armenia

Athanasios Alexiou
Department of Science and Engineering, Novel Global Community Educational Foundation, Hebersham, Australia

Abdel Rahman Mohammad Said Al-Tawaha
Department of Biological Sciences, Al Hussein bin Talal University, Jordan

Acknowledgment

We would like to express our sincere gratitude to the Science Committee of the Republic of Armenia for their generous support and funding of grant 21AG-4C075 and 23PostDoc-4D007, as well as to the Ministry of Science and Higher Education of the Russian Federation for their support and funding of grant FENW-2023-0008. We also acknowledge the support provided by the Strategic Academic Leadership Program of the Southern Federal University ("Priority 2030") grant program. All of these grants are essential in bringing to fruition the edited book titled "Harnessing NanoOmics and Nanozymes for Sustainable Agriculture". This book will be published by IGI, Global USA. Their commitment to advancing research and knowledge in the fields of nanotechnology, omics, enzymology, and sustainable agriculture has been instrumental in making this endeavor a reality. Their support has enabled us to gather and disseminate significant insights, fostering progress and innovation in these critical domains. We express our deep appreciation for their vision and dedication, which have not only enriched the content of this book but have also contributed to the global discourse on sustainable agriculture. We look forward to continued collaboration and the positive impact it will undoubtedly have on the future of agriculture and scientific research.

Section 1

Nano-Omics and Nano-Enzymes: An Introduction

Chapter 1
Impact of Nano-Enzyme and Nanomics for Sustainable Agriculture:
Current Status and Future Prospective

Shivangi Singh
Sardar Vallabhbhai Patel University of Agriculture and Technology, India

Omkar Singh
https://orcid.org/0000-0003-1971-185X
Sardar Vallabhbhai Patel University of Agriculture and Technology, India

Sakshi Singh
Tilak Dhari P.G. College, India

Abhishek Singh
https://orcid.org/0000-0002-0049-6910
Yerevan State University, Armenia

Vishnu D. Rajput
https://orcid.org/0000-0002-6802-4805
Southern Federal University, Russia

Karen Ghazaryan
https://orcid.org/0000-0002-8973-2595
Yerevan State University, Armenia

Vaishali Singh
Sardar Vallabhbhai Patel University of Agriculture and Technology, India

Athanasios Alexiou
Funogen, Athens, Greece

Abdel Rahman Mohammad Said Al-Tawaha
Al Hussein bin Talal University, Jordan

Aleksandr Yesayan
https://orcid.org/0009-0007-2407-3609
Yerevan State University, Armenia

Armine David Chakhmakhchyan
Research Institute of Biology, Yerevan State University, Armenia

Hassan El-Ramady
Soil and Water Dept., Faculty of Agriculture, Kafrelsheikh University, Egypt

ABSTRACT

Nano-enzymes and nanomics have emerged as promising tools for sustainable agriculture, offering novel solutions to enhance crop productivity, protect the environment, and promote food security. This

DOI: 10.4018/979-8-3693-1890-4.ch001

chapter discusses the applications of nano-enzymes and nanomics in agriculture, including crop protection, improving soil fertility, precision farming, and post-harvest management. It also gives an overview of their current state and future prospects. The chapter also discusses the drawbacks and restrictions of various technologies, such as toxicity and safety risks, cost-effectiveness and scaling issues, and regulatory obstacles. Strong instruments for sustainable agriculture could be developed by integrating nanotechnology with other cutting-edge fields like biotechnology and the internet of things (IoT).

1. INTRODUCTION

Nanotechnology has emerged as a revolutionary field that has the potential to transform various sectors, including agriculture. In the field of nanotechnology, nano-enzymes and nanomics have garnered considerable interest. Nano-enzymes are synthetic enzymes that imitate the catalytic functions of natural enzymes while offering distinct features such exceptional stability, cost-effectiveness, and simple handling (Wang et al., 2018; Rajput et al., 2021). Nanomaterials like metal nanoparticles, carbon nanotubes, and graphene are commonly used to produce these enzymes (Huang et al., 2019). Nanomics is the utilization of nanotechnology in the omics discipline, encompassing genomes, proteomics, and metabolomics (Prasad et al., 2017). Nanomics utilizes nanomaterials and nanodevices to investigate and control biological systems at the molecular scale (Chaudhary et al., 2021; Singh et al., 2024a).

Agriculture serves as essential for ensuring food security and sustaining the livelihoods of millions of people globally. The growing global population, climate change, and environmental degradation are major obstacles to the sustainability of agricultural systems (FAO, 2021; Singh et al., 2023). Sustainable agriculture strives to fulfil present and future food and fiber requirements while reducing adverse effects on the environment and guaranteeing the enduring sustainability of agricultural methods (Velten et al., 2015). Sustainable agriculture includes practices including conservation tillage, integrated pest management, precision farming, and the use of bio-based inputs (Khoshnevisan et al., 2021). Implementing sustainable agriculture practices is essential for meeting the United Nations' Sustainable Development Goals (SDGs), namely SDG 2 (Zero Hunger) and SDG 12 (Responsible Consumption and Production) (United Nations, 2021).

Nanotechnology has great potential to help solve the agricultural sector's difficulties and encourage sustainable practices. Nanomaterials and nanodevices have diverse uses in agriculture such as crop protection, soil fertility improvement, precision farming, and post-harvest management (Fraceto et al., 2018; Singh et al., 2022; Shahi et al., 2024). Nanotechnology can enhance input efficiency, decrease environmental impact, and improve crop yields and quality (Chhipa, 2019; Singh et al., 2023; Singh et al., 2024). Nano-fertilizers can transport nutrients directly to plants, reducing nutrient wastage and enhancing soil quality (Dimkpa & Bindraban, 2018). Nano-pesticides can provide regulated release and precise delivery of active components, leading to decreased pesticide usage and minimizing negative environmental effects (Kah et al., 2018; Singh et al., 2023c). Nanotechnology can facilitate the creation of nano-biosensors for monitoring crop health, soil conditions, and environmental factors in real-time, which can help improve precision agriculture methods (Prasad et al., 2017).

Nano-enzymes and nanomics integration in agriculture has the potential to transform different aspects of crop production and management. Nano-enzymes can be used to create new methods for protecting crops, like breaking down pesticide residues and improving plant defense mechanisms (Wang et al.,

2018). Nanotechnology methods, like nano-biosensors and nano-delivery systems, can facilitate the early identification of plant diseases, precise application of agrochemicals, and monitoring of plant stress responses at the molecular level (Chaudhary et al., 2021). Nano-enzymes and nanomics in agriculture can help create sustainable and resilient agricultural systems to address climate change challenges and provide food security for the expanding global population. Yet, implementing nanotechnology in agriculture also prompts worries about the possible hazards and unknown factors linked to the utilization of nanomaterials. The issues involve the possible toxicity of nanomaterials to plants, animals, and humans, the environmental destiny and behavior of nanomaterials, and the socio-economic consequences of nanotechnology implementation (Kah et al., 2018). It is crucial to perform thorough risk assessments and establish suitable regulatory frameworks to guarantee the safe and responsible application of nanotechnology in agriculture.

2. NANO-ENZYMES IN AGRICULTURE

2.1 Types of Nano-Enzymes Used in Agriculture

Nano-enzymes are a promising tool in agriculture due to their unique characteristics and catalytic activity. Artificial enzymes created using nanomaterials display enzyme-like traits, including strong stability, specificity, and catalytic efficiency (Wu et al., 2019; Kolesnikov et al., 2024). Numerous nano-enzymes, such as metal nanoparticles, metal oxide nanoparticles, and carbon-based nanomaterials, have been studied for agricultural applications (Huang et al., 2019).

Metal nanoparticles including gold nanoparticles (AuNPs) and silver nanoparticles (AgNPs) have been extensively researched for their exceptional catalytic capabilities and biocompatibility, making them suitable as nano-enzymes (Deng et al., 2019). AuNPs have peroxidase-like activity, which can be used for breaking down pesticide residues and detecting plant infections (Li et al., 2018). AgNPs also have antibacterial characteristics and have been used to create nano-pesticides and nano-fungicides (Khot et al., 2012).

Metal oxide nanoparticles including cerium oxide nanoparticles (CeO2NPs) and iron oxide nanoparticles (Fe3O4NPs) have been studied as nano-enzymes in agriculture. CeO2NPs exhibit catalase-like and superoxide dismutase-like properties, aiding in the removal of reactive oxygen species (ROS) and safeguarding plants against oxidative stress (Walkey et al., 2015). Fe3O4NPs have been used to create magnetic nano-biosensors for detecting plant illnesses and monitoring plant stress responses (Xianyu et al., 2018).

Carbon-based nanomaterials including carbon nanotubes (CNTs) and graphene oxide (GO) have demonstrated promise as nano-enzymes in agriculture. CNTs have demonstrated peroxidase-like activity and have been utilized for detecting plant hormones and creating nano-biosensors (Khoshnevisan et al., 2021). Graphene oxide (GO) has been utilized as a nano-carrier to transport agrochemicals and plant growth regulators due to its large surface area and exceptional adsorption characteristics (Kabiri et al., 2017).

Figure 1. A) Classification of nanoparticles (NZs) and B) synthesis of nanoparticles

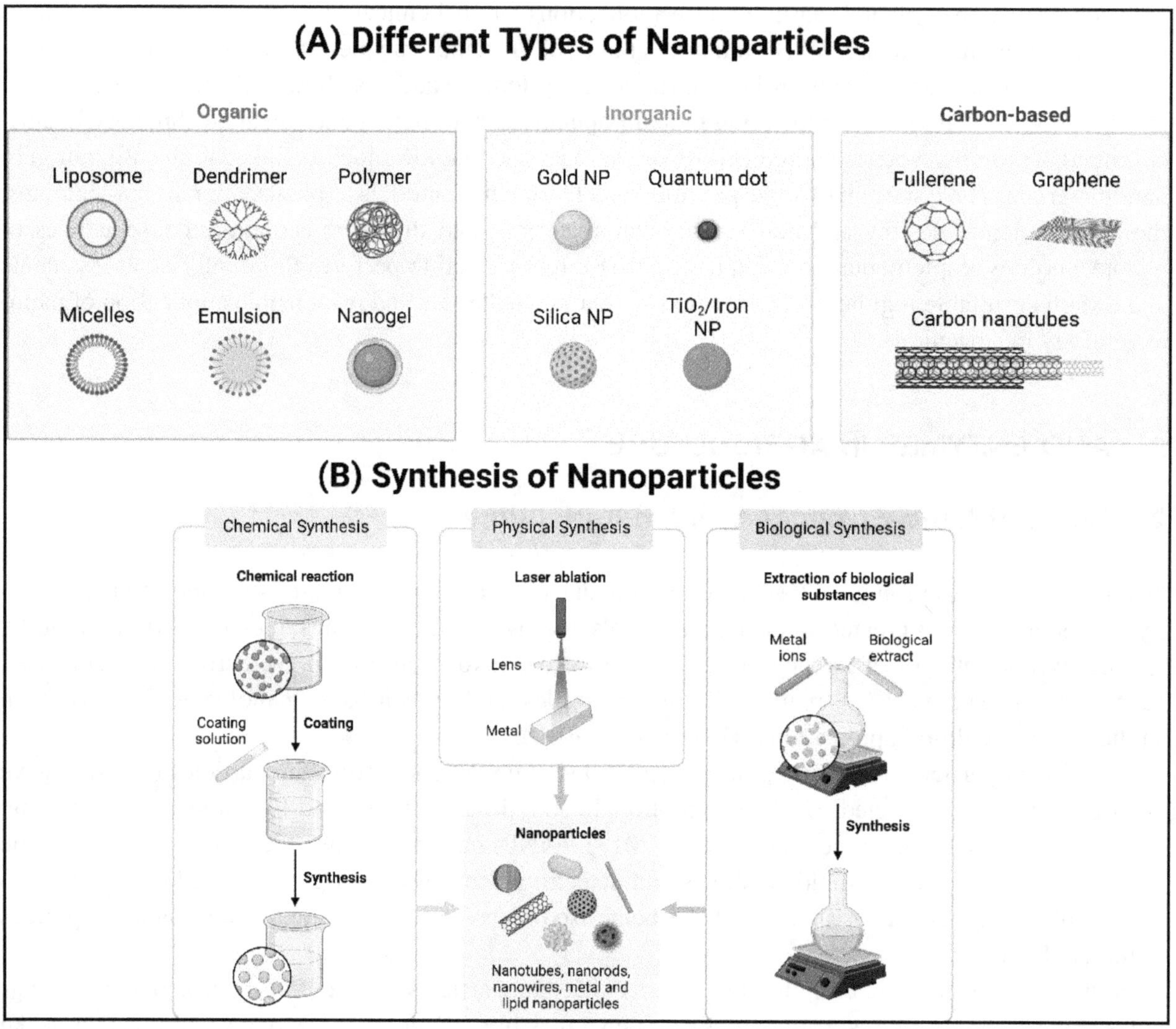

2.2 Mechanism of Action of Nano-Enzymes

Nano-enzymes display catalytic capabilities akin to natural enzymes, although their mode of operation varies from that of biological enzymes. Nano-enzymes exhibit catalytic activity due to their distinctive physicochemical characteristics, including high surface area, surface charge, and the existence of reactive functional groups (Singh et al., 2019). Nano-enzymes function through either intrinsic catalytic activity or synergistic catalytic activity, as classified by Wei & Wang in 2013.

Intrinsic catalytic activity is the natural capacity of nanomaterials to catalyze particular reactions without requiring any modifications or functionalization. Catalytic activity is commonly seen in metal and metal oxide nanoparticles, where the surface atoms or ions function as catalytic sites (Wu et al., 2019). AuNPs have inherent peroxidase-like activity because of the existence of Au (III) ions on their

Figure 2. The general illustration of nanozymes (NZs) and their mechanism of action (Gorgzadeh et al., 2024)

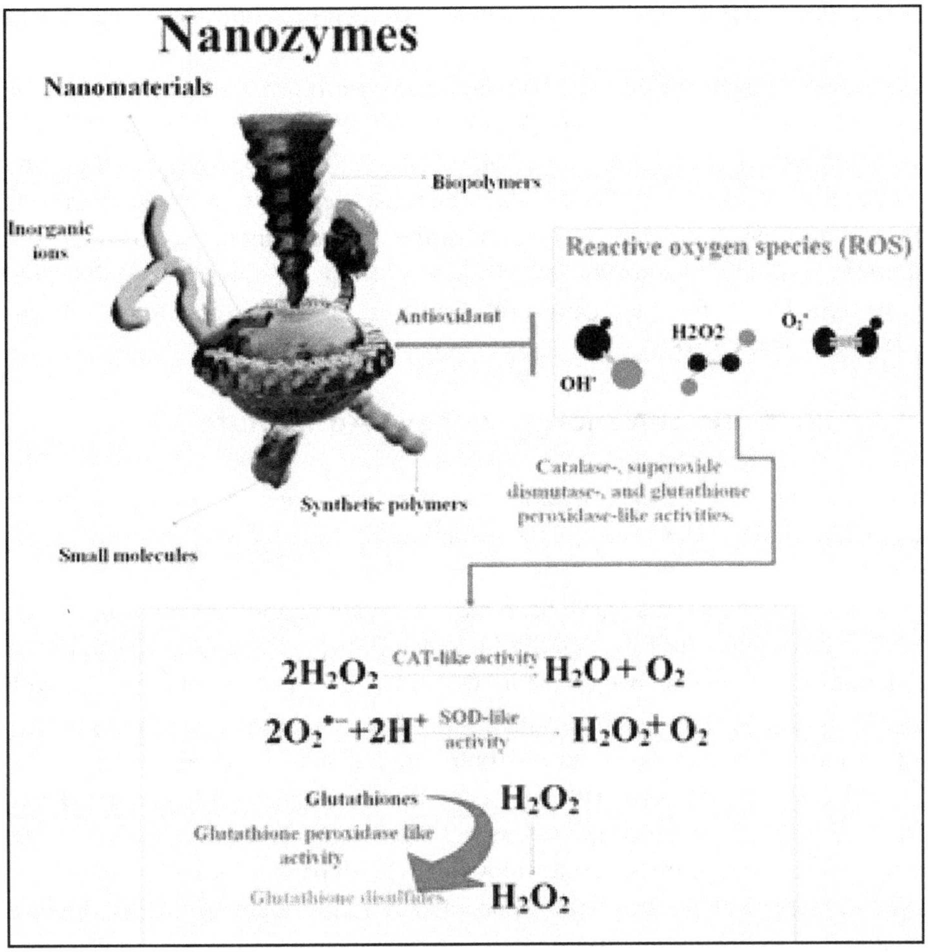

surface, enabling them to catalyze the oxidation of substrates in the presence of hydrogen peroxide (H_2O_2) (Li et al., 2018).

Synergistic catalytic activity combines nanomaterials with enzymes, aptamers, or organic compounds to improve their catalytic efficacy. The nanomaterial acts as a scaffold for immobilizing functional moieties, enhancing the stability, specificity, and catalytic effectiveness of the nano-enzyme system (Huang et al., 2019). For instance, Xianyu et al. (2018) showed that immobilizing glucose oxidase (GOx) on AuNPs improved the enzyme's stability and sensitivity for detecting glucose in plant extracts.

2.3 Advantages of Using Nano-Enzymes in Agriculture

Nano-enzymes have numerous benefits compared to conventional enzymes in agricultural uses. One major benefit is their exceptional stability under severe environmental circumstances, like extreme temperatures, pH levels, and ionic strengths (Wu et al., 2019). Stability is essential for creating strong and dependable nano-enzyme systems for use in the field (Huang et al., 2019). Nano-enzymes have the benefit of being more cost-effective and simpler to produce than natural enzymes. Nano-enzymes can be created by straightforward and scalable techniques such chemical reduction, hydrothermal synthesis, and sol-gel processing (Singh et al., 2019). This enhances their economic feasibility for extensive agricultural use (Fraceto et al., 2018).

Nano-enzymes have excellent catalytic efficiency and specificity due to their extensive surface area and the active sites present on their surface (Wei & Wang, 2013). The great catalytic effectiveness of nano-enzymes enables the use of lower doses compared to natural enzymes, resulting in financial and environmental benefits (Walkey et al., 2015). Nano-enzymes can be readily customized and adapted for particular agricultural uses. Nano-enzymes can have their surface modified to enhance their catalytic activity, selectivity, and biocompatibility (Huang et al., 2019). The ability to build and functionalize nano-enzyme systems allows for the creation of customized solutions to address various agricultural requirements (Fraceto et al., 2018).

2.4 Current Applications of Nano-Enzymes in Agriculture

Nano-enzymes have found various applications in agriculture, ranging from crop protection to post-harvest management. Some of the key applications are discussed below:

- *Crop protection*: Nano-enzymes have been used to create nano-pesticides and nano-fungicides to manage plant pests and illnesses. AgNPs have demonstrated potent antibacterial effects against several plant diseases such as bacteria, fungus, and viruses (Khot et al., 2012). AuNPs modified with plant-based antimicrobial peptides have been shown to hinder the growth of fungal infections including Fusarium graminearum and Phytophthora infestans (Li et al., 2018).
- *Soil fertility enhancement*: Nano-enzymes have been studied to enhance soil fertility and plant nutrition. Fe3O4NPs have been utilized as nano-fertilizers to gradually release iron in soils, a crucial element for plant growth and development (Dimkpa & Bindraban, 2018). CeO2NPs have been found to increase the absorption of phosphorus and nitrogen by plants, leading to enhanced growth and yield (Walkey et al., 2015).
- *Seed treatment*: Nano-enzymes have been used to enhance the germination, vitality, and resilience of seeds against both living and non-living environmental factors. CNTs modified with plant growth regulators including gibberellic acid and indole-3-acetic acid have been proven to enhance seed germination and seedling growth in several crops (Khoshnevisan et al., 2021).
- *Post-harvest management*: Nano-enzymes are used to preserve and ensure the quality of agricultural products throughout post-harvest storage and processing. AgNPs have been added to packaging materials to inhibit microbial spoilage and prolong the shelf life of fruits and vegetables (Kabiri et al., 2017). Nano-enzyme-based biosensors have been created to detect pesticide residues and pollutants in food products, enhancing food safety and quality (Xianyu et al., 2018).

3. NANOMICS IN AGRICULTURE

3.1 Definition and Concept of Nanomics

Nanomics is a developing area that combines nanotechnology with other omics technologies like genomics, proteomics, and metabolomics to investigate biological systems at the nanoscale level (Chaudhary et al., 2021). Nanomics in agriculture refers to using nanomaterials and nanodevices to analyze, manipulate, and control plant systems at the molecular level. Nanomics seeks to offer an in-depth understanding of plant biology, from gene expression to metabolic pathways, to facilitate the creation of precise agriculture and sustainable agricultural production methods (Kashyap et al., 2021).

3.1 Nanomics Techniques Used in Agriculture

- *Nano-biosensors*: Nano-biosensors are analytical tools that use nanomaterials along with biological recognition elements like enzymes, antibodies, or aptamers to detect specific target substances with high sensitivity and selectivity (Jain, 2021). Nano-biosensors have been created in agriculture to detect plant diseases, pesticide residues, and plant stress markers (Khanna et al., 2019). A graphene-based nano-biosensor has been developed to detect Xylella fastidiosa, a bacterial pathogen that causes severe illnesses in several crops (Carsanba et al., 2021).
- *Nano-delivery systems*: Nano-delivery systems are designed nanomaterials that can encase and carry bioactive substances, including fertilizers, insecticides, or plant growth regulators, to precise locations in plants (Chhipa, 2019). These technologies can enhance the efficiency and accuracy of pesticide application, hence decreasing the environmental footprint and boosting crop output (Fraceto et al., 2018). Chitosan nanoparticles have been utilized to control the release of nitrogen fertilizer in wheat, leading to enhanced nitrogen use efficiency and grain yield (Pereira et al., 2017).
- *Nano-fertilizers*: Nano-fertilizers are nanomaterials that can give critical nutrients to plants in a more effective and sustainable manner compared to conventional fertilizers (Dimkpa & Bindraban, 2018; Singh et al., 2023; Singh, O. 2023). Nano-fertilizers can be tailored to deliver nutrients slowly and gradually, complementing the nutrient uptake pattern of plants and minimizing nutrient losses (Kashyap et al., 2021). Iron oxide nanoparticles have been used as nano-fertilizers to apply iron to soybean leaves, leading to increased iron absorption and better plant growth (Nadi et al., 2019).
- *Nano-pesticides*: Nano-pesticides are formulations based on nanomaterials that efficiently manage plant pests and diseases, reducing the environmental impact of traditional pesticides (Khoshnevisan et al., 2021). Nano-pesticides enhance the solubility, stability, and precise delivery of active chemicals, leading to decreased pesticide usage and minimizing harm to unintended organisms (Fraceto et al., 2018). Silver nanoparticles are extensively researched for their broad-spectrum antibacterial properties against different plant diseases, making them suitable as nano-pesticides (Kumari et al., 2020).

3.3 Advantages of Using Nanomics in Agriculture

Nanomics has several benefits compared to conventional methods in agriculture. Nanomics allows for the creation of precise and sensitive methods to identify and diagnose plant diseases early, enabling timely and focused interventions (Kashyap et al., 2021). Nano-biosensors can identify plant infections and stress markers at extremely low levels, aiding in precision agriculture and minimizing crop damage (Jain, 2021). Nanomics can enhance the effectiveness and sustainability of crop nutrition and protection methods. Nano-fertilizers and nano-pesticides offer precise delivery of nutrients and active substances, respectively, which can decrease input requirements and lower environmental pollution levels (Dimkpa & Bindraban, 2018; Khoshnevisan et al., 2021). Nano-delivery systems can improve the bioavailability and uptake of nutrients and agrochemicals by plants, therefore enhancing agricultural output and quality (Chhipa, 2019). Nanomics can facilitate the advancement of intelligent and adaptive agricultural systems. Nano-biosensors can be combined with wireless sensor networks and Internet of Things (IoT) platforms to monitor crop health, soil conditions, and environmental data in real-time (Jain, 2021). This integration can help with data-driven decision-making and optimization of agricultural operations, resulting in enhanced resource usage efficiency and less environmental impact (Kashyap et al., 2021).

3.4 Current Applications of Nanomics in Agriculture

- *Precision farming*: Nanomics has enabled precision farming, utilizing modern technologies for precise crop management at particular sites (Jain, 2021). Nano-biosensors can be used in the field to track soil moisture, nutrient levels, and plant stress indicators, allowing for precise irrigation, fertilization, and pest control (Kashyap et al., 2021). A nano-biosensor using carbon nanotubes has been created to detect soil nitrate levels in real-time, aiding in precise nitrogen fertilization for maize (Inani et al., 2021).
- *Crop monitoring and management:* Nanomics approaches have been used to monitor and manage different crops. Nano-biosensors have been employed for the timely identification of viral diseases in tomatoes, fungal diseases in wheat, and bacterial diseases in citrus. Nano-biosensors provide quick and on-location detection of plant illnesses, enabling prompt and specific disease control methods.
- *Soil health monitoring*: Nanomics has been used to evaluate and track soil health. Nano-biosensors have been created to identify soil pollutants such heavy metals and pesticide residues that might harm soil quality and plant development (Jain, 2021). A gold nanoparticle-based colorimetric nano-biosensor has been developed to detect arsenic in soil samples, offering a quick and straightforward method for assessing soil pollution (Kalluri et al., 2021).
- *Food safety and quality control*: Nanomics has found uses in guaranteeing food safety and quality control along the agricultural value chain. Nano-biosensors have been created for the detection of foodborne infections, mycotoxins, and chemical pollutants in agricultural produce (Jain, 2021). For example, a graphene-based electrochemical nano-biosensor has been described for the sensitive detection of Escherichia coli O157:H7 in fresh produce, leading to the prevention of foodborne infections (Nguyen et al., 2021).

4. CHALLENGES AND LIMITATIONS

4.1 Environmental Concerns

Nanomaterials being used in agriculture have sparked worries about their possible environmental effects. Nanomaterials can undergo transformations such aggregation, disintegration, and interactions with other environmental components after being released into the environment. These changes can impact their fate, transport, and toxicity (Bundschuh et al., 2018). The distinctive characteristics of nanomaterials, namely their small size and large surface area, can increase their reactivity and bioavailability, thereby resulting in unforeseen effects on ecosystems (Grillo et al., 2021).

One significant environmental issue is the buildup of nanomaterials in soil and their effects on soil health. Research has demonstrated that specific nanomaterials, like silver nanoparticles, can change soil microbial communities and impact soil enzymatic activities, which play a vital role in nutrient cycling and soil fertility (Xu et al., 2021). Nanomaterials can be absorbed by plants and moved to various plant components, including edible tissues, which raises concerns regarding their introduction into the food chain (Tripathi et al., 2017). Nanomaterials have the potential to be hazardous to non-target creatures such beneficial insects, aquatic life, and wildlife, posing an environmental problem. Nanomaterials released from agricultural applications can unintentionally expose organisms to them, resulting in negative impacts on their growth, reproduction, and survival (Grillo et al., 2021). Studies have shown that some nano-pesticides are hazardous to honeybees, who play a crucial role as pollinators in agricultural ecosystems (Chaudhary et al., 2021).

It is essential to perform thorough environmental risk evaluations of nanomaterials used in agriculture to address these environmental concerns (Bundschuh et al., 2018). This entails examining the behavior, movement, and alteration of nanomaterials in diverse environmental settings as soil, water, and air, and evaluating their harmful effects on different organisms (Grillo et al., 2021). Additionally, creating safer and more eco-friendly nanomaterials, including biodegradable and biocompatible ones, can reduce any environmental hazards (Khoshnevisan et al., 2021).

4.2 Toxicity and Safety Issues

One important problem in agriculture is the potential toxicity and safety risks related to using nanoparticles. Nanomaterials can have distinct interactions with biological systems, and their toxicity can be influenced by parameters like size, shape, surface charge, and chemical composition (Fraceto et al., 2018). Nanomaterials' small size enables them to quickly enter cell membranes and interact with biological components, which may result in hazardous effects (Usman et al., 2020).

The primary worry is the possible toxicity of nanomaterials to plants. Some studies have found that nanomaterials can have positive effects on plant growth and productivity, whereas others have demonstrated that specific nanomaterials can cause oxidative stress, DNA damage, and changes in gene expression in plants (Ma et al., 2017). Zinc oxide nanoparticles at high concentrations have been shown to hinder seed germination and root formation in wheat (Singh et al., 2018). Another issue is the possible harmful effects of nanoparticles on humans and animals. Ingesting agricultural products with nanomaterials can result in their buildup in the body (Chaudhary et al., 2021). Research has demonstrated that specific nanomaterials, such titanium dioxide nanoparticles, can induce oxidative stress and inflammation in human cells (Usman et al., 2020). Inhalation of nanoparticles during their use in agriculture can endanger farmworkers and adjacent communities (Fraceto et al., 2018).

Rigorous toxicological studies are required to evaluate the potential health risks of nanomaterials used in agriculture in order to address concerns regarding toxicity and safety (Usman et al., 2020). This involves conducting in vitro and in vivo research to assess the immediate and prolonged harmful effects of nanomaterials, as well as their ability to accumulate in organisms and increase in concentration down the food chain (Ma et al., 2017). Standardized standards for safety assessment of nanomaterials in agriculture can aid in promoting their responsible and safe utilization (Chaudhary et al., 2021).

4.3 Regulatory Challenges

Regulating nanomaterials in agriculture is challenging because of their distinct properties and wide range of uses. The existing regulations for agricultural goods and inputs may not be sufficient to manage the particular hazards and uncertainties linked to nanomaterials (Amenta et al., 2015). The absence of uniform definitions, terminology, and categorization systems for nanomaterials might result in discrepancies in their regulation among various countries and regions (Kah et al., 2018). One significant regulatory problem involves doing thorough risk assessments and safety evaluations of nanoparticles utilized in agriculture. Current risk assessment methods and recommendations may not be appropriate for nanomaterials because of their unique physicochemical characteristics and interactions in biological systems (Grillo et al., 2021). Regulators encounter difficulties in establishing the suitable techniques and standards for evaluating the safety of nanomaterials, given their possible environmental and health hazards (Kookana et al., 2014). Transparency and public involvement are necessary regulatory requirements when making decisions about using nanomaterials in agriculture. Public perception and adoption of nanotechnology in food and agriculture can greatly impact the regulatory environment (Chaudhry et al., 2017). Regulators must guarantee clear and widespread communication of information regarding the advantages, dangers, and uncertainties linked to nanomaterials to facilitate educated public involvement in the regulatory procedure (Amenta et al., 2015). International coordination and harmonization of regulatory frameworks for nanomaterials in agriculture are required to tackle these regulatory problems (Kah et al., 2018). This entails creating established processes for characterizing, assessing risks, and evaluating the safety of nanomaterials. It also involves setting explicit criteria for labeling and traceability, as mentioned by Amenta et al. (2015). Engaging many stakeholders such as researchers, industry, policymakers, and the public is essential for creating efficient and transparent regulatory systems (Chaudhry et al., 2017).

4.4 Cost-Effectiveness and Scalability

The cost-effectiveness and scalability of nanomaterials in agriculture are crucial factors for their broad implementation and commercial success. Nanomaterial fabrication typically requires intricate and costly methods such chemical synthesis, physical vapor deposition, and biological synthesis (Chaudhary et al., 2021). The expensive production costs can hinder the availability and accessibility of nanomaterial-based goods for farmers, especially in underdeveloped nations (Chhipa, 2019). The scalability of producing nanomaterials is a major hurdle. To produce nanomaterials on a large scale, it is necessary to optimize manufacturing methods, quality control procedures, and infrastructure (Khoshnevisan et al., 2021). Inconsistent manufacturing processes and variable nanomaterial quality can impede their widespread manufacture and commercialization (Grillo et al., 2021). Research is focused on finding cost-effective and scalable ways for producing nanomaterials to tackle these difficulties. This involves investigating green synthesis methods, like utilizing plant extracts and microorganisms, to decrease the environmental impact and manufacturing expenses of nanomaterials (Prasad et al., 2017). Furthermore, the advancement of sophisticated manufacturing methods like 3D printing and microfluidics can facilitate the accurate manipulation and mass production of nanomaterials with specific features (Chhipa, 2019). One way to enhance the cost-effectiveness and scalability of nanoparticles in agriculture is by creating multifunctional and intelligent nanomaterials. Nanomaterials have the ability to carry out several tasks like delivering nutrients, controlling pests, and managing diseases. This capability helps in minimizing the requirement for multiple products and applications (Khoshnevisan et al., 2021). Furthermore, combining nanomaterials with sensors and precision agriculture tools can improve their efficiency and economic feasibility (Grillo et al., 2021).

5. FUTURE PERSPECTIVES

5.1 Potential Future Applications of Nano-Enzymes and Nanomics in Agriculture

Nano-enzymes and nanomics in agriculture are advancing quickly, with several possible applications that could transform crop cultivation and agricultural management. One potential field is creating intelligent nano-biosensors capable of real-time detection of plant infections, environmental stressors, and nutrient deficits (Jain et al., 2021). Nano-biosensors can be used with wireless networks and Internet of Things (IoT) platforms to facilitate precision agriculture. This allows farmers to observe crop health and make well-informed choices about irrigation, fertilization, and pest control (Khanna et al., 2019). A possible future use could involve creating nano-vaccines for plants. Nano-vaccines can be created to transport particular antigens or RNA interference (RNAi) molecules to plants, stimulating their immune system to combat viral, bacterial, or fungal diseases (Chaudhary et al., 2021). This method offers a more focused and efficient strategy for safeguarding crops against illnesses, hence decreasing reliance on chemical pesticides (Fraceto et al., 2018).Nano-enzymes can be designed to improve plant growth and production. Nano-enzymes capable of repairing atmospheric nitrogen or solubilizing soil phosphorus could be created to enhance plants' nutrient absorption (Kashyap et al., 2021). Nano-enzymes capable of breaking down lignin or cellulose could enhance the effectiveness of producing biofuels from agricultural waste (Prasad et al., 2017). In nanomics, potential future applications may include utilizing nanopore sequencing and single-molecule detection methods to analyze plant genomes, transcriptomes, and proteomes with exceptional precision (Chhipa, 2019). This could help discover new genes and pathways related to plant growth, development, and stress responses, which could aid in developing more robust and productive crop varieties (Chaudhary et al., 2021).

5.2 Integration of Nanotechnology With Other Emerging Technologies (e.g., Biotechnology, IoT)

Combining nanotechnology with biotechnology and IoT may result in the creation of potent instruments for sustainable agriculture. Biotechnology, including genetic engineering and genome editing methods like CRISPR-Cas9, can be utilized to create crop types with improved characteristics including drought tolerance, disease resistance, and higher nutritional quality (Jain et al., 2021). Nanotechnology could enhance these endeavors by offering precise delivery systems for gene editing components or by facilitating the creation of nano-biosensors to monitor the expression of inserted genes (Kashyap et al., 2021).

IoT can be utilized to develop intelligent farming systems that gather and evaluate up-to-the-minute data on crop development, soil quality, and environmental factors (Khanna et al., 2019). Nano-biosensors could be used in the field to monitor parameters and send the data to cloud-based platforms for analysis and decision-making (Fraceto et al., 2018). Nanotechnology and IoT integration could facilitate precise agricultural operations like adjusting fertilizer and pesticide application rates, optimizing irrigation schedules, and detecting crop stress or disease early (Jain et al., 2021). In addition, combining nanotechnology, biotechnology, and IoT could result in the creation of closed-loop agricultural systems, where the products of one process are used as inputs for another (Chaudhary et al., 2021). Agricultural waste can be converted into biofuels or bio-based materials by utilizing nano-enzymes. These products can be utilized in crop production or as raw materials in other industries (Prasad et al., 2017). Implementing this circular strategy may decrease waste, enhance resource utilization efficiency, and advance sustainability in agriculture (Chhipa, 2019).

5.3 Research and Development Needs

To maximize the benefits of nano-enzymes and nanomics in agriculture, many research and development requirements need to be met. Research on the mechanisms of action, stability, and safety of nano-enzymes and nanomaterials in agricultural systems is essential (Grillo et al., 2021). This involves research on the relationships between nanomaterials and plants, soil, and other environmental elements, as well as the possible harmful effects and environmental destiny of these materials (Pandey et al., 2018). Furthermore, standardized techniques and standards are required for the synthesis, characterisation, and application of nano-enzymes and nanomaterials in agriculture (Kah et al., 2018). This would aid in guaranteeing the reproducibility, quality, and safety of these materials, making it easier to transition them from the laboratory to practical applications (Chaudhary et al., 2021). Furthermore, multidisciplinary cooperation are required among researchers from many domains like nanotechnology, biotechnology, agronomy, and environmental science (Jain et al., 2021). These partnerships may result in the creation of creative and comprehensive solutions for agricultural concerns like climate change, soil degradation, and food security (Prasad et al., 2017). Increased investment in research and development of nano-enzymes and nanomics in agriculture is necessary. This encompasses financial backing for fundamental and practical research, together with assistance for technology transfer and commercialization (Chhipa, 2019). Utilizing public-private partnerships and international collaborations can expedite the advancement and implementation of agricultural technology by pooling resources and knowledge (Grillo et al., 2021).

5.4 Policy and Regulatory Considerations

The use of nano-enzymes and nanomics in agriculture presents significant policy and regulatory implications. A clear and standardized regulatory framework is required for evaluating and approving nanomaterials and nano-enabled products in agriculture (Amenta et al., 2015). The framework should be grounded on scientific knowledge and take into account the distinctive characteristics and possible hazards of nanomaterials, along with their planned applications and exposure situations (Kookana et al., 2014). Furthermore, better risk communication and public participation are necessary for the development and management of nano-enzymes and nanomics in agriculture (Chaudhry et al., 2017). The information provided should be clear and easily understood regarding the advantages, disadvantages, and unknown factors related to these technologies. It should also provide ways for the public to be involved in decision-making processes (Grillo et al., 2021). Policies and incentives are required to encourage the responsible and sustainable development and utilization of nano-enzymes and nanomics in agriculture (Prasad et al., 2017). This may involve providing financial resources for studying the environmental and social effects of new technologies, as well as assisting farmers and industries in implementing optimal practices and stewardship initiatives (Chhipa, 2019).

6. CONCLUSION

Nano-enzymes and nanomics have significant promise to advance sustainable agriculture and tackle difficulties in the agricultural sector. These technologies can enhance food security, environmental protection, and socio-economic development by implementing precision agriculture methods, enhancing crop resilience, and creating a circular economy. Collaborative efforts among researchers, policymakers, and stakeholders are essential to effectively utilize the benefits of nano-enzymes and nanomics. By conducting transdisciplinary research, practicing responsible governance, and forming inclusive collaborations, we can harness the transformative power of nanotechnology in agriculture to create a more sustainable and resilient food system for future generations.

ACKNOWLEDGMENT

Acknowledge the support by the Strategic Academic Leadership Program of the Southern Federal University ("Priority 2030").

REFERENCES

Amenta, V., Aschberger, K., Arena, M., Bouwmeester, H., Moniz, F. B., Brandhoff, P., ... Peters, R. J. (2015). Regulatory aspects of nanotechnology in the agri/feed/food sector in EU and non-EU countries. *Regulatory Toxicology and Pharmacology*, *73*(1), 463–476. doi:10.1016/j.yrtph.2015.06.016 PMID:26169479

Bundschuh, M., Filser, J., Lüderwald, S., McKee, M. S., Metreveli, G., Schaumann, G. E., Schulz, R., & Wagner, S. (2018). Nanoparticles in the environment: Where do we come from, where do we go to? *Environmental Sciences Europe*, *30*(1), 1–17. doi:10.1186/s12302-018-0132-6 PMID:29456907

Carsanba, E., Favaretto, C., Moraes, F. C., Cucinelli Neto, R. P., Borin, E., & Bordin, J. R. (2021). A graphene-based biosensor for early detection of Xylella fastidiosa in citrus. *Scientific Reports*, *11*(1), 1–10. doi:10.1038/s41598-021-87496-1 PMID:33414495

Chaudhary, S., Dheri, G. S., & Brar, B. S. (2021). Nanomics: An emerging tool for sustainable agriculture. *Environmental Nanotechnology, Monitoring & Management, 15*, 100455. doi:10.1016/j.enmm.2021.100455

Chhipa, H. (2019). Applications of nanotechnology in agriculture. In *Nanoscience in Food and Agriculture 5* (pp. 115–142). Springer. doi:10.1007/978-3-030-14024-7_6

Chhipa, H. (2019). Applications of nanotechnology in agriculture. *Angewandte Chemie International Edition, 58*(13), 4318–4322. doi:10.1002/anie.201814313 PMID:30714282

Dimkpa, C. O., & Bindraban, P. S. (2018). Nanofertilizers: New products for the industry? *Journal of Agricultural and Food Chemistry, 66*(26), 6462–6473. doi:10.1021/acs.jafc.7b02150 PMID:28535672

FAO. (2021). The State of Food and Agriculture 2021: Making agrifood systems more resilient to shocks and stresses. Food and Agriculture Organization of the United Nations. https://www.fao.org/3/cb4476en/cb4476en.pdf

Fraceto, L. F., Grillo, R., de Medeiros, G. A., Scognamiglio, V., Rea, G., & Bartolucci, C. (2016). Nanotechnology in agriculture: Which innovation potential does it have? *Frontiers in Environmental Science, 4*, 20. doi:10.3389/fenvs.2016.00020

Gorgzadeh, A., Amiri, P. A., Yasamineh, S., Naser, B. K., & abdulallah, K. (2024). The potential use of nanozyme in aging and age-related diseases. *Biogerontology*. Advance online publication. doi:10.1007/s10522-024-10095-w PMID:38466515

Grillo, R., Fraceto, L. F., & Pereira, A. E. S. (2021). Nanotechnology in agriculture: Challenges and opportunities. In *Nanomaterials for Agriculture and Forestry Applications* (pp. 51–77). Elsevier. doi:10.1016/B978-0-12-822840-4.00003-2

Huang, Y., Ren, J., & Qu, X. (2019). Nanozymes: Classification, catalytic mechanisms, activity regulation, and applications. *Chemical Reviews, 119*(6), 4357–4412. doi:10.1021/acs.chemrev.8b00672 PMID:30801188

Inani, H., Jain, R. K., & Jain, P. K. (2021). Development of carbon nanotube-based biosensor for real-time monitoring of soil nitrate levels in maize. *Journal of Hazardous Materials, 408*, 124957. doi:10.1016/j.jhazmat.2020.124957

Jain, A., Ranjan, S., Dasgupta, N., & Ramalingam, C. (2018). Nanomaterials in food and agriculture: An overview on their safety concerns and regulatory issues. *Critical Reviews in Food Science and Nutrition, 58*(2), 297–317. doi:10.1080/10408398.2016.1160363 PMID:27052385

Jain, P. K. (2021). Nanomics-based agriculture for sustainable crop production. In *Nanomaterials for Agriculture and Forestry Applications* (pp. 1–25). Elsevier. doi:10.1016/B978-0-12-822840-4.00001-9

Kabiri, S., Degryse, F., Tran, D. N., da Silva, R. C., McLaughlin, M. J., & Losic, D. (2017). Graphene oxide: A new carrier for slow release of plant micronutrients. *ACS Applied Materials & Interfaces, 9*(49), 43325–43335. doi:10.1021/acsami.7b07890 PMID:29160685

Kah, M., Kookana, R. S., Gogos, A., & Bucheli, T. D. (2018). A critical evaluation of nanopesticides and nanofertilizers against their conventional analogues. *Nature Nanotechnology*, *13*(8), 677–684. doi:10.1038/s41565-018-0131-1 PMID:29736032

Kah, M., Tufenkji, N., & White, J. C. (2019). Nano-enabled strategies to enhance crop nutrition and protection. *Nature Nanotechnology*, *14*(6), 532–540. doi:10.1038/s41565-019-0439-5 PMID:31168071

Kalluri, R., Narayanan, J., & Thangavel, K. (2021). A gold nanoparticle-based colorimetric biosensor for rapid detection of arsenic in soil samples. *Journal of Hazardous Materials*, *403*, 123875. doi:10.1016/j.jhazmat.2020.123875

Kashyap, P. L., Prasanna, R., & Kumar, R. (2021). Nanomics for sustainable agriculture. In *Nanomaterials for Agriculture and Forestry Applications* (pp. 27–50). Elsevier. doi:10.1016/B978-0-12-822840-4.00002-0

Khanna, V. K., Kumar, A., Jain, P. K., & Bhalla, V. (2019). Nanomaterials-based biosensors for food contaminant detection. *Environmental Chemistry Letters*, *17*(2), 845–859. doi:10.1007/s10311-018-00833-z

Khoshnevisan, B., Shafiei, H., Rajaeifar, M. A., & Tabatabaei, M. (2021). Nano-based smart pesticide formulations for sustainable agricultural development. *Environmental Science and Pollution Research International*, *28*(16), 19839–19856. doi:10.1007/s11356-021-12592-8

Khot, L. R., Sankaran, S., Maja, J. M., Ehsani, R., & Schuster, E. W. (2012). Applications of nanomaterials in agricultural production and crop protection: A review. *Crop Protection (Guildford, Surrey)*, *35*, 64–70. doi:10.1016/j.cropro.2012.01.007

Kolesnikov, S., Timoshenko, A., Varduni, V., Minnikova, T., Tsepina, N., Mandzhieva, S., & Rajput, V. D. (2024). *Comparative estimation of ecotoxicity of nano- and microparticles of CuO, ZnO, NiO, Co3O4 by the integral biological indicator of soil state*. Nanotechnology for Environmental Engineering. doi:10.1007/s41204-024-00358-9

Kookana, R. S., Boxall, A. B., Reeves, P. T., Ashauer, R., Beulke, S., Chaudhry, Q., Cornelis, G., Fernandes, T. F., Gan, J., Kah, M., Lynch, I., Ranville, J., Sinclair, C., Spurgeon, D., Tiede, K., & van den Brink, P. J. (2014). Nanopesticides: Guiding principles for regulatory evaluation of environmental risks. *Journal of Agricultural and Food Chemistry*, *62*(19), 4227–4240. doi:10.1021/jf500232f PMID:24754346

Kumari, A., Singla, R., Guliani, A., & Walia, S. (2020). Nanopesticides: Opportunities and challenges for crop protection. In *Nanopesticides* (pp. 151–171). Springer. doi:10.1007/978-3-030-44873-8_8

Li, Z., Liu, X., Jiang, M., & Guo, L. (2018). Specific detection of bioavailable phenanthrene by a whole cell bioreporter in the presence of copper. *Environmental Pollution*, *235*, 381–389. doi:10.1016/j.envpol.2017.12.077

Ma, X., Geiser-Lee, J., Deng, Y., & Kolmakov, A. (2017). Interactions between engineered nanoparticles (ENPs) and plants: Phytotoxicity, uptake and accumulation. *The Science of the Total Environment*, *408*(16), 3053–3061. doi:10.1016/j.scitotenv.2010.03.031 PMID:20435342

Nadi, E., Aynehband, A., & Mojaddam, M. (2019). Effect of nano-iron fertilizer on yield and yield components of chickpea (Cicer arietinum L.) under water deficit stress. *Journal of Crop Science and Biotechnology*, *22*(3), 249–256. doi:10.1007/s12892-018-0152-0

Nguyen, T. M. H., Le, T. T. H., Vu, T. T., Nguyen, T. H., Nguyen, V. H., & Tran, L. D. (2021). A graphene-based electrochemical biosensor for rapid detection of Escherichia coli O157:H7 in fresh produce. *Sensors and Actuators. B, Chemical, 329*, 129196. doi:10.1016/j.snb.2020.129196

Pandey, G., Maheshwari, D. K., & Samreen, S. (2018). Nanotechnology in agro-food industry. In *Nanotechnology* (pp. 147–190). Springer. doi:10.1007/978-981-10-8780-8_6

Pereira, A. E. S., Sandoval-Herrera, I. E., Zavala-Betancourt, S. A., Oliveira, H. C., Ledezma-Pérez, A. S., Romero, J., & Fraceto, L. F. (2017). γ-Polyglutamic acid/chitosan nanoparticles for the plant growth regulator gibberellic acid: Characterization and evaluation of biological activity. *Carbohydrate Polymers, 157*, 1862–1873. doi:10.1016/j.carbpol.2016.11.073 PMID:27987906

Prasad, R., Bhattacharyya, A., & Nguyen, Q. D. (2017). Nanotechnology in sustainable agriculture: Recent developments, challenges, and perspectives. *Frontiers in Microbiology, 8*, 1014. doi:10.3389/fmicb.2017.01014 PMID:28676790

Rajput, V. D., Singh, A., Minkina, T., Rawat, S., Mandzhieva, S., Sushkova, S., & Upadhyay, S. K. (2021). Nano-Enabled Products: Challenges and Opportunities for Sustainable Agriculture. *Plants, 10*(12), 2727. doi:10.3390/plants10122727 PMID:34961197

Shafique, S., Shafique, S., Ahmed, A., & Shabbir, R. (2020). A nano-biosensor for early detection and management of Tomato leaf curl New Delhi virus. *Scientia Horticulturae, 265*, 109211. doi:10.1016/j.scienta.2020.109211

Shahi, U. P., Singh, O., Singh, V. K., Shivangi, P. K., Singh, R., & Singh, V. K. (2024). Nanotechnology in rice farming: Optimizing nutrient management with nanofertilizers. In Sustainable Agriculture: Nanotechnology and Biotechnology for Crop Production and Protection. Walter de Gruyter GmbH. DOI: doi:10.1515/9783111234694-003

Singh, A., Hussain, I., Singh, N. B., & Singh, H. (2018). Uptake, translocation and impact of green synthesized nanoceria on growth and antioxidant enzymes activity of Solanum lycopersicum L. *Ecotoxicology and Environmental Safety, 158*, 345–353. doi:10.1016/j.ecoenv.2018.04.039 PMID:31284122

Singh, A., Rajput, V., Varshney, A., Sharma, R., Ghazaryan, K., Minkina, T., . . . El-Ramady, H. (2024b). Revolutionizing Crop Production: Nanoscale Wonders - Current Applications, Advances, and Future Frontiers. Egyptian Journal of Soil Science, 64(1). doi:10.21608/ejss.2023.246354.1684

Singh, A., Rajput, V. D., Pandey, D., Sharma, R., Ghazaryan, K., & Minkina, T. (2023c). Nano Zinc-Enabled Strategies in Crops for Combatting Zinc Malnutrition in Human Health. *Frontiers in Bioscience (Landmark Edition), 28*(8), 158. doi:10.31083/j.fbl2808158 PMID:37664935

Singh, A., Sharma, R., Rawat, S., Singh, A. K., Rajput, V. D., Fedorov, Y., . . . Chaplygin, V. (2022). Nanomaterial-plant interaction: Views on the pros and cons. Toxicity of Nanoparticles in Plants, 47–68. doi:10.1016/B978-0-323-90774-3.00015-5

Singh, J., Dutta, T., Kim, K. H., Rawat, M., Samddar, P., & Kumar, P. (2018). 'Green' synthesis of metals and their oxide nanoparticles: Applications for environmental remediation. *Journal of Nanobiotechnology, 16*(1), 1–24. doi:10.1186/s12951-018-0408-4 PMID:30373622

Singh, O. (2023). Nanotechnology for sustainability and food security in agriculture: A nanopriming long story in short. In A. Singh, V. Rajput, K. Ghazaryan, S. Gupta, & T. Minkina (Eds.), *Nanopriming Approach to Sustainable Agriculture* (pp. 315–339). IGI Global. doi:10.4018/978-1-6684-7232-3.ch014

Singh, O., Shahi, U. P., Shivangi, Singh, V. K., Singh, P. K., Singh, R., & Singh, A. (2024). Enhancing crop nutrition with zinc nanoparticles: A novel approach to cereal crop micronutrient management. In Sustainable Agriculture: Nanotechnology and Biotechnology for Crop Production and Protection. Walter de Gruyter GmbH. DOI: doi:10.1515/9783111234694-006

Singh, O., Shahi, U.P., Dhyani, B.P., Kumar, S., Vivek, S., & Sengar, R.S., Shivangi, and Singh, A. (2023). Effect of zinc oxide nanoparticles application on growth and yield of basmati rice (Oryza sativa) in alkaline soil. *Indian Journal of Agronomy*, *68*(3), 1–6.

Singh, R. P., Singh, A., Srivastava, S., & Kumar, R. (2018). Gold nanoparticle-based colorimetric sensor for detection of Fusarium graminearum in wheat. *Analytical Methods*, *10*(20), 2374–2380. doi:10.1039/C8AY00473A

Singh, S., Singh, O., & Shahi, U. P. (2023). Nanotechnology: A New Innovation in Biofortification and Food Safety. In A. Singh, V. Rajput, K. Ghazaryan, S. Gupta, & T. Minkina (Eds.), *Nanopriming Approach to Sustainable Agriculture* (pp. 340–351). IGI Global. doi:10.4018/978-1-6684-7232-3.ch015

Singh, Y. K., Singh, B. V., Katiyar, D., Saikanth, D. R. K., Kumar, K., Singh, O., Verma, S., & Kumar, P. (2023). Efficacy of nano fertilizers on yield, attributes and economics of wheat. *International Journal of Environment and Climate Change*, *13*(7), 291–297. doi:10.9734/ijecc/2023/v13i71879

Tripathi, D. K., Shweta, Singh, S., Singh, S., Pandey, R., Singh, V. P., Sharma, N. C., Prasad, S. M., Dubey, N. K., & Chauhan, D. K. (2017). An overview on manufactured nanoparticles in plants: Uptake, translocation, accumulation and phytotoxicity. *Plant Physiology and Biochemistry*, *110*, 2–12. doi:10.1016/j.plaphy.2016.07.030 PMID:27601425

United Nations. (2021). The Sustainable Development Goals Report 2021. United Nations. https://unstats.un.org/sdgs/report/2021/

Usman, M., Farooq, M., Wakeel, A., Nawaz, A., Cheema, S. A., Rehman, H. U., Ashraf, I., & Sanaullah, M. (2020). Nanotechnology in agriculture: Current status, challenges and future opportunities. *The Science of the Total Environment*, *721*, 137778. doi:10.1016/j.scitotenv.2020.137778 PMID:32179352

Velten, S., Leventon, J., Jager, N., & Newig, J. (2015). What is sustainable agriculture? A systematic review. *Sustainability (Basel)*, *7*(6), 7833–7865. doi:10.3390/su7067833

Walkey, C., Das, S., Seal, S., Erlichman, J., Heckman, K., Ghibelli, L., Traversa, E., McGinnis, J. F., & Self, W. T. (2015). Catalytic properties and biomedical applications of cerium oxide nanoparticles. *Environmental Science. Nano*, *2*(1), 33–53. doi:10.1039/C4EN00138A PMID:26207185

Wang, X., Guo, J., Feng, J., & Shi, Z. (2018). Nanozyme-based environmental sensing and remediation. *NanoImpact*, *10*, 21–30. doi:10.1016/j.impact.2018.01.003

Wei, H., & Wang, E. (2013). Nanomaterials with enzyme-like characteristics (nanozymes): Next-generation artificial enzymes. *Chemical Society Reviews*, *42*(14), 6060–6093. doi:10.1039/c3cs35486e PMID:23740388

Wu, J., Wang, X., Wang, Q., Lou, Z., Li, S., Zhu, Y., Qin, L., & Wei, H. (2019). Nanomaterials with enzyme-like characteristics (nanozymes): Next-generation artificial enzymes (II). *Chemical Society Reviews*, *48*(4), 1004–1076. doi:10.1039/C8CS00457A PMID:30534770

Xianyu, Y., Wang, Z., & Jiang, X. (2018). A plasmonic nanosensor for immunoassay via enzyme-triggered click chemistry. *ACS Nano*, *8*(12), 12741–12747. doi:10.1021/nn505857g PMID:25423357

Xu, C., Peng, C., Sun, L., Zhang, S., Huang, H., Chen, Y., & Shi, J. (2021). Distinctive effects of TiO2 and CuO nanoparticles on soil microbes and their community structures in flooded paddy soil. *Soil Biology & Biochemistry*, *43*(8), 1406–1414. doi:10.1016/j.soilbio.2011.03.010

Zaghloul, S. M., El-Shishtawy, R. M., Atia, A. A., Mansour, N. A., & El-Sherbiny, I. M. (2021). Graphene oxide-based biosensor for detection of Candidatus Liberibacter asiaticus bacterium causing citrus greening disease. *Talanta*, *223*, 121714. doi:10.1016/j.talanta.2020.121714

Chapter 2
Nano-Omics Approaches for Studying Plant Growth, Yield, and Nutritional Value

Anaytullah Siddique
Department of Agronomy, School of Agriculture, Lovely Professional University, Phagwara, India

Priyanka Aley
Department of Agronomy, School of Agriculture, Lovely Professional University, Phagwara, India

ABSTRACT

The global population's exponential growth poses a challenge, demanding innovative solutions for sustainable crop yield and nutritional enhancement. Nano-omics, rooted in nanotechnology, revolutionizes agriculture. Nanoparticles (NPs) penetrate plant tissues for precise nutrient delivery, catalyzing peroxidase and catalase synthesis. This scavenges reactive oxygen species, reducing lipid peroxidation, preserving cell membrane integrity, enhancing water and nutrient uptake, and stimulating carbohydrate partitioning. Omics techniques (genomics, transcriptomics, proteomics, and metabolomics) explore molecular interactions. Genomic analysis reveals altered gene expression related to growth, yield, and stress response. Transcriptomic studies show modulation of transcription factors and pathways by nano-zymes. Proteomic investigations focus on protein profile changes induced by nano-zymes, deepening understanding of plant responses. It unveils the interplay between nano-zymes and plants, providing insights into mechanisms for augmented crop growth and nutritional enhancement.

INTRODUCTION

As the world's population continues to grow, the pressure on the agriculture sector to produce enough food while maintaining crop growth, yield, and nutritional quality is increasing (Fischer R A et al., 2018). Unfortunately, environmental changes have made it challenging to meet these goals (Steiner A et al., 2020; Singh et al., 2022). In response, many advanced techniques have been developed to address these challenges, such as breeding techniques, upgraded fertilizers, precision farming, and nano-technology

DOI: 10.4018/979-8-3693-1890-4.ch002

(Usman M et al., 2020). Among these techniques, nano-omics stands out as the most advanced technology, offering new tools and materials on a nanoscale that advance crop productivity and sustainability. By using particles and materials that are incredibly small, nano-omics enables the precision delivery of nutrients, pesticides, and genetic materials to plants, reducing waste and minimizing environmental impact (Pandohee J et al., 2023).

The application of nanotechnology in agriculture involves using nanomaterials to enhance various aspects of crop production, resource management, and crop protection (Singh et al., 2023a). These nanomaterials have unique properties that make them useful for agriculture, such as increased surface area and reactivity (Liu C et al., 2021). Nanotechnology can be used to create precise solutions that address long-standing agricultural problems, promoting sustainable farming practices (Singh et al., 2023b). For example, nano-fertilizers and smart delivery systems can improve nutrient management, while nano-enabled sensors and monitoring devices can provide real-time data on soil health, crop growth, and environmental factors (Shafi A et al., 2020). Nanotechnology has also led to transformative innovations in crop protection and pest management, such as nano-based pesticides, insecticides, and antimicrobial agents that offer targeted delivery and reduced ecological impact (Ghouri M Z et al., 2020).

Additionally, nanotechnology can enhance crop resilience against abiotic stresses like drought or salinity, through the development of stress-tolerant materials and precision delivery systems (Afzal S et al., 2022). Bio-engineered nanoparticles (NPs) emerged as an apparent perspective that elicits the immunity of plants by scavenging the reactive oxygen species (ROS) (Singh et al., 2023c). Finally, nanotechnology plays a role in improving the nutritional quality of crops, such as through nano-fortification and biofortification techniques that enhance the bioavailability of essential nutrients in food crops. It has emerged as a promising tool for addressing various challenges in agriculture, including crop production, resource management, and crop protection (Singh et al., 2023d). However, its integration raises concerns about safety, environmental impact, and ethical considerations. To ensure safe, sustainable, responsible deployment, safety assessments, and regulatory oversight are necessary (Ditta, A et al., 2015).

Finally, nano omics, the study of the interactions between nanoparticles and biological systems at the molecular level, is crucial for understanding the impact of nanotechnology on plant metabolism, gene expression, and protein synthesis, and ensuring its safe and sustainable application in agriculture (Majumdar S et al., 2021). While the integration of nanotechnology in agriculture raises concerns, its potential to transform farming practices and ensure global food security and sustainability cannot be denied. Responsible deployment, safety assessments, and regulatory oversight are essential to harness the full potential of nanotechnology while minimizing its potential risks (Hull M et al., 2018).

NANO-OMICS IN AGRICULTURE

The application of nanotechnology has been around for quite some time and has rapidly expanded into various areas such as medicine, electronics, energy production, and catalysts (Nasrollahzadeh M et al., 2019). Unfortunately, achieving the United Nations' goal of "Zero hunger" has become increasingly difficult due to exponential population growth and climate shifts (Salinas M et al., 2020). However, the emerging field of Nanoomics is studying the interactions between nanoparticles and biological systems at a molecular level, which could have significant impacts on addressing global challenges such as hunger (Nguyen N H et al., 2023). In agriculture, understanding the impact of nanotechnology on plant biology is essential for developing innovative solutions that promote sustainable, efficient, and resilient farming practices (Rajput et al., 2024). The United Nations has set an agenda to achieve the goal of sustainability by 2030, with the primary targets being to develop crop resilience, efficient use of resources through precision farming, and minimize food wastage by recycling processes. Nanomaterials are used in precision farming to increase crop productivity and quality (Salinas M et al., 2020 and Ditta A et al., 2015).

The use of nano-base agrochemicals, such as nano-fertilizers, nano-herbicides, nano-insecticides, nano-pesticides, and nano-fungicides, can significantly benefit agriculture by ensuring a controlled release of agrochemicals at optimum amounts (Prasad M et al., 2018; Singh et al., 2023e). This method can help reduce the extra burden of fertilizers while also suppressing weed populations, insect attacks, and pathogens. There are several formulations based on Nano-Omics that have been introduced in agriculture, such as nano clay, nano Zn EDTA, nano Fe EDTA, nCu, nMn, nMoO3, and nanotubes, which enhance plant growth or initiate the absorption of macro and micronutrients (Vardumyan et al., 2024) . Moreover, nano clay is used as an adsorbent, and nano zeolite is used to ameliorate heavy metal contamination while also acting as a carrier of nutrients in the soil (Kolbert Z et al., 2022).

The study of omics has revolutionized the way we understand the functioning of organisms. With the help of advanced analytical methods and bioinformatics tools, we can now examine biomolecules with greater precision and sensitivity, leading to a deeper understanding of molecular pathways in plants and their responses to various stresses. Systems biology has played a key role in this field, enabling us to comprehend the function of genes, proteins, and metabolites more comprehensively (Debnath, M et al., 2011).

NANOTECHNOLOGY-BASED AGROCHEMICALS

It's interesting to note that in modern agriculture, nanomaterial-based agrochemicals are being used to improve crop productivity and quality. These agrochemicals come in various forms such as nano-fertilizers, nano-pesticides, nano-fungicides, and even nano-biosensors (Singh et al., 2023f). They can be applied either to the soil or directly on the plant's foliage, with the latter being more responsive. However, both methods have their advantages and disadvantages (Vannini et al., 2014).

Nanoparticle-Based Fertilizer

It is well-known that chemical fertilizers have played a major role in Indian agriculture since the Green Revolution. However, the pros and cons of their usage have been a topic of discussion among scientists at regular intervals. The overuse of chemical fertilizers has resulted in several negative impacts, including soil degradation, soil salinization, water pollution, and deviating pH and EC levels. Additionally, the natural microflora and fauna in the soil are reduced, making it challenging to achieve sustainability Mirzajani et al., 2014). Nanoparticle-based fertilizers have emerged as a potential alternative to chemical fertilizers to support crop growth, development, and yield. These fertilizers can be applied to the soil or used as foliar application to protect plants from mineral element deficiencies (Nakabayashi et al., 2019; Nhan et al., 2015). However, the long-term effects of nanoparticle-based fertilizers on soil health, crop productivity, and environmental sustainability are yet to be fully understood. Therefore, more studies are needed to evaluate the safety and effectiveness of these fertilizers before they can be widely adopted in agriculture. The following nano-fertilizers are under consideration by the scientific community to make them more useful in agriculture.

Figure 1. Flow diagram of omics technique and its application in the plant

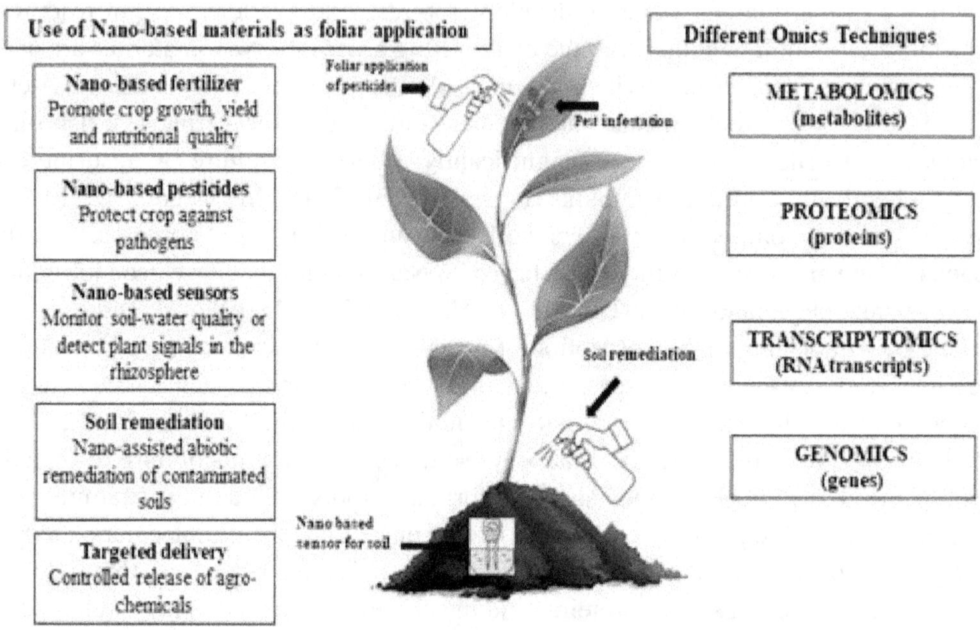

Table 1. Effect of NPS-based nanofertilizers in different crops

SR. NO.	POPULAR NANOFERTILIZERS	MODE OF APPLICATION	RESPONSE IN CROPS	REFERENCE
1	ZnO	Foliar spray	Increased biomass via enhancing nutrient concentration in *Guar*	Raliya R et al., 2023
2	AgNPs	Foliar spray	Reduced bacterial disease spread over Cowpea	Vanti G L et al., 2018
3	CuO	Foliar spray	Reduced disease spread over Tomato	Giannousi K et al., 2013
4	N, P, K NPs	Foliar spray	Improved biological and seed yield in Chickpea	Drostkar E et al., 2016
5	TiO_2	Seed treatment	Promoted seed germination and seedling growth in Wheat	Feizi H et al., 2012
6	B NPs	Foliar application	Enhanced morphological growth in Green beans	Franco-Lagos C L et al., 2023
7	Fe_3O_4	Hydroponics	Increased chlorophyll, total soluble protein, number of chloroplasts in Barley	Tombuloglu H et al., 2019
8	Al_2O_3	Seed treatment	Accelerated the physiological and biochemical parameters in Egyptian Roselle	Abdel Latef A A H et al., 2020
9	S NPs	Soil application	Improved oil, crude protein, methionine, cysteine, and total free amino acid in groundnut	Thirunavukkarasu M et al., 2018
10	Fe and Mn NPs	Seed priming	Improved antioxidant system in rice seedling	Khepar V et al., 2023

Nanoparticle-Based Insecticides and Pesticides

It's interesting to note that insect and pest management in agriculture has historically relied heavily on organic chemicals. However, the residual and harmful effects of these chemicals on soil, the environment, and human health are becoming increasingly apparent. That's where nanoparticle (NPs) based insecticides and pesticides come in (Nie D et al., 2023) These agrochemicals have the potential to be much more beneficial than traditional insecticides and pesticides for controlling insects and pests over agricultural fields. Because of their nano size, NPs can easily penetrate and translocate within the plant, allowing them to reach the target cell quickly (Vurro M et al.,2019). NPs-based nanoemulsion materials can be made with either water or oil and are available in both capsulated and non-capsulated forms. The primary component of nano-based herbicides is biodegradable polymeric material, which may increase the efficiency of herbicides (He J et al., 2023). A variety of materials are used to synthesize nanoparticle-based (NPs) insecticides and pesticides, including synthetic materials, plant-based materials, and microbial content. It is absolutely correct that the use of NPs-based insecticides and pesticides reduces the amount of chemicals that are required to be applied to plants compared to conventional chemicals. This means that the risk of soil and water pollution is significantly reduced, as well as the risk of any harmful effects on human health. By using fewer chemicals, the impact on the environment is minimized, which is a significant benefit of NPs-based insecticides and pesticides (Rani A et al., 2020). Overall, this innovative development in the world of agriculture has the potential to revolutionize the way we approach pest management while also ensuring that our environment and health are protected.

THE MECHANISM BEHIND THE UPTAKE AND TRANSLOCATION OF NPs

Nowadays, nanoparticle (NPs) based research is in focus among the scientific community because of its potential to address the issues related to crops. The production or yield is one of the most common constraints in agriculture that is linked to several factors like soil fertility, adverse environmental conditions, obstacles in nutrient uptake, and translocation (Kumar S 2020). Since the green revolution in India, the use of chemical-based fertilizers, insecticides, pesticides, and fungicides has been in practice even though it is useful to maintain the current level of yield and protect the crops from insects, pests, and diseases but due to unorganized way of practice make them dangerous in many

Figure 2. Flow diagram of capsulated form of NP-based agrochemicals

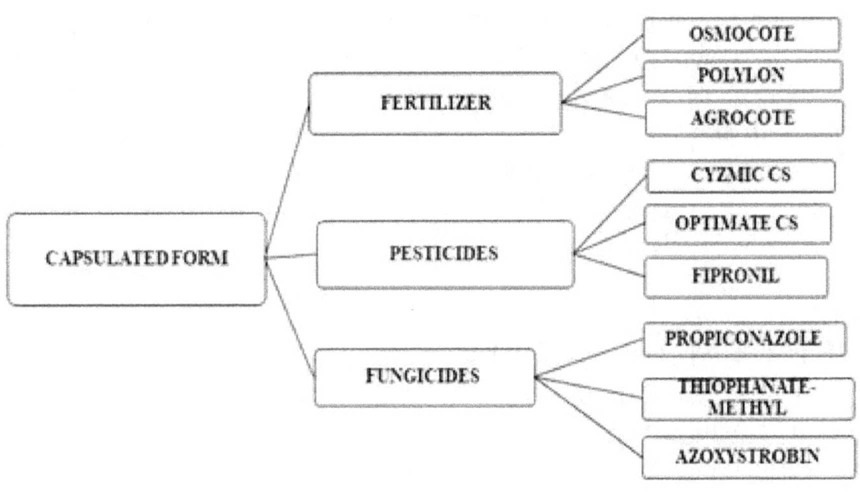

Figure 3. Flow diagram of incapsulated form of NP-based agrochemicals

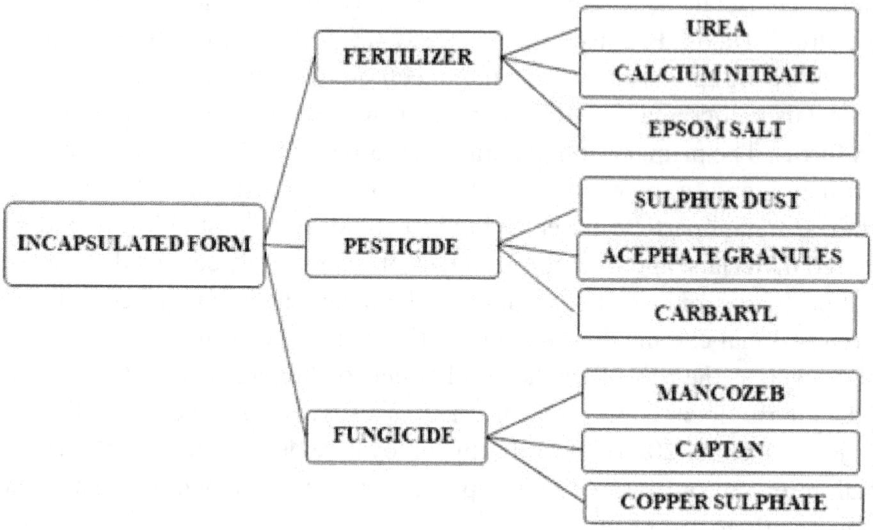

ways (Patil H J et al., 2016). However, the use of NPs based Agrochemicals like nanofertilizers, insecticides, pesticides, and fungicides makes them more easily and faster responsive due to nano size. The application of NPs at pre-optimized rates maintains the plant population by enhancing seed germination, seedling establishment, and vegetative and reproductive growth. Due to changes in soil ecological conditions, the capacity of nutrient uptake via root may reduced hence the translocation and portioning of the nutrient may affected (Aqeel U et al., 2022). Nowadays, there are two popular techniques in use for the application of nano-based agrochemicals either used as seed priming treatments or as a foliar application over the plants directly it depends on the conditions and situations of the plants even though both practices have their own cons pron (Hassanisaadi M et al., 2022). The uptake and translocation of nanoparticles (NPs) based- agrochemicals in plants include complex processes influenced by various factors such as properties of nanoparticles, plant species, and environmental conditions.

Nanoparticles enter plant roots through mechanisms like osmotic pressure and capillary forces, utilizing cell wall pores or plasmodesmata connections. Inside plant cells, entry involves binding to proteins, ion channels, aquaporins, or endocytosis, creating new pores. Transport within the plant can happen through apoplast or symplast, facilitated by plasmodesmata. Entry through the cell wall is size-dependent, with smaller nanoparticles passing easily, while larger ones may use stomata, hydathodes, or flower stigma. Stomatal pores are employed for particles exceeding 40 nm in size, according to the study (Maurel C et al., 2015). Here are some key mechanisms involved in the uptake and translocation ions and molecules.

NUTRIENT UPTAKE AT THE ROOT LEVEL

Endocytosis: This is a highly specific form of endocytosis where receptors on the cell surface bind to specific ligands, triggering the formation of vesicles containing the ligands. This is crucial for the internalization of specific substances, such as hormones or nutrients. Thus nanoparticles can be taken up by plant roots through endocytosis.

Passive diffusion: It is a very common mechanism in the plant to absorb nutrients through the root. It involves the movement of nutrients across cell membranes without the expenditure of energy by the plant. Passive diffusion relies on the concentration gradient of substances, allowing them to move from areas of higher concentration to areas of lower concentration. Thus nanoparticles can also enter plant roots through passive diffusion, depending on their size, surface charge, and the presence of transporters.

TRANSLOCATION BETWEEN ROO-TO-SHOOT:

Vascular transport: Once NPs inside the roots, they can move through the plant's vascular system, including the xylem and phloem. Xylem transport is typically responsible for the upward movement of nanoparticles from roots to shoots.

Size-dependent translocation: Nanoparticles of different sizes may have varying degrees of mobility within the plant. The smaller size of nanoparticles (NPs) may be more easily translocated through the plant's vascular system compared to greater size.

INTERCELLULAR TRANSLOCATION

It is also known as cell-to-cell transport wherein nanoparticles may move from one plant cell to another through plasmodesmata present on the cell wall, which are channels connecting adjacent plant cells. This intercellular transport contributes to the distribution of nanoparticles within the plant. However, the intercellular translocation of NPs preffer two different pathway within the cells.

Symplastic movement: It is a kind of translocation of substances that move through the interconnected cytoplasm of plant cells via plasmodesmata, which are channels that connect the cytoplasm of adjacent cells wherein the translocation of nutrients, ions, and signaling molecules, pass through plasmodesmata, traveling from one cell to another without crossing cell membranes.

Apoplastic Translocation: It is a kind of translocation of substances that the movement occurs through the intercellular spaces of cell walls. In this method, substances or molecules can move through the cell walls without entering the cytoplasm of the cells which is also known as the movement of extracellular spaces. However, to reach the symplastic pathway, substances must eventually cross a plasma membrane.

Overall the nanoparticles undergo an intricate sequence of events to infiltrate the cell membrane and cell wall of the root epidermis, ultimately reaching the plant's vascular bundle (xylem), stele, and finally the leaves. Moving through the intact cell membrane involves the nanoparticles passing through size-specific pores. Before reaching the stele, nanoparticles passively integrate through the apoplast of the endodermis. The uptake mechanism primarily relies on active transport and encompasses various cellular processes such as signaling, recycling, and regulation of the plasma membrane. Nanoparticles gain entry into the plant root through various mechanisms, including osmotic pressure, capillary forces, and pores on the cell wall facilitated by symplastic routes.

Figure 4. Absorption and translocation of NP-based nanoparticles through roots in the plant

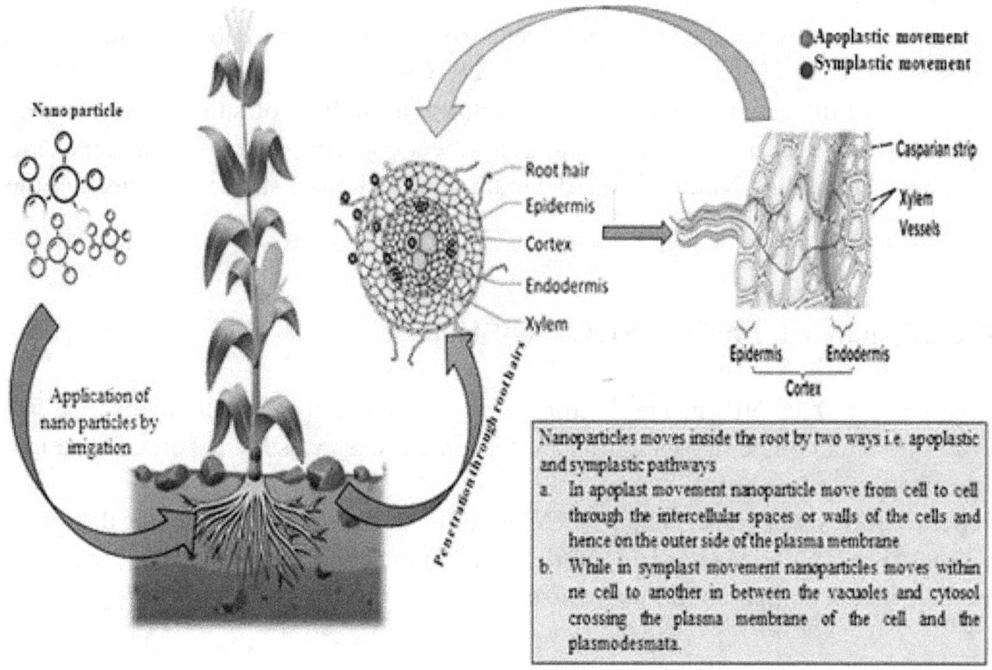

Figure 5. Absorption and translocation of NP-based nanoparticles through stomata in the plant

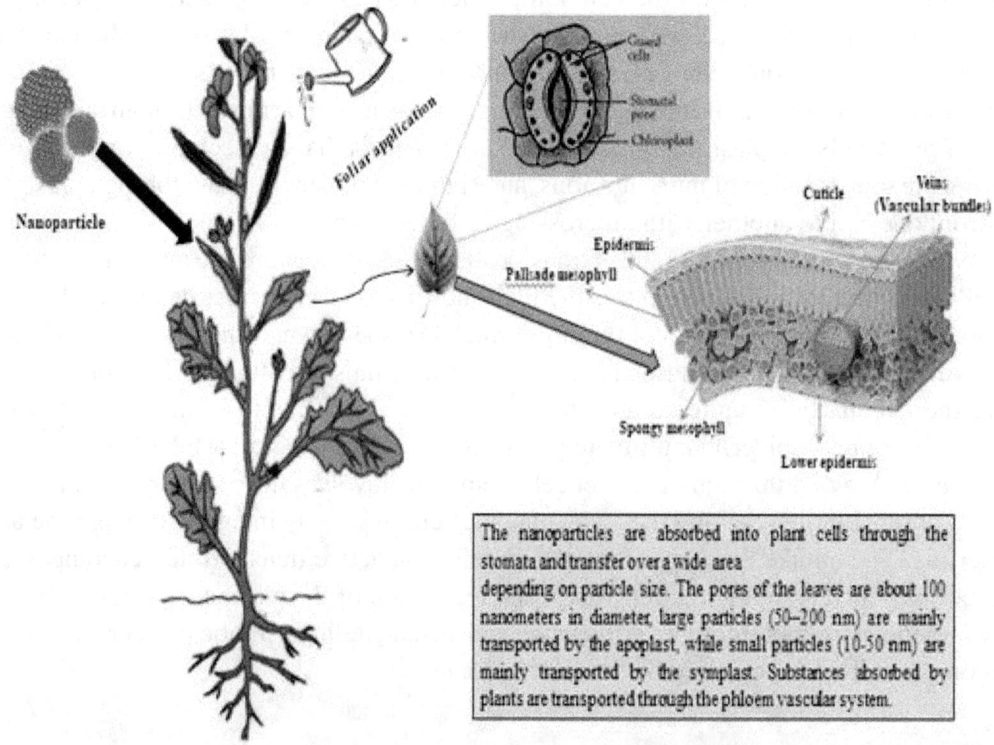

NANO-OMICS AT MOLECULAR LEVEL

To optimize effectiveness and productivity, it is essential to comprehend the interactions between nanomaterials and biological networks in plants. Recently, the study of these interactions has evolved from traditional methods to the latest ones to advanced approaches known as omics (Cembrowska-Lech D et al., 2023). Nowadays, this transition is achievable due to the increased sensitivity and precision of our tools in examining and comprehending biological processes. The term "omics" refers to analyzing all the biomolecules and their connected process to make them more remunerative at the cellular or tissue level in an organism without bias (Yerasu S R et al., 2022). This involves the investigation based on genomics, transcriptomics, proteomics, or metabolomics. By using these omics approaches, scientists are figuring out how genes, proteins, and metabolites in plants respond to different stresses. This helps us understand complex traits in agriculture and how crops react to exposure to nanomaterials (Raza A et al., 2022). Omics techniques are really sensitive, which means they can detect, measure, and differentiate changes in cells and molecules when an organism is exposed to different forms of particles, like ionic, nano, or bulk forms, even at lower and more realistic doses in the environment (Sahoo J P et al., 2020).

In plants, the most commonly used technique among omics methods is transcriptomics, which helps to identify predictive biomarkers of nanomaterial toxicity, linked to observable traits. However, this approach, starting from the bottom with genes, has limitations. It doesn't fully account for uncertainties due to processes that happen after gene activity, like alterations in protein and metabolite levels caused by stimuli. The higher levels of organization in an organism, such as the metabolome or proteome, are not solely determined by genes that influence the function of genes in response to stimuli (Rodziewicz P et al., 2014). Metabolomic analyses play a crucial role in annotating genes or proteins and addressing gaps in our understanding of plant metabolism. Unlike other techniques, metabolomics is not limited to data from specific plant species, making it applicable to various plants. To better understand the interaction between nanomaterials and plants, it is recommended to consider a more holistic approach that links observable traits with the metabolome, complementing other omics methods like proteomics and transcriptomics. This review focuses on groundbreaking studies in plant metabolomics and proteomics to uncover insights into the interaction between nanomaterials and plants. Various omics tools are briefly discussed, and the paper explores new analytical platforms used in metabolomics and proteomics when studying plant responses to nanomaterials. Finally, the review suggests potential weak points in plant biological pathways when dealing with nanomaterials and how integrating data from different omics methods can address key concerns and open up broader applications of nanotechnology in agriculture.

METABOLOMICS AND ITS USES OVER THE PLANTS

It represents the ultimate outcome of cellular processes, revealing an organism's response to external stimuli. Plant metabolic databases contain information on numerous compounds involved in thousands of pathways across a wide variety of 350 plant species (Baghalian K et al., 2014). Out of all, around 200,000 metabolites, can be broadly categorized into two major parts: primary and secondary metabolites (Pandita D et al., 2021). Primary metabolites like carbohydrates, amino acids, vitamins, organic acids, and fatty acids are essential for plant growth while secondary metabolites, on the other hand, are created from primary metabolites and serve as adaptation and defense in plants. Secondary metabolites such as polyketides, terpenoids, steroids, phenylpropanoids, alkaloids, and glucosinolates, each have its pathways

(Mohiuddin AK 2019). Plants respond to substances like engineered nanomaterials (ENMs), and changes in their physiology are driven by molecular events and reflected in the levels of metabolites. These metabolites play roles in interconnected biological pathways like glycolysis, TCA cycle (Tricoboxylic acid), gluconeogenesis, amino acid, secondary metabolite, nitrogen metabolism, and fatty acid metabolism.

Plants release metabolites from their roots as signals to defend against or adapt to stress, also influencing soil chemistry and nutrient availability (Kumar P et al., 2019). Overview of metabolomics of plants and their rhizosphere provides us a momentary glimpse of the complete array of metabolites linked to a plant, offering a direct representation of the current functional condition of the cells.The significance of assessing metabolite levels in plants reacting to nanomaterials, the nanotoxicology community has taken a proactive approach. Nevertheless, numerous investigations have concentrated on the overall content within categories of metabolites, such as sugars, phenolics, flavonoids, chlorophylls, non-enzymatic antioxidants, and lignin, often utilizing less sensitive biochemical assays (Ali SS et al., 2021). To overcome these challenges and recognize the vital role of affected metabolites in response to ENMs, it becomes crucial to enhance the analytical coverage of the plant metabolome.

OBSTACLES DURING THE APPLICATION OF METABOLOMICS

Metabolomic analysis can be split into two main catagories that is untargeted and targeted while the choice among them depends on the goals of the study. Several key factors differentiate untargeted from targeted metabolomics:

1- Degree of sample preparation needed
2- Number of detected metabolites
3- Level of quantification
4- Extent of data processing after acquisition

Untargeted metabolomics adopts a discovery-oriented strategy, exploring the complete spectrum of metabolites within an organism (Rinschen MM et al., 2019). In investigations concerning nanotoxicity, untargeted metabolomics emerges as a valuable tool for unveiling previously unknown metabolites of importance and identifying potential markers while it enables researchers to generate hypotheses regarding the biological pathways implicated in the response to exposures to nanomaterials (Hew P S et al., 2024). Untargeted metabolomics faces practical hurdles due to the wide-ranging composition and physicochemical properties of plant metabolites, making simultaneous extraction and identification challenging. Moreover, the significant variability in molecular weight, spanning from femto- or picomolar concentrations for hormones to millimolar concentrations for organic acids and sugars, adds complexity, particularly in detecting less abundant metabolites, leading to sensitivity and accuracy challenges. In contrast, targeted metabolomics concentrates on specific classes of chemically characterized and biologically annotated metabolites (Allwood J W et al., 2021). It adopts a hypothesis-driven approach, analyzing a predefined set of known metabolites with heightened selectivity and sensitivity to address specific biochemical inquiries. The utilization of calibration and isotopically labeled internal standards facilitates the precise quantitation of the metabolites under investigation.

NANO-PARTICLES USED IN METABOLOMICS

Nano-Based Copper Particles (nCu)

Various research studies have investigated the reactions of diverse crop species to Copper-based Engineered Nanomaterials (ENMs) employing untargeted metabolomics. As an example, exposing plant roots to nano-Copper (nCu) concentrations between 10 to 20 mg L^{-1} for a week resulted in alterations in the accumulation of metabolites in the leaves (Keller A A et al., 2017). In this groundbreaking metabolomic investigation employing 1H-NMR, elevated levels of secondary metabolites associated with defense response and cell signaling, including 4-aminobutyrate (GABA), acetylglucosamine, and phenyl lactate, were identified (Majumdar S et al., 2021). Conversely, metabolites related to amino acid metabolism (Phenylalanine (Phe), Tyrosine (Tyr), Tryptophan (Trp), Arginine (Arg), Proline (Pro)), fatty acid biosynthesis, riboflavin metabolism, and flavonoid biosynthesis were suppressed. Moreover, the application of nano-Copper (nCu) ranging from 200 to 800 mg kg^{-1} induced changes in the metabolite profile of cucumber fruits, thereby affecting their nutritional quality (Zhao L et al., 2016). Analysis through GC-MS and 1H-NMR identified 107 and 10 metabolites, respectively, providing complementary platforms for drawing conclusions (Ahmad F et al., 2022). A plant treated with nCu disrupted 15 biological pathways in the fruits, with significant impacts on galactose metabolism and the tricarboxylic acid (TCA) cycle. Additionally, entire essential amino acid including proline, α-linolenic metabolism, and alkaloid biosynthesis were affected. An accumulation of proline and benzoic acid in the fruits suggests a response to oxidative stress (Zhao L et al., 2016).

Nano-Based Silver Particle (nAg)

The utilization of nanoengineered silver particles in metabolomics research involves studying the impact of silver nanoparticles (AgNPs) on the metabolic profile of biological systems. This approach allows for a detailed examination of how these nanoparticles influence the composition and concentration of various metabolites within organisms (Zhang J L et al., 2018). Researchers may employ advanced metabolomic techniques, such as mass spectrometry (MS) or nuclear magnetic resonance (NMR), to analyze the changes in metabolites when exposed to nano-based silver particles. This could include investigating alterations in amino acid profiles, lipid compositions, and other small molecules that play crucial roles in cellular functions. Metabolomic studies reveal that the detrimental effects observed in plants exposed to nano-silver (nAg) are primarily attributed to the release of silver ions (Ag^{+}ions) (Soria, NGC et al., 2017). Indeed, the exposure of Engineered Nanomaterials (ENMs), particularly nano-silver (nAg), not only affects the physiological processes in plants but also exerts an influence on the plant's metabolic profile. This influence can lead to the modulation of specific metabolites, potentially impacting pathways associated with fatty acids, amino acids, and secondary metabolites (Romero N et al., 2021). Additionally, the exposure to nAg can regulate the plant's stress response, influence nutrient homeostasis, and induce changes in gene expression. Understanding these multifaceted effects is crucial for comprehensively assessing the impact of nano-silver on plant health and function.

Nano-Based Iron Particle (nFe)

In the agricultural context, nanoparticles are under investigation in metabolomics studies to comprehensively comprehend their impact on plants, soil, and the overall agricultural ecosystem. Specifically, iron nanoparticles (nFe) are subjects of metabolomics research, exploring their potential to enhance nutrient availability in soils (Verma K K et al., 2022). The understanding of how nFe influences plant metabolism, nutrient uptake, and related aspects is pivotal for refining and optimizing agricultural practices. Presently, nFe is extensively utilized in plant metabolomic profiling, focusing on nutrient uptake, translocation, iron homeostasis, stress response, biofortification, as well as bioavailability and toxicity assessment (Heikal Y M et al., 2023). The precise and targeted delivery of nano-sized iron particles offers a methodical approach to fortification, ensuring optimal nutrient delivery to specific plant tissues (Shirsat S et al., 2023). This precision enhances the bioavailability of iron within the plant, making it more readily absorbable by the human body upon consumption (Blanco-Rojo R et al., 2019). As a critical component of chlorophyll, responsible for photosynthesis, improved iron availability through nano-sized iron particles (nFe) has the potential to boost photosynthetic efficiency, positively influencing metabolic processes dependent on this vital energy-capturing mechanism (Jafari A et al., 2022).

Research has demonstrated that exposure to nFe induces changes in the plant metabolome, leading to alterations in the concentrations of key metabolites, including amino acids, organic acids, and secondary metabolites crucial for plant metabolism (Yan L et al., 2020). Despite the promising potential of nFe in improving plant metabolism, ongoing research is necessary to fully comprehend the underlying mechanisms and to ensure the safe and responsible integration of nanotechnology in agriculture. Consideration of factors such as dosage, application methods, and plant species is imperative for optimizing the benefits of nFe for plant metabolic processes.

Nano-Based Zinc Particle (nZn)

Nano-based zinc particles (nZn) play a dual role in influencing plants, operating at both the biochemical and molecular levels and impacting various physiological processes. This includes enhancing nutrient uptake, antioxidant defense mechanisms, enzyme activation, stress response, photosynthesis, cellular signaling, and gene expression (Karnwal A et al., 2023). Nano-based zinc particles improves the availability of zinc in the soil, facilitating nutrient uptake and translocation across cell membranes (Lv W et al., 2023). Additionally, it influences the expression of genes associated with zinc transporters and uptake proteins, thereby modulating the molecular processes involved in these essential physiological functions (Hara T et al., 2017). In the realm of chlorophyll synthesis, growth, development, and stress responses, nZn affects antioxidant pathways and the synthesis of enzymes that counter ROS (Zia-ur-Rehman M et al., 2023). A comprehensive understanding of nZn's dual impact at both biochemical and molecular levels is vital for harnessing its potential benefits in enhancing plant growth, development, and stress tolerance. Ongoing research aims to uncover specific mechanisms, ensuring the responsible and effective application of nano-based zinc particles in agriculture.

Nano-Carbon Materials (CBNs)

Carbon-based nanomaterials (CBNs) have unique structural and electronic properties, and their applications have been explored in various fields, including electronics, materials science, and medicine while gained attention for their potential applications in agriculture due to their unique properties and versatile nature (Li Z et al., 2019). Carbon-based nanomaterials (CBNs) exhibit great potential as carriers for fertilizers owing to their stable molecular arrangement, even distribution, and low toxicity when applied (Aacharya R et al., 2020). A notable example is the use of graphene oxide nanoparticles, which prove to be efficient transporters for trace elements. Carbon-based nanomaterials (CBNs) that have been widely investigated in agriculture include.

1. Carbon Nanotubes

Research has delved into the potential of Carbon Nanotubes in enhancing plant growth and development. Investigations have scrutinized the influence of CNTs on seed germination, root elongation, and the overall growth of plants. It act as nanocarriers, facilitating the delivery of nutrients or other bioactive compounds to plants (Mathew S et al., 2021).

2. Graphene and Graphene Oxide (GO)

Extensive research has been conducted on graphene, which is a single layer of carbon atoms arranged in a hexagonal lattice, and its oxide counterpart, graphene oxide, to explore their potential applications in agriculture. These materials exhibit the capability to improve soil water retention, enhance nutrient delivery, and possess antimicrobial properties that could be advantageous for the health of plants (Zhao L et al., 2022).

3. Fullerences

Spherical carbon molecules known as fullerenes possess unique properties. The most prevalent fullerene, C60, has 60 carbon atoms arranged in a structure resembling a soccer ball, with 12 pentagons and 20 hexagons (Mohan Gokhale M et al., 2015). It has wider applications in materials science, nanotechnology, and potential applications in medicine and agriculture. .Although research on the use of fullerenes in agriculture is at a nascent stage, there is potential for them to enhance the efficiency of photosynthesis and act as carriers for delivering agricultural chemicals (Khamitova K K et al., 2019).

4. Activated Carbon Nanomaterials

Activated carbon nanomaterials, have the potential to be utilized in agriculture for various purposes. These applications encompass improving soil properties, enhancing water retention, increasing nutrient availability, aiding in carbon sequestration, and participating in environmental remediation through the adsorption of pollutants. it can contribute to soil enhancement by positively impacting its structure, aeration, and water-holding capacity (Mukherjee A et al., 2016). Their porous nature allows for increased water retention, particularly beneficial in regions prone to drought. Functionalized forms can act as carriers for nutrients, protecting them from leaching and ensuring a controlled release to enhance nutrient availability for plants. Additionally, their strong adsorption capabilities make them valuable for removing pollutants from the soil, preventing their uptake by plants and reducing environmental contamination.

5. Carbon Quantum Dots

Carbon Quantum Dots (CQDs), small carbon nanoparticles with unique quantum properties (particle size < 10 nm), have versatile applications (Devi P et al., 2019). They have been studied for their potential as:

a- Carbon Quantum Dots act as an catalyst to enhance the plant growth growth development via regulating the metabolic process.
b- It is utilized as a detection sensors for several purposes such as to detect nutrient in the soil and harmful compound present in food material.
c- CQDs explored for their ability to serve as carriers for delivering drugs to plants.

CQDs are known for their distinctive properties, including:

a- Fluorescence and Photoluminescence: Valuable for bioimaging and biosensing.
b- Electrochemical Activity: Suitable for use in electrochemical sensors.
c- Biocompatibility and Low Toxicity: Safe for biological applications.
d- High Solubility: Easily dispersible in various solvents.
e- Chemical Stability: Durable under different environmental conditions.
f- Surface Functionality: Allowing for easy modifications.

These properties of Carbon Quantum Dots make them more efficient for the use in wide areas such as biotechnology, medicine, sensing, environmental science and agriculture.

Silica Nanoparticles

Silica nanoparticles refer to nanoscale particles composed predominantly of silicon dioxide (SiO_2), which is similar to the compound found in quartz and sand. It exhibit unique properties due to their small size and high surface area. Silica nanoparticles typically have dimensions in the nanometer range, often occurs less than 100 nanometers (Hoang C V et al., 2022). They can exist in various shapes, including spherical, rod-shaped, or more complex structures. It has a wide range of application from biomedical to agriculture. Their smaller size allows them to navigate biological barriers, and their surface can be functionalized for targeted delivery. It als act as a catalyst supports in various chemical reactions, enhancing reaction rates and selectivity. Moreover, Silica nanoparticles may be employed in agriculture for enhancing nutrient delivery to plants or as carriers for pesticides. Here are some useful application in agriculture to boost the crop growth (Wang L et al., 2022).

a- Silica Nanoparticles as Fertilizer Additives: It can be incorporated into fertilizers or soil amendments to enhance nutrient uptake by plants.
b- Increased Resistance to Abiotic Stress: Silica nanoparticles have been shown to enhance plant resistance to various abiotic stresses such as drought, salinity, and metal toxicity by reducing the rate of transpiration and improving water retention in the soil (Wang L et al., 2022; Ghazaryan et al., 2023).

c- **Enhance Pest and Disease Resistance:** It induce systemic acquired resistance, making plants more resilient to various pathogens and reducing the need for chemical like pesticides and fungicides (Goswami P et al., 2022).

d- **Improve Water Use Efficiency [WUE]:** It improves WUE in the plant in plants by reducing transpiration rates, helping plants retain water and thrive in water-limited conditions (Alsaeedi A et al., 2022).

e- **Strengthening Plant Cell Walls:** Silica based nanoparticles, when deposited in the cell walls of plants, contribute to heightened structural rigidity. This reinforcement enhances the overall strength of the plant, rendering it more resistant to lodging or collapsing under its own weight. The strengthened cell walls provide additional support to the plant structure, contributing to improved stability and resilience (Sarkar M M et al., 2022).

f- **Seed Treatment:** Coating seeds with silica nanoparticles can enhance germination rates and seedling growth while the application can promote an early boost to the development of crops (Goswami P et al., 2022).

g- **Enhanced Antioxidant Defense:** Drought stress often leads the generation of reactive oxygen species (ROS) in plants, causing oxidative damage. Silica nanoparticles have antioxidant properties and can stimulate the activity of antioxidant enzymes, helping plants counteract oxidative stress and maintain cellular integrity (Sharf-Eldin A A et al., 2023).

h- **Upregulation of Stress-Responsive Genes:** Silica nanoparticles may trigger the expression of stress-responsive genes in plants. This activation of specific genes involved in stress tolerance pathways can improve a plant's ability to cope with drought conditions (Sharf-Eldin A A et al., 2023).

It is important to note that the specific mechanisms may vary from plant to plant, and the efficacy of silica nanoparticles in improving entire set of functions including drought resistance may depend on several factors such as concentration, application method, and the timing of application.

CONCLUSION

A Nano-Omics-based investigation holds significant promise concerning the growth, yield, and nutritional value of crops. This research involves the application of nanotechnology within omics disciplines, including genomics, transcriptomics, proteomics, and metabolomics. Over the past few decades, crop yields have experienced a substantial decline due to various factors, with adverse climatic conditions such as heat, cold, salinity, drought, and heavy metal stress being prominent contributors. Concurrently, the global population has seen a sharp increase. In this context, Nano-Omics-based studies offer potential solutions to mitigate challenges related to plant growth, yield, and nutritional value. In contemporary times, a diverse array of nanoparticles such as ZnO, TiO2, AgNO3, Al2O3, and Fe3O4 are readily available. These nanoparticles can be utilized in various forms, catering to specific requirements, playing a versatile role in plant fulfillment, either as fertilizers or as pesticides and fungicides. Due to their nano size, practical applications of nanoparticles are feasible through both soil and foliar applications. Nano-Omics-based manipulations initiate activities at the nanoscale, aiming to enhance plant productivity and nutritional quality through biochemical and molecular processes. These processes include moisture retention, reduction in transpiration rates, acceleration of nutrient use efficiency (NUE), suppression of reactive oxygen species (ROS) production, reinforcement of cell walls, and upregulation of genes.

Overall the integration of Nano-Omics into agriculture provides an opportunity for researchers and practitioners to revolutionize how challenges in crop production are addressed. This integration can lead to increased efficiency, sustainability, and resilience in response to evolving environmental conditions and growing food demands.

ACKNOWLEDGMENT

I am sincerely grateful for the significant contributions that have shaped this chapter, and I want to acknowledge the co-authors for their invaluable input. Additionally, I extend my appreciation to Lovely Professional University for providing an excellent environment that facilitated the compilation of this work. This milestone wouldn't have been possible without the support and resources provided by the university.

REFERENCES

Aacharya, R., & Chhipa, H. (2020). Nanocarbon fertilizers: Implications of carbon nanomaterials in sustainable agriculture production. In *Carbon Nanomaterials for Agri-Food and Environmental Applications* (pp. 297–321). Elsevier. doi:10.1016/B978-0-12-819786-8.00015-3

Abdel Latef, A. A. H., Zaid, A., Abu Alhmad, M. F., & Abdelfattah, K. E. (2020). The impact of priming with Al_2O_3 nanoparticles on growth, pigments, osmolytes, and antioxidant enzymes of Egyptian Roselle (*Hibiscus sabdariffa* L.) cultivar. *Agronomy (Basel)*, *10*(5), 681. doi:10.3390/agronomy10050681

Afzal, S., Singh, M. P., Chaudhary, N., & Singh, N. K. (2022). Application of nanoparticles in developing resilience against abiotic stress in rice plant (*Oryza sativa* L.). In *Plant Perspectives to Global Climate Changes* (pp. 151–172). Academic Press. doi:10.1016/B978-0-323-85665-2.00014-5

Ahmad, F., & Nadeem, H. (2022). Mass Spectroscopy as an analytical tool to harness the production of secondary plant metabolites: The way forward for drug discovery. In *Gene, Drug, and Tissue Engineering* (pp. 77–103). Springer US.

Ali, S. S., Al-Tohamy, R., Koutra, E., Moawad, M. S., Kornaros, M., Mustafa, A. M., & Sun, J. (2021). Nanobiotechnological advancements in agriculture and food industry: Applications, nanotoxicity, and future perspectives. *The Science of the Total Environment*, *792*, 148359. doi:10.1016/j.scitotenv.2021.148359 PMID:34147795

Allwood, J. W., Williams, A., Uthe, H., van Dam, N. M., Mur, L. A., Grant, M. R., & Pétriacq, P. (2021). Unravelling plant responses to stress—The importance of targeted and untargeted metabolomics. *Metabolites*, *11*(8), 558. doi:10.3390/metabo11080558 PMID:34436499

Alsaeedi, A., Elgarawani, M. M., Alshaal, T., & Elhawat, N. (2022). Silicon-and nanosilicon-mediated drought and waterlogging stress tolerance in plants. In *Silicon and Nano-silicon in Environmental Stress Management and Crop Quality Improvement* (pp. 121–152). Academic Press. doi:10.1016/B978-0-323-91225-9.00005-4

Aqeel, U., Aftab, T., Khan, M. M. A., Naeem, M., & Khan, M. N. (2022). A comprehensive review of impacts of diverse nanoparticles on growth, development and physiological adjustments in plants under changing environment. *Chemosphere, 291*, 132672. doi:10.1016/j.chemosphere.2021.132672 PMID:34756946

Baghalian, K., Hajirezaei, M. R., & Schreiber, F. (2014). Plant metabolic modeling: Achieving new insight into metabolism and metabolic engineering. *The Plant Cell, 26*(10), 3847–3866. doi:10.1105/tpc.114.130328 PMID:25344492

Blanco-Rojo, R., & Vaquero, M. P. (2019). Iron bioavailability from food fortification to precision nutrition. A review. *Innovative Food Science & Emerging Technologies, 51*, 126–138. doi:10.1016/j.ifset.2018.04.015

Cembrowska-Lech, D., Krzemińska, A., Miller, T., Nowakowska, A., Adamski, C., Radaczyńska, M., Mikiciuk, M., & Mikiciuk, M. (2023). An integrated multi-omics and artificial intelligence framework for advance plant phenotyping in horticulture. *Biology (Basel), 12*(10), 1298. doi:10.3390/biology12101298 PMID:37887008

Debnath, M., Pandey, M., & Bisen, P. S. (2011). An omics approach to understand the plant abiotic stress. *OMICS: A Journal of Integrative Biology, 15*(11), 739–762. doi:10.1089/omi.2010.0146 PMID:22122668

Devi, P., Saini, S., & Kim, K. H. (2019). The advanced role of carbon quantum dots in nanomedical applications. *Biosensors & Bioelectronics, 141*, 111158. doi:10.1016/j.bios.2019.02.059 PMID:31323605

Ditta, A., Arshad, M., & Ibrahim, M. (2015). Nanoparticles in sustainable agricultural crop production: applications and perspectives. *Nanotechnology and plant sciences: nanoparticles and their impact on plants*, 55-75.

Drostkar, E., Talebi, R., & Kanouni, H. (2016). Foliar application of Fe, Zn and NPK nano-fertilizers on seed yield and morphological traits in chickpea under rainfed condition. *Journal of Resources and Ecology, 4*, 221–228.

Feizi, H., Rezvani Moghaddam, P., Shahtahmassebi, N., & Fotovat, A. (2012). Impact of bulk and nano-sized titanium dioxide (TiO_2) on wheat seed germination and seedling growth. *Biological Trace Element Research, 146*(1), 101–106. doi:10.1007/s12011-011-9222-7 PMID:21979242

Fischer, R. A., & Connor, D. J. (2018). Issues for cropping and agricultural science in the next 20 years. *Field Crops Research, 222*, 121–142. doi:10.1016/j.fcr.2018.03.008

Franco-Lagos, C. L., Sanchez, E., Palacio-Marquez, A., Perez-Alvarez, S., Terrazas-Gomez, M., Villalobos-Cano, O., & Ramirez-Estrada, C. A. (2023). Efficacy of the application of boron nanofertilizer on biomass, yield, nitrogen assimilation and photosynthetic activity in green beans. *Notulae Botanicae Horti Agrobotanici Cluj-Napoca, 51*(1), 12795–12795. doi:10.15835/nbha51112795

Ghazaryan, K., Harutyunyan, A., Khachatryan, H., Singh, A., Minkina, T., Rajput, V., & Movsesyan, H. (2023). Study of tolerance and phytodesalination potential of wheat, oat, emmer, and barley for sustainable saline agriculture. *Applied Ecology and Environmental Research, 21*(5), 4853–4882. doi:10.15666/aeer/2105_48534882

Ghouri, M. Z., Khan, Z., Khan, S. H., Ismail, M., Aftab, S. O., Sultan, Q., & Ahmad, A. (2020). Nanotechnology: Transformation of agriculture and food security. *Bioscience*, *3*, 19.

Giannousi, K., Avramidis, I., & Dendrnou-Samara, C. (2013). Synthesis, characterization and evaluation of copper based nanoparticles as agrochemicals against Phytopthora infestans. *RSC Advances*, *3*(44), 21743–21752. doi:10.1039/c3ra42118j

Goswami, P., Mathur, J., & Srivastava, N. (2022). Silica nanoparticles as novel sustainable approach for plant growth and crop protection. *Heliyon*, *8*(7), e09908. doi:10.1016/j.heliyon.2022.e09908 PMID:35847613

Hara, T., Takeda, T. A., Takagishi, T., Fukue, K., Kambe, T., & Fukada, T. (2017). Physiological roles of zinc transporters: Molecular and genetic importance in zinc homeostasis. *The Journal of Physiological Sciences; JPS*, *67*(2), 283–301. doi:10.1007/s12576-017-0521-4 PMID:28130681

Hassanisaadi, M., Barani, M., Rahdar, A., Heidary, M., Thysiadou, A., & Kyzas, G. Z. (2022). Role of agrochemical-based nanomaterials in plants: Biotic and abiotic stress with germination improvement of seeds. *Plant Growth Regulation*, *97*(2), 375–418. doi:10.1007/s10725-021-00782-w

He, J., Li, J., Gao, Y., He, X., & Hao, G. (2023). Nano-based Smart Formulations: A potential solution to the Hazardous Effects of Pesticide on the Environment. *Journal of Hazardous Materials*, *456*, 131599. doi:10.1016/j.jhazmat.2023.131599 PMID:37210783

Heikal, Y. M., El-Esawi, M. A., El-Ballat, E. M., & Abdel-Aziz, H. M. (2023). Applications of nanoparticles for mitigating salinity and drought stress in plants: An overview on the physiological, biochemical and molecular genetic aspects. *New Zealand Journal of Crop and Horticultural Science*, *51*(3), 297–327. doi:10.1080/01140671.2021.2016870

Hew, P. S., Selamat, J., Jambari, N. N., Murugesu, S., Sanny, M., Khatib, A., & Sukor, R. (2024). Quality and safety of food product–current assessment, issues, and metabolomics as a way forward. *Food Chemistry Advances*, *4*, 100632. doi:10.1016/j.focha.2024.100632

Hoang, C. V., Thoai, D. N., Cam, N. T. D., Phuong, T. T. T., Lieu, N. T., Hien, T. T. T., Nhiem, D. N., Pham, T.-D., Tung, M. H. T., Tran, N. T. T., Mechler, A., & Vo, Q. V. (2022). Large-scale synthesis of Nanosilica from silica sand for plant stimulant applications. *ACS Omega*, *7*(45), 41687–41695. doi:10.1021/acsomega.2c05760 PMID:36406494

Hull, M., & Bowman, D. (Eds.). (2018). *Nanotechnology environmental health and safety: risks, regulation, and management*. William Andrew.

Jafari, A., & Hatami, M. (2022). Foliar-applied nanoscale zero-valent iron (nZVI) and iron oxide (Fe_3O_4) induce differential responses in growth, physiology, antioxidative defense and biochemical indices in Leonurus cardiaca L. *Environmental Research*, *215*, 114254. doi:10.1016/j.envres.2022.114254 PMID:36096173

Karnwal, A., Dohroo, A., & Malik, T. (2023). Unveiling the potential of bioinoculants and nanoparticles in sustainable agriculture for enhanced plant growth and food security. *BioMed Research International*, *2023*, 1–23. Advance online publication. doi:10.1155/2023/6911851 PMID:38075309

Keller, A. A., Adeleye, A. S., Conway, J. R., Garner, K. L., Zhao, L., Cherr, G. N., & Zuverza-Mena, N. (2017). Comparative environmental fate and toxicity of copper nanomaterials. *NanoImpact*, *7*, 28–40. doi:10.1016/j.impact.2017.05.003

Khamitova, K. K., Kayupov, B. A., Yegemova, S. S., Gabdullin, M. T., Abdullin, K. A., Ismailov, D. V., & Kerimbekov, D. S. (2019). The use of fullerenes as a biologically active molecule. *International Journal of Nanotechnology*, *16*(1-3), 100–108. doi:10.1504/IJNT.2019.102396

Khepar, V., Ahuja, R., Sidhu, A., & Samota, M. K. (2023). Nano-sulfides of Fe and Mn efficiently augmented the growth, antioxidant defense system, and metal assimilation in rice seedlings. *ACS Omega*, *8*(33), 30231–30238. doi:10.1021/acsomega.3c03012 PMID:37636944

Kolbert, Z., Szőllősi, R., Rónavári, A., & Molnár, Á. (2022). Nanoforms of essential metals: From hormetic phytoeffects to agricultural potential. *Journal of Experimental Botany*, *73*(6), 1825–1840. doi:10.1093/jxb/erab547 PMID:34922354

Kumar, P., & Dubey, K. K. (2019). Citric acid cycle regulation: Back bone for secondary metabolite production. In *New and Future Developments in Microbial Biotechnology and Bioengineering* (pp. 165–181). Elsevier. doi:10.1016/B978-0-444-63504-4.00013-X

Kumar, S. (2020). Abiotic stresses and their effects on plant growth, yield and nutritional quality of agricultural produce. *International Journal of Food Science and Agriculture*, *4*(4), 367–378. doi:10.26855/ijfsa.2020.12.002

Li, Z., Wang, L., Li, Y., Feng, Y., & Feng, W. (2019). Carbon-based functional nanomaterials: Preparation, properties and applications. *Composites Science and Technology*, *179*, 10–40. doi:10.1016/j.compscitech.2019.04.028

Liu, C., Zhou, H., & Zhou, J. (2021). The applications of nanotechnology in crop production. *Molecules (Basel, Switzerland)*, *26*(23), 7070. doi:10.3390/molecules26237070 PMID:34885650

Lv, W., Geng, H., Zhou, B., Chen, H., Yuan, R., Ma, C., & Wang, F. (2022). The behavior, transport, and positive regulation mechanism of ZnO nanoparticles in a plant-soil-microbe environment. *Environmental Pollution*, *315*, 120368. doi:10.1016/j.envpol.2022.120368 PMID:36216179

Majumdar, S., & Keller, A. A. (2021). Omics to address the opportunities and challenges of nanotechnology in agriculture. *Critical Reviews in Environmental Science and Technology*, *51*(22), 2595–2636. doi:10.1080/10643389.2020.1785264

Mathew, S., Tiwari, D. K., & Tripathi, D. (2021). Interaction of carbon nanotubes with plant system: A review. *Carbon Letters*, *31*(2), 167–176. doi:10.1007/s42823-020-00195-1

Maurel, C., Boursiac, Y., Luu, D. T., Santoni, V., Shahzad, Z., & Verdoucq, L. (2015). Aquaporins in plants. *Physiological Reviews*, *95*(4), 1321–1358. doi:10.1152/physrev.00008.2015 PMID:26336033

Mirzajani, F., Askari, H., Hamzelou, S., Schober, Y., Rompp, A., Ghassempour, A., & Spengler, B. (2014). Proteomics study of silver nanoparticles toxicity on *Oryza sativa* L. *Ecotoxicology and Environmental Safety*, *108*, 335–339. doi:10.1016/j.ecoenv.2014.07.013 PMID:25124680

Mohan Gokhale, M., & Ravindra Somani, R. (2015). Fullerenes: Chemistry and its applications. *Mini-Reviews in Organic Chemistry*, *12*(4), 355–366. doi:10.2174/1570193X12666150930224428

Mohiuddin, A. K. (2019). Chemistry of secondary metabolites. *Annals of Clinical Toxicology*, *2*(1), 1014.

Mukherjee, A., Majumdar, S., Servin, A. D., Pagano, L., Dhankher, O. P., & White, J. C. (2016). Carbon nanomaterials in agriculture: A critical review. *Frontiers in Plant Science*, *7*, 172. doi:10.3389/fpls.2016.00172 PMID:26941751

Nakabayashi, R., Hashimoto, K., Toyooka, K., & Saito, K. (2019). Keeping the shape of plant tissue for visualizing metabolite features in segmentation and correlation analysis of imaging mass spectrometry in Asparagus officinalis. Metabolomics. *Metabolomics*, *15*(2), 24. doi:10.1007/s11306-019-1486-5 PMID:30830429

Nasrollahzadeh, M., Sajadi, S. M., Sajjadi, M., & Issaabadi, Z. (2019). Applications of nanotechnology in daily life. *Interface Science and Technology, 28*, 113-143.

Nguyen, N. H., & Falagan-Lotsch, P. (2023). Mechanistic Insights into the Biological Effects of Engineered Nanomaterials: A Focus on Gold Nanoparticles. *International Journal of Molecular Sciences*, *24*(4), 4109. doi:10.3390/ijms24044109 PMID:36835521

Nhan, L. V., Ma, C., Rui, Y., Liu, S., Li, X., Xing, B., & Liu, L. (2015). Phytotoxic\ mechanism of nanoparticles: Destruction of chloroplasts and vascular bundles and alteration of nutrient absorption. *Scientific Reports*, *5*(1), 11618. doi:10.1038/srep11618 PMID:26108166

Nie, D., Li, J., Xie, Q., Ai, L., Zhu, C., Wu, Y., & Tan, W. (2023). Nanoparticles: A Potential and Effective Method to Control Insect-Borne Diseases. *Bioinorganic Chemistry and Applications*, *2023*, 1–13. Advance online publication. doi:10.1155/2023/5898160 PMID:37213220

Pandita, D., & Pandita, A. (2021). Secondary metabolites in medicinal and aromatic plants (MAPs): potent molecules in nature's arsenal to fight human diseases. *Medicinal and Aromatic Plants: Healthcare and Industrial Applications*, 41-84.

Pandohee, J., Basu, R., Dasgupta, S., Sundarrajan, P., Shaikh, N., Patel, N., & Noor, A. (2023). Applications of multi-omics approaches for food and nutritional security. In *Sustainable Agriculture in the Era of the OMICs Revolution* (pp. 103–118). Springer International Publishing. doi:10.1007/978-3-031-15568-0_5

Patil, H. J., & Solanki, M. K. (2016). Microbial inoculant: modern era of fertilizers and pesticides. Microbial inoculants in sustainable agricultural productivity: vol. 1: research perspectives, 319-343.

Prasad, M., Lambe, U. P., Brar, B., Shah, I., Manimegalai, J., Ranjan, K., & Prasad, G. (2018). Nanotherapeutics: An insight into healthcare and multi-dimensional applications in medical sector of the modern world. *Biomedicine and Pharmacotherapy*, *97*, 1521–1537. doi:10.1016/j.biopha.2017.11.026 PMID:29793315

Rajput, P., Kumar, P., Priya, A., Kumari, S., Shiade, S. R. G., Rajput, V. D., Fathi, A., Pradhan, A., Sarfraz, R., Sushkova, S., Mandzhieva, S., Minkina, T., Soldatov, A., Wong, M. H., & Rensing, C. (2024). Nanomaterials and biochar mediated remediation of emerging contaminants. *The Science of the Total Environment*, *916*, 170064. doi:10.1016/j.scitotenv.2024.170064 PMID:38242481

Raliya, R., & Tarafdar, J. C. (2013). ZnO nanoparticle biosynthesis and its effect on phosphorous-mobilizing enzyme secretion and gum contents in clusterbean (*Cyamopsis tetragonoloba* L.). *Agricultural Research*, *2*(1), 48–57. doi:10.1007/s40003-012-0049-z

Rani, A., Rani, K., Tokas, J., Anamika, Singh, A., Kumar, R., & Kumar, S. (2020). Nanomaterials for agriculture input use efficiency. *Resources use efficiency in agriculture*, 137-175.

Raza, A. (2022). Metabolomics: A systems biology approach for enhancing heat stress tolerance in plants. *Plant Cell Reports*, *41*(3), 741–763. doi:10.1007/s00299-020-02635-8 PMID:33251564

Rinschen, M. M., Ivanisevic, J., Giera, M., & Siuzdak, G. (2019). Identification of bioactive metabolites using activity metabolomics. *Nature Reviews. Molecular Cell Biology*, *20*(6), 353–367. doi:10.1038/s41580-019-0108-4 PMID:30814649

Rodziewicz, P., Swarcewicz, B., Chmielewska, K., Wojakowska, A., & Stobiecki, M. (2014). Influence of abiotic stresses on plant proteome and metabolome changes. *Acta Physiologiae Plantarum*, *36*(1), 1–19. doi:10.1007/s11738-013-1402-y

Romero, N., Castro, G. R., & Gagneten, A. M. (2021). Ecotoxicologic effects of silver nanoparticles on freshwater nontarget species. In *New Trends in Removal of Heavy Metals from Industrial Wastewater* (pp. 705–733). Elsevier. doi:10.1016/B978-0-12-822965-1.00028-3

Sahoo, J. P., Behera, L., Sharma, S. S., Praveena, J., Nayak, S. K., & Samal, K. C. (2020). Omics studies and systems biology perspective towards abiotic stress response in plants. *American Journal of Plant Sciences*, *11*(12), 2172–2194. doi:10.4236/ajps.2020.1112152

Salinas, M., & Baudet, C. (2020). The Challenges of Agriculture: Feeding the World of Tomorrow, on a Transitioning and Endangered Earth. *OIDA International Journal of Sustainable Development*, *13*(5), 21–58.

Sarkar, M. M., Mathur, P., Mitsui, T., & Roy, S. (2022). A review on functionalized silica nanoparticle amendment on plant growth and development under stress. *Plant Growth Regulation*, *98*(3), 421–437. doi:10.1007/s10725-022-00891-0

Shafi, A., Qadir, J., Sabir, S., Zain Khan, M., & Rahman, M. M. (2020). Nanoagriculture: a holistic approach for sustainable development of agriculture. Handbook of Nanomaterials and Nanocomposites for Energy and Environmental Applications, 1-16.

Sharf-Eldin, A. A., Alwutayd, K. M., El-Yazied, A. A., El-Beltagi, H. S., Alharbi, B. M., Eisa, M. A., & Ibrahim, M. F. (2023). Response of maize seedlings to silicon dioxide nanoparticles (SiO2NPs) under drought stress. *Plants*, *12*(14), 2592. doi:10.3390/plants12142592 PMID:37514207

Shirsat, S., & K, S. (2023). Iron oxide nanoparticles as iron micronutrient fertilizer—Opportunities and limitations. *Journal of Plant Nutrition and Soil Science*, jpln.202300203. Advance online publication. doi:10.1002/jpln.202300203

Singh, A., Agrawal, S., Rajput, V. D., Ghazaryan, K., Movsesyan, H. S., Minkina, T., . . . Gupta, S. K. (2023c). Seed Nanopriming. *Nanopriming Approach to Sustainable Agriculture*, 290–313. doi:10.4018/978-1-6684-7232-3.ch013

Singh, A., Rajput, V. D., Pandey, D., Sharma, R., Ghazaryan, K., & Minkina, T. (2023a). Nano Zinc-Enabled Strategies in Crops for Combatting Zinc Malnutrition in Human Health. *Frontiers in Bioscience (Landmark Edition)*, *28*(8), 158. doi:10.31083/j.fbl2808158 PMID:37664935

Singh, A., Rajput, V. D., Sharma, R., Ghazaryan, K., & Minkina, T. (2023d). Salinity stress and nanoparticles: Insights into antioxidative enzymatic resistance, signaling, and defense mechanisms. *Environmental Research*, *235*, 116585. doi:10.1016/j.envres.2023.116585 PMID:37437867

Singh, A., Rajput, V. D., Varshney, A., Ghazaryan, K., & Minkina, T. (2023e). Small Tech, Big Impact: Agri-nanotechnology Journey to Optimize Crop Protection and Production for Sustainable Agriculture. *Plant Stress*, *10*, 100253. doi:10.1016/j.stress.2023.100253

Singh, A., Rawat, S., Rajput, V. D., Ghazaryan, K., Minkina, T., Al Tawaha, A. R. M., . . . Sandhu, K. S. (2023b). Nanoparticle-Based Priming-Enhanced Agricultural Practices. *Nanopriming Approach to Sustainable Agriculture*, 142–157. doi:10.4018/978-1-6684-7232-3.ch006

Singh, A., Sharma, R., Rajput, V. D., Ghazaryan, K., Minkina, T., Al Tawaha, A. R. M., & Varshney, A. (2023f). Green Synthesis of Nanofertilizers and Their Application for Crop Production. *Nanotechnology in the Life Sciences*, 205–231. doi:10.1007/978-3-031-41329-2_8

Singh, A., Sharma, R., Rawat, S., Singh, A. K., Rajput, V. D., Fedorov, Y., . . . Chaplygin, V. (2022). Nanomaterial-plant interaction: Views on the pros and cons. *Toxicity of Nanoparticles in Plants*, 47–68. doi:10.1016/B978-0-323-90774-3.00015-5

Soria, N. G. C., Montes, A., Bisson, M. A., Atilla-Gokcumen, G. E., & Aga, D. S. (2017). Mass spectrometry-based metabolomics to assess uptake of silver nanoparticles by Arabidopsis thaliana. *Environmental Science. Nano*, *4*(10), 1944–1953. doi:10.1039/C7EN00555E

Steiner, A., Aguilar, G., Bomba, K., Bonilla, J. P., Campbell, A., Echeverria, R., & Zebiak, S. (2020). Actions to transform food systems under climate change. https://hdl.handle.net/10568/108489

Thirunavukkarasu, M., Subramanian, K. S., Kannan, P., & Balaji, T. (2018). Response of nano-sulphur to the groundnut. *International Journal of Chemical Studies*, *6*(3), 2067–2072.

Tombuloglu, H., Slimani, Y., Tombuloglu, G., Almessiere, M., & Baykal, A. (2019). Uptake and translocation of magnetite (Fe3O4) nanoparticles and its impact on photosynthetic genes in barley (Hordeum vulgare L.). *Chemosphere*, *226*, 110–122. doi:10.1016/j.chemosphere.2019.03.075 PMID:30925403

Usman, M., Farooq, M., Wakeel, A., Nawaz, A., Cheema, S. A., Rehman, H., Ashraf, I., & Sanaullah, M. (2020). Nanotechnology in agriculture: Current status, challenges and future opportunities. *The Science of the Total Environment*, *721*, 137778. doi:10.1016/j.scitotenv.2020.137778 PMID:32179352

Vannini, C., Domingo, G., Onelli, E., De Mattia, F., Bruni, I., Marsoni, M., & Bracale, M. (2014). Phytotoxic and genotoxic effects of silver nanoparticles exposure on germinating wheat seedlings. *Journal of Plant Physiology*, *171*(13), 1142–1148. doi:10.1016/j.jplph.2014.05.002 PMID:24973586

Vanti, G. L., Nargund, V. B., N, B. K., Vanarchi, R., Kurjogi, M., Mulla, S. I., Tubaki, S., & Patil, R. R. (2018). Synthesis of *Gossypium hirsutum*-derived silver nanoparticles and their antibacterial efficacy against plant pathogens. *Applied Organometallic Chemistry*, *33*(1), e4630. doi:10.1002/aoc.4630

Vardumyan, H., Singh, A., Rajput, V., Minkina, T., El-Ramady, H., & Ghazaryan, K. (2024). Additive-Mediated Phytoextraction of Copper-Contaminated Soils Using Medicago lupulina L. *Egyptian Journal of Soil Science*, *64*(2), 599–618. doi:10.21608/ejss.2024.266169.1714

Verma, K. K., Song, X. P., Joshi, A., Tian, D. D., Rajput, V. D., Singh, M., & Li, Y. R. (2022). Recent trends in nano-fertilizers for sustainable agriculture under climate change for global food security. *Nanomaterials (Basel, Switzerland)*, *12*(1), 173. doi:10.3390/nano12010173 PMID:35010126

Vurro, M., Miguel-Rojas, C., & Pérez-de-Luque, A. (2019). Safe nanotechnologies for increasing the effectiveness of environmentally friendly natural agrochemicals. *Pest Management Science*, *75*(9), 2403–2412. doi:10.1002/ps.5348 PMID:30672106

Wang, L., Ning, C., Pan, T., & Cai, K. (2022). Role of silica nanoparticles in abiotic and biotic stress tolerance in plants: A review. *International Journal of Molecular Sciences*, *23*(4), 1947. doi:10.3390/ijms23041947 PMID:35216062

Yan, L., Li, P., Zhao, X., Ji, R., & Zhao, L. (2020). Physiological and metabolic responses of maize (*Zea mays*) plants to Fe_3O_4 nanoparticles. *The Science of the Total Environment*, *718*, 137400. doi:10.1016/j.scitotenv.2020.137400 PMID:32105936

Yerasu, S. R., Reddy, B. R., Singh, D. P., & Singh, J. (2022). Omics studies for vegetable improvement. In *Omics in Horticultural Crops* (pp. 215–238). Academic Press., doi:10.1016/B978-0-323-89905-5.00018-5

Zhang, J. L., Zhou, Z. P., Pei, Y., Xiang, Q. Q., Chang, X. X., Ling, J., & Chen, L. Q. (2018). Metabolic profiling of silver nanoparticle toxicity in Microcystis aeruginosa. *Environmental Science. Nano*, *5*(11), 2519–2530. doi:10.1039/C8EN00738A

Zhao, L., Huang, Y., Zhou, H., Adeleye, A. S., Wang, H., Ortiz, C., & Keller, A. A. (2016). GC-TOF-MS based metabolomics and ICP-MS based metallomics of cucumber (Cucumis sativus) fruits reveal alteration of metabolites profile and biological pathway disruption induced by nano copper. *Environmental Science. Nano*, *3*(5), 1114–1123. doi:10.1039/C6EN00093B

Zhao, L., Wang, W., Fu, X., Liu, A., Cao, J., & Liu, J. (2022). Graphene oxide, a novel nanomaterial as soil water retention agent, dramatically enhances drought stress tolerance in soybean plants. *Frontiers in Plant Science*, *13*, 810905. doi:10.3389/fpls.2022.810905 PMID:35242153

Zia-ur-Rehman, M., Anayatullah, S., Irfan, E., Hussain, S. M., Rizwan, M., Sohail, M. I., Jafir, M., Ahmad, T., Usman, M., & Alharby, H. F. (2023). Nanoparticles assisted regulation of oxidative stress and antioxidant enzyme system in plants under salt stress: A review. *Chemosphere*, *314*, 137649. doi:10.1016/j.chemosphere.2022.137649 PMID:36587917

Section 2

Role of Nanotechnological Aspects, Nano-Omics, and Nano-Enzymes in Ensuring Soil Improvement for Agricultural Sustainability

Chapter 3
Enhancing Nutrient Uptake With Nano Fertilizers and Soil Amendments

Bhavana Tomar
G.D. Goenka University, India

Tirunima Patle
https://orcid.org/0000-0003-3268-6118
Shri Vaishnav Vidyapeeth Vishwavidyalaya, India

Sneh Singh Parihar
ITM University, Gwalior, India

Prashant Singh
https://orcid.org/0009-0008-9667-3598
Mandsaur University, India

Shiv Singh Tomar
G.D. Goenka University, India

ABSTRACT

Traditional fertilizers often lead to nutrient runoff, leaching, and soil degradation, exacerbating environmental issues such as water pollution and greenhouse gas emissions. In this chapter's introduction, the authors highlight the pressing challenges facing global agriculture, including rapid population growth, climate change, and resource constraints, necessitating innovative solutions for sustainable food production. Nanotechnology emerges as a promising avenue to address these challenges, offering tailored approaches to nutrient management and soil conditioning. Nano-fertilizers, nanocarriers, and nanostructures present novel solutions by leveraging the unique properties of nanoparticles. These nanoparticles provide controlled release mechanisms, improved solubility, and targeted delivery, optimizing nutrient uptake efficiency and minimizing environmental impact. The chapter delves into the formulation and composition of nano-fertilizers, exploring advanced techniques such as nanoencapsulation and stimuli-responsive materials. Mechanisms of nutrient release and uptake by plants are elucidated, showcasing how nanoparticles interact with plant roots, modify rhizosphere chemistry, and enhance mycorrhizal associations. Moreover, the role of nanoparticles in promoting nutrient uptake, particularly micronutrients, underscores their potential to improve crop growth and development, even in adverse conditions like saline soils. Furthermore, nano-enhanced soil amendments offer a transformative approach to soil management, contributing to soil health, structure, and fertility. By integrating nanotechnology into agricultural practices, researchers aim to foster more efficient, sustainable, and environmentally friendly farming methods. Ultimately, the chapter sets the stage for exploring the innovative potential of nanotechnology in revolutionizing agriculture and ensuring food security in the face of mounting global challenges.

DOI: 10.4018/979-8-3693-1890-4.ch003

INTRODUCTION

Agriculture, the backbone of human civilization, has undergone remarkable transformations over the centuries. From traditional farming methods to the Green Revolution, technological advancements have played a pivotal role in enhancing crop yields and ensuring food security (Yadav et al.,2023). As we stand on the precipice of the 21st century, the challenges facing global agriculture are more complex than ever. The global farming community is facing an unprecedented convergence of challenges, from rapid population growth, climate change, soil degradation, and resource constraints, demands an innovative solution to sustainably meet the world's growing food needs. Nutrient management is a pivotal component of crop production, exerting a profound influence on plant growth, development, and overall yield. Essential nutrients, including nitrogen, phosphorus, potassium, and micronutrients, play a critical role in sustaining healthy crops (Khatri and Bhateria.,2023). However, conventional nutrient management faces challenges, including issues like leaching, volatilization, and fixation, which compromise the efficiency of nutrient delivery to plants and also leads to soil degradation, and environmental pollution (Gade et al.,2023). Imbalances in nutrient availability can adversely affect crop performance, impacting both quality and quantity of yields. Furthermore, environmental concerns, such as water pollution and greenhouse gas emissions, are associated with the excessive use of traditional fertilizers (Sekhon.,2014).

In this context, nanotechnology rises to the occasion as a game-changer, offering a tailored response to the intricate problems that have long hindered agricultural progress (Singh et al., 2022). The infusion of nanotechnology has ushered in a transformative era, marking a significant departure from conventional practices. It can be a promising solution to these challenges by providing innovative approaches to nutrient management in agriculture. Nano-fertilizers, with their tailored formulations, offer controlled and sustained release of nutrients, addressing the limitations of conventional fertilizers (Ozer et al.,2014). Nanoparticles play a crucial role in enhancing nutrient uptake efficiency by facilitating better absorption through plant roots, thereby contributing to improved crop yields. Nano-sized materials, specifically designed for agricultural applications, have shown immense potential in enhancing nutrient uptake, improving soil health, and ultimately boosting crop yields and addressing the escalating challenges faced by the global farming community. Nano fertilizers and soil amendments, purpose-built for agricultural challenges, stand as avant-garde tools that not only address the immediate concerns of nutrient management and soil health but also hold the key to fostering a more sustainable and resilient agricultural future (Nandhakumar, 2023).

Nanotechnology, operating at the nanoscale, involves the manipulation and control of materials with dimensions less than 100 nanometers. This discipline offers unique properties and behaviors at this scale, and in agriculture, it holds promise for revolutionizing various aspects of crop production. Nanotechnology applications in agriculture include the development and utilization of nanomaterials, such as nanoparticles, nanocarriers, and nanostructures (kumar et al., 2023; Rajput et al., 2021).

Nanozymes are nanostructures with enzyme-like catalytic activities. They possess several advantages over natural enzymes, such as higher stability, reusability, and tunable catalytic properties. Nanozymes have been extensively explored for various applications in biomedical, environmental, and agricultural fields. In the context of plant nutrition, nanozymes can play a significant role in improving nutrient uptake efficiency. They can be used to enhance the availability and accessibility of essential nutrients for plant growth and development. Nanozymes can catalyze reactions that convert nutrients into forms that are more easily absorbed by plants or facilitate nutrient transport across cell membranes.

For example, nanozymes can be employed to enhance the uptake of micronutrients like iron and zinc, which are essential for plant metabolism. By catalyzing the conversion of insoluble forms of these nutrients into soluble forms, nanozymes can increase their bioavailability to plants, thereby promoting better growth and productivity. Furthermore, nanozymes can also be engineered to target specific nutrient deficiencies in plants, allowing for tailored nutrient supplementation strategies. This targeted approach can help optimize nutrient uptake and utilization, leading to improved crop yields and overall plant health.

The precision afforded by nanotechnology extends to the realm of precision agriculture, where targeted delivery of nutrients aligns with the specific requirements of crops. This precision not only optimizes resource utilization but also minimizes environmental impact (Oleg et al., 2017). In essence, nanotechnology in agriculture represents a transformative pathway towards more efficient, sustainable, and environmentally friendly farming practices (Singh et al., 2023a). It holds the potential to usher in a new era of precision-driven agriculture that meets the global demand for increased food production while minimizing adverse environmental effects. Nano fertilizers and soil amendments herald a new era of in agriculture, reshaping the conventional landscape of nutrient management and soil conditioning(Cicek and Nadaroglu.,2015). Unlike their traditional counterparts, these Nano-formulations are not bound by a one-size-fits-all approach. Instead, they epitomize the essence of nanotechnology's gift to agriculture by offering an unprecedented level of precision in delivering nutrients and enhancing soil quality(Mukhopadhyay.,2014). As we delve deeper into this chapter, the innovative potential of Nano fertilizers and soil amendments unfolds, promising a future where agriculture not only adapts to change but thrives amidst it, propelled by the precision and efficacy of Nano-technological solutions(Srilatha.,2014).

Types of Nanoparticles Used in Agriculture

Nano-fertilizers are designed to address the limitations of conventional fertilizers. These nanoparticles encapsulate or carry essential nutrients, allowing for controlled release and improved efficiency in nutrient utilization by plants. They offer targeted nutrient delivery, minimizing wastage and environmental impact. (Gogos et al., 2012)

- Nanocarriers serve as delivery vehicles for various agricultural inputs, such as pesticides, herbicides, and growth regulators. These nanoparticles can encapsulate active ingredients, protecting them from degradation and ensuring precise delivery to the intended target, reducing the overall usage of agrochemicals (Singh et al., 2023 b, c).
- Nanostructures, including nanoclays and nanocomposites, are employed to enhance soil properties. These nanoparticles improve soil structure, water retention, and nutrient availability, contributing to sustainable soil management practices (Khodakovskaya et al., 2009).

Properties That Make Nanoparticles Effective

Nanoparticles in agriculture, including nano-fertilizers and nanocarriers, offer a sophisticated approach to nutrient management (Singh et al., 2023d). Their unique properties, such as high surface area, controlled release, improved solubility, and targeted delivery, make them effective tools for enhancing nutrient delivery to plants and improving overall agricultural sustainability (Rico et al., 2011).

Nanoparticles possess an exceptionally high surface area compared to their larger counterparts. This characteristic enhances the contact area with plant roots, facilitating efficient nutrient absorption (Singh et al., 2024a). Nano-fertilizers and nanocarriers can be engineered to provide controlled and sustained release of nutrients. This controlled release aligns with the specific needs of plants, optimizing nutrient uptake and minimizing nutrient losses through leaching or volatilization. Many nanoparticles exhibit improved solubility characteristics, enhancing the availability of nutrients in the soil. (Wang et al., 2016) This increased solubility contributes to better nutrient absorption by plant roots. They allow for targeted delivery of nutrients to specific plant tissues. This targeted approach minimizes off-target effects and ensures that nutrients reach the areas where they are most needed, maximizing their impact on crop growth and development. Nanocarriers provide protection to encapsulated active ingredients, such as pesticides or growth regulators, from environmental factors. This protection ensures the stability and efficacy of these inputs, leading to improved pest control and crop protection.

Formulation and Composition of Nano-Fertilizers

Research in the field of nano-fertilizers highlights their potential in optimizing nutrient management through advanced formulations, controlled release mechanisms, and improved nutrient uptake efficiency compared to traditional fertilizers (Singh et al., 2024b). Nano-fertilizers can be composed of various nanoparticles, such as metal and metal oxide nanoparticles, carbon-based nanoparticles, and polymer-based nanoparticles. (Ramiyan et al., 2023). For example, zinc oxide nanoparticles have been explored for their potential as carriers for zinc nutrients (Gauba et al., 2023). Nano NPK formulations are designed to enhance the delivery of essential nutrients. Studies have investigated the formulation of nano NPK fertilizers using techniques like co-precipitation and sol-gel methods (Prasad et al., 2014).

Encapsulation of nutrients in nano-fertilizers involves advanced techniques. Nanoencapsulation methods, such as coacervation and layer-by-layer assembly, have been employed to protect nutrients from leaching and environmental degradation (Sinha et al., 2019). Smart nano-fertilizers utilize responsive materials that release nutrients in response to specific environmental cues. Stimuli-responsive polymer-based nanomaterials have been explored for controlled nutrient release (Jain et al., 2016).

Mechanisms of Nutrient Release and Uptake by Plants

The gradual and sustained delivery of nutrients leads to controlled release of nano particle. Nanoparticles can control the release rate, influencing nutrient availability for plant uptake (Marchiole et al., 2020). Facilitates enhanced nutrient uptake is by the unique properties of nanoparticles. For instance, engineered nanoparticles have been shown to improve phosphorus uptake by modifying the rhizosphere chemistry (Wang et al., 2016). Nano-particles interact with plant roots at the nanoscale, affecting root morphology and nutrient absorption. Studies have investigated the role of nano-fertilizers in promoting root growth and nutrient uptake (de la Rosa et al., 2019). Nano-fertilizers exhibit higher nutrient use efficiency compared to traditional fertilizers. Research has shown improved nitrogen use efficiency with nano-fertilizers in comparison to conventional nitrogen sources (Javed et al., 2022. Nano-fertilizers contribute to reduced environmental impact by minimizing nutrient runoff and leaching. Studies have highlighted the potential of nano-fertilizers in reducing nutrient losses and environmental pollution (Gogos et al., 2012). Assessing the long-term benefits of nano-fertilizers. While initial costs may be higher, the extended nutrient release and efficiency may result in economic benefits over time (Iqbal et al., 2019). Nano-fertilizers show adaptability to various soil conditions, ensuring improved nutrient availability. Investigations into the interaction between nano-fertilizers and different soil types contribute to understanding their adaptability (Fatima et al., 2021).

Role of Nanostructures in Enhancing Nutrient Uptake

Enhanced nutrient uptake mechanisms through nanotechnology involves the precise delivery of nutrients, improved solubility and bioavailability, facilitated nutrient uptake pathways, and increased nutrient use efficiency. The interaction between nanoparticles and plant roots encompasses modifications in root surfaces, rhizosphere chemistry, promotion of mycorrhizal associations, and the cellular uptake of nanoparticles. The nanostructures can encapsulate nutrients, protecting them from environmental factors and ensuring a controlled release. This controlled release provides a sustained and targeted supply of nutrients, optimizing plant uptake (Sun et al., 2023). The nano-ZnO significantly enhanced the nutrient uptake efficiency of maize plants compared to traditional zinc fertilizers. The nanoparticles facilitated better absorption through plant roots, resulting in increased zinc content in the plant tissues this demonstrate the potential of nano-fertilizers to improve nutrient uptake, particularly micronutrients, leading to enhanced plant growth and development. (Rizwan et al. 2018). Nanoparticles can enhance the solubility and bioavailability of certain nutrients. For instance, engineered nanoparticles can alter the chemical form of nutrients in the soil, making them more accessible to plant roots. This improved solubility increases the chances of nutrient uptake by plants, especially in soils with low nutrient availability (Liu et al., 2021). The nanoscale size and unique properties of nanoparticles enable them to interact directly with plant cell membranes. This interaction facilitates nutrient uptake pathways, potentially enhancing the absorption of essential elements. Nanoparticles may play a role in promoting the efficiency of nutrient absorption through roots (Khodakovskaya et al., 2012). The silica nanoparticles were applied as a nano-fertilizer, significantly increased wheat yields compared to traditional fertilizers. The nanoparticles improved water retention in the soil and enhanced nutrient availability, enabling better crop performance in saline environments. Nano-fertilizers demonstrates the potential to address specific challenges such as saline soils, contributing to increased crop productivity in adverse conditions (Wang et al., 2017). Nano-fertilizers contribute to increased nutrient use efficiency by delivering nutrients directly to the targeted plant tissues. This targeted delivery minimizes losses through leaching or volatilization, ensuring that a higher proportion of applied nutrients is utilized by plants. The improved efficiency results in enhanced plant growth and development (Rizwan et al., 2020, Frontiers in Plant Science). Nano-fertilizers with controlled release mechanisms contribute to efficient nutrient utilization, reducing environmental impact and improving crop productivity. The crops treated with nano-urea showed improved nitrogen use efficiency and higher yields compared to those treated with conventional urea nano-urea exhibited a slower and sustained release of nitrogen over time. (Prasad et al., 2016).

Interaction Between Nanoparticles and Plant Roots

Nanoparticles interact with the root surface, leading to modifications in root morphology and structure. This interaction can increase the surface area of roots, creating more sites for nutrient absorption. Enhanced root structures contribute to improved nutrient uptake by plants, particularly in nutrient-deficient soils (Wang et al., 2015). The application of nano-fertilizers containing iron nanoparticles significantly increased iron uptake by rice plants cultivation highlight the potential of nano particles in addressing specific nutrient deficiencies and improving crop yields in real-world agricultural in Southeast Asia (Singh et al., 2021). Nanoparticles influence the chemical environment in the rhizosphere, the region around plant roots. This modification in rhizosphere chemistry can alter the availability of nutrients, making them more accessible to plant roots. The improved nutrient environment in the rhizosphere enhances nutrient uptake by plants (Raliya et al., 2016). Nano-fertilizers can positively influence nutrient uptake, leading to improved crop quality and nutritional content, the nano-hydroxyapatite enhanced phosphorus uptake by the plants, leading to improved growth, yield, and nutritional quality of tomatoes compared to conventional phosphorus fertilizers (Jha et al., 2023). Mycorrhizal associations in the rhizosphere, fostering symbiotic relationships between plants and mycorrhizal fungi also get influence by nanoparticles. Mycorrhizal associations enhance nutrient absorption, especially phosphorus, as these fungi can facilitate nutrient transport to plant roots. Nano-fertilizers may play a role in promoting and optimizing these beneficial associations (Bhardwaj et al., 2022). Nanoparticles can be taken up by plant cells through endocytosis, a cellular process where the cell engulfs particles. This cellular uptake mechanism allows nanoparticles to enter plant cells, potentially influencing plant physiology. Understanding the intricate details of nanoparticle endocytosis provides insights into how nanoparticles interact with plant cells at the molecular level (Giraldo et al., 2014).

Nano-Enhanced Soil Amendments and Quality

The quest for sustainable agriculture has led scientists and researchers to explore innovative solutions, and one such frontier is the integration of nanotechnology into soil management practices. Nano-enhanced soil amendments represent a groundbreaking approach to improve soil quality and foster enhanced agricultural productivity, meticulously integrated into soil amendments, contribute to the enhancement of soil health, structure, and overall fertility (Patle et al., 2023; Ghazaryan et al., 2023).

One of the key challenges in agriculture is optimizing nutrient delivery to plants. Nano-enhanced soil amendments provide a platform for precisely targeted nutrient delivery. Nano-fertilizers, for instance, can be engineered to release nutrients gradually, ensuring a sustained supply to plants (Patle et al.,2023) The increased surface area of nano-sized particles also enhances nutrient bioavailability, allowing plants to absorb essential elements more efficiently. Nano-enhanced soil amendments offer innovative solutions for improving soil structure, water retention, and nutrient availability (Rajput et al 2022). The introduction of nanostructures into the soil has the potential to address various soil-related challenges, promoting sustainable agricultural practices and contributing to environmental conservation. It involves the incorporation of nanomaterials into the soil to improve its properties. Nanostructures, such as nanoclays and nanocomposites, are designed to address specific soil-related challenges and enhance overall soil quality (Hasan et al., 2020). applications of nano materials various aspects of soil management, including soil structure improvement, water retention enhancement, and nutrient availability optimization. At the molecular level, Nano-sized materials introduce a level of precision previously unattainable. By interacting with soil particles, nanostructures work to mitigate issues such as compaction, promoting aeration and water infiltration. This molecular approach not only enhances the physical properties of the soil but also creates a favorable environment for root development and nutrient absorption.

These applications contribute to sustainable agricultural practices and environmental stewardship (Rizwan et al., 2021). They contribute to soil structure modification by influencing the arrangement of soil particles. For example, the incorporation of certain nanoparticles can alter soil aggregation, leading to improved soil structure and aeration (Wang et al., 2016). Nanoparticles can enhance water retention in the soil. This is particularly important in water-scarce regions where improving the soil's water-holding capacity can contribute to more efficient water use in agriculture. Nano-enhanced soil amendments act as water reservoirs, releasing moisture gradually to meet plant needs (Gogos et al., 2015). It can influence nutrient availability by interacting with soil components. For instance, nanocomposites may facilitate the controlled release of nutrients, ensuring a sustained supply for plant uptake. This contributes to optimized nutrient availability, positively impacting crop growth and yield (Lombi et al., 2017). It may positively influence microbial communities in the soil. Some nanoparticles can act as carriers for beneficial microbes or stimulate microbial activity, contributing to improved nutrient cycling and overall soil health (de la Rosa et al., 2021). Nanostructures can enhance the stability of soil aggregates, leading to a reduction in soil erosion. This is particularly relevant in areas prone to erosion, where maintaining soil structure is crucial for preventing nutrient loss and preserving arable land (Raliya et al., 2016). Beyond conventional soil enhancement, this subsection delves into the pivotal role of nanostructures in soil remediation and pollution control. Nano-sized materials exhibit unique adsorption and catalytic properties, making them effective agents in mitigating soil contaminants. From heavy metals to organic pollutants, nanostructures offer a promising avenue for cleaning contaminated soils, restoring them to a state conducive to healthy plant growth and sustainable agriculture (Ghazaryan et al., 2018). Soil contamination, whether from heavy metals or pollutants, poses a significant threat to agricultural sustainability (Vardumyan et al., 2024). Nano-enhanced soil amendments emerge as promising solutions for remediation efforts, showcasing the potential to address and mitigate the risks associated with soil pollution.(Rizwan et al., 2021) investigates the capability of nano-sized particles to adsorb or immobilize contaminants in soil. The study emphasizes how these nanomaterials can prevent the uptake of contaminants by plants, thereby reducing the risk of soil pollution. The nano-enhanced amendments in providing a protective barrier against the adverse effects of soil contamination. Nano-sized particles not only act as protective barriers but also play a crucial role in facilitating the degradation of organic pollutants in the soil. Studies, such as the one by Mehta et al., 2021, delve into the mechanisms by which nanomaterials contribute to the breakdown of organic pollutants, thereby contributing to the creation of cleaner and healthier soils. This dual functionality of nanomaterials holds promise for comprehensive soil remediation strategies. Nano-sized particles, exemplified by iron oxide nanoparticles, exhibit exceptional adsorption capabilities. In the study conducted by (Li et al., 2016) these particles effectively adsorbed heavy metals like lead and cadmium, reducing their bioavailability and mitigating the risk of soil pollution. Additionally, nano-enhanced processes involving nanoparticles like titanium dioxide, as explored by (Waghmode et al.,2019) contribute to the efficient degradation of organic pollutants in soil. This dual functionality of nanomaterials in adsorption and pollutant degradation holds promise for creating cleaner and healthier soils. The integration of nano-enhanced soil amendments into modern agricultural practices represents a paradigm shift in soil management. The need for sustainable and efficient agriculture has never been more critical, and nano-enhanced solutions offer a promising avenue to address the challenges of soil degradation, nutrient management, and crop quality. As we move towards a future where food security and environmental sustainability are paramount, embracing the potential of nano-enhanced soil amendments becomes imperative for shaping resilient and productive agricultural systems.

Pros and Cons Associated With Application of Nanotechnology in Agriculture

The benefits of nanotechnology in agriculture include increased nutrient use efficiency, a reduction in environmental impact through minimized runoff and pollution, and the promotion of sustainable agricultural practices. Nanotechnology enables the precise delivery of nutrients to plants through nano-fertilizers. This precision ensures that nutrients are released in a controlled manner, aligning with the specific needs of plants at different growth stages. This targeted approach enhances nutrient use efficiency by minimizing wastage and optimizing plant uptake (Rani et al., 2020).it often incorporate controlled release mechanisms, allowing for a sustained supply of nutrients over an extended period. This controlled release ensures that nutrients are available to plants when they are most needed, contributing to improved nutrient use efficiency compared to traditional fertilizers (Singh et al., 2018). The by nano-fertilizers minimizes the risk of nutrient losses through leaching or volatilization through controlled release of nutrients. This reduction in nutrient losses enhances the overall efficiency of nutrient utilization by plants, leading to improved crop yields (Ali et al., 2022). It also contribute to a reduction in environmental impact by minimizing nutrient runoff. The controlled release and targeted delivery of nutrients prevent excess nutrients from entering water bodies, thereby mitigating the risk of water pollution (Gogos et al., 2012). They may contribute to lower greenhouse gas emissions through optimized nutrient use efficiency. By reducing the need for frequent fertilizer applications, nano-fertilizers have the potential to decrease the environmental footprint associated with fertilizer production and application (Kah et al., 2018). The environmental benefits extend to promoting sustainable agricultural practices. Nano-fertilizers, with their reduced environmental impact, align with the goals of sustainable agriculture by minimizing negative effects on ecosystems and biodiversity (Lombi et al., 2017).

However, potential challenges and concerns revolve around issues such as toxicity. One of the primary concerns associated with nanotechnology in agriculture is the potential toxicity of nanoparticles. Some nanoparticles may have adverse effects on soil organisms, beneficial insects, and even plants. Understanding the ecotoxicity of nano-fertilizers is crucial for assessing their safety and environmental impact (Kah et al., 2019). The fate and transport of nanoparticles in the environment are complex processes. Nano-fertilizers may undergo transformations in the soil, affecting their behavior and potential interactions with organisms. Studying the environmental fate of nanoparticles is essential for predicting their long-term impact (Rico et al., 2011). As the regulatory landscape for nano-fertilizers is evolving, and there may be challenges in establishing standardized regulations. Ensuring the safe and responsible use of nano-fertilizers requires collaboration between scientists, regulators, and industry stakeholders (Garg et al., 2023).

Challenges and Regulatory Considerations for Nanotechnology in Agriculture

The integration of nanotechnology into agriculture introduces a set of intricate challenges, particularly regarding the environmental impact and safety of nanomaterials. While short-term benefits in nutrient uptake and crop yield improvement are evident, understanding the long-term effects of nanotechnology on soil health is imperative. Questions about potential soil contamination, alterations in microbial diversity, and the persistence of nanomaterials over time necessitate a holistic understanding (Rico et al., 2011).

Nano-sized particles, owing to their unique properties, raise concerns about unforeseen risks, such as bioaccumulation and adverse effects on non-target organisms. The fate, transport, and overall impact of nanoparticles on the environment are critical areas that demand meticulous investigation which sheds light on the potential risks associated with nanomaterials, emphasizing the importance of addressing questions related to their environmental fate and impact and to evaluate the potential accumulation of nanoparticles in soil and water, their effects on non-target organisms, and their long-term environmental impact (Klaine et al., 2012). This multifaceted approach is essential to assess the overall environmental implications of nanotechnology in agriculture. Addressing these concerns ensures that the transformative potential of nanotechnology is realized in a manner that is environmentally responsible and sustainable for future agricultural practices. The ZnO NPs enhanced nutrient availability and crop yields in the short term, concerns arose regarding their long-term effects on soil health The accumulation of ZnO NPs in the soil led to alterations in microbial diversity, potentially influencing nutrient cycling and soil ecosystem dynamics ehich the complexity of balancing short-term agricultural benefits with the need for a comprehensive understanding of the enduring environmental consequences associated with nanotechnology applications in agriculture. (Wang et al., 2018)

The incorporation of nanotechnology into agriculture presents practical challenges, particularly concerning economic feasibility and scalability. The intricate processes involved in the production of nano-materials frequently result in elevated costs, posing economic barriers for farmers, especially in resource-constrained regions. Achieving a balance between innovation and economic feasibility is paramount for facilitating broad adoption across diverse agricultural settings (Shapira et al., 2015)

The evolving regulatory landscape for nanotechnology in agriculture presents challenges in terms of standardization. Clear guidelines for the production, use, and disposal of nano fertilizers and soil amendments are essential to ensure safety. However, the lack of standardized testing protocols and regulatory frameworks hinders widespread adoption, necessitating efforts to navigate these complexities (Younis et al., 2012). Navigating the economic viability and scalability of nanotechnology in agriculture is exemplified the application of nano-fertilizers in rice cultivation by comparing their costs and benefits with traditional fertilizers and highlighted their long-term economic benefits through enhanced nutrient use efficiency and increased crop yields. This the intricate dynamics of economic considerations in the adoption of nanotechnology in agriculture, emphasizing the potential for sustainable and economically viable practices (Younas et al.,2020, Pandey et al.2018).

Current Regulations and Guidelines

- **United States:** In the United States, the Environmental Protection Agency (EPA), the Food and Drug Administration (FDA), and the U.S. Department of Agriculture (USDA) play roles in regulating nanotechnology applications in agriculture. The EPA assesses nanopesticides, the FDA evaluates nanomaterials in food products, and the USDA monitors various aspects, including the release of genetically modified organisms.
- **European Union:** The European Union (EU) has established regulations for nanomaterials in agriculture. The EU's REACH regulation (Registration, Evaluation, Authorization, and Restriction of Chemicals) applies to nanomaterials, ensuring their safety and proper documentation before entering the market.
- Nanotechnology in agriculture in India, the Department of Science and Technology (DST), Ministry of Environment, Forest and Climate Change (MoEFCC), and the Department of Biotechnology (DBT) are key entities involved in overseeing nanotechnology applications. MoEFCC deals with the environmental clearances, DBT provides the information on nanobiotechnology and research initiatives. Additionally, the Food Safety and Standards Authority of India (FSSAI), the Indian Council of Medical Research, and institutions like the National Institute of Nutrition (NIN) (https://www.nin.res.in/) and Indian Institutes of Technology (IITs)also work collaboratively.
- **Other Countries:** Various countries worldwide have started incorporating nanotechnology into their agricultural practices. Countries like Canada, Australia, and China have established specific

regulations or guidelines to address the use of nanomaterials in agriculture, focusing on risk assessment and product safety.

Ethical Considerations in Nanotechnology for Agriculture

Ensuring the ethical use of nanotechnology in agriculture involves a rigorous examination, as exemplified in a comprehensive study by Macnaghten et al. (2010, Public Understanding of Science). The study underscores the significance of thorough risk assessments to guarantee the safety of both human health and the environment. Macnaghten et al.'s research emphasizes the critical need for assessing potential risks associated with the release of nanoparticles into the environment and their broader impact on ecosystems. Ethical considerations extend beyond risk assessments to transparency and consumer awareness. emphasize transparency in labeling nanomaterial-containing agricultural products. Providing clear information enables consumers to make informed choices about the products they use in agriculture, aligning with ethical principles. Moreover, ethical dimensions in nanotechnology for agriculture encompass evaluating the socioeconomic impacts (Bhattacharya et al.,2023). Assessing potential disparities in access to nanotechnology-related benefits, ensuring equitable distribution of resources, and considering implications for small-scale farmers are integral components of ethical decision-making. This contributes to the broader understanding of the ethical landscape surrounding nanotechnology in agriculture, emphasizing the need for comprehensive evaluations beyond technical aspects(Khan 2014).

Global Collaboration and Standardization in Nanotechnology for Agriculture

The imperative for global collaboration and standardization in the ethical application of nanotechnology in agriculture is underscored by initiatives such as those championed by the Organisation for Economic Co-operation and Development (OECD). Their efforts focus on harmonizing regulatory approaches and facilitating the global exchange of information, as highlighted in OECD's nanotechnology program. International collaboration is a cornerstone of ethical considerations in nanotechnology, as explored in a study by (Hull et al., 2018). The study emphasizes the importance of harmonizing efforts on a global scale to address the ethical challenges associated with nanotechnology in agriculture. Simultaneously, concerted efforts are underway to establish international standards for nanotechnology in agriculture. Standardization, as exemplified by initiatives like those undertaken by the International Organization for Standardization (ISO), seeks to ensure consistency in testing methodologies, risk assessment procedures, and regulatory frameworks across diverse regions. This pursuit of standardization is crucial for fostering ethical practices in the deployment of nanotechnology globally (Soltani et al., 2019).

Public Engagement and Awareness in Nanotechnology for Agriculture

The ethical utilization of nanotechnology in agriculture underscores the significance of public engagement and awareness. Stakeholder involvement, including farmers, consumers, and environmental organizations, is paramount in decision-making processes related to nanotechnology applications, as highlighted in a study by (Macnaghten et al., 2010, Public Understanding of Science). Engaging stakeholders fosters awareness, ensures diverse perspectives are considered, and helps address concerns associated with nanotechnology in agriculture (Mukherjee et al., 2019). Education and communication play pivotal roles in promoting public understanding of nanotechnology applications in agriculture (Davies et al., 2019) Ethical practices necessitate transparent communication and the dissemination of accurate information to consumers about the benefits, potential risks, and safety measures associated with nanotechnology applications. This aligns with efforts advocated by organizations like the European Commission, which emphasize the importance of clear communication regarding nanomaterial-containing agricultural products.

CONCLUSION

The exploration of nanotechnology in agriculture signifies a transformative journey with profound implications for the future of nutrient management in crop production. This synthesis of knowledge underscores the groundbreaking potential of nanotechnology, heralding a revolution in agricultural practices. Key findings include the precision achieved in nutrient delivery through nano-fertilizers, utilizing controlled release mechanisms to significantly enhance nutrient uptake efficiency. Furthermore, nano-enhanced soil amendments, nanotechnology contribute significantly to augmenting soil structure, water retention, and nutrient availability, addressing crucial challenges in modern agriculture. The positive impact of nano-fertilizers on increased crop yields, as highlighted in case studies and research findings, underscores their pivotal role in fostering sustainable and productive agricultural practices. Ethical and environmental considerations are paramount, demanding stringent risk assessments, transparent practices, and global collaboration to mitigate potential adverse effects. Emerging trends, including nano sensors for precision agriculture and genetic engineering integration with nanotechnology, signify the future trajectory of this field. These findings hold significant implications for agricultural practices. The advent of nano sensors and nano-fertilizers heralds an era of precise, data-driven agriculture, empowering farmers to tailor nutrient applications based on real-time soil and plant conditions. Nano-enhanced soil amendments emerge as pivotal contributors to sustainable soil management, positively impacting soil health, water conservation, and reducing environmental footprints. The economic viability of nanotechnology in agriculture assumes paramount importance, with ongoing research aiming to comprehend the scalability and cost-effectiveness of nano-enabled products, imperative for their widespread adoption. Stakeholder engagement and ethical considerations emerge as guiding principles for the responsible deployment of nanotechnology in agriculture. Public awareness, transparent practices, and inclusive decision-making processes are essential for ensuring ethical and sustainable practices. The transformative potential of nanotechnology in agriculture, however, necessitates a thoughtful navigation of challenges encompassing long-term environmental impact, rigorous risk assessment, economic viability, stakeholder engagement, and robust regulatory frameworks. As the field continues to advance, sustained research efforts and collaborative endeavors will be indispensable in fully realizing the benefits of nanotechnology while ensuring its ethical and sustainable integration into global agriculture. In conclusion, the odyssey into nanotechnology for agriculture unveils not merely technological strides but also underscores the imperative of a holistic and responsible approach, presenting a trajectory towards a more resilient, productive, and sustainable future for global agriculture.

REFERENCES

Ali, B., Saleem, M. H., Ali, S., Shahid, M., Sagir, M., Tahir, M. B., Qureshi, K. A., Jaremko, M., Selim, S., Hussain, A., Rizwan, M., Ishaq, W., & Rehman, M. (2022). Mitigation of salinity stress in barley genotypes with variable salt tolerance by application of zinc oxide nanoparticles. *Frontiers in Plant Science, 13*, 973782. doi:10.3389/fpls.2022.973782 PMID:36072329

Bhardwaj, A. K., Arya, G., Kumar, R., Hamed, L., Pirasteh-Anosheh, H., Jasrotia, P., Kashyap, P. L., & Singh, G. P. (2022). Switching to nanonutrients for sustaining agroecosystems and environment: The challenges and benefits in moving up from ionic to particle feeding. *Journal of Nanobiotechnology, 20*(1), 19. doi:10.1186/s12951-021-01177-9 PMID:34983548

Bhattacharya, B., Roy, P., Bhattacharya, S., Prasad, B., & Mandal, A. K. (2023). Nanotechnology and sustainable development: Overcoming the obstacles by adopting ethical practices for future farming. In Engineered Nanomaterials for Sustainable Agricultural Production, Soil Improvement and Stress Management (pp. 431-445). Academic Press.

Cicek, S., & Nadaroglu, H. (2015). The use of nanotechnology in the agriculture. *Advances in Nano Research, 3*(4), 207–223. doi:10.12989/anr.2015.3.4.207

Davies, S. R., Kearnes, M., & Macnaghten, P. (2019). Nanotechnology and public engagement: a new kind of (social) science? In *Nano meets macro* (pp. 473–500). Jenny Stanford Publishing. doi:10.1201/9780429067150-32

de la Rosa, G., Vázquez-Núñez, E., Molina-Guerrero, C., Serafín-Muñoz, A. H., & Vera-Reyes, I. (2021). Interactions of nanomaterials and plants at the cellular level: Current knowledge and relevant gaps. *Nanotechnology for Environmental Engineering, 6*(1), 1–19. doi:10.1007/s41204-020-00100-1

Fatima, F., Hashim, A., & Anees, S. (2021). Efficacy of nanoparticles as nanofertilizer production: A review. *Environmental Science and Pollution Research International, 28*(2), 1292–1303. doi:10.1007/s11356-020-11218-9 PMID:33070292

Gade, A., Ingle, P., Nimbalkar, U., Rai, M., Raut, R., Vedpathak, M., Jagtap, P., & Abd-Elsalam, K. A. (2023). Nanofertilizers: The Next Generation of Agrochemicals for Long-Term Impact on Sustainability in Farming Systems. *Agrochemicals, 2*(2), 257–278. doi:10.3390/agrochemicals2020017

Garg, D., Sridhar, K., Stephen Inbaraj, B., Chawla, P., Tripathi, M., & Sharma, M. (2023). Nano-biofertilizer formulations for agriculture: A systematic review on recent advances and prospective applications. *Bioengineering (Basel, Switzerland), 10*(9), 1010. doi:10.3390/bioengineering10091010 PMID:37760112

Gauba, A., Hari, S. K., Ramamoorthy, V., Vellasamy, S., Govindan, G., & Arasu, M. V. (2023). The versatility of green synthesized zinc oxide nanoparticles in sustainable agriculture: A review on metal-microbe interaction that rewards agriculture. *Physiological and Molecular Plant Pathology, 125*, 102023. doi:10.1016/j.pmpp.2023.102023

Giraldo, J. P., Landry, M. P., Faltermeier, S. M., McNicholas, T. P., Iverson, N. M., Boghossian, A. A., Reuel, N. F., Hilmer, A. J., Sen, F., Brew, J. A., & Strano, M. S. (2014). Plant nanobionics approach to augment photosynthesis and biochemical sensing. *Nature Materials, 13*(4), 400–408. doi:10.1038/nmat3890 PMID:24633343

Gogos, A., Knauer, K., & Bucheli, T. D. (2012). Nanomaterials in plant protection and fertilization: Current state, foreseen applications, and research priorities. *Journal of Agricultural and Food Chemistry, 60*(39), 9781–9792. doi:10.1021/jf302154y PMID:22963545

Hasan, M. K., Shopan, J., & Ahammed, G. J. (2020). Nanomaterials and soil health for agricultural crop production: current status and future prospects. *Nanomaterials for agriculture and forestry applications*, 289-312.

Hull, M., & Bowman, D. (Eds.). (2018). *Nanotechnology environmental health and safety: risks, regulation, and management*. William Andrew.

Iqbal, M., Umar, S., & Mahmooduzzafar. (2019). Nano-fertilization to enhance nutrient use efficiency and productivity of crop plants. *Nanomaterials and plant potential*, 473-505.

Jain, S., Cherukupalli, S. K., Mahmood, A., Gorantla, S., Rapalli, V. K., Dubey, S. K., & Singhvi, G. (2019). Emerging nanoparticulate systems: Preparation techniques and stimuli responsive release characteristics. *Journal of Applied Pharmaceutical Science, 9*(8), 130–143. doi:10.7324/JAPS.2019.90817

Javed, T., Singhal, R. K., Shabbir, R., Shah, A. N., Kumar, P., Jinger, D., ... Siuta, D. (2022). Recent advances in agronomic and physio-molecular approaches for improving nitrogen use efficiency in crop plants. *Frontiers in Plant Science, 13*, 877544. doi:10.3389/fpls.2022.877544 PMID:35574130

Jha, A., Pathania, D., Damathia, B., Raizada, P., Rustagi, S., Singh, P., ... Chaudhary, V. (2023). Panorama of biogenic nano-fertilizers: A road to sustainable agriculture. *Environmental Research, 235*, 116456. doi:10.1016/j.envres.2023.116456 PMID:37343760

Kah, M., Navarro, D., Kookana, R. S., Kirby, J. K., Santra, S., Ozcan, A., & Kabiri, S. (2019). Impact of (nano) formulations on the distribution and wash-off of copper pesticides and fertilisers applied on citrus leaves. *Environmental Chemistry, 16*(6), 401–410. doi:10.1071/EN18279

Khan, A. S. (2014, May). Ethics and nanotechnology. In *2014 IEEE International Symposium on Ethics in Science, Technology and Engineering* (pp. 1-14). IEEE.

Khatri, A., & Bhateria, R. (2023). Efficacy of Nanofertilizers Over Chemical Fertilizers in Boosting Agronomic Production. *Nature Environment and Pollution Technology, 22*(2), 767–776. doi:10.46488/NEPT.2023.v22i02.019

Khodakovskaya, M., Dervishi, E., Mahmood, M., Xu, Y., Li, Z., Watanabe, F., & Biris, A. S. (2009). Carbon nanotubes are able to penetrate plant seed coat and dramatically affect seed germination and plant growth. *ACS Nano, 3*(10), 3221–3227. doi:10.1021/nn900887m PMID:19772305

Khodakovskaya, M. V., De Silva, K., Biris, A. S., Dervishi, E., & Villagarcia, H. (2012). Carbon nanotubes induce growth enhancement of tobacco cells. *ACS Nano, 6*(3), 2128–2135. doi:10.1021/nn204643g PMID:22360840

Klaine, S. J., Koelmans, A. A., Horne, N., Carley, S., Handy, R. D., Kapustka, L., Nowack, B., & von der Kammer, F. (2012). Paradigms to assess the environmental impact of manufactured nanomaterials. *Environmental Toxicology and Chemistry, 31*(1), 3–14. doi:10.1002/etc.733 PMID:22162122

Kumar, N., Samota, S. R., Venkatesh, K., & Tripathi, S. C. (2023). Global trends in use of nano-fertilizers for crop production: Advantages and constraints–A review. *Soil & Tillage Research, 228*, 105645. doi:10.1016/j.still.2023.105645

Li, Q., Chen, X., Zhuang, J., & Chen, X. (2016). Decontaminating soil organic pollutants with manufactured nanoparticles. *Environmental Science and Pollution Research International, 23*(12), 11533–11548. doi:10.1007/s11356-016-6255-7 PMID:26906002

Liu, X., Wei, Z., Ma, Y., Liu, J., & Liu, F. (2021). Effects of biochar amendment and reduced irrigation on growth, physiology, water-use efficiency and nutrients uptake of tobacco (Nicotiana tabacum L.) on two different soil types. *The Science of the Total Environment, 770*, 144769. doi:10.1016/j.scitotenv.2020.144769 PMID:33736368

Lombi, E., Donner, E., Dusinska, M., & Wickson, F. (2019). A One Health approach to managing the applications and implications of nanotechnologies in agriculture. *Nature Nanotechnology, 14*(6), 523–531. doi:10.1038/s41565-019-0460-8 PMID:31168074

Macnaghten, P., Kearnes, M. B., & Wynne, B. (2005). Nanotechnology, governance, and public deliberation: What role for the social sciences? *Science Communication, 27*(2), 268–291. doi:10.1177/1075547005281531

Marchiol, L., Iafisco, M., Fellet, G., & Adamiano, A. (2020). Nanotechnology support the next agricultural revolution: Perspectives to enhancement of nutrient use efficiency. *Advances in Agronomy, 161*, 27–116. doi:10.1016/bs.agron.2019.12.001

Mehta, M., Sharma, M., Pathania, K., Jena, P. K., & Bhushan, I. (2021). Degradation of synthetic dyes using nanoparticles: A mini-review. *Environmental Science and Pollution Research International, 28*(36), 49434–49446. doi:10.1007/s11356-021-15470-5 PMID:34350572

Mukherjee, A., Maity, A., Pramanik, P., Shubha, K., Joshi, D. C., & Wani, S. H. (2019). Public perception about use of nanotechnology in agriculture. In *Advances in phytonanotechnology* (pp. 405–418). Academic Press. doi:10.1016/B978-0-12-815322-2.00019-5

Mukhopadhyay, S. S. (2014). Nanotechnology in agriculture: Prospects and constraints. *Nanotechnology, Science and Applications*, 63–71. doi:10.2147/NSA.S39409 PMID:25187699

Nandhakumar, M. R., Muthukrishnan, R., Kumar, T. P., Sureshkumar, P., & Nandini, S. (2023). Nano Fertilizer in Crop Production: The Changing Scenario. *International Journal of Environment and Climate Change, 13*(6), 158–170. doi:10.9734/ijecc/2023/v13i61811

Oleg, F. (2017). *Dmitry, Beilin*. Nanotechnology in Agriculture. doi:10.1201/9781315229287-7

Özer, E. A., Özcan, M., & Didin, M. (2014). Nanotechnology in food and agriculture industry. *Food processing: Strategies for quality assessment*, 477-497.

Pandey, G. (2018). Challenges and future prospects of agri-nanotechnology for sustainable agriculture in India. *Environmental Technology & Innovation, 11*, 299–307. doi:10.1016/j.eti.2018.06.012

Qiao, A., Kalevaru, V. N., Radnik, J., Duvel, A., Heitjans, P., Kumar, A. S. H., Prasad, P. S. S., Lingaiah, N., & Martin, A. (2014). Oxidative dehydrogenation of ethane to ethylene over V2O5/Al2O3 catalysts: Effect of source of alumina on the catalytic performance. *Industrial & Engineering Chemistry Research*, *53*(49), 18711–18721. doi:10.1021/ie5008344

Rajput, V. D., Minkina, T., Ahmed, B., Singh, V. K., Mandzhieva, S., Sushkova, S., Bauer, T., Verma, K. K., Shan, S., van Hullebusch, E. D., & Wang, B. (2022). Nano-biochar: A novel solution for sustainable agriculture and environmental remediation. *Environmental Research*, *210*, 112891. doi:10.1016/j.envres.2022.112891 PMID:35183514

Raliya, R., Tarafdar, J. C., & Biswas, P. (2016). Enhancing the mobilization of native phosphorus in the mung bean rhizosphere using ZnO nanoparticles synthesized by soil fungi. *Journal of Agricultural and Food Chemistry*, *64*(16), 3111–3118. doi:10.1021/acs.jafc.5b05224 PMID:27054413

Ramaiyan, S., Vijayakumar, P., Balasubramanian, R. A. B., & Nallathambi, M. (2023). Influence of nano-zinc oxide and fortified rice residue compost on rice productivity, zinc biofortification, zinc use efficiency, soil quality, zinc fractions and profitability in different rice production systems. *Journal of Plant Nutrition*, *46*(17), 1–22. doi:10.1080/01904167.2023.2212722

Rani, A., Rani, K., Tokas, J., Anamika, Singh, A., Kumar, R., ... Kumar, S. (2020). Nanomaterials for agriculture input use efficiency. *Resources use efficiency in agriculture*, 137-175.

Rico, C. M., Majumdar, S., Duarte-Gardea, M., Peralta-Videa, J. R., & Gardea-Torresdey, J. L. (2011). Interaction of nanoparticles with edible plants and their possible implications in the food chain. *Journal of Agricultural and Food Chemistry*, *59*(8), 3485–3498. doi:10.1021/jf104517j PMID:21405020

Rizwan, M., Ali, S., Rehman, M. Z., Riaz, M., Adrees, M., Hussain, A., Zahir, Z. A., & Rinklebe, J. (2021). Effects of nanoparticles on trace element uptake and toxicity in plants: A review. *Ecotoxicology and Environmental Safety*, *221*, 112437. doi:10.1016/j.ecoenv.2021.112437 PMID:34153540

Sekhon, B. S. (2014). Nanotechnology in agri-food production: An overview. *Nanotechnology, Science and Applications*, 31–53. doi:10.2147/NSA.S39406 PMID:24966671

Shapira, P., & Youtie, J. (2015). The economic contributions of nanotechnology to green and sustainable growth. *Green processes for nanotechnology: From inorganic to bioinspired nanomaterials*, 409-434.

Singh, A., Kumar, A., Jaswal, A., Singh, M., & Gaikwad, D. S. (2018). Nutrient use efficiency concept and interventions for improving nitrogen use efficiency. *Plant Archives*, *18*(1), 1015–1023.

Singh, A., Margaryan, G., Harutyunyan, A., Movsesyan, H., Khachatryan, H., Rajput, V., . . . Ghazaryan, K. (2024b). Advancing Agricultural Resilience in Ararat Plain, Armenia: Utilizing Biogenic Nanoparticles and Biochar under Saline Environments to Optimize Food Security and Foster European Trade. *Egyptian Journal of Soil Science, 64*(2). doi:10.21608/ejss.2024.257000.1703

Singh, A., Rajput, V. D., Pandey, D., Sharma, R., Ghazaryan, K., & Minkina, T. (2023b). Nano Zinc-Enabled Strategies in Crops for Combatting Zinc Malnutrition in Human Health. *Frontiers in Bioscience (Landmark Edition)*, *28*(8), 158. doi:10.31083/j.fbl2808158 PMID:37664935

Singh, A., Rajput, V. D., Sharma, R., Ghazaryan, K., & Minkina, T. (2023c). Salinity stress and nanoparticles: Insights into antioxidative enzymatic resistance, signaling, and defense mechanisms. *Environmental Research*, *235*, 116585. doi:10.1016/j.envres.2023.116585 PMID:37437867

Singh, A., Rajput, V. D., Tawaha, A. R. M., Al Zoubi, O. M., Habeeb, T., Rawat, S., Ghazaryan, K., & Minkina, T. (2023a). A Review on Crop Responses to Nanofertilizers for Mitigation of Multiple Environmental Stresses. *Ecological Engineering & Environmental Technology*, *24*(7), 280–296. doi:10.12912/27197050/169313

Singh, A., Rajput, V. D., Varshney, A., Ghazaryan, K., & Minkina, T. (2023d). Small Tech, Big Impact: Agri-nanotechnology Journey to Optimize Crop Protection and Production for Sustainable Agriculture. *Plant Stress*, *10*, 100253. doi:10.1016/j.stress.2023.100253

Singh, A., Sengar, R. S., Rajput, V. D., Al-Ghzawi, A. L., Shahi, U. P., Ghazaryan, K., Minkina, T., Tawaha, A. R. M. A., Al Zoubi, O. M., & Habeeb, T. (2024a). Impact of Salinity Stress and Zinc Oxide Nanoparticles on Macro and Micronutrient Assimilation: Unraveling the Link between Environmental Factors and Nutrient Uptake. *Journal of Ecological Engineering*, *25*(2), 1–9. doi:10.12911/22998993/172947

Singh, A., Sharma, R., Rawat, S., Singh, A. K., Rajput, V. D., Fedorov, Y., . . . Chaplygin, V. (2022). Nanomaterial-plant interaction: Views on the pros and cons. *Toxicity of Nanoparticles in Plants*, 47–68. doi:10.1016/B978-0-323-90774-3.00015-5

Singh, H., Sharma, A., Bhardwaj, S. K., Arya, S. K., Bhardwaj, N., & Khatri, M. (2021). Recent advances in the applications of nano-agrochemicals for sustainable agricultural development. *Environmental Science. Processes & Impacts*, *23*(2), 213–239. doi:10.1039/D0EM00404A PMID:33447834

Sinha, T., Bhagwatwar, P., Krishnamoorthy, C., & Chidambaram, R. (2019). Polymer based micro-and nanoencapsulation of agrochemicals. *Polymers for agri-food applications*, 5-28.

Soltani, A. M., & Pouypouy, H. (2019). Standardization and regulations of nanotechnology and recent government policies across the world on nanomaterials. In *Advances in phytonanotechnology* (pp. 419–446). Academic Press. doi:10.1016/B978-0-12-815322-2.00020-1

Srilatha, B. (2011). Nanotechnology in agriculture. *Journal of Nanomedicine & Nanotechnology*, *2*(7).

Sun, M., Zhao, C., Shang, H., Hao, Y., Han, L., Qian, K., White, J. C., Ma, C., & Xing, B. (2023). ZnO quantum dots outperform nanoscale and bulk particles for enhancing tomato (Solanum lycopersicum) growth and nutritional values. *The Science of the Total Environment*, *857*, 159330. doi:10.1016/j.scitotenv.2022.159330 PMID:36228785

Waghmode, M. S., Gunjal, A. B., Mulla, J. A., Patil, N. N., & Nawani, N. N. (2019). Studies on the titanium dioxide nanoparticles: Biosynthesis, applications and remediation. *SN Applied Sciences*, *1*(4), 310. doi:10.1007/s42452-019-0337-3

Wang, P., Lombi, E., Menzies, N. W., Zhao, F. J., & Kopittke, P. M. (2018). Engineered silver nanoparticles in terrestrial environments: A meta-analysis shows that the overall environmental risk is small. *Environmental Science. Nano*, *5*(11), 2531–2544. doi:10.1039/C8EN00486B

Wang, P., Lombi, E., Zhao, F. J., & Kopittke, P. M. (2016). Nanotechnology: A new opportunity in plant sciences. *Trends in Plant Science*, *21*(8), 699–712. doi:10.1016/j.tplants.2016.04.005 PMID:27130471

Yadav, A., Yadav, K., & Abd-Elsalam, K. A. (2023). Nanofertilizers: Types, Delivery and Advantages in Agricultural Sustainability. *Agrochemicals*, *2*(2), 296–336. doi:10.3390/agrochemicals2020019

Younas, A., Yousaf, Z., Riaz, N., Rashid, M., Razzaq, Z., Tanveer, M., & Huang, S. (2020). Role of nanotechnology for enhanced rice production. *Nutrient dynamics for sustainable crop production*, 315-350.

Younis, S. A., Kim, K. H., Shaheen, S. M., Antoniadis, V., Tsang, Y. F., Rinklebe, J., Deep, A., & Brown, R. J. (2021). Advancements of nanotechnologies in crop promotion and soil fertility: Benefits, life cycle assessment, and legislation policies. *Renewable & Sustainable Energy Reviews*, *152*, 111686. doi:10.1016/j.rser.2021.111686

Chapter 4
Nanotechnology for Soil Remediation and Land Restoration

Muhammad Majeed
https://orcid.org/0000-0001-7663-2563
University of Gujrat, Pakistan

Wajeeha Noor
https://orcid.org/0000-0003-0728-0765
University of Central Punjab, Lahore, Pakistan

Arshad Rasool
https://orcid.org/0000-0001-9043-1207
Minhaj University, Lahore, Pakistan

Muhammad Saleem
University of Veterinary and Animal Sciences, Ravi, Pakistan

Iqra Bashir
University of Veterinary and Animal Sciences, Ravi, Pakistan

Rabia Fatima
Clarkson University, USA

Allah Bakhsh Gulshan
Ghazi University, Pakistan

Dwaipayan Sinha
https://orcid.org/0000-0001-7870-8998
Government General Degree College, Mohanpur, India

Satarupa Nath
Rishi Bankim Chandra College for Women, Naihati, India

Shumaila Hassan
University of Agriculture, Faisalabad, Pakistan

Zehra Batool
University of Baltistan Skardu, Pakistan

Shahid Abbas
Ghazi University, Pakistan

Arslan Rafique
University of Gujrat, Pakistan

ABSTRACT

Nanotechnological techniques are promising to standard spectroscopy methods as they are more durable, sensitive, targeted, and affordable. Nanotechnology has shown significant new promising approaches for efficient nutrient management to ensure sustainable agriculture. The combined approach of nano-technology and phytotechnology called "nano-phytoremediation" is used to reduce toxic environments

DOI: 10.4018/979-8-3693-1890-4.ch004

and restore these degraded resources. Nanotechnology also has a significant impact on nutrient uptake in precision agriculture. It's an eco-friendly and cost-effective method for managing and cleaning up soil conditions. Nanoremediation employs nanomaterials to enhance adsorption and sequestration of pollutants and has emerged as a cutting-edge approach with promising in situ and ex situ applications. Applications based on nanotechnology are easy to use, inexpensive, and suggest improved remediation and treatment techniques that could significantly lower soil contamination.

1. INTRODUCTION

Human-caused environmental degradation poses a challenge to a clean, green, and sustainable environment which results in air, water and soil pollution (Loiseau et al., 2023). Land degradation & soil contamination is a serious global challenge & threat to productivity as it has already affected approximately 6 billion hectares of soil globally. The agriculture land deterioration has been caused majorly by soil erosion, salt damage, declining soil fertility; mining, deforestation, improper fertilizer use, and heavy metal contamination in the soil (Ahmed et al., 2021; Lamichhane et al., 2024).

Soil remediation and land restoration are crucial processes to ensure ecological health and sustainable land usage (Singh et al., 2022). Contaminated soil poses significant risks to agriculture and also has an adverse impact on human health as shown in figure above (Loiseau et al., 2023). Though remediating soil contamination and deterioration is a huge challenge due to the increase of released toxins, with the development of new technologies, efficient pollutant removal has also been made progress recently. Effective methods for quickening the detoxification of a contaminated site include bioremediation, phytoremediation and biostimulation (Kumari et al., 2023). There are several land restoration techniques that can help recover degraded soils and reduce the financial losses that land degradation causes to agriculture and human health worldwide. This global soil resource degradation negatively affects human well-being, food security, and agricultural productivity (Rajput et al., 2021). Therefore, it needs to be addressed immediately. Remediation soil treatment with nanotechnology may offer a long-term way to restore the degraded soil resources. Nanotechnology can also be potentially applied in improving phytoremediation and bioremediation, as well as in the remediation of soil contaminated with heavy metals, pesticides, and their residues, as well as persistent organic pollutants (POPs) (S. Ali et al., 2023; Kumar et al., 2023; Vardumyan et al., 2024).

Nanotechnology based soil and land restoration has shown great potential to achieve the goal of sustainable agriculture (Singh et al., 2023a). Nanotechnology has been utilized in agriculture for crop protection and production with a focus on nanopesticides, nanomaterials, nanofertilizers, nanobiosensors, and nano-enabled remediation techniques for contaminated soils (Bhalla et al., 2022).

Nanotechnological approaches accelerate the clearance of hazardous substances from contaminated soils and are deeply rooted in biological processes (Singh et al., 2023b, c). Numerous approaches, including the use of rhizobacteria, genetic engineering, and chemical additions, have been tried to improve the efficacy of phytoremediation (Azeem et al., 2023). In addition to their beneficial effects on eliminating the contaminants, NMs may have both positive and negative interactions with biotic and abiotic elements (Singh et al., 2024a). For this reason, a great deal of work has been done to assess the synergistic effects of using NMs in conjunction with bioremediation techniques and to clarify the physical, chemical, and biological interactions that may occur in soil or water (Mubarok et al., 2024). Another major obstacle to the in situ applications of Multifunctional nanocomposites is the limited mixing and subsurface transit of nanoparticles. This challenge could be mitigated using engineered approaches for conjunction of nanomaterials with other technologies which could be further developed and tested (Liao et al., 2023).

2. NANOMATERIALS FOR CONTAMINANT DETECTION

The accumulation of heavy metals and hazardous compounds in soil leads to soil contamination that present a threat to humanity in the form of numerous diseases as the crops ingested. To mitigate this challenge, researchers have developed a new generation of nanosensors that will detect heavy trace metals in soil (Swain et al., 2023). Nanosensors have been used for early detection to monitor and identify asymptomatic plant illnesses, nutritional deficiencies, soil health, and water requirements (Singh et al., 2024b). This real time detection of environmental pollutants has greatly benefited (Manimegalai et al., 2023).

The contaminant's detection is carried out using a variety of nanotechnology-based techniques, such as surface increase scattering, surface plasmon resonance, and unique forms of nanoparticles like silicon oxide, zinc oxide, and silver. Significant interest has also been shown in biosensing technology for pollution treatment and detection (Fadiji et al., 2023). Although the advanced spectroscopic monitoring techniques are effective, their use in smart agriculture is limited by their high instrumental and operational costs, ongoing human supervision, and portability concerns. Chemiresistive nanosensors provide a number of advantages, including high sensitivity, quick real-time detection, labor-free and economical operation, and consistent response to fluctuating humidity and meteorological circumstances (Lo Presti et al., 2023).

Nanomaterials and nanosensors have also replaced spectroscopy methods for detecting pesticides. They not only track the required concentration and growth progress but also analyze the interaction with microbial populations (Ahmad et al., 2023). Nanobiosensors have been also utilized to monitor soil conditions for pesticides, herbicides, insecticides, diseases, fertilizers, and crop growth. They are used to heavy metals like Hg2+. As3+, and Cu2+. These technological advances have the ability to solve many of the issues (Banerjee et al., 2024).

In 2013, Da Silva and coworkers created an innovative nanosensor called a bionanosensor which has the ability to detect organophosphate pesticides. They developed AChE (acetylcholinesterase) biosensor using multilayer films incorporating MWCNTs, chitosan (CS), and AChE liposome bioreactors (ALB). Following developments in the field of nanotechnology, Xiong and group also shared their work on organophosphorus pesticide detection using enzyme-functionalized nanostructured biosensors in 2018 (Rastogi et al., 2019; Dehghanian et al., 2023). According to researched data, it has been evident that bionanosensors enable faster and more accurate detection of viruses and bacteria, improving food safety for consumers. The monitoring of agricultural pollutants could benefit from it as these advanced nanosensors can detect insects, illnesses, infections, toxins, and pollutants rapidly (Nguyen et al., 2023).

2.1. Nanoparticles in Soil Decontamination

Nanoparticles have been promisingly utilized for soil Decontamination to restore Environmental sustainability (Singh et al., 2023d). With the advancements and development in the research area of nanotechnology, more innovative ways to use NPs effectively are emerging continuously (Liu et al., 2023; Arif et al., 2021). They are being employed for removal of heavy metals, degradation of organic compounds and accelerating the remediation processes. For these purposes nZVI, Fe_3O_4 NPs, TiO_2 NPs and Ag NPs with other Nps have been used in several research studies to mitigate the soil contamination.

3. NANOREMEDIATION STRATEGIES FOR CONTAMINATED SOIL

Contaminated soil poses a significant environmental challenge, necessitating innovative solutions for effective remediation (Ghazaryan et al., 2018). Traditional remediation methods often have limitations therefore switching to applying innovative nanoremediation as it showed 80% cost reduction with more efficient results. Innovative adaptable nanomaterials could make it easier to implement strategies for concurrent agricultural output and soil remediation, but a comprehensive and mechanistic evaluation of safety and efficacy is required (Tyagi et al., 2023). It deploys advanced technologies called NEMs (Nanoengineered materials) which includes nanosensing and nanophotocatalysis using nanocatalysts and nanoadsorbents (Ganie et al., 2021). The enhanced Adsorption and Sequestration of Pollutants is majorly unique due to the high surface area to volume ratio property of nanoparticles (NPs). Additionally, NPs exhibit high affinity towards target pollutants, further increasing selectivity and efficiency (New et al., 2023).

Nanoremediation strategies offer a powerful tool for tackling soil contamination. For this purpose, Metal oxide nanoparticles, Carbon-based nanomaterials, biochar based nanocomposites, Calcium peroxide, ACNPs, ZVI NPs, TiO_2 NPs, and Fe_2O_3 NPs showcase remarkable efficacy in remediating a diverse range of soil contaminants both in situ and ex situ (S. Ali et al., 2023; A. V. Singh et al., 2023)

In 2021, Liu and group published research where they developed a novel approach to soil restoration in paddy soil contaminated with pentachlorophenol (PCP). They utilized a nanoscale zero-valent iron (nZVI) approach in conjunction for safe rice cultivation. Results clearly depict the efficacy as rice quality and yield both were improved by this co-treatment (A. V. Singh et al., 2023).

Similarly, nZVI nanoremediation strategy was applied to brownfield soil which was highly contaminated with As and Hg. But results show the potential of effective nanoremediation as it reduced the toxicity of contaminants up to 80% (Li et al., 2023). Nanomaterials have revolutionized soil decontamination with cost-effective in situ applications, like nZVI and Nano-absorbents, providing dynamic, targeted remediation without excavation. Ex situ approaches, such as nano reagent washing and phytoremediation with nanofertilizers, offer precise control, high treatment rates, and reduced environmental risks (Chandra et al., 2023; Biswas et al., 2023)

3.2. Nanoparticle-Assisted Phytoremediation

Several Nanoparticles are employed to sustainably increase the efficiency of phytoremediation (Arora et al., 2023). Nano-phyto-remediation involves applying nanoremediation along with nanoparticles or nanomaterials to remediate soil. In order to restore soil fertility, research on both above- and below-ground microorganisms is essential since these organisms may be linked to plant immunology and physiological processes. Therefore it is important to analyze the plant-microbe-nanoparticle interaction (Sukul et al., 2023).

Plants may interact with the NPs that are present in soil and these NPs then make their way up the food chain, where they may pose a health risk to people. The figure 8 depicts potential interactions along the soil plant microbe NPs nexus (Hua et al., 2023; Daniel et al., 2023). Phytoremediation typically involves the collaboration of plants and microbes. Plants uptake contaminants through their roots, while microbes contribute to contaminant degradation and nutrient mobilization in the rhizosphere. NPs can significantly enhance nutrient delivery, microbial activity, plant defense and root growth. Consequently, nano phytoremediation improves plant-based remediation efficiency (Arora et al., 2023).

4. NANOTECHNOLOGY FOR SOIL STRUCTURE IMPROVEMENT

Nanotechnology offers a promising approach to address soil degradation and enhance soil structure, leading to improved soil health and productivity. This section would discuss the transformative impact of nanoparticles on soil aggregation and structure along with nano-enabled soil amendments (Ansari, 2023).

4.1. Nanoparticles for Soil Aggregation and Structure

Nano-biotechnology aids in the creation of innovative, inexpensive, and environmentally friendly nano-biofertilizers with the potential qualities of both nanomaterials and biofertilizers. The effectiveness of nano-biofertilizer in enhancing agricultural productivity has been established by its role in plant and soil systems. They work in concert to provide increased retention of soil moisture and vital plant nutrients (Kumari et al., 2023; Chehelgerdi et al., 2023).

A recent article of 2023 discussed the impact of NPs on soil porosity & quality. According to that, Zhang did research which concluded improved soil quality & reduced metal bioavailability by remediation of heavy metals in soil using nZVI (zero-valent iron nanoparticles). These nanoparticles aid in the transformation of hazardous, mobile metal ions into more stable, less mobile forms through reduction processes, lowering the possibility of plant uptake of the metals and their subsequent introduction into the food chain. This strategy has the potential to improve soil quality and agricultural output in contaminated areas in addition to addressing environmental issues (Elshayb, Nada, Farroh, et al., 2022; Sinha et al., 2023a).

A study has also been done to analyze the impact of NPs on water retention in 2019. It utilized the co-polymerization of AA monomer and AM monomer employing $Ca(OH)_2$ nanoparticles as cross-linkers results in a unique PAM/PAA/$Ca(OH)_2$ super-absorbent hydrogel (SAH). Resultantly, the improved swelling ratio (SR) and water retention capability has been observed due to the SAH network that made it easier for water molecules to enter the hydrogel network, which can increase water absorbency (Lv et al., 2019). Another research has been done to study the relation of NPs and water retention. The graph in figure 11 clearly depicts the highest values of efficient water when BC+ZnO NPs both were applied as the green bar topped in both of the seasons (Elshayb et al., 2022; Naveed et al., 2024a).

Similarly, another research was done to assess the impact of NPs (TNP, SNP, and ZNP) on root penetration showed how endodontically treated roots' ability to withstand breakage. It was concluded that the endodontically treated roots' resistance to breakage and continue to grow (Periakaruppan et al., 2023; Sangeeta Dash et al., 2023). NPs have been employed to increase crop yields, enhance resource efficiency, and improve nutrient delivery. A recent study showed that applying iron oxide nanoparticles enhanced the efficiency of the plants' uptake of nutrients. This tactic not only increases the plants' ability to utilize nutrients, but it also reduces the amount of nutrients that could be lost to their surroundings (Majeed, Hayat Bhatti, et al., 2020).

4.2. Nano-Enabled Soil Amendments for Fertility

The section will discuss the nano-enabled soil amendments which use nanoparticles (NPs) to control nutrient release, improve soil conditioning, and enhance plant nutrient uptake (Majeed, Bhatti, & Amjad, 2021). The potential for improving the efficiency of nutrient and pesticide AI consumption s enormous with these nano-enabled controlled-release fertilizers and insecticides. To obtain high-efficiency and safe nanopesticides and fertilizers, a comprehensive design strategy is advised for adoption. The figure 13 shows how environmental stimuli activate the release of source in a controlled environment. (Hassan et al., 2022; Malik et al., 2023).

Global food production must increase by 50% by 2050 to meet the demands of the growing population, nearing 9.8 billion (Fujita et al., 2023). Nanotechnology emerges as a key player in addressing this challenge, holding promise for a more robust, efficient, and sustainable agricultural system. Nano-pesticides, nano-herbicides, and nano-bactericides, alongside nano-delivery systems, aim to enhance agrochemical efficiency, minimize runoff, promote plant growth, and combat crop diseases. In the realm of nano-enabled soil amendments for fertility, key focuses include controlled nutrient release and sustainable soil conditioning for agriculture, signaling a pivotal step in agricultural innovation (B. K. Singh et al., 2023)

5. NANOMATERIALS IN EROSION CONTROL

Soil erosion poses a significant threat to global land productivity and food security (Ghazaryan et al., 2023). It occurs when the detachment and transport of soil particles through wind or water which carry away the fertile soil (Aziz et al., 2023; Naveed et al., 2024b). The slow erosion of exposed topsoil, or desertification, as a result of salinization, compaction, and loss of organic matter and structure is a worry for the environment since it reduces the amount of fertile land and, in turn, soil production (Rehman et al., 2023).

5.1. Nanofibers and Nanoparticles for Erosion Prevention

In order to maintain a sustainable agriculture ensuring food security, it is crucial to control this. Nanomaterials, along with Nano fibers and nanoparticles, with their unique properties and versatility, offer promising solutions for tackling this critical challenge (Muhammad et al., 2024). Several nanomaterial such as ENMs (engineered nanomaterials) and MENMs (Metal-based nanomaterials) function as SAPs (superabsorbent polymers), dust suppressants, and nano mulching to enhance erosion barriers, soil structure & vegetation (as shown in figure 14). These innovative nanomaterial usages pave the way for sustainable land management practices to protect our precious soil resources for future generations (Haq et al., 2022).

Recently in 2021, a research study was done to investigate the potential application of nanofiber (NF). The reduced soil and water loss clearly verify the usage of NF which suggests that NF has high absorbance capacity as well as water retention in soil and additionally reduces soil loss by increased adhesion properties. Hence, it can be effectively utilized for revegetation and land restoration (Arshad et al., 2022).

5.2. Nanoremediation and Microbial Interactions

Remediation when coupled with nanotechnology is termed as nanoremediation. It holds the potential for decontamination. Nanobioremediation uses NPs and living organisms to remediate contaminated pollutants. Therefore a proper understanding of the interaction between them is essential as microbial interactions play a crucial role in enhancing nanoremediation efficiency and they interact with NPs in several ways (Hassan, Haq, Majeed, et al., 2022). In 2018, Tan and coworkers shared the interaction linkages among contaminants, living organisms, environmental factors and nanomaterials (NM) as shown in figure 15. The specific interactions between NPs and microorganisms depend on the type of NP, the contaminant level, and the environmental conditions Naveed et al., 2024a.

5.3. Nanomaterials and Their Impact on Microbial Communities

The introduction of nanomaterials into soil and water ecosystems has the potential to significantly impact microbial communities, both positively and negatively. While nanomaterials can enhance microbial-assisted degradation of contaminants and promote soil fertility, they can also disrupt microbial biodiversity and alter ecosystem functioning. In order to maximize the benefits and minimize the risks, it is crucial to conduct thorough risk assessment, optimize NP design and synthesis, and monitor microbial responses (Hussain et al., 2023a). Nanomaterials are playing an increasingly important role in microbial-assisted degradation of contaminants and maintaining soil biodiversity during remediation. Reportedly, microbes have been employed as catalyst extraction to facilitate nanomaterials in remediation. As it could be beneficial or adversely affecting depending upon the interaction & mechanisms (as shown in figure 16). Therefore, several novel ways has been shared to mitigate the toxicity and maximize the benefits (Bashir et al., 2023; Tariq, et al., 2023; Jamil et al., 2022).

6. MONITORING AND ASSESSMENT USING NANOTECHNOLOGY

Nanotechnology offers unprecedented capabilities for detecting, analyzing, and monitoring environmental contaminants. Monitoring and assessing soil using nanotechnology is crucial. Soil health depends on various properties and real-time assessment is required for its management. For this, nanomaterials based nanobiosensors & engineered NPs have been proven useful for detection and removal of soil contaminants which can share real-time monitoring for soil health (Altug et al., 2022; Haq et al., 2022b).

6.1. Nanoscale Tools for Assessing Soil Health

Soil microbial activity monitoring is an essential for assessing soil health, identifying potential problems, and informing sustainable land management practices. Emerging nanotools revolutionize soil microbial activity monitoring using nanosensors, nano biomarkers and nanoprobe that aim to provide a more comprehensive and accurate understanding of soil health (Muhammad, Waheed, et al., 2024; Rohilla et al., 2021). The developed nanosensors ensure soil health by real time monitoring and measuring the soil conditions including diseases, toxicity and nutrient level (Bashir et al., 2023b; Rimpika et al., 2023). Besides the nanotools that have been discussed earlier, a novel approach "WSN

(Wireless sensor networks)" is also being employed in precision agriculture (PA). Such networks are of high importance for estimating irrigation needs and monitoring soil during PA. The system consists of configuring multiple types of sensors and wifi nodes above and below the ground (Sinha et al., 2023b). Moreover, Nanoscale imaging is remarkably essential in soil structure analysis. It aids in developing and deploying the suitable strategy for soil remediation and enhancing yield. In 2021, an integrated approach FIB- SEM (focused ion beam - scanning electron microscope) was utilized to study the soil structure for characterization of soil health. This imaging gave high resolution results levelling up the traditional XCT (X-ray computer tomography) technique but with nanoscale soil structure for deep analysis (Waheed et al., 2023).

7. ENVIRONMENTAL IMPACT AND SAFETY CONSIDERATIONS

The persistence of nanomaterials in the environment poses ecotoxicological risks within food chains. Therefore, safe and effective implementation of nanotechnology for soil decontamination requires a multi-faceted approach for environmental protection and human health (Ameen et al., 2021). By adhering to some guidelines we can ensure safe application of nanoparticles in soils. These guidelines include proper proceeding at some checkpoints like thorough risk assessment and monitoring, optimization of NP designs and synthesis, implementing controlled applications of NP, employing strategic biosurveillance and following standardized protocols. The establishment of these guidelines and assessment can aid to promote sustainable environmental safety by reducing potential risks and toxicological impact of nanotechnology (H. Ali et al., 2023; Hussain et al., 2023b). Nanoparticles' presence gives rise to several potential environmental and health risks due to bioaccumulation, biodegadation and bioconcentration which needs development of strategies for its mitigation (Elshayb, Nada, Sadek, et al., 2022). If these NPs are inhaled, ingested, or form contact (through skin), they can create several public health problems. Therefore, implication of safety conditions must be followed through legalised developed regulation after their implementation (Balasubramaniam et al., 2023).

8. FUTURE DIRECTIONS AND RESEARCH PRIORITIES

Emerging nanotechnology approaches hold immense potential for tackling a diverse range of soil contaminants to improve soil health. For a sustainable future, interdisciplinary collaborations between soil scientists, engineers, chemists, and toxicologists are crucial for developing and implementing effective nanoremediation strategies (Pandey et al., 2023). In order to advance soil remediation, research should focus on optimizing nanoparticle design, developing targeted delivery systems, evaluating long-term environmental impacts, and establishing clear regulatory frameworks for sustainable nanoremediation practices (Tassadduq et al., 2022).

The global nanomaterials market is poised for significant growth, with projections indicating a compound annual growth rate of 29.5% between 2017 and 2020, reaching an estimated value of $75.8 billion. Currently, there are nearly 400 companies actively involved in nanotechnology research and development worldwide, and this number is expected to surpass 1,000 by 2027 (Waheed et al., 2022; Majeed et al., 2023a). Recently in 2021, a novel nano enabled based system has been developed using benign Fe-oxalate capped metal oxide nanomaterials. These OCIMnox and OCICuox ensure reduced toxicity and increased nutrient uptake as nanofertilizers and show great potential for further usage as innovatively synthesized materials (Ullah et al., 2022; Majeed et al., 2023b).

The detection of microbes, contaminants and toxic NPs in soil is crucial due to its impact on plant growth and human health. Advanced nanomaterial-based biosensors have emerged as valuable tools for real-time detection (Sinha et al., 2023c). However, further research is needed to enhance the sensitivity and specificity of these sensors for large-scale applications in assessing soil microbes. Future studies should focus on exploring the efficiency of single and multiple species detection, combining detection methods with bacterial beneficial molecule identification, and developing innovative, flexible nanosensors. Understanding nanosensor behavior, trafficking, toxicity, and efficiency within microbes is crucial for the development of selective nanosensors that can unravel intricate microbial signaling mechanisms and provide unprecedented insights into microbial responses and functions in the microbiome under diverse environmental conditions (Majeed, Tariq, et al., 2021).

Significant advancements have been seen in nanotechnology to develop nanoformulations. It has emerged as a promising solution to address the limitations of conventional pesticides. Current trends and research focuses on the use of nanoparticles as pesticides for plant protection. The researchers emphasize cutting-edge approaches utilizing nanostructures as active components and biopesticides (Majeed, Bhatti, Pieroni, et al., 2021). Nanotechnology is revolutionizing agriculture, offering innovative solutions to address pressing challenges like food security, sustainability, and the achievement of 2030 Sustainable Development Goals (SDGs) as shown in figure 20 (Majeed, Khan, et al., 2022). Nanofertilizers, nanopesticides, and nanoremediation hold promise for transforming conventional agriculture into a more efficient, resilient, and sustainable system. Novel research directions, such as targeted delivery, multifunctionality, responsiveness, and nanobiosensors, are being explored to enhance the efficacy and sustainability of nanotechnology-based applications (Majeed, Lu, et al., 2022).

According to BCC research, the increasing adoption of nanoparticles (NPs) for environmental remediation is evident in the market recently. The global market for nanotechnology in environmental remediation has grown steadily, reaching \$23.4 billion in 2014 and projected to reach \$123.13 billion by 2022. While initial costs of NPs were high, they have significantly decreased, making them more viable for remediation applications (Majeed et al., 2021b; Majeed et al., 2023; Bhattacharya et al., 2023).

The NPs immobilization is an emerging methodology which has been utilized to mitigate the soil remediation challenges. It has been proven beneficial strategy as it prevents aggregation, enhances stability and reduces toxicology with controlled release (Khan et al., 2022). This approach uses various different materials as support such as zeolites, resins and chitosan etc. Additionally, polyelectrolytes have also shown great potential as delivery agents by offering targeted and improved transport as well as long term activity (Majeed, Bhatti, et al., 2020).

In soil science, NanoSIMS has also emerged as a revolutionary tool. It has provided insights of the soil processes at the micro level. The ability to analyze elemental at high resolution has given rise to understand nutrient cycle, microbes-NP interaction and soil processes. Despite its remarkable capabilities, NanoSIMS has certain limitations which need to be addressed in future to make this technique more effective and reliable. The limitations include time consumption while sample preparation, interpretation of data and requirements of sample size. NanoSIMS can also be coupled with other analytical approaches to mitigate the limitations. The combinations of techniques also open the doors for more understanding of soil health, processes and properties (Khoja et al., 2022; Naveed et al., 2024c).

CONCLUSION

In a nutshell, this chapter highlights the promising potential of nanotechnology in soil remediation and land restoration. It undoubtedly offers innovative and cost-effective solutions to address soil decontamination. The chapter has discussed various nanotechnology based applications which includes nanosensors for detection of contaminants, nanoparticles for decontamination of soil, nanomaterials for analyzing the soil structure, and nanofibers for the purpose of erosion control. The study also focuses on the significance of interdisciplinary collaborations and future research along with safety considerations for environmental protection.

REFERENCES

Ahmad, N., Naeem, M., Ali, H., Alabbosh, K. F., Hussain, H., Khan, I., Siddiqui, S. A., Khan, A. A., & Iqbal, B. (2023). From challenges to solutions: The impact of melatonin on abiotic stress synergies in horticultural plants via redox regulation and epigenetic signaling. *Scientia Horticulturae*, *321*, 112369. doi:10.1016/j.scienta.2023.112369

Ahmed, S., Khan, F. S. A., Mubarak, N. M., Khalid, M., Tan, Y. H., Mazari, S. A., Karri, R. R., & Abdullah, E. C. (2021). Emerging pollutants and their removal using visible-light responsive photocatalysis – A comprehensive review. *Journal of Environmental Chemical Engineering*, *9*(6), 106643. doi:10.1016/j.jece.2021.106643

Ali, H., Muhammad, Z., Majeed, M., Aziz, R., Khan, A., Mangrio, W. M., Abdo, H. G., Almohamad, H., & Al Dughairi, A. A. (2023). Vegetation diversity pattern during spring season in relation to topographic and edaphic variables in sub-tropical zone. *Botanical Studies*, *64*(1), 25. Advance online publication. doi:10.1186/s40529-023-00398-5 PMID:37716923

Ali, S., Tyagi, A., & Bae, H. (2023). ROS interplay between plant growth and stress biology: Challenges and future perspectives. *Plant Physiology and Biochemistry*, *203*, 108032. doi:10.1016/j.plaphy.2023.108032 PMID:37757722

Altug, H., Oh, S. H., Maier, S. A., & Homola, J. (2022). Advances and applications of nanophotonic biosensors. In Nature Nanotechnology (Vol. 17, Issue 1, pp. 5–16). Nature Research. doi:10.1038/s41565-021-01045-5

Ansari, M. A. (2023). Nanotechnology in Food and Plant Science: Challenges and Future Prospects. In Plants (Vol. 12, Issue 13). Multidisciplinary Digital Publishing Institute (MDPI). doi:10.3390/plants12132565

Arif, U., Bhatti, K. H., Ajaib, M., Wagay, N. A., Majeed, M., Zeb, J., Hameed, A., & Kiani, J. (2021). Ethnobotanical indigenous knowledge of Tehsil Charhoi, District Kotli, Azad Jammu and Kashmir, Pakistan. *Ethnobotany Research and Applications*, *22*. Advance online publication. doi:10.32859/era.22.50.1-24

Arora, D., Arora, A., Bala, R., Panghal, V., & Kumar, S. (2023). Enhancement in Phytoremediation Efficiency of Tagetus erecta with the Application of Nano-scale Zero Valent Iron (nZVI) for the Restoration of Lead Contaminated Soil: An Approach Toward Sustainability. *Water, Air, and Soil Pollution*, *234*(8), 535. Advance online publication. doi:10.1007/s11270-023-06540-0

Arshad, F., Waheed, M., Harun, N., Fatima, K., Khan, B. A., Fatima, K., Abbas, Z., Jabeen, S., & Majeed, M. (2022). Indigenous farmer's perception about fodder and foraging species of Semi-arid lowlands of Pakistan: A case study of District Kasur, Pakistan. *Taiwania*, *67*(4), 510–523. doi:10.6165/tai.2022.67.510

Azeem, M., Pirjan, K., Qasim, M., Mahmood, A., Javed, T., Muhammad, H., Yang, S., Dong, R., Ali, B., & Rahimi, M. (2023). Salinity stress improves antioxidant potential by modulating physio-biochemical responses in Moringa oleifera Lam. *Scientific Reports*, *13*(1), 2895. Advance online publication. doi:10.1038/s41598-023-29954-6 PMID:36807545

Aziz, A., Anwar, M. M., Majeed, M., Fatima, S., Mehdi, S. S., Mangrio, W. M., Elbouzidi, A., Abdullah, M., Shaukat, S., Zahid, N., Mahmoud, E. A., Casini, R., Yessoufou, K., & Elansary, H. O. (2023). Quantifying Landscape and Social Amenities as Ecosystem Services in Rapidly Changing Peri-Urban Landscape. *Land (Basel)*, *12*(2), 477. Advance online publication. doi:10.3390/land12020477

Balasubramaniam, T., Shen, G., Esmaeili, N., & Zhang, H. (2023). Plants' Response Mechanisms to Salinity Stress. In Plants (Vol. 12, Issue 12). MDPI. doi:10.3390/plants12122253

Banerjee, S., Mondal, S., Islam, J., Sarkar, R., Saha, B., & Sen, A. (2024). Rhizospheric nano-remediation salvages arsenic genotoxicity: Zinc-oxide nanoparticles articulate better oxidative stress management, reduce arsenic uptake, and increase yield in Pisum sativum (L.). *The Science of the Total Environment*, *913*, 169493. doi:10.1016/j.scitotenv.2023.169493 PMID:38151134

Bashir, S. M., Altaf, M., Hussain, T., Umair, M., Majeed, M., Mangrio, W. M., Khan, A. M., Gulshan, A. B., Hamed, M. H., Ashraf, S., Amjad, M. S., Bussmann, R. W., Abbasi, A. M., Casini, R., Alataway, A., Dewidar, A. Z., Al-Yafrsi, M., Amin, M. H., & Elansary, H. O. (2023b). Vernacular Taxonomy, Cultural and Ethnopharmacological Applications of Avian and Mammalian Species in the Vicinity of Ayubia National Park, Himalayan Region. *Biology (Basel)*, *12*(4), 609. Advance online publication. doi:10.3390/biology12040609 PMID:37106809

Bhalla, G., Bhalla, B., Kumar, V., & Sharma, A. (2022). Bioremediation and phytoremediation of pesticides residues from contaminated water: a novel approach. *Pesticides Remediation Technologies from Water and Wastewater*, 339–363. doi:10.1016/B978-0-323-90893-1.00016-7

Bhattacharya, R., & Bhattacharya, R. **Majeed, M.,** Bhandari, S., Aziz, R., Sinha, D., ... & Niyogi, S. S. Environmental Rehabilitation of Industrial Waste Dumping Site. In *Biohydrometallurgical Processes* (pp. 44–89). CRC Press. doi:10.1201/9781003451457-4

Biswas, P., Polash, S. A., Dey, D., Kaium, M. A., Mahmud, A. R., Yasmin, F., Baral, S. K., Islam, M. A., Rahaman, T. I., Abdullah, A., Ema, T. I., Khan, D. A., Bibi, S., Chopra, H., Kamel, M., Najda, A., Fouda, M. M. A., Rehan, U. S. M., Mheidat, M., ... Hasan, M. N. (2023). Advanced implications of nanotechnology in disease control and environmental perspectives. *Biomedicine and Pharmacotherapy*, *158*, 114172. doi:10.1016/j.biopha.2022.114172 PMID:36916399

Chandra, J., Hasan, N., Nasir, N., Wahab, S., Thanikachalama, P. V., Sahebkar, A., Ahmad, F. J., & Kesharwani, P. (2023). Nanotechnology-empowered strategies in treatment of skin cancer. *Environmental Research*, *235*, 116649. doi:10.1016/j.envres.2023.116649 PMID:37451568

Chehelgerdi, M., Chehelgerdi, M., Allela, O. Q. B., Pecho, R. D. C., Jayasankar, N., Rao, D. P., Thamaraikani, T., Vasanthan, M., Viktor, P., Lakshmaiya, N., Saadh, M. J., Amajd, A., Abo-Zaid, M. A., Castillo-Acobo, R. Y., Ismail, A. H., Amin, A. H., & Akhavan-Sigari, R. (2023). Progressing nanotechnology to improve targeted cancer treatment: overcoming hurdles in its clinical implementation. In Molecular Cancer (Vol. 22, Issue 1). BioMed Central Ltd. doi:10.1186/s12943-023-01865-0

Daniel, A. I., Keyster, M., & Klein, A. (2023). Biogenic zinc oxide nanoparticles: A viable agricultural tool to control plant pathogenic fungi and its potential effects on soil and plants. *The Science of the Total Environment*, *897*, 165483. doi:10.1016/j.scitotenv.2023.165483 PMID:37442458

Dehghanian, Z., Asgari Lajayer, B., Biglari Quchan Atigh, Z., Nayeri, S., Ahmadabadi, M., Taghipour, L., Senapathi, V., Astatkie, T., & Price, G. W. (2023). Micro (nano) plastics uptake, toxicity and detoxification in plants: Challenges and prospects. *Ecotoxicology and Environmental Safety*, *268*, 115676. doi:10.1016/j.ecoenv.2023.115676 PMID:37979355

Elshayb, O. M., Nada, A. M., Farroh, K. Y., Al-Huqail, A. A., Aljabri, M., Binothman, N., & Seleiman, M. F. (2022). Utilizing Urea–Chitosan Nanohybrid for Minimizing Synthetic Urea Application and Maximizing Oryza sativa L. Productivity and N Uptake. *Agriculture*, *12*(7), 944. Advance online publication. doi:10.3390/agriculture12070944

Elshayb, O. M., Nada, A. M., Sadek, A. H., Ismail, S. H., Shami, A., Alharbi, B. M., Alhammad, B. A., & Seleiman, M. F. (2022). The Integrative Effects of Biochar and ZnO Nanoparticles for Enhancing Rice Productivity and Water Use Efficiency under Irrigation Deficit Conditions. *Plants*, *11*(11), 1416. Advance online publication. doi:10.3390/plants11111416 PMID:35684189

Fadiji, A. E., Yadav, A. N., Santoyo, G., & Babalola, O. O. (2023). Understanding the plant-microbe interactions in environments exposed to abiotic stresses: An overview. *Microbiological Research*, *271*, 127368. doi:10.1016/j.micres.2023.127368 PMID:36965460

Fujita, H., Ohta, S., Nakamura, N., Somiya, M., & Horie, M. (2023). Progress of Endogenous and Exogenous Nanoparticles for Cancer Therapy and Diagnostics. In Genes (Vol. 14, Issue 2). MDPI. doi:10.3390/genes14020259

Ghazaryan, K. A., Harutyunyan, A. S., Khachatryan, H. E., Singh, A., Minkina, T. M., Rajput, V. D., & Movsesyan, H. S. (2023). Study of tolerance and phytodesalination potential of wheat, oat, emmer, and barley for sustainable saline agriculture. *Applied Ecology and Environmental Research*, *21*(5), 4853–4882. doi:10.15666/aeer/2105_48534882

Ghazaryan, K. A., Movsesyan, H. S., Khachatryan, H. E., & Ghazaryan, N. P. (2018). Geochemistry of Potentially Toxic Trace Elements in Soils of Mining Area: A Case Study from Zangezur Copper and Molybdenum Combine, Armenia. *Bulletin of Environmental Contamination and Toxicology*, *101*(6), 732–737. doi:10.1007/s00128-018-2443-0 PMID:30229278

Haq, S. M., Tariq, A., Li, Q., Yaqoob, U., Majeed, M., Hassan, M., Fatima, S., Kumar, M., Bussmann, R. W., Moazzam, M. F. U., & Aslam, M. (2022a). Influence of Edaphic Properties in Determining Forest Community Patterns of the Zabarwan Mountain Range in the Kashmir Himalayas. *Forests*, *13*(8), 1214. Advance online publication. doi:10.3390/f13081214

Hassan, M., Haq, S. M., Ahmad, R., Majeed, M., Sahito, H. A., Shirani, M., Mubeen, I., Aziz, M. A., Pieroni, A., Bussmann, R. W., Alataway, A., Dewidar, A. Z., Al-Yafrsi, M., Elansary, H. O., & Yessoufou, K. (2022). Traditional Use of Wild and Domestic Fauna among Different Ethnic Groups in the Western Himalayas—A Cross Cultural Analysis. *Animals (Basel)*, *12*(17), 2276. Advance online publication. doi:10.3390/ani12172276 PMID:36077997

Hassan, M., Haq, S. M., Majeed, M., Umair, M., Sahito, H. A., Shirani, M., Waheed, M., Aziz, R., Ahmad, R., Bussmann, R. W., Alataway, A., Dewidar, A. Z., El-Abedin, T. K. Z., Al-Yafrsi, M., Elansary, H. O., & Yessoufou, K. (2022). Traditional Food and Medicine: Ethno-Traditional Usage of Fish Fauna across the Valley of Kashmir: A Western Himalayan Region. *Diversity*, *14*(6), 455. Advance online publication. doi:10.3390/d14060455

Hua, Z., Ma, S., Ouyang, Z., Liu, P., Qiang, H., & Guo, X. (2023). The review of nanoplastics in plants: Detection, analysis, uptake, migration and risk. *Trends in Analytical Chemistry*, *158*, 116889. doi:10.1016/j.trac.2022.116889

Hussain, S., Raza, A., Abdo, H. G., Mubeen, M., Tariq, A., Nasim, W., Majeed, M., Almohamad, H., & Al Dughairi, A. A. (2023b). Relation of land surface temperature with different vegetation indices using multi-temporal remote sensing data in Sahiwal region, Pakistan. *Geoscience Letters*, *10*(1), 33. Advance online publication. doi:10.1186/s40562-023-00287-6

Jamil, M. D., Waheed, M., Akhtar, S., Bangash, N., Chaudhari, S. K., Majeed, M., Hussain, M., Ali, K., & Jones, D. A. (2022). Invasive Plants Diversity, Ecological Status, and Distribution Pattern in Relation to Edaphic Factors in Different Habitat Types of District Mandi Bahauddin, Punjab, Pakistan. *Sustainability (Switzerland), 14*(20). doi:10.3390/su142013312

Khan, A. M., Li, Q., Saqib, Z., Khan, N., Habib, T., Khalid, N., Majeed, M., & Tariq, A. (2022). MaxEnt Modelling and Impact of Climate Change on Habitat Suitability Variations of Economically Important Chilgoza Pine (Pinus gerardiana Wall.) in South Asia. *Forests*, *13*(5), 715. Advance online publication. doi:10.3390/f13050715

Khoja, A. A., Haq, S. M., Majeed, M., Hassan, M., Waheed, M., Yaqoob, U., Bussmann, R. W., Alataway, A., Dewidar, A. Z., Al-Yafrsi, M., Elansary, H. O., Yessoufou, K., & Zaman, W. (2022). Diversity, Ecological and Traditional Knowledge of Pteridophytes in the Western Himalayas. *Diversity*, *14*(8), 628. Advance online publication. doi:10.3390/d14080628

Kumar, R., Kumar, M., & Luthra, G. (2023). Fundamental approaches and applications of nanotechnology: A mini review. *Materials Today: Proceedings*. Advance online publication. doi:10.1016/j.matpr.2022.12.172

Kumari, R., Suman, K., Karmakar, S., Mishra, V., Lakra, S. G., Saurav, G. K., & Mahto, B. K. (2023). Regulation and safety measures for nanotechnology-based agri-products. In Frontiers in Genome Editing (Vol. 5). Frontiers Media SA. doi:10.3389/fgeed.2023.1200987

Lamichhane, J. R., Barbetti, M. J., Chilvers, M. I., Pandey, A. K., & Steinberg, C. (2024). Exploiting root exudates to manage soil-borne disease complexes in a changing climate. In Trends in Microbiology (Vol. 32, Issue 1, pp. 27–37). Elsevier Ltd. doi:10.1016/j.tim.2023.07.011

Li, S., Chen, L., & Fu, Y. (2023). Nanotechnology-based ocular drug delivery systems: recent advances and future prospects. In Journal of Nanobiotechnology (Vol. 21, Issue 1). BioMed Central Ltd. doi:10.1186/s12951-023-01992-2

Liao, Z., Fan, J., Lai, Z., Bai, Z., Wang, H., Cheng, M., Zhang, F., & Li, Z. (2023). Response network and regulatory measures of plant-soil-rhizosphere environment to drought stress. In *Advances in Agronomy* (Vol. 180, pp. 93–196). Academic Press Inc. doi:10.1016/bs.agron.2023.03.002

Liu, J., Li, W., Chen, H., Li, S., Yang, L., Peng, K., Cai, C., & Huang, X. (2023). Applications of functional nanoparticle–stabilized surfactant foam in petroleum-contaminated soil remediation. *Journal of Hazardous Materials*, *443*, 130267. doi:10.1016/j.jhazmat.2022.130267 PMID:36444047

Lo Presti, D., Di Tocco, J., Massaroni, C., Cimini, S., De Gara, L., Singh, S., Raucci, A., Manganiello, G., Woo, S. L., Schena, E., & Cinti, S. (2023). Current understanding, challenges and perspective on portable systems applied to plant monitoring and precision agriculture. *Biosensors & Bioelectronics*, *222*, 115005. doi:10.1016/j.bios.2022.115005 PMID:36527829

Loiseau, B., Carrière, S. D., Jougnot, D., Singha, K., Mary, B., Delpierre, N., Guérin, R., & Martin-StPaul, N. K. (2023). The geophysical toolbox applied to forest ecosystems – A review. *The Science of the Total Environment*, *899*, 165503. doi:10.1016/j.scitotenv.2023.165503 PMID:37454861

Majeed, M., Tariq, A., Anwar, M. M., Khan, A. M., Arshad, F., Mumtaz, F., Farhan, M., Zhang, L., Zafar, A., Aziz, M., Abbasi, S., Rahman, G., Hussain, S., Waheed, M., Fatima, K., & Shaukat, S. (2021). Monitoring of Land Use–Land Cover Change and Potential Causal Factors of Climate Change in Jhelum District, Punjab, Pakistan, through GIS and Multi-Temporal Satellite Data (pp. 129-147). https://www.mdpi.com/books/book/6906-land-use-land-cover-and-natural-hazards-interactions-changes-and-impacts

Majeed, M., Bhatti, K. H., & Amjad, M. S. (2021). Impact of climatic variations on the flowering phenology of plant species in jhelum district, punjab, pakistan. *Applied Ecology and Environmental Research*, *19*(5), 3343–3376. doi:10.15666/aeer/1905_33433376

Majeed, M., Bhatti, K. H., Amjad, M. S., Abbasi, A. M., Bussmann, R. W., Nawaz, F., Rashid, A., Mehmood, A., Mahmood, M., Khan, W. M., & Ahmad, K. S. (2020). Ethno-veterinary uses of poaceae in Punjab, Pakistan. *PLoS ONE, 15*(11 November). doi:10.1371/journal.pone.0241705

Majeed, M., Bhatti, K. H., Pieroni, A., Sõukand, R., Bussmann, R. W., Khan, A. M., Chaudhari, S. K., Aziz, M. A., & Amjad, M. S. (2021). Gathered wild food plants among diverse religious groups in Jhelum District, Punjab, Pakistan. *Foods*, *10*(3), 594. Advance online publication. doi:10.3390/foods10030594 PMID:33799901

MajeedM.Hayat BhattiK.Shoaib AmjadM.Mehmood AbbasiA.RashidA.Nawaz Muhammad Nawaz ShareefF.MehmoodA.MahmoodM. (2020). *Ethno-veterinary practices of Poaceae taxa in Punjab, Pakistan*. doi:10.21203/rs.3.rs-15532/v1

Majeed, M., Khan, A. M., Habib, T., Anwar, M. M., Sahito, H. A., Khan, N., & Ali, K. (2022). Vegetation analysis and environmental indicators of an arid tropical forest ecosystem of Pakistan. *Ecological Indicators*, *142*, 109291. Advance online publication. doi:10.1016/j.ecolind.2022.109291

Majeed, M., Lu, L., Anwar, M. M., Tariq, A., Qin, S., El-Hefnawy, M. E., El-Sharnouby, M., Li, Q., & Alasmari, A. (2023). Prediction of flash flood susceptibility using integrating analytic hierarchy process (AHP) and frequency ratio (FR) algorithms. *Frontiers in Environmental Science*, *10*, 1037547. Advance online publication. doi:10.3389/fenvs.2022.1037547

Majeed, M., Lu, L., Haq, S. M., Waheed, M., Sahito, H. A., Fatima, S., Aziz, R., Bussmann, R. W., Tariq, A., Ullah, I., & Aslam, M. (2022). Spatiotemporal Distribution Patterns of Climbers along an Abiotic Gradient in Jhelum District, Punjab, Pakistan. *Forests*, *13*(8), 1244. Advance online publication. doi:10.3390/f13081244

Majeed, M., Muhammad, M., Hussain, T., Ali, M., Naz, T., Nawaz, S., & Abbas, G. (2023a). Nanopriming-Based Management of Biotic Stresses for Sustainable Agriculture. In *Nanopriming Approach to Sustainable Agriculture* (pp. 263–289). IGI Global. doi:10.4018/978-1-6684-7232-3.ch012

Majeed, M., Muhammad, M., Nawaz, S., Naz, T., Iqbal, M. M., Zahid, N., ... Mehboob, M. (2023b). Nanopriming for Crop Management for Sustainable Agriculture: An Overview. *Nanopriming Approach to Sustainable Agriculture*, 110-141. doi:10.4018/978-1-6684-7232-3.ch005

Majeed, M., Tariq, A., Anwar, M. M., Khan, A. M., Arshad, F., Mumtaz, F., Farhan, M., Zhang, L., Zafar, A., Aziz, M., Abbasi, S., Rahman, G., Hussain, S., Waheed, M., Fatima, K., & Shaukat, S. (2021). Monitoring of land use–Land cover change and potential causal factors of climate change in Jhelum district, Punjab, Pakistan, through GIS and multi-temporal satellite data. *Land (Basel)*, *10*(10), 1026. Advance online publication. doi:10.3390/land10101026

Malik, S., Muhammad, K., & Waheed, Y. (2023). Nanotechnology: A Revolution in Modern Industry. In Molecules (Vol. 28, Issue 2). MDPI. doi:10.3390/molecules28020661

Manimegalai, S., Vickram, S., Deena, S. R., Rohini, K., Thanigaivel, S., Manikandan, S., Subbaiya, R., Karmegam, N., Kim, W., & Govarthanan, M. (2023). Carbon-based nanomaterial intervention and efficient removal of various contaminants from effluents – A review. *Chemosphere*, *312*, 137319. doi:10.1016/j.chemosphere.2022.137319 PMID:36410505

Mubarok, S., Nuraini, A., Hamdani, J. S., Suminar, E., Kusumiyati, K., Budiarto, R., Lestari, F. W., Rahmat, B. P. N., & Ezura, H. (2024). Antioxidative response of parthenocarpic tomato, iaa9-3 and iaa9-5, under heat stress condition. *Plant Physiology and Biochemistry*, *207*, 108333. doi:10.1016/j.plaphy.2024.108333 PMID:38181640

Muhammad, M., Basit, A., Majeed, M., Shah, A. A., Ullah, I., Mohamed, H. I., Khan, A., & Ghanaim, A. M. (2024). Bacterial pigments and their applications. In *Bacterial Secondary Metabolites* (pp. 277–298). Elsevier. doi:10.1016/B978-0-323-95251-4.00019-3

Muhammad, M., Waheed, A., Wahab, A., Majeed, M., Nazim, M., Liu, Y. H., Li, L., & Li, W. J. (2024). Soil salinity and drought tolerance: An evaluation of plant growth, productivity, microbial diversity, and amelioration strategies. In *Plant Stress* (Vol. 11). Elsevier B.V. doi:10.1016/j.stress.2023.100319

Naveed, M. Muhammad Majeed, Khizra Jabeen, Nimra Hanif, Rida Naveed, Sania Saleem & Nida Khan (2024). Plant Disease Diagnosis with Artificial Intelligence (AI). Microbial Data Intelligence and Computational Techniques for Sustainable Computing (pp 217–234). Springer. doi:10.1007/978-981-99-9621-6_15

Naveed, Zahid, Fatima, Saleem, Majeed, Abid, Javed, Wazir, & Qasim. (2024). Disease Diagnosis with Artificial Intelligence (AI). In Microbial Data Intelligence and Computational Techniques for Sustainable Computing (pp 235–258). Springer. doi:10.1007/978-981-99-9621-6_16

Naveed, Memon, Abdullah, Makhdoom, Azeem, Mehmood, Salahuddin, Rajpoot, & Majeed. (2024). Use Cases and Future Aspects of Intelligent Techniques in Microbial Data Analysis. In Microbial Data Intelligence and Computational Techniques for Sustainable Computing (pp 259–280). Springer. doi:10.1007/978-981-99-9621-6_17

New, W. X., Ogbezode, J. E., & Gani, P. (2023). Nanoparticles in Soil Remediation: Challenges and Opportunities. *Industrial and Domestic Waste Management*, *3*(2). Advance online publication. doi:10.53623/idwm.v3i2.357

Nguyen, N. T. T., Nguyen, L. M., Nguyen, T. T. T., Tran, U. P. N., Nguyen, D. T. C., & Van Tran, T. (2023). A critical review on the bio-mediated green synthesis and multiple applications of magnesium oxide nanoparticles. *Chemosphere*, *312*, 137301. doi:10.1016/j.chemosphere.2022.137301 PMID:36410506

Pandey, P., Tripathi, A., Dwivedi, S., Lal, K., & Jhang, T. (2023). Deciphering the mechanisms, hormonal signaling, and potential applications of endophytic microbes to mediate stress tolerance in medicinal plants. In Frontiers in Plant Science (Vol. 14). Frontiers Media SA. doi:10.3389/fpls.2023.1250020

Periakaruppan, R., Vanathi, P., & Vidhya, D. (2023). Management of soil nutrient deficiency by nano-metal oxides. *Nanometal Oxides in Horticulture and Agronomy*, 291–320. doi:10.1016/B978-0-323-91809-1.00003-2

Rajput, V. D., Singh, A., Minkina, T., Rawat, S., Mandzhieva, S., Sushkova, S., Shuvaeva, V., Nazarenko, O., Rajput, P., Komariah, Verma, K. K., Singh, A. K., Rao, M., & Upadhyay, S. K. (2021). Nano-Enabled Products: Challenges and Opportunities for Sustainable Agriculture. *Plants*, *10*(12), 2727. doi:10.3390/plants10122727 PMID:34961197

Rehman, A., Aziz, A., Anwar, M. M., Majeed, M., Albanai, J. A., Almohamad, H., & Abdo, H. G. (2023). Quantifying the impacts of urbanization on urban green, evidences from Maga City, Lahore Pakistan. *Discover Sustainability*, *4*(1), 48. Advance online publication. doi:10.1007/s43621-023-00169-z

Rimpika, A., Anushi, Manasa, S., Sharma, S., Thakur, A., Shilpa, & Sood, A. (2023). An Overview of Precision Farming. *International Journal of Environment and Climate Change*, *13*(12), 441–456. doi:10.9734/ijecc/2023/v13i123701

Rohilla, D., Chaudhary, S., & Umar, A. (2021). An Overview of Advanced Nanomaterials for Sensor Applications. *Engineering and Science*, *16*, 47–70. doi:10.30919/es8d552

Sangeeta Dash, E., Yadav, A. L., Bora, R., & Kumar Pandey, A. (2023). *Research Trends in Agriculture Science* (Vol. I). www.bhumipublishing.com

Singh, A., Rajput, V., Varshney, A., Sharma, R., Ghazaryan, K., Minkina, T., . . . El-Ramady, H. (2024b). Revolutionizing Crop Production: Nanoscale Wonders - Current Applications, Advances, and Future Frontiers. *Egyptian Journal of Soil Science, 64*(1). doi:10.21608/ejss.2023.246354.1684

Singh, A., Rajput, V. D., Tawaha, A. R. M., Al Zoubi, O. M., Habeeb, T., Rawat, S., Ghazaryan, K., & Minkina, T. (2023a). A Review on Crop Responses to Nanofertilizers for Mitigation of Multiple Environmental Stresses. *Ecological Engineering & Environmental Technology*, *24*(7), 280–296. doi:10.12912/27197050/169313

Singh, A., Rajput, V. D., Tawaha, A. R. M., Al Zoubi, O. M., Habeeb, T., Rawat, S., Ghazaryan, K., & Minkina, T. (2023b). A Review on Crop Responses to Nanofertilizers for Mitigation of Multiple Environmental Stresses. *Ecological Engineering & Environmental Technology*, *24*(7), 280–296. doi:10.12912/27197050/169313

Singh, A., Rajput, V. D., Varshney, A., Ghazaryan, K., & Minkina, T. (2023c). Small Tech, Big Impact: Agri-nanotechnology Journey to Optimize Crop Protection and Production for Sustainable Agriculture. *Plant Stress*, *10*, 100253. doi:10.1016/j.stress.2023.100253

Singh, A., Sengar, R. S., Rajput, V. D., Al-Ghzawi, A. L., Shahi, U. P., Ghazaryan, K., Minkina, T., Tawaha, A. R. M. A., Al Zoubi, O. M., & Habeeb, T. (2024a). Impact of Salinity Stress and Zinc Oxide Nanoparticles on Macro and Micronutrient Assimilation: Unraveling the Link between Environmental Factors and Nutrient Uptake. *Journal of Ecological Engineering*, *25*(2), 1–9. doi:10.12911/22998993/172947

Singh, A., Sharma, R., Rajput, V. D., Ghazaryan, K., Minkina, T., Al Tawaha, A. R. M., & Varshney, A. (2023d). Green Synthesis of Nanofertilizers and Their Application for Crop Production. *Nanotechnology in the Life Sciences*, 205–231. doi:10.1007/978-3-031-41329-2_8

Singh, A., Sharma, R., Rawat, S., Singh, A. K., Rajput, V. D., Fedorov, Y., . . . Chaplygin, V. (2022). Nanomaterial-plant interaction: Views on the pros and cons. *Toxicity of Nanoparticles in Plants*, 47–68. doi:10.1016/B978-0-323-90774-3.00015-5

Singh, A. V., Katz, A., Maharjan, R. S., Gadicherla, A. K., Richter, M. H., Heyda, J., del Pino, P., Laux, P., & Luch, A. (2023). Coronavirus-mimicking nanoparticles (CorNPs) in artificial saliva droplets and nanoaerosols: Influence of shape and environmental factors on particokinetics/particle aerodynamics. *The Science of the Total Environment*, *860*, 160503. doi:10.1016/j.scitotenv.2022.160503 PMID:36442637

Singh, B. K., Delgado-Baquerizo, M., Egidi, E., Guirado, E., Leach, J. E., Liu, H., & Trivedi, P. (2023). Climate change impacts on plant pathogens, food security and paths forward. In Nature Reviews Microbiology (Vol. 21, Issue 10, pp. 640–656). Nature Research. doi:10.1038/s41579-023-00900-7

Singh, S., Chakma, S., Alawa, B., Kalyanasundaram, M., & Diwan, V. (2023). Identification, characterization, and implications of microplastics in soil – A case study of Bhopal, central India. *Journal of Hazardous Materials Advances*, *9*, 100225. doi:10.1016/j.hazadv.2022.100225

Sinha, D., Maurya, A. K., Abdi, G., Majeed, M., Agarwal, R., Mukherjee, R., Ganguly, S., Aziz, R., Bhatia, M., Majgaonkar, A., Seal, S., Das, M., Banerjee, S., Chowdhury, S., Adeyemi, S. B., & Chen, J. T. (2023a). Integrated Genomic Selection for Accelerating Breeding Programs of Climate-Smart Cereals. In Genes (Vol. 14, Issue 7). Multidisciplinary Digital Publishing Institute (MDPI). doi:10.3390/genes14071484

Sukul, U., Das, K., Chen, J. S., Sharma, R. K., Dey, G., Banerjee, P., Taharia, M., Lee, C. I., Maity, J. P., Lin, P. Y., & Chen, C. Y. (2023). Insight interactions of engineered nanoparticles with aquatic higher plants for phytoaccumulation, phytotoxicity, and phytoremediation applications: A review. *Aquatic Toxicology (Amsterdam, Netherlands)*, *264*, 106713. doi:10.1016/j.aquatox.2023.106713 PMID:37866164

Swain, R., Sahoo, S., Behera, M., & Rout, G. R. (2023). Instigating prevalent abiotic stress resilience in crop by exogenous application of phytohormones and nutrient. In *Frontiers in Plant Science* (Vol. 14). Frontiers Media S.A., doi:10.3389/fpls.2023.1104874

Tariq, A., Mumtaz, F., Majeed, M., & Zeng, X. (2023a). Spatio-temporal assessment of land use land cover based on trajectories and cellular automata Markov modelling and its impact on land surface temperature of Lahore district Pakistan. *Environmental Monitoring and Assessment, 195*(1). doi:10.1007/s10661-022-10738-w

Tassadduq, S. S., Akhtar, S., Waheed, M., Bangash, N., Nayab, D. E., Majeed, M., Abbasi, S., Muhammad, M., Alataway, A., Dewidar, A. Z., Elansary, H. O., & Yessoufou, K. (2022). Ecological Distribution Patterns of Wild Grasses and Abiotic Factors. *Sustainability (Basel)*, *14*(18), 11117. Advance online publication. doi:10.3390/su141811117

Tyagi, A., Ali, S., Ramakrishna, G., Singh, A., Park, S., Mahmoudi, H., & Bae, H. (2023). Revisiting the Role of Polyamines in Plant Growth and Abiotic Stress Resilience: Mechanisms, Crosstalk, and Future Perspectives. In Journal of Plant Growth Regulation (Vol. 42, Issue 8, pp. 5074–5098). Springer. doi:10.1007/s00344-022-10847-3

Ullah, I., Aslam, B., Shah, S. H. I. A., Tariq, A., Qin, S., Majeed, M., & Havenith, H. B. (2022). An Integrated Approach of Machine Learning, Remote Sensing, and GIS Data for the Landslide Susceptibility Mapping. *Land (Basel)*, *11*(8), 1265. Advance online publication. doi:10.3390/land11081265

Waheed, M., Arshad, F., Majeed, M., Fatima, S., Mukhtar, N., Aziz, R., Mangrio, W. M., Almohamad, H., Al Dughairi, A. A., Al-Mutiry, M., & Abdo, H. G. (2022). Community Structure and Distribution Pattern of Woody Vegetation in Response to Soil properties in Semi-Arid Lowland District Kasur Punjab, Pakistan. *Land (Basel)*, *11*(12), 2145. Advance online publication. doi:10.3390/land11122145

Waheed, M., Arshad, F., Majeed, M., Haq, S. M., Aziz, R., Bussmann, R. W., Ali, K., Subhan, F., Jones, D. A., & Zaitouny, A. (2023). *Potential distribution of a noxious weed (Solanum viarum Du-nal), current status, and future invasion risk based on MaxEnt modeling*. Geology, Ecology, and Landscapes. do i:10.1080/24749508.2023.2179752

Chapter 5
Precision Farming With Nanoscale Sensors:
Monitoring Soil, Water, and Crop Health

Jyoti Rani
Chaudhary Devi Lal University, India

Arshad Rasool
Minhaj University, Lahore, Pakistan

Hassan Sher
University of Swat, Pakistan

Rabia Fatima
https://orcid.org/0000-0001-9043-1207
Clarkson University, USA

Javeed Hussain
University of Kotli, Pakistan

Muhammad Majeed
https://orcid.org/0000-0001-7663-2563
University of Gujrat, Pakistan

Faisal Hussain
https://orcid.org/0000-0001-5571-6186
Ghazi University, Pakistan

Olim K. Khojimatov
Tashkent Botanical Garden of the Academy of Sciences of the Republic of Uzbekistan, Uzbekistan

Khadim Hussain
Islamia University, Bahawalpur, Pakistan

ABSTRACT

In the demesne of sustainable agriculture, diverse scientific disciplines contribute significantly, with nanotechnology taking center stage. Recent strides in precision agriculture with nanotechnology primarily emphasize real-time, environmentally conscious sensing of crop, water, and soil parameters to optimize crop yield. Nanotechnology's focus extends to the development of innovative nanomaterials, enhancing soil nutrient availability, seed germination, and seedling growth while concurrently addressing biotic and abiotic stress. Furthermore, the application of diverse nano-based fertilizers and pesticides directly onto crops at reduced dosages emerges as a key strategy, mitigating risks to both human health and the environment. The exploration of advanced nanodiagnostic systems and nanobiosensors plays a pivotal role in monitoring plant diseases, soil quality parameters, and stress, thereby offering a substantial advantage in precision agriculture. Nanosensors are composed of nanoscale particles such as nanoscale wires, carbon nanotubes, thin films, and nanoparticles.

DOI: 10.4018/979-8-3693-1890-4.ch005

Figure 1. Overview of nanosensor and its uses in precision agriculture

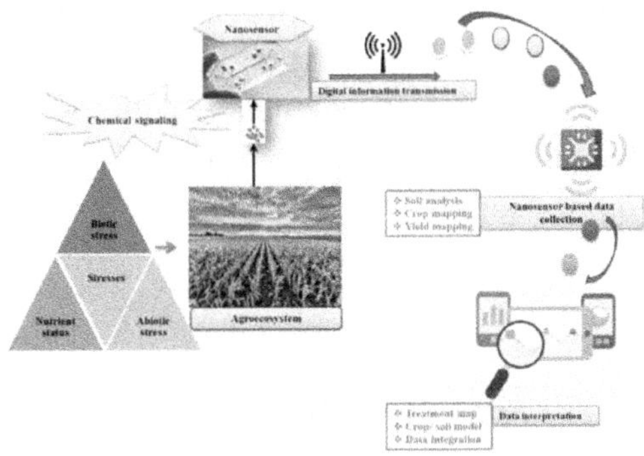

1. INTRODUCTION

To enhance agriculture for increased productivity, economic growth, and environmental sustainability, a departure from traditional methods is imperative (Rajput et al., 2021). Farmers, driven by yield concerns, have increasingly relied on agrochemicals. However, expanding agricultural land due to population growth has led to heightened agrochemical usage, causing pollution in air, water, and soil (Muhammad, Waheed, et al., 2024). Recognizing the harmful impact, the scientific community aims to enhance conventional farming for environmental protection. Global awareness is pressuring farmers to adopt modern methods, with precision agriculture standing out for its reduced agrochemical usage, offering crop-specific treatments to increase yields (Sangeeta Dash et al., 2023). Nanotechnology has emerged as a key approach, providing innovative solutions for environmentally friendly practices in precision farming. Nanomaterials, with unique properties, serve as the foundational building blocks. Precision agriculture incorporates various nano-based tools like nanosensors, nanofertilizers (NFs), nanoherbicides, nanoscale carriers, and nanopesticides, reducing reliance on diverse agrochemicals (Singh et al., 2022). Integrating nanotechnology has the potential to preserve soil and water health without compromising crop productivity (Goyal et al., 2023). Nano-biosensors play a crucial role in precise monitoring of crop growth and soil properties in precision agriculture.

Contemporary agriculture is undergoing a transformative shift to address feeding a growing global population while mitigating environmental impact. Precision farming, a departure from conventional methods, involves meticulous management guided by real-time data and site-specific conditions (Ghazaryan et al., 2018). Through information and communication technologies, precision farming optimizes crop yields, reduces environmental impact, and elevates overall farm sustainability (Sangeeta Dash et al., 2023). Nanoscale sensors in agriculture operate at molecular and atomic levels, revolutionizing various aspects. They offer novel capabilities for monitoring critical parameters like soil conditions, water quality, and crop health with unprecedented precision. The integration of nanoscale sensors into precision farming systems fundamentally transforms comprehension and management of agricultural processes (Yadav et al., 2023).

This integrated approach emphasizes the essential role of various nanosensors in water, soil, and plant systems. Exploring the application of biosynthesized nanoparticles in advanced agriculture provides a nuanced understanding of their potential. The chapter offers a comprehensive overview of nanosensors, addressing challenges and risks in real-time, emphasizing their key role in mitigating the environmental impact of conventional agrochemicals while ensuring increased agricultural productivity. As agriculture evolves, integrating cutting-edge technologies becomes essential to meet the demands of a growing global population while safeguarding the environment for future generations. These sensors detect changes in conductance when semiconducting carbon nanotubes interact with specific chemicals. The comprehensive compilation presented in this chapter delves into the multifaceted applications and roles of nanosensors in precision agriculture and crop protection. In summary, nanosensors emerge as indispensable tools, significantly contributing to crop protection and advancing the principles of precision agriculture.

2. NANOSCALE SENSORS: FUNDAMENTALS AND TYPES

This section introduces nanoscale sensors in precision farming, emphasizing their compact dimensions (around 100 nm) and unique properties in detecting events at the nanoscale. Integrated into precision agriculture, nanosensors provide real-time insights into soil, water, and plant health, facilitating timely interventions and optimizing practices for sustainability (Ullah et al., 2022). This integration marks a transition toward enhanced agricultural management and ecological sustainability.

2.1 Principles of Nanoscale Sensors

Nanoscale sensors, operating within the realm of nanotechnology, manipulate matter at the molecular or atomic scale. These minute sensors, measured in nanometers, leverage unique properties at this scale for highly sensitive and precise measurements. Functioning based on principles like quantum effects and size-dependent properties, nanosensors distinguish themselves from traditional sensing technologies (Altug et al., 2022). Comprising key components such as a sensor, analytic, transducer, detector, and a feedback path, nanosensors prove effective in monitoring electrical changes to measure levels at the single-molecule scale (Figure 2). Real-world applications demonstrate the capabilities of nanosensors, such as carbon nanotube sensors detecting ammonia and responding with enhanced conductivity when interacting with water vapor (Rohilla et al., 2021). This dynamic field, still in its early stages, holds promise for diverse applications, relying on multidisciplinary approaches.

2.2 Different Types of Nanosensors

Nanosensors, influential in diverse sectors like food, medicine, agriculture, and the environment, exhibit predominant classifications as electrochemical, photonic, and colorimetric devices (Pramanik et al., 2020) (Table 1, Figure 3). Electrochemical nanosensors rely on alterations in resistance within nanomaterials due to analyte binding, particularly in one-dimensional materials like nanowires and nanotubes. Chemical nanosensors identify reduced electrical conductivity resulting from molecular attachment to nanomaterials, while physical chemical nanosensors detect conductivity changes through physical manipulation, often utilizing capacitors. Surface-enhanced Raman spectroscopy nanosensors, plasmonic nanosensors, and electromagnetic nanosensors exemplify this category.

Photonic nanosensors, employing Bragg grating color changes in response to chemical stimuli, quantify sample concentration. Colorimetric nanosensors, using visible color changes induced by analyte presence, find applications in detecting heavy metals and hazardous gases, offering portable alternatives for environmental monitoring. Gold nanoparticles, for instance, play a crucial role in swiftly identifying

Figure 2. Key components of a nanosensor

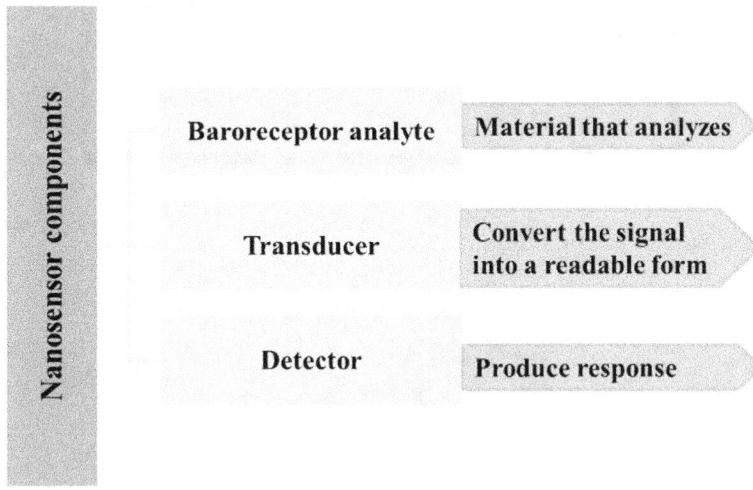

pollutants in the field, contributing to environmental protection. The transformative impact of nanosensors across these domains underscores their significance and the ongoing advancements in their applications.

2.2.1 Nanosensors for Soil Monitoring

Nanosensors designed for soil monitoring are essential in precision farming, offering real-time information on soil composition, moisture, and nutrient levels. These tiny sensors enable accurate assessment of soil health, empowering farmers to optimize fertilization for improved crop yield. Biomonitoring, involving real-time data collection on elements like fertilizers and pathogens, is critical. Ideal nanobiosensors possess traits such as quick response, stability during storage, portability, affordability, biocompatibility, and accuracy. Operating at the atomic scale, nanobiosensors show remarkable sensitivity, enabling precise virus detection at low concentrations. An instance is the graphene-based nanosensor for soil nitrogen measurement, integrated into a wireless network for timely nutrient information. Another example is

Table 1. Classification of nanosensors on the basis of stimuli for their activation

Stimuli	Features
Biological	Biomass (identities, concentrations, states)
Chemical	Components (identities, concentrations, states)
Electrical	Charge, current, potential, dielectric constant, conductivity
Magnetic	Magnetic field, flux, permeability
Mechanical	Position, acceleration, stress, strain, force, pressure, mass, density, viscosity, moment, torque acoustic wave amplitude, phase, polarization, velocity
Optical	Absorbance, reflectance, fluorescence, luminescence, refractive index, light scattering
Thermal	Temperature, flux, thermal conductivity, specific heat

Figure 3. Key words

Electrochemical sensors: Analyze mass and electrochemical transduction mechanism
Optical nanosensor: Measure energy, delay time, amplitude and decay phase

Figure 4. Types of nanosensor and its applications

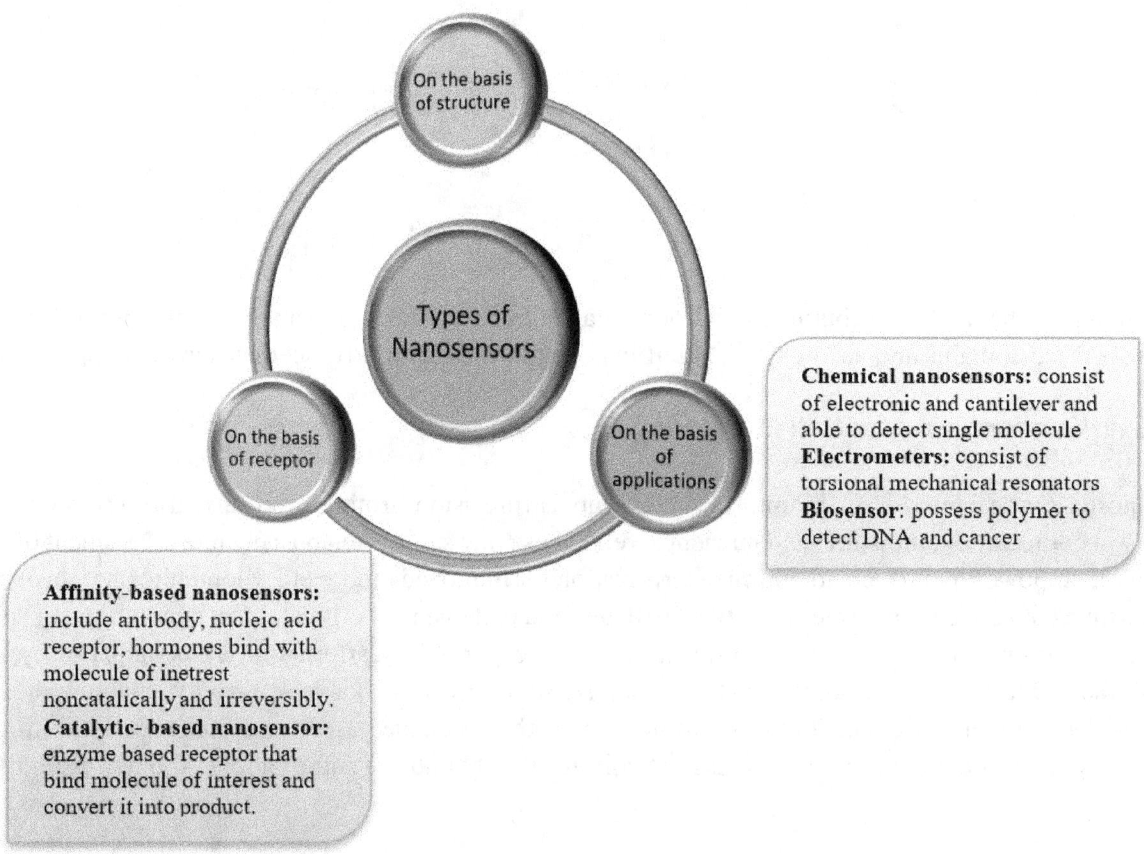

the graphene oxide-based sensor detecting humidity changes from 0.1% to 90%. These advancements enhance agricultural decision-making through accessible and immediate soil insights (Sinha et al., 2023a).

2.2.2 Nanosensors for Water Quality Assessment

Global climate changes are causing water scarcity on land, leading to increased droughts and irregular rainfall patterns. This poses a significant challenge as clean water is essential for crucial activities such as agriculture, industry, and human consumption. Agricultural chemical use further compounds the issue, polluting underground water with contaminants like germs, salts, metals, chemicals, medicines, and even radioactive substances (Ali et al., 2023). Traditionally, spectrometry has been employed to assess water pollution by analyzing light, but its limitations in cost and real-time responses hinder on-the-spot applications. However, the emergence of nanosensors, especially miniature ones, presents a revolutionary solution. These nanosensors offer quick, real-time analysis, portability, high sensitivity, and user-friendly operation, making them increasingly popular for water quality assessment.

Scientists are leveraging nanosensors to detect water pollutants efficiently. Nanostructured porous silicon sensors identify various metal ions, while carbon nanotube-based sensors exhibit high sensitivity to glucose. Some use special nanoalloys to detect diverse organic matters simultaneously, and pH-based sensors, employing dyes for color changes, assess water acidity. Intrinsically conducting polymers like polyaniline (PANI) are utilized for their sensing capabilities (Waheed et al., 2023; Naveed et al., 2024a). These nanosensors play a crucial role in improving water quality for agricultural purposes, accurately detecting pollutants, assessing pH levels, and ensuring overall water suitability. Integrating these tiny sensors into water systems not only enhances accuracy but also aids farmers in optimizing crop irrigation practices.

2.2.3 Nanosensors for Crop Health Monitoring

Precision farming linked small sensors and wireless networks for real-time monitoring and control, enabling farmers to make informed decisions for enhanced productivity. Nanosensors, such as those employing special materials for biosensing, play a crucial role in assessing soil moisture, water levels, nutrient content, and plant diseases. These sensors provide valuable insights into the health of crops, detecting stress, illness, and nutrient deficiencies (Bashir et al., 2023a). The use of atomic force microscope-based nanosensors is noteworthy, as they can identify specific chemicals like metsulfuron-methyl, influencing plant growth. Nanosensors excel in swiftly and accurately identifying bacteria and viruses, ensuring food safety. Employing advanced tools like spectroscopy and imaging aids in disease detection, stress assessment, and nutrient-related issues. Nanosensors contribute to early disease warnings by analyzing volatile organic compounds (VOCs) released during plant stress, with examples including gold nanoparticle-based sensors detecting bacterial infections (Hussain et al., 2023a; Naveed et al., 2024b).

Nanosensors are effective in identifying drought-induced stress by detecting changes in chemical releases, offering early indications of plant stress. Scientists utilize super tiny diagnostic systems with nanosensors to detect plant diseases promptly, employing materials like metal nanoparticles and quantum dots. Quantum dots, acting as miniature crystals, exhibit high sensitivity in identifying plant diseases. Various methods, including electrochemical techniques, special electrodes with gold and copper, and carbon tubes with enzymes, enable nanosensors to identify harmful chemicals and pesticides (Rimpika et al., 2023). Nanosensors with carbon tubes prove useful in detecting gases like ammonia and sulfur dioxide, monitoring their impact on crops and overall health. Fluorescence spectroscopy and gold nanoparticles exhibit heightened sensitivity in identifying harmful agents.

Advanced tools permit scientists to comprehend the interaction between plant tissues and cells with harmful bugs (Ullah et al., 2022). Probing plant tissues measures pressure and chemical concentrations in water vessels, aiding in understanding bug movement and growth within the plant. This non-invasive approach surpasses traditional methods that damaged plants during bacterial detection, allowing for the study of bug dynamics in new tissues, previously inaccessible. Overall, nanosensors and advanced tools revolutionize precision farming, offering unprecedented insights for sustainable and efficient crop management (Majeed, Tariq, et al., 2021).

3. SOIL MONITORING WITH NANOSCALE SENSORS

Nanoscale sensors are transforming soil monitoring by providing unmatched precision and real-time data crucial for optimizing agricultural practices. Embedded within the soil matrix, these sensors adeptly capture and transmit information on vital soil characteristics such as pH, nutrient levels, and moisture content. The capability of nanoscale sensors to operate at the micro-scale, offering high spatial resolution, contributes to their effectiveness in soil monitoring. This allows for precise mapping of soil composition variations between fields, facilitating targeted interventions (Majeed, Tariq, et al., 2022). The continuous monitoring capability of nanosensors enhances resource efficiency by enabling timely adjustments to fertilization and irrigation schedules. Precision farming takes a revolutionary turn when nanoscale sensors are used for soil monitoring. With these sensors, farmers can work with unmatched precision and gain comprehensive insights into critical soil parameters. This capacity facilitates the adoption of sustainable agricultural practices, optimizes resource efficiency, and allows for well-informed decision-making (Sinha et al., 2023b).

3.1 Nanosensors for Soil Mapping

3.1.1 Monitoring Soil Nutrients

Nanosensors, utilizing materials such as carbon nanotubes, graphene, and nanoclays, offer real-time soil nutrient monitoring by detecting and quantifying potassium, phosphorus, and nitrogen (Yadav et al., 2023). These sensors, binding with specific nutrients, enable farmers to make informed decisions on crop management and fertilizer application, leading to reduced waste, increased efficiency, and environmental mitigation. For instance, a graphene-based nanosensor integrated into a wireless network provides real-time insights into soil nitrogen content, aiding farmers in precise decision-making (Zain et al., 2024).

3.1.2 Microbial Activity Monitoring

Nanosensors, transcending conventional nutrient assessments, play a pivotal role in advancing soil health through the monitoring of microbial activity (Chaupoo et al., 2023). In real-time, these sensors track the dynamic shifts within microbial communities, providing invaluable insights into the intricate biological processes occurring in the soil. This granular level of detail equips farmers with the knowledge needed to make informed decisions about soil management, nurturing enhanced crop health and bolstering overall productivity. The integration of nanosensors in agricultural practices represents a transformative approach, where technology meets the intricacies of soil biology, paving the way for sustainable and optimized farming strategies (Hassan, Haq, Ahmad, et al., 2022).

3.2 Soil Moisture Sensors at the Nanoscale

Regular assessment of soil moisture and texture is vital for successful crop production, with variations impacting yields spatially and temporally. Traditional methods for estimating soil moisture in agriculture, often inaccurate and requiring frequent calibration, are being replaced by humidity-based nanosensors (Wang, 2020). These sensors utilize electrical transduction with a hygroscopic probe, altering dielectric characteristics upon water absorption. Polymers, ceramics, and composite-based nanosensors bring advantages such as enhanced stability, prolonged durability, and improved adaptability in various agricultural applications. Their integration could significantly enhance precision in soil temperature and moisture measurements. Many nanosensor devices with wireless communication systems offer economical, user-friendly, and real-time data. Notable examples encompass carbon nanotube and graphene-based nanosensors, like the graphene oxide-based sensor capable of detecting humidity changes as low as 0.1% (Hooshmand et al., 2023).

3.2.1 Impact on Irrigation Management and Water Conservation

Global water scarcity, intensified by climate change-induced droughts and erratic monsoons, poses challenges for securing clean water for agriculture, industry, and human consumption. Agrochemical usage exacerbates groundwater pollution with pathogens, salts, metals, and pharmaceutical compounds (Waheed et al., 2022). Addressing this crisis requires innovative and sustainable water treatment technologies, especially for recycling contaminated water. In the face of adverse climate conditions, precision and site-specific irrigation management emerge as viable solutions to enhance crop productivity. Leveraging advanced technologies such as GPS, GIS, and automated machine guidance allows for judicious water application. Integrating low-flying drones or high-resolution imaging satellites aids in measuring plant or soil water content, enabling precise irrigation in affected areas. Despite its potential in reducing water usage, precision irrigation encounters challenges like cloud interference and high data processing demands (Majeed, Bhatti, Pieroni, et al., 2021).

The fusion of remote sensing technology and crop simulation models improves agricultural management. Additionally, applying nanotechnology in microirrigation enhances water quality through improved filtering methods. Nanoparticle-based biosensors enable real-time detection, measurement, and removal of water-based contaminants, contributing to nanofiltration membrane applications that can desalinate water and reduce filter and membrane clogging risks (Sinha et al., 2023c; Naveed et al., 2024c). The integration of nanoscale sensors into precision agriculture significantly aids water conservation. Accurate measurements of soil moisture and temperature facilitate targeted and efficient irrigation, conserving water resources and enhancing crop performance. This synergy of nanotechnology and precision agriculture underscores the potential for sustainable farming practices prioritizing resource efficiency, as outlined in Table 2 for bionanosensors in precision agriculture.

Table 2. Key bionenosensor types and applications in precision agriculture

Bionanosensor Type	Role of Nanosensor	Application	Material Type	Reference
Electrochemical biosensors	Pesticide biosensors	Monitoring of soil pesticides and herbicides	enzymes, conducting polymers	John et al., 2021; Jang et al., 2016; Kashyap, Kumar, 2021
Potentiometric biosensors	Nutrient biosensors	Precise monitoring of soil pesticides and herbicides	ion-selective electrodes, polymers	
Fluorescence-based biosensors	Plant disease biosensors	Monitoring plant disease and stress	quantum dots, fluorescent proteins	
Soil moisture sensors	Irrigation biosensors	Monitoring irrigation	ceramics, metal oxides	
Chemiresistor sensors	Environmental biosensor	Monitoring of soil quality parameters	polymers, metal oxides	
Capacitive humidity sensors	Moisture sensors	Monitoring soil humidity	ceramics, polymers	

4. WATER QUALITY ASSESSMENT USING NANOSCALE SENSORS

Nanosensors have stepped up as an invaluable technology, working on a nanoscale size for quick detection of analytes such as heavy metals, pesticides, and herbicides at trace levels with high sensitivity and selectivity, which is crucial in environmental monitoring (Majeed, Khan, et al., 2022; Majeed et al., 2021b). Mainly, two approaches can be used to accomplish the usage of nanosensors in water quality control (Majeed, Lu, et al., 2022).

4.1. Nanosensors for Detecting Pollutants in Water

Nanosensors comprise a recognition element and a physical transducer. The recognition element interacts with the analyte, inducing a detectable signal via the transducer that can be processed and amplified, allowing for the detection and quantification of contaminants in water (Majeed et al., 2023).

4.1.1 Heavy Metal Detection

"Heavy metals" are naturally occurring elements that have high atomic mass and density at least five times greater than water. Due to their non-biodegradable nature, they keep accumulating in the environment and increase toxicity in water. The use of water contaminated with heavy metals can pose detrimental effects on agriculture, humans, and other living organisms (Saleh & Hassan, n.d.). The permissible concentration of heavy metals in drinking water is regulated by various standards and guidelines specific to each country. Different nanosensors have been developed for the rapid detection of highly toxic heavy metals in water even at trace levels (Majeed, Bhatti, et al., 2020) (Figure 4).

4.1.2 Pesticide Monitoring

Pesticides are any substances that are incorporated into the environment to control or destroy unwanted life forms (Khan et al., 2022). They are poorly soluble in water and become noxious due to prolonged persistence and bioaccumulation in the environment owing to their chemical stability and high lipophilicity. According to the European Union, the maximum permissible concentration of any pesticide in drinking water is 0.1 μg/L and for all pesticides combined is 0.5 μg/L. Nanosensors help to identify and monitor pollutants in water sources which plays a crucial role in safeguarding the environment (Majeed, Bhatti, & Amjad, 2021).

4.2. Nanotechnology in Assessing Water pH and Salinity

4.2.1. Importance for Crop Health

Salinity and the pH of water are important parameters in environmental monitoring and water quality assessment (Singh et al., 2023 a, b; Tariq et al., 2023). Nanosensors made up of nanomaterials such as nanocomposites and quantum dots, can be used for pH sensing and salinity in water providing a sensitive and accurate measurement. Their proper management can help maintain optimal growing conditions for crops, promoting healthy growth and productivity (Arif et al., 2021; Ghazaryan et al., 2023).

Figure 5. Schematic illustration of the detection process of nanobiosensors

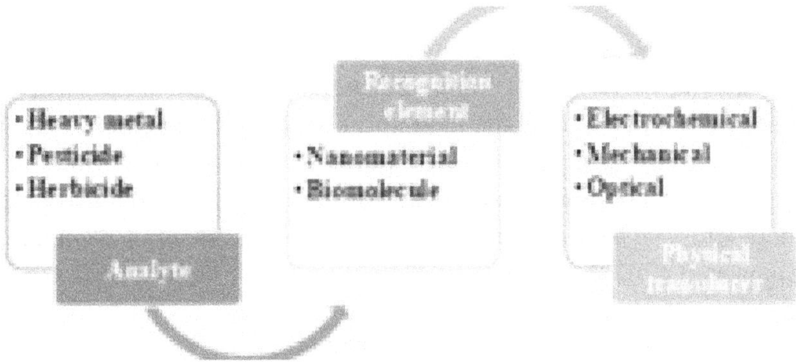

4.2.2. Precision Irrigation Based on Water Quality

Water quality plays a crucial role in precision irrigation, as it affects the health and productivity of crops. Nanosensors can be used for real-time monitoring of water quality parameters such as pH, electrical conductivity, nutrient levels, salinity, and contaminants ensuring that irrigation water meets quality standards (Majeed, Hayat Bhatti, et al., 2020). By integrating nanosensors with precision irrigation systems, farmers can optimize water usage, minimize water wastage, and ensure that crops receive the right amount of water and nutrients for optimal growth and yield based on the specific needs of crops and soil conditions (Singh et al., 2023c). It supports sustainable farming practices and contributes to the long-term viability of the agriculture industry (Aziz et al., 2023).

5. CROP HEALTH MONITORING WITH NANOSCALE SENSORS

Human health and agribusiness both depend on crop plants and their better productivity but continuous manual monitoring of crop plants has been a challenging task for decades (Singh et al., 2023d). However, budding nanotechnology such as nanosensors is now replacing manual monitoring for early detection of plant stress by enabling real-time monitoring of plant health via electronic devices (Rehman et al., 2023).

5.1. Early Disease Detection Using Nanosensors

5.1.1. Monitoring Plant Pathogens at the Molecular Level

Nanosensors appeal to smart agriculture since they facilitate early diagnosis of plant diseases and enable low limits of detection. It is now possible to identify the various components of DNA by using nanosensors, which are sensitive enough to detect single molecules (Singh et al., 2024b) These nanosensors can recognize and identify the DNA sequences of many fungal pathogens as well as viral diseases which offers insightful information to optimize agricultural productivity and maintain plant health. Moreover, another method of printing conductive circuits directly onto the leaf with micrometer precision may facilitate engineering and observation of new plant functions (Muhammad, Basit, et al., 2024).

5.1.2. Preventive Measures Based on Nanoscale Data

Identification of diseases and stress is challenging for manual plant health monitoring due to the asymptomatic or delayed onset of symptoms. By continuously monitoring plant health at the molecular level, preventive measures can be taken to mitigate the impact of stressors on plants. Nanosensors can detect various targets of plant diseases, including DNA, proteins, and volatile organic compounds (VOCs) (Haq et al., 2022) . They can provide valuable data on plant stress patterns and help in understanding the communication of stress signals within plants. Prompt management of plant stress is the key to efficient farming (Khoja et al., 2022).

5.2. Nanosensors for Assessing Crop Stress

5.2.1. Temperature and Humidity Monitoring

Nanosensors with nanomaterials having high surface area and fast diffusion properties can be effectively utilized in agriculture to provide real-time monitoring of environmental conditions such as temperature and humidity. These nanosensors can be used as plant wearable devices that detect plants' microclimate and wirelessly gather temperature and humidity data. The electrochemically produced graphene quantum dots have demonstrated a significant level of sensitivity to humidity within two to three minutes of response time. Metal-based and carbon-based nanomaterials have demonstrated temperature sensitivity and have been utilized for temperature monitoring in different applications (Haq et al., 2022b).

5.2.2. Impact on Overall Crop Yield

Nanosensors are emerging and powerful tools that can monitor plant health in terms of growth and development; allowing for early detection and prevention of crop diseases and contamination. The most commonly used nanosensors in the agricultural industry are electrochemical and piezoelectric. Fluorescence resonance energy transfer (FRET), surface-enhanced Raman scattering (SERS), and corona-phase molecule identification are examples of typical nanosensor detection techniques (Bashir et al., 2023b). The use of nanosensors in agriculture has the potential to revolutionize crop management practices by enabling precise and targeted interventions, such as adjusting irrigation and ventilation systems, to create optimal growing conditions for crops and contribute to increased crop yield and sustainability (Hussain et al., 2023b).

6. INTEGRATION OF NANOSCALE SENSORS IN PRECISION FARMING SYSTEMS

The incorporation of nanoscale sensors in precision farming systems assists farmers in increasing crop productivity through improved fertilization supervision and minimizing inputs through the use of agricultural resources such as water, nutrients, and agrochemicals (Hassan, Haq, Majeed, et al., 2022)

6.1. Data Collection and Analytics

In agriculture, remote sensing techniques like RGB and thermal imaging play a key role in monitoring crop health and detecting diseases early. Real-time field monitoring, facilitated by wireless sensors within Wireless Sensor Networks (WSNs), supports precision agriculture by measuring crucial parameters like humidity, soil moisture, and climate. This approach contributes significantly to improving crop yield (Badino et al., n.d.).

6.1.1 Remote Sensing Applications

Remote sensing helps us keep an eye on field changes and adjust water use in different areas. By using adjustable irrigation systems like center pivots, we can handle water shortages or excess caused by severe weather, aiming to improve crop yields while saving nutrients and water. Throughout the growing season, multiple remote sensing images help us check soil moisture, crop water needs, and identify crop stress. We use this info to decide when and how much to irrigate, making water use more efficient (Arshad et al., 2022). In agriculture, remote sensing is like a real-time camera capturing how light bounces off plants, showing us if they're healthy. Whether from satellites, aircraft, or up close, we use this light data to check chlorophyll levels, leaf size, plant coverage, and other important crop details (Bhatia et al., 2024). For plant health, scientists use "vegetative analysis" with computer programs and models to assess data from instruments that detect plants from afar. The Normalized Difference Vegetation Index (NDVI) is a key tool, helping us monitor and ensure healthy plant growth.

Plant diseases can cause big losses, so it's crucial to detect them early. Traditional methods like field scouting take a lot of time and can be mistake-prone. Remote sensing, using techniques like RGB and thermal imaging, is a better way to monitor diseases, especially in the early stages. Soil mapping at different levels, local to national, is essential for managing natural resources sustainably. Detailed maps covering crop conditions and soil variations help us use variable-rate technology effectively, adjusting crop inputs based on the land's needs (Späti et al., 2021).

6.1.2 Real-Time Monitoring for Decision-Making

Precision agriculture relies on real-time field monitoring, accomplished through the deployment of wireless sensors in Wireless Sensor Networks (WSNs) using a systematic approach. In Wireless Sensor Networks (WSNs), sensor nodes sense, gather, and measure data about the environment. Thousands of these nodes work collaboratively without the need for extensive infrastructure to monitor the area and transmit data to users. Sensors in agriculture play crucial roles in measuring various parameters like humidity, soil moisture, climate, intrusion detection, and water level, contributing to increased crop yield.

These sensors are instrumental in collecting meteorological, soil, and crop data, surveying land for improved farming practices, assessing crop resource needs, identifying urgent requirements for irrigation and fertilizers, and safeguarding fields from intruders.

6.2. Automation and Control Systems

6.2.1 Smart Farming Technologies

Automated farming, also known as smart farming or precision agriculture, employs cutting-edge technologies to enhance agricultural processes. It aims to optimize operations like planting, irrigation, fertilization, pest management, crop observation, and harvesting (Singh et al., 2024a). Utilizing sensors and data collection, these systems gather information on soil conditions, weather, and crop health. Advanced algorithms and machine learning analyze this data to provide insights for informed crop management decision (Hooshmand et al., 2023).

Harvesting robots, designed for autonomous fruit, vegetable, and grain harvesting, navigate fields and perform precise harvesting tasks. Automated irrigation systems use sensors and meteorological data to optimize water distribution, increasing efficiency and reducing waste. Drones equipped with cameras and sensors fly over fields, collecting data on crop growth, health, and nutrient levels, helping farmers detect issues early on. Devices with sensors and algorithms automate seed planting, strategically improving crop yield and germination rates. Weed-removing robots, guided by computer vision and machine learning, target weeds without harming crops, reducing the need for manual labor and herbicides.

6.2.2 Integration With Existing Precision Farming Practices

Agriculture is advancing by integrating technologies like robotics, sensors, drones, data analytics, and artificial intelligence (Goyal et al., 2023). Machine vision (MV), also known as agro-vision, interprets pixel images to offer quick, reliable, and non-destructive data for monitoring cultivation processes. It enhances machines processing images, and new deep-learning techniques are creating intelligent robots for real-time field applications and crop health analysis using multispectral imagery. Regular smartphones can now use MV systems for crop monitoring (Tassadduq et al., 2022).

The Internet of Things (IoT) connects smart devices using sensors for real-time data flow, benefitting industries like agriculture. With about 20 billion connected devices, IoT improves process and service management, and there's consideration of combining it with blockchain for security. Artificial Intelligence (AI) is increasingly integrated into automation systems and agricultural machinery. AI's machine learning algorithms handle large datasets, providing insights for predicting market trends, identifying plant diseases, and optimizing irrigation schedules based on weather forecasts, supporting informed decision-making by farmers (Ullah et al., 2022). Autonomous vehicles, equipped with sensors, AI, and GPS, are used in agriculture for tasks like planting, spraying, and harvesting. These self-driving vehicles navigate fields, perform precise tasks, and operate without human intervention, reducing the potential for errors and increasing operational efficiency.

7. BENEFITS AND CHALLENGES OF PRECISION FARMING WITH NANOSCALE SENSORS

Nanoscale sensors in precision farming offer increased crop productivity and sustainability by precisely monitoring resource usage. Yet, integrating complex sensor networks and addressing data management challenges pose obstacles for widespread adoption. Achieving successful implementation hinges on balancing the benefits with resolving these challenges.

7.1. Increased Efficiency and Resource Optimization

Precision farming employs cutting-edge technologies, sensors, and analytics to optimize agricultural productivity and resource use through real-time data gathering. The integration of IoT, satellite-based remote sensing, and AI enables data-driven decision-making, enhancing monitoring and resource allocation. Precision farming benefits farmers by improving resource management, increasing crop yields, reducing costs, and promoting environmental sustainability. Future integration of blockchain, robotics, and drones holds potential for further innovation in agricultural resource management, ultimately revolutionizing modern agriculture by addressing challenges and fostering sustainability.

7.2. Environmental Sustainability and Reduced Inputs

Precision farming maximizes input utilization and machinery efficiency, enhancing farm productivity, profitability, and environmental sustainability (Singh et al., 2024c). Leveraging cutting-edge technologies and data analytics streamlines farming practices, while navigational aids minimize fossil fuel and input usage (Majeed, Bhatti, Pieroni, et al., 2021). Environmentally, precision farming reduces chemical runoff and water pollution through optimized fertilizer and pesticide application. Techniques like UAV-based spraying and reduced tillage preserve soil structure and enhance crop production. Precision farming supports biodiversity, addresses water scarcity with efficient irrigation, and optimizes inputs, reducing greenhouse gas emissions and contributing to climate change mitigation. It plays a vital role in ensuring food security and a robust global food system for future generations (Majeed, Lu, et al., 2022).

7.3. Challenges in Implementation and Adoption

Implementing automation in agriculture poses challenges, particularly the substantial upfront costs associated with acquiring machinery, sensors, and control systems. Small-scale farmers or those with limited financial resources may find these costs prohibitive. Integrating automation technologies with existing farm infrastructure and equipment is complex and may require additional investments or modifications for compatibility. Automated farming systems, relying on technologies like robotics, sensors, and data analytics, face challenges in consistent and reliable operation, especially in diverse and harsh agricultural environments. Maintenance of these systems demands specialized skills, creating challenges for farmers and farmworkers, particularly in regions with limited access to training resources (Sinha et al., 2023c).

The diversity of tasks and operations in agriculture requires automation systems to adjust to varying crops, field conditions, and farming practices, necessitating flexibility and customizability. The extensive data generated by automation tools poses challenges in terms of management, analysis, and interpretation, especially for farmers lacking expertise. Concerns about data privacy and security add complexity, emphasizing the importance of safeguarding sensitive farm data. Addressing these challenges requires collaborative efforts among farmers, researchers, technology providers, policymakers, and stakeholders, ensuring the unlocking of automation's full potential while mitigating drawbacks (Majeed et al., 2023).

7.4. Future Directions and Emerging Technologies

7.4.1 Advances in Nanosensor Technology

Biosensors, incorporating biological elements like cells and proteins, generate signals for analysis (Bhatia et al., 2024). Nanosensors play a crucial role in autonomous monitoring in agriculture, strategically positioned to collaborate with GPS for accurate crop development tracking. This provides vital data on optimal planting and harvesting times, enhancing agronomic techniques. Nanotechnology facilitates real-time monitoring of crop growth, soil fertility, moisture, temperature, nutrient capacity, pathogens, and diseases. Nanosensors precisely measure soil parameters like pH, nutrients, pesticides, and moisture, aiding in nitrogen uptake estimation and pathogen detection. This promotes sustainable farming by optimizing input use. Nano-chips, a type of microarray, enable high-sensitivity detection of single nucleotide changes in bacteria and viruses (Hooshmand et al., 2023).

7.4.2 Integration With Other Emerging Technologies

Integrating automation technologies with the Internet of Things (IoT) and blockchain enables the development of a connected and transparent agricultural ecosystem. IoT sensors play a key role by delivering real-time data on diverse parameters, while blockchain ensures traceability and instills trust throughout the entire supply chain, spanning from the farm to the consumer's table (Goyal et al., 2023).

7.4.3 Potential for Addressing Global Agricultural Challenges

Agriculture faces significant challenges in the 21st century, requiring innovative solutions to enhance productivity sustainably, adapt to climate change, and support development in agriculture-dependent developing countries. Precision farming, vertical farming, and the integration of AI and machine learning offer potential remedies to increase crop yields and minimize environmental impact. Increased funding for research and development is essential to create tools and methods enabling farmers to adapt to changing conditions and boost productivity. Precision agriculture, utilizing advanced technology and data analysis, emerges as a crucial strategy to address global agricultural challenges, feeding a growing population while mitigating environmental impacts (Muhammad, Waheed, et al., 2024). Overcoming challenges involves bridging the knowledge gap between farmers and technology, government support for infrastructure development, and addressing issues with professional agricultural sensors. The agricultural sector requires high-quality IoT-powered sensors, and wireless power transfer options can enhance their longevity. Cross-technology communication, advancements in machine vision, and 6G mobile networks play crucial roles in progress. Clever hydroponics and breeding technology improvements contribute to a more productive and sustainable agricultural future, necessitating efforts in education, infrastructure, sensor technology, and creative agricultural methods (Hussain et al., 2023a).

CONCLUSION

In conclusion, the integration of nanotechnology in precision agriculture holds immense promise for revolutionizing farming practices. The combination of nanosensors and advanced Io T tools not only enhances input management and conservation but also has the potential to significantly boost crop yields compared to conventional methods. This transformative shift, underpinned by the unique capabilities of nanotechnology, underscores its pivotal role in reshaping the agricultural landscape. The empowerment of farmers with rapid decision-making tools, coupled with the efficient delivery of crucial data on crop growth and soil conditions, further accentuates the impact of nanotechnology in precision agriculture. As nanotechnology progresses, promising advancements in disease diagnosis, treatment speed, and plant nutrient uptake are poised to contribute to a more resilient and efficient agricultural ecosystem. However, it is crucial to acknowledge that the interaction between nanoparticles and soil remains not fully understood, necessitating ongoing research to comprehend their impact on plant nutrition under real-world field conditions. Despite this, the trajectory of precision farming, propelled by continuous innovation and collaboration, suggests a pivotal role in ensuring sustainable agriculture, food security, and a resilient global food system. In the evolving landscape of precision farming, driven by nanosensor technologies, the potential to shape a future marked by food security and environmental consciousness is evident. Through collaborative efforts and ongoing innovation, precision farming emerges as a key player in securing a more sustainable and food-secure future for generations to come.

REFERENCES

Ali, H., Muhammad, Z., Majeed, M., Aziz, R., Khan, A., Mangrio, W. M., Abdo, H. G., Almohamad, H., & Al Dughairi, A. A. (2023). Vegetation diversity pattern during spring season in relation to topographic and edaphic variables in sub-tropical zone. *Botanical Studies*, *64*(1), 25. Advance online publication. doi:10.1186/s40529-023-00398-5 PMID:37716923

Altug, H., Oh, S. H., Maier, S. A., & Homola, J. (2022). Advances and applications of nanophotonic biosensors. In Nature Nanotechnology (Vol. 17, Issue 1, pp. 5–16). Nature Research. doi:10.1038/s41565-021-01045-5

Arif, U., Bhatti, K. H., Ajaib, M., Wagay, N. A., Majeed, M., Zeb, J., Hameed, A., & Kiani, J. (2021). Ethnobotanical indigenous knowledge of Tehsil Charhoi, District Kotli, Azad Jammu and Kashmir, Pakistan. *Ethnobotany Research and Applications*, *22*. Advance online publication. doi:10.32859/era.22.50.1-24

Arshad, F., Waheed, M., Harun, N., Fatima, K., Khan, B. A., Fatima, K., Abbas, Z., Jabeen, S., & Majeed, M. (2022). Indigenous farmer's perception about fodder and foraging species of Semi-arid lowlands of Pakistan: A case study of District Kasur, Pakistan. *Taiwania*, *67*(4), 510–523. doi:10.6165/tai.2022.67.510

Aziz, A., Anwar, M. M., Majeed, M., Fatima, S., Mehdi, S. S., Mangrio, W. M., Elbouzidi, A., Abdullah, M., Shaukat, S., Zahid, N., Mahmoud, E. A., Casini, R., Yessoufou, K., & Elansary, H. O. (2023). Quantifying Landscape and Social Amenities as Ecosystem Services in Rapidly Changing Peri-Urban Landscape. *Land (Basel)*, *12*(2), 477. Advance online publication. doi:10.3390/land12020477

Badino, P., Gras, A., Johanna Groot, M., Shoaib Amjad, M., & Morales-de la Nuez, A. (n.d.). *Quantitative ethnoveterinary study on plant resource utilization by indigenous communities in high-altitude regions*. Academic Press.

Bashir, S. M., Altaf, M., Hussain, T., Umair, M., Majeed, M., Mangrio, W. M., Khan, A. M., Gulshan, A. B., Hamed, M. H., Ashraf, S., Amjad, M. S., Bussmann, R. W., Abbasi, A. M., Casini, R., Alataway, A., Dewidar, A. Z., Al-Yafrsi, M., Amin, M. H., & Elansary, H. O. (2023a). Vernacular Taxonomy, Cultural and Ethnopharmacological Applications of Avian and Mammalian Species in the Vicinity of Ayubia National Park, Himalayan Region. *Biology (Basel), 12*(4), 609. Advance online publication. doi:10.3390/biology12040609 PMID:37106809

Bhatia, D., Paul, S., Acharjee, T., & Ramachairy, S. S. (2024). Biosensors and their widespread impact on human health. *Sensors International, 5*, 100257. Advance online publication. doi:10.1016/j.sintl.2023.100257

Bhattacharya, R., Bhattacharya, R., Majeed, M., Bhandari, S., Aziz, R., & Sinha, D. Niyogi, S. S. Environmental Rehabilitation of Industrial Waste Dumping Site. In *Biohydrometallurgical Processes* (pp. 44–89). CRC Press. doi:10.1201/9781003451457-4

Chaupoo, A. S., Dhakad, A., Beese, S., Kumar, A., Kumar, P., Pandey, S. K., & Kurdekar, A. K. (2023). Transforming Horticulture: The Influence of Nanotechnology on Crop Enhancement. *International Journal of Environment and Climate Change, 13*(11), 3381–3400. doi:10.9734/ijecc/2023/v13i113512

Ghazaryan, K. A., Harutyunyan, A. S., Khachatryan, H. E., Singh, A., Minkina, T. M., Rajput, V. D., & Movsesyan, H. S. (2023). Study of tolerance and phytodesalination potential of wheat, oat, emmer, and barley for sustainable saline agriculture. *Applied Ecology and Environmental Research, 21*(5), 4853–4882. doi:10.15666/aeer/2105_48534882

Ghazaryan, K. A., Movsesyan, H. S., Khachatryan, H. E., & Ghazaryan, N. P. (2018). Geochemistry of Potentially Toxic Trace Elements in Soils of Mining Area: A Case Study from Zangezur Copper and Molybdenum Combine, Armenia. *Bulletin of Environmental Contamination and Toxicology, 101*(6), 732–737. doi:10.1007/s00128-018-2443-0 PMID:30229278

Goyal, V., Rani, D., Mehrotra, S., Deng, C., & Wang, Y. (2023). Unlocking the Potential of Nano-Enabled Precision Agriculture for Efficient and Sustainable Farming. In Plants (Vol. 12, Issue 21). Multidisciplinary Digital Publishing Institute (MDPI). doi:10.3390/plants12213744

Haq, S. M., Tariq, A., Li, Q., Yaqoob, U., Majeed, M., Hassan, M., Fatima, S., Kumar, M., Bussmann, R. W., Moazzam, M. F. U., & Aslam, M. (2022a). Influence of Edaphic Properties in Determining Forest Community Patterns of the Zabarwan Mountain Range in the Kashmir Himalayas. *Forests, 13*(8), 1214. Advance online publication. doi:10.3390/f13081214

Hassan, M., Haq, S. M., Ahmad, R., Majeed, M., Sahito, H. A., Shirani, M., Mubeen, I., Aziz, M. A., Pieroni, A., Bussmann, R. W., Alataway, A., Dewidar, A. Z., Al-Yafrsi, M., Elansary, H. O., & Yessoufou, K. (2022). Traditional Use of Wild and Domestic Fauna among Different Ethnic Groups in the Western Himalayas—A Cross Cultural Analysis. *Animals (Basel), 12*(17), 2276. Advance online publication. doi:10.3390/ani12172276 PMID:36077997

Hassan, M., Haq, S. M., Majeed, M., Umair, M., Sahito, H. A., Shirani, M., Waheed, M., Aziz, R., Ahmad, R., Bussmann, R. W., Alataway, A., Dewidar, A. Z., El-Abedin, T. K. Z., Al-Yafrsi, M., Elansary, H. O., & Yessoufou, K. (2022). Traditional Food and Medicine: Ethno-Traditional Usage of Fish Fauna across the Valley of Kashmir: A Western Himalayan Region. *Diversity*, *14*(6), 455. Advance online publication. doi:10.3390/d14060455

Hooshmand, S., Kassanos, P., Keshavarz, M., Duru, P., Kayalan, C. I., Kale, İ., & Bayazit, M. K. (2023). Wearable Nano-Based Gas Sensors for Environmental Monitoring and Encountered Challenges in Optimization. In Sensors (Basel, Switzerland) (Vol. 23, Issue 20). doi:10.3390/s23208648

Hussain, S., Raza, A., Abdo, H. G., Mubeen, M., Tariq, A., Nasim, W., Majeed, M., Almohamad, H., & Al Dughairi, A. A. (2023a). Relation of land surface temperature with different vegetation indices using multi-temporal remote sensing data in Sahiwal region, Pakistan. *Geoscience Letters*, *10*(1), 33. Advance online publication. doi:10.1186/s40562-023-00287-6

Jamil, M. D., Waheed, M., Akhtar, S., Bangash, N., Chaudhari, S. K., Majeed, M., Hussain, M., Ali, K., & Jones, D. A. (2022). Invasive Plants Diversity, Ecological Status, and Distribution Pattern in Relation to Edaphic Factors in Different Habitat Types of District Mandi Bahauddin, Punjab, Pakistan. *Sustainability (Switzerland), 14*(20). https://doi.org/ doi:10.3390/su142013312

Khan, A. M., Li, Q., Saqib, Z., Khan, N., Habib, T., Khalid, N., Majeed, M., & Tariq, A. (2022). MaxEnt Modelling and Impact of Climate Change on Habitat Suitability Variations of Economically Important Chilgoza Pine (Pinus gerardiana Wall.) in South Asia. *Forests*, *13*(5), 715. Advance online publication. doi:10.3390/f13050715

Khoja, A. A., Haq, S. M., Majeed, M., Hassan, M., Waheed, M., Yaqoob, U., Bussmann, R. W., Alataway, A., Dewidar, A. Z., Al-Yafrsi, M., Elansary, H. O., Yessoufou, K., & Zaman, W. (2022). Diversity, Ecological and Traditional Knowledge of Pteridophytes in the Western Himalayas. *Diversity*, *14*(8), 628. Advance online publication. doi:10.3390/d14080628

Majeed, M., Bhatti, K. H., & Amjad, M. S. (2021). Impact of climatic variations on the flowering phenology of plant species in jhelum district, punjab, pakistan. *Applied Ecology and Environmental Research*, *19*(5), 3343–3376. doi:10.15666/aeer/1905_33433376

Majeed, M., Bhatti, K. H., Amjad, M. S., Abbasi, A. M., Bussmann, R. W., Nawaz, F., Rashid, A., Mehmood, A., Mahmood, M., Khan, W. M., & Ahmad, K. S. (2020). Ethno-veterinary uses of poaceae in Punjab, Pakistan. *PLoS ONE, 15*(11). doi:10.1371/journal.pone.0241705

Majeed, M., Bhatti, K. H., Pieroni, A., Sõukand, R., Bussmann, R. W., Khan, A. M., Chaudhari, S. K., Aziz, M. A., & Amjad, M. S. (2021). Gathered wild food plants among diverse religious groups in Jhelum District, Punjab, Pakistan. *Foods*, *10*(3), 594. Advance online publication. doi:10.3390/foods10030594 PMID:33799901

MajeedM.Hayat BhattiK.Shoaib AmjadM.Mehmood AbbasiA.RashidA.Nawaz Muhammad Nawaz ShareefF.MehmoodA.MahmoodM. (2020). *Ethno-veterinary practices of Poaceae taxa in Punjab, Pakistan*. doi:10.21203/rs.3.rs-15532/v1

Majeed, M., Khan, A. M., Habib, T., Anwar, M. M., Sahito, H. A., Khan, N., & Ali, K. (2022). Vegetation analysis and environmental indicators of an arid tropical forest ecosystem of Pakistan. *Ecological Indicators*, *142*, 109291. Advance online publication. doi:10.1016/j.ecolind.2022.109291

Majeed, M., Lu, L., Anwar, M. M., Tariq, A., Qin, S., El-Hefnawy, M. E., El-Sharnouby, M., Li, Q., & Alasmari, A. (2023). Prediction of flash flood susceptibility using integrating analytic hierarchy process (AHP) and frequency ratio (FR) algorithms. *Frontiers in Environmental Science*, *10*, 1037547. Advance online publication. doi:10.3389/fenvs.2022.1037547

Majeed, M., Lu, L., Haq, S. M., Waheed, M., Sahito, H. A., Fatima, S., Aziz, R., Bussmann, R. W., Tariq, A., Ullah, I., & Aslam, M. (2022). Spatiotemporal Distribution Patterns of Climbers along an Abiotic Gradient in Jhelum District, Punjab, Pakistan. *Forests*, *13*(8), 1244. Advance online publication. doi:10.3390/f13081244

Majeed, M., Tariq, A., Anwar, M. M., Khan, A. M., Arshad, F., Mumtaz, F., Farhan, M., Zhang, L., Zafar, A., Aziz, M., Abbasi, S., Rahman, G., Hussain, S., Waheed, M., Fatima, K., & Shaukat, S. (2021). Monitoring of land use–Land cover change and potential causal factors of climate change in Jhelum district, Punjab, Pakistan, through GIS and multi-temporal satellite data. *Land (Basel)*, *10*(10), 1026. Advance online publication. doi:10.3390/land10101026

Majeed, M., Tariq, A., Anwar, M. M., Khan, A. M., Arshad, F., Mumtaz, F., Farhan, M., Zhang, L., Zafar, A., Aziz, M., Abbasi, S., Rahman, G., Hussain, S., Waheed, M., Fatima, K., & Shaukat, S. (2021). Monitoring of Land Use–Land Cover Change and Potential Causal Factors of Climate Change in Jhelum District, Punjab, Pakistan, through GIS and Multi-Temporal Satellite Data; (pp. 129-147). https://www.mdpi.com/books/book/6906-land-use-land-cover-and-natural-hazards-interactions-changes-and-impacts

Majeed, M., Tariq, A., Haq, S. M., Waheed, M., Anwar, M. M., Li, Q., Aslam, M., Abbasi, S., Mousa, B. G., & Jamil, A. (2022). A Detailed Ecological Exploration of the Distribution Patterns of Wild Poaceae from the Jhelum District (Punjab), Pakistan. *Sustainability (Basel)*, *14*(7), 3786. Advance online publication. doi:10.3390/su14073786

Muhammad, M., Basit, A., Majeed, M., Shah, A. A., Ullah, I., Mohamed, H. I., Khan, A., & Ghanaim, A. M. (2024). Bacterial pigments and their applications. In *Bacterial Secondary Metabolites* (pp. 277–298). Elsevier., doi:10.1016/B978-0-323-95251-4.00019-3

Muhammad, M., Waheed, A., Wahab, A., Majeed, M., Nazim, M., Liu, Y. H., Li, L., & Li, W. J. (2024). Soil salinity and drought tolerance: An evaluation of plant growth, productivity, microbial diversity, and amelioration strategies. In *Plant Stress* (Vol. 11). Elsevier B.V., doi:10.1016/j.stress.2023.100319

Naveed, Majeed, Jabeen, Hanif, Naveed, Saleem, & Khan. (2024). Plant Disease Diagnosis with Artificial Intelligence (AI). Microbial Data Intelligence and Computational Techniques for Sustainable Computing (pp 217–234). Springer. doi:10.1007/978-981-99-9621-6_15

Naveed, Zahid, Fatima, Saleem, Majeed, Abid, Javed, Wazir, & Qasim. (2024). Disease Diagnosis with Artificial Intelligence (AI). In Microbial Data Intelligence and Computational Techniques for Sustainable Computing (pp 235–258). Springer. https://doi.org/ doi:10.1007/978-981-99-9621-6_16

Naveed, Memon, Abdullah, Makhdoom, Azeem, Mehmood, Salahuddin, Rajpoot, & Majeed. (2024). Use Cases and Future Aspects of Intelligent Techniques in Microbial Data Analysis. In Microbial Data Intelligence and Computational Techniques for Sustainable Computing (pp 259–280). Springer. doi:10.1007/978-981-99-9621-6_17

Pramanik, P. K. D., Solanki, A., Debnath, A., Nayyar, A., El-Sappagh, S., & Kwak, K. S. (2020). Advancing Modern Healthcare with Nanotechnology, Nanobiosensors, and Internet of Nano Things: Taxonomies, Applications, Architecture, and Challenges. *IEEE Access : Practical Innovations, Open Solutions*, *8*, 65230–65266. doi:10.1109/ACCESS.2020.2984269

Rajput, V. D., Singh, A., Minkina, T., Rawat, S., Mandzhieva, S., Sushkova, S., Shuvaeva, V., Nazarenko, O., Rajput, P., Komariah, Verma, K. K., Singh, A. K., Rao, M., & Upadhyay, S. K. (2021). Nano-Enabled Products: Challenges and Opportunities for Sustainable Agriculture. *Plants*, *10*(12), 2727. doi:10.3390/plants10122727 PMID:34961197

Rehman, A., Aziz, A., Anwar, M. M., Majeed, M., Albanai, J. A., Almohamad, H., & Abdo, H. G. (2023). Quantifying the impacts of urbanization on urban green, evidences from Maga City, Lahore Pakistan. *Discover Sustainability*, *4*(1), 48. Advance online publication. doi:10.1007/s43621-023-00169-z

Rimpika, A., Anushi, Manasa, S., Sharma, S., Thakur, A., Shilpa, & Sood, A. (2023). An Overview of Precision Farming. *International Journal of Environment and Climate Change*, *13*(12), 441–456. doi:10.9734/ijecc/2023/v13i123701

Rohilla, D., Chaudhary, S., & Umar, A. (2021). An Overview of Advanced Nanomaterials for Sensor Applications. *Engineering and Science*, *16*, 47–70. doi:10.30919/es8d552

Saleh, H. M., & Hassan, A. I. (n.d.). *Environmental Impact and Remediation of Heavy Metals*. Academic Press.

Sangeeta Dash, E., Yadav, A. L., Bora, R., & Kumar Pandey, A. (2023). *Research Trends in Agriculture Science* (Vol. I). www.bhumipublishing.com

Singh, A., Margaryan, G., Harutyunyan, A., S. Movsesyan, H., Khachatryan, H., Rajput, V., . . . Ghazaryan, K. (2024c). Advancing Agricultural Resilience in Ararat Plain, Armenia: Utilizing Biogenic Nanoparticles and Biochar under Saline Environments to Optimize Food Security and Foster European Trade. *Egyptian Journal of Soil Science, 64*(2). doi:10.21608/ejss.2024.257000.1703

Singh, A., Rajput, V., Varshney, A., Sharma, R., Ghazaryan, K., Minkina, T., . . . El-Ramady, H. (2024b). Revolutionizing Crop Production: Nanoscale Wonders - Current Applications, Advances, and Future Frontiers. *Egyptian Journal of Soil Science, 64*(1). doi:10.21608/ejss.2023.246354.1684

Singh, A., Rajput, V. D., Pandey, D., Sharma, R., Ghazaryan, K., & Minkina, T. (2023b). Nano Zinc-Enabled Strategies in Crops for Combatting Zinc Malnutrition in Human Health. *Frontiers in Bioscience (Landmark Edition)*, *28*(8), 158. doi:10.31083/j.fbl2808158 PMID:37664935

Singh, A., Rajput, V. D., Sharma, R., Ghazaryan, K., & Minkina, T. (2023d). Salinity stress and nanoparticles: Insights into antioxidative enzymatic resistance, signaling, and defense mechanisms. *Environmental Research*, *235*, 116585. doi:10.1016/j.envres.2023.116585 PMID:37437867

Singh, A., Rajput, V. D., Tawaha, A. R. M., Al Zoubi, O. M., Habeeb, T., Rawat, S., Ghazaryan, K., & Minkina, T. (2023c). A Review on Crop Responses to Nanofertilizers for Mitigation of Multiple Environmental Stresses. *Ecological Engineering & Environmental Technology*, *24*(7), 280–296. doi:10.12912/27197050/169313

Singh, A., Rajput, V. D., Varshney, A., Ghazaryan, K., & Minkina, T. (2023a). Small Tech, Big Impact: Agri-nanotechnology Journey to Optimize Crop Protection and Production for Sustainable Agriculture. *Plant Stress*, *10*, 100253. doi:10.1016/j.stress.2023.100253

Singh, A., Sengar, R. S., Rajput, V. D., Al-Ghzawi, A. L., Shahi, U. P., Ghazaryan, K., Minkina, T., Tawaha, A. R. M. A., Al Zoubi, O. M., & Habeeb, T. (2024a). Impact of Salinity Stress and Zinc Oxide Nanoparticles on Macro and Micronutrient Assimilation: Unraveling the Link between Environmental Factors and Nutrient Uptake. *Journal of Ecological Engineering*, *25*(2), 1–9. doi:10.12911/22998993/172947

Singh, A., Sharma, R., Rawat, S., Singh, A. K., Rajput, V. D., Fedorov, Y., . . . Chaplygin, V. (2022). Nanomaterial-plant interaction: Views on the pros and cons. *Toxicity of Nanoparticles in Plants*, 47–68. doi:10.1016/B978-0-323-90774-3.00015-5

Sinha, D., Maurya, A. K., Abdi, G., Majeed, M., Agarwal, R., Mukherjee, R., Ganguly, S., Aziz, R., Bhatia, M., Majgaonkar, A., Seal, S., Das, M., Banerjee, S., Chowdhury, S., Adeyemi, S. B., & Chen, J. T. (2023). Integrated Genomic Selection for Accelerating Breeding Programs of Climate-Smart Cereals. In Genes (Vol. 14, Issue 7). Multidisciplinary Digital Publishing Institute (MDPI). doi:10.3390/genes14071484

Späti, K., Huber, R., & Finger, R. (2021). Benefits of Increasing Information Accuracy in Variable Rate Technologies. *Ecological Economics*, *185*, 107047. Advance online publication. doi:10.1016/j.ecolecon.2021.107047

Tariq, A., Mumtaz, F., Majeed, M., & Zeng, X. (2023a). Spatio-temporal assessment of land use land cover based on trajectories and cellular automata Markov modelling and its impact on land surface temperature of Lahore district Pakistan. *Environmental Monitoring and Assessment, 195*(1). doi:10.1007/s10661-022-10738-w

Tassadduq, S. S., Akhtar, S., Waheed, M., Bangash, N., Nayab, D. E., Majeed, M., Abbasi, S., Muhammad, M., Alataway, A., Dewidar, A. Z., Elansary, H. O., & Yessoufou, K. (2022). Ecological Distribution Patterns of Wild Grasses and Abiotic Factors. *Sustainability (Basel)*, *14*(18), 11117. Advance online publication. doi:10.3390/su141811117

Ullah, I., Aslam, B., Shah, S. H. I. A., Tariq, A., Qin, S., Majeed, M., & Havenith, H. B. (2022). An Integrated Approach of Machine Learning, Remote Sensing, and GIS Data for the Landslide Susceptibility Mapping. *Land (Basel)*, *11*(8), 1265. Advance online publication. doi:10.3390/land11081265

Waheed, M., Arshad, F., Majeed, M., Fatima, S., Mukhtar, N., Aziz, R., Mangrio, W. M., Almohamad, H., Al Dughairi, A. A., Al-Mutiry, M., & Abdo, H. G. (2022). Community Structure and Distribution Pattern of Woody Vegetation in Response to Soil properties in Semi-Arid Lowland District Kasur Punjab, Pakistan. *Land (Basel)*, *11*(12), 2145. Advance online publication. doi:10.3390/land11122145

Waheed, M., Arshad, F., Majeed, M., Haq, S. M., Aziz, R., Bussmann, R. W., Ali, K., Subhan, F., Jones, D. A., & Zaitouny, A. (2023). *Potential distribution of a noxious weed (Solanum viarum Du-nal), current status, and future invasion risk based on MaxEnt modeling*. Geology, Ecology, and Landscapes. doi:10.1080/24749508.2023.2179752

Wang, S. (2020). A Comprehensive Review of Wearable Applications and Material Construction. *Ozean Journal of Applied Sciences*, *10*(06), 364–408. doi:10.4236/ojapps.2020.106027

Yadav, A., Yadav, K., Ahmad, R., & Abd-Elsalam, K. A. (2023). Emerging Frontiers in Nanotechnology for Precision Agriculture: Advancements, Hurdles and Prospects. *Agrochemicals*, *2*(2), 220–256. doi:10.3390/agrochemicals2020016

Zain, M., Ma, H., Ur Rahman, S., Nuruzzaman, M., Chaudhary, S., Azeem, I., Mehmood, F., Duan, A., & Sun, C. (2024). Nanotechnology in precision agriculture: Advancing towards sustainable crop production. In Plant Physiology and Biochemistry (Vol. 206). Elsevier Masson s.r.l. doi:10.1016/j.plaphy.2023.108244

Chapter 6
Nanomaterials for Sustainable Pest and Disease Management in Agriculture

Ayesha Shafqat
School of Botany, Minhaj University, Lahore, Pakistan

Arshad Rasool
School of Botany, Minhaj University, Lahore, Pakistan

Rabia Fatima
https://orcid.org/0000-0001-9043-1207
Clarkson University, USA

Muhammad Majeed
https://orcid.org/0000-0001-7663-2563
University of Gujrat, Pakistan

Khursheed Ur Rahman
Hazara University, Pakistan

Altaf Hussain Narejo
University of Karachi, Pakistan

Muhammad Ishtiaq
University of Bhimber, Pakistan

Shazia Kousar
University of Baltistan, Pakistan

Afshan Farid
https://orcid.org/0009-0009-7237-4689
PMAS Arid Agriculture University, Pakistan

ABSTRACT

The escalating challenges in global agriculture necessitate the exploration of environmentally sustainable alternatives for pest and disease management. Nanomaterials, characterized by their unique properties at the nanoscale, offer a promising avenue for addressing these challenges in a sustainable manner. This chapter reviews recent developments in the application of nanomaterials for sustainable pest and disease management in agriculture. The authors examine the diverse range of nanomaterials, including nanoparticles, nanocomposites, and nanostructured coatings, and their efficacy against various agricultural pests and pathogens. Nanomaterials demonstrate targeted delivery mechanisms, increased stability, and prolonged effectiveness, minimizing the ecological impact associated with conventional agrochemicals. Consideration is given to the interactions between nanomaterials and plants, as well as their effects on soil microbiota, to assess their overall impact on ecosystem health.

DOI: 10.4018/979-8-3693-1890-4.ch006

1. INTRODUCTION

Plant disease has been a main consideration impacting food creation and human cultural improvement more than millennia. All through the early rural period, the event of plant sickness pestilences and bugs was viewed as a discipline from the divine beings and plain plant infection the executives' approaches were very restricted (Späti et al., 2021) . Every year, 20%-40% of yields are disappeared because of plant bugs and microbes. Existing plant sickness the executives depends prevalently on poisonous pesticides that are possibly hurtful to people and the climate (Singh et al., 2023). Plant infection results from complex communications among biotic and abiotic factors including hosts, microbes, and conditions, to which ought to be added vectors for certain illnesses and human exercises that change the collaboration purposefully or inadvertently through horticultural practices, for example, trimming frameworks, obstruction quality arrangement and use of pesticides (Lamichhane et al., 2024).Furthermore, the potential of nanomaterials in countering resistance development in pests and pathogens is discussed, presenting a sustainable alternative to current agricultural practices. Environmental and economic aspects of adopting nanomaterial-based strategies are deliberated, along with challenges related to toxicity, regulatory frameworks, and scalability. This concludes by highlighting the integration of nanomaterials into precision agriculture and smart farming systems, underscoring their potential to revolutionize pest and disease management, fostering a more resilient and sustainable agricultural sector. The adoption of nanomaterials holds promise for achieving effective control while minimizing the ecological footprint associated with traditional pest and disease management practices.

Insect pests and vectors of creature and plant illnesses are a significant requirement to the improvement of horticultural efficiency, and a consistent danger to food security and vocations, especially in less evolved nations (Pandey et al., 2023). Given commonly low yields and the general absence of huge food saves, when sickness plagues happened food deficiencies could undoubtedly create bringing about grievous impacts on human culture -, for example, the Irish Starvation brought about by potato late curse during the 1840s and the 1943 Bengal starvation brought about by rice earthy brown spot significant requirements to crop productivity by directly harming the harvest and by causing plant focuses on that decline yields. Around the world, pre-collect and post-gather misfortunes to bugs, weeds, and plant microorganisms are assessed at 45% (Pandey et al., 2023); additional losses can be attributed to vertebrate pests. These misfortunes happen even with the utilization of synthetic pesticides and safe assortments. Neither establishing more useful assortments nor putting an expanded land region under creation can address the issue for food if yield and post-collect misfortunes because of nuisances stay at the ongoing level(Wang, 2020).

2. SUSTAINABLE AGRICULTURE PRACTICES

Despite the fact that there is no bound together settled upon meaning of sustainable farming, we can say that the possibility of manageable horticulture drought to be in concurrence with the meaning of feasible turn of events, Therefore, and as outlined in the UN's 1987 Brundtland report (WCED, 1987), sustainable agriculture ought to have the option to meet the ongoing requirements of society without compromising the capacity of people in the future to address their own issues. It ought to consider natural, social, and financial maintainability, which encompass the III focal pillars of reasonable turn of sustainable development (Sangeeta Dash et al., 2023a). It is generally difficult to understand what each term alludes to - whether it is a methodology, a training, a bunch of related rehearses, a movement or instrument, and so on. This means that all of these methods generally had, Clear principles, Environmental, economic, and social objectives, had developed as approaches by their own doing throughout some time (for example agroecology or manageable heightening) or were high on the strategy plan from their beginning . (for example carbon cultivating). Frequently these methodologies might be relevant inside an assortment of creation types and conditions, or think about the entire ranch/framework in their plan. Generally speaking they have professionals, and once in a while a market/name previously connected with them (e.g. organic farming).

Even though they could have fluctuated with regards to scope (more "overall, for example, agroecology or practical intensification" focused, for example, permaculture or maybe high nature esteem farming), a key binding together trademar (Yadav et al., 2023)k is the way that they are decisions that might be made by ranchers and will to a great extent decide how they deal with their homestead in the long term.

3. PURPOSE AND SIGNIFICANCE OF NANOMATERIAL'S IN PEST AND DISEASE MANAGEMENT

Nanotechnology get its origin from Greek word for "dwarf," "nano," is the origin of the word "nano." Nanotechnology has resulted in the creation of novel ideas and agricultural products that have enormous potential for addressing the aforementioned issues (Rajput et al., 2021). Nanotechnology has considerably progressed in medication and pharmacology, yet has gotten similarly less interest for horticultural applications (Altug et al., 2022). The use of nanotechnology in agriculture is currently being explored in plant hormone delivery, seed germination, water management, and transfer of target genes, nanobarcoding, nanosensors, and controlled release of agrichemicals (Rimpika et al., 2023). Nanoparticles have been engineered by material scientists to have the desired properties, like shape, pore size, and surface properties, so that they can be used as protectants or for precise and targeted delivery of an active, like a pesticide, through adsorption, encapsulation, or conjugation (Goyal et al., 2023).

The nanoscale utilization of agrochemicals in farming as nanofertilizers, nanopesticides, nanosensors, and nanoformulations has changed customary agro-rehearses (Singh et al., 2022). In crop security, various sorts of nanoformulations, like nanoemulsion, nanosuspension, and nanoparticulate, are recommended to handle various issues concerned to agriculture (Rohilla et al., 2021; Singh et al., 2023a). As agricultural nanotechnology develops, the impending to deliver a new generation of pesticides and other actives for plant disease management also increased. The utilization of nanoparticles to safeguard plants can happen by means of two unique instruments: (a) nanoparticles themselves, which protect crops, or b) nanoparticles, which act as carriers for existing pesticides or other actives, like double-stranded RNA (dsRNA). These nanoparticles can be sprayed on seeds, foliar tissue, or roots, or they can be drenched or soaked.

Nanoparticles, as transporters, can give a few advantages, similar to

(I) upgraded timeframe of realistic usability,
(II) further developed dissolvability of inadequately water-solvent pesticides,
(III) decreased poisonousness, and
(IV) supporting site-explicit take-up into the objective nuisance (Malik et al., 2023).

An increase in the effectiveness of the activity and stability of the nanopesticides under environmental pressures (UV and rain), resulting in a significant reduction in the number of applications, a decrease in their toxicity, and a reduction in their costs, are two additional potential benefits to reduce toxicity effect This schematic shows different nanomaterials as either protectants or transporters for actives like insect sprays, fungicides, herbicides, or RNA-impedance particles, focusing on a great many bugs and microorganisms. It likewise features the possible advantages of nanomaterial applications, for example, further developed timeframe of realistic usability, target site-explicit take-up, and expanded dissolvability, while diminishing soil filtering and harmfulness (Fujita et al., 2023) Figure 1.

3.1. Nanomaterials in Pest Management

Pesticides are unavoidable in cultivation or farming industry to overhaul crop yield. In abundance of 90% of bug harms used are lost due to drift, depleting in soil, degradation process (Photolysis, hydrolysis), and microbial activities (Lamichhane et al., 2024). Metal nanoparticles are convincing against plant microorganisms, bugs, and irritations in various assessments. In this manner, nanoparticles can be used to make new bug toxic substances, pesticides, and bug repellant definitions (Bhatia et al., 2024). Nanoformulation of standard pesticide with polymers or metal nanoparticles is a significantly mentioning area of pesticide industry. Nanoencapsulation of pesticides is helpful in controlled and slow appearance of dynamic trimmings by control in the outside shell of nanocapsule, which conveyances low measurements over a postpone period and lessens excess run-off of undesirable pesticide (Victoria et al., 2024).

Site-targeted delivery and immovability of active ingredient is another advantage of nanotechnology in plant conservation and protection. Correspondingly, nanoemulsion with water or oil rallies the solubility and competence of pesticide against different pests (Periakaruppan et al., 2023). Different types of nanopesticide have been synthesized and are summarized in Table 1. Silver nanoparticle (Ag) is broad-spectrum active agent against phytopathogen as Biploaris sorokinniana, Botrytis cinerea, etc. Nanomaterials like SiO2, TiO2, and ZnO are expanding their presence in pesticides to safeguard the plants from contagious, bacterial, and bug bothers. Au, Ag and Cu nanoparticles are being utilized as bio-nanosensors and electrical-nanosensors to recognize the irritation (Al-Sahli et al., 2024). Nanoparticles are one of the viable natural insect poisons to tackle the recent concern of ecological contamination. Epitome of green pesticides by utilizing nanoparticle and settle dynamic fixings which will diminish risks Table 1.

Figure 1. Nanomaterials as protectants or carriers to provide crop protection

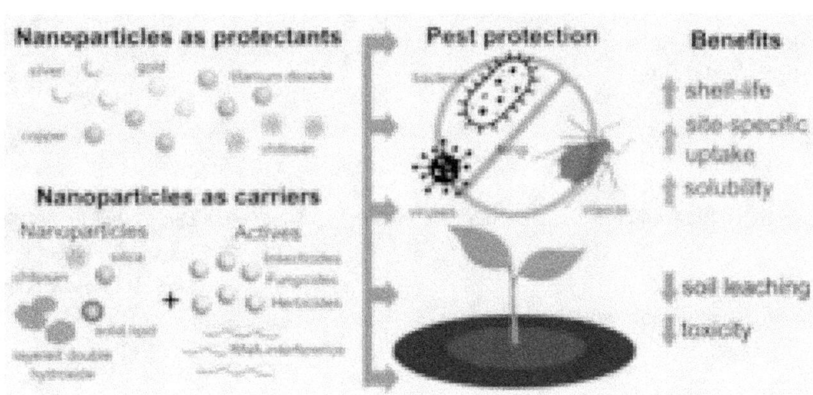

The degree of nanotoxicity in the climate predominantly relies upon focus (under 100 ppm), size (in excess of 20 nm) and organization of the nanoparticle (Sinha et al., 2023a; Singh et al., 2023b). Pesticides are the basis to defend against major biological disasters in agriculture and are important to ensure national food security. Pesticides are inevitable in agriculture to enhance crop yield. More than 90 percent of insecticides used are lost due to drift, leaching in soil, degradation process (Photolysis, hydrolysis), and microbial activities (Waheed et al., 2023).

4. TYPES OF NANO PESTICIDES

4.1. Target Released Nano Pesticides

A controlled pesticide delivery or brilliant conveyance arrangement of pesticides is utilized ashore by choosing a reasonable course to manage the target pest this approach limits pesticide utilization and bit by bit accomplishes more viable use of pesticide. This framework permits just straightforward, slow discharge as opposed to a responsive delivery given ecological condition like light, temperature, soil, pH, humidity, and enzyme (Hussain et al., 2023a).

4.2. Light Sensitive Nano Pesticides

Luminophore and light-delicate materials are covered with pesticides and presented to light. The physical and synthetic properties of these transporter materials are changed which brought about the breakdown and arrival of pesticides. A few light-delicate particles can change over light into heat energy which can make light-touchy pesticides (Yadav et al., 2023).

Table 1. Advancement in nanopesticide development: different types of nanoformulations of conventional pesticides and nanomaterial showed significant impact on pest control Environ Chem Lett (2017) 15:15–22

	Pesticide	Crop disease	Pathogen	Impact
(A) Nanoformulation				
Cellulose/silica nanocomposites	Tebuconazole	Leaf senescence and decay	Rust and moulds	50% improvement in release rate
Porous hollow silica	Validamycin	–	–	Loading capacity improved up to 36%
Calcium carbonate	Validamycin	–	*Rhizoctonia solani*	Controlled release up to 2 weeks
TiO_2	With Zn and Ag	Spot of tomato	*Xanthomonas perforans*	High photocatalytic activity and bacterial reduction
Porous hollow silica	2,4-Dichlorophenoxyacetate	–	–	Improved herbicide activity period
Lignin	Diuron	–	–	Improved controlled release rate

4.3. pH-Sensitive Nano Pesticides

The pH-sensitive carrier material is covered with pesticides, nano pesticides are two sorts. The ionization or deionization of monoacid for the most part happens inside a pH scope of 4-8, for which protons are acknowledged at low pH and pesticide is delivered at unbiased and base pH (Zhou et al., 2023).

4.4. Enzyme Responsive Pesticide

When a pest interacts with plants, a series of changes occur including changes in plant enzymes. Enzyme-responsive polymers can react physically and chemically to such stimulation. Protein responsive polymer can respond genuinely and artificially to such excitement. These polymers can be utilized to cover the pesticides. At the point when Lepidopteran bugs assault the harvest, change in plant catalysts causes the polymer container materials to break under the activity of cutinase and discharge the pesticide (Arif et al., 2021).

4.5. Nano Pheromone and Nanoparapheromone

They are naturally occurring volatile compounds can be used to eradicate pest control approaches. They are delicate to wind, rain, and unsteady owing to photo-oxidation, auto-oxidation, and isomerization (Pramanik et al., 2020).

4.6. Nano Encapsulation

Nano capsulation is a process in which pesticides material is incorporated in shell, structure orientation is formed, and now a days beneficial bacteria are coated with nano materials and provide protection against pests (Rohilla et al., 2021).

4.7. Nanoparticles for Insect Control

Insects are tracked down in a wide range of conditions and possess minimal multiple thirds of the known types of creatures on the planet. Bugs benefited from a wide range of plants including crop plants, backwoods trees, therapeutic plants and weeds (Altug et al., 2022). They additionally plague the food and other put away items in godowns, receptacles, capacity designs and bundles making an enormous measure of misfortune the put away food and furthermore crumbling of food quality. Insect cause injury to plants and put away items either straightforwardly or by implication in their endeavors to get food. Bugs that cause under 5% harm are not considered as vermin. The bugs which cause harm somewhere in the range of 5 and 10% are called minor vermin and those that cause harm over 10% are considered as significant nuisances that overall bug bothers caused an expected 14% misfortune, plant microbes cause a 13% misfortune and weeds a 13% misfortune (Payne et al., 2023).

5. CHEMICAL PESTICIDES AND NANOTECHNOLOGY

The rising utilization of pesticides adversely affected farmers, buyers, nontarget life forms and the climate. A test to adjust between crop creation and ecological insurance can be accomplished by taking on nanotechnology (Sangeeta Dash et al., 2023b). These are synthetically alterable particles comprising of a huge surface-to-volume proportion, and their actual properties make them sufficiently fit to target living beings Nanoparticles are compelling nanocides against bugs and vermin. Various sorts of nanoparticles, viz. nano silica, silver, aluminum oxide, zinc oxide, titanium dioxide and oil in water (nanoemulsions), can be really utilized as nanocides (Grillo et al., 2021; Singh et al., 2023c). Nanotubes announced with aluminosilicate can connect to establish surfaces and hair of bugs through which they go into the bug body and influence their body capabilities Silica-based nanoformulation was tried against red flour scarab T. castaneum and gave successful control of the nuisance Aluminum oxide and silicon dioxide nanopesticides act by restricting to the insects cuticle, trailed by physico-sorption of waxes and lipids, leading to insects death (Zhou et al., 2023).

5.1. Entrance Pathway of Nanoparticles Into Insect's Bodies

Nanoparticles take entry into insets through physical contact, ingestion, and inward breath. In actual contact nanoparticles enter the exoskeleton, restricting the nanomaterials to sulfur from protein or phosphorus from DNA in intracellular space, prompting denature of organelles and catalyst brought about cellular fraction and cell death. The take-up, movement, and accumulation of nanoparticles rely upon plant species, age, development, and natural condition. They go into cells by a dispersion interaction. In plants ZnO2+, Cu2+, Al2+, and Au2+ are moved through the underground root growth, and Fe3O4 goes into the foliar part of the plant. Plant cell walls go about as an obstruction for section of any outer specialist including nanoparticle into the plant system (Chaupoo et al., 2023; Singh et al., 2023d).

Cell plasma film permits under 5 to 20 nm nanoparticles. Catalyst responsive pesticide when a nuisance cooperates with plants, a progression of changes happen remembering changes for plant proteins. Protein responsive polymer can respond actually and artificially to such feeling. These polymers can be utilized to cover the pesticides. At the point when Lepidopteran bugs assault the yield, change in plant catalysts causes the polymer container materials to break under the activity of cutinase and discharge the pesticide (Zain et al., 2024).

There is an opportunity for growth of pore or enlistment of new cell wall pore. The nanoparticle goes into cells by endocytosis with assistance of depression like construction that structures around the nanoparticle by the plasma membrane (Bashir et al., 2023a). They may similarly cross the layer using introduced transport carrier protein or through molecule channels. When nanoparticles are applied on the leaf surface, they enter through stomata opening and subsequently moved to various tissue. Internal breath of nanoparticles prompts absorption in cells through phagocytosis which causes disease and changes the activity of metabolic characteristics and diminishes lipid, protein, and glucose levels. Reduced food utilization in the long run prompts the death of the bug (Ali et al., 2023) Inward breath likewise modified the movement of sensory system catalysts and layer potential.eg. Zn NPS and TiO2 NPS are tie to acetylcholine esterase and β carboxyl esterase and influence their movement. The method of activity of these nanoparticles is like the organo phosphorous and carbonate concentration in insect spray (Tassadduq et al., 2022).

5.2. Nanomaterials for Nematode Management

Nematodes are ubiquitous in nature, and a large number of the species are parasitic to plants, causing huge financial misfortunes in different yields. Plant-parasitic nematodes (phyto-parasitic) attack the foundations of plants and secure themselves to redirect supplements from the plant toward their own development (Khider et al., 2023). Nematodes are multicellular creatures in the phylum Ecdysozoa, that is to say, creatures that can shed their fingernail skin. Plant-parasitic nematodes incorporate around 15% of the total number of known nematodes species, what rate encompasses generally.

The existence pattern of a nematode has six phases, which incorporate the egg stage, J1 or first stage hatchlings, J2 or second stage hatchlings, J3 or third-stage hatchlings, J4 or fourth-stage hatchlings, and grown-up stage. The initial four phases are the youthful stages and are known as adolescent stages. The female lays 200 to 300 eggs in the dirt or in plant tissues, separately or in bunches as an egg mass, that trapdoor out into hatchlings which are like grown-ups apparently (Khider et al., 2023). Various type of nematodes are witnessed in agricultural land.

5.3. Ectoparasitic

Ectoparasitic nematodes benefits from an external perspective of plant tissue and utilize their stylet to take care of from plant roots. Nematodes that utilization this technique can brush on various plants, making it more straightforward for them to switch has, yet their additional portability makes them truly vulnerable to environmental vacillations and hunters. This sort of nematodes can have incredibly lengthy stylets, which permit them to take care of profound inside the plant root. One of the case of this kind is *Tylenchorhynchus claytoni* (Payne et al., 2023).

5.4. Semi-Endoparasitic

Nematodes that feed as semi-endoparasites can somewhat enter the plant and feed sooner or later in their life cycle. The styles of these nematodes infiltrate the upper tissues of the root and permit the invading nematodes to form a constant feeding cell. The best example of this type is *Rotylenchulus reniformis* ("Society of Nematologists 1961," 2023).

5.5. Sedentary Endoparasites

The most damaging nematodes in the world have a sedentary endoparasitic lifestyle. The two main nematodes in this group are the cyst nematodes (Heterodera and Globodera) and the root-knot nematodes (Meloidogyne). In these nematodes the J2 invades the plant near the tip of a root and migrates throught the tissue to the developing vascular cells (Sangeeta Dash et al., 2023b).

5.6. Stem and Bulb Nematodes

Stem and bulb nematodes attack the upper and lower portions of plants. They use water movies to relocate up the stem of the plant and subsequently are really harming under wet circumstances. The irresistible phases of these nematodes is the fourth-stage adolescent. This stage frequently enters arising plant tissues subterranean, yet can creep up stems in a film of water and enter the shoots through buds, petioles, or stomata.

5.7. Seed Gall Nematodes

This kind of nematodes has a special component, it relocates as a second-instar hatchling in water movies to the plant passes on to benefit from it as an ectoparasite. This instrument cause deformity of the plant leaves. At the point when the swarmed plants starting to deliver blossoms, the second-instar hatchlings infiltrate the new blossoms and feeds on the recently shaped seeds. This sort of nematode is scattered alongside the seed during planting and collect (Aziz et al., 2023).

5.8. Foliar Nematodes

This type of nematodes is related to the *Aphelenchoides* genus. The grown-up nematodes move through the water film from the stems to the leaves of their host plant. After the grown-up of this nematode arrives at the plant leaves, it infiltrates the leaves through regular pores, for example, stomata. The horrendous taking care of by grown-up nematodes harms the plant cells and the leaves become dry. At the point when the temperature diminishes in the colder time of year, these grown-ups harbor in the dead leaves until the temperature increments and the weather conditions is optimum (Hassan, Haq, Ahmad, et al., 2022).

5.9. Control of Plant-Parasitic Nematodes

Substance control was extremely successful against bug however after some time the viability of the nematocide utilized diminished. These variables incorporate the accompanying. Soil fumigants are applied to the dirts as fluid, emulsions, or granules yet the dynamic fixing spread in the dirt as a gas. Halogenated hydrocarbons called DD, EDB, MBr, and DBCP are generally utilized soil fumigents. Utilizing bug sprays, large numbers of the organophosphorus subsidiaries and carbamates additionally have nematocidal properties. Phorate, fensulfothion, thionazin, dichlohenation, disulfoton, methomyl, and vydate have been tried broadly everywhere. A portion of these mixtures have become disliked in view of their high mammalian harmfulness and perseverance in the dirt (Majeed, Bhatti, & Amjad, 2021).

Conventional escalated cultivating strategies depend to a great extent on the utilization of nematocides, reliance on this approach should reduce as natural and wellbeing worries around these harmful substance increment. Natural limitations on nematocidal use for controlling plant-parasitic nematodes have produced expanding interest in organic control measures (Majeed et al., 2020).

5.10. Nanotechnology and Plant-Parasitic Nematodes

Plant pathogenic nematodes are infinitesimal yet powerful as yield misfortunes of $85 billion have been assessed overall every year. Yearly assessed crop misfortunes because of nematodes in India have been worked on a mission to be around 242.1 billion. (Rehman et al., 2023). Broad exploration in created nations and in excess of 70 non-industrial nations have affirmed the worry of disastrous nature of PPNs and it has become extremely important to deal with these to increment crop efficiency. Plants pathogenic nematodes are minute, worm formed creatures present inside plant materials or rhizospheric soil itself. Out of complete known types of nematodes, just 10% are plants pathogenic in nature and these feed on plants prompting the decrease in the development of harvests and yield proficiency (Muhammad, Basit, et al., 2024).

Utilization of nanoparticles in plant sickness the board is a novel and extravagant methodology that might demonstrate extremely viable in future with the advancement of use part of nanotechnology. A nematicide is a synthetic used to kill plant parasitic nematodes. These are expansive range poisons having high instability or different properties advancing relocating through a soil for example Aldicarb (Table 2). Normal Regular nematicide in neem cake root exudates of marigold (Tagetes) in likewise observed to be nematocidal Paeciliomyces organisms is additionally Neamalophagoui parasites, superheated steam is likewise utilized.

The nanotechnology has the enormous likely possibility of utilizations in horticulture and plants sickness the executives specifically direct use of nanoparticles to soil or seed or foliage in the easiest to shield plants from various sorts of microbe attack by stifling there as by the compound strategies (Muhammad, Waheed, et al., 2024; Naveed et al., 2024a).

Nanotechnology extraordinarily affects organic sciences and more nanomaterials are being utilized in medication, drug store, and agribusiness. Numerous nano-items or nanomaterials were utilized against plant-parasitic nematodes, for example, nano-silver and nano-sulfur. These days, the fast improvement of nanotechnology presents a better approach to work on the presentation of customary pesticide definitions through the development of nanotechnology-based horticulture framework, for example, drug conveys and a controllable medication focusing on and delivering framework (Bashir et al., 2023a).

Nanobiotechnology holds the commitment of controlled delivery and site-designated conveyance of agrochemicals (Singh et al., 2023; Naveed et al., 2024b). A plenty of substance, physical and organic strategies keeps on developing prompting the creation of respectable metal nanoparticles. Plant parasitic nematodes are major farming irritations causing crop misfortunes worth many billions bucks every year around the world. Customary control measures rely on profoundly harmful nematicides. In principle, seed treatment for nematode control is ideal, however is generally ineffective because of poor rhizosphere conveyance. Dynamic elements of different nanoparticles have likewise shown proof of being possibly

Table 2. List of chemicals used as nematicides

Crop	Nematode pest	Nematicide	Application rate[2]	Application techniques
Potato	*Globodera* sp.	Aldi carb	2.24-3.36	Incorporated in row
		Oxamyl	4.0-5.5	-
		Carbofuran	4.0-5.5	-
Tomato, cucurbits	*Meloidogyne* spp.	Aldi carb	3.36	Incorporated in 30 cm bands
		Ethoprophos	0.9-2.9	Incorporated in bands
		Oxamyl	0.6-1.2	Incorporated in bands
		Fenamiphos	1.6-3.3	Incorporated in bands
	Meloidogyne spp.	Dazomet	30-50g/m²	Incorporated in bands and irrigated time interval before planting
Citrus	*Tylenchulus semipenetrans*	Fenamiphos	10.8-21.6	Annual treatment applied along drip-line
	Tylenchulus semipenetrans	Aldi carb	5.5-11.0	Annual treatment applied along drip-line
Grape	*Xiphinema index*	Aldi carb	10-May	In band for nursery use
	Meloidogyne spp.	fenamiphos (e.c. formulation)	10	In band for nursery usein band for nursery use
Banana	*Radophilus similis* and/or	Carbofuran	2-4 ga.i.per plant	Applied around plant 2-3 times per year
	Helicotylenchus multicinctus and/or	Ethoprophos	2-4 ga.i.per plant	Applied around plant 2-3 times per year
	Pratylenchus spp.	Fenamiphos	2-4 ga.i.per plant	Applied around plant 2-3 times per year
	Meloidogyne spp.	Isazofos	2-4 ga.i.per plant	Applied around plant 2-3 times per year
		Ebufos	2-4 ga.i.per plant	Applied around plant 2-3 times per year

powerful nematicides, which makes these nanoparticles a reasonable respectable source to control nematode disease in plants (Sangeeta Dash et al., 2023a).

AgNp is notable to have hostile to nematicide properties and in this manner utilized in the field of medication, and are presently getting consideration in the fields of farming and cultivation.

The Natural control specialists assume a significant part in controlling the play bothers like nematodes, weeds, bug, and parasites. Gold, Silver NP have nematicidal action that might give an option in contrast to high-take a chance with engineered nematicides or conflicting natural control specialists (Figure 2). High recurrence (every other week) and high application dosagesof Gold, Silver NP might be expected to accomplish viable field viability for root-hitch nematodes. Consolidating AgNP with a water system framework, for example, fertigation or tank-blend with viable synthetics that supplement the AgNP nematicidal impact might build materialness of AgNP. Further comprehension of the component in the nematicidal activity of Silver NP additionally warrants improvement of Gold, Silver NP adequacy (Yadav et al., 2023) (Table 3).

6. NANOPARTICLES FOR FUNGAL DISEASES

Crop cultivators persistently fight with parasitic sicknesses influencing their yields. Plant pathogenic parasites should be controlled assuming buyer interest in created nations for premium quality and different food varieties is to be met; while top notch oats, products of the soil are a sign of financial development in emerging nations. Business farming depends vigorously upon high contributions of substance pesticides to safeguard crops against microorganism sand bothers. The opposition of bug sand plant microorganisms against pesticides and fungicides is quickly turning into a difficult issue (Dorjee et al., 2023;Naveed et al., 2024c; Majeed et al., 2024c).

Organisms are answerable for over 70% of all significant harvest illnesses and critical yield misfortune is observed in a few yield animal types like rice, wheat, barley, cotton, ground nut and grapevine. In agribusiness, yearly yield misfortunes due to pre-and post-gather parasitic illness surpass 200 billion euros.. Almost one fourth of food crops overall are impacted by mycotoxin, aflatoxins, ergottoxins, Fusarium toxins. Although most of contagious species are saprophytes, various them are parasites, with around15, 000 of them, causing sickness in plants, the greater part having a place with the Ascomycetes

Figure 2. Nematodes attack on plants vs. nano particles application

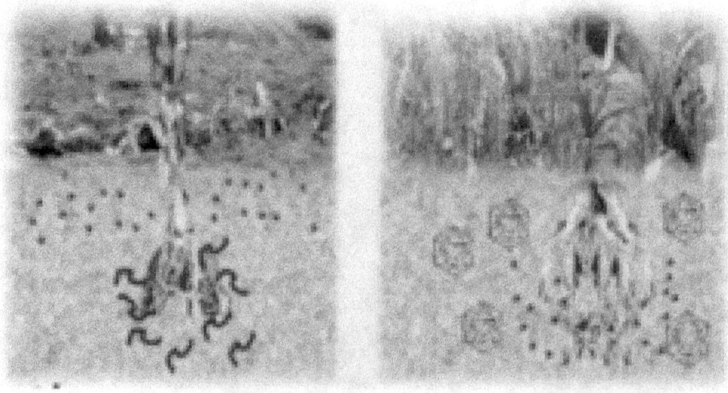

and Basidiomycetes Contagious plant illnesses are by and large dealt with the utilizations of compound fungicides. Compound control has been seen as exceptionally successful for a few parasitic illnesses, yet it leaves cut off unambiguous impacts that obliterate valuable creatures alongside microbes. Fungicide opposition (absence of responsiveness) is a significant reason for unfortunate infectious prevention in parasitic microorganisms (Li et al., 2023; Jamil et al., 2022). The advancement of fungicide opposition is affected by complex collaborations of elements, for example, the method of activity of the fungicide the science of the microorganism, fungicide use design, and the editing framework. Understanding the science of fungicide obstruction, how it creates, and how it tends to be managed urgent for guaranteeing maintainable infectious prevention with fungicides. A few microemulsion details are accessible available, including plant development controllers and foundational fungicides, for wide range infectious prevention in various sorts of target plants (Khider et al., 2023).

It is vital to comprehend parasitic science to foster more successful treatment and anticipation techniques for contagious diseases. Contagious microorganisms can be partitioned into two significant gatherings: filamentous growths and yeasts. The majority of the essential microorganisms are filamentous parasites, while the greater part of the deft microbes (which cause contaminations when a patient's resistant protections are compromised) are yeasts. The fundamental science of filamentous organisms and their lifecycle is displayed in Fig. 1. Filamentous parasites are for the most part made out of lengthened, extended hyphae with cell walls, and normally replicate physically or abiogenetically by creating a large number of spores (Li et al., 2023).

Parasites have fostered a few obstruction systems remembering modifications for drug targets, change in cell pathways (like adjustment in sterol biosynthesis), decrease in the intercellular convergence of target compound, overexpression of the antifungal medication target, enactment of stress reaction flagging, and overexpression of efflux pump proteins (Ahmed et al., 2023).

The limited number of antifungal specialists, the critical harmfulness of these specialists towards mammalian cells, the rise of safe species, and the insurance given by biofilms towards customary antimycotic specialists plainly show the need to foster new antifungal specialists to successfully treat and control safe contagious diseases. Moreover, organisms have their natural opposition components to antifungal medications, including biofilm arrangement, contrasts in cell porousness, and a few systems covering with those ensnared in gained obstruction including objective contrariness, stress reaction signaling, and efflux pump proteins overexpression.

Nanoparticles are promising cutting edge antimicrobial specialists (Grillo et al., 2021). Their antibacterial properties have been broadly contemplated, yet their utilization as antifungal specialists has gotten less consideration (Saleh & Hassan, n.d.). Nonetheless, nanoparticles have numerous properties that could make them helpful antimycotic specialists. For example, nanoparticles are exceptionally tuneable, they have high unambiguous surface region, and their special physical and synthetic properties permit them to go after microorganisms by means of a few distinct mechanism (Hooshmand et al., 2023). Our gathering and others have as of late represented that going after microorganisms with nanoparticles that apply a few synchronous components of activity can be a powerful methodology to forestall or fundamentally postpone the beginning of antibacterial opposition. Both natural and inorganic nanoparticles with antimycotic properties have been created, and each kind of nanoparticle has novel advantages and restrictions (Zain et al., 2024).

Figure 3. The mechanisms of fungal acquired and intrinsic resistance which can be developed through several mechanisms

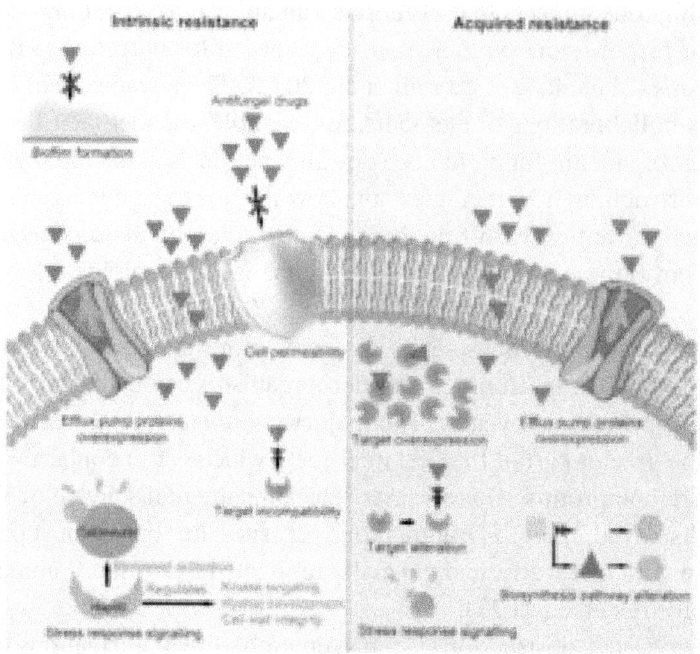

7. NANOMATERIAL FOR BACTERIAL DISEASES

Microscopic organisms focusing on nanomaterials have been broadly utilized in the conclusion and treatment of bacterial irresistible sicknesses. These nanomaterials show incredible potential as antimicrobial specialists because of their wide range antibacterial limit and somewhat low poisonousness. As of late, nanomaterials have worked on the exact recognition of microorganisms, gave restorative systems against nosocomial diseases and worked with the conveyance of antigenic protein immunizations that actuate humoral and cell invulnerability. Biomaterial inserts, which have customarily been blocked by bacterial colonization, benefit from their capacity to keep microbes from framing biofilms and spreading into contiguous tissues (Majeed, Lu, et al., 2022).

Nanomaterial-based treatments address promising instruments to battle against bacterial contaminations that are hard to treat, for instance, those described by biofilm development or antimicrobial opposition. Au NPs have been generally utilized as biosensors for microbes location. Thiolation of the illusory phages coordinated on different bacterial microbes caused conglomeration of AuNPs, bringing about a noticeable colorimetric response before around 100 cells of the objective microscopic organisms (Haq et al., 2022).

The Advancement of Natural, Inorganic and Half and half NPs for Irresistible diseases Microorganisms focusing on nanomaterials have been broadly used to treat bacterial, contagious, viral and parasitic irresistible illnesses lately (Waheed et al., 2022), and their applications incorporate direct bactericidal impacts, embodiment of probiotics, and exemplification of bactericidal specialists. Inorganic nanomaterials with direct bactericidal impacts incorporate calcium silicate (CaSi), silver (Ag), copper (Cu), and Gold (Au) NPs. Among the inorganic nanomaterials, Ag NPs have been one of the most generally utilized antimicrobial materials because of their wide range antibacterial limit and low affinity to induce drug obstruction.

Ag NPs had a better antimicrobial impact in both gram-positive and gram-negative microbes, for instance methicillin-safe Staphylococcus aureus (MRSA), vancomycin-safe Enterococcus (VRE), Pseudomonas aeruginosa, and Klebsiella pneumoniae (Majeed, Tariq, et al., 2022). Au NPs have been displayed to diminish the impact of lipopolysaccharide (LPS), a significant poison in gram-negative microbes that causes fulminant resistant reactions through its impact on dendritic cells and safe responses (Majeed, Tariq, et al., 2021). Change metal oxides and chalcogenide (TMO&C) nanomaterials have been portrayed as having brilliant physiochemical, electronic, and optical properties. They have a utilitarian design get together, demonstrating their advantages in battling bacterial contaminations (Sinha et al., 2023b). Additionally, attractive iron oxide NPs (MIONPs) display further developed adherence to Staphylococcus and extracellular polymeric substances, consequently expanding bacterial biofilm destruction through the embodiment of antimicrobial agents (Ullah et al., 2022) Figure 4.

8. POSSIBLE POTENTIAL RISKS ASSOCIATED WITH NANO MATERIALS

Numerous researchers think about nanotechnology as the following legitimate move toward science, incorporating designing with science, science, medication, and physical science (Majeed, Bhatti, Pieroni, et al., 2021). When they arrive at the climate, the effect of NPs is straightforwardly connected with their capacity to gather in organic entities, hurting them. Organic entities are normally presented to nanomaterials, however it doesn't mean they are harmless; to be sure, they can have an exceptionally destructive impact contingent upon the conditions. Plus, while normally started NPs are typically joined over the long run and structure higher size materials, made NPs will generally persevere because of the utilization of surfactants and stabilizers. Thus, an assessment of what the work of these materials could mean for the climate is required (Sinha et al., 2023c).

Figure 4. Impact on non-target organisms and environment

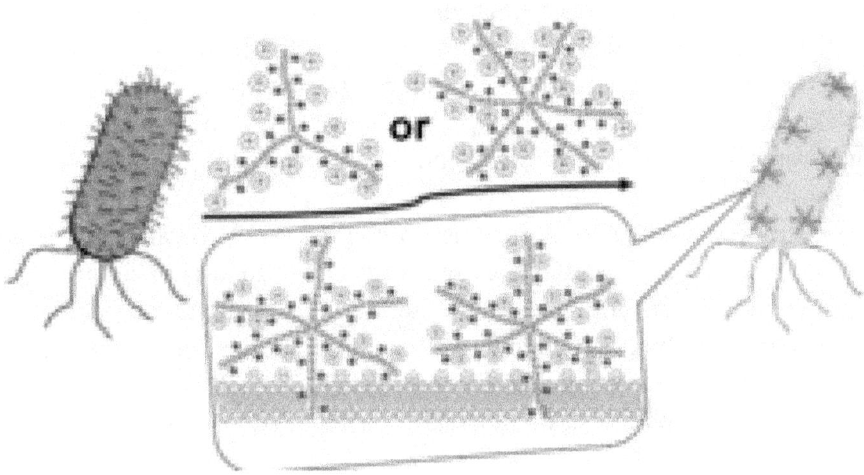

Nanotechnology is supposed to be the premise of a significant number of the vitally mechanical developments of the 21st 100 years. Given the rising rates for nanomaterial creation, the potential for their delivery in the climate and ensuing consequences for biological system wellbeing is turning into a rising worry that should be addressed (Sinha et al., 2023c). Fabricated nanomaterials will enter the climate through deliberate deliveries as well as inadvertent deliveries like air discharges and strong or fluid waste streams from creation offices (Majeed, Khan, et al., 2022). Furthermore, nanomaterials in paints, textures, and individual and medical services items, including sunscreens and beauty care products, enter the climate corresponding to their utilization. Radiated nanomaterials will eventually store ashore and water surface. Nanomaterials arriving at in the land can possibly defile soil, and relocate into surface and ground waters. Particles in strong squanders, squander water effluents, direct releases, or unintentional spillages can be moved to amphibian frameworks by wind or water spillover. The greatest delivery in the climate can emerge out of spillages related with the transportation of fabricated nanomaterials from creation offices to other assembling locales, purposeful deliveries for ecological applications, and diffuse deliveries related with wear and disintegration from general use (Majeed et al., 2023).

9. HEALTH CONCERNS ASSOCIATED WITH AGRICULTURAL NANOMATERIAL

Because of the tiny size of designed nanomaterials, inward breath openness might possibly happen to airborne particles made out of nanomaterials covering a size range from a couple of nanometers to a few micrometers in width. Nanomaterials might agglomerate into bigger particles or longer fiber chains, which might change their properties and may affect their conduct in the indoor and outside conditions as well as their possible openness and passage into the human body (Panda et al., 2023). They can store in the respiratory framework and have nanostructure-affected poisonousness because of high surface region, high surface movement, surprising morphology, little widths, or corruption into more modest particles after statement. Particles framed from the corruption or comminuting of nanomaterials may likewise introduce an expected gamble in the event that they show nanostructure-subordinate natural movement.. Nanoparticles have high statement efficiencies in the lungs of solid people, and, surprisingly, higher efficiencies in people with asthma or constant obstructive pneumonic illnesses (Li et al., 2023).

When breathed in, nanoparticles store dispersedly upon the alveolar surface, probable prompting a dissipated chemo-attractant signal and bringing about lower acknowledgment and alveolar macrophage reactions (Khoja et al., 2022) have revealed that the testimony of 20 nm particles is 2.7 times more noteworthy than 100 nm particles and 4.3 times more noteworthy than 200 nm particles (Haq et al., 2022b) have shown that higher affidavit efficiencies happen in patients with asthma or constant obstructive pneumonic sickness than in sound subjects, perhaps because of diminished leeway capacity. They observed that there was under 25% leeway of 50-and 100-nm particles during the initial 24 hrs. after inward breath.

Skin can be presented to strong nanoscale particles through either deliberate or nonintentional means. The external skin comprises of a 10 μm thick, extreme layer of dead keratinized cells (layer corneum) that is hard to pass for particles, ionic mixtures, and water-dissolvable mixtures. Deliberate dermal openness to nanoscale materials might incorporate the utilization of moisturizers, creams, wound dressing, cleansers and socks containing silver nanomaterials (Haq et al., 2022b). Nano TiO2 and ZnO materials can be uncovered as a sunscreen part or stringy materials covered with nanoscale substances for water or stain repellent properties. Nonintentional openness could include dermal contact with human substances created during nanomaterial production or burning.

It stays muddled whether various sorts of nanomaterials will enter the skin and have toxicological effects incorporates skin or other organ cytotoxicity, through amassing in skin or digestion to significantly more modest particles or due to photoactivated nanoparticles. The states of purpose may likewise influence the type of the compound (Bashir et al., 2023b). During average customer use, the particles might be delivered in one specific structure, however under additional upsetting circumstances, the structure might change. For instance, materials are exposed to washing, drying, and pressing. During washing, warm or high temp water alongside detergent. may build the arrival of nanomaterials. Drying materials will expose the nanofibers to intensity and fomentation. Pressing applies a lot of intensity, strain, and scraped spot. Capable improvement of any new materials expects that dangers to wellbeing and the overall climate related with the turn of events, creation, use and removal of these materials are tended to.This is important to safeguard laborers associated with creation and utilization of these materials, the general population, and the environment. Notwithstanding, it additionally assists with illuminating general society about the improvement regarding these new, possibly valuable, materials. The revealed potential dangers of nanomaterials are summed up in Table 3.

Pulmonary bioassay studies with nanoscale and fine-quartz particles in rats: Toxicity is not dependent upon particle size but on surface characteristics.

9.1. Nano Material and Environment

To assess the discharge of NPs into the climate, various models and approaches have been utilized, whose procedures fluctuate significantly both in conceptualization and in information assortment; from additional general models to models that are utilized for a particular situation of a decided nanomate-

Table 3. Nano material and their possible risks

Nanomaterials	Possible Risks
Carbon nanomaterials, silica nanoparticle	Pulmonary inflammation, granulomas, and fibrosis
Carbon, silver and gold nanomaterials	Distribution into other organs including the central nervous system
Quantum dots, carbon and TiO_2 nanoparticles	Skin penetration
MnO_2, TiO_2, and carbon nanoparticles	May enter brain through nasal epithelium olfactory neurons
TiO_2, Al_2O_3, carbon black, Co, and Ni nanoparticles	May be more toxic than micron sized particles

rial (Hussain et al., 2023b). Most generally utilized models are Material Stream Investigation (MFA), which are utilized to foresee the item outflows to specialized compartments (landfills, reusing plants, water medicines plants, and so on) and, from that point, to the climate; and the Natural Destiny Models (EFM) (Bhattacharya et al., 2023), which depict the vehicle and conveyance through the different ecological compartments (water, earth, air). At the end of the day, MFA centers around discharges created during the existence pattern of the item and where they are at last saved, while EFM portrays how NPs, whenever outflow has occurred, are shipped, changed and corrupted between the different ecological compartments .

The different EFM models can be arranged by whether they are steady or variable models over the long run, or whether the changes of the NPs are subject to their properties, natural circumstances, or both. The manner in which an EFM model is proposed can radically influence the outcomes. The absolute most wonderful models are SB4N model and RedNano model, the two of which incorporate the assessment of change processes because of arrangement; and the Stream model, which explicitly ponder likely compound changes of Ag and ZnO NPs (Hassan, Haq, Majeed, et al., 2022).

When inside the cell, the harmfulness and harm caused to various cell compartments (Figure 5) will rely upon the kind of NPs. For instance, carbon nanotubes typically produce harmfulness in mitochondria (Arshad et al., 2022), while Ag NPs for the most part tie to the layer, altering its porousness and morphology (Arshad et al., 2022). Lysosomes are organelles where NPs end to be discharged out from the cell, yet this gathering can be a wellspring of poisonousness as well. Cores harm happens when NPs can diffuse through the atomic layer and create aggravations in the typical capability of the Golgi complex. The greater part of the poisonous impacts created by NPs include the development of ROS (Figure 5).

Figure 5. Toxic effects on the different intracellular objectives
Published online 2020 Dec 31. doi: 10.3390/ma14010166

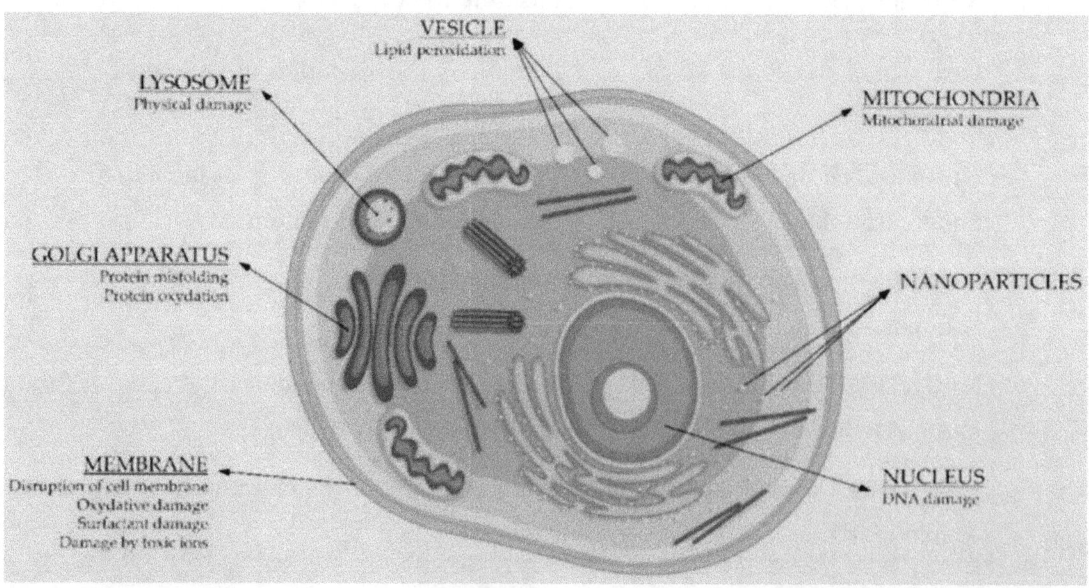

9.2. Plants, Environment, and NPs

Plants are a vital gathering of living creatures with regards to concentrating on any wellspring of ecological effect, because of their connection with each natural compartment that can contain NPs like soil, water, and air . Besides, they are a fundamental piece of the established pecking order, and assume a significant part in moving NPs to creatures and people. The most normal systems plants use to ingest NPs are through leaves, blossoms surface, roots or plant harmed regions. Most of plant harmfulness reads up generally are for human utilization, for example, corn, wheat or soy are centered around boundary investigation like the pace of seed germination, root development or nitrogen obsession (Nguyen et al., 2023) . Concentrates on show that the convergences of NPs in the soil are more raised than in water or air. Subsequently, plants seems to be the primary source of nano particles of trophic levels in the food chain (Karnwal et al., 2023).

9.3. Future Directions

Regardless of the promising development of nanotechnology in varied fields, its agricultural applications have not been translated to meet global needs primarily due to shallow awareness and biosafety concerns (Banerjee et al., 2024).The foremost reason for scarcity of commercial development of nanotechnology in agriculture is geographically limited existence of legislative framework, regulatory guidelines and negative public opinion. The growing challenge of global food security and climate change strongly underline the commercial applications of nanotechnology-based products for agricultural sector. Therefore, there is an urgent need to thoroughly assess the risk assessment and risk management factors associated with application of nanoparticles in agricultural sector, prior to implementation of regulatory guidelines (Yuvaraj et al., 2023).

Number of distributions and effectively investigated models show serious areas of strength for exceptionally in this field and significant certainty that nanopesticide-based details, like nanoemulsions, nanodispesions, and NPs have a splendid future and potential for creating more secure and more successful substance pesticide definitions for bug control, which possibly could bring about progressive changes in this field (Dehghanian et al., 2023). A few administrative bodies viz. US Food and Medication Organization (USFDA), Association for Financial Collaboration and Improvement (OECD), Worldwide Standard Association (ISO) are embraced difficulties toward this path subsequently offices including USFDA explicitly upholds regulation toward the utilization of nanocides (1370. On soil while, ISO and OECD just give rules and ideas to the administrative bodies.

9.4. Recommendations for Future Research

Promising utilizations of nanotechnology in all areas can't be ignored. Be that as it may, at the same time vulnerability and negative discernment versus nanotechnological mediations in farming area should be treated in a serious way. Consequently, there is need to put forth broad attempts in sending and working on the cutting edge explores in light of perceived information holes (Ahmed et al., 2023). Approving the passable degree of nanoparticles portion inside wellbeing limits should be investigated and explained. This could be accomplished by endeavoring fixation subordinate concentrate in regular soil framework to comprehend the exact dynamic and non-poisonous portion of nanoparticles (Khider et al., 2023).

A comprehension of the transgene rational and trophic chain move impacts of nanoparticles applications on plants should be incorporated to acquire exhaustive information on nontoxicity. Curiously, choice of reasonable level along with contemplating Trans generational and trophic chain move impacts could give satisfactory security appraisals. A reasonable outline of the dirt physico-synthetic qualities of the rural fields where nanoparticles are to be applied may help in lessening their gamble toward plant and soil biota. Changing soil climate in a manner to alter the destiny, transport and bioavailability of nanoparticles to lessen their resulting poisonousness could enormously accomplish their protected and valuable applications in farming. For instance, high level soil the executives rehearses for further developing soil conditions could help with decreasing vehicle, bioavailability and further poisonousness of nanoparticles with huge positive effect in agroecosystem(Dorjee et al., 2023; Tariq et al., 2023). At last, and in particular, we firmly suggest the consideration of biosynthesized nanoparticles as requirements for significant and top to bottom explores. Recovering the climate agreeable methodology of green blend of nanoparticles, it is accepted that biosynthesized nanoparticles may have somewhat lesser or no harmfulness and subsequently future explores should exactly zero in on their practical utility.

REFERENCES

Ahmed, T., Luo, J., Noman, M., Ijaz, M., Wang, X., Masood, H. A., Manzoor, N., Wang, Y., & Li, B. (2023). Microbe-mediated nanoparticle intervention for the management of plant diseases. *Crop Health*, *1*(1), 3. Advance online publication. doi:10.1007/s44297-023-00006-9

Al-Sahli, S. A., Al-Otibi, F., Alharbi, R. I., Amina, M., & Al Musayeib, N. M. (2024). Silver nanoparticles improve the fungicidal properties of Rhazya stricta decne aqueous extract against plant pathogens. *Scientific Reports*, *14*(1), 1297. doi:10.1038/s41598-024-51855-5 PMID:38221517

Ali, H., Muhammad, Z., Majeed, M., Aziz, R., Khan, A., Mangrio, W. M., Abdo, H. G., Almohamad, H., & Al Dughairi, A. A. (2023). Vegetation diversity pattern during spring season in relation to topographic and edaphic variables in sub-tropical zone. *Botanical Studies*, *64*(1), 25. Advance online publication. doi:10.1186/s40529-023-00398-5 PMID:37716923

Altug, H., Oh, S. H., Maier, S. A., & Homola, J. (2022). Advances and applications of nanophotonic biosensors. In Nature Nanotechnology (Vol. 17, Issue 1, pp. 5–16). Nature Research. doi:10.1038/s41565-021-01045-5

Arif, U., Bhatti, K. H., Ajaib, M., Wagay, N. A., Majeed, M., Zeb, J., Hameed, A., & Kiani, J. (2021). Ethnobotanical indigenous knowledge of Tehsil Charhoi, District Kotli, Azad Jammu and Kashmir, Pakistan. *Ethnobotany Research and Applications*, *22*. Advance online publication. doi:10.32859/era.22.50.1-24

Arshad, F., Waheed, M., Harun, N., Fatima, K., Khan, B. A., Fatima, K., Abbas, Z., Jabeen, S., & Majeed, M. (2022). Indigenous farmer's perception about fodder and foraging species of Semi-arid lowlands of Pakistan: A case study of District Kasur, Pakistan. *Taiwania*, *67*(4), 510–523. doi:10.6165/tai.2022.67.510

Aziz, A., Anwar, M. M., Majeed, M., Fatima, S., Mehdi, S. S., Mangrio, W. M., Elbouzidi, A., Abdullah, M., Shaukat, S., Zahid, N., Mahmoud, E. A., Casini, R., Yessoufou, K., & Elansary, H. O. (2023). Quantifying Landscape and Social Amenities as Ecosystem Services in Rapidly Changing Peri-Urban Landscape. *Land (Basel)*, *12*(2), 477. Advance online publication. doi:10.3390/land12020477

Banerjee, S., Mondal, S., Islam, J., Sarkar, R., Saha, B., & Sen, A. (2024). Rhizospheric nano-remediation salvages arsenic genotoxicity: Zinc-oxide nanoparticles articulate better oxidative stress management, reduce arsenic uptake, and increase yield in Pisum sativum (L.). *The Science of the Total Environment, 913*, 169493. doi:10.1016/j.scitotenv.2023.169493 PMID:38151134

Bashir, S. M., Altaf, M., Hussain, T., Umair, M., Majeed, M., Mangrio, W. M., Khan, A. M., Gulshan, A. B., Hamed, M. H., Ashraf, S., Amjad, M. S., Bussmann, R. W., Abbasi, A. M., Casini, R., Alataway, A., Dewidar, A. Z., Al-Yafrsi, M., Amin, M. H., & Elansary, H. O. (2023a). Vernacular Taxonomy, Cultural and Ethnopharmacological Applications of Avian and Mammalian Species in the Vicinity of Ayubia National Park, Himalayan Region. *Biology (Basel), 12*(4), 609. Advance online publication. doi:10.3390/biology12040609 PMID:37106809

Bhatia, D., Paul, S., Acharjee, T., & Ramachairy, S. S. (2024). Biosensors and their widespread impact on human health. *Sensors International, 5*, 100257. Advance online publication. doi:10.1016/j.sintl.2023.100257

Bhattacharya, R., Bhattacharya, R., Majeed, M., Bhandari, S., Aziz, R., Sinha, D., & Niyogi, S. S. (2023). Environmental Rehabilitation of Industrial Waste Dumping Site. In *Biohydrometallurgical Processes* (pp. 44–89). CRC Press. doi:10.1201/9781003451457-4

Chaupoo, A. S., Dhakad, A., Beese, S., Kumar, A., Kumar, P., Pandey, S. K., & Kurdekar, A. K. (2023). Transforming Horticulture: The Influence of Nanotechnology on Crop Enhancement. *International Journal of Environment and Climate Change, 13*(11), 3381–3400. doi:10.9734/ijecc/2023/v13i113512

Dehghanian, Z., Asgari Lajayer, B., Biglari Quchan Atigh, Z., Nayeri, S., Ahmadabadi, M., Taghipour, L., Senapathi, V., Astatkie, T., & Price, G. W. (2023). Micro (nano) plastics uptake, toxicity and detoxification in plants: Challenges and prospects. *Ecotoxicology and Environmental Safety, 268*, 115676. doi:10.1016/j.ecoenv.2023.115676 PMID:37979355

Dorjee, L., Gogoi, R., Kamil, D., Kumar, R., Mondal, T. K., Pattanayak, S., & Gurung, B. (2023). Essential oil-grafted copper nanoparticles as a potential next-generation fungicide for holistic disease management in maize. *Frontiers in Microbiology, 14*, 1204512. Advance online publication. doi:10.3389/fmicb.2023.1204512 PMID:37485521

Fujita, H., Ohta, S., Nakamura, N., Somiya, M., & Horie, M. (2023). Progress of Endogenous and Exogenous Nanoparticles for Cancer Therapy and Diagnostics. In Genes (Vol. 14, Issue 2). MDPI. doi:10.3390/genes14020259

Goyal, V., Rani, D., Mehrotra, S., Deng, C., & Wang, Y. (2023). Unlocking the Potential of Nano-Enabled Precision Agriculture for Efficient and Sustainable Farming. In Plants (Vol. 12, Issue 21). Multidisciplinary Digital Publishing Institute (MDPI). doi:10.3390/plants12213744

Grillo, R., Fraceto, L. F., Amorim, M. J. B., Scott-Fordsmand, J. J., Schoonjans, R., & Chaudhry, Q. (2021). Ecotoxicological and regulatory aspects of environmental sustainability of nanopesticides. *Journal of Hazardous Materials, 404*, 124148. doi:10.1016/j.jhazmat.2020.124148 PMID:33059255

Haq, S. M., Tariq, A., Li, Q., Yaqoob, U., Majeed, M., Hassan, M., Fatima, S., Kumar, M., Bussmann, R. W., Moazzam, M. F. U., & Aslam, M. (2022a). Influence of Edaphic Properties in Determining Forest Community Patterns of the Zabarwan Mountain Range in the Kashmir Himalayas. *Forests*, *13*(8), 1214. Advance online publication. doi:10.3390/f13081214

Hassan, M., Haq, S. M., Ahmad, R., Majeed, M., Sahito, H. A., Shirani, M., Mubeen, I., Aziz, M. A., Pieroni, A., Bussmann, R. W., Alataway, A., Dewidar, A. Z., Al-Yafrsi, M., Elansary, H. O., & Yessoufou, K. (2022). Traditional Use of Wild and Domestic Fauna among Different Ethnic Groups in the Western Himalayas—A Cross Cultural Analysis. *Animals (Basel)*, *12*(17), 2276. Advance online publication. doi:10.3390/ani12172276 PMID:36077997

Hassan, M., Haq, S. M., Majeed, M., Umair, M., Sahito, H. A., Shirani, M., Waheed, M., Aziz, R., Ahmad, R., Bussmann, R. W., Alataway, A., Dewidar, A. Z., El-Abedin, T. K. Z., Al-Yafrsi, M., Elansary, H. O., & Yessoufou, K. (2022). Traditional Food and Medicine: Ethno-Traditional Usage of Fish Fauna across the Valley of Kashmir: A Western Himalayan Region. *Diversity*, *14*(6), 455. Advance online publication. doi:10.3390/d14060455

Hooshmand, S., Kassanos, P., Keshavarz, M., Duru, P., Kayalan, C. I., Kale, İ., & Bayazit, M. K. (2023). Wearable Nano-Based Gas Sensors for Environmental Monitoring and Encountered Challenges in Optimization. In Sensors (Basel, Switzerland) (Vol. 23, Issue 20). doi:10.3390/s23208648

Hussain, S., Raza, A., Abdo, H. G., Mubeen, M., Tariq, A., Nasim, W., Majeed, M., Almohamad, H., & Al Dughairi, A. A. (2023a). Relation of land surface temperature with different vegetation indices using multi-temporal remote sensing data in Sahiwal region, Pakistan. *Geoscience Letters*, *10*(1), 33. Advance online publication. doi:10.1186/s40562-023-00287-6

Jamil, M. D., Waheed, M., Akhtar, S., Bangash, N., Chaudhari, S. K., Majeed, M., Hussain, M., Ali, K., & Jones, D. A. (2022). Invasive Plants Diversity, Ecological Status, and Distribution Pattern in Relation to Edaphic Factors in Different Habitat Types of District Mandi Bahauddin, Punjab, Pakistan. *Sustainability (Basel)*, *14*(20), 13312. Advance online publication. doi:10.3390/su142013312

Karnwal, A., Dohroo, A., & Malik, T. (2023). Unveiling the Potential of Bioinoculants and Nanoparticles in Sustainable Agriculture for Enhanced Plant Growth and Food Security. In *BioMed Research International* (Vol. 2023). Hindawi Limited. doi:10.1155/2023/6911851

Khoja, A. A., Haq, S. M., Majeed, M., Hassan, M., Waheed, M., Yaqoob, U., Bussmann, R. W., Alataway, A., Dewidar, A. Z., Al-Yafrsi, M., Elansary, H. O., Yessoufou, K., & Zaman, W. (2022). Diversity, Ecological and Traditional Knowledge of Pteridophytes in the Western Himalayas. *Diversity*, *14*(8), 628. Advance online publication. doi:10.3390/d14080628

Lamichhane, J. R., Barbetti, M. J., Chilvers, M. I., Pandey, A. K., & Steinberg, C. (2024). Exploiting root exudates to manage soil-borne disease complexes in a changing climate. In Trends in Microbiology (Vol. 32, Issue 1, pp. 27–37). Elsevier Ltd. doi:10.1016/j.tim.2023.07.011

Li, Y., Zhang, P., Li, M., Shakoor, N., Adeel, M., Zhou, P., Guo, M., Jiang, Y., Zhao, W., Lou, B. Z., & Rui, Y. (2023). Application and mechanisms of metal-based nanoparticles in the control of bacterial and fungal crop diseases. In Pest Management Science (Vol. 79, Issue 1, pp. 21–36). John Wiley and Sons Ltd. doi:10.1002/ps.7218

Majeed, M., Bhatti, K. H., & Amjad, M. S. (2021). Impact of climatic variations on the flowering phenology of plant species in jhelum district, punjab, pakistan. *Applied Ecology and Environmental Research, 19*(5), 3343–3376. doi:10.15666/aeer/1905_33433376

Majeed, M., Bhatti, K. H., Pieroni, A., Sõukand, R., Bussmann, R. W., Khan, A. M., Chaudhari, S. K., Aziz, M. A., & Amjad, M. S. (2021). Gathered wild food plants among diverse religious groups in Jhelum District, Punjab, Pakistan. *Foods, 10*(3), 594. Advance online publication. doi:10.3390/foods10030594 PMID:33799901

MajeedM.Hayat BhattiK.Shoaib AmjadM.Mehmood AbbasiA.RashidA.Nawaz Muhammad Nawaz ShareefF.MehmoodA.MahmoodM. (2020). *Ethno-veterinary practices of Poaceae taxa in Punjab, Pakistan*. doi:10.21203/rs.3.rs-15532/v1

Majeed, M., Khan, A. M., Habib, T., Anwar, M. M., Sahito, H. A., Khan, N., & Ali, K. (2022). Vegetation analysis and environmental indicators of an arid tropical forest ecosystem of Pakistan. *Ecological Indicators, 142*, 109291. Advance online publication. doi:10.1016/j.ecolind.2022.109291

Majeed, M., Lu, L., Anwar, M. M., Tariq, A., Qin, S., El-Hefnawy, M. E., El-Sharnouby, M., Li, Q., & Alasmari, A. (2023). Prediction of flash flood susceptibility using integrating analytic hierarchy process (AHP) and frequency ratio (FR) algorithms. *Frontiers in Environmental Science, 10*, 1037547. Advance online publication. doi:10.3389/fenvs.2022.1037547

Majeed, M., Lu, L., Haq, S. M., Waheed, M., Sahito, H. A., Fatima, S., Aziz, R., Bussmann, R. W., Tariq, A., Ullah, I., & Aslam, M. (2022). Spatiotemporal Distribution Patterns of Climbers along an Abiotic Gradient in Jhelum District, Punjab, Pakistan. *Forests, 13*(8), 1244. Advance online publication. doi:10.3390/f13081244

Majeed, M., Tariq, A., Anwar, M. M., Khan, A. M., Arshad, F., Mumtaz, F., Farhan, M., Zhang, L., Zafar, A., Aziz, M., Abbasi, S., Rahman, G., Hussain, S., Waheed, M., Fatima, K., & Shaukat, S. (2021). Monitoring of land use–Land cover change and potential causal factors of climate change in Jhelum district, Punjab, Pakistan, through GIS and multi-temporal satellite data. *Land (Basel), 10*(10), 1026. Advance online publication. doi:10.3390/land10101026

Majeed, M., Tariq, A., Anwar, M. M., Khan, A. M., Arshad, F., Mumtaz, F., Farhan, M., Zhang, L., Zafar, A., Aziz, M., Abbasi, S., Rahman, G., Hussain, S., Waheed, M., Fatima, K., & Shaukat, S. (2021). Monitoring of Land Use–Land Cover Change and Potential Causal Factors of Climate Change in Jhelum District, Punjab, Pakistan, through GIS and Multi-Temporal Satellite Data; (pp. 129-147). https://www.mdpi.com/books/book/6906-land-use-land-cover-and-natural-hazards-interactions-changes-and-impacts

Majeed, M., Tariq, A., Haq, S. M., Waheed, M., Anwar, M. M., Li, Q., Aslam, M., Abbasi, S., Mousa, B. G., & Jamil, A. (2022). A Detailed Ecological Exploration of the Distribution Patterns of Wild Poaceae from the Jhelum District (Punjab), Pakistan. *Sustainability (Basel), 14*(7), 3786. Advance online publication. doi:10.3390/su14073786

Malik, S., Muhammad, K., & Waheed, Y. (2023). Nanotechnology: A Revolution in Modern Industry. In Molecules (Vol. 28, Issue 2). MDPI. doi:10.3390/molecules28020661

Muhammad, M., Basit, A., Majeed, M., Shah, A. A., Ullah, I., Mohamed, H. I., Khan, A., & Ghanaim, A. M. (2024). Bacterial pigments and their applications. In *Bacterial Secondary Metabolites* (pp. 277–298). Elsevier. doi:10.1016/B978-0-323-95251-4.00019-3

Muhammad, M., Waheed, A., Wahab, A., Majeed, M., Nazim, M., Liu, Y. H., Li, L., & Li, W. J. (2024). Soil salinity and drought tolerance: An evaluation of plant growth, productivity, microbial diversity, and amelioration strategies. In *Plant Stress* (Vol. 11). Elsevier B.V. doi:10.1016/j.stress.2023.100319

Naveed, Majeed, Jabeen, Hanif, Naveed, Saleem, & Khan. (2024a). Plant Disease Diagnosis with Artificial Intelligence (AI). Microbial Data Intelligence and Computational Techniques for Sustainable Computing (pp 217–234). Springer. doi:10.1007/978-981-99-9621-6_15

Naveed, Zahid, Fatima, Saleem, Majeed, Abid, Javed, Wazir, & Qasim. (2024b). Disease Diagnosis with Artificial Intelligence (AI). In Microbial Data Intelligence and Computational Techniques for Sustainable Computing (pp 235–258). Springer. doi:10.1007/978-981-99-9621-6_16

Naveed, Memon, Abdullah, Makhdoom, Azeem, Mehmood, Salahuddin, Rajpoot, & Majeed. (2024c). Use Cases and Future Aspects of Intelligent Techniques in Microbial Data Analysis. In Microbial Data Intelligence and Computational Techniques for Sustainable Computing (pp 259–280). Springer. doi:10.1007/978-981-99-9621-6_17

Nguyen, N. T. T., Nguyen, L. M., Nguyen, T. T. T., Tran, U. P. N., Nguyen, D. T. C., & Van Tran, T. (2023). A critical review on the bio-mediated green synthesis and multiple applications of magnesium oxide nanoparticles. *Chemosphere*, *312*, 137301. doi:10.1016/j.chemosphere.2022.137301 PMID:36410506

PandaP.ChakrobortyS.BabuS.KrishnaN. (2023). *The Use of Silver Nanoparticles in Environmental Remediation*. doi:10.20944/preprints202301.0330.v1

Pandey, P., Tripathi, A., Dwivedi, S., Lal, K., & Jhang, T. (2023). Deciphering the mechanisms, hormonal signaling, and potential applications of endophytic microbes to mediate stress tolerance in medicinal plants. In Frontiers in Plant Science (Vol. 14). Frontiers Media SA. doi:10.3389/fpls.2023.1250020

Payne, P. R., Finlay-Smits, S., Small, B., Cave, V., & Kean, J. (2023). What's that bug? Community participation in biosecurity in Mount Maunganui, New Zealand. *Biological Invasions*, *25*(2), 593–610. doi:10.1007/s10530-022-02937-w

Periakaruppan, R., Vanathi, P., & Vidhya, D. (2023). Management of soil nutrient deficiency by nanometal oxides. *Nanometal Oxides in Horticulture and Agronomy*, 291–320. doi:10.1016/B978-0-323-91809-1.00003-2

Pramanik, P. K. D., Solanki, A., Debnath, A., Nayyar, A., El-Sappagh, S., & Kwak, K. S. (2020). Advancing Modern Healthcare with Nanotechnology, Nanobiosensors, and Internet of Nano Things: Taxonomies, Applications, Architecture, and Challenges. *IEEE Access : Practical Innovations, Open Solutions*, *8*, 65230–65266. doi:10.1109/ACCESS.2020.2984269

Rajput, V. D., Singh, A., Minkina, T., Rawat, S., Mandzhieva, S., Sushkova, S., Shuvaeva, V., Nazarenko, O., Rajput, P., Komariah, Verma, K. K., Singh, A. K., Rao, M., & Upadhyay, S. K. (2021). Nano-Enabled Products: Challenges and Opportunities for Sustainable Agriculture. *Plants*, *10*(12), 2727. doi:10.3390/plants10122727 PMID:34961197

Rehman, A., Aziz, A., Anwar, M. M., Majeed, M., Albanai, J. A., Almohamad, H., & Abdo, H. G. (2023). Quantifying the impacts of urbanization on urban green, evidences from Maga City, Lahore Pakistan. *Discover Sustainability*, *4*(1), 48. Advance online publication. doi:10.1007/s43621-023-00169-z

Rimpika, A., Anushi, Manasa, S., Sharma, S., Thakur, A., Shilpa, & Sood, A.Anusha K. N. (2023). An Overview of Precision Farming. *International Journal of Environment and Climate Change*, *13*(12), 441–456. doi:10.9734/ijecc/2023/v13i123701

Rohilla, D., Chaudhary, S., & Umar, A. (2021). An Overview of Advanced Nanomaterials for Sensor Applications. *Engineering and Science*, *16*, 47–70. doi:10.30919/es8d552

Saleh, H. M., & Hassan, A. I. (n.d.). *Environmental Impact and Remediation of Heavy Metals*. Academic Press.

Sangeeta Dash, E., Yadav, A. L., Bora, R., & Kumar Pandey, A. (2023). *Research Trends in Agriculture Science* (Vol. I). www.bhumipublishing.com

Singh, A., Agrawal, S., Rajput, V. D., Ghazaryan, K., Movsesyan, H. S., Minkina, T., . . . Gupta, S. K. (2023d). Seed Nanopriming. *Nanopriming Approach to Sustainable Agriculture*, 290–313. doi:10.4018/978-1-6684-7232-3.ch013

Singh, A., Rajput, V. D., Pandey, D., Sharma, R., Ghazaryan, K., & Minkina, T. (2023c). Nano Zinc-Enabled Strategies in Crops for Combatting Zinc Malnutrition in Human Health. *Frontiers in Bioscience (Landmark Edition)*, *28*(8), 158. doi:10.31083/j.fbl2808158 PMID:37664935

Singh, A., Rajput, V. D., Sharma, R., Ghazaryan, K., & Minkina, T. (2023f). Salinity stress and nanoparticles: Insights into antioxidative enzymatic resistance, signaling, and defense mechanisms. *Environmental Research*, *235*, 116585. doi:10.1016/j.envres.2023.116585 PMID:37437867

Singh, A., Rajput, V. D., Varshney, A., Ghazaryan, K., & Minkina, T. (2023a). Small Tech, Big Impact: Agri-nanotechnology Journey to Optimize Crop Protection and Production for Sustainable Agriculture. *Plant Stress*, *10*, 100253. doi:10.1016/j.stress.2023.100253

Singh, A., Rawat, S., Rajput, V. D., Ghazaryan, K., Minkina, T., Al Tawaha, A. R. M., . . . Sandhu, K. S. (2023b). Nanoparticle-Based Priming-Enhanced Agricultural Practices. *Nanopriming Approach to Sustainable Agriculture*, 142–157. doi:10.4018/978-1-6684-7232-3.ch006

Singh, A., Sharma, R., Rajput, V. D., Ghazaryan, K., Minkina, T., Al Tawaha, A. R. M., & Varshney, A. (2023e). Green Synthesis of Nanofertilizers and Their Application for Crop Production. *Nanotechnology in the Life Sciences*, 205–231. doi:10.1007/978-3-031-41329-2_8

Singh, A., Sharma, R., Rawat, S., Singh, A. K., Rajput, V. D., Fedorov, Y., . . . Chaplygin, V. (2022). Nanomaterial-plant interaction: Views on the pros and cons. *Toxicity of Nanoparticles in Plants*, 47–68. doi:10.1016/B978-0-323-90774-3.00015-5

Sinha, D., Maurya, A. K., Abdi, G., Majeed, M., Agarwal, R., Mukherjee, R., Ganguly, S., Aziz, R., Bhatia, M., Majgaonkar, A., Seal, S., Das, M., Banerjee, S., Chowdhury, S., Adeyemi, S. B., & Chen, J. T. (2023a). Integrated Genomic Selection for Accelerating Breeding Programs of Climate-Smart Cereals. In Genes (Vol. 14, Issue 7). Multidisciplinary Digital Publishing Institute (MDPI). doi:10.3390/genes14071484

Society of Nematologists. (1961). *Journal of Nematology*, *55*(1), 20230047. Advance online publication. doi:10.2478/jofnem-2023-0047

Späti, K., Huber, R., & Finger, R. (2021). Benefits of Increasing Information Accuracy in Variable Rate Technologies. *Ecological Economics*, *185*, 107047. Advance online publication. doi:10.1016/j.ecolecon.2021.107047

Tariq, A., Mumtaz, F., Majeed, M., & Zeng, X. (2023). Spatio-temporal assessment of land use land cover based on trajectories and cellular automata Markov modelling and its impact on land surface temperature of Lahore district Pakistan. *Environmental Monitoring and Assessment*, *195*(1), 114. Advance online publication. doi:10.1007/s10661-022-10738-w PMID:36385403

Tassadduq, S. S., Akhtar, S., Waheed, M., Bangash, N., Nayab, D. E., Majeed, M., Abbasi, S., Muhammad, M., Alataway, A., Dewidar, A. Z., Elansary, H. O., & Yessoufou, K. (2022). Ecological Distribution Patterns of Wild Grasses and Abiotic Factors. *Sustainability (Basel)*, *14*(18), 11117. Advance online publication. doi:10.3390/su141811117

Ullah, I., Aslam, B., Shah, S. H. I. A., Tariq, A., Qin, S., Majeed, M., & Havenith, H. B. (2022). An Integrated Approach of Machine Learning, Remote Sensing, and GIS Data for the Landslide Susceptibility Mapping. *Land (Basel)*, *11*(8), 1265. Advance online publication. doi:10.3390/land11081265

Victoria, J., Tripathi, S., Prakash, V., Tiwari, K., Mahra, S., Sharma, A., Rana, S., Kandhol, N., Sahi, S., Tripathi, D. K., & Sharma, S. (2024). Encapsulated nanopesticides application in plant protection: Quo vadis? *Plant Physiology and Biochemistry*, *206*, 108225. doi:10.1016/j.plaphy.2023.108225 PMID:38147708

Waheed, M., Arshad, F., Majeed, M., Fatima, S., Mukhtar, N., Aziz, R., Mangrio, W. M., Almohamad, H., Al Dughairi, A. A., Al-Mutiry, M., & Abdo, H. G. (2022). Community Structure and Distribution Pattern of Woody Vegetation in Response to Soil properties in Semi-Arid Lowland District Kasur Punjab, Pakistan. *Land (Basel)*, *11*(12), 2145. Advance online publication. doi:10.3390/land11122145

Waheed, M., Arshad, F., Majeed, M., Haq, S. M., Aziz, R., Bussmann, R. W., Ali, K., Subhan, F., Jones, D. A., & Zaitouny, A. (2023). *Potential distribution of a noxious weed (Solanum viarum Du-nal), current status, and future invasion risk based on MaxEnt modeling*. Geology, Ecology, and Landscapes. doi:10.1080/24749508.2023.2179752

Wang, S. (2020). A Comprehensive Review of Wearable Applications and Material Construction. *Ozean Journal of Applied Sciences*, *10*(06), 364–408. doi:10.4236/ojapps.2020.106027

Yadav, A., Yadav, K., Ahmad, R., & Abd-Elsalam, K. A. (2023). Emerging Frontiers in Nanotechnology for Precision Agriculture: Advancements, Hurdles and Prospects. *Agrochemicals*, *2*(2), 220–256. doi:10.3390/agrochemicals2020016

Yuvaraj, M., Sathya Priya, R., Jagathjothi, N., Saranya, M., Suganthi, N., Sharmila, R., Cyriac, J., Anitha, R., & Subramanian, K. S. (2023). Silicon nanoparticles (SiNPs): Challenges and perspectives for sustainable agriculture. *Physiological and Molecular Plant Pathology*, *128*, 102161. doi:10.1016/j.pmpp.2023.102161

Zain, M., Ma, H., Ur Rahman, S., Nuruzzaman, M., Chaudhary, S., Azeem, I., Mehmood, F., Duan, A., & Sun, C. (2024). Nanotechnology in precision agriculture: Advancing towards sustainable crop production. In Plant Physiology and Biochemistry (Vol. 206). Elsevier Masson s.r.l. doi:10.1016/j.plaphy.2023.108244

Zhou, J., Liu, G., Guo, Z., Wang, M., Qi, C., Chen, G., Huang, X., Yan, S., & Xu, D. (2023). Stimuli-responsive pesticide carriers based on porous nanomaterials: A review. *Chemical Engineering Journal*, *455*, 140167. doi:10.1016/j.cej.2022.140167

Chapter 7
Transformative Nanotechnology Applications in Soil Remediation and Ecosystem Restoration

Mohamed Nouri
https://orcid.org/0000-0002-5751-8949
Sultan Moulay Slimane University, Morocco

ABSTRACT

The soil's composition, structure, and nutrients have altered throughout time. These changes are natural and anthropogenic. Economic, environmental, and social benefits, including ecosystem services, come from restoring degraded land and exploiting marginal land. Most physiochemical remedies for damaged land are difficult, costly, and time-consuming. Nano-remediation and nano-restoration are novel, efficient, cost-effective, eco-friendly, and dependable for toxin remediation and risk reduction. The high surface area/volume ratio, increased reactivity, customizable physical qualities, and adaptability of nanoscale entities make them attractive for soil remediation. Different nanomaterials (NMs) and nanotools can clean up pollutants. Both foreign chemicals and polluted location affect these methods. Decontaminating soil contaminants with nanoscale entities reduces their detrimental effects on humans, plants, and animals. It also discusses nanoparticles (NPs) and ex- and in-situ cleanup. The authors discuss nanoscale item uses in soil remediation and land restoration in this chapter.

INTRODUCTION

Soil is heavily affected by environmental pollution and can be described as a widespread problem. The presence of hazardous compounds in soil leads to soil pollution, which in turn disrupts the biological, chemical, and physical soil characteristics. This ultimately results in negative alterations to soil fertility and the ecosystem (Rajput et al., 2021).

Soil pollution is a significant worldwide issue; the soil's quality has deteriorated due to the human actions: waste disposal, mining, emissions from foundries and electronic industries, and the use of wastewater or sludge for irrigation, lead to the introduction of organic contaminants, like pesticides,

DOI: 10.4018/979-8-3693-1890-4.ch007

polycyclic aromatic hydrocarbons (PACs), polychlorinated biphenyls (PCBs), and polycyclic aromatic hydrocarbons (PAHs), inorganic contaminants, including heavy metals (HMs) and metalloids, as well as a wide range of other impurities, are frequently found in soil (Ganie et al., 2021; Vardumyan et al., 2024). Because they are resistant to chemical and biological breakdown, organic contaminants and inorganic ions accumulate in soil over time, where they cause harm (Ganie et al., 2021). HMs, particularly when present in high concentrations, can disrupt various physiological processes in organisms. This includes modifying enzyme specificity, impairing cell function, and causing damage to the membrane of cell and DNA structure (Fajardo et al., 2020). An assessment of the mobility and bioavailability of these contaminants can provide insights into their harmful impacts. Therefore, in order to decrease their harmfulness, it is imperative to decrease their accessibility to plants and other organisms. Immobilization is a practice that can successfully decrease the toxicity of metals in the soil. It does this by introducing a likely agent to the soil via ion exchange, adsorption, or precipitation, so reducing the amount of bioavailable and mobile fraction of these contaminants (Ganie et al., 2021).

Various methods, both ex-situ and in-situ, have been utilized for soil remediation over the past few years (Ghazaryan et al., 2018). These methods include vitrification, solidification, photocatalysis, extraction, and biological technologies. These approaches differ in their efficacy, expense, time consumption, and potential for secondary contamination when applied in the field. Presently, there is a growing fascination in the utilization of NMs due to its novel technique in rehabilitating soils that have been damaged by HMs (Fajardo et al., 2020).

Although significant efforts have been made to reduce the impacts of human activities, the need for environmental cleanup still remains and must be diminished. Given this perspective, nano-remediation is gaining more and more interest as an innovative technique that might completely transform the approach to handling and resolving dangerous waste and environmental pollution, while also having the ability to tackle environmental issues (Arsenov et al., 2023). Extensive study investigates the unique features of NMs. Consequently, they demonstrate extensive capacity and present multiple opportunities in various industries (Arsenov et al., 2023).

Extensive research is now being conducted to develop and construct effective and dependable methods to break down or convert environmental toxins that are of concern. The utilization of NMs in environmental remediation has gathered substantial interest owed to their distinct impute: cost-effectiveness, sensitivity, exceptional electrical capabilities, high surface / volume ratio, and superior catalytic activities (Ganie et al., 2021).

Although NPs possess evident benefits, they are not immune to certain drawbacks, among which are potential hazards to human and environmental health, ease of aggregation, susceptibility to geochemical changes, and a propensity for quick passivation (A. Singh et al., 2022; Zhu et al., 2019). At now, there is a lack of knowledge and evidence about the hazards linked to the negative impacts of NPs on native microbial populations and ecosystem function at different stages of the cleanup process (Marcon et al., 2021). Due to a lack of understanding, cautious and protective regulatory measures have been put in place (Bardos et al., 2018).

This chapter provides an overview of how nanotechnology is used to address the issue of contaminated soil containing both organic and/or inorganic pollutants, with a focus on remediation and restoration. In addition, the sustainability assessment was conducted to address the issue of contamination legacy, with particular attention to the implications on health and the environment.

NPs SYNTHESIS

NMs have gathered significant consideration due to their intriguing characteristics, including a high surface/volume ratio, extremely small size, diverse morphologies, forms, architecture, high reactivity, and level of functionalization (A. Singh et al., 2023; A. Singh, et al., 2023). The distinctive characteristics of these substances make them suitable for application in diverse fields of environmental study. Nevertheless, the physical structure of NPs, including their form and size, significantly impacts their fascinating characteristics. This has led to enormous research and the production of NPs using many techniques. Modifying the substrate and precursor concentration, medium pH, and synthesis temperature can change the NPs' properties (Arsenov et al., 2023). The accurate and finely-tuned production of NPs is the essential initial step for advance application. NMs can be categorized into various groups according to their synthesis method (biological, chemical, or physical), material composition, functionality, nano-scale size, surface morphology, phase composition, shape, origin/source, physiochemical and magnetic properties, and crystallinity (Arsenov et al., 2023). NM synthesis can be categorized into two methods: (i) top-down and (ii) bottom-up nanofabrication methodes. The top-down strategy concern the fragmentation of bulk materials into smaller bits using physical processes (A. Singh, et al., 2023). This process leads to the creation of significant quantities of NPs. However, the ability to assemble these NPs may be limited to specific applications depending on the physical and optical characteristics of the NPs. Additionally, there are some constraints associated with this method, such as the inherent flaws in the resulting NPs or the susceptibility of NPs to potential harm. In addition, the time needed for material fabrication, which entails expensive instrumentalization and expenditures, hinders the practicality of a large-scale use (Ekrami et al., 2022). Conversely, the bottom-up strategy is all about self-assembly and is now trending. Atoms or molecules are gathered and then formed into different NPs. The production of these NPs is influenced by chemical principles, which involve the addition of a capping agent, typically a surfactant. This surfactant helps control the growth of the NPs, ensuring they reach the required size (Ekrami et al., 2022). The technique of bottom-up encompasses a range of chemical methods. The primary challenge of this technique lies in the inconsistency of homogeneity in NPs, particularly in their sizes, which can significantly impact their properties (Decarolis et al., 2020). Notwithstanding these disadvantages, the bottom-up synthesis approach is more popular and produces better outcomes than the top-down strategy. Applications involving large-scale operations and industrial synthesis are ideal for this class. Its decreased price makes it more appealing for use in a variety of sectors going forward (Arsenov et al., 2023).

Irrespective of the synthesis process, NPs can be fabricated using traditional approaches that involve multiple artificial paths based on distinct chemical and physical mechanisms (A. Singh, et al., 2024). While traditional technologies offer numerous benefits in the production of high-quality NMs, they also have inherent limitations. The conventional techniques used for NP synthesis are characterized by their time-consuming nature, high cost, and energy-intensive processes. Additionally, these methods include the use of harmful chemicals during production, which hinders their practicality and economic viability (Y. Wang et al., 2019). For instance, the synthesis of NMs by physical techniques often necessitates a significant amount of time for the manufacturing of NPs. Moreover, this process is quite costly, which restricts their widespread application. In addition, chemical techniques for synthesizing NPs include the employment of diverse hazardous compounds, such as organic solvents, which are caustic, combustible, and raise environmental apprehensions. Concurrently, these techniques produce substantial quantities of harmful by-products and refuse, giving rise to apprehensions regarding environmental integrity and human well-being (Patil & Chandrasekaran, 2020). Therefore, the development of sustainable substitute solutions has become inevitable. Currently, the new engineered materials incorporate the biological-mediated production of NMs. The emergence of novel, eco-friendly techniques for NP synthesis is generating considerable attention in the ongoing study and development of nano-remediation. These technologies provide numerous advantages in terms of their positive impact on the natural world and the environment. In essence, the ecologically compatible production of NMs can be achieved by implementing regulated, cleaned-up, controlled, and remediated methods (Ying et al., 2022). Their biodegradable nature, utilization of renewable and sustainable resources in manufacture, and significant potential for application in environmental remediation procedures are the factors contributing to this connection (Arsenov et al., 2023).

A novel strategy for synthesising NPs is developing with the aim of resolving the toxicity problem, a major obstacle in the conventional NP production process. In order to create NPs that are safe for the environment, this method merges nanotechnology with green chemistry. The general plan for synthesising NPs using a biological technique is simple. Green, safe, trustworthy, non-hazardous, efficient, and cost-effective NP fabrication is at the heart of this sustainable approach. Further, to lessen the environmental impact of traditional synthesis methods, it is recommended to use green synthesis of NPs as a crucial methodology. The use of potentially harmful solvents is commonplace in these processes, endangering both people and the planet. Moreover, they produce a great deal of trash and hazardous by-products (Ying et al., 2022). Natural resources, including important phytochemicals and biological components, are used as reducing and stabilizing agents in this NM synthesis technique. The produced NPs are more stable as a result than those made using conventional synthesis techniques. Various living things, including plants, algae, bacteria, and fungi, can be employed to achieve this goal (J. Singh et al., 2018).

NMs Surface functionalization is frequently utilized to enhance the reliability, sensitivity, practicality, and selectivity of a system. Therefore, a biofunctionalized NM exhibits the distinct characteristics of various integrated element inside a individual condensed process. These would possess a multipurpose capability, allowing them to execute numerous applications (Basak et al., 2020).

THE MECHANISM UNDERLYING NANO-REMEDIATION

The primary focus of contemporary science is to discover innovative and effective techniques for converting or breaking down environmental pollutants. The employment of NMs for environmental remediation has attracted significant interest due to its remarkable attributes like as electrical properties, high surface area/volume ratio, sensitivity, cost-effectiveness, and improved catalytic capabilities (Fei et al., 2022).

NM-based nano-remediation is cutting-edge. Nano-remediation could tackle 21st-century pollution challenges, polluted land management, and environmental balance. It offers new ways to quickly and effectively remove pollutants from contaminated areas. Nanostructures in technologies can reduce the total cost of pollution remediation, shorten the clean-up time, eliminate the necessity to treat and dispose of polluted materials, and decrease contaminant concentrations to nearly zero, all on-site (Ganie et al., 2021).

NMs can move through tiny soil openings or stay in groundwater. This allows NPs to have longer reactivity, better dispersion, and reach farther places than bigger particles (A. Singh, Sengar, et al., 2024). Nano-remediation uses NMs' strong reactivity and wide surface area to remove organohalogenated compounds, hydrocarbons, and HMs. Utilizing NMs allows dirty soil remediation while lowering chemical use during purification (Visentin et al., 2021).

Nanotechnology can be used on numerous soil types to disintegrate, remove, or render harmless hazardous substances in contaminated situations. Global study focuses on NMs for environmental remediation. NMs are successful and cost-effective, giving them a viable choice for treating hazardous contaminants at a lesser cost than earlier methods. There are benefits to using NMs instead of physico-chemical approaches for environmental cleanup. The method is promising for eliminating organic and inorganic hazardous contaminants due to its reduced labor needs, financial efficiency, and environmental and social sustainability (Arsenov et al., 2023).

The nano-remediation materials possess three distinct modes of action (Marcon et al., 2021):

1. The *physical process* in which pollutants are adsorbed and immobilized on the particles surface due to their extensive reactivity and surface area.
2. The *process of detoxification* triggers and/or accelerates the initial chemical decomposition that results in the conversion of toxic elements into less damaging substances. Oxidation-reduction processes are the primary mechanisms.
3. The *bio-cooperative* way involves particles that enhance the breakdown of contaminants by making them more accessible and transforming them into species that can be more easily remediated.

The diverse nature of nano-remediators in addressing environmental contaminants is depicted in Figure 1.

Given its multimodal nature and the wide range of nanostructures, any type of contamination has the potential to be effectively treated (Marcon et al., 2021).

NMs, despite their diminutive dimensions, provide a cost-efficient remedy for tackling some issues linked to conventional site remediation. Nano-remediation can be employed in both in-situ and ex-situ employments, mirroring traditional procedures. The in-situ strategy involves remediating toxins at their original place, whereas the ex-situ approach involves transporting them to a different site for remediation (Ali et al., 2023).

DESTINY OF NPs IN THE SOIL

The restorative fate of NPs and their connection with soil constituents have been the subject of little research. The complexity of soil systems, which contain heterogeneous liquid, gas, and soil phases, is one cause. The biggest problems occur in unsaturated soils, when water only partially fills the gaps between soil particles (Marcon et al., 2021).

Many factors govern soil NPs' reactivity, stability, mobility, and toxicity. Physical, chemical, and biological changes can occur in soils once NPs are introduced. Interactions with inorganic and organic soil elements and the environment often cause these modifications. Dissolved humic components can be solidified into particles by adjusting the slurry's pH and ionic strength. Humic acids can have their functional groups separated, altering the charge of some NPs. The rate of NP deposition and the strength of repulsive forces between particles are both affected by the magnitude of these changes (Marcon et al., 2021). The key physical process that happens directly after NPs enter the soil system is aggregation. Aggregation increases aggregate size and decreases surface area, affecting behavior and mobility in porous media. Aggregation can be hetero- or homo-aggregated. The same NPs aggregate in homo-

Figure 1. Nano-remediation processes for decontamination procedures

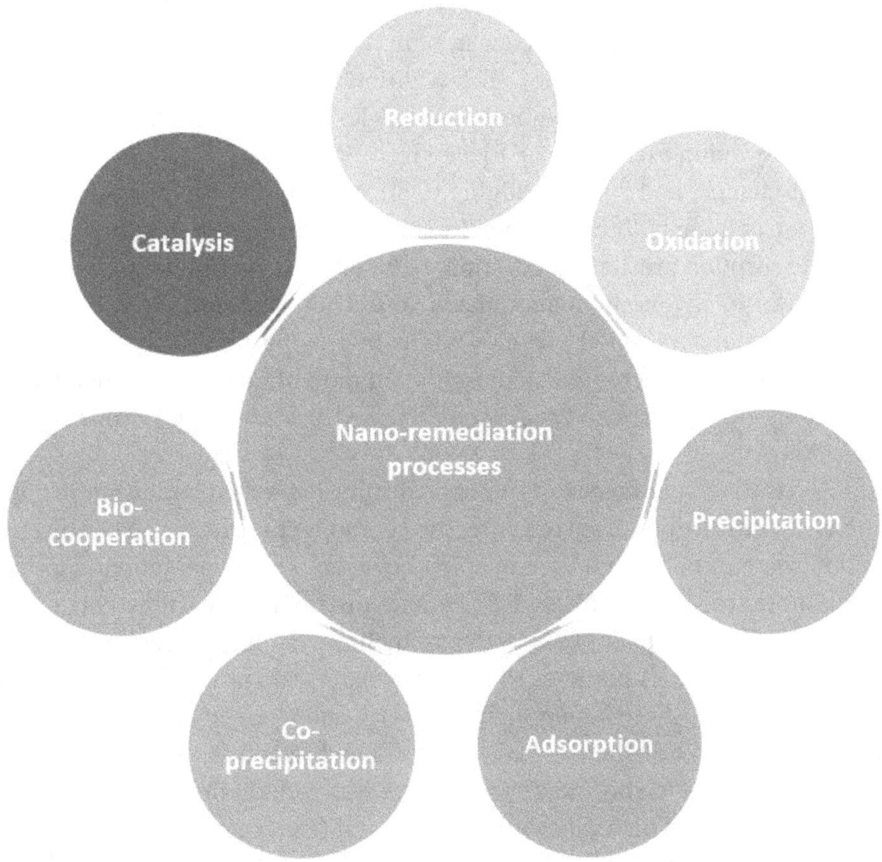

aggregation, while ambient particles and NPs, such natural colloids, aggregate in hetero-aggregation (Ahmed et al., 2021).

Through stability and absorption, soil organic matter (SOM) controls NP behavior and fate. SOM absorption reduces surface-active area, improving NP effects. SOM can affect NP stability and mobility differently depending on their properties, which may cause conflict. SOM also improves NP electrostatic stability and bridging flocculation, improving aggregations and cohesion (Ahmed et al., 2021).

If the forces that attach NP to soil particles are minimized, transportation will occur. Groundwaters will be less accessible to NP transport due to hetero- and homo-aggregation and deposition. NPs may interact with soil components or become trapped in soil pores after injection. Importantly, NPs that firmly bind to soil components will become permanent and block soil pores. Smaller NPs that can slip into soil particle gaps may move further before becoming stuck in the soil structure (Fazeli Sangani et al., 2018). Unsurprisingly, the concentration of particles has a substantial impact on their destiny and movement. Regardless of the spectrum of sizes, the ability to move at minimal concentrations (< 0.1 mg/L) of polystyrene sulfonate modified nZVI was shown by (Phenrat et al., 2009) in water-soaked sand. The opposite is true for larger and more Fe^0-containing particles, which are more likely to agglomerate due to the increased strength of the magnetic forces between them. Further studies using sand columns in the lab have shown that coatings have a positive effect on particle mobility. Coatings made of polymers are used to combat the immobility of bare nZVI. These coatings cause the particles to become electrosterically stabilized, which prevents them from aggregating (Marcon et al., 2021).

It is vital to note that these alterations occur in both directions, as humic compounds also undergo changes upon interaction with nanoparticles. The addition of TiO_2 NPs alters the natural structure of humic acid extracted from peat soil, as demonstrated by (K. Yang et al., 2009). The adsorption mechanisms involved are likely anion exchange due to electrostatic contacts, hydrogen bonding, hydrophobic interactions, ligand exchange, and cation bridging. Collectively, these interactions have a noteworthy effect on the ability of pollutants to be retained in soils that have been treated with nanoparticles, as evidenced by specific sorption profiles. In addition, (Chang & Kang, 2009) performed laboratory tests to assess the breakdown of pyrene in contaminated soil. They observed significant alterations in pH and oxidation-reduction potential (ORP) leveles while employing nZVI (Chang & Kang, 2009). The pH rose from 4.8 to 8.5 when particles with a mass-to-soil ratio of up to 10% were added. Concurrently, the ORP dropped from +400 to -700 mV, suggesting that Fe^0 had introduced strong reducing conditions (Chang & Kang, 2009).

Transporting nZVI to the contaminated site without forming large clusters that might impede subsurface migration is crucial. Additionally, before reaching the contaminant, particles can't react with oxygen or water. This could mean that the particles are unable to convert a chemical into an inert substance. This reference provides more information about nZVI's size distribution, particle concentration, and zeta potential (Marcon et al., 2021). A major issue arises from this ever-changing state of affairs: the remediation process itself causes the nZVI particles to lose their nanoscale properties, specifically their ability to disperse, while simultaneously reducing their reactivity and tendency to clump. This highlights the critical importance of hydraulic characteristics of the soil and injection technology. The ideal situation would be for the injected nZVI to stay put after delivery; this would allow for a gradual transformation into iron minerals, which should naturally settle into the soil (Zhao et al., 2016).

Natural resources like soil are essential. Human activities have polluted the soil with many organic and inorganic pollutants, degrading it. Mixed contaminated soils, which are polluted with organic chemicals and metals, are difficult to remediate since they must be reduced simultaneously.

REMEDIATION OF SOIL INORGANIC POLLUTANTS USING NMs

The existence of perilous inorganic pollutants in soil is a noteworthy concern due to their elevated toxicity and persistent presence in the ecosystem. Typically, HMs have the ability to become part of plant systems and can pollute the food chain, so presenting a significant risk to the quality of food and the safety of human health.

There has been an increasing demand for pollution-free environments in recent decades, with a special focus on the soil system. HM contamination poses a significant risk to the maintenance of soil systems. This is mostly owing to its various sources, toxicity, inability to break down naturally, and attitude to store all time (Fajardo et al., 2020). HMs, have a negative impact on various physiological processes. They can cause changes in the specificity of enzymes, disrupt cellular function, and harm the cell membrane and DNA structure. Consequently, the extensive contamination of HMs in the soil system is presently a highly significant environmental issue (Fajardo et al., 2020).

Many in-situ and ex-situ approaches have been utilized to remove HMs from polluted soils in recent decades. Many traditional physicochemical techniques and biological remediation are used (Ahmed et al., 2021). Conventional physicochemical remediation uses harmful chemicals that enter the environment untreated. In contaminated areas, biological remediation uses plants and microorganisms to reduce HM toxicity. However, these approaches are expensive, time-consuming, insufficient for microorganisms, limited in bioavailability of pollutants on geographical and temporal scales, and inefficient in real-world field conditions (Ahmed et al., 2021). Cost-effectiveness, operational expenses, time restrictions, location of the contaminated site, soil contaminants, public approval, and execution readiness determine the efficacy of conventional remediation procedures. Thus, the widespread use of HMs immobilization techniques requires new, environmentally safe, non-toxic, affordable, and reusable adsorbents. Nano-remediation, which uses artificial NPs to clean the environment, has emerged to solve these issues. Nano-remediation can also decrease cleanup time, improve long-term efficacy, and decrease hazardous material concentrations to negligible levels (Ahmed et al., 2021).

Therefore, there is still a requirement for effective on-site stabilizing solutions to guarantee a decrease in the bio-accessibility of metals and metalloids. Nano-remediation has emerged as a potential method for detoxifying soils polluted with HM, and has been the topic of recent investigation in the field of soil remediation. nZVI is the most often employed type of NPs in the field of nano-remediation, specifically those consisting of Fe (Fajardo et al., 2020).

The examination of (Li et al., 2021) demonstrates that the use of nZVI is highly successful in immobilizing As in a soil that has been influenced by smelting activities. As a result of the treatment, the solubility and leachability of As are greatly reduced. The fixation of nZVI is crucial for the efficient restoration of soil biological function and the promotion of the establishment and growth of pioneering plants. Specifically, a concentration of 0.2% nZVI demonstrated the highest efficacy in enhancing soil dehydrogenase activity and decreasing the availability of As in the rhizosphere of ryegrass and alfalfa. The results demonstrate that using nZVI in soils with high levels of As contamination is an effective method for expediting ecological restoration (Li et al., 2021).

In addition, various types of NPs (including metallic oxide, metallic, polymeric, carbonaceous, and nanocomposites) have been effectively utilized for the purpose of removing HMs. For instance, (Camilli et al., 2014) documented that the utilization of carbon nanotube (CNTs) nano-sponges effectively eliminated the harmful organic solvent dichlorobenzene and Cd from the soil. In general, nano-sponges containing iron and sulfur have been found to improve the effectiveness of absorbing Cd and other pollutants from soil. In a similar vein, (Matos et al., 2017) examined the impact of CNTs on soil remediation, specifically focusing on the immobilization of HM ions (Zn^{2+}, Cu^{2+}, Pb^{2+}, and Ni^{2+}). The study found that the addition of a modest quantity of carbon nanotubes (CNTs) greatly enhances the ability to adsorb pollutants and consequently increases the ability to immobilize HMs in the polluted soils. (Gong et al., 2018) conducted a study where they examined the use of $Fe_3(PO_4)_2$ NPs to treat soil. This treatment effectively reduced the mobility of Cd by 70%. This examination additionally proposed that the production of cadmium phosphate is the primary factor contributing to the decreased bioaccumulation of Cd in soil (Ahmed et al., 2021).

(Qiao et al., 2017) found that the presence of biochar-supported Fe NPs reduced the availability of Cd in plants. Therefore, the efficiency of immobilizing Cd was determined to be 81% after a duration of 28 days in the experiment. In a separate investigation, (Su et al., 2016) proposed that the usage of nZVI ion has proven effective in treating soil contaminated with Cr by completely immobilizing Cr(VI) with a 100% success rate. Furthermore, there has been a notable augmentation in plant growth accompanied by a reduction in the harmful effects of Cr (VI) on plants. In a similar vein, (Noman et al., 2020) recovered that the existence of Cu NPs resulted in a lessening in the movement of Cd from soil to wheat plants, leading to enhanced plant development and nutritional levels. In a separate investigation, (Manzoor et al., 2021) shown that the presence of FeO NPs improved the adverse effects of Cd stress on wheat plants by enhancing plant growth, chlorophyll, and antioxidants levels. In order to enhance the implementation of nano-remediation applications, researchers should collaborate with area policymakers to facilitate the advancement of innovative research aimed at identifying sustainable nano-solutions (Ahmed et al., 2021).

An experimentation was effected to assess the effectiveness of utilizing Fe-based NPs in a magnetic separation process as a pre-treatment to improve the recovery of As from soil contaminated with a high concentration of this element (Diego Baragaño et al., 2021). The results propose that the application of NPs to soils prior to treatment can enhance the efficacy of magnetic separation, particularly when there is a notable disparity in magnetic characteristics between the soil and the NPs. Therefore, soils that are lacking in ferri- and ferro-magnetic minerals would be excellent choices for remediation using a two-step process involving pre-treatment with NPs followed by magnetic soil washing. Moreover, the integration of nano-remediation with soil washing procedures presents a synergistic treatment strategy that effectively decreases the quantity of contaminated soil while enhancing its stability (Diego Baragaño et al., 2021).

The ability of two iron-based NMs, specifically goethite nanospheres (nGoethite) and nZVI, to prevent the movement of As in a contaminated soil was assessed and compared by (D. Baragaño et al., 2020b). Hence, the findings demonstrate that both nZVI and nGoethite possess the ability to efficiently immobilize As in this specific contaminated site. Utilizing lower concentrations of nGoethite shows potential as an effective method for remediating soils polluted with As. Nevertheless, the toxicity of the soil was amplified at elevated doses, perhaps as a result of a significant increase in electrical conductivity. Both forms of NP showed a small increase in the availability of Fe (D. Baragaño et al., 2020b).

Additionally, soils that have been poisoned with As provide a significant worldwide environmental issue that need careful attention. Nanocomposites (nZVI, iron oxide NP, TiO_2, nano-silica, FeS NPs, bimetallic NPs, etc.) have demonstrated exceptional efficacy and remarkable capability in eliminating or immobilizing As in soils. An analysis of the latest research shows that a significant amount of research has been conducted using nZVI and iron oxide NPs. The processes involved in immobilizing As and reducing its toxicity in soil include precipitation, adsorption, and oxidation of As (III) to As (V). The primary mechanism by which arsenic is immobilized in soil is adsorption. Nevertheless, the behavior of As in nanocomposites can be influenced by various factors in the soil system, including pH, organic matter, mineral composition, redox potential, and physical qualities such as texture and porosity (Alidokht et al., 2021).

When applying NMs to multi-contaminated soils, it is important to exercise attention to avoid the mobilization and immobilization of antagonist inorganic contaminants. This is particularly relevant for certain elements, such as As/Sb and Pb/Zn. Prior to the application of NMs, it is imperative to evaluate the soil pH, the potential long-term release of NMs, and the type of munitions utilized in the locations under study. Regardless, employing multiple treatments of NMs may be the optimal approach to prevent the mobilization of metals in soils contaminated by various metals (Rodríguez-Seijo et al., 2020).

An assessment was conducted on a phytonanoremediation technique that involved the combination of *Avicennia germinans* and nZVI. The evaluation spanned a duration of 5 months and utilized polluted soil. The developed phytonanoremediation system proved to be an effective tool for assessing the interaction between *A. germinans* and nZVI, both alone and in combination, in order to remediate Cd, Pb, and As in soil. The findings of this work indicate that the interaction between nZVI and *A. germinans* presents a new and effective method for increasing the bioavailability of Cd^{2+} and Pb^{2+} and facilitating the translocation of these pollutants into the stems and leaves of mangroves (Soto Hidalgo et al., 2020).

(Gong et al., 2017) utilized starch-stabilized nZVI (S-nZVI) to mitigate the buildup of Cd and oxidative damage in *Boehmeria nivea* (L.). The findings reveal that the presence of S-nZVI at a low concentration increases the effectiveness of phytoremediation.

Recent advances in nano-remediation have immobilized HMs and metalloids. The use of nZVI to stabilize and reduce metal availability is common. Needle-like nZVI increases iron particle surface area. Therefore, this size-dependent trait increases responsiveness. Multiple investigations have shown that nZVI effectively prevents As migration in soils, including in real life. In this scenario, nZVI structural differences affect particle reactivity and aging. Furthermore, nZVI toxicity risks must be considered. This issue requires consideration of NP dose, exposure duration, oxidation rate, and Fe availability (D. Baragaño et al., 2020b).

To address these challenges, alternative NPs utilizing iron oxides have been employed for the goal of environmental cleanup. They have demonstrated superior stability compared to nZVI when employed for the removal of metals from water or soil. Iron oxy-hydroxide, specifically α-FeOOH, is a naturally occurring oxide mineral called goethite. It effectively captures and isolates contaminants, such as arsenic, by sorption mechanisms. Synthetic goethite has been successfully created and utilized for the purpose of immobilizing As (D. Baragaño et al., 2020b).

Recent nano-remediation of soil contaminated with multiple HMs reduced Zn and Pb mobility, especially in the near term. However, after 120 days of treatment, this strategy was ineffective and the soil's toxicity was similar. Thus, deploying nZVI on a big scale raises questions about its efficacy and cost in real life. Especially with older soil that may have lower pollutant availability due to matrix component interactions. Additionally, its pros and negatives must be carefully considered. The detrimental impacts of nano-remediation on living things must be considered. Since nZVI and the pollutants being targeted can have mixed effects, their stability can fluctuate over time (Fajardo et al., 2020).

REMEDIATION OF SOIL ORGANIC POLLUTANTS USING NMs

The occurrence of soil contamination by petroleum hydrocarbons has been extensively documented, especially in nations with abundant oil resources. Hydrocarbon pollution has diminished the accessibility of cultivable land in the majority of impacted populations. TPHs, total petroleum hydrocarbons, are commonly discharged into the environment due to occurrences such as industrial accidents, oil spills, and pipeline vandalism. As a result, they have a substantial impact on environmental pollution as organic pollutants. Consequently, both domestic and global groups have advocated for the restoration of unclean soil (Bolade et al., 2021).

Nano-biosorbents have two potential classes of adsorption sites for capturing PAH from hydrocarbon-polluted areas (Basak et al., 2020): (i) sequestration on the outside areas of NPs, and (ii) attachment within the nanopores inside the nano-biosorbents. Each NP's outer surface contains evenly distributed hydrophobic regions that are effective in attracting organic contaminants. However, these hydrophobic interfaces only provided partial clarification of the interaction behavior between the NPs and organic contaminant (Basak et al., 2020).

In addition, (H. Wang et al., 2015) proved a method for removing PAH (benzo[a]pyrene) from polluted soil surfaces utilizing silica NPs that are coated with lipid bi-layers of *Pseudomonas aeruginosa*. The nano-bioconjugate material was extremely small in size, but had a significant surface area. The interactions between each particle, as well as with other organic particle in the solid environment, were determined by the characteristics and aggregation situations present on their surface. As shown by (H. Wang et al., 2015), the lipid molecule bio-functionalizing silica NPs has a dual function. It adsorbs or sequesters the PAH, making it a remediation agent; concurrently, it keeps the NPs colloidally stable, which is critical for getting them to the source of the pollution (H. Wang et al., 2015).

The study of (Bolade et al., 2021) revealed that sodium persulphate as an oxidizing agent effectively enhanced the breakdown of TPH in polluted soil. The ideal conditions for the complete breakdown of TPH were found to be a pH of 6.0 and an oxidant dosage of 0.74 M, achieved within a timeframe of 5 days and 14 hours. The bioactive components derived from *Azadirachta indica* can serve as diminishing and stabilizing agents during the creation of NPs. Biosynthesized iron NPs (FeNP (2:1)) were utilized to degrade TPH in soil and activate sodium persulphate for the same objective. The utilization of NPs-activated persulphate effectively decomposed approximately 99% of total TPH present in the soil within a duration of 5 days and 14 hours (Bolade et al., 2021).

(Romeh, 2021) utilized Ficus-derived zerovalent iron NPs immobilized on adsorbents (F-Fe^0 ad) as an environmentally friendly nanotechnology approach. They employed *Plantago major* L. for phytoremediation purposes, specifically to eliminate chlorpyrifos. The combination of *P. major* L. and F-Fe^0 fixed on adsorbents demonstrated a significantly higher efficiency in removing chlorpyrifos, achieving nearly 100% removal. In comparison, *P. major* L. alone achieved a removal rate of 43.76%, while F-Fe^0 alone achieved a removal rate of 81.69%.

(Atta et al., 2020) utilized FeNPs coated with natural compounds derived from myrrh to remediate crude oil in the existence of *Eichhornia crassipes*. The findings demonstrated that the inclusion of MNPs (magnetite NPs) with myrrh reserves significantly enhanced the remediation process by up to 95%. Additionally, the bioaccumulation of sulfur in *E. crassipes* exhibited a positive correlation with the amount of oil contamination.

Finally, the introduction of non-functionalized MNPs to a soil contaminated with both As and PAHs resulted in a decrease in the availability of As and a significant immobilization of organic molecules. Higher concentrations of NPs (over 1%) resulted in increased immobilization of As, with some cases achieving nearly complete immobilization (D. Baragaño et al., 2020a). Moreover, following the application of MNP, there was no reduction of As(V) to As(III), which is the most poisonous and easily movable form. In addition, this remediation approach also resulted in the immobilization of organic molecules (D. Baragaño et al., 2020a). Even at the lowest NP dose, the concentration of TPH fell, and the concentrations of PAHs were lowered as a result of sorption processes. It is worth mentioning that the immobilization of PAH was more significant for compounds with a higher molecular weight (D. Baragaño et al., 2020a).

THE DIFFICULTIES ASSOCIATED WITH USING SUSTAINABLE REMEDIATION TECHNOLOGIES

The selection and implementation of a suitable sustainable remediation technology depend on several factors that can impact the efficiency and suitability of the remediation process. There is a dearth of conclusive scientific information about the ways in which these elements, which encompass both biotic and abiotic concerns, as well as scientific and non-scientific aspects, influence the effectiveness and practical implementation of the remediation procedure. Soil contamination can have long-term effects, thus it's important to know all the characteristics (Figure 2) before choosing and predicting a remediation technology. Furthermore, to fully implement the principles of sustainable remediation, it is crucial to approach contaminated area rehabilitation as a novel mindset, wherein the incorporation of environmental, economic, and social factors is regarded as a fundamental component in the decision-making process. Governments, academics, and industry are all crucial in advancing sustainable remediation and addressing current challenges. This is necessary for both emerging and industrialized countries to successfully include sustainability into their remediation efforts (Ambaye et al., 2022).

ASSESSING THE ECONOMIC VIABILITY OF NMs

Given the growing interest in cost reduction for nanotechnology applications in various biological sectors, it is important to investigate the challenges and opportunities associated with producing more affordable NMs using sustainable and abundant raw materials (Mpongwana & Rathilal, 2022). Given the current circumstances, intensifying the investigation of biological resources may be the most effective means to further reduce the expenses associated with NMs (Ahmad et al., 2023).

There are numerous environmental contamination sites where agricultural trash is disposed of. The utilization of various waste materials to create NMs has significant consequences in environmental remediation, namely in the areas of photodegradation and soil decontamination. These procedures are both environmentally sustainable and economically efficient. The implementation of green chemistry can significantly improve the economic viability of NMs by utilizing natural agents to achieve superior results at lower prices. This can be achieved by efficient scaling and adhering to a zero-waste policy (Aswathi et al., 2022). Environmental remediation can effectively restore the ecosystem by utilizing a two-pronged approach: transforming biodegradable trash into valuable nanotechnology-based goods and deploying them for diverse ecological remediation goals. Furthermore, the biowastes containing biomolecules, which serve as precursors and agents for modifying substances in environmentally friendly synthesis, are readily accessible. Therefore, the sustainable production of NMs utilizing this source is economically viable, particularly when conducted on a large-scale industrial level (Ahmad et al., 2023).

Many green approaches utilize a wide range of biological entities to produce various NMs, with the goal of meeting the increasing need for environmental sustainability. Additionally, bio-based NPs have a positive impact on the environment since they decrease waste and enhance the value of NP production

Figure 2. Parameters controlling sustainable remediation technology

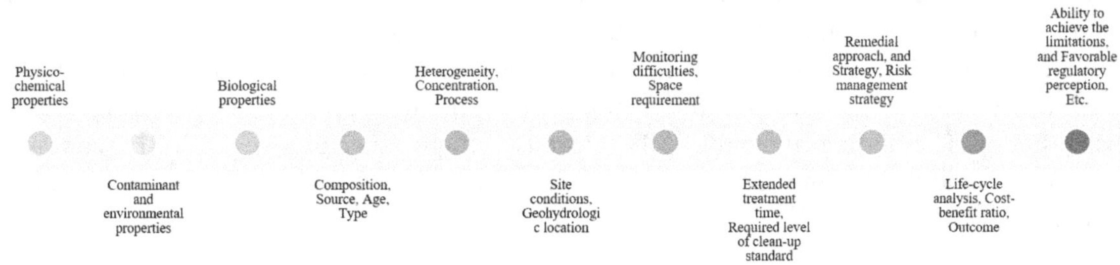

(Chauhan et al., 2020). This method of synthesizing NPs is preferred because it aligns with the principles of "green chemistry" and "sustainable development". The essential principles of "green synthesis" can be elucidated by various factors, including waste prevention or minimization, reduction of byproducts or contamination, utilization of less toxic or non-toxic substances, as well as the incorporation of renewable feedstock and energy (J. Singh et al., 2018). Furthermore, the concept of "sustainable development" entails the careful management of present demands while ensuring that the needs of future generations are not compromised (Pal et al., 2019).

NANO-REMEDIATORS' ECOTOXICITY

The employment of NPs to eliminate contaminants from the environment has shown promising results. However, field-scale applications have been hampered by concerns about the NPs' potential harm to living organisms. Large-scale NP production can pollute the environment by depositing NPs and other contaminants. NPs are harmful to humans, microorganisms, plants, and animals (Kumar Das et al., 2022). NMs have been found to cause significant health consequences when individuals are exposed to them by ingestion, inhalation, or skin contact (Ahmad et al., 2023).

In addition, NPs are widely recognized for their ability to cause many detrimental effects on living beings, such as DNA damage, disruption of membrane integrity, and interference with cellular uptake, among other impacts. The principal harmful mechanism of NPs is the generation of ROS, which leads to oxidative stress and affects lipids, proteins, and nucleic acids (Fajardo et al., 2020).

The process of creating and altering NMs through physical and chemical means is frequently unsustainable since it requires significant amounts of energy and involves the use of potentially harmful compounds. To address this issue, researchers have turned to isolated extracts from plants and microorganisms as a more environmentally friendly approach, known as green synthesis. The biomolecules present in various plant tissues, including fruits, roots, seeds, flowers, and leaves, serve as agents for synthesis and coating. They play a crucial role in decreasing and stabilizing processes, hence reducing the demand on synthetic chemical substances for modifications. In contrast, the process of capping NPs with various substances such as proteins, enzymes, minerals, carbohydrates, vitamins, lipids, and phytochemical ingredients results in the development of distinct photochemical and photophysical properties. This capping mechanism effectively inhibits the aggregation, instability, and toxicity of the NPs (Ahmad et al., 2023).

NPs based on nZVI are widely recognized as environmentally benign materials and have been considerably utilized in the rehabilitation of soil. Nevertheless, the actual likelihood of simultaneous exposure to nZVI and pollutants, particularly the mechanism by which their harmful effects interact, remains largely uncertain. The study of (M. Yang et al., 2023) was discovered that the toxicity of TCEP (tris(2-chloroethyl) phosphate) to earthworms was enhanced by nZVI through iron harming, rather than the traditional "Trojan horse effect". Earthworms' food intake and growth were indirectly inhibited by TCEP, which primarily induced neurotoxicity. Because of this, nZVI exacerbated TCEP-induced malnutrition by making intestinal damage worse and blocking energy metabolism directly. The findings unequivocally showed the importance of using nZVI-based remediation technology and conducting toxicological assessments in newly discovered contaminations (M. Yang et al., 2023).

(Liang et al., 2017) found that earthworms' intestinal damage was worsened by the simultaneous presence of increased amounts of nZVI and decabromodiphenyl ether (BDE209). The growth and reproduction of earthworms were hindered by both nZVI and dioxin-like 3,3′,4,4′-tetrachloro-biphenyls (PCB77). The synergistic impact seen may be attributed to the harm caused to the earthworm epidermis (Zhang et al., 2021). In general, there is a dearth of sufficient study on the combined hazardous effects of NMs and contaminants. It is imperative to determine the harmful interaction mechanism in order to guarantee the secure implementation of nano-remediation technology (M. Yang et al., 2023).

The previous research has determined that the combined toxicity of NMs and contaminants is a result of the in vivo adsorption-release-migration mechanism, commonly referred to as the "Trojan horse effect" (Bhagat et al., 2021). For instance, the addition of 1.0 mg/L carbon nanotubes, acting as a potent carrier, intensified the acute toxicity of carbofuran to *Nile tilapia*. This resulted in a decrease in the 96-hour LC50 value from 2.46 to 0.50 mg/L (Campos-Garcia et al., 2015). While the Trojan horse effect could contribute to the combined toxicity of nZVI and organophosphorus flame retardant (OPFR), it is vital to note that the adsorption capabilities of metallic NPs like nZVI are significantly inferior to those of carbon-based NMs. Hence, the Trojan horse phenomenon alone is inadequate to account for the combined toxicity, and it is plausible that specific biological pathways are implicated (Hou et al., 2022).

Soil invertebrates are particularly vulnerable to the harmful effects of nZVI and other pollutants when iron poisoning is present. Evidence utilizing modern data analysis methods, such as multi-omics, is urgently needed due to the lack of knowledge regarding the biology level interaction mechanism between nZVI and pollutants (M. Yang et al., 2023).

However, the inclusion of nZVI proved to be the most efficient factor in decreasing soil lindane concentrations. The most significant decrease in the quantities of extractable Zn occurred when *Brassica napus*, nZVI, and organic modifications were present simultaneously. The concurrent application of the three actions (rhizoremediation using *B. napus*, nano-remediation employing nZVI, and bio-stimulation through organic amendments) resulted in a substantial increase in soil microbial biomass, biodiversity, and activity. Consequently, this led to a notable enhancement in soil health. The analysis determined that employing a combination of remediation procedures can effectively address the constraints inherent in any individual strategy (Hidalgo et al., 2022).

Regarding toxicity, MNPs differed from other iron-based NPs in that they did not have a major impact on the availability of Fe, did not alter the chemical form of As, and actually lowered the toxicity to plants when the dose exceeded 1%. This reduction in toxicity was due to the immobilization of both As and organic contaminants (D. Baragaño et al., 2020a). These findings expand the potential application of MNPs, not just for immobilization techniques, but also for hybrid remediation approaches. One example is to adopt a strategy that takes advantage of the magnetic characteristics of MNPs in combination with soil washing. This approach can effectively recover contaminants that have been adsorbed onto the surfaces of the NPs (D. Baragaño et al., 2020a).

Based on the positive outcomes achieved, such as the effective immobilization of pollutants and the decrease of soil toxicity, along with the suitable characteristics of the magnetite NPs employed, the utilization of MNPs in soil nano-remediation appears as a viable option for addressing simultaneous pollution by As and PAHs, either individually or in conjunction with other remediation methods (D. Baragaño et al., 2020a).

OBSTACLES AND VIEWPOINTS

NP remediation is a growing method for addressing global water and soil pollution. Due to their advanced surface properties, nano-engineered metallic NPs, nano-membranes, nano-adsorbents, nano-sensors, and enhanced photocatalysts show promise in sustainable and environmental remediation. Application materials with nano-scale components have different characteristics. High surface / volume ratio, chemical stability, adjustable absorption properties, strong adsorption capability, practical surface arrangements, and cost efficiency make these resources effective and efficient in protecting and monitoring the environment. Nano-remediation technologies have several uses, but environmental issues prevent their widespread implementation. Many nano-remediation methods are only suited for labs, limiting their commercialization. Commercialize nano-remediation techniques to ensure accurate use for human and environmental health (Fei et al., 2022).

NMs can remediate soil, but they have restrictions and must meet certain parameters. First, move the NMs to the right soil site. The given NMs must also respond well to the source medium pollutants. NMs must also efficiently interact with target pollutants and start the chemical reaction. NM modification agents should be environmentally and biota-friendly. Particle stabilization can improve NP delivery, but in low-permeability soils, the distance is still limited. When directly injected into contaminated soil, biogeochemical and physical soil factors such hydraulic conductivity, oxygen level, mineral constitutions, presence of other components, response time, temperature, and pH can affect NM transport. Most NM altering agents were poisonous and might harm organisms in soil (Ahmad et al., 2023).

The employment of NPs has been confirmed to be a beneficial method for addressing contamination in soil, as previously noted. The potential of nano-remediation technology is promising, but, additional research is required to evaluate any future hazards (Ali et al., 2023).

Here are some important factors to take into account (Ali et al., 2023):

a. Independent investigation is essential to know the stability and mobility of NMs.
b. To increase NM adoption, production costs must be reduced.
c. Nano-remediation should prioritize NM recycling and reuse.
d. Assess NMs' durability and stability in different climatic conditions.
e. Remediation NMs' biosafety and environmental impact must be assessed immediately.
f. To improve the reliability and applicability of nano-remediation for contaminant removal findings, future studies should use actual environmental conditions rather than laboratory simulations.
g. The approval by the authorities, meaning the establishment of particular rules through assessment by governmental bodies, as well as by society.

Understanding these parameters will help develop safe and effective nano-remediation methods to reduce pollution risks (Ali et al., 2023).

CONCLUSION

Environmental contaminants can be remedied using NPs. Researchers are exploring using nanotechnology-based remediation strategies for environmental cleanup since they outperform others. NPs' quick loss of reactivity, inclination to clump, susceptibility to geochemical variables, and potential health and environmental concerns if misused limit their employment. At a specific area, contamination hazards must be weighed against nano-remediation concerns, which include unknown and long-term effects and uncertainties.

To analyze potential dangerous responses, one must first understand an NM's "mode of action". Further studies should examine the likelihood of NPs entering matrices using customized coatings. Additionally, nano-kinetic studies should be conducted to understand and control the physicochemical processes that determine NP fate. This might be useful for academics, developers, and scientists trying to figure out how these novel materials behave in the real world. To improve results, academic research and lab- and field-scale trials should collaborate. However, economic reality and private interests must be considered. Many groups are using NPs to remediate polluted soils, but researchers and governments must address unanswered questions to make this technique reliable. This should be supplemented by the development of new prospective NMs with unique and complementary properties to present remediation approaches.

REFERENCES

Ahmad, M. A., Adeel, M., Shakoor, N., Javed, R., Ishfaq, M., Peng, Y., Zain, M., Azeem, I., Ali, I., Usman, M., Wu, Z., Gohari, G., Xu, M., Rui, Y., Zhang, Z., White, J. C., & Deng, X. (2023). Modifying engineered nanomaterials to produce next generation agents for environmental remediation. *The Science of the Total Environment*, *894*, 164861. doi:10.1016/j.scitotenv.2023.164861 PMID:37343875

Ahmed, T., Noman, M., Ijaz, M., Ali, S., Rizwan, M., Ijaz, U., Hameed, A., Ahmad, U., Wang, Y., Sun, G., & Li, B. (2021). Current trends and future prospective in nanoremediation of heavy metals contaminated soils: A way forward towards sustainable agriculture. *Ecotoxicology and Environmental Safety*, *227*, 112888. doi:10.1016/j.ecoenv.2021.112888 PMID:34649136

Ali, Q., Zia, M. A., Kamran, M., Shabaan, M., Zulfiqar, U., Ahmad, M., Iqbal, R., & Maqsood, M. F. (2023). Nanoremediation for heavy metal contamination: A review. *Hybrid Advances*, *4*, 100091. doi:10.1016/j.hybadv.2023.100091

Alidokht, L., Anastopoulos, I., Ntarlagiannis, D., Soupios, P., Tawabini, B., Kalderis, D., & Khataee, A. (2021). Recent advances in the application of nanomaterials for the remediation of arsenic-contaminated water and soil. *Journal of Environmental Chemical Engineering*, *9*(4), 105533. doi:10.1016/j.jece.2021.105533

Ambaye, T. G., Chebbi, A., Formicola, F., Prasad, S., Gomez, F. H., Franzetti, A., & Vaccari, M. (2022). Remediation of soil polluted with petroleum hydrocarbons and its reuse for agriculture: Recent progress, challenges, and perspectives. *Chemosphere*, *293*, 133572. doi:10.1016/j.chemosphere.2022.133572 PMID:35016966

Arsenov, D., Beljin, J., Jović, D., Maletić, S., Borišev, M., & Borišev, I. (2023). Nanomaterials as endorsed environmental remediation tools for the next generation: Eco-safety and sustainability. *Journal of Geochemical Exploration*, *253*, 107283. doi:10.1016/j.gexplo.2023.107283

Aswathi, V. P., Meera, S., Maria, C. G. A. & Nidhin, M. (2022). Green synthesis of nanoparticles from biodegradable waste extracts and their applications: a critical review. *Nanotechnology for Environmental Engineering, 8*(2), 377–397. doi:10.1007/s41204-022-00276-8

Atta, A. M., Mohamed, N. H., Hegazy, A. K., Moustafa, Y. M., Mohamed, R. R., Safwat, G. & Diab, A. A. (2020). Green Technology for Remediation of Water Polluted with Petroleum Crude Oil: Using of Eichhornia crassipes (Mart.) Solms Combined with Magnetic Nanoparticles Capped with Myrrh Resources of Saudi Arabia. *Nanomaterials, 10*(2), 262. doi:10.3390/nano10020262

Baragaño, D., Alonso, J., Gallego, J. R., Lobo, M. C., & Gil-Díaz, M. (2020a). Magnetite nanoparticles for the remediation of soils co-contaminated with As and PAHs. *Chemical Engineering Journal, 399*, 125809. doi:10.1016/j.cej.2020.125809

Baragaño, D., Alonso, J., Gallego, J. R., Lobo, M. C., & Gil-Díaz, M. (2020b). Zero valent iron and goethite nanoparticles as new promising remediation techniques for As-polluted soils. *Chemosphere, 238*, 124624. doi:10.1016/j.chemosphere.2019.124624 PMID:31472353

Baragaño, D., Gallego, J. L. R., María Menéndez-Aguado, J., Marina, M. A., & Sierra, C.Baragaño. (2021). As sorption onto Fe-based nanoparticles and recovery from soils by means of wet high intensity magnetic separation. *Chemical Engineering Journal, 408*, 127325. doi:10.1016/j.cej.2020.127325

Bardos, P., Merly, C., Kvapil, P., & Koschitzky, H. P. (2018). Status of nanoremediation and its potential for future deployment: Risk-benefit and benchmarking appraisals. *Remediation (New York, N.Y.), 28*(3), 43–56. doi:10.1002/rem.21559

Basak, G., Hazra, C., & Sen, R. (2020). Biofunctionalized nanomaterials for in situ clean-up of hydrocarbon contamination: A quantum jump in global bioremediation research. *Journal of Environmental Management, 256*, 109913. doi:10.1016/j.jenvman.2019.109913 PMID:31818738

Bhagat, J., Nishimura, N., & Shimada, Y. (2021). Toxicological interactions of microplastics/nanoplastics and environmental contaminants: Current knowledge and future perspectives. *Journal of Hazardous Materials, 405*, 123913. doi:10.1016/j.jhazmat.2020.123913 PMID:33127190

Bolade, O. P., Akinsiku, A. A., Oluwafemi, O. S., Williams, A. B., & Benson, N. U. (2021). Biogenic iron oxide nanoparticles and activated sodium persulphate for hydrocarbon remediation in contaminated soil. *Environmental Technology & Innovation, 23*, 101719. doi:10.1016/j.eti.2021.101719

Camilli, L., Pisani, C., Gautron, E., Scarselli, M., Castrucci, P., D'Orazio, F., Passacantando, M., Moscone, D., & De Crescenzi, M. (2014). A three-dimensional carbon nanotube network for water treatment. *Nanotechnology, 25*(6), 065701. doi:10.1088/0957-4484/25/6/065701 PMID:24434944

Campos-Garcia, J., Martinez, D. S. T., Alves, O. L., Leonardo, A. F. G., & Barbieri, E. (2015). Ecotoxicological effects of carbofuran and oxidised multiwalled carbon nanotubes on the freshwater fish Nile tilapia: Nanotubes enhance pesticide ecotoxicity. *Ecotoxicology and Environmental Safety, 111*, 131–137. doi:10.1016/j.ecoenv.2014.10.005 PMID:25450925

Chang, M. C., & Kang, H. Y. (2009). Remediation of pyrene-contaminated soil by synthesized nanoscale zero-valent iron particles. *Journal of Environmental Science and Health. Part A, Toxic/Hazardous Substances & Environmental Engineering*, *44*(6), 576–582. doi:10.1080/10934520902784609 PMID:19337920

Chauhan, A. K., Kataria, N., & Garg, V. K. (2020). Green fabrication of ZnO nanoparticles using Eucalyptus spp. leaves extract and their application in wastewater remediation. *Chemosphere*, *247*, 125803. doi:10.1016/j.chemosphere.2019.125803 PMID:31972482

Decarolis, D., Odarchenko, Y., Herbert, J. J., Qiu, C., Longo, A., & Beale, A. M. (2020). Identification of the key steps in the self-assembly of homogeneous gold metal nanoparticles produced using inverse micelles. *Physical Chemistry Chemical Physics*, *22*(34), 18824–18834. doi:10.1039/C9CP03473K PMID:31475258

Ekrami, E., Pouresmaieli, M., Hashemiyoon, E., Noorbakhsh, N., & Mahmoudifard, M. (2022). Nanotechnology: A sustainable solution for heavy metals remediation. *Environmental Nanotechnology, Monitoring & Management*, *18*, 100718. doi:10.1016/j.enmm.2022.100718

Fajardo, C., Sánchez-Fortún, S., Costa, G., Nande, M., Botías, P., García-Cantalejo, J., Mengs, G., & Martín, M. (2020). Evaluation of nanoremediation strategy in a Pb, Zn and Cd contaminated soil. *The Science of the Total Environment*, *706*, 136041. doi:10.1016/j.scitotenv.2019.136041 PMID:31855644

Fazeli Sangani, M., Owens, G., & Fotovat, A. (2018). Transport of engineered nanoparticles in soils and aquifers. doi:10.1139/er-2018-0022

Fei, L., Bilal, M., Qamar, S. A., Imran, H. M., Riasat, A., Jahangeer, M., Ghafoor, M., Ali, N., & Iqbal, H. M. N. (2022). Nano-remediation technologies for the sustainable mitigation of persistent organic pollutants. *Environmental Research*, *211*, 113060. doi:10.1016/j.envres.2022.113060 PMID:35283076

Ganie, A. S., Bano, S., Khan, N., Sultana, S., Rehman, Z., Rahman, M. M., Sabir, S., Coulon, F., & Khan, M. Z. (2021). Nanoremediation technologies for sustainable remediation of contaminated environments: Recent advances and challenges. *Chemosphere*, *275*, 130065. doi:10.1016/j.chemosphere.2021.130065 PMID:33652279

Ghazaryan, K. A., Movsesyan, H. S., Khachatryan, H. E., & Ghazaryan, N. P. (2018). Geochemistry of Potentially Toxic Trace Elements in Soils of Mining Area: A Case Study from Zangezur Copper and Molybdenum Combine, Armenia. *Bulletin of Environmental Contamination and Toxicology*, *101*(6), 732–737. doi:10.1007/s00128-018-2443-0 PMID:30229278

Gong, X., Huang, D., Liu, Y., Peng, Z., Zeng, G., Xu, P., Cheng, M., Wang, R., & Wan, J. (2018). Remediation of contaminated soils by biotechnology with nanomaterials: Bio-behavior, applications, and perspectives. *Critical Reviews in Biotechnology*, *38*(3), 455–468. doi:10.1080/07388551.2017.1368446 PMID:28903604

Gong, X., Huang, D., Liu, Y., Zeng, G., Wang, R., Wan, J., Zhang, C., Cheng, M., Qin, X., & Xue, W. (2017). Stabilized Nanoscale Zerovalent Iron Mediated Cadmium Accumulation and Oxidative Damage of Boehmeria nivea (L.) Gaudich Cultivated in Cadmium Contaminated Sediments. *Environmental Science & Technology*, *51*(19), 11308–11316. doi:10.1021/acs.est.7b03164 PMID:28850225

Hidalgo, J., Anza, M., Epelde, L., Becerril, J. M., & Garbisu, C. (2022). Zero-valent iron nanoparticles and organic amendment assisted rhizoremediation of mixed contaminated soil using Brassica napus. *Environmental Technology & Innovation*, *28*, 102621. doi:10.1016/j.eti.2022.102621

Hou, J., Hu, C., Li, P., & Lin, D. (2022). Multidimensional bioresponses in nematodes contribute to the antagonistic toxic interaction between pentachlorophenol and TiO2 nanoparticles in soil. *Journal of Hazardous Materials*, *424*, 127587. doi:10.1016/j.jhazmat.2021.127587 PMID:34740505

Kumar Das, P., Mohanty, C., Krishna Purohit, G., Mishra, S., & Palo, S. (2022). Nanoparticle assisted environmental remediation: Applications, toxicological implications and recommendations for a sustainable environment. *Environmental Nanotechnology, Monitoring & Management*, *18*, 100679. doi:10.1016/j.enmm.2022.100679

Li, L., Wang, X., Zhou, Y., & Xing, B. (2021). Effectiveness and limitation of A-nZVI for restoration of a highly As-contaminated soil. *Journal of Cleaner Production*, *284*, 124691. doi:10.1016/j.jclepro.2020.124691

Liang, J., Xia, X., Zaman, W. Q., Zhang, W., Lin, K., Hu, S., & Lin, Z. (2017). Bioaccumulation and toxic effects of decabromodiphenyl ether in the presence of nanoscale zero-valent iron in an earthworm–soil system. *Chemosphere*, *169*, 78–88. doi:10.1016/j.chemosphere.2016.11.012 PMID:27863305

Manzoor, N., Ahmed, T., Noman, M., Shahid, M., Nazir, M. M., Ali, L., Alnusaire, T. S., Li, B., Schulin, R., & Wang, G. (2021). Iron oxide nanoparticles ameliorated the cadmium and salinity stresses in wheat plants, facilitating photosynthetic pigments and restricting cadmium uptake. *The Science of the Total Environment*, *769*, 145221. doi:10.1016/j.scitotenv.2021.145221 PMID:33736258

Marcon, L., Oliveras, J., & Puntes, V. F. (2021). In situ nanoremediation of soils and groundwaters from the nanoparticle's standpoint: A review. *The Science of the Total Environment*, *791*, 148324. doi:10.1016/j.scitotenv.2021.148324 PMID:34412401

Matos, M. P. S. R., Correia, A. A. S., & Rasteiro, M. G. (2017). Application of carbon nanotubes to immobilize heavy metals in contaminated soils. *Journal of Nanoparticle Research*, *19*(4), 1–11. doi:10.1007/s11051-017-3830-x

Mpongwana, N. & Rathilal, S. (2022). A Review of the Techno-Economic Feasibility of Nanoparticle Application for Wastewater Treatment. *Water, 14*(10), 1550. doi:10.3390/w14101550

Noman, M., Ahmed, T., Hussain, S., Niazi, M. B. K., Shahid, M., & Song, F. (2020). Biogenic copper nanoparticles synthesized by using a copper-resistant strain Shigella flexneri SNT22 reduced the translocation of cadmium from soil to wheat plants. *Journal of Hazardous Materials*, *398*, 123175. doi:10.1016/j.jhazmat.2020.123175 PMID:32768848

Pal, G., Rai, P. & Pandey, A. (2019). Green synthesis of nanoparticles: A greener approach for a cleaner future. *Green Synthesis, Characterization and Applications of Nanoparticles*, 1–26. doi:10.1016/B978-0-08-102579-6.00001-0

Patil, S. & Chandrasekaran, R. (2020). Biogenic nanoparticles: a comprehensive perspective in synthesis, characterization, application and its challenges. *Journal of Genetic Engineering and Biotechnology, 18*(1), 1–23. doi:10.1186/s43141-020-00081-3

Phenrat, T., Kim, H. J., Fagerlund, F., Illangasekare, T., Tilton, R. D., & Lowry, G. V. (2009). Particle size distribution, concentration, and magnetic attraction affect transport of polymer-modified Fe0 nanoparticles in sand columns. *Environmental Science & Technology*, *43*(13), 5079–5085. doi:10.1021/es900171v PMID:19673310

Qiao, Y., Wu, J., Xu, Y., Fang, Z., Zheng, L., Cheng, W., Tsang, E. P., Fang, J., & Zhao, D. (2017). Remediation of cadmium in soil by biochar-supported iron phosphate nanoparticles. *Ecological Engineering*, *106*, 515–522. doi:10.1016/j.ecoleng.2017.06.023

Rajput, V. D., Singh, A., Minkina, T., Rawat, S., Mandzhieva, S., Sushkova, S., Shuvaeva, V., Nazarenko, O., Rajput, P., Komariah, Verma, K. K., Singh, A. K., Rao, M., & Upadhyay, S. K. (2021). Nano-enabled products: Challenges and opportunities for sustainable agriculture. *Plants*, *10*(12), 2727. Advance online publication. doi:10.3390/plants10122727 PMID:34961197

Rodríguez-Seijo, A., Vega, F. A., & Arenas-Lago, D. (2020). Assessment of iron-based and calcium-phosphate nanomaterials for immobilisation of potentially toxic elements in soils from a shooting range berm. *Journal of Environmental Management*, *267*, 110640. doi:10.1016/j.jenvman.2020.110640 PMID:32421672

Romeh, A. A. (2021). Synergistic effect of Ficus-zero valent iron supported on adsorbents and Plantago major for chlorpyrifos phytoremediation from water. *International Journal of Phytoremediation*, *23*(2), 151–161. doi:10.1080/15226514.2020.1803201 PMID:32772548

Singh, A., Rajput, V. D., Pandey, D., Sharma, R., Ghazaryan, K. & Minkina, T. (2023). Nano Zinc-Enabled Strategies in Crops for Combatting Zinc Malnutrition in Human Health. *Frontiers in Bioscience-Landmark, 28*(8), 158. doi:10.31083/j.fbl2808158

Singh, A., Rajput, V. D., Sharma, R., Ghazaryan, K., & Minkina, T. (2023). Salinity stress and nanoparticles: Insights into antioxidative enzymatic resistance, signaling, and defense mechanisms. *Environmental Research*, *235*, 116585. doi:10.1016/j.envres.2023.116585 PMID:37437867

Singh, A., Rajput, V. D., Varshney, A., Ghazaryan, K., & Minkina, T. (2023). Small Tech, Big Impact: Agri-nanotechnology Journey to Optimize Crop Protection and Production for Sustainable Agriculture. *Plant Stress*, *10*, 100253. doi:10.1016/j.stress.2023.100253

Singh, A., Rajput, V. D., Varshney, A., Sharma, R., Ghazaryan, K., Minkina, T., Alexiou, A. & El-Ramady, H. (2024). Revolutionizing Crop Production: Nanoscale Wonders - Current Applications, Advances, and Future Frontiers. *Egyptian Journal of Soil Science, 64*(1). doi:10.21608/ejss.2023.246354.1684

Singh, A., Sengar, R. S., Rajput, V. D., Al-Ghzawi, A. L., Shahi, U. P., Ghazaryan, K., Minkina, T., Al Tawaha, A. R. M., Al Zoubi, O. M., & Habeeb, T. (2024). Impact of Salinity Stress and Zinc Oxide Nanoparticles on Macro and Micronutrient Assimilation: Unraveling the Link between Environmental Factors and Nutrient Uptake. *Journal of Ecological Engineering*, *25*(2), 1–9. doi:10.12911/22998993/172947

Singh, A., Sharma, R., Rawat, S., Singh, A. K., Rajput, V. D., Fedorov, Y., Minkina, T. & Chaplygin, V. (2022). Nanomaterial-plant interaction: Views on the pros and cons. *Toxicity of Nanoparticles in Plants: An Evaluation of Cyto/Morpho-Physiological, Biochemical and Molecular Responses*, 47–68. doi:10.1016/B978-0-323-90774-3.00015-5

Singh, J., Dutta, T., Kim, K. H., Rawat, M., Samddar, P. & Kumar, P. (2018). 'Green' synthesis of metals and their oxide nanoparticles: applications for environmental remediation. *Journal of Nanobiotechnology, 16*(1), 1–24. doi:10.1186/s12951-018-0408-4

Soto Hidalgo, K. T., Carrión-Huertas, P. J., Kinch, R. T., Betancourt, L. E., & Cabrera, C. R. (2020). Phytonanoremediation by Avicennia Germinans (black mangrove) and Nano Zero Valent Iron for Heavy Metal Uptake from Cienaga Las Cucharillas Wetland Soils. *Environmental Nanotechnology, Monitoring & Management*, *14*, 100363. doi:10.1016/j.enmm.2020.100363

Su, H., Fang, Z., Tsang, P. E., Zheng, L., Cheng, W., Fang, J., & Zhao, D. (2016). Remediation of hexavalent chromium contaminated soil by biochar-supported zero-valent iron nanoparticles. *Journal of Hazardous Materials*, *318*, 533–540. doi:10.1016/j.jhazmat.2016.07.039 PMID:27469041

Vardumyan, H., Singh, A., Rajput, V. D., Minkina, T., Ragab El-Ramady, H., & Ghazaryan, K. (2024). Additive-Mediated Phytoextraction of Copper-Contaminated Soils Using Medicago lupulina L. *Egyptian Journal of Soil Science*, *64*(2), 599–618. doi:10.21608/ejss.2024.266169.1714

Visentin, C., Trentin, A. W. da S., Braun, A. B., & Thomé, A. (2021). Nano scale zero valent iron production methods applied to contaminated sites remediation: An overview of production and environmental aspects. *Journal of Hazardous Materials*, *410*, 124614. doi:10.1016/j.jhazmat.2020.124614 PMID:33246810

Wang, H., Kim, B., & Wunder, S. L. (2015). Nanoparticle-supported lipid bilayers as an in situ remediation strategy for hydrophobic organic contaminants in soils. *Environmental Science & Technology*, *49*(1), 529–536. doi:10.1021/es504832n PMID:25454259

Wang, Y., O'Connor, D., Shen, Z., Lo, I. M. C., Tsang, D. C. W., Pehkonen, S., Pu, S., & Hou, D. (2019). Green synthesis of nanoparticles for the remediation of contaminated waters and soils: Constituents, synthesizing methods, and influencing factors. *Journal of Cleaner Production*, *226*, 540–549. doi:10.1016/j.jclepro.2019.04.128

Yang, K., Lin, D., & Xing, B. (2009). Interactions of humic acid with nanosized inorganic oxides. *Langmuir*, *25*(6), 3571–3576. doi:10.1021/la803701b PMID:19708146

Yang, M., Wu, X., He, C., Zhang, J., Hou, J., & Lin, D. (2023). nZVI-induced iron poisoning aggravated the toxicity of TCEP to earthworm in soil. *Environmental Pollution*, *317*, 120785. doi:10.1016/j.envpol.2022.120785 PMID:36460191

Ying, S., Guan, Z., Ofoegbu, P. C., Clubb, P., Rico, C., He, F., & Hong, J. (2022). Green synthesis of nanoparticles: Current developments and limitations. *Environmental Technology & Innovation*, *26*, 102336. doi:10.1016/j.eti.2022.102336

Zhang, F., He, M., Zhang, C., Lin, D., & Zhang, J. (2021). Combined toxic effects of dioxin-like PCB77 with Fe-based nanoparticles in earthworm Eisenia fetida. *The Science of the Total Environment*, *766*, 144347. doi:10.1016/j.scitotenv.2020.144347 PMID:33418254

Zhao, X., Liu, W., Cai, Z., Han, B., Qian, T., & Zhao, D. (2016). An overview of preparation and applications of stabilized zero-valent iron nanoparticles for soil and groundwater remediation. *Water Research*, *100*, 245–266. doi:10.1016/j.watres.2016.05.019 PMID:27206054

Zhu, Y., Liu, X., Hu, Y., Wang, R., Chen, M., Wu, J., Wang, Y., Kang, S., Sun, Y., & Zhu, M. (2019). Behavior, remediation effect and toxicity of nanomaterials in water environments. *Environmental Research*, *174*, 54–60. doi:10.1016/j.envres.2019.04.014 PMID:31029942

Section 3

Role of Nanotechnological Aspects, Nano-Omics, and Nano-Enzymes for Crop Production and Protection

Chapter 8
Nanotechnology for Improved Crop Resilience in Challenging Environments

Arshad Rasool
Minhaj University, Lahore, Pakistan

Milka Tariq
University of Agriculture, Faisalabad, Pakistan

Muhammad Asif
Punjab University, Lahore, Pakistan

Ghulam Mujtaba Shah
Hazara University, Mansehra, Pakistan

Rabia Fatima
https://orcid.org/0000-0001-9043-1207
Clarkson University, USA

Muhammad Majeed
https://orcid.org/0000-0001-7663-2563
University of Gujrat, Pakistan

Robina Aziz
Government College, Women University, Sialkot, Pakistan

Jehanzeb Khan
https://orcid.org/0000-0003-3517-2761
Government Postgraduat College, Kohat, Pakistan

Khadim Hussain
Islamia University, Bahawalpur, Pakistan

Abduraimov Ozodbek
Academy Sciences, Uzbekistan

ABSTRACT

The global agricultural landscape faces unprecedented challenges, including climate change, soil degradation, and water scarcity, necessitating innovative approaches to enhance crop resilience. Nanotechnology has emerged as a promising avenue for addressing these challenges by offering precise tools to manipulate and engineer materials at the nanoscale. This chapter explores the application of nanotechnology to enhance crop resilience in challenging environments. Nanomaterials, such as nanoparticles and nanocomposites, exhibit unique physicochemical properties that can positively impact plant growth, stress response, and overall crop performance. Engineered nanomaterials have demonstrated the ability to

DOI: 10.4018/979-8-3693-1890-4.ch008

enhance nutrient uptake, mitigate abiotic stressors, and bolster plant defense mechanisms. Additionally, nanoscale delivery systems enable targeted and controlled release of agrochemicals, optimizing their efficacy while minimizing environmental impact. This chapter highlights recent advancements in nanotechnology-driven strategies to improve crop resilience, encompassing nanoscale nutrient delivery, stress-responsive nanomaterials, and precision agriculture technologies. Furthermore, the potential risks and ethical considerations associated with nanotechnology in agriculture are discussed. Overall, the integration of nanotechnology holds great promise for sustainable agriculture, offering novel solutions to ensure food security and mitigate the impact of challenging environmental conditions on global crop production.

1. INTRODUCTION

The food production and distribution chain are under tremendous pressure due to various factors, including soaring climate change, population growth, water scarcity, and soil contamination (Rajput et al., 2021). The diminishing arable land because of urbanization and industrialization further compounds the issue, making food security the biggest concern of the century (Cheng et al., 2023). To meet global food demand, more productive and stress resilient system is crucial. Although the crop production system has improved significantly since the green revolution, but it cannot be sustained without technological advancements. Furthermore, the inefficient utilization of agrochemicals has jeopardized sustainable crop production (Singh et al., 2022). The use of synthetic fertilizers is detrimental to plant sustainability due to low nutrient efficacy, increased environmental risks, higher production costs, and reduced farmer profitability. Traditional breeding methods have indeed played a significant role in enhancing agricultural traits and crop productivity over time. However, these methods can be time-consuming due to approaches such as pure line selection, backcrossing, and single seed descent, and they may face limitations in introducing new traits between species, which can hinder the rapid development of improved varieties. Additionally, there is growing recognition that current agricultural practices need to evolve towards more sustainable and efficient methods to address environmental concerns and ensure long-term ecological balance. Therefore, a technological revolution in making stress resilient crop management system is necessary to sustainably meet global food demands in the future (Banerjee et al., 2024).

Nanotechnology, a rapidly advancing interdisciplinary field, has emerged as a promising avenue for addressing the challenges facing global agriculture, particularly in the context of climate change and environmental stressors (Singh et al., 2023a). The application of nanotechnology in agriculture, known as nanobiotechnology, holds immense potential for enhancing crop resilience in adverse conditions. By leveraging nanoscale materials and devices, researchers aim to develop innovative solutions to mitigate the impact of abiotic and biotic stresses on crops. These stressors include drought, salinity, extreme temperatures, and various pathogens that threaten agricultural productivity. The unique physicochemical properties exhibited by nanomaterials enable precise manipulation of biological processes at the molecular level, offering opportunities to optimize plant growth, nutrient utilization, and defense mechanisms. This scientific exploration of nanotechnology for improved crop resilience not only signifies a paradigm shift in sustainable agriculture but also underscores the urgency of adopting novel strategies to ensure food security in the face of evolving environmental challenges (Dehghanian et al., 2023).

Figure 1. The prospective applications of engineered nanomaterials (ENMs) in sustainable agriculture aim to enhance the stress resilience of crop plants (Science Direct)

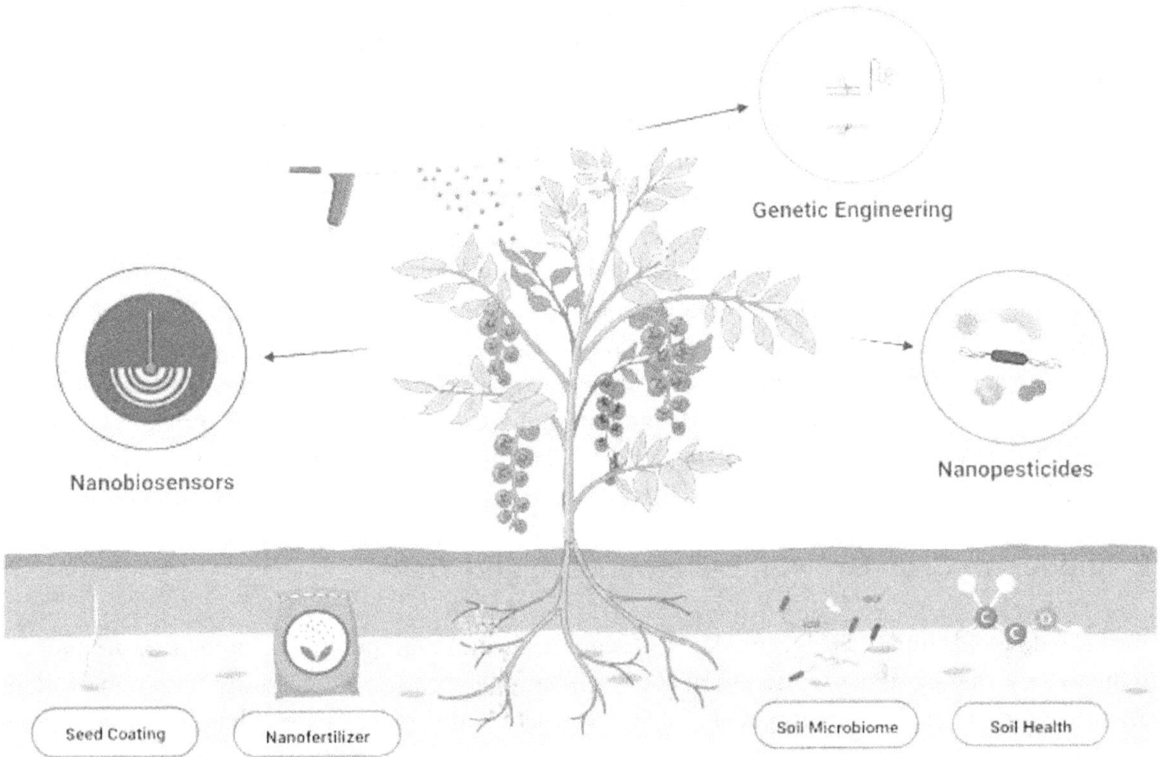

2. UNDERSTANDING CHALLENGING ENVIRONMENTS

Arid zones, characterized by persistent water shortages, high evaporation, and frequent droughts, cover a significant portion of the Earth's land area. Various classification systems exist, including those based on winter temperatures, length of growing period, and precipitation-to-evapotranspiration ratio. Arid regions occupy 10.6% of the Earth's land area, with an additional 15.2% in semi-arid zones. Notably, more than a third of the global population resides in dryland regions (Majeed, Bhatti, & Amjad, 2021).

2.1. Characteristics of Arid Ecosystems

Arid environments exhibit higher evaporation than precipitation, leading to water scarcity, climate variability, and high wind velocities. Aridity results from factors such as stable air masses, absence of storm systems, and limited convective currents. Low precipitation and high temperatures contribute to elevated evaporation rates, leading to aridic and xeric soil moisture regimes. Water scarcity poses a major constraint on agricultural production in arid regions, affecting soil fertility and hindering plant growth (Majeed, Hayat Bhatti, et al., 2020; Singh et al., 2023b). Managing water scarcity in arid regions requires innovative approaches. Efficient irrigation systems, wastewater reuse, and careful crop selection are essential. Implementing more efficient irrigation practices could save significant water resources. However, challenges such as pollutants, soil quality decline, and system flaws accompany wastewater reuse. Crop selection plays a crucial role, with certain crops demonstrating better tolerance to arid conditions. Regional decision tools are imperative for optimal agricultural management in arid areas, considering water availability and inconsistent rainfall. In addressing water scarcity challenges, measures like plastic film mulching and the use of superabsorbent polymers (SAPs) have gained attention. Plastic film mulching proves valuable for enhancing crop yields and water use efficiency, although its impact may vary with geographical and climatic factors (Arif et al., 2021). SAPs contribute to soil moisture retention and nutrient storage, positively affecting agricultural productivity in arid soils.

Soil organic carbon (SOC) and soil organic matter (SOM) play critical roles in arid ecosystems, influencing soil structure, drought resilience, and crop yields. Arid regions often exhibit low organic carbon stocks but significant inorganic carbon content, mainly in the form of soil carbonates. Vegetative coverings, appropriate crop rotations, and deep-rooted plant species contribute to SOC improvement, aiding soil structure and stability. Soil salinization poses a severe threat to agricultural sustainability in arid regions. Various causes, including poor drainage, irrigation with salt-rich water, and climate-related factors, contribute to soil salinization. Mitigation strategies include agronomic practices, improved drainage, cultivation of salt-resistant plants, and mechanical extraction of salts (Hassan, Haq, Ahmad, et al., 2022).

2.2. Impact on Crop Growth, Development, and Yield

In the face of global climate change, marked by escalating average temperatures, intensified droughts, and heightened weather extremes, agricultural ecosystems on the brink are combating with unprecedented challenges (Aziz et al., 2023; Tariq et al., 2023). Nowhere is this more evident than in the Middle East, where the pressing issues of food insecurity are compounded by a rapidly expanding population. The

Figure 2. Graphical representation of arid soil (MDPI)

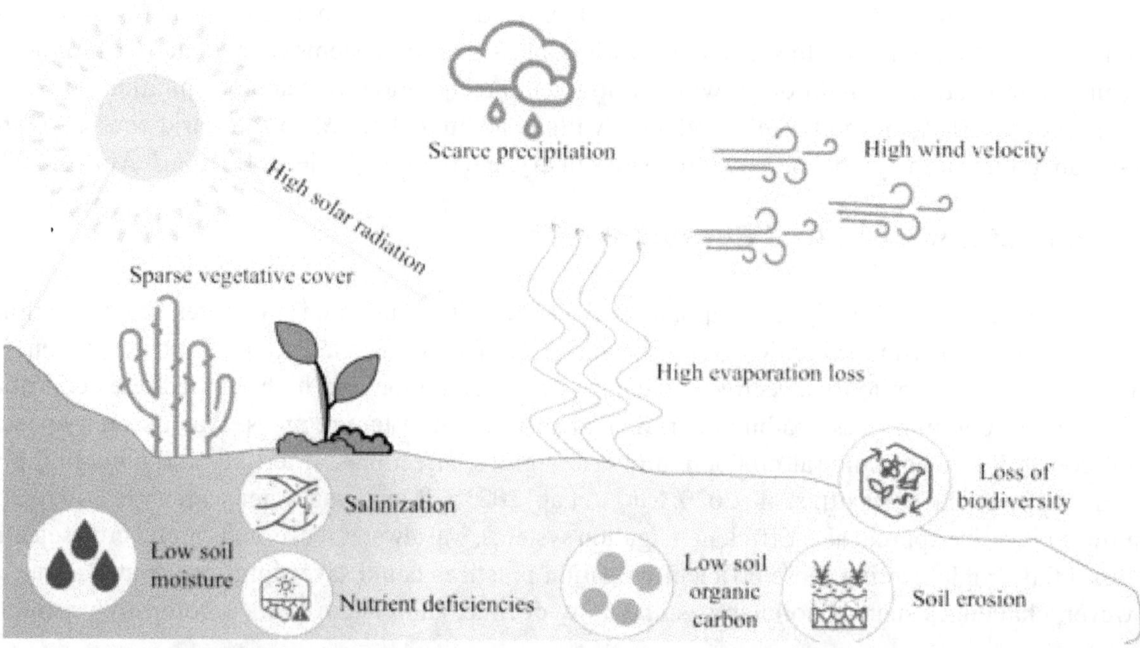

inadequacy of the world's current major crops to provide essential calories, proteins, fats, and nutrients is glaring, further accentuated by the burdens of political and economic instability. In this complex scenario, the imperative for new, locally adapted crops becomes evident. Halophytes emerge as a promising avenue for addressing the nutritional needs of a burgeoning human population, provided their growth and reproductive performance can be tailored to meet the demands of this evolving world. As we navigate this new agricultural landscape, the exploration of halophytes holds potential as a transformative solution to feed and sustain populations amidst these challenging circumstances (Haq et al., 2022a; Naveed et al., 2024a).

In the dynamic interaction between plants and their environment, various stresses pose challenges to growth and development, both in natural and agricultural settings. Foremost among these challenges is drought, a particularly severe environmental stressor that significantly impacts plant productivity. Given that approximately 80–95% of a plant's fresh biomass comprises water, the crucial role of water in numerous physiological processes, spanning growth, development, and metabolism, cannot be overstated. Recognized as a primary environmental stress, especially in drought-prone regions, drought is identified as the foremost threat to global food security, historically catalyzing significant famines (Muhammad, Waheed, et al., 2024). Agricultural ramifications of drought are further intensified by the depletion of water resources and escalating food demand driven by a burgeoning world population. The unpredictable nature of drought hinges on factors like irregular rainfall distribution, evapotranspiration, and water-holding capacity in the rhizosphere. Additionally, the occurrence of physiological drought or pseudo-drought, where plants struggle to uptake water despite adequate soil moisture, adds another layer of complexity to the challenge. Underlying the visible effects of drought stress are disruptions across molecular, biochemical, physiological, morphological, and ecological dimensions of plant traits and processes (Fadiji et al., 2023a). In water deficit environments, plant yield and quality suffer adverse consequences. The impact varies based on growth stages, age, plant species, and the severity and duration of drought. However, plants exhibit a remarkable ability to adapt by modulating resource utilization and adjusting their growth patterns in response to adverse environmental conditions, including drought (Lo Presti et al., 2023).

3. NANOMATERIALS FOR ENHANCED WATER MANAGEMENT

The role of nanomaterials in water treatment and purification has emerged as a transformative frontier, offering a promising alternative to conventional technologies that are often costly and time-consuming. This paradigm shift is particularly advantageous for developing nations like India and Bangladesh, where the feasibility and cost-effectiveness of implementing new removal processes are crucial factors. Qualitatively, nanomaterials can replace traditional counterparts, known for their high raw material requirements, energy intensity, or environmental harm. Embracing green chemistry principles in nanoparticle production further contributes to reduced waste generation and safer chemical syntheses (Yuvaraj et al., 2023). Yet, to substantiate these qualitative advantages, more quantitative data is essential. The ongoing debate surrounding the safety of nanoparticles and their potential environmental impact adds complexity to the discourse. While nanotechnology holds fervent hope for providing clean water to developing countries efficiently and sustainably, the Janus face of nanomaterials introduces challenges (Singh et al., 2023c; Naveed et al., 2024b). The catalytic activity of nanoparticles, beneficial for pollutant degradation, may trigger toxic responses when taken up by cells (Manimegalai et al., 2023). Despite these concerns, the cost-reducing potential of nanotechnology positions it as a viable and effective long-term solution for contaminant removal from water. Nanoparticles, serving as potent sorbents, separation media, catalysts for photochemical degradation, and components in nanofiltration membranes, exemplify the diverse roles nanomaterials can play in revolutionizing water treatment methodologies (Lin et al., 2023).

3.1. Nanoscale Sensors for Soil Moisture Monitoring

In the domain of contemporary agriculture, nanotechnology has asserted its prominence, emerging as a pivotal tool with transformative influence. Its applications span a spectrum of agricultural domains, offering multifaceted benefits to enhance productivity. Nanotechnology exhibits the potential to augment agricultural production significantly by optimizing nutrient utilization through the application of nanoformulations in fertilizers (Manimegalai et al., 2023). Beyond this, its impact extends to crop enhancement, the detection and treatment of diseases, and the elucidation of host-parasite interactions at the molecular level, facilitated by nanosensors. The scope of nanotechnology encompasses diverse aspects such as plant disease diagnostics, the removal of contaminants from soil and water, postharvest management of vegetables and flowers, and the reclamation of salt-affected soils. Nanobiosensors, a distinctive facet of this technology, prove instrumental in sensing pathogens, fertilizers, moisture levels, and soil pH. This not only aims to minimize the application of plant protection products but also strives to reduce nutrient loss, ultimately fostering enhanced crop yields through judicious nutrient management (Lo Presti et al., 2023).

Soil moisture monitoring, a critical aspect of precision agriculture, involves the measurement of either soil water potential or soil water content and has found extensive application in irrigation scheduling (Lin et al., 2023). Understanding the dynamics of soil moisture within the plant root zone is pivotal for effective irrigation management, influencing both the volume of irrigation water applied and plant water uptake. Various methods exist for assessing soil moisture levels, ranging from direct measurements through gravimetric sampling to indirect methods utilizing electromagnetic properties, heat conductivity, neutron

Figure 3. Monitoring strategy of soil using different nanoscale sensors

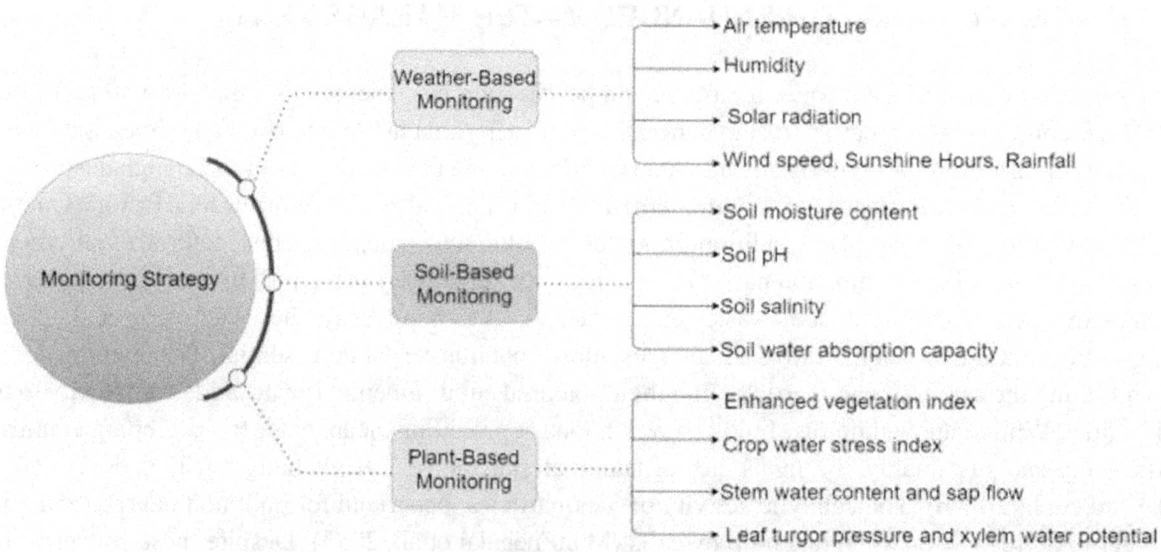

count, water potential, and electrical resistance. In the era of rapid advancements in microcomputer and communication technology, a diverse array of soil sensors, including ground-based, aerial, and satellite moisture sensors, are gaining prominence in the toolkit for precision irrigation. These sensors, with their minimal footprint in the field and the ability to operate at multiple depths, offer a comprehensive understanding of soil moisture dynamics. Deploying sensors at different depths enhances accuracy and provides insights into changes in soil water content in response to irrigation and crop water use (Ahmad et al., 2023).

Soil sensors contribute a wealth of data on the physical, chemical, and mechanical properties of the soil, utilizing measurements such as optical, radiometric, mechanical, acoustic, electrical, electromagnetic, pneumatic, or electrochemical parameters. This wealth of information aids in determining crucial parameters, including the Maximum Allowable Depletion (MAD) (Li et al., 2020; Majeed et al., 2021b). Comprehensive insights into the functionality and characteristics of soil moisture sensors can be found in the works of (Karnwal et al., 2023; Naveed et al., 2024c), and (Nguyen et al., 2023), providing a valuable resource for researchers and practitioners in the field of precision agriculture.

In the area of smart irrigation systems, effective monitoring plays a pivotal role in optimizing water use efficiency by scrutinizing factors that directly impact crop growth and development. The paradigm of monitoring within the context of smart irrigation extends to the real-time collection of data pertaining to soil conditions, plant health, and weather parameters, encompassing the entire cropped area. This approach is facilitated through the utilization of advanced communication technologies (Panda et al., 2023). The establishment of a real-time monitoring system necessitates the seamless integration of sensors with a wireless sensor communication network or an Internet of Things (IoT) framework. Wireless networks assume a paramount role in enabling real-time monitoring in smart irrigation, owing to their inherent capabilities in sensing, processing, and transmitting data, as emphasized by

Figure 4. Soil moisture sensors require specific conditions to function properly

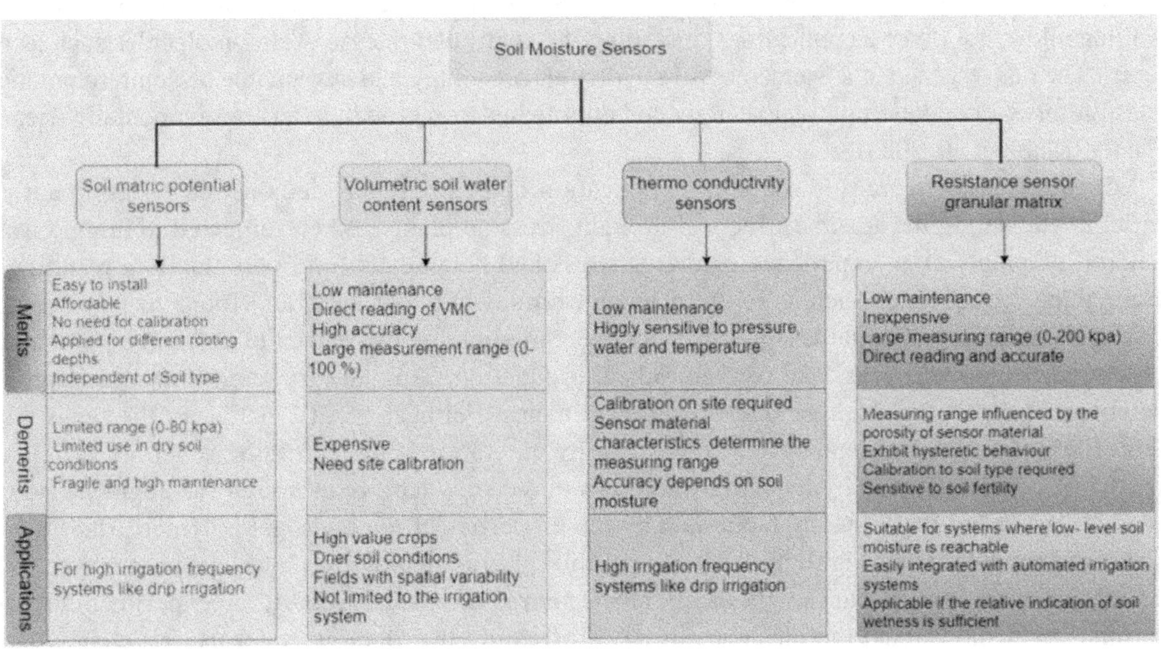

Hamami and Nassereddine (2020). Within the wireless sensor network, diverse sensor nodes are interconnected through a wireless connection module, creating a robust framework for comprehensive and instantaneous data collection and analysis, thereby contributing to the enhancement of smart irrigation practices (Dorjee et al., 2023).

3.2. Nanoparticles for Water Retention in Soil

The utilization of water-absorbing polymers, commonly referred to as hydrogels, in agriculture traces back to the early 1980s. These crosslinked hydrophilic water-soluble polymers exhibit the remarkable ability to absorb substantial amounts of water, up to a hundred times their dry weight, and release it under mechanical stress. In the agricultural context, hydrogels, often termed superabsorbent polymers (SAP), play a crucial role in enhancing soil water absorption in arid regions by dispersing absorbed water into the soil. Beyond agriculture, SAPs find applications in diverse fields such as hygienic biosensor products, wound dressing, regenerative medicines, tissue engineering, food additives, drug delivery systems, artificial snow, coal dewatering, sealing, biomedical applications, pharmaceuticals, and barrier materials for regulating biological adhesions and diagnostics separation of biomolecules or cells. Key to these applications are the distinctive water absorbency and retention capabilities of hydrogels, positioning SAPs as versatile entities, particularly in agriculture (Khider et al., 2023; Li et al., 2023). Hydrogels, when used as soil conditioners, enhance nutrient and water retention capabilities, serving as agents for slow-release fertilizers. Their application improves soil density, structure, and permeability, leading to enhanced water infiltration and reduced run-off and erosion. This, in turn, positively impacts crop productivity and feasible yield. The controlled release of stored water from hydrogels during soil drying, facilitated by a diffusion mechanism, ensures prolonged soil moisture. Mixing hydrogels with soil increases soil porosity, fostering better oxygen circulation to plant roots. A noteworthy development is the patented formulation by Sannino et al. featuring cellulose-based polyelectrolyte hydrogels that are both biocompatible and biodegradable, offering adjustable swelling abilities. These hydrogels can absorb significant amounts of water or aqueous solutions, and their unique memory of shape after swelling allows for diverse applications, including the controlled release of tiny molecules such as nutrients. The convergence of advancements in hydrogel technology and sustainable agriculture practices holds promise for addressing water scarcity and optimizing crop growth in an environmentally friendly manner (Malik et al., 2023).

Mitigating water scarcity through the application of hydrogel has demonstrated significant potential in alleviating the adverse effects of drought stress on plants. When subjected to limited water availability, plants often experience oxidative stress and heightened lipid peroxidation, resulting in observable consequences such as reduced height, diminished leaf area, and foliar matrix damage. Hydrogel emerges as a valuable solution, effectively minimizing the impact of drought on plants and subsequently reducing stress and oxygen radical production. This, in turn, contributes to improved growth and enhanced yield, even in challenging climates (Fujita et al., 2023). One of the key mechanisms through which hydrogel aids in water scarcity mitigation is by increasing water use efficiency (WUE). WUE is improved by reducing water loss in the soil, achieved through the augmentation of the soil's water-holding capacity facilitated by the properties of the hydrogel polymer. Abd El-Aziz et al. substantiate this by highlighting the capability of polymers within hydrogels to absorb, hold, and release water in accordance with the plant's water requirements. Notably, an experiment focused on tomatoes demonstrated the influence of an eco-friendly hydrogel on water use efficiency under various irrigation levels. The application of hydrogel resulted in a noteworthy reduction in the amount of water required for tomato irrigation compared to the control treatment without hydrogel. These findings underscore the positive impact of hydrogel treatments on the productivity of tomato plants and, consequently, on the values of water use efficiency. The use of natural and eco-friendly hydrogels holds promise in reducing water loss and enhancing water use efficiency, thus sustaining productivity during periods of drought stress. This reaffirms the potential of hydrogel technology as a viable and environmentally friendly approach to address water scarcity challenges in agriculture (Pramanik et al., 2020; Rohilla et al., 2021).

Drought stress affects approximately half of the global arid and semi-arid regions, significantly impacting plant photosynthesis, growth, and physio-biochemical processes. The disruption of these essential processes leads to inhibited plant growth, development, and eventual yield loss. Researchers have identified numerous harsh impacts of drought stress, including oxidative stress induced by excessive reactive oxygen species (ROS) production, nutrition imbalance, altered cell membrane integrity, and disturbances in biomolecules like DNA, proteins, and lipids. Additionally, imbalanced photosynthetic efficiency, reduced turgor pressure, and altered leaf gas exchange rates contribute to a significant loss in total biomass and productivity under water stress conditions (Banerjee et al., 2024; Muhammad, Waheed, et al., 2024). Drought stress adversely affects various morphological characteristics of plants, such as seed germination, plant height, relative root length, root diameter, total biomass of leaves and roots, number of leaves per plant, and number of branches per plant, impacting a wide range of crops. Physiologically, crops under drought conditions experience partial stomatal closure and increased photorespiration due to carbon metabolism imbalance (Sangeeta Dash et al., 2023). Elevated production of ROS during stress further damages chloroplasts through oxidation, collectively limiting photosynthate production and ultimately reducing agricultural productivity. In response to these challenges, plants activate natural defense systems involving morphological, physiological, and biochemical adaptations. These adaptations include leaf rolling, altered leaf angle, development of a deep root system, drought-resistant epigenetic phenotypic plasticity, and gene activation (Liao et al., 2023).

4. NANOTECHNOLOGY FOR IMPROVED NUTRIENT UPTAKE

Addressing challenges like food security necessitates the refinement of fertilizer particles through advanced nanoscience technology to optimize their usage, minimize losses, and improve efficiency in nutrient uptake (Singh et al., 2023d). Nanotechnology, characterized by its innovative and interdisciplinary approach, involves the design, development, and application of materials and devices at the molecular level in the nanometer scale (1 to 100 nanometers) (Swain et al., 2023). Nanoparticles exhibit small size, a large surface-to-volume ratio, heightened chemical reactivity, enhanced solubility, and unique magnetic and optical properties. Nanofertilizers, a prominent focus of research and fertilizer industries, stand out due to their controlled and targeted delivery mechanisms (Singh et al., 2024a). These nanofertilizers have the potential to enhance crop productivity by positively influencing seed germination, seedling growth, photosynthetic activity, energy metabolism, protein synthesis, and enzymatic functions. Beyond nanofertilizers, nanotechnology-based soil-binding products play a crucial role in improving soil water management and preventing soil erosion (Singh et al., 2024b). The potential applications of nanotechnology in plant nutrient management and soil improvement are promising, as discussed in this chapter. The assessment of advanced research in this field underscores the potential for nanotechnology to revolutionize agricultural practices, providing sustainable solutions to enhance food security and mitigate challenges associated with soil and nutrient management (Fadiji et al., 2023b).

The field of nano-nutrition for crop plants is evolving, and various NPs have demonstrated success as both macro- and micronutrients (Singh et al., 2024c). These nanoparticles contribute to enhancing soil fertility, optimizing nutrient uptake by crop plants, and ultimately improving overall productivity. However, it is crucial to address concerns regarding the potential toxicity of these NPs in crop plants. Achieving a balance necessitates a more focused approach on dose optimization to ensure normal growth and productivity. Additionally, there is a growing emphasis on synthesizing NPs derived from organic or natural sources that are easily biodegradable (Grillo et al., 2021).

4.1. Nanofertilizers for Efficient Nutrient Delivery

In recent years, nanotechnology has emerged as a transformative force in plant science and agriculture. Advancements in nanotechnology have facilitated the large-scale production of nanoparticles of physiologically important metals (Vardumyan et al., 2024). These nanoparticles are now incorporated into fertilizer formulations to enhance plant cell uptake and minimize nutrient loss. Nanoparticles possess attributes such as high surface area, sorption capacity, and controlled-release kinetics, transforming them into "smart delivery systems." Nanostructured fertilizers exhibit the potential to increase nutrient use efficiency through mechanisms like targeted delivery and slow or controlled release (Zhou et al., 2023). They can precisely release active ingredients in response to environmental triggers and biological demands. Recent laboratory-scale investigations have reported that nano-fertilizers hold the promise of improving crop productivity by accelerating seed germination, enhancing seedling growth, boosting photosynthetic activity, optimizing nitrogen metabolism, and facilitating carbohydrate and protein synthesis. However, given the nascent nature of this technology, ethical and safety concerns surrounding the use of nanoparticles in plant productivity are numerous and must be meticulously evaluated before widespread adoption of these so-called nano-fertilizers in agricultural fields. A judicious and thorough assessment of these ethical and safety considerations is imperative to ensure responsible and sustainable utilization of nanotechnology in agriculture (Payne et al., 2023; Jamilet al., 2023).

4.2. Nanomaterials for Enhancing Nutrient Absorption

While nano-based fertilizers have demonstrated superior effectiveness compared to their conventional counterparts in addressing plant productivity and developmental issues, their widespread application in agricultural fields raises concerns about nutrient loss. The potential failure to reach targeted sites may lead to nutrient fixation in the soil or contribute to water pollution through leaching (Ghazaryan et al., 2018; Al-Sahli et al., 2024). This, in turn, can result in a significant decline in soil fertility, negatively impacting crucial soil parameters and overall crop productivity. A recent addition to the realm of nanofertilizers, known as slow-release fertilizers, is evolving as a promising solution. In the context of sustainable agriculture, slow-release nanofertilizers utilize nanocarriers to achieve the targeted delivery of active nutrient molecules in a controlled manner. The robust and diffusion-based controlled release properties of these nanocarriers enable the precise delivery of agrochemicals to plants under field conditions. Various nano-based carriers, such as mesoporous silica, carbon-based nanomaterials, nanoclays, nanohydroxyapatite, and polymer-based nanomaterials, serve as effective delivery systems for a range of agrochemicals, including micronutrients, macronutrients, and biofertilizers. The development of slow-release nanofertilizers represents a significant stride toward sustainable agriculture, promising efficient nutrient delivery while minimizing adverse environmental impacts (Periakaruppan et al., 2023).

In the domain of smart agricultural practices, the development of a Smart Target Delivery System, based on environmental triggers and biological demands, stands as a technological breakthrough. Progress in the commercial production of physiologically and agronomically essential metal core nanoparticles (NPs) for fertilizer formulations marks a significant stride in curbing nutrient losses and enhancing plant use-efficiency. These advancements qualify as 'smart delivery systems,' offering highly controlled, specific, remotely coordinated, and targeted approaches that circumvent biological barriers. Nano fertilizers, when applied to plants or soil, hold promise for improving fertilizer uptake and performance. Notably, nano fertilizers exhibit slower release rates, aiding soil fertility and nutritional balance by reducing runoff into groundwater and mitigating the risks of toxicity due to over-application. Zeolites, possessing nano-porous properties, demonstrate high selectivity for plant nutrients, ensuring slow and controlled nutrient release triggered by various factors, thereby minimizing nutrient loss and ecological risks. Nano fertilizers designed for slow release and targeted delivery, such as urea-coated hydroxyapatite NPs, have proven effective in significantly enhancing crop yields with reduced urea amounts compared to conventional fertilizers (Victoria et al., 2024).

In biologically synthesized NPs, the protein corona (PC), a dynamic and complex biomolecular coating formed around NPs in physiological environments, plays a pivotal role. The PC influences NP uptake, bio-distribution inside cells, and cellular functions. Protein-coating NPs can achieve targeted delivery in a plant model, as demonstrated by gold NPs coated with antibodies accurately targeting specific plant structures. The composition of PC, influenced by biophysiochemical properties and factors like ionic strength, temperature, pH, and density, critically shapes physiological interactions between NPs and cells in the plant system. The Vroman effect, a competitive protein adsorption process on the NP surface, governs the formation and evolution of the PC over time, impacting NP dissemination and bio-dispersion. The size, surface charge, and curvature of NPs play key roles in protein adsorption and PC development, with implications for NP stability, homo-aggregation, and interaction with cell membranes and receptors. The dynamic nature of the PC, including the transition from a 'soft corona' to a 'hard corona,' further underscores the complexity of NP interactions in biological media. Overall, understanding and engineering the properties of NPs and their PC are crucial for optimizing their performance in agricultural applications, ensuring effective nutrient delivery while considering the dynamic nature of biological systems (Pandey et al., 2023).

5. NANOMATERIALS AND STRESS TOLERANCE

Nanotechnology emerges as a promising avenue with diverse applications, notably in agriculture, offering solutions to the challenges faced by many in developing countries, including food shortages caused by environmental adversities such as global warming, drought, floods, extreme climatic conditions, salinity, and soil fertility loss. Abiotic stresses like global warming, often accompanied by drought stress, and cold stress leading to membrane rigidification and solute leakage, impact the sustainable growth and production of crops. Salinity stress hampers water and nutrient uptake, disrupting growth and yield, while flooding produces toxic compounds and gases harmful to crop plants (Singh et al., 2023; Ghazaryan et al., 2023). The common denominator across these stresses is the enhanced generation of reactive oxygen species (ROS), causing cellular damage crucial to plant survival. Addressing these challenges requires an improved and sustainable farming technology, along with cultivars resistant to various hazards. In this context, nanomaterials have emerged as crucial tools for enhancing the growth and productivity of crops facing abiotic stresses. Nanoparticles, with their high surface energy and surface/volume ratio, exhibit enhanced bioavailability and bioactivity compared to their standard or bulk forms. They demonstrate the ability to penetrate plant cells, influencing key events in the plant life cycle such as seed germination, seedling growth, root formation, photosynthesis, flowering, and ultimately, yield. While nanomaterials showcase beneficial effects on plant systems under abiotic stress, it's important to acknowledge potential toxicity concerns (Bhatia et al., 2024). This chapter delves into modern strategies adopted for abiotic stress mitigation in plants, leveraging the potential of nanomaterials to maximize crop yield. The exploration of nanotechnology in agriculture offers a nuanced understanding of both its advantages and potential challenges, contributing to the ongoing discourse on sustainable and resilient farming practices (Hooshmand et al., 2023).

5.1. Nanoparticles for Heat and Cold Stress Mitigation

Abiotic stresses inflict various disorders on plants, including osmotic stress in cells, retardation in cell development, diminished photosynthetic activity, seed dormancy, delayed reproduction, and ultimately, a negative impact on yield (Chaupoo et al., 2023). Water-deficit stress, the most prevalent abiotic stress in nature, causes substantial damage to crop plants. The severity of water deficit stress stems from the reduction in plant relative water content, leading to osmotic and oxidative stress. Salinity stress produces similar effects as drought stress, exacerbating the challenges faced by plants. Drought and salinity are among the most menacing global abiotic stresses, necessitating a cascade of morphological, physiological, and molecular changes in plants (Balasubramaniam et al., 2023). To survive these stresses, plants need to undergo osmotic adjustment, reactive oxygen species (ROS) detoxification, stomata closure, and cellular signaling. Additionally, high temperatures can impact various aspects of plant physiology, including hormone production, nutrient uptake, stomatal conductance, transpiration rate, photosynthetic activities, enzymatic activity, antioxidant levels, membrane stability index, and ROS production. Similarly, chilling stress in plants can affect tissue water content, membrane fluidity, and chlorophyll content (Azeem et al., 2023). These abiotic stresses create a complex web of challenges for plants, necessitating a comprehensive understanding of their responses at the cellular and molecular levels for the development of effective mitigation strategies (Azeem et al., 2023; Mubarok et al., 2024).

5.2. Nanotechnology for Salinity Stress

Saline soil, characterized by a high concentration of soluble salts with an electrical conductivity (ECe) value of 4 dS/mL or higher, poses significant challenges to plant growth. The presence of salinity makes it difficult for plant roots to absorb water, posing hazards to plant health. Salinity-resistant plants exhibit various adaptations at morphological, biochemical, and physiological levels to sustain their life cycles. Projections indicate that 50% of cultivated agricultural lands will face salt stress by 2050. Salts like NaCl and Na2SO4 are primary contributors to soil salinity in agricultural lands (Balasubramaniam et al., 2023). Germination and early seedling stages are particularly vulnerable to soil salinity, as it disrupts ionic and osmotic equilibrium, leading to physiological drought in plants. Salinity induces various cellular and metabolic changes, including disruptions in cellular growth and expansion, plant membrane instability, ion toxicity, altered metabolism, inhibited seed germination, and reduced photosynthesis. In addition to these effects, salinity stress results in a decrease in osmotic potential, leading to reduced uptake of nutrients and water by salinity-stressed roots (Rehman et al., 2023). The closure of stomata induced by salinity inhibits CO2 fixation, causing damage to photosynthetic pigments and adversely affecting photosynthetic processes and electron carriers. Hormonal and nutrient imbalances are also observed in plants under salinity stress. The hormonal levels of abscisic acid (ABA) and indole-3-acetic acid (IAA) become disrupted (Wu et al., 2005). Furthermore, salts like NaCl increase the concentrations of Na+ and Cl- ions, leading to ionic stress and competition with essential nutrients such as K+, Ca2+, and Mg2+, resulting in nutrient deficiency in plants (Hassan, Haq, Majeed, et al., 2022). These negative implications of NaCl salt contribute to a decline in photosynthetic activity, the generation of reactive oxygen species (ROS), and programmed cell death (Arshad et al., 2022). Understanding and addressing the multifaceted impacts of salinity stress on crops are crucial for developing effective strategies to mitigate its effects on agricultural productivity(Muhammad, Basit, et al., 2024).

6. NANOGENOMICS FOR CROP IMPROVEMENT

Efforts to enhance crop yields sustainably while minimizing the use of pesticides and fertilizers are at the forefront of agricultural research. Traditional plant breeding methods, although effective, are often laborious, time-consuming, and present challenges. To overcome these limitations, novel plant breeding technologies equipped with advanced capabilities, such as gene knockout/in, epigenetic modifications, and the generation of heritable targeted mutations, are crucial. Recent developments in biotechnology, particularly third-generation genome editing (GE) techniques like CRISPR/Cas9, along with genome sequencing, plant-based synthetic biology, and bioengineering, have opened new avenues for efficient crop improvement. Traditional plant breeding involves complex backcrossing and selection processes, making it time-consuming and less effective. Mutation breeding, while effective, is also time-consuming and prone to random mutations. Transgenic breeding, although successful, faces public acceptance challenges due to the use of genetically modified organisms (GMOs). In contrast, third-generation GE methods, particularly CRISPR/Cas9, offer a cost-effective, stable, time-saving, and less labor-intensive solution. Importantly, these methods can bypass GMO regulations as they do not involve the insertion of foreign DNA (Hussain et al., 2023a).

In parallel, speed breeding, which extends the photoperiod and enhances flowering time, allows for multiple generations per year, significantly accelerating the breeding process. Combining speed breeding with GE technologies holds promise for developing elite organic varieties within 1–2 years, a substantial improvement compared to traditional methods. Additionally, nanotechnology has emerged as a valuable tool for improving the delivery of CRISPR components into plant cells. Nanoparticles (NPs) coated with DNA, RNA, or ribonucleoproteins (RNPs) can be employed to achieve efficient and targeted delivery of CRISPR reagents into meristematic cells (Bashir et al., 2023a). GE techniques, categorized into three generations, with CRISPR/Cas9 being the third generation, have revolutionized precise genome modifications. Overcoming challenges related to plant transformation, nanotechnology provides a robust, efficient, and species-independent method for delivering biomodifier-conjugated complexes. The use of nanotechnology in conjunction with CRISPR techniques offers opportunities for overcoming barriers to genetically engineered plant production. Nanomaterial engineering is a cutting-edge technology with the potential to develop crops for sustainable farming systems. Nanobiotechnology techniques, coupled with advanced genetic editing tools, provide precise and efficient gene selection, reducing the time required for removing unwanted genes and facilitating access to essential genes from large plantations.

6.1. Nanoscale Tools for Studying Plant Genes in Challenging Environments

In the past decade, nanotechnology-based sensing has garnered significant attention for detecting various contaminants in water, with detection levels reaching parts per billion (ppb) for metals and organic pollutants in both laboratory and field settings (Haq et al., 2022b). Ongoing advancements in nanotechnology are expected to yield new sensing technologies in the next decade, with applications extending beyond water sensing. Agricultural and food products often face challenges related to perishability, prompting extensive research in nanotechnology to enhance shelf-life, freshness, quality, and safety. The demand for traceability in the food industry has risen due to concerns about food safety and authenticity. Traceability ensures the ability to track food products throughout production, processing, distribution, and consumption. Advanced traceability systems are crucial for efficient product recalls during food safety outbreaks. Addressing essential technical challenges, nanotechnology-based tracing devices, such as nucleic acid engineered nanobio barcodes, developed by Luo and his group at Cornell University, offer a comprehensive solution. These devices not only validate the origin of agricultural products but also provide information about the product's journey, aiding informed decision-making processes (Khoja et al., 2022).

Challenges in implementing new technologies, particularly managing costs, must be addressed for widespread adoption. The minimal use of materials in precise nanomanufacturing may contribute to cost reduction, but economic considerations for specific nanotechnology applications need careful examination. In the realm of clean energy, nanotechnology plays a pivotal role, particularly in the development of inexpensive, safe, and renewable energy sources. Nanotechnology-enhanced solar cells, a prominent example of "green nanotechnology," are actively researched for their potential application in agriculture (Majeed, Bhatti, et al., 2020). Solar energy conversion, energy storage, and nanotechnology-based solar-thermal energy conversion systems are areas of active exploration. Researchers worldwide are focused on reducing costs in photocatalysts and energy materials to enhance the economic feasibility of these technologies. Another avenue of nanotechnology's contribution to clean energy involves the conversion of biomass into fuels, chemical intermediates, and specialty chemicals.

Nanostructures, with their large surface area per unit volume and precise control over composition and structure, hold promise as catalysts for biomass conversion. As biomass gains importance as an industrial feedstock, the development of economically feasible catalysts becomes critical, and nanotechnology offers inherent advantages in this domain. The pursuit of harnessing solar energy and biomass conversion through nanotechnology is expected to persist and intensify in the coming years, driven by the grand challenge of providing sustainable and accessible energy solutions for humanity (Khan et al., 2022).

7. INTEGRATION OF NANOTECHNOLOGY IN CROP MANAGEMENT

Nanotechnology has emerged as a valuable tool in enhancing various facets of agricultural production, with a particular focus on seed germination and plant growth to improve crop yields and quality. The intricate process of seed germination, influenced by environmental factors, genetic traits, and soil parameters, has seen positive effects from the application of nanoparticles (NPs), including Carbon

Nanotubes (CNTs), silicon dioxide (SiO_2) NPs, ZnO NPs, titanium dioxide (TiO_2) NPs, and gold (Au) NPs across diverse crops such as tomato, wheat, rice, pearl millet, soybean, barley, and maize. Studies have demonstrated that NPs play a crucial role in augmenting seed germination rates by influencing antioxidant enzyme activities, water utilization rates, and oxygen levels (Khoja et al., 2022; Majeed et al., 2023). For instance, Au NPs significantly enhanced the germination rate of pearl millet, while ZnO NPs positively impacted wheat germination by increasing antioxidant enzyme activity. TiO_2 NPs have proven beneficial in promoting seed germination, reducing germination time in crops like tomato and wheatgrass. Non-metallic NPs, such as multiwalled carbon nanotubes (MWCNTs), have also shown promise in stimulating seed germination in soybean, barley, and corn by enhancing water assimilation capabilities. Furthermore, various metallic and non-metallic NPs, including Ag NPs, ZnO NPs, TiO_2 NPs, silica NPs, and MWCNTs, have demonstrated the ability to enhance the growth, photosynthesis, and yield of crops like spinach, cotton, maize, soybean, and barley. These NPs primarily act by mediating antioxidant enzyme activity, influencing plant cell morphology, and altering protein and organic compound content. Silica NPs, for example, promote maize growth, influencing cell wall thickness and protein content (Majeed, Lu, et al., 2022).

Additionally, NPs can induce gene expression related to nutrient assimilation and growth regulation, influencing the transcriptome of treated seedlings. NPs also have the potential to regulate plant hormone balance, as seen in the foliar application of Ag NPs inducing auxin-related gene expression in common bean varieties. However, it is essential to acknowledge that NPs, owing to their structural and surface reactivity properties, can induce intracellular oxidative stress and genetic damage, potentially leading to reduced crop yields and physiological disorders. The concentration of NPs applied is crucial, as high concentrations can result in adverse effects. Metallic NPs, in particular, may exhibit toxicity, impacting plant growth and development. Careful consideration and determination of safe NP doses for different crop species are imperative for sustainable and effective agricultural applications (Majeed, Khan, et al., 2022).

Nanobiotechnology, particularly the use of nanomaterials to facilitate the transfer of biomolecules such as DNA and RNA, stands out as a promising avenue within the realm of genetic modifications. This approach encompasses various applications, including transgene expression, genome editing, and gene silencing (Sinha et al., 2023a). Overcoming the challenges posed by the plant cell wall's physical and chemical properties is a crucial aspect of this technology. Pollen, characterized by a chemically inert cell wall, proves to be an excellent target for transient gene expression. Notably, imidazolium-coated Single-Walled Carbon Nanotubes (SWCNTs) demonstrated high efficiency in delivering plasmid DNA encoding green fluorescent protein (GFP) into oil palm pollen, achieving both successful delivery and robust GFP activity. Similarly, chitosan-coated SWCNTs enabled the transformation of a DNA plasmid into chloroplasts across various plant species, resulting in elevated transient expression levels.

In addition to transgene expression, recent years have witnessed a surge in studies exploring nanomaterial-based gene silencing and genome editing (Majeed, Bhatti, Pieroni, et al., 2021). Nanomaterials offer specific delivery of genetically engineered plasmids, presenting innovative approaches for rapid genome modifications in plants. For example, research demonstrated the successful transfer of DNA-CNT conjugates into multiple plant species, including tobacco, arugula, cotton, and wheat. Platforms utilizing CNT-mediated Small Interfering RNA (siRNA) delivery showcased high silencing efficiency in plant cells, emphasizing effective intracellular transferability. Polyethylenimine-coated Gold Nanoparticles (PEI-AuNPs) effectively delivered siRNA into intact plant cells, resulting in a substantial decrease (at least 76%) in the expression of the target gene. While nanobiotechnology in genome modification holds tremendous promise, its stability necessitates further investigation. Additionally, potential ecological implications and impacts on other species underscore the need for comprehensive research to address these concerns and ensure the responsible advancement of this transformative technology. Continued exploration and refinement will be crucial for harnessing the full potential of nanobiotechnology-assisted genome modification in plants (Majeed, Tariq, Anwar, et al., 2021).

7.1. Precision Agriculture Using Nanoscale Data

Precision agriculture (PA) is undergoing a transformative evolution driven by advancements in Internet of Things (IoT) technologies. This evolution is geared towards achieving key features that enhance efficiency and increase crop yields (Majeed, Tariq, et al., 2022). These features encompass data metrics for monitoring, decision-making, crop protection and management, waste reduction, operational cost efficiency, crop variability, data trend analysis for various use cases, and pests control. The core components shaping and evaluating Key Performance Indicators (KPIs) in precision agriculture rely on technological factors such as sensors, mobility, real-time monitoring, connectivity, and the ease of deploying applications (Majeed, Tariq, et al., 2022; Sinha et al., 2023b). With the agricultural sector already embracing the next industrial revolution, farmers are incorporating innovative technologies to meet growing food demands. This integration empowers farmers with timely and informed decisions, optimizing the use of inputs in the right place and at the right time. The introduction of new technologies provides additional tools to refine and enhance decision-making in precision agriculture. The widespread digitization of agri-food production facilitates the precise utilization of inputs, leading to positive environmental impacts. The overall performance of precision agriculture is elevated with the adoption of new technologies.

Machine Learning (ML) plays a pivotal role in optimizing various aspects, including error correction, security, traffic prediction, data processing, and resource management. Low latency, a crucial requirement for precision agriculture, can be achieved through local computation in the network, employing edge or fog computing. Sensor nodes, traditionally constrained in energy and computation, benefit from lightweight processing at the edge, ensuring interoperability and isolating nodes from the core network. The data generated by sensors in precision agriculture is voluminous, underscoring the significance of big data for real-time analysis and context awareness. Blockchain technology enhances precision agriculture solutions, providing efficiency, speed, and security. Emerging technologies like Software Defined Network (SDN) and Artificial Intelligence (AI) contribute to real-time network virtualization, improving resource and energy utilization and enhancing application responsiveness (Ullah et al., 2022).

Motivated by the pressing need for food amid declining natural resources and the widespread deployment of IoT devices, precision agriculture becomes crucial in addressing agricultural challenges. Current trends indicate a shortfall in achieving Zero Hunger by 2030, with the potential number of people affected by hunger exceeding 840 million. Agriculture stands out as a critical and dominant application of IoT, with specific requirements varying across crop types and application-specific use cases. The integration of AI, ML, nanotechnology, energy-efficient frameworks, and sensor network development is transforming the entire precision agriculture landscape. Energy harvesting and remote sensing revolutionize network management, design, and architecture within the precision agriculture framework. Blockchain and big data contribute to rapid analysis and diagnosis with enhanced privacy features (Waheed et al., 2022).

7.2. Challenges and Ethical Considerations

Precision farming emerges as a superior option for mitigating production costs and maximizing productivity, addressing the limitations inherent in traditional agricultural practices. A range of cutting-edge techniques enhances precision farming, leveraging nanotechnology processes that provide precise control at the nanometer scale. Nanotechnology proves instrumental in disease detection and treatment, offering the potential to minimize damage to non-target plant tissues and enhance plant nutrient capabilities. The application of machine-centric biological nanotechnology holds transformative potential, capable of yielding remarkable results when implemented judiciously. However, the realization of benefits from nanotechnology and molecular biology in agriculture faces challenges, particularly the existence of unfilled positions due to a shortage of highly skilled personnel.

While nanotechnology introduces active ingredients with nanoscale dimensions, resulting in increased efficacy and penetration, the full spectrum of its impacts, both positive and negative, is not fully understood and may only become apparent over time. As technology advances exponentially, the adoption of nanotechnology raises concerns about job obsolescence, potentially rendering certain roles obsolete. This shift prompts ethical considerations, particularly regarding accountability in the use of nanotechnology, including the involvement of mechanical nanobots. Addressing these ethical questions becomes essential to ensure responsible and sustainable integration of nanotechnology in precision farming, balancing its potential benefits with the need for careful oversight and ethical considerations in its implementation (Tassadduq et al., 2022).

7.3. Future Directions

The rapid advancements in nanotechnology have ushered in a new era of possibilities, particularly in the application of engineered nanoparticles with diameters below 100 nm. This has opened up vast potential in disease diagnosis, drug delivery, gene therapy, and medical imaging. However, the comprehensive exploration of the potential health risks associated with human exposure to nanoparticles has often been overlooked. Engineered nanoparticles can instigate the generation of reactive oxygen species (ROS) and contribute to inflammatory-mediated diseases, underscoring the imperative for in-depth studies to assess exposure risks (Hussain et al., 2023b). The manipulation of nanoscale particles for specific applications introduces challenges, as the absence of established exposure standards raises concerns about potential adverse effects on the human body. The extended persistence of nanoparticles due to their small size, despite a higher surface-to-volume ratio that enhances reactivity, poses potential risks upon entry into the biochemical system. The life cycle of nanomaterials, from production and application to waste disposal, entails implicit effects on both human life and the environment. Beyond human health considerations, the impact on agriculture and aquatic ecosystems raises alarms, warranting a comprehensive examination of the repercussions (Ali et al., 2023). Societal apprehensions, including concerns about unregulated weapons manufacturing and privacy infringements through advanced surveillance systems, have contributed to a negative public perception of nanotechnology. Drawing parallels with the challenges faced by genetically modified organisms (GMOs), it becomes evident that prioritizing prudence is paramount. To navigate the ethical, health, and safety concerns surrounding nanotechnology, the establishment of regulatory bodies and the implementation of universal policies emerge as essential measures. Fostering awareness and conducting rigorous research are crucial components of a proactive approach to identify and mitigate potential hazards, ensuring that the promising field of nanotechnology evolves responsibly and sustainably . (Waheed et al., 2023).

The future trajectory of nanotechnologies in agriculture holds great promise, particularly in enhancing crop resilience to diverse environmental challenges. Emerging nanotechnologies dedicated to bolstering crop resilience are poised to revolutionize the agricultural landscape. Collaborative efforts among scientists, researchers, and agricultural stakeholders are crucial to navigating this uncharted territory and harnessing the full potential of nanotechnological interventions (Altug et al., 2022; Sinha et al., 2023c). Research priorities in this domain are pivotal for understanding the nuanced interactions between nanomaterials and plants, deciphering the optimal delivery mechanisms, and ensuring the long-term ecological sustainability of these applications. The integration of nanotechnologies into agriculture demands a multidisciplinary approach, merging expertise in nanoscience, agronomy, and environmental science. Ongoing endeavors to unravel the molecular and physiological responses of crops to nanomaterials pave the way for precision agriculture and sustainable food production. As we venture into this realm, careful consideration of ethical, environmental, and regulatory aspects remains paramount to steer these advancements toward a future where nanotechnologies contribute significantly to global food security while minimizing potential risks (Bashir et al., 2023b; Bhattacharya et al., 2023).

CONCLUSION

The imperative to increase agricultural production as a linchpin for achieving sustainable development goals, including eradicating poverty and hunger, conserving natural resources, and addressing climate change, has propelled nanotechnology into the forefront of modern farming practices. Nanotechnology stands out as a potent tool for fostering "precision farming," offering efficient solutions for sustainable agriculture, optimized nutrient utilization, and environmental preservation. Notably, the advent of nanotechnology-enabled products, exemplified by "smart nano-delivery capsules," presents novel avenues for controlled and gradual release of various agents such as pesticides, herbicides, fertilizers, and micronutrients, tailored to the specific needs of plants triggered by environmental stimuli. This comprehensive review delves into the fundamental properties, application advantages, and future trajectories of nanotechnology in agricultural economics, with a focus on promoting crop growth and soil reclamation aligned with sustainable development goals. The analysis encompasses a thorough examination of obstacles associated with the use of agri-nanotech products, considering their life cycle assessment in soil and plants (Rimpika et al., 2023). The review places particular emphasis on assessing the impact of diverse nanotech products on crop growth in comparison to traditional market technologies. Furthermore, it explores the merits of nanobiochar and nanobubbles as promising natural nanotech products to enhance crop physiology, mitigate environmental pollution, and facilitate carbon sequestration in the soil. The discussion extends to the technical intricacies of converting agricultural biomass wastes into cost-effective biofuels, while also addressing regulatory frameworks, legislative policies, and economic considerations to fortify public awareness and acceptance of nanotechnologies in agro-environmental domains. By elucidating the opportunities and challenges inherent in these areas, the review provides insights to actively propel the development and implementation of nano-enabled products in agriculture and associated industries (Goyal et al., 2023). The significance of nanotechnology in bolstering crop resilience lies in its capacity to address challenges such as climate change, resource depletion, and environmental pollution. The scientific evidence presented highlights the positive impacts of nanomaterials on seed germination, plant growth, and overall crop yield. As a call to action, the widespread adoption of nanotechnology in agriculture is advocated, emphasizing the need for collaborative efforts among researchers, policymakers, and practitioners to overcome existing barriers. Furthermore, there is an imperative for continued research to unravel the full potential of nanotechnology in addressing emerging agricultural challenges, ensuring food security, and achieving sustainable development goals. By embracing nanotechnology, agriculture can stride towards a more efficient, resilient, and environmentally conscious future (Yadav et al., 2023).

REFERENCES

Ahmad, N., Naeem, M., Ali, H., Alabbosh, K. F., Hussain, H., Khan, I., Siddiqui, S. A., Khan, A. A., & Iqbal, B. (2023). From challenges to solutions: The impact of melatonin on abiotic stress synergies in horticultural plants via redox regulation and epigenetic signaling. *Scientia Horticulturae*, *321*, 112369. doi:10.1016/j.scienta.2023.112369

Al-Sahli, S. A., Al-Otibi, F., Alharbi, R. I., Amina, M., & Al Musayeib, N. M. (2024). Silver nanoparticles improve the fungicidal properties of Rhazya stricta decne aqueous extract against plant pathogens. *Scientific Reports*, *14*(1), 1297. doi:10.1038/s41598-024-51855-5 PMID:38221517

Ali, H., Muhammad, Z., Majeed, M., Aziz, R., Khan, A., Mangrio, W. M., Abdo, H. G., Almohamad, H., & Al Dughairi, A. A. (2023). Vegetation diversity pattern during spring season in relation to topographic and edaphic variables in sub-tropical zone. *Botanical Studies*, *64*(1), 25. Advance online publication. doi:10.1186/s40529-023-00398-5 PMID:37716923

Altug, H., Oh, S. H., Maier, S. A., & Homola, J. (2022). Advances and applications of nanophotonic biosensors. In Nature Nanotechnology (Vol. 17, Issue 1, pp. 5–16). Nature Research. doi:10.1038/s41565-021-01045-5

Arif, U., Bhatti, K. H., Ajaib, M., Wagay, N. A., Majeed, M., Zeb, J., Hameed, A., & Kiani, J. (2021). Ethnobotanical indigenous knowledge of Tehsil Charhoi, District Kotli, Azad Jammu and Kashmir, Pakistan. *Ethnobotany Research and Applications*, *22*. Advance online publication. doi:10.32859/era.22.50.1-24

Arshad, F., Waheed, M., Harun, N., Fatima, K., Khan, B. A., Fatima, K., Abbas, Z., Jabeen, S., & Majeed, M. (2022). Indigenous farmer's perception about fodder and foraging species of Semi-arid lowlands of Pakistan: A case study of District Kasur, Pakistan. *Taiwania*, *67*(4), 510–523. doi:10.6165/tai.2022.67.510

Azeem, M., Pirjan, K., Qasim, M., Mahmood, A., Javed, T., Muhammad, H., Yang, S., Dong, R., Ali, B., & Rahimi, M. (2023). Salinity stress improves antioxidant potential by modulating physio-biochemical responses in Moringa oleifera Lam. *Scientific Reports*, *13*(1), 2895. Advance online publication. doi:10.1038/s41598-023-29954-6 PMID:36807545

Aziz, A., Anwar, M. M., Majeed, M., Fatima, S., Mehdi, S. S., Mangrio, W. M., Elbouzidi, A., Abdullah, M., Shaukat, S., Zahid, N., Mahmoud, E. A., Casini, R., Yessoufou, K., & Elansary, H. O. (2023). Quantifying Landscape and Social Amenities as Ecosystem Services in Rapidly Changing Peri-Urban Landscape. *Land (Basel)*, *12*(2), 477. Advance online publication. doi:10.3390/land12020477

Balasubramaniam, T., Shen, G., Esmaeili, N., & Zhang, H. (2023). Plants' Response Mechanisms to Salinity Stress. In Plants (Vol. 12, Issue 12). MDPI. doi:10.3390/plants12122253

Banerjee, S., Mondal, S., Islam, J., Sarkar, R., Saha, B., & Sen, A. (2024). Rhizospheric nano-remediation salvages arsenic genotoxicity: Zinc-oxide nanoparticles articulate better oxidative stress management, reduce arsenic uptake, and increase yield in Pisum sativum (L.). *The Science of the Total Environment*, *913*, 169493. doi:10.1016/j.scitotenv.2023.169493 PMID:38151134

Bashir, S. M., Altaf, M., Hussain, T., Umair, M., Majeed, M., Mangrio, W. M., Khan, A. M., Gulshan, A. B., Hamed, M. H., Ashraf, S., Amjad, M. S., Bussmann, R. W., Abbasi, A. M., Casini, R., Alataway, A., Dewidar, A. Z., Al-Yafrsi, M., Amin, M. H., & Elansary, H. O. (2023a). Vernacular Taxonomy, Cultural and Ethnopharmacological Applications of Avian and Mammalian Species in the Vicinity of Ayubia National Park, Himalayan Region. *Biology (Basel)*, *12*(4), 609. Advance online publication. doi:10.3390/biology12040609 PMID:37106809

Bhatia, D., Paul, S., Acharjee, T., & Ramachairy, S. S. (2024). Biosensors and their widespread impact on human health. *Sensors International*, *5*, 100257. Advance online publication. doi:10.1016/j.sintl.2023.100257

Bhattacharya, R., Bhattacharya, R., Majeed, M., Bhandari, S., Aziz, R., Sinha, D., & Niyogi, S. S. (2023). Environmental Rehabilitation of Industrial Waste Dumping Site. In *Biohydrometallurgical Processes* (pp. 44–89). CRC Press. doi:10.1201/9781003451457-4

Chaupoo, A. S., Dhakad, A., Beese, S., Kumar, A., Kumar, P., Pandey, S. K., & Kurdekar, A. K. (2023). Transforming Horticulture: The Influence of Nanotechnology on Crop Enhancement. *International Journal of Environment and Climate Change*, *13*(11), 3381–3400. doi:10.9734/ijecc/2023/v13i113512

Cheng, C., Fu, J., Su, H., & Ren, L. (2023). Recent Advancements in Agriculture Robots: Benefits and Challenges. In Machines (Vol. 11, Issue 1). MDPI. doi:10.3390/machines11010048

Dehghanian, Z., Asgari Lajayer, B., Biglari Quchan Atigh, Z., Nayeri, S., Ahmadabadi, M., Taghipour, L., Senapathi, V., Astatkie, T., & Price, G. W. (2023). Micro (nano) plastics uptake, toxicity and detoxification in plants: Challenges and prospects. *Ecotoxicology and Environmental Safety*, *268*, 115676. doi:10.1016/j.ecoenv.2023.115676 PMID:37979355

Dorjee, L., Gogoi, R., Kamil, D., Kumar, R., Mondal, T. K., Pattanayak, S., & Gurung, B. (2023). Essential oil-grafted copper nanoparticles as a potential next-generation fungicide for holistic disease management in maize. *Frontiers in Microbiology*, *14*, 1204512. Advance online publication. doi:10.3389/fmicb.2023.1204512 PMID:37485521

Fadiji, A. E., Yadav, A. N., Santoyo, G., & Babalola, O. O. (2023). Understanding the plant-microbe interactions in environments exposed to abiotic stresses: An overview. *Microbiological Research*, *271*, 127368. doi:10.1016/j.micres.2023.127368 PMID:36965460

Fujita, H., Ohta, S., Nakamura, N., Somiya, M., & Horie, M. (2023). Progress of Endogenous and Exogenous Nanoparticles for Cancer Therapy and Diagnostics. In Genes (Vol. 14, Issue 2). MDPI. doi:10.3390/genes14020259

Ghazaryan, K. A., Harutyunyan, A. S., Khachatryan, H. E., Singh, A., Minkina, T. M., Rajput, V. D., & Movsesyan, H. S. (2023). Study of tolerance and phytodesalination potential of wheat, oat, emmer, and barley for sustainable saline agriculture. *Applied Ecology and Environmental Research*, *21*(5), 4853–4882. doi:10.15666/aeer/2105_48534882

Ghazaryan, K. A., Movsesyan, H. S., Khachatryan, H. E., & Ghazaryan, N. P. (2018). Geochemistry of Potentially Toxic Trace Elements in Soils of Mining Area: A Case Study from Zangezur Copper and Molybdenum Combine, Armenia. *Bulletin of Environmental Contamination and Toxicology*, *101*(6), 732–737. doi:10.1007/s00128-018-2443-0 PMID:30229278

Goyal, V., Rani, D., Mehrotra, S., Deng, C., & Wang, Y. (2023). Unlocking the Potential of Nano-Enabled Precision Agriculture for Efficient and Sustainable Farming. In Plants (Vol. 12, Issue 21). Multidisciplinary Digital Publishing Institute (MDPI). doi:10.3390/plants12213744

Grillo, R., Fraceto, L. F., Amorim, M. J. B., Scott-Fordsmand, J. J., Schoonjans, R., & Chaudhry, Q. (2021). Ecotoxicological and regulatory aspects of environmental sustainability of nanopesticides. *Journal of Hazardous Materials*, *404*, 124148. doi:10.1016/j.jhazmat.2020.124148 PMID:33059255

Haq, S. M., Tariq, A., Li, Q., Yaqoob, U., Majeed, M., Hassan, M., Fatima, S., Kumar, M., Bussmann, R. W., Moazzam, M. F. U., & Aslam, M. (2022a). Influence of Edaphic Properties in Determining Forest Community Patterns of the Zabarwan Mountain Range in the Kashmir Himalayas. *Forests*, *13*(8), 1214. Advance online publication. doi:10.3390/f13081214

Hassan, M., Haq, S. M., Ahmad, R., Majeed, M., Sahito, H. A., Shirani, M., Mubeen, I., Aziz, M. A., Pieroni, A., Bussmann, R. W., Alataway, A., Dewidar, A. Z., Al-Yafrsi, M., Elansary, H. O., & Yessoufou, K. (2022). Traditional Use of Wild and Domestic Fauna among Different Ethnic Groups in the Western Himalayas—A Cross Cultural Analysis. *Animals (Basel)*, *12*(17), 2276. Advance online publication. doi:10.3390/ani12172276 PMID:36077997

Hassan, M., Haq, S. M., Majeed, M., Umair, M., Sahito, H. A., Shirani, M., Waheed, M., Aziz, R., Ahmad, R., Bussmann, R. W., Alataway, A., Dewidar, A. Z., El-Abedin, T. K. Z., Al-Yafrsi, M., Elansary, H. O., & Yessoufou, K. (2022). Traditional Food and Medicine: Ethno-Traditional Usage of Fish Fauna across the Valley of Kashmir: A Western Himalayan Region. *Diversity*, *14*(6), 455. Advance online publication. doi:10.3390/d14060455

Hooshmand, S., Kassanos, P., Keshavarz, M., Duru, P., Kayalan, C. I., Kale, İ., & Bayazit, M. K. (2023). Wearable Nano-Based Gas Sensors for Environmental Monitoring and Encountered Challenges in Optimization. In Sensors (Basel, Switzerland) (Vol. 23, Issue 20). doi:10.3390/s23208648

Hussain, S., Raza, A., Abdo, H. G., Mubeen, M., Tariq, A., Nasim, W., Majeed, M., Almohamad, H., & Al Dughairi, A. A. (2023a). Relation of land surface temperature with different vegetation indices using multi-temporal remote sensing data in Sahiwal region, Pakistan. *Geoscience Letters*, *10*(1), 33. Advance online publication. doi:10.1186/s40562-023-00287-6

Jamil, M. D., Waheed, M., Akhtar, S., Bangash, N., Chaudhari, S. K., Majeed, M., Hussain, M., Ali, K., & Jones, D. A. (2022). Invasive Plants Diversity, Ecological Status, and Distribution Pattern in Relation to Edaphic Factors in Different Habitat Types of District Mandi Bahauddin, Punjab, Pakistan. *Sustainability (Basel)*, *14*(20), 13312. Advance online publication. doi:10.3390/su142013312

Karnwal, A., Dohroo, A., & Malik, T. (2023). Unveiling the Potential of Bioinoculants and Nanoparticles in Sustainable Agriculture for Enhanced Plant Growth and Food Security. In *BioMed Research International* (Vol. 2023). Hindawi Limited., doi:10.1155/2023/6911851

Khan, A. M., Li, Q., Saqib, Z., Khan, N., Habib, T., Khalid, N., Majeed, M., & Tariq, A. (2022). MaxEnt Modelling and Impact of Climate Change on Habitat Suitability Variations of Economically Important Chilgoza Pine (Pinus gerardiana Wall.) in South Asia. *Forests*, *13*(5), 715. Advance online publication. doi:10.3390/f13050715

Khoja, A. A., Haq, S. M., Majeed, M., Hassan, M., Waheed, M., Yaqoob, U., Bussmann, R. W., Alataway, A., Dewidar, A. Z., Al-Yafrsi, M., Elansary, H. O., Yessoufou, K., & Zaman, W. (2022). Diversity, Ecological and Traditional Knowledge of Pteridophytes in the Western Himalayas. *Diversity*, *14*(8), 628. Advance online publication. doi:10.3390/d14080628

Li, Y., Zhang, P., Li, M., Shakoor, N., Adeel, M., Zhou, P., Guo, M., Jiang, Y., Zhao, W., Lou, B. Z., & Rui, Y. (2023). Application and mechanisms of metal-based nanoparticles in the control of bacterial and fungal crop diseases. In Pest Management Science (Vol. 79, Issue 1, pp. 21–36). John Wiley and Sons Ltd. doi:10.1002/ps.7218

Liao, Z., Fan, J., Lai, Z., Bai, Z., Wang, H., Cheng, M., Zhang, F., & Li, Z. (2023). Response network and regulatory measures of plant-soil-rhizosphere environment to drought stress. In *Advances in Agronomy* (Vol. 180, pp. 93–196). Academic Press Inc. doi:10.1016/bs.agron.2023.03.002

Lin, H., Jie, B., Ye, J., Zhai, Y., Luo, Z., Shao, G., Chen, R., Zhang, X., & Yang, Y. (2023). Recent advance of macroscopic metal-organic frameworks for water treatment: A review. *Surfaces and Interfaces*, *36*, 102564. doi:10.1016/j.surfin.2022.102564

Lo Presti, D., Di Tocco, J., Massaroni, C., Cimini, S., De Gara, L., Singh, S., Raucci, A., Manganiello, G., Woo, S. L., Schena, E., & Cinti, S. (2023). Current understanding, challenges and perspective on portable systems applied to plant monitoring and precision agriculture. *Biosensors & Bioelectronics*, *222*, 115005. doi:10.1016/j.bios.2022.115005 PMID:36527829

Majeed, M., Bhatti, K. H., & Amjad, M. S. (2021). Impact of climatic variations on the flowering phenology of plant species in jhelum district, punjab, pakistan. *Applied Ecology and Environmental Research*, *19*(5), 3343–3376. doi:10.15666/aeer/1905_33433376

Majeed, M., Bhatti, K. H., Amjad, M. S., Abbasi, A. M., Bussmann, R. W., Nawaz, F., Rashid, A., Mehmood, A., Mahmood, M., Khan, W. M., & Ahmad, K. S. (2020). Ethno-veterinary uses of poaceae in Punjab, Pakistan. *PLoS ONE, 15*(11). doi:10.1371/journal.pone.0241705

Majeed, M., Bhatti, K. H., Pieroni, A., Sõukand, R., Bussmann, R. W., Khan, A. M., Chaudhari, S. K., Aziz, M. A., & Amjad, M. S. (2021). Gathered wild food plants among diverse religious groups in Jhelum District, Punjab, Pakistan. *Foods*, *10*(3), 594. Advance online publication. doi:10.3390/foods10030594 PMID:33799901

MajeedM.Hayat BhattiK.Shoaib AmjadM.Mehmood AbbasiA.RashidA.Nawaz Muhammad Nawaz ShareefF.MehmoodA.MahmoodM. (2020). *Ethno-veterinary practices of Poaceae taxa in Punjab, Pakistan*. doi:10.21203/rs.3.rs-15532/v1

Majeed, M., Khan, A. M., Habib, T., Anwar, M. M., Sahito, H. A., Khan, N., & Ali, K. (2022). Vegetation analysis and environmental indicators of an arid tropical forest ecosystem of Pakistan. *Ecological Indicators*, *142*, 109291. Advance online publication. doi:10.1016/j.ecolind.2022.109291

Majeed, M., Lu, L., Anwar, M. M., Tariq, A., Qin, S., El-Hefnawy, M. E., El-Sharnouby, M., Li, Q., & Alasmari, A. (2023). Prediction of flash flood susceptibility using integrating analytic hierarchy process (AHP) and frequency ratio (FR) algorithms. *Frontiers in Environmental Science*, *10*, 1037547. Advance online publication. doi:10.3389/fenvs.2022.1037547

Majeed, M., Lu, L., Haq, S. M., Waheed, M., Sahito, H. A., Fatima, S., Aziz, R., Bussmann, R. W., Tariq, A., Ullah, I., & Aslam, M. (2022). Spatiotemporal Distribution Patterns of Climbers along an Abiotic Gradient in Jhelum District, Punjab, Pakistan. *Forests*, *13*(8), 1244. Advance online publication. doi:10.3390/f13081244

Majeed, M., Tariq, A., Anwar, M. M., Khan, A. M., Arshad, F., Mumtaz, F., Farhan, M., Zhang, L., Zafar, A., Aziz, M., Abbasi, S., Rahman, G., Hussain, S., Waheed, M., Fatima, K., & Shaukat, S. (2021). Monitoring of land use–Land cover change and potential causal factors of climate change in Jhelum district, Punjab, Pakistan, through GIS and multi-temporal satellite data. *Land (Basel)*, *10*(10), 1026. Advance online publication. doi:10.3390/land10101026

Majeed, M., Tariq, A., Anwar, M. M., Khan, A. M., Arshad, F., Mumtaz, F., Farhan, M., Zhang, L., Zafar, A., Aziz, M., Abbasi, S., Rahman, G., Hussain, S., Waheed, M., Fatima, K., & Shaukat, S. (2021). Monitoring of Land Use–Land Cover Change and Potential Causal Factors of Climate Change in Jhelum District, Punjab, Pakistan, through GIS and Multi-Temporal Satellite Data (pp. 129-147).https://www.mdpi.com/books/book/6906-land-use-land-cover-and-natural-hazards-interactions-changes-and-impacts

Majeed, M., Tariq, A., Haq, S. M., Waheed, M., Anwar, M. M., Li, Q., Aslam, M., Abbasi, S., Mousa, B. G., & Jamil, A. (2022). A Detailed Ecological Exploration of the Distribution Patterns of Wild Poaceae from the Jhelum District (Punjab), Pakistan. *Sustainability (Basel)*, *14*(7), 3786. Advance online publication. doi:10.3390/su14073786

Malik, S., Muhammad, K., & Waheed, Y. (2023). Nanotechnology: A Revolution in Modern Industry. In Molecules (Vol. 28, Issue 2). MDPI. doi:10.3390/molecules28020661

Manimegalai, S., Vickram, S., Deena, S. R., Rohini, K., Thanigaivel, S., Manikandan, S., Subbaiya, R., Karmegam, N., Kim, W., & Govarthanan, M. (2023). Carbon-based nanomaterial intervention and efficient removal of various contaminants from effluents – A review. *Chemosphere*, *312*, 137319. doi:10.1016/j.chemosphere.2022.137319 PMID:36410505

Mubarok, S., Nuraini, A., Hamdani, J. S., Suminar, E., Kusumiyati, K., Budiarto, R., Lestari, F. W., Rahmat, B. P. N., & Ezura, H. (2024). Antioxidative response of parthenocarpic tomato, iaa9-3 and iaa9-5, under heat stress condition. *Plant Physiology and Biochemistry*, *207*, 108333. doi:10.1016/j.plaphy.2024.108333 PMID:38181640

Muhammad, M., Basit, A., Majeed, M., Shah, A. A., Ullah, I., Mohamed, H. I., Khan, A., & Ghanaim, A. M. (2024). Bacterial pigments and their applications. In *Bacterial Secondary Metabolites* (pp. 277–298). Elsevier., doi:10.1016/B978-0-323-95251-4.00019-3

Muhammad, M., Waheed, A., Wahab, A., Majeed, M., Nazim, M., Liu, Y. H., Li, L., & Li, W. J. (2024). Soil salinity and drought tolerance: An evaluation of plant growth, productivity, microbial diversity, and amelioration strategies. In *Plant Stress* (Vol. 11). Elsevier B.V. doi:10.1016/j.stress.2023.100319

Naveed, Majeed, Jabeen, Hanif, Naveed, Saleem, & Khan. (2024a). Plant Disease Diagnosis with Artificial Intelligence (AI). Microbial Data Intelligence and Computational Techniques for Sustainable Computing (pp 217–234). Springer. doi:10.1007/978-981-99-9621-6_15

Naveed, Zahid, Fatima, Saleem, Majeed, Abid, Javed, Wazir, & Qasim. (2024b). Disease Diagnosis with Artificial Intelligence (AI). In Microbial Data Intelligence and Computational Techniques for Sustainable Computing (pp 235–258). Springer. doi:10.1007/978-981-99-9621-6_16

Naveed, Memon, Abdullah, Makhdoom, Azeem, Mehmood, Salahuddin, Rajpoot, & Majeed. (2024c). Use Cases and Future Aspects of Intelligent Techniques in Microbial Data Analysis. In Microbial Data Intelligence and Computational Techniques for Sustainable Computing (pp 259–280). Springer. doi:10.1007/978-981-99-9621-6_17

Nguyen, N. T. T., Nguyen, L. M., Nguyen, T. T. T., Tran, U. P. N., Nguyen, D. T. C., & Van Tran, T. (2023). A critical review on the bio-mediated green synthesis and multiple applications of magnesium oxide nanoparticles. *Chemosphere*, *312*, 137301. doi:10.1016/j.chemosphere.2022.137301 PMID:36410506

PandaP.ChakrobortyS.BabuS.KrishnaN. (2023). *The Use of Silver Nanoparticles in Environmental Remediation*. doi:10.20944/preprints202301.0330.v1

Pandey, P., Tripathi, A., Dwivedi, S., Lal, K., & Jhang, T. (2023). Deciphering the mechanisms, hormonal signaling, and potential applications of endophytic microbes to mediate stress tolerance in medicinal plants. In Frontiers in Plant Science (Vol. 14). Frontiers Media SA. doi:10.3389/fpls.2023.1250020

Payne, P. R., Finlay-Smits, S., Small, B., Cave, V., & Kean, J. (2023). What's that bug? Community participation in biosecurity in Mount Maunganui, New Zealand. *Biological Invasions*, *25*(2), 593–610. doi:10.1007/s10530-022-02937-w

Periakaruppan, R., Vanathi, P., & Vidhya, D. (2023). Management of soil nutrient deficiency by nanometal oxides. *Nanometal Oxides in Horticulture and Agronomy*, 291–320. doi:10.1016/B978-0-323-91809-1.00003-2

Pramanik, P. K. D., Solanki, A., Debnath, A., Nayyar, A., El-Sappagh, S., & Kwak, K. S. (2020). Advancing Modern Healthcare with Nanotechnology, Nanobiosensors, and Internet of Nano Things: Taxonomies, Applications, Architecture, and Challenges. *IEEE Access : Practical Innovations, Open Solutions*, *8*, 65230–65266. doi:10.1109/ACCESS.2020.2984269

Rajput, V. D., Singh, A., Minkina, T., Rawat, S., Mandzhieva, S., Sushkova, S., Shuvaeva, V., Nazarenko, O., Rajput, P., Komariah, Verma, K. K., Singh, A. K., Rao, M., & Upadhyay, S. K. (2021). Nano-Enabled Products: Challenges and Opportunities for Sustainable Agriculture. *Plants*, *10*(12), 2727. doi:10.3390/plants10122727 PMID:34961197

Rehman, A., Aziz, A., Anwar, M. M., Majeed, M., Albanai, J. A., Almohamad, H., & Abdo, H. G. (2023). Quantifying the impacts of urbanization on urban green, evidences from Maga City, Lahore Pakistan. *Discover Sustainability*, *4*(1), 48. Advance online publication. doi:10.1007/s43621-023-00169-z

Rimpika, A., Anushi, Manasa, S., Sharma, S., Thakur, A., Shilpa, & Sood, A.. (2023). An Overview of Precision Farming. *International Journal of Environment and Climate Change, 13*(12), 441–456. doi:10.9734/ijecc/2023/v13i123701

Rohilla, D., Chaudhary, S., & Umar, A. (2021). An Overview of Advanced Nanomaterials for Sensor Applications. *Engineering and Science, 16*, 47–70. doi:10.30919/es8d552

Sangeeta Dash, E., Yadav, A. L., Bora, R., & Kumar Pandey, A. (2023). *Research Trends in Agriculture Science* (Vol. I). www.bhumipublishing.com

Singh, A., Margaryan, G., Harutyunyan, A., S. Movsesyan, H., Khachatryan, H., Rajput, V., . . . Ghazaryan, K. (2024b). Advancing Agricultural Resilience in Ararat Plain, Armenia: Utilizing Biogenic Nanoparticles and Biochar under Saline Environments to Optimize Food Security and Foster European Trade. *Egyptian Journal of Soil Science, 64*(2). doi:10.21608/ejss.2024.257000.1703

Singh, A., Rajput, V., Varshney, A., Sharma, R., Ghazaryan, K., Minkina, T., . . . El-Ramady, H. (2024a). Revolutionizing Crop Production: Nanoscale Wonders - Current Applications, Advances, and Future Frontiers. *Egyptian Journal of Soil Science, 64*(1). https://doi.org/ doi:10.21608/ejss.2023.246354.1684

Singh, A., Rajput, V. D., Pandey, D., Sharma, R., Ghazaryan, K., & Minkina, T. (2023b). Nano Zinc-Enabled Strategies in Crops for Combatting Zinc Malnutrition in Human Health. *Frontiers in Bioscience (Landmark Edition), 28*(8), 158. doi:10.31083/j.fbl2808158 PMID:37664935

Singh, A., Rajput, V. D., Sharma, R., Ghazaryan, K., & Minkina, T. (2023a). Salinity stress and nanoparticles: Insights into antioxidative enzymatic resistance, signaling, and defense mechanisms. *Environmental Research, 235*, 116585. doi:10.1016/j.envres.2023.116585 PMID:37437867

Singh, A., Rajput, V. D., Tawaha, A. R. M., Al Zoubi, O. M., Habeeb, T., Rawat, S., Ghazaryan, K., & Minkina, T. (2023d). A Review on Crop Responses to Nanofertilizers for Mitigation of Multiple Environmental Stresses. *Ecological Engineering & Environmental Technology, 24*(7), 280–296. doi:10.12912/27197050/169313

Singh, A., Rajput, V. D., Varshney, A., Ghazaryan, K., & Minkina, T. (2023c). Small Tech, Big Impact: Agri-nanotechnology Journey to Optimize Crop Protection and Production for Sustainable Agriculture. *Plant Stress, 10*, 100253. doi:10.1016/j.stress.2023.100253

Singh, A., Sengar, R. S., Rajput, V. D., Al-Ghzawi, A. L., Shahi, U. P., Ghazaryan, K., Minkina, T., Tawaha, A. R. M. A., Al Zoubi, O. M., & Habeeb, T. (2024c). Impact of Salinity Stress and Zinc Oxide Nanoparticles on Macro and Micronutrient Assimilation: Unraveling the Link between Environmental Factors and Nutrient Uptake. *Journal of Ecological Engineering, 25*(2), 1–9. doi:10.12911/22998993/172947

Singh, A., Sharma, R., Rawat, S., Singh, A. K., Rajput, V. D., Fedorov, Y., . . . Chaplygin, V. (2022). Nanomaterial-plant interaction: Views on the pros and cons. *Toxicity of Nanoparticles in Plants*, 47–68. doi:10.1016/B978-0-323-90774-3.00015-5

Singh, B. K., Delgado-Baquerizo, M., Egidi, E., Guirado, E., Leach, J. E., Liu, H., & Trivedi, P. (2023). Climate change impacts on plant pathogens, food security and paths forward. In Nature Reviews Microbiology (Vol. 21, Issue 10, pp. 640–656). Nature Research. doi:10.1038/s41579-023-00900-7

Sinha, D., Maurya, A. K., Abdi, G., Majeed, M., Agarwal, R., Mukherjee, R., Ganguly, S., Aziz, R., Bhatia, M., Majgaonkar, A., Seal, S., Das, M., Banerjee, S., Chowdhury, S., Adeyemi, S. B., & Chen, J. T. (2023c). Integrated Genomic Selection for Accelerating Breeding Programs of Climate-Smart Cereals. In Genes (Vol. 14, Issue 7). Multidisciplinary Digital Publishing Institute (MDPI). doi:10.3390/genes14071484

Swain, R., Sahoo, S., Behera, M., & Rout, G. R. (2023). Instigating prevalent abiotic stress resilience in crop by exogenous application of phytohormones and nutrient. In *Frontiers in Plant Science* (Vol. 14). Frontiers Media S.A. doi:10.3389/fpls.2023.1104874

Tariq, A., Mumtaz, F., Majeed, M., & Zeng, X. (2023). Spatio-temporal assessment of land use land cover based on trajectories and cellular automata Markov modelling and its impact on land surface temperature of Lahore district Pakistan. *Environmental Monitoring and Assessment*, *195*(1), 114. Advance online publication. doi:10.1007/s10661-022-10738-w PMID:36385403

Tassadduq, S. S., Akhtar, S., Waheed, M., Bangash, N., Nayab, D. E., Majeed, M., Abbasi, S., Muhammad, M., Alataway, A., Dewidar, A. Z., Elansary, H. O., & Yessoufou, K. (2022). Ecological Distribution Patterns of Wild Grasses and Abiotic Factors. *Sustainability (Basel)*, *14*(18), 11117. Advance online publication. doi:10.3390/su141811117

Ullah, I., Aslam, B., Shah, S. H. I. A., Tariq, A., Qin, S., Majeed, M., & Havenith, H. B. (2022). An Integrated Approach of Machine Learning, Remote Sensing, and GIS Data for the Landslide Susceptibility Mapping. *Land (Basel)*, *11*(8), 1265. Advance online publication. doi:10.3390/land11081265

Vardumyan, H., Singh, A., Rajput, V., Minkina, T., El-Ramady, H., & Ghazaryan, K. (2024). Additive-Mediated Phytoextraction of Copper-Contaminated Soils Using Medicago lupulina L. *Egyptian Journal of Soil Science*, *64*(2), 599–618. doi:10.21608/ejss.2024.266169.1714

Victoria, J., Tripathi, S., Prakash, V., Tiwari, K., Mahra, S., Sharma, A., Rana, S., Kandhol, N., Sahi, S., Tripathi, D. K., & Sharma, S. (2024). Encapsulated nanopesticides application in plant protection: Quo vadis? *Plant Physiology and Biochemistry*, *206*, 108225. doi:10.1016/j.plaphy.2023.108225 PMID:38147708

Waheed, M., Arshad, F., Majeed, M., Fatima, S., Mukhtar, N., Aziz, R., Mangrio, W. M., Almohamad, H., Al Dughairi, A. A., Al-Mutiry, M., & Abdo, H. G. (2022). Community Structure and Distribution Pattern of Woody Vegetation in Response to Soil properties in Semi-Arid Lowland District Kasur Punjab, Pakistan. *Land (Basel)*, *11*(12), 2145. Advance online publication. doi:10.3390/land11122145

Waheed, M., Arshad, F., Majeed, M., Haq, S. M., Aziz, R., Bussmann, R. W., Ali, K., Subhan, F., Jones, D. A., & Zaitouny, A. (2023). *Potential distribution of a noxious weed (Solanum viarum Du-nal), current status, and future invasion risk based on MaxEnt modeling*. Geology, Ecology, and Landscapes. doi:10.1080/24749508.2023.2179752

Wang, S. (2020). A Comprehensive Review of Wearable Applications and Material Construction. *Ozean Journal of Applied Sciences*, *10*(6), 364–408. doi:10.4236/ojapps.2020.106027

Yadav, A., Yadav, K., Ahmad, R., & Abd-Elsalam, K. A. (2023). Emerging Frontiers in Nanotechnology for Precision Agriculture: Advancements, Hurdles and Prospects. *Agrochemicals*, *2*(2), 220–256. doi:10.3390/agrochemicals2020016

Yuvaraj, M., Sathya Priya, R., Jagathjothi, N., Saranya, M., Suganthi, N., Sharmila, R., Cyriac, J., Anitha, R., & Subramanian, K. S. (2023). Silicon nanoparticles (SiNPs): Challenges and perspectives for sustainable agriculture. *Physiological and Molecular Plant Pathology*, *128*, 102161. doi:10.1016/j.pmpp.2023.102161

Zhou, J., Liu, G., Guo, Z., Wang, M., Qi, C., Chen, G., Huang, X., Yan, S., & Xu, D. (2023). Stimuli-responsive pesticide carriers based on porous nanomaterials: A review. *Chemical Engineering Journal*, *455*, 140167. doi:10.1016/j.cej.2022.140167

Chapter 9
Nanotechnology and Agricultural Sustainability:
Environmental Impacts and Benefits

Mona Kumari
G.D. Goenka University, India

Bhavana Tomar
G.D. Goenka University, India

Prashant Singh
https://orcid.org/0009-0008-9667-3598
Mandsaur University, India

Shiv Singh Tomar
G.D. Goenka University, India

Tirunima Patle
https://orcid.org/0000-0003-3268-6118
Shri Vaishnav Vidyapeeth Vishwavidyalaya, India

Sneh Singh Parihar
ITM University, Gwalior, India

ABSTRACT

The intricate interplay between finite natural resources—such as land, water, and soil—and the world's expanding population highlights the need for agricultural expansion to take operational, environmental, and economic sustainability into account. Beyond addressing poverty and hunger, agricultural progress requires overcoming existing challenges. Sustainable growth in agriculture relies on embracing cutting-edge techniques, notably the integration of nanotechnology. Nanotechnology, characterized by its tiny dimensions, plays a crucial role in coordinating agricultural processes. Numerous potential advantages of this cutting-edge technology include decreased agricultural inputs, increased nutrient absorption from the soil at the nanoscale, and improved food safety and quality. Nonetheless, there are obstacles to overcome in the fields of agriculture, food, and natural resources when applying nanotechnology, such as concerns about vulnerability, sustainability, human health, and general well-being. Nanotechnology emerges as an essential tool for strengthening the agricultural and food industry through the use of innovative nano instruments. These advancements enable rapid disease diagnostics, boost plants' nutrient absorption capabilities, and address various critical concerns. In order to properly monitor product and nutrient levels while boosting productivity without compromising soil and water integrity, there is interest in using nanotechnology in agriculture for specialized applications such as nano fertilizers and nano insecticides. Moreover, nanotechnology proves crucial in safeguarding against insect pests and

DOI: 10.4018/979-8-3693-1890-4.ch009

microbial diseases, serving as sensors to monitor soil health and crop quality, fostering significant progress in agriculture. The strategic application of nanotechnology offers a glimmer of hope for resolving modern issues like climate change, food security, and sustainability. Researchers, leading agricultural innovation, actively explore the transformative potential of nanotechnology, recognizing its pivotal role in reshaping agricultural practices. It is essential to acknowledge that the field is continually evolving, and the dynamic nature of information requires ongoing exploration and adaptation in the pursuit of sustainable and resilient agricultural systems.

INTRODUCTION

Agriculture, as a pivotal and stable sector, serves as the primary source of raw materials for the food and feed industries. The problems associated with finite natural resources—such as soil, water, and land—as well as the world's population growth underscore the critical need for agricultural expansion that is both environmentally sound and economically feasible. Economic growth-induced variations in agricultural nutrient balances underscore the significance of soil fertility development, particularly in developing countries. The imperative for agricultural development becomes even more apparent when addressing poverty and hunger, requiring proactive measures (Prasad et al., 2017).

The intersection of agricultural poverty and nutritional processes emphasizes the adoption of new technologies to enhance agricultural output (Smit and Yunlong, 1994). New knowledge is the foundation of current paradigms in food and nutritional security, and agricultural growth is closely related to many other areas, including social inclusion, health, energy, climate change, ecological processes, natural resources, and governance. According to Thornhill et al. (2016), sustainable food production depends on environmental performance and the active participation of food chain ecosystems as agriculture strives for recovery. The catalyst role of nanotechnology in providing a range of nanomaterials, including nanopesticides and nanofertilizers, emerges as a promising avenue (Altabbaa et al., 2023).

Adopting cutting-edge methods like nanotechnology is unquestionably necessary for agriculture to thrive sustainably. Glancing back in time, Feynman's 1959 talk titled "Plenty of Room at the Bottom" signaled the beginning of nanoprocessing, and in 1974, Professor Norio Tanaguchi coined the word nanotechnology (Bulovic et al., 2004). Subsequently, nanotechnology witnessed significant development, with recent instruments enabling the precise consideration and isolation of nanomaterials (Gibney, 2015; Bonnell and Huey, 2001). The world market size of nanotechnology, projected to grow from approximately US$ 110.6 billion in 2002 to US$ 891.1 billion in 2015, underscores its expanding influence (Prasad et al., 2017).

The demand for nanomaterials, particularly in packaging, has increased in the food and agriculture sector, with Europe, Asia, and North America contributing significantly. Nanotechnology, at one billionth of a meter (nm), exhibits altered properties due to changes in atoms, developing magnetic power (Rajput et al.,2022). Particularly in biological processes, the increased surface area that comes from nanoparticles' reduced size leads to enhanced reactivity (Pokropivny et al., 2007; Prasad, 2014). The versatile nature of nanotechnology is further enhanced by its unique electrical, plasmonic, and optical features associated with quantum confinement effects (Sun, 2007; Aziz et al., 2015; Prasad et al., 2016).

Biological host molecule detection, monitoring, and diagnosis technologies that are quick, dependable, and affordable have become more important to the agricultural industry in the twenty-first century. Producing nanomaterials from plant systems through green nanotechnology has become a safe and effective substitute that lowers greenhouse gas emissions and waste (Prasad, 2014; Kandasamy and Prema, 2015).

Even while agrochemicals like fertilizers and pesticides are essential to contemporary agriculture, there are some drawbacks to using them, including the possibility of water contamination and residues on food products. High-tech farming methods could be revolutionized by the creation of a system with engineered smart nanotools, minimizing environmental effect and increasing yields in terms of quality and quantity (Sekhon, 2014; Liu and Lal, 2015). Biosensors have great promise as they utilize the benefits of nanotechnology. According to Fraceto et al. (2016) and Sertova (2015), nanomaterials enable the downsizing of smart devices into tiny, compact forms, such as nanosensors, which are very important for biochemical analysis. The role of nanotechnology in detecting mycotoxins in various foods is particularly noteworthy (Sertova, 2015).

Advancements in nanotechnology, characterized by refined instruments for isolating and considering nanomaterials, peaked in 2016, with around 14,000 documents and 2,707 matching patents listed in global patent databases (Gibney, 2015; Prasad et al., 2017). Given these insights, further research is imperative to elucidate the applications of nanotechnology, ensuring the desired quality in agricultural production.

2. NANOPARTICLES AND THEIR ROLES

a) Carbon Nanotubes

With a resemblance to a coiled two-dimensional graphene sheet, carbon nanotubes are a unique type of carbon. They are primarily classified as either single-walled or multi-walled nanotubes (SWNTs or MWNTs), and both have an amazing tensile strength of about 200 GPa. These tubes find applications in reinforcing composites and nanoelectromechanical systems. Structurally, the nanotube system consists of seamlessly wrapped graphitic layers forming cylinders. Phthalamic nanoparticles (NPS) or quantum dots (QDs) intended for plant protein labeling have been incorporated recently (Chahiye et al., 2014).

The diverse properties of carbon nanotubes, encompassing mechanical, electronic, thermal, and optical aspects, intricately link to their geometrical dimensions, especially their diameter, which is typically around 1mm for most SWNTs (Singh et al.,2024) . This diameter strongly correlates with synthesis techniques, the combination of q & a bonds, and electron orbital rehybridization. By utilizing these qualities, opportunities arise to create a range of nanodevices with special conductivity, optical, and thermal qualities that can be used in farming and promote sustainable farming practices (Raliya et al., 2013). Carbon nanotubes can serve as carriers for delivering agrochemicals and other substances, reducing the environmental release of chemicals and minimizing harm to the plant's tissues (Hazirostamlo et al., 2015).

b) Quantum Dots

The features of semiconductor quantum dots (QDs) include high quantum yield, molar extinction coefficients, large effective excitation, and broad absorption spectra with narrow, symmetric fluorescence spectra covering the ultraviolet to near-infrared. They are ideal candidates for quantum confinement of materials and charge carriers, have a size-tunable band energy, and show strong resistance to photochemical degradation (Androvitsanlas et al., 2016). Additionally, QDs work as photocatalysts in the chemical reaction of water to hydrogen catalyzed by light, providing a pathway toward solar fuel production

(Konstantatos and Sargent, 2009). Little doses of quantum dots (QDs) have been shown to be cytotoxic to seeds and seedlings. QDs can also be utilized in plant systems for live imaging, which enables the verification of established physiological processes (Das et al., 2015).

c) Nanorods

Nanorods represent multifunctional Plasmonic substances capable of effectively coupling with sensing phenomena and regulating energy size tunably. These substances can be used with MEMS to produce particular field reactions (Bulovic et al., 2004). Significant physiological alterations and phytotoxicity have been demonstrated by gold nanorods in watermelon plants, especially at high concentrations (Wan et al., 2014). Furthermore, Nima et al. (2014) report that they exhibit the capacity to convey the auxin growth regulator 2,4-D, which has a substantial impact on the regulation of tobacco cell culture growth.

d) Microsand Momcat Encapsulation

According to Rodriguez et al. (2016), encapsulation is the process of enveloping an object in a coating or embedding it in a homogeneous or heterogeneous matrix, producing capsules with a variety of advantageous characteristics. According to Ozdemir and Kemerliy (2016), the benefits of encapsulating techniques include controlled release, precise targeting, and protection of materials or objects from unfavorable surroundings. Regarding the dimensions and form of the capsule, various encapsulation technologies are given. A macroscale capsule is produced by macro-encapsulation or coating, whereas a micro- and nanoscale particle is produced by micro- and nanoencapsulation (Ozdemir and Kemerli, 2016). Within a chamber with a polymeric membrane enclosing an inner liquid core, chemicals are contained within vesicular systems known as nanocapsules. The transport of drugs, protection, enhanced bioavailability of food ingredients or nutraceuticals, food fortification, and the self-healing of different materials are all areas of great interest for micro and nanoparticles (NPs). Given the effectiveness of nanoencapsulating several medications, including anti-inflammatory chemicals and pesticides, these technologies have a great deal of promise in the field of agriculture (Haolong et al., 2011).

e) Nanoemulsion

Nanoscale droplets, which are smaller than 100 nm, make up nanoemulsions in an oil/water mixture (Anton and Vandamme, 2011). Nanoemulsions and microemulsions might not differ fundamentally, but their physical characteristics can change greatly from those of emulsions made on a larger scale (Gupta et al., 2016). Elastic modulus, Laplace pressure, and the surface area to volume ratio are all significantly impacted by the size of the droplets. Notably, Nanoemulsions show noticeably greater values for these parameters compared to ordinary emulsions. Additionally, most nanoemulsions are optically transparent, providing technical advantages for incorporation into beverages. The formulation of nanoemulsions necessitates high energy, requiring specialized devices capable of generating severe shear strain, like high pressure homogenizers or ultrasonic generators. A "low-energy" technique for creating nanoemulsions was presented by Tadros et al. (2004). It involved the interaction of two liquid phases: an aqueous phase, which includes pure water, and a homogeneous phase that included a lipophilic phase, hydrophilic surfactant, and possibly a solvent, polymers, or drug (Gupta et al., 2016). This approach seems to be the easiest and doesn't call for any energy-intensive special gadgets(Vardumyan et al.,2024). Because it can regulate materials to increase productivity, nanotechnology is essential to the advancement of sustainable agricultural development (Gruere, 2012; Mukhopadhyay, 2014). Additionally, it contributes to monitoring water quality, identifying suitable pesticides, and maintaining soil health—essential components for sustainable agriculture. The multifaceted attributes and applications of nanomaterials make it challenging to provide a comprehensive a risk assessment for their health and the environment (Parsad et al., 2014). The degree of particle aggregation, shape, charge, surface structure, behavior, and chemical composition of manufactured nanoparticles (NPs) all influence their toxicity (Ion et al., 2010). In order to achieve sustainable development, research on nanotechnology in agriculture is currently crucial. Interesting applications in the field of agri-food have been reported for nanotubes, fullerenes, biosensors, controlled delivery systems, and nanofiltration (Sabir et al., 2014). Apart from aiding in the direct degradation of waste and dangerous substances, nanotechnology furthermore aims to enhance the efficiency of the microbes already involved in this procedure. To finish these procedures, three typical methods are used: phytoremediation (plants), mycoremediation (fungi and mushrooms), and bioremediation (beneficial bacteria). Therefore, by using microorganisms to effectively and environmentally remove heavy metals from soil and water, agricultural bioremediation supports sustainable remediation technologies and addresses and restores the natural state of soil (Dixit et al., 2015).

Agriculture With Nanotechnology for Sustainable Development

Through the control of nutrients, nanotechnology is essential for increasing agricultural output and making a substantial contribution to the field's sustainable development. It actively monitors water quality and pesticides, aiding in the overall health of agricultural ecosystems. However, evaluating the health and environmental risks associated with nanomaterials poses challenges due to their diverse attributes and activities. Engineered nanoparticles exhibit varying levels of toxicity influenced by factors such as chemical makeup, form, charge, surface structure, behavior, and the way particles clump or separate(Singh et al.,2022).

Nanotechnology research is now essential to the agricultural industry, as seen by applications in fields such as agri-food, where nanotubes, fullerenes, biosensors, controlled delivery systems, and nanofiltration are notable examples. Resource management, plant medicine delivery, and maintaining soil fertility have all benefited from this approach. Risk assessment, food processing, packaging systems, biomass, and agricultural waste are all included in its scope. Environmental monitoring, especially for soil and water contamination, heavily relies on nanosensors, which include biosensors, electrochemical sensors, optical sensors, and nano-detection devices(Rajput et al.,2021).

In addition to directly accelerating the breakdown of waste and hazardous materials, nanomaterials also improve the effectiveness of the microbes that break down trash. Bioremediation strategies, incorporating living organisms like beneficial microbes, plants (phytoremediation), fungi, and mushrooms (mycoremediation), contribute to breaking down or removing toxins from agricultural soil and water(Singh et al.,2023). This approach aids in sustainable remediation technologies, effectively removing heavy metals through microorganisms. In this light, the investigation of nano-nano interactions as a means of eliminating harmful elements from agricultural soil and advancing sustainability is an interesting topic.

Nano Fertilizers: Nano fertilizers have just been introduced to the market, especially for use in agriculture. However, the formulation of major chemical agricultural fertilizers is yet to be shaped by the advancements in nanotechnology.

Table 1. Some commercial products of nano fertilizers

Commercial Product	Content	Company
Nano – GroTM	PGRS and immunity enhancers	Agro- nanotechnology Corp, FL, US.
Nano-Green	Extract of Crops	Nano-green Science, Inc. India
Nano-Ag Answer	Microorganisms	Urth Agriculture, Ca, Us
Biozar Nano Fertilizer	Multiple organic materials	JU Agric Sciences, Ret. Ltd. Janakpuri, New Delhi, India
TAG Nano (NPK, Phos, Zinc, Cal, etc) Fertilizer	Protein- lactis gluconate, humic acid	Tropical Agrosystem, India (P) Ltd., India.

(b) Nanopesticides

The incorporation of nanomaterials for the enhancement and protection of food production stands as a swiftly advancing frontier. In the realm of agriculture, where the impact of insect pests on landscapes and crops is considerable, nanomaterials emerge as potential transformative agents for efficient management of host pathogens and pests (Khota et al., 2012). A paradigm shift has occurred with the advent of nano-encapsulated pesticide formulations, which offer features such slow-release mechanisms that enhance stability, permeability, solubility, and specificity (Bhattacharyya et al., 2016). These benefits stem from encapsulating active ingredients, safeguarding them against premature degradation, and enhancing their efficacy in pest control over an extended period(Ghazaryan et al.,2023). The development of nano-encapsulated pesticides not only decreases the volume of pesticides but also reduces human exposure, promoting an environmentally conscious approach to crop production and protection (Nuruzzanan et al., 2016).

According to Grillo et al. (2016), there is thus a pressing need to develop a safe and effective pesticide delivery method in order to increase world food production while reducing harmful environmental effects on biological systems. Notably, certain chemical companies, like Switzerland's Syngenta, openly endorse nano-scale pesticides, marketing them as "micro-encapsulated pesticides." Syngenta's products, like Penncap-M, Subdue MAXX, Osprey's Chyella, and Karate ZEON alongside micro-encapsulated pesticides from BASF, demonstrate a proactive approach in addressing the nanoscale dimension (Gouin, 2004). In Australia, Syngenta also promotes products like Prino MAXX, Banner MAXX, Subdue MAXX, indicating a global interest in the adoption of nanopesticides(Ghazaryan et al.,2018).

The Impact of Nanoparticles on Ecotoxicology

Engineered nanoparticles (NPs) are becoming more and more prevalent in the environment as a result of the development of nanotechnologies. The toxicity of regularly used industrial NPs is being thoroughly studied in order to protect plant life and human health from potential negative consequences (Du et al., 2017). Metal nanoparticles (NPs) exhibit antibacterial, anticandidal, and antifungal properties, but their toxicity is dependent on parameters such as their solubility and binding selectivity to biological sites (Patra and Baek, 2017). Mold > yeast > Gram-negative > Gram-positive microorganisms exhibit a sensitivity hierarchy to the nanotoxic effects of nanoparticles, which are closely linked to the features of the targeted cell wall. According to (Hoffmann et al.,2007), some nanoparticles (NPs) exhibit photochemical activity, which enables them to directly transfer electrons from oxygen to create superoxide radicals when exposed to light.

Nanotechnology's Influence on the Food Industry

Nanoscale biosensors are pivotal in pathogen detection and diagnosis, while nanotechnology introduces innovative solutions like nanoscale filtration systems to enhance food texture modification. In the realm of nanotechnology, biosensors interact with food surfaces, preserving the gloss and colors of food. In addition to serving as tag sensors, magnetic nanocomposites are useful for controlled release, intelligent packaging, nano-additives, and nano-coding paper and plastic materials for identity and authenticity. Food quality is monitored and controlled, portable DNA/protein chips, in-situ sensors, label and packaging sensing, and nutraceutical administration are all components of quality food testing (Singh et al.,2023).

Impact on Food Processing

Food manufacturing makes extensive use of nanotechnology, utilizing nanocarrier systems to transport nutrients and supplements as well as integrating organic nanosized additives. Foods like milk and meat include nanoscale proteins, and some nutrients—particularly vitamins—are encapsulated to be effectively absorbed by the digestive system and delivered into the bloodstream. These NPs are used to fortify some foods and drinks without changing their flavor or appearance. NP emulsions improve the consistency and smoothness of ice cream, as seen by developments from businesses that supply omega-3 fatty acids, such as KD Pharma.

Packaging and Labeling of Food

Ensuring freshness, safety, quality, and flavor are critical in the food industry. Product packaging and labeling are requirements for producers all the way through the supply chain. A major problem is to build intelligent packaging that offers useful information. Packaging materials that detect food oxidation have recently been introduced; these materials are incorporated with nanosensors. As oxidation occurs, nanoparticle-based sensors indicate color changes, offering insights into the nature of packaged foods. Successful applications include packaging for milk and meat (Bumbudsanpharoke and Ko, 2015). Nanostructured nanoparticles serve as effective barriers, preventing gas diffusion and are employed in beverage bottles, minimizing CO2 loss, reducing packaging weight, extending shelf life, and providing other advantages.

Impact on the Growth of Cultivated Plants, Ecotoxicology, and Sustainability

Both fields and aquatic habitats are favorable to the growth of cultivated plants, which are vital to the food chain. Notably, plants like Lepidium sativum and Pisum sativum can accumulate particular iron oxide nanoparticles (Fe3O4-NPs), a magnetic form of iron ore, highlighting the function of nanoparticles in the natural ecosystem (Abbas et al., 2016). Understanding the ecotoxicological aspects of agricultural fields containing nanomaterials is crucial. If nanomaterials in agricultural plants prove non-toxic, it could potentially lead to increased agricultural crop production.

Nano Biosensors

Nano biosensors are greatly influenced by the characteristics of nanoscale materials, both physically and chemically. Sagadevan and Periasamy (2014) stress the importance of integrating nanomaterials and utilizing advanced signal transduction technologies to enhance biosensor sensitivity and overall performance. The field of nanobiosensors is growing quickly in response to the increasing need for quick, precise, and affordable biosensor systems in a variety of industries, such as healthcare, agriculture, food and beverage, process industries, genomic analysis, environmental monitoring, defense, and security. As of right now, research and development on nanotechnology-based biosensors is just getting started (Fogel and Limson, 2016).

Positive Impacts

Nanotechnology brings about positive effects by strengthening materials and devices, improving the efficiency of monitoring devices, minimizing energy consumption, reducing material waste, mitigating toxicity, remediating environmental pollution, and promoting renewable energy production. These positive outcomes are clearly seen in the usage of nanomaterials and nanoparticles, which lead to notable resource conservation and improved productivity in manufacturing and energy-related applications. Energy consumption, pollution, and greenhouse gas emissions are just a few of the issues that nanotechnology has the capacity to address. As a result, it has the potential to have positive effects on the economy, society, and environment (Singh et al.,2023).

A number of global issues, such as the availability of alternative energy, protecting human rights to clean water, protecting wildlife, cleaning up brownfields, and lowering the prevalence of disease worldwide, could be helped by nanotechnology. Significant environmental benefits are possible thanks to it, including improved pollution detection and removal, more efficient and cleaner industrial processes, high-precision manufacturing to minimize waste production, abundant and clean power from efficient solar cells, reduced dependency on large industrial plants, and environmental damage repair.

Energy Consumption: The integration of graphene into coating materials, requiring only a single layer, eliminates the need for multifunctional film coatings. Applications include applying graphene-based coatings to wind turbine blades or airplane bodies, resulting in weight savings and increased efficiency.

Cost Saving on Materials: Advances in nanotechnology can decrease costs in alternative energy methods, such as hybrid automobiles.

Reduced Waste on Raw Materials: The simultaneous usage of raw materials can be more effectively achieved by conducting large-scale testing on a smaller scale. Chemical processes can be made more efficient and productive by using catalysts or reagents at the nanoscale.

Environmental Monitoring and Protection: Advanced nanotechnology has led to the creation of detectors for faster and more accurate nuclear leak detection, as seen at the Fukushima Daiichi Nuclear Power Plant. These detectors, considered among the best in Washington, can detect even trace amounts of radiation.

Affordable and Clean Energy: Nanotechnology-based prototype solar panels turn sunlight into power more efficiently than conventional designs. Batteries use nanotechnology, which may lead to advancements in fuel cell catalysts and materials for storing hydrogen. Nanotechnology can improve energy generation, conversion, and storage for fuel cells, solar cells, thermo-to-electric systems, biomass energy, hydrogen storage, secondary batteries, supercapacitors, and thermal storage fluids by increasing surface area and decreasing storage unit weight.

Ensuring the Human Right to Clean Water: Nanotechnology provides cost-effective water purification through rapid, low-cost impurity detection. Magnetic interactions using ultra-small rust can aid in removing arsenic from drinking water.

Pollution Reduction and Environmental Progress: Nanotechnology facilitates the production of lighter cars and machinery that require less fuel, as well as the development of alternative fuel and energy sources. It also enables materials for detecting and cleaning up environmental contaminants. The University College of Dublin (UCD) Center for Bio-Nano Interactions (CBNI) studies the impact of nanoparticles dispersed in the environmental milieu, examining how this interaction affects organic stability, dispersibility, environmental fate, and behavior.

Global Disease Burden Reduction: Advances in healthcare through enhanced diagnosis and treatment will contribute to a global improvement in personal well-being.

Economic Crisis Mitigation: It is anticipated that investments in nanotechnology will boost economic growth and assist the emergence of auxiliary sectors including product marketing, waste disposal and recycling, and intellectual property and liability disputes (Singh et al.,2024).

Possible Negative Effects

Nanoparticles, which have larger surface areas compared to bulk materials, may pose increased risks to both human health and the environment, surpassing those associated with their larger counterparts. The uncertainty surrounding the specific nanotechnology products or nanomaterials, including their shape, size, and chemical compositions, contributes to various adverse environmental effects. Notably, carbon nanotubes exhibit health effects similar to asbestos when interacting with lung epithelia (Poland et al., 2009). However, the consequences of Carbon Nano Tubes (CNTs) at low concentrations in soils, sediments, and natural waters remain unclear.

The variability in the shape and size of nanomaterials plays a crucial role in determining their toxicity, further complicated by the lack of standardized information and characterization methods for airborne nanoparticles, hindering environmental protection efforts. The chemical structure is a critical factor in nanomaterial toxicity, with even slight changes in chemical functional groups significantly altering their properties. A comprehensive risk assessment, covering exposure risk, toxicological analysis, transport risk, persistence risk, transformation risk, and recyclability, is essential at every stage of nanotechnology.

As nanoscale materials continue to decrease in size, detecting toxic nanoparticles in waste becomes increasingly challenging, posing a potential environmental threat. Various mechanisms through which nanoparticles or nanomaterials can become toxic and impact the surrounding environment include hydrophobic and hydrophilic nanoparticles. Ongoing research explores the use of $TiO2$ powder coatings to mitigate weathering effects on composite materials.

Environmental Consequences and Uncertainties

The risks associated with current consumer and industrial applications of nanotechnology remain uncertain and unquantified. CBNI's research aims to comprehend the potential impact of nanoparticles on ecosystems, spanning plants, animals, and microorganisms. Environmental effects and risks linked to nanotechnology exhibit limitations and inconsistencies. Nanotechnology has the potential to cause significant environmental harm due to its high energy requirements for creating nanoparticles, which in turn requires a significant amount of energy; dispersing hazardous and persistent nanomaterials that harm the environment; reducing rates of recovery and recycling; unclear environmental consequences at other stages of the life cycle; and raising concerns about a technical and labor shortage.

Graphene, celebrated for its exceptional properties benefiting the environment and the economy, also poses potential environmental risks. The toxicity of graphene remains unknown, complicating its removal from waste. Interactions between graphene and materials/biological systems in the environment may yield unexpected outcomes. Although polymer nanocomposites benefit from graphene's thermal conductivity and fire retardancy, contamination of graphene with other materials during the manufacturing process can provide a fire risk.

Future Perspectives

Looking ahead to a future where living and non-living components coexist harmoniously to ensure food stability, several important considerations come to the forefront:

- Conducting a thorough life-cycle analysis of nanofoods/nanomaterials is crucial, accompanied by the establishment of a comprehensive database. International collaboration is essential to formulate policies, ideas, and regulations guiding the responsible application of this knowledge.
- The synergy of biosensors with nanotechnology can play a pivotal role in managing insect pests and maintaining the quality of agricultural food products. Intelligent food packaging, incorporating nanosensors, provides consumers with real-time information about the status of specific food items.
- Attaining excellence in the application of these technologies requires the creation of new communication channels and discussions on databases. Involving various stakeholders, including consumers, researchers, authorities, and industrial sectors, is vital for deliberating on the impacts of this technology on human life, science, and the economy.
- Nanotechnology is poised to open innovative and economically viable approaches to ensuring human safety and nutrition.

CONCLUSION

Recognizing agriculture's pivotal role as the primary source of sustenance for both humans and animals, there is an urgent need to integrate contemporary knowledge into agricultural practices. Embracing agricultural modernization and adopting new technologies, particularly nanotechnology, has the potential to significantly enhance agricultural production. This involves utilizing nanoformulations for agrochemicals, such as nanofertilizers and nanopesticides, to optimize crop yield.

Incorporating nanosensors in crop protection facilitates early disease detection and identification of agrochemical residues. The recent use of nanomaterials provides an environmentally friendly solution to mitigate the impact of plant pathogens.

Innovative methods, like the green synthesis of nanoparticles using plant extracts, offer notable advantages in terms of simplicity, convenience, and eco-friendliness, requiring minimal reaction time. Nanomaterials prepared through these environmentally sustainable approaches have the potential to elevate agriculture by improving processes such as fertilization, plant growth regulation, targeted delivery of pesticides to specific sites, wastewater treatment, and enhanced nutrient absorption in plants. Moreover, these methods contribute to minimizing the use of harmful chemicals that can pollute the environment. The creation of an extensive database, an alarm system, and international collaboration for regulation and legislation are essential to fully realizing the advantages of nanotechnological processes in agriculture.

REFERENCES

Abbas, S. S., Haneef, M., Lohani, M., Tabassum, H., & Khan, A. F. (2016). Nanomaterials used as a plants growth enhancer: An update. *International Journal of Pharmaceutical Sciences Review and Research*, *5*, 17–23.

Altabbaa, S., Mann, N. A., Chauhan, N., Utkarsh, K., Thakur, N., & Mahmoud, G. A. E. (2023). Era connecting nanotechnology with agricultural sustainability: Issues and challenges. *Nanotechnology for Environmental Engineering*, *8*(2), 481–498. doi:10.1007/s41204-022-00289-3

Androvitsaneas, P., Young, A. B., Schneider, C., Maier, S., Kamp, M., Höfling, S., Knauer, S., Harbord, E., Hu, C. Y., Rarity, J. G., & Oulton, R. (2016). Charged quantum dot micropillar system for deterministic light-matter interactions. *Physical Review. B*, *93*(24), 241409. doi:10.1103/PhysRevB.93.241409

Anton, N., & Vandamme, T. F. (2011). Nanoemulsions and micro emulsions: Clarification. *International Journal of Pharmaceutics*, *377*, 142–147. doi:10.1016/j.ijpharm.2009.05.014 PMID:19454306

Aziz, N., Faraz, M., Pandey, R., Shakir, M., Fatma, T., Varma, A., Barman, I., & Prasad, R. (2015). Facile algae-derived route to biogenic silver nanoparticles: Synthesis, antibacterial, and photocatalytic properties. *Langmuir*, *31*(42), 11605–11612. doi:10.1021/acs.langmuir.5b03081 PMID:26447769

Berekaa, M. M. (2015). Nanotechnology in food industry; advances in food processing, packaging and food safety. *International Journal of Current Microbiology and Applied Sciences*, *4*(5), 345–357.

. Bhattacharyya, A., Duraisamy, P., Govindarajan, M., Buhroo, A. A., & Prasad, R. (2016). Nanobiofungicides: emerging trend in insect pest control. *Advances and applications through fungal nanobiotechnology*, 307-319.

Bonnell, D. A., & Huey, B. D. (2001). Basic principles of scanning probe microscopy. *Scanning Probe Microscopy & Spectroscopy: Theory, Techniques, and Applications*, 7-42.

Bulovic, V., Mandell, A., & Perlman, A. (2004). Molecular memory device. *US 20050116256 A*, *1*, 388.

Bulovic, V., Mandell, A., & Perlman, A. (2004). *U.S. Patent No. 6,781,868*. Washington, DC: U.S. Patent and Trademark Office.

Bumbudsanpharoke, N., & Ko, S. (2015). Nano-food packaging: An overview of market, migration research, and safety regulations. *Journal of Food Science*, *80*(5), R910–R923. doi:10.1111/1750-3841.12861 PMID:25881665

Campbell, B. M., Thornton, P., Zougmoré, R., Van Asten, P., & Lipper, L. (2014). Sustainable intensification: What is its role in climate smart agriculture? *Current Opinion in Environmental Sustainability*, *8*, 39–43. doi:10.1016/j.cosust.2014.07.002

Chahine, N. O., Collette, N. M., Thomas, C. B., Genetos, D. C., & Loots, G. G. (2014). Nanocomposite scaffold for chondrocyte growth and cartilage tissue engineering: Effects of carbon nanotube surface functionalization. *Tissue Engineering. Part A*, *20*(17-18), 2305–2315. doi:10.1089/ten.tea.2013.0328 PMID:24593020

Das, S., Wolfson, B. P., Tetard, L., Tharkur, J., Bazata, J., & Santra, S. (2015). Effect of N-acetyl cysteine coated CdS: Mn/ZnS quantum dots on seed germination and seedling growth of snow pea (Pisum sativum L.): imaging and spectroscopic studies. *Environmental Science. Nano*, *2*(2), 203–212. doi:10.1039/C4EN00198B

Dixit, R., Wasiullah, X., Malaviya, D., Pandiyan, K., Singh, U. B., Sahu, A., Shukla, R., Singh, B., Rai, J., Sharma, P., Lade, H., & Paul, D. (2015). Bioremediation of heavy metals from soil and aquatic environment: An overview of principles and criteria of fundamental processes. *Sustainability (Basel)*, *7*(2), 2189–2212. doi:10.3390/su7022189

Dixit, R., Wasiullah, X., Malaviya, D., Pandiyan, K., Singh, U. B., Sahu, A., Shukla, R., Singh, B., Rai, J., Sharma, P., Lade, H., & Paul, D. (2015). Bioremediation of heavy metals from soil and aquatic environment: An overview of principles and criteria of fundamental processes. *Sustainability (Basel)*, *7*(2), 2189–2212. doi:10.3390/su7022189

Du, W., Tan, W., Peralta-Videa, J. R., Gardea-Torresdey, J. L., Ji, R., Yin, Y., & Guo, H. (2017). Interaction of metal oxide nanoparticles with higher terrestrial plants: Physiological and biochemical aspects. *Plant Physiology and Biochemistry*, *110*, 210–225. doi:10.1016/j.plaphy.2016.04.024 PMID:27137632

Feynman, R. P. (1996). *No ordinary genius: the illustrated Richard Feynman*. WW Norton & Company.

Floros, J. D., Newsome, R., Fisher, W., Barbosa-Cánovas, G. V., Chen, H., Dunne, C. P., German, J. B., Hall, R. L., Heldman, D. R., Karwe, M. V., Knabel, S. J., Labuza, T. P., Lund, D. B., Newell-McGloughlin, M., Robinson, J. L., Sebranek, J. G., Shewfelt, R. L., Tracy, W. F., Weaver, C. M., & Ziegler, G. R. (2010). Feeding the world today and tomorrow: the importance of food science and technology: an IFT scientific review. *Comprehensive Reviews in Food Science and Food Safety*, *9*(5), 572–599. doi:10.1111/j.1541-4337.2010.00127.x PMID:33467827

Fogel, R., & Limson, J. (2016). Developing biosensors in developing countries: South Africa as a case study. *Biosensors (Basel)*, *6*(1), 5. doi:10.3390/bios6010005 PMID:26848700

Fraceto, L. F., Grillo, R., de Medeiros, G. A., Scognamiglio, V., Rea, G., & Bartolucci, C. (2016). Nanotechnology in agriculture: Which innovation potential does it have? *Frontiers in Environmental Science*, *4*, 20. doi:10.3389/fenvs.2016.00020

Ghazaryan, K., Harutyunyan, A., Khachatryan, H., Singh, A., Minkina, T., Rajput, V., & Movsesyan, H. (2023). Study of tolerance and phytodesalination potential of wheat, oat, emmer, and barley for sustainable saline agriculture. *Applied Ecology and Environmental Research*, *21*(5).

Ghazaryan, K. A., Movsesyan, H. S., Khachatryan, H. E., & Ghazaryan, N. P. (2018). Geochemistry of potentially toxic trace elements in soils of mining area: A case study from Zangezur Copper and Molybdenum Combine, Armenia. *Bulletin of Environmental Contamination and Toxicology*, *101*(6), 732–737. doi:10.1007/s00128-018-2443-0 PMID:30229278

Gibney, E., Cami, J., & Hedberg, J. (2015). Buckyballs in space solve 100-year-old riddle. *Nature*, *523*, 296–297.

Gouin, S. (2004). Microencapsulation: Industrial appraisal of existing technologies and trends. *Trends in Food Science & Technology*, *15*(7-8), 330–347. doi:10.1016/j.tifs.2003.10.005

Grillo, R., Abhilash, P. C., & Fraceto, L. F. (2016). Nanotechnology applied to bio-encapsulation of pesticides. *Journal of Nanoscience and Nanotechnology*, *16*(1), 1231–1234. doi:10.1166/jnn.2016.12332 PMID:27398594

Gruère, G. P. (2012). Implications of nanotechnology growth in food and agriculture in OECD countries. *Food Policy*, *37*(2), 191–198. doi:10.1016/j.foodpol.2012.01.001

Gruère, G. P. (2012). Implications of nanotechnology growth in food and agriculture in OECD countries. *Food Policy*, *37*(2), 191–198. doi:10.1016/j.foodpol.2012.01.001

Gupta, A., Eral, H. B., Hatton, T. A., & Doyle, P. S. (2016). Nanoemulsions: Formation, properties and applications. *Soft Matter*, *12*(11), 2826–2841. doi:10.1039/C5SM02958A PMID:26924445

Hajirostamlo, B., Mirsaeedghazi, N., Arefnia, M., Ali, M., & Shariati, E. A. F. (2015). The role of research and development in agriculture and its dependent concepts in agriculture. *Science and Engineering*, *4*(1), 79–81.

Hoffmann, M., Hotze, E. M., & Wiesner, M. R. (2007). Reactive oxygen species generation on nanoparticulate material. Mcgraw-Hill Professional.

Ion, A. C., Ion, I., Culetu, A., & Gherase, D. (2010). Carbon-based nanomaterials. Environmental applications. Romania.

Kah, M., & Hofmann, T. (2014). Nanopesticide research: Current trends and future priorities. *Environment International*, *63*, 224–235. doi:10.1016/j.envint.2013.11.015 PMID:24333990

Kandasamy, S., & Prema, R. S. (2015). Methods of synthesis of nano particles and its applications. Academic Press.

Khot, L. R., Sankaran, S., Maja, J. M., Ehsani, R., & Schuster, E. W. (2012). Applications of nanomaterials in agricultural production and crop protection: A review. *Crop Protection (Guildford, Surrey)*, *35*, 64–70. doi:10.1016/j.cropro.2012.01.007

Konstantatos, G., & Sargent, E. H. (2009). Solution-processed quantum dot photodetectors. *Proceedings of the IEEE*, *97*(10), 1666–1683. doi:10.1109/JPROC.2009.2025612

Li, H., Yang, Y., Wang, Y., Wang, C., Li, W., & Wu, L. (2011). Self-assembly and ion-trapping properties of inorganic nanocapsule-surfactant hybrid spheres. *Soft Matter*, *7*(6), 2668–2673. doi:10.1039/c0sm01044h

Liu, R., & Lal, R. (2015). Potentials of engineered nanoparticles as fertilizers for increasing agronomic productions. *The Science of the Total Environment*, *514*, 131–139. doi:10.1016/j.scitotenv.2015.01.104 PMID:25659311

Mukhopadhyay, S. S. (2014). Nanotechnology in agriculture: Prospects and constraints. *Nanotechnology, Science and Applications*, 63–71. doi:10.2147/NSA.S39409 PMID:25187699

Nima, Z. A., Lahiani, M. H., Watanabe, F., Xu, Y., Khodakovskaya, M. V., & Biris, A. S. (2014). Plasmonically active nanorods for delivery of bio-active agents and high-sensitivity SERS detection in planta. *RSC Advances*, *4*(110), 64985–64993. doi:10.1039/C4RA10358K

Nuruzzaman, M. D., Rahman, M. M., Liu, Y., & Naidu, R. (2016). Nanoencapsulation, nano-guard for pesticides: A new window for safe application. *Journal of Agricultural and Food Chemistry*, *64*(7), 1447–1483. doi:10.1021/acs.jafc.5b05214 PMID:26730488

Ozdemir, M., & Kemerli, T. (2016). Innovative applications of micro and nanoencapsulation in food packaging. *Encapsulation and controlled release technologies in food systems*, 333-378.

Patra, J. K., & Baek, K. H. (2017). Antibacterial activity and synergistic antibacterial potential of biosynthesized silver nanoparticles against foodborne pathogenic bacteria along with its anticandidal and antioxidant effects. *Frontiers in Microbiology*, *8*, 167. doi:10.3389/fmicb.2017.00167 PMID:28261161

Pokropivny, V., Lohmus, R., Hussainova, I., Pokropivny, A., & Vlassov, S. (2007). *Introduction to nanomaterials and nanotechnology*. Tartu University Press.

Prasad, R. (2014). Synthesis of silver nanoparticles in photosynthetic plants. *Journal of Nanoparticles*.

Prasad, R., Kumar, V., & Prasad, K. S. (2014). Nanotechnology in sustainable agriculture: Present concerns and future aspects. *African Journal of Biotechnology*, *13*(6), 705–713. doi:10.5897/AJBX2013.13554

Prasad, R., Pandey, R., & Barman, I. (2016). Engineering tailored nanoparticles with microbes: Quo vadis? *Wiley Interdisciplinary Reviews. Nanomedicine and Nanobiotechnology*, *8*(2), 316–330. doi:10.1002/wnan.1363 PMID:26271947

Rajput, V. D., Kumari, A., Minkina, T., Barakhov, A., Singh, S., Mandzhieva, S. S., Sushkova, S., Ranjan, A., Rajput, P., & Garg, M. C. (2023). A practical evaluation on integrated role of biochar and nanomaterials in soil remediation processes. *Environmental Geochemistry and Health*, *45*(12), 9435–9449. doi:10.1007/s10653-022-01375-w PMID:36070110

Rajput, V. D., Singh, A., Minkina, T., Rawat, S., Mandzhieva, S., Sushkova, S., Shuvaeva, V., Nazarenko, O., Rajput, P., Komariah, Verma, K. K., Singh, A. K., Rao, M., & Upadhyay, S. K. (2021). Nano-enabled products: Challenges and opportunities for sustainable agriculture. *Plants*, *10*(12), 2727. doi:10.3390/plants10122727 PMID:34961197

Raliya, R., Tarafdar, J. C., Gulecha, K., Choudhary, K., Ram, R., Mal, P., & Saran, R. P. (2013). Scope of nanoscience and nanotechnology in agriculture. *Journal of Applied Biology and Biotechnology, 1*(3), 41-44.

Rodríguez, J., Martín, M. J., Ruiz, M. A., & Clares, B. (2016). Current encapsulation strategies for bioactive oils: From alimentary to pharmaceutical perspectives. *Food Research International*, *83*, 41–59. doi:10.1016/j.foodres.2016.01.032

Sabir, S., Arshad, M., & Chaudhari, S. K. (2014). Zinc oxide nanoparticles for revolutionizing agriculture: Synthesis and applications. *TheScientificWorldJournal*, *2014*, 2014. doi:10.1155/2014/925494 PMID:25436235

Sagadevan, S., & Periasamy, M. (2014). Recent trends in nanobiosensors and their applications-a review. *Reviews on Advanced Materials Science*, *36*, 62–69.

Sekhon, B. S. (2014). Nanotechnology in agri-food production: An overview. *Nanotechnology, Science and Applications*, 31–53. doi:10.2147/NSA.S39406 PMID:24966671

Sertova, N. M. (2015). Application of nanotechnology in detection of mycotoxins and in agricultural sector. *Journal of Central European Agriculture*, *16*(2), 117–130. doi:10.5513/JCEA01/16.2.1597

Singh, A., Agrawal, S., Rajput, V. D., Ghazaryan, K., Movsesyan, H. S., Minkina, T., . . . Gupta, S. K. (2023). Seed Nanopriming: An Innovative Approach for Upregulating Seed Germination and Plant Growth Under Salinity Stress. In Nanopriming Approach to Sustainable Agriculture (pp. 290-313). IGI Global.

Singh, A., Margaryan, G., Harutyunyan, A., & Movsesyan, S, H., Khachatryan, H., Rajput, V. D., ... & Ghazaryan, K. (. (2024). Advancing Agricultural Resilience in Ararat Plain, Armenia: Utilizing Biogenic Nanoparticles and Biochar under Saline Environments to Optimize Food Security and Foster European Trade. *Egyptian Journal of Soil Science*, *64*(2). Advance online publication. doi:10.21608/ejss.2024.257000.1703

. Singh, A., Rajput, V. D., Al Tawaha, A. R. M., Al Zoubi, O. M., Habeeb, T., Rawat, S., ... Minkina, T. (2023). A review on crop responses to nanofertilizers for mitigation of multiple environmental stresses. *Ecological Engineering & Environmental Technology, 24*.

Singh, A., Rajput, V. D., Pandey, D., Sharma, R., Ghazaryan, K., & Minkina, T. (2023). Nano Zinc-Enabled Strategies in Crops for Combatting Zinc Malnutrition in Human Health. *Frontiers in Bioscience (Landmark Edition)*, *28*(8), 158. doi:10.31083/j.fbl2808158 PMID:37664935

Singh, A., Rajput, V. D., Sharma, R., Ghazaryan, K., & Minkina, T. (2023). Salinity stress and nanoparticles: Insights into antioxidative enzymatic resistance, signaling, and defense mechanisms. *Environmental Research*, *235*, 116585. doi:10.1016/j.envres.2023.116585 PMID:37437867

Singh, A., Rajput, V. D., Varshney, A., Ghazaryan, K., & Minkina, T. (2023). Small tech, big impact: Agri-nanotechnology journey to optimize crop protection and production for sustainable agriculture. *Plant Stress*, *10*, 100253. doi:10.1016/j.stress.2023.100253

Singh, A., Rawat, S., Rajput, V. D., Ghazaryan, K., Minkina, T., Al Tawaha, A. R. M., . . . Sandhu, K. S. (2023). Nanoparticle-Based Priming-Enhanced Agricultural Practices: Promoting Sustainable Plant Growth and Development. In Nanopriming Approach to Sustainable Agriculture (pp. 142-157). IGI Global.

Singh, A., Rawat, S., Rajput, V. D., Ghazaryan, K., Minkina, T., Al Tawaha, A. R. M., . . . Sandhu, K. S. (2023). Nanoparticle-Based Priming-Enhanced Agricultural Practices: Promoting Sustainable Plant Growth and Development. In Nanopriming Approach to Sustainable Agriculture (pp. 142-157). IGI Global.

Singh, A., Rawat, S., Rajput, V. D., Ghazaryan, K., Minkina, T., Al Tawaha, A. R. M., & Varshney, A. (2023). Impact of Nanofertilizers for the Mitigation of Multiple Environmental Stresses. In *Nanofertilizers for Sustainable Agroecosystems: Recent Advances and Future Trends* (pp. 431–454). Springer Nature Switzerland.

Singh, A., Sengar, R. S., Rajput, V. D., Al-Ghzawi, A. L., Shahi, U. P., Ghazaryan, K., Minkina, T., Tawaha, A. R. M. A., Al Zoubi, O. M., & Habeeb, T. (2024). Impact of Salinity Stress and Zinc Oxide Nanoparticles on Macro and Micronutrient Assimilation: Unraveling the Link between Environmental Factors and Nutrient Uptake. *Journal of Ecological Engineering*, *25*(2), 1–9. doi:10.12911/22998993/172947

Singh, A., Sharma, R., Rawat, S., Singh, A. K., Rajput, V. D., Fedorov, Y., & Chaplygin, V. (2022). Nanomaterial-plant interaction: Views on the pros and cons. In *Toxicity of nanoparticles in plants* (pp. 47–68). Academic Press. doi:10.1016/B978-0-323-90774-3.00015-5

Sun, C. Q. (2007). Size dependence of nanostructures: Impact of bond order deficiency. *Progress in Solid State Chemistry*, *35*(1), 1–159. doi:10.1016/j.progsolidstchem.2006.03.001

Tadros, T., Izquierdo, P., Esquena, J., & Solans, C. (2004). Formation and stability of nano-emulsions. *Advances in Colloid and Interface Science*, *108*, 303–318. doi:10.1016/j.cis.2003.10.023 PMID:15072948

Thornhill, S., Vargyas, E., Fitzgerald, T., & Chisholm, N. (2016). Household food security and biofuel feedstock production in rural Mozambique and Tanzania. *Food Security*, *8*(5), 953–971. doi:10.1007/s12571-016-0603-9

Vardumyan, H., Singh, A., Rajput, V. D., Minkina, T., El-Ramady, H., & Ghazaryan, K. (2024). Additive-Mediated Phytoextraction of Copper-Contaminated Soils Using Medicago lupulina L. *Egyptian Journal of Soil Science*, *64*(2), 599–618. doi:10.21608/ejss.2024.266169.1714

Vidotti, M., Carvalhal, R. F., Mendes, R. K., Ferreira, D., & Kubota, L. T. (2011). Biosensors based on gold nanostructures. *Journal of the Brazilian Chemical Society*, *22*(1), 3–20. doi:10.1590/S0103-50532011000100002

Viswanathan, S., & Radecki, J. (2008). Nanomaterials in electrochemical biosensors for food analysis-a review. *Polish Journal of Food and Nutrition Sciences*, *58*(2).

Wan, Y., Li, J., Ren, H., Huang, J., & Yuan, H. (2014). Physiological investigation of gold nanorods toward watermelon. *Journal of Nanoscience and Nanotechnology*, *14*(8), 6089–6094. doi:10.1166/jnn.2014.8853 PMID:25936063

Yang, X., Liang, Q., Chen, Y., & Wang, B. (2019). Alteration of methanogenic archaeon by ethanol contribute to the enhancement of biogenic methane production of lignite. *Frontiers in Microbiology*, *10*, 2323. doi:10.3389/fmicb.2019.02323 PMID:31649654

Yunlong, C., & Smit, B. (1994). Sustainability in agriculture: A general review. *Agriculture, Ecosystems & Environment*, *49*(3), 299–307. doi:10.1016/0167-8809(94)90059-0

Chapter 10
Nanotechnology Solutions for Sustainable Pest and Disease Control for Sustainable Agriculture and Food Security

Prashant Singh
https://orcid.org/0009-0008-9667-3598
Mandsaur University, India

Bhavana Tomar
G.D. Goenka University, India

Tirunima Patle
https://orcid.org/0000-0003-3268-6118
Shri Vaishnav Vidyapeeth Vishwavidyalaya, India

Sneh Singh Parihar
ITM University, Gwalior, India

Shiv Singh Tomar
G.D. Goenka University, India

Dayashankar Singh Singh
Mandsaur University, India

ABSTRACT

Each year, a significant portion of crops, ranging from 20% to 40%, succumb to plant pests and pathogens, resulting in substantial agricultural losses. Traditional methods of managing plant diseases heavily rely on the application of toxic pesticides, posing potential hazards to both human health and the environment. Nanotechnology presents a promising avenue for addressing these challenges by offering various advantages over conventional pesticides. These include mitigating toxicity, enhancing shelf-life, and improving the solubility of pesticides that are poorly water-soluble, thereby potentially yielding positive environmental outcomes. This review delves into two primary approaches for leveraging nanoparticles in plant disease management: employing nanoparticles independently as protective agents or utilizing them as carriers, often termed as 'magic bullets,' for delivering a range of substances such as herbicides, insecticides, fungicides, fertilizers, and RNA-interference molecules or genes directly to specific cellular organelles within plants. Nanoparticles encapsulate active compounds with high stability and biodegradability, shielding them from degradation by external factors or the host plant itself. Moreover, they minimize inadvertent dispersion into the soil, consequently reducing the need for multiple active compounds in plant treatments and thereby lowering environmental impacts. Additionally, nanoparticles

DOI: 10.4018/979-8-3693-1890-4.ch010

can be functionalized with biomolecules like antibodies or aptamers to ensure target selectivity and specificity. Despite these benefits, there remain certain challenges associated with the use of nano devices for plant protection. Foremost among these is the insufficient research on the potential toxicity of certain nanomaterials, such as nano silver and nano gold, to plants, animals, and ecosystems. Accumulation of nanomaterials in plant and animal tissues could potentially enter the food chain, necessitating rigorous safety assessments and consumer education efforts. Nonetheless, the adoption of non-toxic materials, such as starch, chitin, or nano clays, as alternatives to metals, can mitigate such risks.

INTRODUCTION

Plant diseases and pests severely reduce food yields, resulting in losses that are estimated to be between 20% and 40% of global production each year. The present pest control method makes extensive use of pesticides, including herbicides, fungicides, and insecticides (Flood, J. 2010). Pesticides harm creatures that are not their intended targets because pest populations to resurrect, and cause resistance to develop—all despite their availability, speed of action, and dependability (Stephenson, G.R. 2003).

Additionally, about 90% of pesticides that are applied are lost during or after application, which has sparked interest in creating environmentally friendly, affordable pesticides. With the ability to completely transform civilization, nanotechnology offers exciting new opportunities in the food, medical, and agricultural sectors. Nanotechnology in crop protection allows for the controlled release of encapsulated pesticides, fertilizers, and other agrochemicals to combat pests and diseases. Nanosensors enable the early detection of pollutants and plant diseases, including pesticide residues (Ghormade et al., 2011; Lin et al., 2020).

The application of nanoparticles in crop protection encourages the creation of effective plant disease management strategies, providing fresh approaches to the problems stated above. Insect pests account for a substantial 14% crop loss worldwide, while plant infections are estimated to cause losses of up to 13%, or US $2,000 billion annually (Pimentel, 2009). By making it possible to administer pesticides, herbicides, and fertilizers safely and effectively at lower concentrations, nanomaterials lessen the negative impacts on pollinating insects and human health (Kuzma and VerHage, 2006; Mousavi and Rezaei, 2011).

Nano pesticide formulations allow for gradual release and improve the solubility of poorly soluble active chemicals, hence reducing toxicity and increasing efficacy (Kah et al., 2012). Due to their strong reactivity at the nanoscale, pesticide-loaded nanoparticles can be triggered to release slowly in response to environmental cues (Lauterwasser.C. 2005), providing enhanced crop protection (Debnath et al., 2011). Although nanotechnology has advanced significantly in pharmacology and medicine, its applications in agriculture are still largely untapped (Sinha, K. et al. 2017; Balaure, P.C. et al. 2017).

Nanozymes are materials that possess inherent enzyme-like characteristics. With comparable catalytic kinetics and mechanisms, they can selectively catalyze natural enzyme substrates under physiological conditions. When compared to natural enzymes, nanozymes have a number of distinct benefits, such as easy mass production, cheap cost, great stability, and customizable activity. Furthermore, nanozymes represent a novel class of synthetic enzymes that possess not only the catalytic activity like that of an enzyme but also the distinct physicochemical characteristics of nanomaterials, including superparamagnetism, photothermal properties, and fluorescence. Nanozymes have been extensively developed for in vivo disease monitoring and treatment, as well as for in vitro detection, thanks to their unique combination of physicochemical features and enzyme-like catalytic capabilities.

For instance, the majority of the research on using nanozymes to identify and detect fungi has, thus far, only focused on the identification of mycotoxins and other chemical substances linked to fungal infections. There is therefore room to investigate the use of nanozymes in the direct identification of fungi in food, particularly in the agricultural production of those commodities. Due to their inability to germinate, a large number of fungi that are present in seeds seriously impair their utility. Moreover, mushrooms yield mycotoxins, which, if ingested, can seriously harm people's health.

Numerous applications of agricultural nanotechnology exist, including increased seed germination, genetic transfer, plant hormone delivery, water management, nanobarcoding, controlled release of agrichemicals, and nanosensors (Hayles, J.; Johnson, L.; Worthley, C.; Losic, D. 2017). To accurately and precisely administer pesticides, material scientists use conjugation, encapsulation, or adsorption to generate customized nanoparticles (Khandelwal et al. 2016). As agricultural nanotechnology develops, there will be a huge increase in the ability to produce a new generation of insecticides and other actives for managing plant diseases.

There are two ways in which nanoparticles can be used to protect plants:

(a) Crop protection can be obtained from the nanoparticles themselves.
(b) Using nanoparticles as delivery systems for already-available pesticides or other active components, such as double-stranded RNA (dsRNA), which can be absorbed in water or sprayed on seeds, foliar tissue, or roots. When used as transporters, nanoparticles can provide several benefits, including
    (i) Extended shelf life.
    (ii) Pesticides that are not very soluble in water become more soluble.
    (iii) Decreased toxicity.
    (iv) Increasing the target pest's site-specific absorption.

Insect pest management and correct chemical adsorption by plants are made possible by the delayed and effective release of agrochemicals, such as pesticides or insecticides, to a specific host plant through nano encapsulation (Scrinis and Lyons, 2007). Several research have demonstrated the dependability, affordability, and effectiveness of using various metal nanoparticles as insect pest controllers (Stadler et al., 2010; Barik et al., 2008; Goswami et al., 2010). They also allow for uniform and incredibly small droplet sizes, which makes efficient delivery systems for industrial applications possible (Forgiarini et al., 2001; Lee and Tadros, 1982). Additionally, nanoparticles are a viable and effective substitute for conventional pesticides, especially when it comes to fighting pests that are resistant to pesticides.

Using nanoparticles for direct protection as well as dsRNA carriers for RNA interference (RNAi)-mediated protection and pesticide, fungicide, and herbicide delivery are the topics of this review of recent developments in plant disease control. Notwithstanding the potential of nanotechnology, its limited commercial implementations have hindered its advancement in agricultural applications. Only a limited percentage of research on agrochemicals loaded with nanoparticles has involved field testing or addressed environmental issues. For agricultural nanotechnology to advance, it is imperative that further studies concentrate on assessing crop plants, target pests, and conduct both short- and long-term field trials.

Figure 1. Activity of nanoparticles in plant disease management

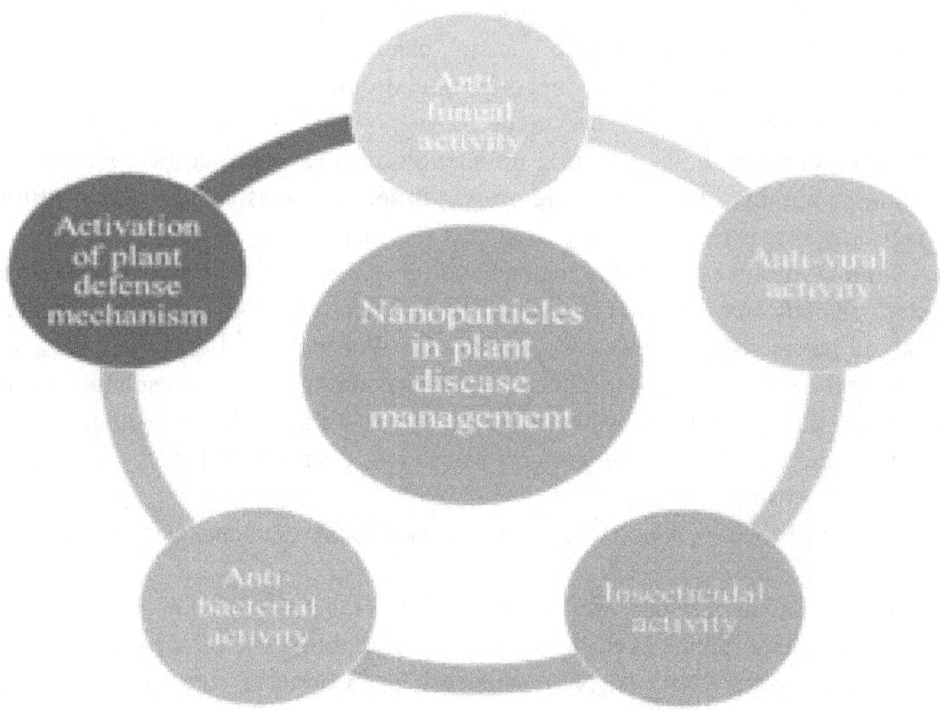

PLANT DISEASE DIAGNOSIS USING NANO PARTICLES

Traditionally, plant disease diagnosis has been based on human observers' visual evaluation of the condition followed by microscopic analysis. Microbiological diagnostics examines characteristics of pathogens, including fruiting bodies, mycelium structure, spore form, color, and arrangement. Pathogens can then be cultured on particular culture media and separated for additional examination (Fang and Ramasamy, 2015). Unfortunately, these processes—which are frequently carried out in industries and research laboratories—are time-consuming, expensive, personnel-intensive, and facility-dependent. This hinders their broad application in the diagnosis of disease, particularly in underdeveloped nations.

Diagnostic techniques include microbiology, molecular biology, and serology are also used to identify plant diseases. Many direct methods are available for monitoring plant diseases, including enzyme-linked immunosorbent assay (ELISA), immunofluorescence, flow cytometry, fluorescence in situ hybridization, polymerase chain reaction (PCR) for nucleic acid detection, and immunofluorescence. In addition, indirect methods such thermography, gas chromatography, fluorescence imaging, and hyperspectral methods are employed (Mc Cartney et al. 2003).

For example, the development of better fungicides and resistant crop types is aided by the use of molecular techniques to identify phytopathogens and study fungicide resistance in wheat. Additionally, phytopathogen populations and their interactions within plants are studied using molecular approaches. New sensors based on nanotechnology are being developed that offer fast, accurate, and affordable plant pathogen identification. Farmers can now detect infections, volatile substances, chemical residues in crops, and environmental changes with the help of these sensors. Metal, metal oxide, and carbon nanoparticles are often utilized nanomaterials in agriculture. Agriculture could undergo a transformation when cutting-edge tactics and nanosensing technologies are combined. A few different kinds of nanosensors are described here (Omanovic-Miklicanina and Maksimovic, 2016).

Diagnosis of Plant Disease Through Metal NPs

According to (Singh et al. 2017), metal nanoparticles (NPs) are remarkable due to their high melting temperatures, hardness, catalytic activity, and distinctive colors. They are able to detect infections with exceptional precision and higher sensitivity than conventional detection limits due to their raised surface area to volume ratio. A less complicated and more affordable option to enzyme tests is the electrochemical detection of metal nanoparticles (NPs). Consequently, in the diagnosis of phytopathogens, metal nanoparticles are gradually replacing enzyme labeling techniques. Nucleic acids with silver (Ag), gold (Au), zinc (ZnS), lead (PbS), and cadmium (CdS) concentrations are all highly useful for sample detection. Because of their higher surface atom fraction, greater surface area, and stronger antibacterial qualities than bulk silver, Ag NPs are the most studied and applied of these in bio-systems.

Diagnosis of Plant Disease Through Metal Oxide NPs

Plant science finds various applications for metal oxides due to their high density of edge surface sites. This benefit is especially noticeable when using metal oxide nanoparticles (NPs) as solid-state gas detectors in commercial, industrial, and residential environments (Sun et al., 2012). These instruments can distinguish volatile metabolites peculiar to a given pathogen with the right adjustments. Moreover, metal oxide nanoparticles (NPs) are useful sensors for identifying volatile organic chemicals due to their affordability, conductivity, and malleability for shape (Fang and Ramasamy, 2015). ZnONPs are utilized, for instance, in the creation of biosensors and gas sensors (Sabir et al., 2014). According to Fang et al. (2014), plants afflicted with the pathogenic fungus Phytophthora cactorum have been known to release volatile p-ethylguaiacol when they are linked to carbon electrodes. TiO2NPs and SnO2NPs have also been utilized to detect this release.

Diagnosis of Plant Disease Through Magnetic NPs

Since the size of magnetic nanoparticles (NPs) is similar to that of the magnetic domain, they have special properties. They behave in two ways: they are super paramagnets and single-domain ferromagnets. Although magnetic nanoparticles have long been used in biomedical science, little is known about their potential use in plant pathology. Using carbon-coated magnetic nanoparticles, researchers have tracked the path, deposition, and movement of NPs within plant cells (González-Melendi et al. 2008). Furthermore, a brand-new NP immunoassay has been created to measure mycotoxin levels in plants in real time. This test, according to Mak et al. (2010), employs magnetic nanotags attached to a spin valve sensor surface that has been immobilized using capture antibodies. Furthermore, superparamagnetic NPs have been used to produce a reliable and quick ELISA technique that dramatically shortens coating, enzyme blocking, and competition periods when compared to traditional ELISA (Radoi et al. 2008).

Diagnosis of Plant Disease Through Carbon NPs

One of the most common atoms on Earth, carbon is the essential component of many metabolic reactions. Because of its unique benefits over other materials, carbon nanoparticles (NPs) are the most widely used artificial nanomaterial. Carbon nanotubes (CNTs) are cylindrical carbon allotropes that have been demonstrated to improve plant development through increased water absorption. CNTs are essential parts of biosensors because of their remarkable electrochemical qualities, which are their flexibility in binding to many chemical species (Balasubramanian and Burghard, 2006), high length-to-diameter ratio, and capacity to promote rapid electron transfer kinetics (Yu et al., 2003). By taking use of these characteristics, CNTs are being used to diagnose plant diseases. They are able to identify plant metabolites that are important for evaluating phytopathological diseases, as changes in the metabolism of aromatic compounds can be an indicator of these diseases. Among volatile metabolites, phenolic chemicals are particularly important for controlling plant diseases. However, electrode corrosion brought on by phenol exposure, which results in the development of dimeric or polymeric oxidation products, makes it difficult to detect phenols with amperometric and potentiometric instruments.

NANO-DIAGNOSTIC SENSORS AND EQUIPMENT

In order to effectively manage polycyclic diseases in agriculture, early crop pathogen detection is crucial. This calls for the use of sensitive gadgets and effective identification methods. Plant illnesses are detected using a variety of techniques, such as detecting metabolites that are produced in response to stress in plants, such as methyl jasmonate, salicylic acid, and jasmononic acid; alternatively, the pathogen might be identified by genetic, serological, or volatile studies. Also employed is host-induced biomarker analysis, which looks at volatiles, proteins, and transcripts (Martinelli et al., 2015). Whether directly or by the detection of substances released by the diseases themselves, nanosensors are essential for the detection of pathogens. For example, plant pathogenic fungus Sclerotinia sclerotiorum has been detected using nanosensors that include CuNPs with a gold electrode. By measuring the rate at which salicylic acid is produced, these nanosensors enable the estimate of according to Wang et al. (2010), the severity of the infection.

Portable Diagnostic Tools

Our constant need for early detection has spurred us to develop cutting-edge diagnostic tools that can find even the tiniest signs of illness. According to (Weigl et al. 2008), these instruments need to have competitive sensors, incorporate on-chip processing for sample preparation, and reduce the number of manipulation steps. Recent technology advancements have made it possible for agriculture to use portable diagnostic tools for plant diseases. A range of portable technologies are currently on the market, such as portable genome sequencers, lateral flow devices (LFDs), nanodiagnostic kits, immunoprinting kits, portable PCR equipment, and loop-mediated isothermal amplification (LAMP-PCR). These are quick, strong, and easy to use tools. Nevertheless, they are mostly in the developmental stage, frequently costly, and not easily available (Khiyami et al., 2014).

Nano FABRICATION IMAGING

Phytopathogen detection is made easier by the viewing of plant tissues and cells made possible by advanced nano diagnostic imaging techniques. With the advent of this new tool, phytopathologists now have much greater access to diagnostic tools that can help with early disease diagnosis (Rosen et al., 2011). Through the manipulation of nanoparticles' (NPs') physical and chemical properties, these instruments have enhanced signal strength, contrast, tissue specificity, and imaging time. The properties of pathogens, their interactions with hosts, and the onset of infection processes are all revealed by means of such imaging techniques. To effectively design compounds to counteract the impacts of infections, a deeper understanding of disease pathways is necessary (Meng et al., 2005).

NANOPARTICLE TYPES FOR MANAGEMENT OF PLANT DISEASES

Nanoparticles as Protectants

When compared to molecules and bulk materials, nanoparticles—which vary in size from 10 to 100 nanometers (nm)—have unique chemical, physical, and biological properties (Yang, W.; Peters, J.I.; Williams, R.O. 2008). These tiny compounds can be applied directly to the roots, foliage, or seeds of plants to offer protection from insects, bacteria, fungi, viruses, and other pests and diseases. Silver, copper, zinc oxide, and titanium dioxide have received special attention due to their antibacterial, antifungal, and antiviral properties (Kah, M.; Hofmann, T. 2014; Gogos, A.; Knauer, K.; Bucheli, T.D. 2012). A brief synopsis and updates from recent literature reviews on the types of nanoparticles now in use are given in this section (Mishra, S.; Singh, H. 2014; Sadeghi, R., Yao, Y.; Kokini, J.L.; Rodriguez, R.J. (2017).

A recent spike in the popularity of silver nanoparticles has been attributed to the utilization of using "green synthesis" techniques, one can use plants, bacteria, fungi, or yeast (Rafique, M.; Sadaf, I.; Rafique, M.S.; Tahir, M.B. 2017). Surprisingly, studies have shown that silver nanoparticles exhibit strong antifungal properties against a range of pathogens, including Botrytis cinerea, Rhizoctonia solani, Macrophomina phaseolina, Alternaria alternata, and Sclerotinia sclerotiorum (Krishnaraj, C. et al. 2012). According to Jain and Kothari (2014), the sun-hemp rosette virus was completely suppressed when silver nanoparticles were sprayed on bean leaves. Furthermore, silver nanoparticles were found to improve resistance against the bean yellow mosaic virus in fava bean plants when administered after infection, outperforming pre-infection or simultaneous application (Elbeshehy et al.). Production, toxicity, and soil interaction are still major obstacles in the way of their great potential for managing plant diseases.

Nanoparticles of gold, copper, titanium dioxide, and silver are also frequently employed. Although the main use of copper and titanium dioxide nanoparticles is in fertilizers, more research should be done to determine how well they may be used to control plant diseases. Fertilizers containing titanium dioxide nanoparticles have demonstrated potential in preventing viral infections and shielding plants from bacterial infections (Sadeghi et al.). Moreover, it has been shown that the introduction of gold nanoparticles via mechanical abrasion can damage Barley yellow mosaic virus particles, giving plants resilience.

Another extensively studied nanoparticle is chitosan, which has negligible toxicity to humans and animals and advantageous biological qualities such biodegradability, biocompatibility, and antibacterial activity (Cota-Arriola, et al. 2013). It has been discovered that chitosan nanoparticles can create viral resistance in a variety of plant tissues, shielding alfalfa, hemp, peanut, potato, and cucumber from

Figure 2. Nanoparticles as protectants

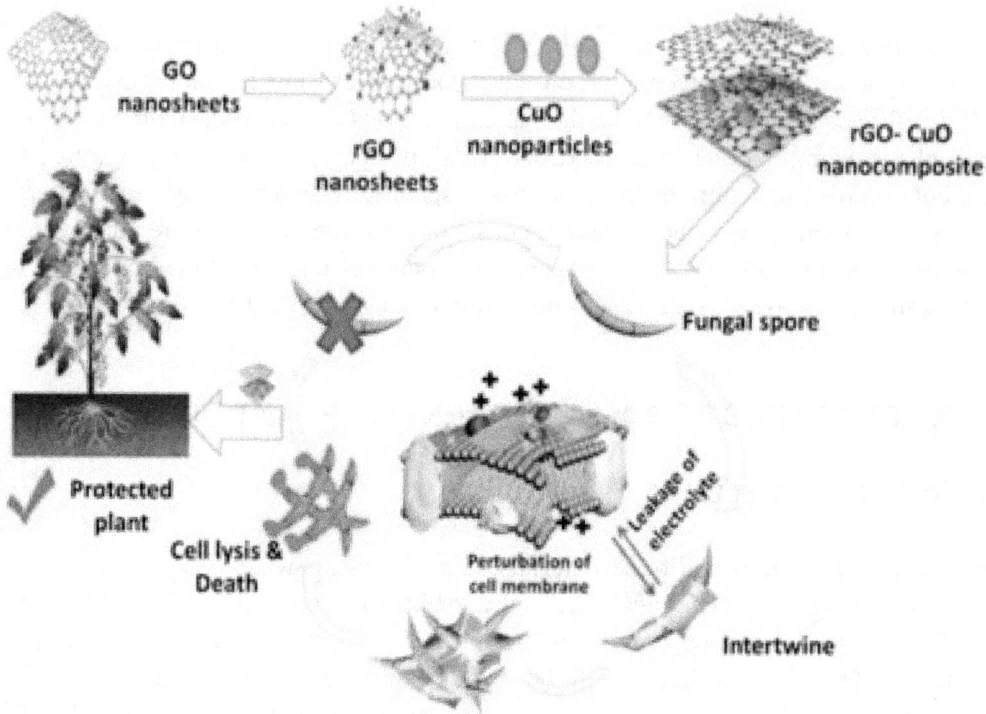

mosaic viruses (Kochkina, Z. et al. 1994). Additionally, they have antibacterial qualities that work well against fungi, such as Fusarium bunch rot in grapes and Fusarium crown root rot in tomatoes, but less well against bacteria (Malerba and Cerana 2016).

Agglutination, breakdown of cell membranes, reduction of H+ ATPase activity, and creation of toxins, microbial growth, mRNA and protein synthesis, and obstruction of nutrition flow are some of the mechanisms that underlie chitosan's antimicrobial activities. Additionally, chitosan has demonstrated effectiveness against a variety of pests, including as cotton leafworms, oleander aphids, root-knot nematodes, and spear psylla nymphs. Chitosan is a highly promising substance for use in agriculture, both on its own and as a carrier.

Nanoparticles That Act as Carrier

In order to facilitate the creation of effective agricultural formulations, nanoparticles operate as carriers for the entrapment, encapsulation, absorption, or attachment of active compounds. Below is a list of common nanoparticles that are employed as carriers for fungicides, herbicides, insecticides, and chemicals that induce RNAi.

Silica nanoparticles- One unique feature of silica nanoparticles is their ease of synthesis, which allows for exact control over their size, shape, and structure. Mody et al. (2014) claim that this makes them superior delivery vehicles. Mesoporous and porous hollow silica nanoparticles (PHSNs) are two examples; both have a spherical structure with porous characteristics. In order to protect the active components and provide prolonged release, insecticides are often encapsulated within PHSNs and MSNs. PHSNs' shell structure provides extra defense against UV deterioration. According to literature, silica nanoparticles are a sensible option for creating pest management solutions because silicon has been used to improve plant resilience against a variety of environmental conditions (Barik, T.; Sahu, B.; Swain, V., 2008).

Chitosan nanoparticles- Because chitosan nanoparticles are hydrophobic, they are only partially soluble in water. As a result, to increase their solubility, they are often mixed with copolymers, both inorganic and organic. Due of its hydroxyl groups and reactive amines, chitosan can undergo graft reactions, modifications, and ionic interactions, all of which improve its characteristics (Li, M.; Huang, Q.; Wu, Y. 2011). Moreover, chitosan adheres strongly to the leaf and stem epidermis, prolonging contact periods and promoting the absorption of advantageous substances.

Solid lipid nanoparticles (SLNs) - At room temperature, lipids that stay solid are used to make emulsions and solid lipid nanoparticles (SLNs). One advantage of employing SLNs is their ability to form a matrix that, in the absence of organic solvents, can capture lipophilic active molecules (Ekambaram et al., 2012). Additionally, SLNs enable the controlled release of different lipophilic components by decreasing the mobility of the active inside the solid matrix. When SLNs are dissolved in water, surfactants are added to stabilize the compound. However, their primary drawbacks are restricted loading efficiency and the potential for the active to escape the structure while it is being stored (Tamjidi et al., 2013).

Layered double hydroxides (LDHs) - Layered double hydroxides (LDHs) are clay-like materials that self-organize into hexagonal sheets (Xu et al., 2006). Active molecules that are confined in the interlayer space comprise the layers of LDHs. LDH nanoparticles can decompose when they come into contact with acidic conditions like dampness and atmospheric carbon dioxide. Plant cell walls can be more easily penetrated by physiologically active substances when using positively charged LDH lactate delaminated nanoparticles (Bao et al., 2016).

Plant disease management typically uses silica, chitosan, solid lipid nanoparticles (SLN), and LDH nanoparticles as carriers. Parts devoted to insecticide, fungicide, herbicide, or RNAi contain in-depth talks on applying these and other less common forms of nanoparticles.

A BRIEF GUIDE TO FUNGICIDES AND INSECTICIDES

This section provides a brief overview of the various pesticide kinds and their classification system in an effort to encourage appropriate use and lower pesticide resistance. There are two types of insecticides: contact (which need direct touch) and systemic (absorbed by the plant). According to their effects on physiological processes such nerve/muscle, development, respiration, midgut, or unknown/non-specific, they can be further divided into at least 55 chemical classes (Sparks, T.C.; Nauen, R. 2015). Pesticide resistance can be avoided by rotating the use of thirty notable mechanisms of action (MoA). These consist of various locations and unidentified groups.

Fungicides work by either penetrating the plant at different systemic levels or by establishing a barrier on it without being absorbed. For fungicides, it is important to have enough spray coverage because there is rarely systemic movement throughout the entire plant. Fungicides are classified into fourteen different MoA categories, which are reflected by the 49 codes issued by the Fungicide Resistance Action Committee (FRAC) (Leadbeater, A.; Gisi, U.; Klein, K.-H. 2012).

Herbicides can be non-selective, affecting all plants, or selective, focusing solely on a limited set of weeds. There are three distinct ways to administer herbicides: sprinkling them on the soil prior to planting, applying them pre-emergence (before weed seedlings emerge from the soil), or applying them post-emergence (after weed seedlings have emerged). There are 23 Herbicide Resistant Action Committee (HRAC) groups for herbicides in an effort to avoid pesticide resistance. It is important to note that rotation within these groups is crucial.

Nanoparticles as Carriers of Insecticides

Numerous research with a variety of nanoparticles and conventional pesticides (27 studies) as well as bioactive compounds with insecticidal properties (13 studies) have been conducted since the early 2000s. The loading of pesticides into nanoparticles was first introduced. These studies have examined a wide variety of essential oils and eight distinct modes of action (MoA) that are not categorized by the Insecticide Resistance Action Committee (IRAC). Lipids (4 studies), chitosan (11 study), and silica (8 studies) were determined to be the most often studied nanoparticle carriers. Spodoptera litura (5 study), Helicoverpa armigera (4 studies), and Tetranychus urticae (4 studies) were the most often targeted pests.

Improving stability, decreasing volatilization, increasing water solubility, and facilitating a gradual release of active molecules were the main goals of these investigations. In the context of low water-soluble insecticides, the conventional practice involves using organic solvents to enhance solubility, thereby increasing the cost and toxicity of the insecticide. However, nanoparticles offer an alternative approach by increasing solubility and subsequently reducing toxicity. Notably, research has successfully loaded low-water-soluble insecticides into modified chitosan and porous silica Cui, H.; Zhao, X.; Cui, B.; Wang, Y.; Sun, C. 2014). None of these research has specifically assessed for decreased environmental toxicity, despite these developments.

A prominent instance is the application of modified chitosan nanoparticles for the purpose of loading the hydrophobic insecticide azadirachtin; research by Lu et al. shows that this compound has sustained drug release in addition to favorably reducing cell proliferation in S. litura ovarian cell lines. Another study carried out in 2015 by Liu et al. discovered that H. armigera larvae treated to dendrimers containing hydrophobic thiamethoxam showed increased absorption and mortality. Surprisingly, Dendrimer nanoparticles containing H. armigera were loaded it showed a marked increase in toxicity, despite the fact that it is normally resistant to thiamethoxam. Similarly, anacardic acid, when intercalated into LDH nanoparticles (Nguyen et al. 2015), shown that the application of direct mustard leaf or S. litura skin treatment resulted in higher mortality rates than the use of anacardic acid alone. These results highlight the possible advantages of using nanoparticles to improve the solubility of active components.

The evaporation or volatilization of the active components is a common cause of pesticide loss after application. Essential oils have insecticidal effects, but they evaporate quickly because they are unstable chemically when exposed to air, light, heat, and moisture content (Lai et al. 2006). In a particular study,

Table 1. Using nanoparticles to deliver insecticides and the intended target

Insecticides	Nanoparticles	Crops	Targeted Pest
Avermectin	Polydopamine	Cucumber	Aphids
Chlorfenapyr	Silica	Brassica chinese	Cotton bollworm
Imidacloprid	Sodium alginate	unspecified	Leafhopper
Azadirachtin	Zinc oxide and chitosan	Groundnut	Groundnut bruchid
Garlic essential oil	PEG	Rice	Red flour beetle

SLN nanoparticle-encapsulated glass vials were sprayed with essential oil of Artemisia arborescens L. The evaporation rate was 45.5%, compared to 80% for the essential oil alone, with an initial burst release that happened after 48 hours. Comparably, red flour beetles (Tribolium castaneum) exposed to garlic essential oil encapsulated in polyethylene glycol (PEG) after 5 months revealed 80% mortality, while essential oil alone produced just 11% mortality.

Another approach to address evaporation issues is the use of nanoparticles to enhance the stability of active molecules, allowing for sustained release and reduced insecticide usage. According to Kumar et al. (2016), imidacloprid sprayed on okra plants that was encapsulated in sodium alginate was equally effective in field tests as imidacloprid applied alone.In a field study conducted by (Song et al. 2012), Brassica chinese was treated with silica-encapsulated chlorfenapyr, and over the course of three days, in comparison to the diamondback moth (Plutella xylostella), the study's control was comparable to or superior.

Furthermore, research has shown that nano-insecticides have the ability to offer prolonged release, which would increase both safety and efficacy. When termites were fed fipronil put into silica-shelled nanoparticles, for example, there was no first burst release, which would have allowed worker termites to return the bait to the nest (Wibowo et al. 2014). Compared to commercial insecticides, this increased the 100% mortality window by 3 days, which allowed for a more thorough colony removal. To evaluate azadirachtin's efficacy over 180 days in groundnut bruchid storage conditions, the medication was affixed onto zinc oxide or chitosan nanoparticles by (Jenne et al. 2018). Against the backdrop of other assessed formulations, neem seed kernel extract loaded into zinc oxide nanoparticles demonstrated a weight reduction of 54.61% in groundnut bruchid. These studies collectively demonstrate the effectiveness of nanoparticle formulations in addressing volatility issues and improving the overall performance of insecticides.

Nanoparticles as Carrier of Fungicides

Commencing in 1997, the initial exploration of nanofungicides involved integrating fungicides into solid wood, as demonstrated by (Laks, P.; Heiden, P.; Liu, Y., 2002). The study examined essential oils that were not included in the fungicide groups and covered nine FRAC groups. Polymer mixes, silica, and chitosan emerged as the predominant nanoparticle carriers investigated. Numerous fungi were employed to assess nanofungicide efficiency, although limited attention was given to plant testing, and toxicity studies were scarce. Nanoparticles were used to solve issues like low water solubility, reduced volatilization, and improved stability for a slow continuous release, much like pesticides.

In order to enhance the poor water-solubility of tebuconazole and boost its adherence to leaf surfaces, Hatfaludi et al. (2004) employed bacterial ghosts that were nanosized and derived from non-denatured empty cell envelopes of Gram-negative bacteria, namely Pectobacterium cypripedii. Among the plants evaluated, fluorescently tagged ghosts without loaded fungicide exhibited the highest adherence to rice leaves (with 55% surviving) when subjected to strong simulated rain in a glasshouse setting (rice, soya, cabbage, cotton, barley, and corn). On the other hand, soy leaves had the lowest adherence rate (10% remaining). Distinct plant performances were seen in connection to rainfall when six plants were treated with either ghost-loaded tebuconazole or two commercial tebuconazole treatments (WP 25 and EW 250) against various fungus. Remarkably, after being cleaned up 24 hours after treatment, ghost-loaded tebuconazole exhibited an efficacy comparable to or higher than WP 25 treatments, although the controls treated with EW 250 were often more effective.

Chitosan–lactide copolymer nanoparticles were treated with an extra low-water-soluble fungicide, pyraclostrobin, at different dosages. After three to five days, the nanofungicide's effectiveness in preventing C. gossypii inhibition was either lower than or equal to that of commercial pyraclostrobin. Seven days after therapy, an increase in inhibition was noticed. In another experiment, lecithin/chitosan was coated with the low-soluble fungicide kaempferol, and the mixture was stored for 60 days on a Petri plate with Fusarium oxysporum showed 67% inhibitory efficiency (Ilk, S.; Saglam, N.; Özgen, 2017).

Chlorothalonil and tebuconazole were loaded into different nanoparticles in a series of research to address the fungicides' limited solubility for use in solid woods. Four investigations looked at the mass loss in southern yellow pine wood over a period of 50–55 days as a result of Gloeophyllum trabeum decay. Because of its size and stability, hydrophobic chlorothalonil enclosed in nanoparticles showed inefficiency. However, minimal fungal-induced degradation of the wood was obtained with an increase in concentration. Tebuconazole coated with nanoparticles dramatically decreased the amount of fungicide administered (from about 2 kg active/m3 to 0.2 kg/m3) while barely affecting degradation. Because of its tiny pit pores, birch wood presented treatment-related issues. When encapsulated using a surfactant-free technique, chlorothalonil plus tebuconazole resulted in a more stable aqueous suspension and smaller median particle sizes, which led to higher uptake into the wood. Using southern pine exposed to G. trabeum and birch wood subjected to T. versicolor, this technique showed the effectiveness of smaller, more stable, surfactant-free nanoparticles against fungal deterioration. Additionally, it exhibited minimal degradation (five percent or less mass loss) at far lower doses for both fungicides than industry norms.

Despite their well-known fungicidal qualities, essential oils evaporate quickly, which prevents their widespread commercial application. In comparison to bulk essential oil components, (Janatova et al. 2015) demonstrated increased antifungal activity 14 days after infection with Aspergillus niger following the effective encapsulation of five different essential oil components into MSN.In a similar vein, Nasseri et al. (2016) employed SLNs to stabilize Zataria multiflora essential oil, providing protection against six types of fungus.

Although little study has been done in this area, leaching—the transfer of chemicals and water through soil—is a significant pesticide control concern. The fungicide metalaxyl was loaded onto mesoporous silica nanoparticles (MSNs) in a study by (Wanyama 2013). Over the course of a 30-day period in soil, the leaching of encapsulated metalaxyl (11.5%) and free metalaxyl (76% release) showed noteworthy variations. Interestingly, encapsulated metalaxyl released at a rate of 47% when exposed to water, highlighting the importance of assessing pesticide behavior in agricultural settings. Using solid lipid or polymeric nanoparticles loaded with tebuconazole and/or carbendazim, (Campos et al. 2015)

Table 2. Using nanoparticles to deliver fungicides and the intended target fungus

Fungicides	Nanoparticles	Crops	Targeted Fungi
Tebuconazole	PVP and PVP copolymer	Southern pine sapwood	G. trabeum
Metalaxyl	MSN	wheat	E. graminis, L. nodorum,
Carbendazim	Polymeric and SLN	Bean seeds	C. gossypii
Ferbam	Gold	Tea	R. stolonifera
Pyrimethanil	MSN	Cucumber	A. parasiticus
Flusilazole	Chitosan–PLA graft copolymer	Wheat and barley	-
essential oil	Chitosan	cucumber	A. niger

investigated the cytotoxicity and observed a reduction in toxicity towards preosteoblast and fibroblast mice cell lines. Moreover, nanoparticles reduced release rates in soil leaching trials in comparison to commercial formulation. The effects of poly (butylene succinate) and poly (lactic acid) shells loaded with azoxystrobin and difenoconazole on zebrafish were studied by (Wang et al. 2018). They found that the shells were less hazardous than other formulations.The noteworthy finding is that validamycin-containing nanosized calcium carbonate produced a delayed release of active molecules. This was initially less effective than validamycin alone, but with time, the efficacy improved. Additionally, using carbendazim-loaded polymeric nanoparticles to inhibit fungal growth against Fusarium oxysporum and Aspergillus parasiticus, (Kumar et al. 2016) demonstrated enhanced fungal suppression on Cucumis sativa, Zea mays, and Lycopersicum esculentum seeds. Using MSNs to administer pyrimethanil, (Zhao et al. 2017) looked at how well cucumber plants absorbed it and calculated the lowest danger of buildup in edible plant portions. Their research greatly advances our knowledge of the dispersion and transfer of insecticides loaded with MSN when applied topically.

Nanoparticles and RNAi for Plant Protection

The identification of the RNAi pathway has enabled the development of innovative and inventive methods for the control of illnesses and pests. A system that has evolved to be conserved in eukaryotes, RNA interference (RNAi) is essential for controlling development, growth, and host protection against transposons and viruses. It's interesting to note that weeds, fungi, viruses, and insects can all be targeted with this technique (Baulcombe, D. 2004).

In plants, RNA interference (RNAi) is started by double-stranded RNA (dsRNA). Subsequently, dsRNA is transformed into small-interfering RNA (siRNA) by DCL-like enzymes. The RNA of pathogens is degraded by these siRNAs when they are integrated into a RISCs (RNA-induced silencing complexes) and control them by base pairing. The pathogenic RNA cannot be used as a translation template as a result of its degradation. Since its discovery, the RNAi pathway has shown to be a potent tool for employing genetic modification to address pests and plant diseases (Robinson et al. 2014).

On the other hand, the usage of genetically modified organisms (GMOs) is controversial and governed by strict laws in many nations. Consequently, there is ongoing research focused on developing new delivery methods for dsRNA.

Nanotechnology in Crop and Crop Protection Industry

In many developing nations, agriculture serves as the cornerstone of their economies. The potential for progress in these countries hinges on their embrace of innovative agricultural technologies geared towards sustainability and efficiency. Consider the field of phytopathology, where it's estimated that plant diseases can slash crop yields by 10–20%, dealing significant blows to the global economy. In order to overcome this obstacle, there's a growing urgency to explore new avenues of disease management. Enter nanotechnology, offering precise delivery mechanisms for pesticides and fertilizers, thereby laying the groundwork for what's now known as "precision farming." This cutting-edge concept revolves around maximizing crop yields while minimizing resource input. By leveraging nanotechnology, precision farming can monitor real-time environmental conditions and tailor interventions accordingly, utilizing a network of wireless sensors and advanced computing systems. This approach, epitomized by the use of "smart dust" composed of sensors and robots, promises targeted and efficient pest control and irrigation management. The integration of nanosensors not only enhances agricultural productivity but also mitigates environmental pollution stemming from traditional farming practices. It's clear that such advancements will empower farmers with better decision-making tools. Case in point, LOFAR_Agro's application of nanosensors to monitor the microclimate of potato crops, offering insights into controlling Phytophthora infection.

Nanotechnology for Controlling Plant Pathogens

More than 70% of the main crop diseases are caused by fungi, which are important plant pathogens (Agrios, 2012). Farmers face ongoing challenges in safeguarding their crops from microbial diseases, often resorting to costly and environmentally harmful agrochemicals. Unfortunately, these practices exacerbate rather than alleviate the situation, contributing to issues like pest resistance on a global scale. It is imperative to reassess traditional agricultural methods and explore innovative substitutes (Ismail et al., 2017; Bhattacharyya et al., 2016).

Agro nanotechnology presents itself as a promising solution to agronomic problems (Sangeetha et al., 2017a). Nanoparticles (NPs) and nanotechnology-based tools offer targeted approaches to controlling plant diseases and combating pathogens. Studies have shown that NPs are effective against a range of bacterial and fungal diseases, including Aspergillus flavus, Bacillus subtilis, and Alternaria alternata (He et al., 2011; Lamsal et al., 2011; Park et al., 2006; Ocsoy et al., 2013; Jo et al., 2009; Kim et al., 2012; Rajiv et al., 2013). Because of their diverse inhibitory effects on phytopathogens, silver nanoparticles (AgNPs) have attracted a lot of attention (Bhaskar et al., 2016; Park et al., 2006; Rabab and EL-Shafey, 2013).

According to Dar and Soytong (2014) and Xue et al. (2014), NPs can be combined in a way that enhances their antimicrobial activity in a synergistic way with biocontrol agents, essential oils, or biopolymer-based substances like cupric and sulfur NPs. This can potentially reduce the total amount of pesticides required and delay the appearance of pest resistance. Furthermore, the incorporation of active components that have the potential to self-inactivate when exposed to sunlight promotes the creation of ecologically benign "nanocides" that have increased pesticide activity as well as residue degradation capabilities. Smart delivery systems, such as nano-dispersed formulations, offer cost-effective and less toxic alternatives to conventional agrochemicals, facilitating mass-scale production of fungicides while promoting targeted crop protection.

LIMITATION OF AGRONANOTECHNOLOGY

When addressing Integrated Pest Management (IPM), it's crucial to explore superior alternatives, especially considering potential future limitations on existing agrochemicals by national regulatory bodies (Srinivasan and Tung, 2015). Nanotechnology presents promising avenues, yet its full integration into agriculture remains pending commercialization. The agricultural community exhibits limited interest in biochemical based on nanotechnology since there aren't enough financial incentives. Various factors contribute to the slow adoption of nanotechnology, including governmental disengagement, a predominant focus on conventional farming methods, minimal competition within the agricultural sector, waning youth enthusiasm, heightened input costs, insufficient farming expertise, and public indifference (Mukhopadhyay, 2014). Agro nanotechnology holds vast potential to enhance farmer livelihoods by revolutionizing agricultural practices. Raising awareness about this science is crucial for bolstering its acceptance and expediting the discovery of new applications in the field.

FUTURE PROSPECTIVES

The potential uses of nanomaterials in agriculture are being investigated more and more, with a focus on reducing the need for plant protection products, minimizing nutrient loss during fertilization, and optimizing nutrient management to enhance yields. Despite the promising benefits, the integration of nanotechnology into agriculture remains relatively limited compared to other sectors. Numerous nanoparticles, nanocapsules, and viral capsids are examples of nanotools and devices that show promise for water treatment, targeted delivery of active chemicals, enhanced plant nutrient absorption, and disease detection and treatment. Recently, powerful substitutes for natural enzymes known as nanozymes have been discovered. As was previously said, despite being in the early phases of research, their application has grown significantly in a variety of biomolecule detection and treatment techniques. Although nanozymes have several benefits, including low cost, high stability, durability, ease of mass production, and long-term storage capacity, there are a number of issues that need to be resolved before they can be used in real-world applications. For instance, ongoing research investigates the application of starch-based nanoparticles as environmentally friendly pesticide and biostimulant delivery methods. For extremely sensitive bio-chemical sensors, novel nanomaterials with distinctive chemical, physical, and mechanical properties—such as fullerenes, nanofibers, and electrochemically active carbon nanotubes—have been produced. While some nano-products tailored for agriculture have entered the market, such as soil-enhancers promoting efficient water distribution and storage, their commercial adoption remains limited due to high development costs. Usually, greater profits in the pharmaceutical or medical industries outweigh these expenditures, which are not yet realized in agriculture. Nonetheless, research persists within the agro-chemical industry to explore potential future advantages and widen the scope of nano-based solutions for agricultural challenges.

CONCLUSION

Over the past few decades, the global human population has surged at an unprecedented rate, leading to rapid industrialization, the loss of arable land, and widespread urbanization. Current agricultural methods such as plant breeding and IPM (Integrated Pest Management) are struggling to keep pace with the demands of feeding billions of people. Thus, there is a pressing need for innovative solutions that can address both current and future food requirements. One promising avenue is agro-nanotechnology, a field that has emerged relatively recently, offering potential solutions to these challenges. By harnessing the power of nanoparticles (NPs), we can potentially reduce reliance on chemical inputs, minimize nutrient loss, and boost crop yields. With regard to the application of nanozyme-based systems for pathogen detection and control, this study provides a thorough summary of current developments. When it comes to improved catalytic activity, high stability, easy preparation, customizable size, and cost-effectiveness, nanozymes are clearly better to their natural enzyme counterparts. Nanozymes are positioned as very promising agents for antibacterial applications and flexible detection medium due to their properties. Although modern nanozymes have impressive antibacterial and detecting capabilities, most of their applications are still in the early stages. This technique is very promising for addressing problems including rising chemical input costs, overuse of pesticides, and groundwater and soil contamination. For instance, zero-valent iron nanoparticles show promise in remediating soil contaminated with pesticides due to their strong affinity for heavy metals and organic compounds. Additionally, iron nanoparticles exhibit excellent soil-binding properties akin to calcium carbonate (CaCO3). Furthermore, to address concerns about environmental sustainability, there should be a greater emphasis on utilizing agricultural waste materials as raw materials for nanoparticle manufacturing.

REFERENCES

Agrios, G. (2012). *Plant Pathology*. Elsevier. Retrieved from http://books.google.ie/books?id= xLdSrKhThNEC&printsec=frontcover&dq=0323144039&hl=&cd=1&source=gbs_api

Balasubramanian, K., & Burghard, M. (2006). Biosensors based on carbon nanotubes. *Analytical and Bioanalytical Chemistry*, *385*(3), 452–468. doi:10.1007/s00216-006-0314-8 PMID:16568294

Balaure, P. C., Gudovan, D., & Gudovan, I. (2017). Nanopesticides: a new paradigm in crop protection. In New pesticides and soil sensors (pp. 129-192). Academic Press.

Bao, W., Wang, J., Wang, Q., O'Hare, D., & Wan, Y. (2016). Layered double hydroxide nanotransporter for molecule delivery to intact plant cells. *Scientific Reports*, *6*(1), 26738. doi:10.1038/srep26738 PMID:27221055

Barik, T. K., Sahu, B., & Swain, V. (2008). Nanosilica—From medicine to pest control. *Parasitology Research*, *103*(2), 253–258. doi:10.1007/s00436-008-0975-7 PMID:18438740

Barik, T. K., Sahu, B., & Swain, V. (2008). Nanosilica—From medicine to pest control. *Parasitology Research*, *103*(2), 253–258. doi:10.1007/s00436-008-0975-7 PMID:18438740

Baulcombe, D. (2004). RNA silencing in plants. *Nature*, *431*(7006), 356–363. doi:10.1038/nature02874 PMID:15372043

Bhaskar, B., Ahammed, S. K., Chaitanya, B. H., Rasheed, V. A., & Prasad, T. N. V. K. V. (2016). Silver nanoparticles: mycogenesis, characterization and its anti-plant pathogenic applications. Int J Res Appl Nat Soc Sci, 4(10), 105-114.

Campos, E. V. R., Oliveira, J. L. D., da Silva, C. M. G., Pascoli, M., Pasquoto, T., Lima, R., Abhilash, P. C., & Fernandes Fraceto, L. (2015). Polymeric and solid lipid nanoparticles for sustained release of carbendazim and tebuconazole in agricultural applications. *Scientific Reports*, *5*(1), 13809. doi:10.1038/srep13809 PMID:26346969

Cota-Arriola, O., Onofre Cortez-Rocha, M., Burgos-Hernández, A., Marina Ezquerra-Brauer, J., & Plascencia-Jatomea, M. (2013). Controlled release matrices and micro/nanoparticles of chitosan with antimicrobial potential: Development of new strategies for microbial control in agriculture. *Journal of the Science of Food and Agriculture*, *93*(7), 1525–1536. doi:10.1002/jsfa.6060 PMID:23512598

Dar, J., & Soytong, K. (2014). Construction and characterization of copolymer nanomaterials loaded with bioactive compounds from Chaetomium species. *Agricultural Technology (Thailand)*, *10*(4), 823–831.

Debnath, N., Das, S., Seth, D., Chandra, R., Bhattacharya, S. C., & Goswami, A. (2011). Entomotoxic effect of silica nanoparticles against Sitophilus oryzae (L.). *Journal of Pest Science*, *84*(1), 99–105. doi:10.1007/s10340-010-0332-3

Dwivedi, S., Saquib, Q., Al-Khedhairy, A. A., & Musarrat, J. (2016). Understanding the role of nanomaterials in agriculture. In D. Singh, H. Singh, & R. Prabha (Eds.), *Microbial Inoculants in Sustainable Agricultural Productivity* (pp. 271–288). Springer. doi:10.1007/978-81-322-2644-4_17

Ekambaram, P. (2012). a. Abdul hasan sathali and k. Priyanka; Solid lipid nanoparticles: A review; sci. Revs. *Chemical Communications (Cambridge)*, *2*(1).

Elamawi, R. M. A., & El-Shafey, R. A. (2013). Inhibition effects of silver nanoparticles against rice blast disease caused by Magnaporthe grisea. *Egyptian Journal of Agricultural Research*, *91*(4), 1271–1283. doi:10.21608/ejar.2013.165104

Fang, Y., & Ramasamy, R. P. (2015). Current and prospective methods for plant disease detection. *Biosensors (Basel)*, *5*(3), 537–561. doi:10.3390/bios5030537 PMID:26287253

Fang, Y., Umasankar, Y., & Ramasamy, R. P. (2014). Electrochemical detection of p-ethylguaiacol, a fungi infected fruit volatile using metal oxide nanoparticles. *Analyst*, *139*(15), 3804–3810. doi:10.1039/C4AN00384E PMID:24895939

Farkas, G. L., & Kiraaly, Z. (1962). Role of phenolic compounds in the physiology of plant diseases and disease resistance. *Journal of Phytopathology*, *44*(2), 105–150. doi:10.1111/j.1439-0434.1962.tb02005.x

Flood, J. (2010). The importance of plant health to food security. *Food Security*, *2*(3), 215–231. doi:10.1007/s12571-010-0072-5

Forgiarini, A., Esquena, J., Gonzalez, C., & Solans, C. (2001). Formation of nano-emulsions by low-energy emulsification methods at constant temperature. *Langmuir*, *17*(7), 2076–2083. doi:10.1021/la001362n

Fu, L., Wang, Z., Dhankher, O. P., & Xing, B. (2020). Nanotechnology as a new sustainable approach for controlling crop diseases and increasing agricultural production. *Journal of Experimental Botany*, *71*(2), 507–519. doi:10.1093/jxb/erz314 PMID:31270541

Ghormade, V., Deshpande, M. V., & Paknikar, K. M. (2011). Perspectives for nano-biotechnology enabled protection and nutrition of plants. *Biotechnology Advances*, *29*(6), 792–803. doi:10.1016/j.biotechadv.2011.06.007 PMID:21729746

Gogos, A., Knauer, K., & Bucheli, T. D. (2012). Nanomaterials in plant protection and fertilization: Current state, foreseen applications, and research priorities. *Journal of Agricultural and Food Chemistry*, *60*(39), 9781–9792. doi:10.1021/jf302154y PMID:22963545

González-Melendi, P., Fernández-Pacheco, R., Coronado, M. J., Corredor, E., Testillano, P. S., Risueño, M. C., & Pérez-de-Luque, A. (2008). Nanoparticles as smart treatment-delivery systems in plants: Assessment of different techniques of microscopy for their visualization in plant tissues. *Annals of Botany*, *101*(1), 187–195. doi:10.1093/aob/mcm283 PMID:17998213

Goswami, A., Roy, I., Sengupta, S., & Debnath, N. (2010). Novel applications of solid and liquid formulations of nanoparticles against insect pests and pathogens. *Thin Solid Films*, *519*(3), 1252–1257. doi:10.1016/j.tsf.2010.08.079

Hatfaludi, T., Liska, M., Zellinger, D., Ousman, J. P., Szostak, M., árpád Ambrus, Jalava, K., & Lubitz, W. (2004). Bacterial ghost technology for pesticide delivery. *Journal of Agricultural and Food Chemistry*, *52*(18), 5627–5634. doi:10.1021/jf049489w PMID:15373403

Hayles, J., Johnson, L., Worthley, C., & Losic, D. (2017). Nanopesticides: a review of current research and perspectives. *New pesticides and soil sensors*, 193-225.

He, L., Liu, Y., Mustapha, A., & Lin, M. (2011). Antifungal activity of zinc oxide nanoparticles against Botrytis cinerea and Penicillium expansum. *Microbiological Research*, *166*(3), 207–215. doi:10.1016/j.micres.2010.03.003 PMID:20630731

Ilk, S., Saglam, N., & Özgen, M. (2017). Kaempferol loaded lecithin/chitosan nanoparticles: Preparation, characterization, and their potential applications as a sustainable antifungal agent. *Artificial Cells, Nanomedicine, and Biotechnology*, *45*(5), 907–916. doi:10.1080/21691401.2016.1192040 PMID:27265551

Ismail, M., Prasad, R., Ibrahim, A. I., & Ahmed, A. I. (2017). Modern prospects of nanotechnology in plant pathology. *Nanotechnology: An Agricultural Paradigm*, 305-317.

Jain, D., & Kothari, S. L. (2014). Green synthesis of silver nanoparticles and their application in plant virus inhibition. *Journal of Mycology and Plant Pathology*, *44*(1), 21–24.

Janatova, A., Bernardos, A., Smid, J., Frankova, A., Lhotka, M., Kourimská, L., & Kloucek, P. (2015). Long-term antifungal activity of volatile essential oil components released from mesoporous silica materials. *Industrial Crops and Products*, *67*, 216–220. doi:10.1016/j.indcrop.2015.01.019

Jenne, M., Kambham, M., Tollamadugu, N. P., Karanam, H. P., Tirupati, M. K., Reddy Balam, R., & Yagireddy, M. (2018). The use of slow releasing nanoparticle encapsulated Azadirachtin formulations for the management of Caryedon serratus O. (groundnut bruchid). *IET Nanobiotechnology / IET*, *12*(7), 963–967. doi:10.1049/iet-nbt.2017.0276 PMID:30247138

Jo, Y. K., Kim, B. H., & Jung, G. (2009). Antifungal activity of silver ions and nanoparticles on phytopathogenic fungi. *Plant Disease*, *93*(10), 1037–1043. doi:10.1094/PDIS-93-10-1037 PMID:30754381

Kah, M., Beulke, S., Tiede, K., & Hofmann, T. (2013). Nanopesticides: State of knowledge, environmental fate, and exposure modeling. *Critical Reviews in Environmental Science and Technology*, *43*(16), 1823–1867. doi:10.1080/10643389.2012.671750

Kah, M., & Hofmann, T. (2014). Nanopesticide research: Current trends and future priorities. *Environment International*, *63*, 224–235. doi:10.1016/j.envint.2013.11.015 PMID:24333990

Khandelwal, N., Barbole, R. S., Banerjee, S. S., Chate, G. P., Biradar, A. V., Khandare, J. J., & Giri, A. P. (2016). Budding trends in integrated pest management using advanced micro-and nano-materials: Challenges and perspectives. *Journal of Environmental Management*, *184*, 157–169. doi:10.1016/j.jenvman.2016.09.071 PMID:27697374

Khiyami, M. A., Almoammar, H., Awad, Y. M., Alghuthaymi, M. A., & Abd-Elsalam, K. A. (2014). Plant pathogen nanodiagnostic techniques: Forthcoming changes? *Biotechnology, Biotechnological Equipment*, *28*(5), 775–785. doi:10.1080/13102818.2014.960739 PMID:26740775

Kim, S. W., Jung, J. H., Lamsal, K., Kim, Y. S., Min, J. S., & Lee, Y. S. (2012). Antifungal effects of silver nanoparticles (AgNPs) against various plant pathogenic fungi. *Mycobiology*, *40*(1), 53–58. doi:10.5941/MYCO.2012.40.1.053 PMID:22783135

Krishnaraj, C., Ramachandran, R., Mohan, K., & Kalaichelvan, P. T. (2012). Optimization for rapid synthesis of silver nanoparticles and its effect on phytopathogenic fungi. *Spectrochimica Acta. Part A: Molecular and Biomolecular Spectroscopy*, *93*, 95–99. doi:10.1016/j.saa.2012.03.002 PMID:22465774

Kumar, D. R., Kumar, P. S., Gandhi, M. R., Al-Dhabi, N. A., Paulraj, M. G., & Ignacimuthu, S. (2016). Delivery of chitosan/dsRNA nanoparticles for silencing of wing development vestigial (vg) gene in Aedes aegypti mosquitoes. *International Journal of Biological Macromolecules*, *86*, 89–95. doi:10.1016/j.ijbiomac.2016.01.030 PMID:26794313

Kuzma, J., & VerHage, P. (2006). *Nanotechnology in agriculture and food production: anticipated applications*. Project on emerging nanotechnologies.

Lai, F., Wissing, S. A., Müller, R. H., & Fadda, A. M. (2006). Artemisia arborescens L essential oil-loaded solid lipid nanoparticles for potential agricultural application: Preparation and characterization. *AAPS PharmSciTech*, *7*(1), E10–E18. doi:10.1208/pt070102 PMID:28290017

Lamsal, K., Kim, S. W., Jung, J. H., Kim, Y. S., Kim, K. S., & Lee, Y. S. (2011). Application of silver nanoparticles for the control of Colletotrichum species in vitro and pepper anthracnose disease in field. *Mycobiology*, *39*(3), 194–199. doi:10.5941/MYCO.2011.39.3.194 PMID:22783103

Lauterwasser, C. (2005). Small sizes that matter: Opportunities and risks of Nanotechnologies. *Report in co-operation with the OECD International Futures Programme.*

Lee, G. W. J., & Tadros, T. F. (1982). Formation and stability of emulsions produced by dilution of emulsifiable concentrates. Part I. An investigation of the dispersion on dilution of emulsifiable concentrates containing cationic and non-ionic surfactants. *Colloids and Surfaces*, *5*(2), 105–115. doi:10.1016/0166-6622(82)80066-8

Li, M., Huang, Q., & Wu, Y. (2011). A novel chitosan-poly (lactide) copolymer and its submicron particles as imidacloprid carriers. *Pest Management Science*, *67*(7), 831–836. doi:10.1002/ps.2120 PMID:21370387

Liu, X., He, B., Xu, Z., Yin, M., Yang, W., Zhang, H., & Shen, J. (2015). A functionalized fluorescent dendrimer as a pesticide nanocarrier: Application in pest control. *Nanoscale*, *7*(2), 445–449. doi:10.1039/C4NR05733C PMID:25429649

Liu, Y., Laks, P., & Heiden, P. (2002). Controlled release of biocides in solid wood. II. Efficacy against Trametes versicolor and Gloeophyllum trabeum wood decay fungi. *Journal of Applied Polymer Science*, *86*(3), 608–614. doi:10.1002/app.10897

Mak, A. C., Osterfeld, S. J., Yu, H., Wang, S. X., Davis, R. W., Jejelowo, O. A., & Pourmand, N. (2010). Sensitive giant magnetoresistive-based immunoassay for multiplex mycotoxin detection. *Biosensors & Bioelectronics*, *25*(7), 1635–1639. doi:10.1016/j.bios.2009.11.028 PMID:20047828

Malerba, M., & Cerana, R. (2016). Chitosan effects on plant systems. *International Journal of Molecular Sciences*, *17*(7), 996. doi:10.3390/ijms17070996 PMID:27347928

Martinelli, F., Scalenghe, R., Davino, S., Panno, S., Scuderi, G., Ruisi, P., & Dandekar, A. M. (2015). Advanced methods of plant disease detection. A review. *Agronomy for Sustainable Development*, *35*(1), 1–25. doi:10.1007/s13593-014-0246-1

McCartney, H. A., Foster, S. J., Fraaije, B. A., & Ward, E. (2003). Molecular diagnostics for fungal plant pathogens. *Pest Management Science: Formerly. Pest Management Science*, *59*(2), 129–142. doi:10.1002/ps.575

Meng, Y., Li, Y., Galvani, C. D., Hao, G., Turner, J. N., Burr, T. J., & Hoch, H. C. (2005). Upstream migration of Xylella fastidiosa via pilus-driven twitching motility. *Journal of Bacteriology*, *187*(16), 5560–5567. doi:10.1128/JB.187.16.5560-5567.2005 PMID:16077100

Mishra, S., & Singh, H. B. (2015). Biosynthesized silver nanoparticles as a nanoweapon against phytopathogens: Exploring their scope and potential in agriculture. *Applied Microbiology and Biotechnology*, *99*(3), 1097–1107. doi:10.1007/s00253-014-6296-0 PMID:25547832

Mody, V. V., Cox, A., Shah, S., Singh, A., Bevins, W., & Parihar, H. (2014). Magnetic nanoparticle drug delivery systems for targeting tumor. *Applied Nanoscience*, *4*(4), 385–392. doi:10.1007/s13204-013-0216-y

Mousavi, S. R., & Rezaei, M. (2011). Nanotechnology in agriculture and food production. *Journal of Applied Environmental and Biological Sciences*, *1*(10), 414–419.

Mukhopadhyay, S. S. (2014). Nanotechnology in agriculture: Prospects and constraints. *Nanotechnology, Science and Applications*, 63–71. doi:10.2147/NSA.S39409 PMID:25187699

Nasseri, M., Golmohammadzadeh, S., Arouiee, H., Jaafari, M. R., & Neamati, H. (2016). Antifungal activity of Zataria multiflora essential oil-loaded solid lipid nanoparticles in-vitro condition. *Iranian Journal of Basic Medical Sciences.*, *19*(11), 1231. PMID:27917280

Ocsoy, I., Paret, M. L., Ocsoy, M. A., Kunwar, S., Chen, T., You, M., & Tan, W. (2013). Nanotechnology in plant disease management: DNA-directed silver nanoparticles on graphene oxide as an antibacterial against Xanthomonas perforans. *ACS Nano*, *7*(10), 8972–8980. doi:10.1021/nn4034794 PMID:24016217

Omanović-Mikličanina, E., & Maksimović, M. (2016). Nanosensors applications in agriculture and food industry. *Bull Chem Technol Bosnia Herzegovina*, *47*, 59–70.

Park, H. J., Kim, S. H., Kim, H. J., & Choi, S. H. (2006). A new composition of nanosized silica-silver for control of various plant diseases. *The Plant Pathology Journal*, *22*(3), 295–302. doi:10.5423/PPJ.2006.22.3.295

Pimentel, D. (1995). Amounts of pesticides reaching target pests: Environmental impacts and ethics. *Journal of Agricultural & Environmental Ethics*, *8*(1), 17–29. doi:10.1007/BF02286399

Prasad, R., Bhattacharyya, A., & Nguyen, Q. D. (2017). Nanotechnology in sustainable agriculture: Recent developments, challenges, and perspectives. *Frontiers in Microbiology*, *8*, 1014. doi:10.3389/fmicb.2017.01014 PMID:28676790

Prasad, R., Kumar, V., & Prasad, K. S. (2014). Nanotechnology in sustainable agriculture: Present concerns and future aspects. *African Journal of Biotechnology*, *13*(6), 705–713. doi:10.5897/AJBX2013.13554

Radoi, A., Targa, M., Prieto-Simon, B., & Marty, J. L. (2008). Enzyme-Linked Immunosorbent Assay (ELISA) based on superparamagnetic nanoparticles for aflatoxin M1 detection. *Talanta*, *77*(1), 138–143. doi:10.1016/j.talanta.2008.05.048 PMID:18804611

Rafique, M., Sadaf, I., Rafique, M. S., & Tahir, M. B. (2017). A review on green synthesis of silver nanoparticles and their applications. *Artificial Cells, Nanomedicine, and Biotechnology*, *45*(7), 1272–1291. doi:10.1080/21691401.2016.1241792 PMID:27825269

Rajiv, P., Rajeshwari, S., & Venckatesh, R. (2013). Bio-Fabrication of zinc oxide nanoparticles using leaf extract of Parthenium hysterophorus L. and its size-dependent antifungal activity against plant fungal pathogens. *Spectrochimica Acta. Part A: Molecular and Biomolecular Spectroscopy*, *112*, 384–387. doi:10.1016/j.saa.2013.04.072 PMID:23686093

Robinson, K. E., Worrall, E. A., & Mitter, N. (2014). Double stranded RNA expression and its topical application for non-transgenic resistance to plant viruses. *Journal of Plant Biochemistry and Biotechnology*, *23*(3), 231–237. doi:10.1007/s13562-014-0260-z

Rosen, J. E., Yoffe, S., Meerasa, A., Verma, M., & Gu, F. X. (2011). Nanotechnology and diagnostic imaging: New advances in contrast agent technology. *Journal of Nanomedicine & Nanotechnology*, *2*(5), 115–126. doi:10.4172/2157-7439.1000115

Sabir, S., Arshad, M., & Chaudhari, S. K. (2014). Zinc oxide nanoparticles for revolutionizing agriculture: Synthesis and applications. *TheScientificWorldJournal*. doi:10.1155/2014/925494 PMID:25436235

Sadeghi, R., Rodriguez, R. J., Yao, Y., & Kokini, J. L. (2017). Advances in nanotechnology as they pertain to food and agriculture: Benefits and risks. *Annual Review of Food Science and Technology*, *8*(1), 467–492. doi:10.1146/annurev-food-041715-033338 PMID:28125343

Sangeetha, J., Thangadurai, D., Hospet, R., Harish, E. R., Purushotham, P., Mujeeb, M. A., & Prasad, R. (2017). Nanoagrotechnology for soil quality, crop performance and environmental management. *Nanotechnology: an agricultural paradigm*, 73-97.

Schofield, C. L., Haines, A. H., Field, R. A., & Russell, D. A. (2006). Silver and gold glyconanoparticles for colorimetric bioassays. *Langmuir*, *22*(15), 6707–6711. doi:10.1021/la060288r PMID:16831017

Scrinis, G., & Lyons, K. (2007). The emerging nano-corporate paradigm: Nanotechnology and the transformation of nature, food and agri-food systems. *International Journal of Sociology of Agriculture and Food*, *15*(2), 22–44.

Singh, P., Kumari, K., Vishvakrma, V. K., Mehrotra, G. K., Chandra, R., Kumar, D., ... & Shahare, V. V. (2017). Metal NPs (Au, Ag, and Cu): Synthesis, stabilization, and their role in green chemistry and drug delivery. *Green Technologies and Environmental Sustainability*, 309-337.

Sinha, K., Ghosh, J., & Sil, P. C. (2017). New pesticides: a cutting-edge view of contributions from nanotechnology for the development of sustainable agricultural pest control. In *New pesticides and soil sensors* (pp. 47–79). Academic Press. doi:10.1016/B978-0-12-804299-1.00003-5

Song, M. R., Cui, S. M., Gao, F., Liu, Y. R., Fan, C. L., Lei, T. Q., & Liu, D. C. (2012). Dispersible silica nanoparticles as carrier for enhanced bioactivity of chlorfenapyr. *Journal of Pesticide Science*, *37*(3), 258–260. doi:10.1584/jpestics.D12-027

Sparks, T. C., & Nauen, R. (2015). IRAC: Mode of action classification and insecticide resistance management. *Pesticide Biochemistry and Physiology*, *121*, 122–128. doi:10.1016/j.pestbp.2014.11.014 PMID:26047120

Srinivasan, C., & Saraswathi, R. (2010). Nano-agriculture—Carbon nanotubes enhance tomato seed germination and plant growth. *Current Science*, *99*(3), 274–275.

Stadler, T., Buteler, M., & Weaver, D. K. (2010). Novel use of nanostructured alumina as an insecticide. *Pest Management Science: Formerly. Pest Management Science*, *66*(6), 577–579. doi:10.1002/ps.1915

Stephenson, G. R., Coats, J. R., & Yamamoto, H. (2001). Pesticide use and world food production: risks and benefits. In *Expert Committee on Weeds Comité d'experts en malherbologie. Proceedings of the 2000 National Meeting* (pp. 9-15). Citeseer.

Sun, Y. F., Liu, S. B., Meng, F. L., Liu, J. Y., Jin, Z., Kong, L. T., & Liu, J. H. (2012). Metal oxide nanostructures and their gas sensing properties: A review. *Sensors (Basel)*, *12*(3), 2610–2631. doi:10.3390/s120302610 PMID:22736968

Tamjidi, F., Shahedi, M., Varshosaz, J., & Nasirpour, A. (2013). Nanostructured lipid carriers (NLC): A potential delivery system for bioactive food molecules. *Innovative Food Science & Emerging Technologies*, *19*, 29–43. doi:10.1016/j.ifset.2013.03.002

Usha Rani, P., Madhusudhanamurthy, J., & Sreedhar, B. (2014). Dynamic adsorption of α-pinene and linalool on silica nanoparticles for enhanced antifeedant activity against agricultural pests. *Journal of Pest Science*, *87*(1), 191–200. doi:10.1007/s10340-013-0538-2

Vandana Ghormade, V. G., Deshpande, M. V., & Paknikar, K. M. (2011). Perspectives for nano-biotechnology enabled protection and nutrition of plants. Academic Press.

Wang, Y., Cui, H., Sun, C., Zhao, X., & Cui, B. (2014). Construction and evaluation of controlled-release delivery system of Abamectin using porous silica nanoparticles as carriers. *Nanoscale Research Letters*, *9*(1), 1–6. doi:10.1186/1556-276X-9-655 PMID:26088998

Wang, Y., Li, C., Wang, Y., Zhang, Y., & Li, X. (2018). Compound pesticide controlled release system based on the mixture of poly (butylene succinate) and PLA. *Journal of Microencapsulation*, *35*(5), 494–503. doi:10.1080/02652048.2018.1538265 PMID:30395751

Wang, Z., Wei, F., Liu, S. Y., Xu, Q., Huang, J. Y., Dong, X. Y., & Chen, H. (2010). Electrocatalytic oxidation of phytohormone salicylic acid at copper nanoparticles-modified gold electrode and its detection in oilseed rape infected with fungal pathogen Sclerotinia sclerotiorum. *Talanta*, *80*(3), 1277–1281. doi:10.1016/j.talanta.2009.09.023 PMID:20006087

Wanyika, H. (2014). Sustained release of fungicide metalaxyl by mesoporous silica nanospheres. In Nanotechnology for Sustainable Development (pp. 321-329). Springer International Publishing.

Weigl, B., Domingo, G., LaBarre, P., & Gerlach, J. (2008). Towards non-and minimally instrumented, microfluidics-based diagnostic devices. *Lab on a Chip*, *8*(12), 1999–2014. doi:10.1039/b811314a PMID:19023463

Wibowo, D., Zhao, C. X., Peters, B. C., & Middelberg, A. P. (2014). Sustained release of fipronil insecticide in vitro and in vivo from biocompatible silica nanocapsules. *Journal of Agricultural and Food Chemistry*, *62*(52), 12504–12511. doi:10.1021/jf504455x PMID:25479362

Xu, L., Cao, L. D., Li, F. M., Wang, X. J., & Huang, Q. L. (2014). Utilization of chitosan-lactide copolymer nanoparticles as controlled release pesticide carrier for pyraclostrobin against Colletotrichum gossypii Southw. *Journal of Dispersion Science and Technology*, *35*(4), 544–550. doi:10.1080/01932691.2013.800455

Xu, Z. P., Stevenson, G. S., Lu, C. Q., Lu, G. Q., Bartlett, P. F., & Gray, P. P. (2006). Stable suspension of layered double hydroxide nanoparticles in aqueous solution. *Journal of the American Chemical Society*, *128*(1), 36–37. doi:10.1021/ja056652a PMID:16390109

Xue, J., Luo, Z., Li, P., Ding, Y., Cui, Y., & Wu, Q. (2014). A residue-free green synergistic antifungal nanotechnology for pesticide thiram by ZnO nanoparticles. *Scientific Reports*, *4*(1), 5408. doi:10.1038/srep05408 PMID:25023938

Yu, X., Chattopadhyay, D., Galeska, I., Papadimitrakopoulos, F., & Rusling, J. F. (2003). Peroxidase activity of enzymes bound to the ends of single-wall carbon nanotube forest electrodes. *Electrochemistry Communications*, *5*(5), 408–411. doi:10.1016/S1388-2481(03)00076-6

Zhao, P., Cao, L., Ma, D., Zhou, Z., Huang, Q., & Pan, C. (2017). Synthesis of pyrimethanil-loaded mesoporous silica nanoparticles and its distribution and dissipation in cucumber plants. *Molecules (Basel, Switzerland)*, *22*(5), 817. doi:10.3390/molecules22050817 PMID:28509872

Section 4

Nanotechnology for Soil and Plant Conservation: Nano-Omics and Non-Enzymes for Human Welfare

Chapter 11
Nanotechnology-Based Soil Improvement and Conservation for Enhancement of Crop Production

Tirunima Patle
https://orcid.org/0000-0003-3268-6118
Shri Vaishnav Institute of Agriculture, Shri Vaishnav Vidyapeeth Vishwavidyalaya, India

Bhavana Tomar
G.D. Goenka University, India

Sneh Singh Parihar
ITM Unversity, Gwalior, India

Shiv Singh Tomar
G.D. Goenka University, India

Prashant Singh
https://orcid.org/0009-0008-9667-3598
Mandsaur University, India

ABSTRACT

The soil degradation and loss of fertile land pose severe threats to global agriculture and environmental sustainability. Soil quality degradation and loss of arable land have become pressing environmental concerns, call for innovative, sustainable, and transformative approaches to soil management, remediation, and restoration. Nanotechnology can be a transformative approach for soil remediation and land restoration. Nanomaterials remarkable properties offers a paradigm shift in soil science and environmental restoration. With their extraordinarily high surface area, tailored reactivity, and adjustable physicochemical characteristics, nanomaterials present a profound opportunity to revolutionize our approach to soil remediation and land rejuvenation. It delves into their intricate synthesis processes, advanced characterization techniques, and the underlying mechanisms governing their actions of nanomaterials and nanocomposites within the soil matrix. Within the realm of soil remediation, nanotechnology emerges as a versatile and potent tool. It excels in adsorbing, immobilizing, and degrading contaminants, including stubborn heavy metals, persistent organic pollutants, and recalcitrant pesticides. Additionally, the controlled-release capabilities of nano fertilizers come to the fore, facilitating precise nutrient delivery to plants, thereby enhancing fertility of soil and promoting plant growth. Apart from contamination reme-

DOI: 10.4018/979-8-3693-1890-4.ch011

diation, nano-materials have immense potential to improve soil structure by enhancing water retention, promoting aeration and facilitating nutrient diffusion. This multifaceted approach extends to the realm of bioremediation, where nanotechnology amplifies microbial-mediated processes, elevating enzymatic activities and accelerating contaminant degradation.

INTRODUCTION

The surge in human activities and technical interventions has significantly intensified the introduction of potentially harmful substances, agrochemicals and surplus of nutrients into soil (Midhat L et al., 2019). Soil, as a fundamental medium for crop growth, plays a vital role in aiding absorption of nutrients by plants (Parikh SJ et al., 2012). While agriculture is vital for human survival, its practices often disrupt natural ecosystems, creating an ongoing conflict between exploiting natural resources for benefits and practicing sustainable land management (Struik, PC et al., 2017 ; Pretty, J et al., 2014). This conflict poses a significant danger to quality of life, livelihoods, and overall sustainable development. (Martin JL et al.,2016) Consequently, preserving soil has become important amid increasing population pressure and the decreasing availability of fertile land due to technological activities (Rajput et al., 2021). Soil degradation and the decline of arable land pose substantial concerns for global agriculture and environmental sustainability (Ghazaryan et al., 2018, 2023) . As soil quality is declining and arable land minimizing, there is an urgent need for innovative, transformative and sustainable approaches for soil management, remediation, and restoration (Vardumyan et al., 2024).

In response to these challenges, nanotechnology emerges as a transformative solution capable of revolutionizing our understanding of soil science and environmental rehabilitation . It acts as a pivotal player in soil remediation, revolutionizing approaches to eliminate contaminants and enhance soil quality and fertility. The unique features of nanomaterials, particularly their high surface area specificity and reactivity causing from a small particle size, make them exceptionally effective in soil remediation processes. This facilitates enhanced adsorption and reaction capacities, allowing for more effective removal of contaminants from the soil matrix (Panpatte DG et al., 2016). Consequently, nanotechnology offers a promising solution to mitigate the adverse impacts of soil contamination arising from various human activities (Singh et al., 2022). Notably, the application of nanomaterials in soil remediation aligns with sustainable land management practices, emphasizing targeted interventions that minimizes ecological disruption (Panpatte DG et al., 2016; Galdames A 2017). The application of nanomaterials for soil remediation has opened new avenues for addressing environmental challenges with increased precision and more efficiency.

Numerous forms of nanomaterials, including carbon nanotubes, magnetic nanoparticles, zerovalent iron nanoparticles, and others, have been explored for their high effectiveness in soil reclamation (Xue W et al., 2018; Kumar A et al.,2019). Each nanomaterial possesses unique properties contributing to its efficacy in remediating specific types of soil contaminants. The following sections of this book chapter is delve into the applications of these nanomaterials in soil remediation, exploring their mechanisms of action, benefits, and challenges. Additionally, understanding the potential ecotoxicological effects of nanomaterials is crucial for ensuring the sustainable and responsible use of these technologies (Klaine SJ et al., 2018; Gottschalk F et al., 2013). This book chapter aims to provide a comprehensive overview of the current state of research on nanotechnology applications in soil remediation, focusing on the functions, fate, mobility, and possible ecotoxicology effects associated with various nanomaterials.

By synthesizing existing knowledge in this field, this review seeks to contribute to the development of informed strategies for the application of nanotechnology in soil remediation and land restoration.

The emergence of nanotechnology presents a global opportunity to effectively remediate or restore polluted soil (Panpatte DG et al., 2016; Galdames A 2017). Recognized for its potential as an environmentally cleaner technology, nanotechnology is praised for mitigating toxicities associated with various metals and metalloids (Klaine SJ et al., 2018; Gottschalk F et al., 2013). Moreover, it has emerged as a promising method for remediating pollutants in diverse environmental matrices, with soils being a prominent focal point (Panpatte DG et al., 2016). In this context, soil remediation stands out as a primary domain where nanotechnological approaches have gained widespread application. Recent explorations of nanoparticles (NPs) for contaminant removal, utilizing mechanisms such as adsorption, redox reactions, precipitation, and co-precipitation, underscore their significant specific surface area (Kumar A et al., 2019).. Nanomaterials, such as nanoparticles or nanocomposites, are engineered for specific soil remediation purposes, and their interactions with soil components are analyzed through omics techniques. Molecular-level investigations explore how nanomaterials influence soil microbial communities, nutrient cycling, and plant-microbe interactions. Understanding these interactions is pivotal for predicting the ecological implications and potential risks associated with the application of nanotechnology in soil remediation.

Overall, the primary goal of this book chapter is to present the current findings and achievements in the application of nanotechnology for soil remediation. with a focus on addressing soil contamination caused by various compounds. The chapter specifically delves into developments from the past decade, witnessing a significant increase in studies exploring nanotechnology for soil remediation. It also considers the potential environmental impact of nanoparticle (NP) use in soil remediation and explore the possibility of merging nano remediation with other soil remediation technologies.

NANOMATERIALS FOR SOIL CONSERVATION

Nanoparticles

Nanoparticles utilized in soil remediation predominantly consist of metal and metal oxide counterparts, exemplified by iron and titanium dioxide (Singh et al., 2023a). The distinctive attributes of these nanomaterials, notably their high surface area and reactivity, underpin their efficacy in addressing soil contaminants (Zhang Y et al., 2018; Dimkpa CO et al., 2011). Environmental application of nanomaterial are represented in fig.no.1. Iron nanoparticles, known for their pronounced redox reactivity, and titanium dioxide nanoparticles, distinguished by photocatalytic prowess, epitomize the forefront of nanoparticle-driven soil remediation. The confluence of heightened surface area, reactivity, and adeptness in redox reactions positions these nanoparticles as versatile entities for orchestrating adsorption, precipitation, and transformative processes crucial for mitigating soil-borne contaminants. The use of metal and metal oxide nanoparticles for soil remediation is nuanced and technologically advanced. The amplified surface area inherent to these nanoparticles augments their adsorption capacities, enabling selective binding with contaminants inherent in the soil matrix (Singh et al., 2023b). This attribute proves instrumental in immobilizing and sequestering diverse contaminants, contributing to their effective removal (Mondal A et al., 2021). Redox reactions, integral to the functionality of metal nanoparticles, facilitate reduction or oxidation of contaminants. Iron nanoparticles, in particular, function as potent electron donors in reductive reactions, effecting the transformation of certain pollutants into less deleterious species. The prowess of nanoparticles to engage in redox reactions broadens their applicability, spanning a gamut of soil contaminants encompassing heavy metals and chlorinated compounds (Govekar S et al., 2018).

Nanocomposites

Nanocomposites consist of nano-sized phases with unique structures and chemistry providing distinct electrical, mechanical, electrochemical, optical, catalytic and structural properties compared to their individual components. Introduction of nanocomposites for soil remediation represents a purposeful integration of different nanomaterials, generating synergistic effects that surpass the capabilities of individual elements. This strategic combination aims to enhance remediation efficiency by capitalizing on the unique attributes of diverse nanoscale entities. The development of nanocomposites is guided by a thorough understanding of the interactions between constituents, aiming to harness collective strengths and enhance reactivity, stability, and functional efficacy (Reginatto C et al., 2020; Khan A et al., 2010).

Application of nanocomposites for soil remediation involves instances where these composite structures exhibit remarkable stability and reactivity. For example, the integration of graphene oxide with iron oxide nanoparticles has shown significant synergy, providing simultaneous adsorption and reduction capabilities for contaminants (Reginatto C et al., 2020). The physicochemical properties, including elution behaviour, eluent concentration, and pH titration studies confirms the nanocomposites maximum adsorption of metal ions. The improved PANI-zicronium titanium phosphate (PZTP) nanocomposite

Figure 1. Environmental application of nanomaterial (Govekar S et al., 2018)

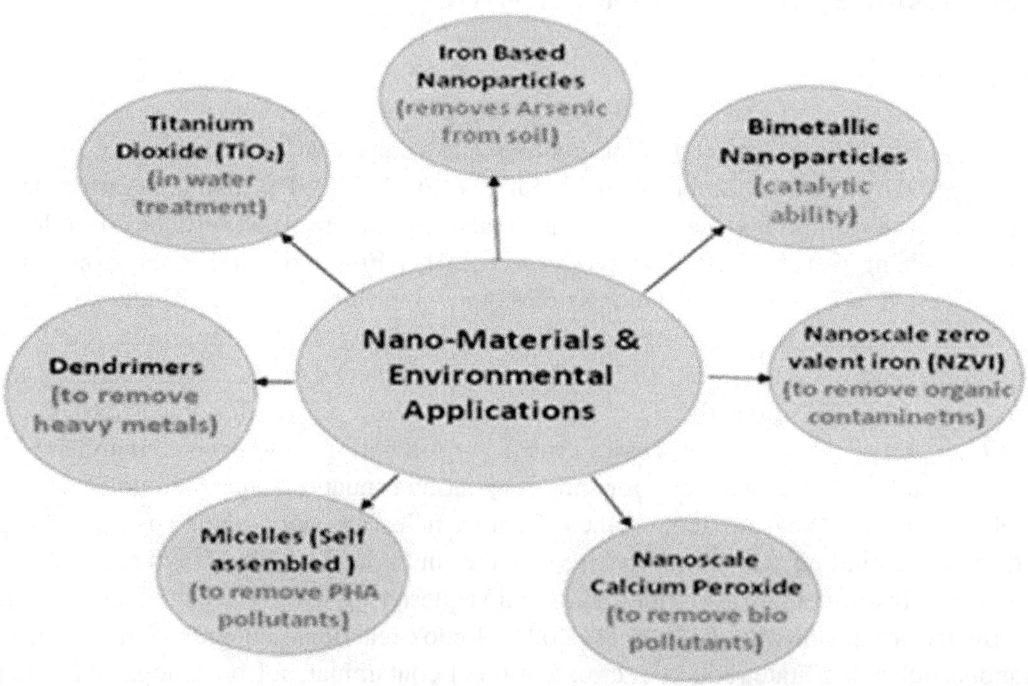

cation exchanger developed using the sol-gel method has a high ion-exchange capacity as well as better chemical and thermal stability. This material effectively removes and recovers Hg2þ and Pb2þ ions from environmental waste, with a detection range of 10^-10 to 10^-1 mol m3. (Khan MZ et al., 2011; Khan AA et al., 2011; Shahadat m et al., 2015).

In the realm of organic pollutant remediation, the use of nanocomposites featuring clay nanoparticles and metal oxide nanoparticles has gained attention. This composite system exhibits superior adsorption and catalytic degradation capacities compared to their individual counterparts (Younis U et al., 2012). The orchestrated interplay within the nanocomposite structure amplifies overall performance, offering a comprehensive solution to the intricate challenges posed by diverse soil contaminants. Photocatalytic experiments on the composite material, which involve the degradation of MB under Xe lamp irradiation, show a higher degradation rate than Degussa P25 or Bi2S3-TiO2. The results suggested that Bi2S3-TiO2 heterojunction/polymer nanocomposite has the potential for photochemical destruction of organic contaminants. (Ma A et al., 2012).

Furthermore, the design and fabrication of nanocomposites often employ advanced techniques such as sol-gel processes, co-precipitation methods, and controlled synthesis approaches. These methodologies ensure precise control over the morphology, structure, and composition of the nanocomposite, further optimizing its remediation functionalities (Reginatto C et al., 2020).

Nanostructured Materials

Nanostructured materials, exemplified by nanotubes and nanofibers, mark a transformative phase in soil remediation strategies, owing to their distinctive attributes, especially the high aspect ratio and surface area. Nanotubes, such as carbon nanotubes, and nanofibers, constructed from various materials including polymers or metals, possess inherent structural advantages that enhance their efficacy in soil remediation (Baalousha M et al., 2016). The elevated aspect ratio amplifies their contact interfaces with contaminants, while the increased surface area enhances adsorption and reactivity.

The applications of nanostructured materials in soil remediation are broad and underscore their multifaceted contributions to the remediation landscape. Due to their high aspect ratio, nanostructured materials can serve as potent platforms for facilitating contaminant degradation. For example, carbon nanotubes exhibit exceptional catalytic activities, promoting the breakdown of organic pollutants through oxidative processes (Baalousha M et al., 2016).

In the realm of enhancing microbial activity in soil, nanostructured materials play a crucial role. They can function as carriers for microbial inoculants, safeguarding them against environmental stresses and precisely delivering them to target zones in the soil. This targeted delivery enhances microbial colonization, fostering enzymatic activities and accelerates the degradation of contaminants (You Y et al., 2017).

Moreover, nanostructured materials contribute to improving soil structure. Nanofibers, when incorporated into soil matrices, enhance water retention, promote aeration, and facilitate nutrient diffusion. The structural modifications induced by nanostructured materials enhance the overall fertility and resilience of the soil ecosystem (Gupta VK et al., 2019). Various types of Nanostructured materials are represented in fig.no.2.

Carbonaceous nanomaterials have distinctive features, comprising high microporosity, substantial surface area, excellent adsorption capabilities and environment friendly nature. Architectures such as fullerene C540, fullerene C60, MWCNTs, SWCNTs, graphene, and activated carbon nanoparticles embody these characteristics. (Matos MP et al., 2017; Marcon L et al., 2017). Furthermore, activation or functionalization of carbon-based nanomaterials provides additional benefits similar to other environmental remediation applications. CNTs have recently gained popularity because of their better adsorption capability when compared to biochar, graphene, graphene oxides, and granular activated carbon. Adsorption is controlled by a variety of variables, including exposure area and surface functional groups such as -COOH and OH. Adsorption capacity can be improved by including functional groups such as -NH2 and -SH, oxidation processes, nonmagnetic metal oxide coating, and grafting of magnetic iron oxides. The increased surface area, high surface-to-volume ratio, and increased reactivity promote flocculation but reduce qualities favorable to nanoremediation. The use of the surfactant poloxamer 407 has successfully stabilized multi-walled carbon nanotubes (Matos MP et al., 2017). CNTs are effective at removing heavy metal ions such as Pb2+, Cu2+, Ni2+, and ZN2+; however, heavy metal immobilization is dependent on pH, organic matter content, and the presence of silt and clay particles. Furthermore, CNTs is successful in remediation of soil contaminated with total petroleum hydrocarbons, crude oil, Cr (VI), Cd, DDT, and hexachlorocyclohexane, contributing to an increase in microbial populations and boosting plant development. (Shan J et al., 2015). CNTs' application approaches include their integration into membrane filtration, separation columns, and aqueous dispersion systems.

Types of Contaminates

Soil contamination from anthropogenic activity includes a wide spectrum of contaminants such as petroleum hydrocarbons, chlorinated solvents, and heavy metals. This contamination exposes the soil to possible dangers to ecosystems and creates significant difficulties to human health and the environment (Galdames A et al., 2017). Understanding of characteristics and sources of these contaminants is crucial for devising effective soil remediation strategies. This section provides an overview about the types of soil contaminants, encompassing heavy metals, organic pollutants, and nutrient imbalances.

Heavy Metals

Heavy metals such as mercury, lead, cadmium, and arsenic grew due to industrial emissions, mining activities, and agricultural inputs (Kumar A et al. 2019; Gottschalk F et al., 2013). These contaminants

Figure 2. Nanostructured materials (Khan Au et al., 2020)

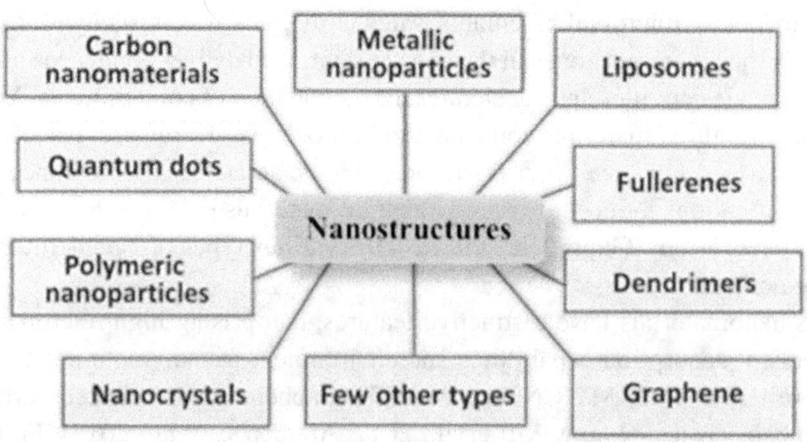

impact the soil, persisting and posing long-term threats to ecosystems and human health (Singh et al., 2023c, d). They cause soil acidification, inhibit plant nutrient uptake, and accumulate in the food chain, all of which have a negative impact on health (Singh et al., 2023e).

Nanotechnology offers innovative heavy metal remediation options. Nanoparticles with a large surface area and reactivity can selectively adsorb or precipitate heavy metals, restricting their mobility in the soil. (Gottschalk F et al., 2013; Khan AU et al., 2020). The use of nanomaterials, such as zerovalent iron nanoparticles, has shown promise in reducing concentrations of heavy metals in contaminated soils. Nanomaterials used for removal of Heavy Metals are presented in fig no.3.

Application of nano zero-valent iron (nZVI) for soil remediation is suitable owing to its minimal environmental disruption, cost effectiveness, swift kinetics and non-toxic nature (Mondal A et al., 2021). (Tian H et al., 2020) reported that, the initial synthesis and utilization of nZVI, where iron nanoscale particles (10 to 100 nm) were synthesized from Fe^{2+} and Fe^{3+}. This innovation proved extremely effective at removing numerous pollutants from water, particularly halogenated organic compounds, which are widely found in soil and groundwater. Their groundbreaking research proved the ability of nZVI to cleanse and convert environmental pollutants such as chlorinated organic solvents, polychlorinated biphenyls, and organochlorine pestiticides.

Furthermore, the writers focused on altering nZVI, which could enhance both process speed and efficiency. (Reginatto C etal., 2020) characterized and evaluated the applicability of polyvinylpyrrolidone (PVP)-enhanced nZVI for remediation of TCE-contaminated soil. The study found that, the size of the prepared PVP-nZVI was around 70 nm, with an isoelectric point around 8.5. The study concluded that, PVP-nZVI technology holds promise for remediating TCE-contaminated soil, achieving a removal efficiency of approximately 84.73%.

The study's findings suggest that PVP-nZVI technology shows potential for addressing soil contaminated with trichloroethylene (TCE). Another study found that Multi-Walled Carbon Nanotubes (MWCNTs) were successful at removing chromium (VI) from soil and groundwater. (Shubair T et al., 2018). A different study conducted by (Xue W et al 2018) explored the immobilization of selenite (Se(IV)) in soil and groundwater using Fe-Me binary oxide nanoparticles [18]. Additionally, the results indicated that the adsorption process of chromium (VI) by Fe/FeS with an S/F molar ratio of 0.207 followed a pseudo-second-order kinetic model, with a sorption capacity of 69.7 mg $g-1$, as simulated by the Langmuir isotherm model (Blundell SP etal., 2021). A regenerable magnetic ligand nanoparticle was developed for the rapid removal of cadmium (Cd) and lead (Pb) from contaminated water (Hua S et al., 2017). The outcomes showcased the excellent performance of the mega-ligand, as both Cd and Pb were swiftly eliminated from contaminated water, with Cd removal taking less than 2 hours and Pb removal in less than 15 minutes. Importantly, the mega-ligand exhibited consistent performance in Cd and Pb removal across varying pH levels (3–10) (Hua S et al., 2017).

Organic Pollutants

Organic pollutants, comprising pesticides, herbicides, and industrial chemicals, pose challenges due to their persistence and toxicity. These contaminants often seep into the soil, affecting both terrestrial and aquatic ecosystems. The prevalence of hazardous persistent organic pollutants (POPs) is increasing daily, posing a continuous threat. The emergence of POPs in groundwater and drinking water presents a substantial risk due to their easy transference from one source to another and subsequent release into new environments (Alharbi et al., 2018). These pollutants linger in the environment for extended

Figure 3. Nanomaterials used for removal of heavy metals (Ahmed T et al., 2021)

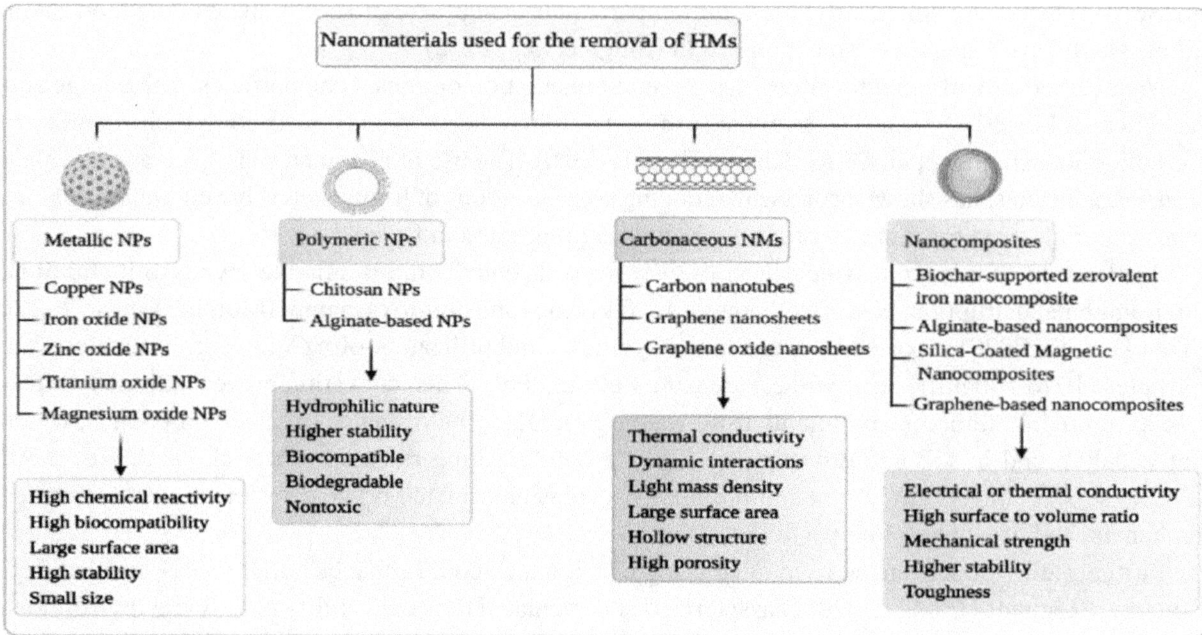

periods, having the potential to accumulate within the human body. Remarkably toxic even at minimal concentrations, POPs can seamlessly traverse the food chain, impacting organisms situated far from the contamination site (El-Shahawani MS et al., 2010). POPs have become a global problem due to their persistent properties, ability to move across vast distances, and proclivity to bioaccumulation.

The Stockholm Convention classifies 22 nonpolar organic compounds as POPs, with organochlorines (OCs) being the largest class. This category contains substances like aldrin and dichlorodiphenyltrichloroethane (DDT), followed by organophosphorus (OPs) insecticides, flame retardants, polyaromatic hydrocarbons (PAHs), and other chemicals such as PCBs. (Fernandez P et al., 2003; Stokke OS et al., 2001-2). Additionally, the byproducts of the aforementioned compounds also fall under the same classification. Traditional remediation methods encounter limitations in effectively addressing these persistent organic pollutants.

Nanotechnology provides a tailored way to remediating organic contaminants. The use of nanomaterials in soil remediation has the potential to improve the efficiency and long-term removal of organic pollutants. Nanomaterials are used in remediation because of their intrinsic properties. They are classified as (i) nanocatalysts, (ii) nanoadsorbents, and (iii) nanomembranes. These three classes are widely used in wastewater treatment (Singh SP et al., 2018). Nanomaterials used to remediate Persistent Organic Pollutants (POPs) are classified as metallic (metal/metal oxides, carbon-based nanomaterials, and nanocomposites) or non-metallic (comprising nanomembranes, polymeric nanomaterials, metal-organic frameworks, etc.) (Khan NA et al., 2019; Kunduru KR et al., 2017). In nano-based remediation of organic pollutants, the most common techniques are oxidation and adsorption, followed by degradation (Khan SH et al., 2020) Nanotechnological methods used to cure POPs and nano materials are represented in fir.no.4 and table no.1.

Figure 4. Nanotechnological methods used to cure POPs (Boulkhessaim et al., 2012)

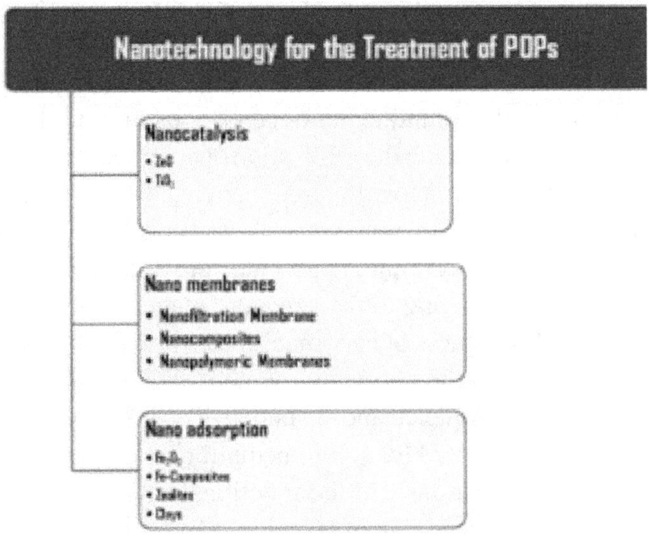

Table 1. Nanomaterials are used for the removal of POPs

Contaminant	Nanomaterials	References
Organic Pollutant	Ag/ZnO, ZnO-Bi, ZnO	(Kong SP et al., 2021)
	Mesoporous silica	(Kong SP et al., 2021)
	Graphene oxide-Ag NP	(Sarno M et al., 2017)
	TiO2	(Abdennouri M et al., 2016)
	ZnO	(Navarro S et al., 2009)
	TiO2 -rGO	(Luna-Sanguino G et al., 2020)
	Palladium and AgNp-embedded-zinc oxide nanostars	(Veerakumar P et al., 2021)
	RE3+-doped nano-TiO2	(Zeng R et al., 2010)
	MoS2/ZnS embedded in N/S doped carbon	(Ahmed T et al., 2020)
	Magnetite and cobalt ferrite-decorated graphene oxide composite.	(Tabasum A et al., 2020)
	CuO and NiO nanoparticles	(Iqbal A et al., 2021)

APPLICATIONS OF NANOTECHNOLOGY IN SOIL REMEDIATION

Contaminant Sequestration

The utilization of nanotechnology in the sequestration of contaminants highlights the diverse mechanisms by which nanomaterials immobilize pollutants. Engineered nanomaterials, with their combination of physical encapsulation and chemical reactions, along with long-term stability, emerge as powerful tools for preventing contaminant leaching in soil remediation efforts.

Immobilization: Nanomaterials take a forefront position in strategies for contaminant sequestration, showcasing their effectiveness in immobilizing contaminants through multifaceted mechanisms. Two key avenues, physical encapsulation, and chemical reactions demonstrate how nanomaterials achieve successful immobilization (Thakur M et al., 2020 ; (Li X et al., 2019). In physical encapsulation, nanomaterials establish a protective shield around contaminants, constraining their mobility within the soil matrix. This encapsulation can occur through the adsorption of contaminants onto the high surface area of nanoparticles or their entrapment within nanostructured matrices. On the other hand, chemical reactions involve transformative interactions between nanomaterials and contaminants, resulting in the formation of stable, less mobile compounds. For example, nanoscale zero-valent iron (nZVI) can undergo redox reactions with certain contaminants, leading to their reduction and immobilization [59].

Long-Term Stability: The effectiveness of nanomaterials in preventing the leaching of contaminants relies on their enduring stability within the soil environment. The unique physicochemical properties of nanomaterials contribute to their resilience and sustained efficacy over time (Thakur M et al., 2020).

Nanomaterials, especially those tailored for soil remediation, undergo engineering with features that bolster their stability. Surface modifications and the functionalization of nanoparticles guarantee their extended presence in the soil, reducing the risk of leaching. Moreover, the creation of stable complexes between nanomaterials and contaminants plays a role in the persistent immobilization of pollutants. Additionally, the intrinsic durability of specific nanomaterials, like graphene-based nanocomposites, secures their longevity in the soil matrix. Their structural robustness and resistance to environmental degradation empower these nanomaterials to uphold their effectiveness in sequestering contaminants over prolonged periods.

The technical aspects of nanotechnology in enhanced bioremediation unfold through the intricate interplay between nanomaterials and soil microbiota. The catalytic and modulatory roles of nanomaterials, as grounded in prior research findings, position them as key facilitators for precision bioremediation interventions in contaminated soil environments. Nanomaterials exert a transformative influence on microbial activity in the soil, serving as catalytic agents to enhance bioremediation processes. The symbiotic interaction between nanomaterials and soil microbes orchestrates a conducive environment for the degradation of organic contaminants, characterized by intricate redox reactions (Wu Y et al., 2019; Khan I et al., 2019).

Nano bioremediation, an economically viable approach, integrates the use of plants and microbes to decompose pollutant compounds, resulting in improved soil quality and reduced pollution levels. The objective of this technique is to eliminate, retain, or reduce pollutants in the soil (Ashraf S et al., 2019; (Chandra R et al., 2017). While bioremediation has previously been boosted via chemical additives or biotechnology, nanotechnology brings a new dimension, substantially elevating the process(Gong X et al., 2018). Nano bioremediation combines nanotechnology with bioremediation on a nanoscale, with contaminants adsorbed, destroyed, or transformed by the unique physicochemical features of NPs. NPs also operate as catalysts, decreasing the activation energy needed to break down compounds (Mehndiratta et al., 2013). The procedure primarily uses carbon- and metal-based NPs, with polymeric NPs such as nanocapsules or nanospheres excelling at eliminating persistent pesticide chemicals and long-chain hydrocarbons (Gong J et al., 2009); Chen C et al., 2017). However, addressing heavy metals (HMs) poses challenges due to their non-biodegradable nature and tendency to enter biological systems and food chains (Chouhan R et al., 2020). Traditional strategies for removing HMs from polluted soils include biosorption and bioaccumulation using plants and bacteria. Recent evidence, however, highlights the extraordinary results of utilizing NPs in HM repair. NPs, either used concurrently or sequentially with certain microorganisms, produce impressive effects, speeding HM removal by functioning as nanocarriers of microbes or microbial biosorbents. (Briffa J et al., 2020; Mishra M et al., 2021; Abdi O et al., 2015). This enhancement of microbial metabolic pathways results in increased enzymatic activities and accelerated rates of microbial degradation. The catalytic abilities of nanomaterials stimulate microbial consortia, presenting a sophisticated approach to enhancing the efficiency of organic pollutant removal in contaminated soils. A fundamental aspect of nanomaterial-driven enhanced bioremediation lies in the modulation of nutrient bioavailability, critically influencing the metabolic vigour of microbial communities (Ayangbenro et al., 2017; Raliya R et al., 2015).

In a recent study that focused on the elimination of Cu, Cd, Cr, and Pb, HM-resistant bacteria, notably B. cereus (PMBL-3) and L. macroides (PMBL-7), showed substantial efficacy. Adding 5 mg L−1 ZnO NPs significantly improved metal removal, resulting in a 60% reduction in Cr, 70% in Cu, and 85% in Pb. This performance exceeded the individual efforts of B. cereus (80% Cr, 60% Cu) and L. macroides (55% Cr, 50% Cu) at neutral pH (Bangano D et al., 2020). At neutral pH, ZnO NPs' negatively charged surface promotes electrostatic interactions with metal cations. Nonetheless, at lower pH values, metal hydroxides precipitated, resulting in competition between hydrogen ions and adsorbents for binding (Xie Yet al., 2011). Another study investigated the use of probiotic bacteria, specifically L. casei and L. fermentum, for Cd absorption from water in conjunction with Se5+ ions and SeNPs. L. casei absorbed more Cd with Se4+ ions (65%) than with Se NPs (55.90%), which can be attributed to Se5+'s increased solubility. Furthermore, L. fermentum was much more efficient in Cd absorption (50.87%) than L. casei (43.78%). The investigation demonstrated that while the percentage of Cd adsorption by L. casei in combination with Se NPs exhibited no significant change, increasing the Se NPs ratio percentage marginally boosted Cd absorption (from 5.49% to 16.54%) in the presence of L. casei (Dong G et al., 2013).

Additionally, nanomaterials serve as precision carriers for microbial inoculants, protecting these entities from environmental stresses. The strategic delivery of microbial populations to targeted soil zones amplifies their colonization potential, enhances enzymatic activities, and maximizes their efficacy in degrading contaminants.

Environmental Impact and Risks of Nanotechnology in Soil Remediation

Understanding the environmental impact and risks associated with nanotechnology in soil remediation necessitates a multidimensional approach. An all-inclusive framework for evaluating the safety and sustainability of nanomaterial applications in remediation strategies can be achieved by considering toxicity, persistence and bioaccumulation.

Toxicity: The assessment of potential nanomaterial toxicity is crucial for determining their safety in soil remediation (Singh et al., 2024a). Various factors like particle size, surface charge and composition, influence the interactions between nanomaterials and soil organisms, plants and humans (Kah M et al., 2018; Nair R et al., 2014). Concerns arise from the size-dependent toxicity of nanoparticles, with smaller particles often displaying greater bioavailability and reactivity. Surface charge, influenced by factors like pH, affects the interactions between nanomaterials and cell membranes, influencing cellular uptake and potential toxicity (Singh et al., 2024b). Furthermore, variations in nanomaterial composition introduce diverse toxicological profiles, necessitating comprehensive studies to establish safety thresholds for specific nanomaterials in different environmental compartments.

Persistence: The environmental persistence of nanomaterials is a critical aspect that may have long-term effects on ecosystems. Aggregation, sedimentation, and transformations all influence the destiny of nanomaterials in the environment. The persistence of these materials raises questions about cumulative impacts (Nowack B et al., 2018).

Aggregation can result in the settling of nanomaterials in sediments, influencing benthic ecosystems. Transformations, such as oxidation or reduction, may alter the physicochemical properties of nanomaterials, impacting their environmental behaviour. Understanding the persistence of nanomaterials in soil is crucial for anticipating their long-term effects on ecosystems and developing strategies to mitigate any adverse consequences.

Bioaccumulation: Examining the potential bioaccumulation of nanomaterials in organisms is crucial for understanding their implications in food chains. Nanomaterials may enter the food web through plant uptake, ingestion by soil organisms, or direct exposure to aquatic organisms (Zuverza-Mena N et al., 2016). The propensity of nano-materials to bioaccumulate varies, influenced by factors like shape, size and surface properties. Nanomaterials may accumulate in plant tissues, potentially reaching concentrations that pose risks to herbivores or organisms higher in the food chain. Comprehensive assessments of bioaccumulation potential are necessary to evaluate the overall risk of nanomaterials in soil ecosystems.

Future directions and challenges in nanotechnology for soil remediation: The future directions of nanotechnology in soil remediation involve the use of environmentally friendly nanomaterials and advanced monitoring technologies. Overcoming challenges related to scalability, cost-effectiveness, and risk assessment protocols, alongside the establishment of a robust regulatory framework, will be crucial for realizing the full potential of nanotechnology in sustainable soil remediation practices. The future of nanotechnology in soil remediation holds promising prospects, with emerging trends focusing on the development of environmentally friendly nanomaterials and nanosensors for real-time monitoring (Jampilek J et al., 2019; Sun Y et al., 202).

Environmentally friendly nanomaterials are designed to minimize ecological impacts while retaining high efficacy in soil remediation. Innovations include the use of biodegradable nanomaterials, green synthesis methods, and the incorporation of sustainable materials. These developments align with the growing emphasis on sustainable practices and the reduction of adverse environmental effects associated with traditional nanomaterials. Nanosensors play a pivotal role in real-time monitoring, providing accurate and timely information on soil conditions and the efficiency of remediation processes. Advancements in sensor technologies, such as miniaturization, enhanced sensitivity, and wireless connectivity, enable continuous monitoring and data-driven decision-making in soil remediation projects.

Despite the advancements, challenges in scalability and cost-effectiveness persist. The translation of laboratory-scale successes to field-scale applications remains a hurdle. The scalability of nanotechnology for large-scale soil remediation projects requires overcoming obstacles related to the production, deployment, and recovery of nanomaterials in complex environmental matrices. Cost considerations, including the synthesis, application, and monitoring expenses, remain significant barriers to widespread adoption.

Developing standardized risk assessment protocols are critical for ensuring the safe and responsible application of nanotechnology for soil remediation [83][84]. The diverse physicochemical properties of nanomaterials necessitate comprehensive risk assessments that consider potential impacts on soil organisms, plants, and human health. Establishing standardized protocols will facilitate a consistent and reliable evaluation of the environmental & health risks associated with several nanomaterials, aiding regulatory agencies, researchers, and industry stakeholders.

The establishment of a robust regulatory framework is imperative to govern the use of nanotechnology in soil remediation. Regulations should address issues of safety, efficacy, and environmental impact, providing guidelines for the responsible development and application of nanomaterials. A harmonized international regulatory framework will help navigate the complexities of nanotechnology, ensuring that innovations adhere to ethical and environmental standards while fostering global collaboration in research and development.

CONCLUSION

Nanotechnology stands as a transformative force in soil remediation, offering precise and efficient tools to tackle complex environmental challenges. The critical role it plays in restoring soil health, mitigating contamination, and promoting sustainable land use necessitates a collective commitment to responsible research, collaboration, and implementation. The unique features of nanomaterials, comprising high reactivity, tailored physicochemical characteristics and exceptional surface area position them as versatile tools for precise and effective remediation strategies. Nano omics and nano enzymes in soil remediation yields significant insights, showcasing Nano Omics capacity to provide a comprehensive molecular understanding of nanomaterial-soil interactions and Nano enzymes enhanced catalytic activities for targeted contaminant degradation.

As we envision a future where soil degradation is countered by advanced technologies, the continued exploration and refinement of nanotechnology underscore our commitment to a resilient and sustainable planet.

REFERENCES

Abdennouri, M., Baâlala, M., Galadi, A., El Makhfouk, M., Bensitel, M., Nohair, K., Barka, N., Boussaoud, A., Barka, N., . . . Barka, N. (2016). Photocatalytic degradation of pesticides by titanium dioxide and titanium pillared purified clays. *Arabian Journal of Chemistry, 9*, S313–S318. https://doi.org/doi:10.1016/j.arabjc.2011.04.005

Abdi, O., & Kazemi, M. (2015). A review study of biosorption of heavy metals and comparison between different biosorbents. *Journal of Materials & Environmental Sciences*, *6*, 1386–1399.

Ahamad, T., Naushad, M., Al-Saeedi, S. I., Almotairi, S., & Alshehri, S. M. (2020). Fabrication of MoS2/ZnS embedded in N/S doped carbon for the photocatalytic degradation of pesticide. *Materials Letters*, *263*, 127271. doi:10.1016/j.matlet.2019.127271

Ahmed, T., Noman, M., Ijaz, M., Ali, S., Rizwan, M., Ijaz, U., Hameed, A., Ahmad, U., Wang, Y., Sun, G., & Li, B. (2021). Current trends and future prospective in nanoremediation of heavy metals contaminated soils: A way forward towards sustainable agriculture. *Ecotoxicology and Environmental Safety*, *227*, 112888. doi:10.1016/j.ecoenv.2021.112888 PMID:34649136

Alharbi, O. M. L., Basheer, A. A., Khattab, R. A., & Ali, I. (2018). Health and environmental effects of persistent organic pollutants. *Journal of Molecular Liquids*, *263*, 442–453. doi:10.1016/j.molliq.2018.05.029

Ashraf, S., Siddiqa, A., Shahida, S., & Qaisar, S. (2019). Titanium-based nanocomposite materials for arsenic removal from water: A review. *Heliyon*, *5*(5), e01577. doi:10.1016/j.heliyon.2019.e01577 PMID:31193228

Ayangbenro, A. S., & Babalola, O. O. (2017). A new strategy for heavy metal polluted environments: A review of microbial biosorbents. *International Journal of Environmental Research and Public Health*, *14*(1), 94. doi:10.3390/ijerph14010094 PMID:28106848

Baalousha, M., Stolpe, B., Lead, J. R., & Colman, B. P. (2016). Nanoparticle dispersion in environmentally important matrices and implications for exposure: A review. *Reviews of Environmental Contamination and Toxicology*, *236*, 53–97.

Baragaño, D., Forján, R., Welte, L., & Gallego, J. L. R. (2020). Nanoremediation of As and metals polluted soils by means of graphene oxide nanoparticles. *Scientific Reports*, *10*(1), 1896. doi:10.1038/s41598-020-58852-4 PMID:32024880

Blundell, S. P., & Owens, G. (2021). Evaluation of enhancement techniques for the dechlorination of DDT by nanoscale zero-valent iron. *Chemosphere*, *264*(1), 128324. doi:10.1016/j.chemosphere.2020.128324 PMID:33022509

Boulkhessaim, S., Gacem, A., Khan, S. H., Amari, A., Yadav, V. K., Harharah, H. N., Elkhaleefa, A. M., Yadav, K. K., Rather, S. U., Ahn, H. J., & Jeon, B. H. (2022). Emerging trends in the remediation of persistent organic pollutants using nanomaterials and related processes: A review. *Nanomaterials (Basel, Switzerland)*, *12*(13), 2148. doi:10.3390/nano12132148 PMID:35807983

Briffa, J., Sinagra, E., & Blundell, R. (2020). Heavy metal pollution in the environment and their toxicological effects on humans. *Heliyon*, *6*(9), e04691. doi:10.1016/j.heliyon.2020.e04691 PMID:32964150

Chandra, R., Kumar, V., Tripathi, S., & Sharma, P. (2017). Phytoremediation of industrial pollutants and life cycle assessment. In *Phytoremediation of environmental pollutants* (pp. 441–470). CRC Press. doi:10.4324/9781315161549-18

Chauhan, R., Yadav, H. O. S., & Sehrawat, N. (2020). Nanobioremediation: A new and a versatile tool for sustainable environmental cleanup-Overview. *Journal of Materials & Environmental Sciences*, *11*, 564–573.

Chen, C., Tsyusko, O. V., McNear, D. H. Jr, Judy, J., Lewis, R. W., & Unrine, J. M. (2017). Effects of biosolids from a wastewater treatment plant receiving manufactured nanomaterials on Medicago truncatula and associated soil microbial communities at low nanomaterial concentrations. *The Science of the Total Environment*, *609*, 799–806. doi:10.1016/j.scitotenv.2017.07.188 PMID:28768212

Dimkpa, C. O., McLean, J. E., Martineau, N., & Britt, D. W. (2011). Interaction of silver nanoparticles with an environmentally beneficial bacterium, Pseudomonas chlororaphis. *Journal of Hazardous Materials*, *199*(1-3), 68–76. doi:10.1016/j.jhazmat.2011.01.118 PMID:21339046

Dong, G., Wang, Y., Gong, L., Wang, M., Wang, H., He, N., Zheng, Y., & Li, Q. (2013). Formation of soluble Cr(III) end-products and nanoparticles during Cr(VI) reduction by bacillus cereus strain XMCr-6. *Biochemical Engineering Journal*, *70*, 166–172. doi:10.1016/j.bej.2012.11.002

El-Shahawi, M. S., Hamza, A., Bashammakh, A. S., & Al-Saggaf, W. T. (2010). An overview on the accumulation, distribution, transformations,toxicity and analytical methods for the monitoring of persistent organic pollutants. *Talanta*, *80*(5), 1587–1597. doi:10.1016/j.talanta.2009.09.055 PMID:20152382

Fernández, P., & Grimalt, J. O. (2003). On the global distribution of persistent organic pollutants. *Chimia*, *57*(9), 514–521. doi:10.2533/000942903777679000

Galdames, A., Mendoza, A., Orueta, M., de Soto García, I. S., Sánchez, M., Virto, I., & Vilas, J. L. (2017). Development of new remediation technologies for contaminated soils based on the application of zero-valent iron nanoparticles and bioremediation with compost. *Resource-Efficient Technologies*, *3*(2), 166–176. doi:10.1016/j.reffit.2017.03.008

Gong, J. L., Wang, B., Zeng, G. M., Yang, C. P., Niu, C. G., Niu, Q. Y., Zhou, W. J., & Liang, Y. (2009). Removal of cationic dyes from aqueous solution using magnetic multi-wall carbon nanotube nanocomposite as adsorbent. *Journal of Hazardous Materials*, *164*(2–3), 1517–1522. doi:10.1016/j.jhazmat.2008.09.072 PMID:18977077

Gong, X., Huang, D., Liu, Y., Peng, Z., Zeng, G., Xu, P., Cheng, M., Wang, R., & Wan, J. (2018). Remediation of contaminated soils by biotechnology with nanomaterials: Bio-behavior, applications, and perspectives. *Critical Reviews in Biotechnology*, *38*(3), 455–468. doi:10.1080/07388551.2017.1368446 PMID:28903604

Gottschalk, F., Sonderer, T., Scholz, R. W., & Nowack, B. (2013). Modeled environmental concentrations of engineered nanomaterials (TiO2, ZnO, Ag, CNT, fullerenes) for different regions. *Environmental Science & Technology*, *46*(21), 11584–11592. PMID:20000512

Govekar, S., Kumar, R., Suresh, R., & Sridharan, K. (2018). Nanotechnology to sustain a clean environment. In Emerging trends of nanotechnology in environment and sustainability: A review-based approach (pp. 3–11). doi:10.1007/978-3-319-71327-4_1

Gupta, V. K., Fakhri, A., Agarwal, S., Al-Salem, S., Atarod, M., Karami, C., Wang, C., & Varjani, S. (2019). A critical review on nanocomposites: From chemical and physical methods of synthesis to properties and applications. *Environmental Chemistry Letters*, *17*(2), 629–654.

Hua, S., Gong, J. L., Zeng, G. M., Yao, F. B., Guo, M., & Ou, X. M. (2017). Remediation of organochlorine pesticides contaminated lake sediment using activated carbon and carbon nanotubes. *Chemosphere*, *177*, 65–76. doi:10.1016/j.chemosphere.2017.02.133 PMID:28284117

Iqbal, A., Haq, A. U., Cerrón-Calle, G. A., Naqvi, S. A. R., Westerhoff, P., & Garcia-Segura, S. (2021). Green synthesis of flower-shaped copper oxide and nickel oxide nanoparticles via Capparis decidua leaf extract for synergic adsorption-photocatalytic degradation of pesticides. *Catalysts*, *11*(7), 806. doi:10.3390/catal11070806

Jampílek, J., Kráτký, M., Konefał, R., & Králík, J. (2019). Nanomaterials for soil remediation: A review. *Environmental Chemistry Letters*, *17*(3), 1263–1270.

Kah, M., Kookana, R. S., Gogos, A., & Bucheli, T. D. (2018). A critical evaluation of nanopesticides and nanofertilizers against their conventional analogues. *Nature Nanotechnology*, *13*(8), 677–684. doi:10.1038/s41565-018-0131-1 PMID:29736032

Khan, A. A., & Paquiza, L. (2011). Characterization and ion-exchange behavior of thermally stable nanocomposite polyaniline zirconium titanium phosphate: Its analytical application in separation of toxic metals. *Desalination*, *265*(1–3), 242–254. doi:10.1016/j.desal.2010.07.058

Khan, A. A., Paquiza, L., & Khan, A. (2010). An advanced nano-composite cation-exchanger polypyrrolezirconium titanium phosphate as a Th(IV)-selective potentiometric sensor: Preparation, characterization and its analytical application. *Journal of Materials Science*, *45*(13), 3610–3625. doi:10.1007/s10853-010-4407-6

Khan, A. U., Khan, M., Cho, M. H., & Khan, M. M. (2020). Selected nanotechnologies and nanostructures for drug delivery, nanomedicine and cure. *Bioprocess and Biosystems Engineering*, *43*(8), 1339–1357. doi:10.1007/s00449-020-02330-8 PMID:32193755

Khan, I., Saeed, K., & Khan, I. (2019). Nanoparticles: Properties, applications and toxicities. *Arabian Journal of Chemistry*, *12*(7), 908–931. doi:10.1016/j.arabjc.2017.05.011

Khan, M. Z., Mondal, P. K., & Sabir, S. (2011a). Bioremediation of 2-chlorophenol containing wastewater by aerobic granules-kinetics and toxicity. *Journal of Hazardous Materials*, *190*(1–3), 222–228. doi:10.1016/j.jhazmat.2011.03.029 PMID:21466913

Khan, N. A., Khan, S. U., Ahmed, S., Farooqi, I. H., Dhingra, A., Hussain, A., & Changani, F. (2019). Applications of nanotechnology in water and wastewater treatment: A review. *Asian Journal of Water, Environment and Pollution*, *16*(4), 81–86. doi:10.3233/AJW190051

Khan, S. H. (2020). Green nanotechnology for the environment and sustainable development. In *Green materials for wastewater treatment* (pp. 13–46). Springer. doi:10.1007/978-3-030-17724-9_2

Khan, S. H., Pathak, B., & Fulekar, M. H. (2020). A study on the influence of metal (Fe, Bi, and Ag) doping on structural, optical, and antimicrobial activity of ZnO nanostructures. *Advanced Composites and Hybrid Materials*, *3*(4), 551–569. doi:10.1007/s42114-020-00174-0

Klaine, S. J., Koelmans, A. A., Horne, N., Carley, S., Handy, R. D., Kapustka, L., Nowack, B., & Reddy, C. M. (2018). Paradigms to assess the environmental impact of manufactured nanomaterials. *Environmental Toxicology and Chemistry*, *37*(7), 1853–1859. PMID:22162122

Kong, X. P., Zhang, B. H., & Wang, J. (2021). Multiple roles of mesoporous silica in safe pesticide application by nanotechnology: A review. *Journal of Agricultural and Food Chemistry*, *69*(24), 6735–6754. doi:10.1021/acs.jafc.1c01091 PMID:34110151

Kumar, A., Bisht, B. S., Joshi, V. D., & Dhewa, T. (2019). Review on bioremediation of polluted environment: A management tool. *International Journal of Environmental Science and Technology*, *16*(7), 3439–3462.

Kunduru, K. R., Nazarkovsky, M., Farah, S., Pawar, R. P., Basu, A., & Domb, A. J. (2017). Nanotechnology for water purification: Applications of nanotechnology methods in wastewater treatment. In A. M. Grumezescu (Ed.), Water purification (pp. 33–74). Academic Press.

Li, X., Zhu, Y., Li, Y., Zhang, D., Zhang, H., & Zhang, Z. (2019). Nanomaterials as novel stabilizers for soil stabilization: Mechanism, ecological risk, and environmental implication. *The Science of the Total Environment*, *646*, 1566–1575.

Luna-Sanguino, G., Ruíz-Delgado, A., Tolosana-Moranchel, A., Pascual, L., Malato, S., Bahamonde, A., & Faraldos, M. (2020). Solar photocatalytic degradation of pesticides over TiO2 -rGO nanocomposites at pilot plant scale. *The Science of the Total Environment*, *737*, 140286. doi:10.1016/j.scitotenv.2020.140286 PMID:32783863

Ma, A., Wei, Y., Zhou, Z., Xu, W., Ren, F., Ma, H., & Wang, J. (2012). Preparation Bi2S3eTiO2 heterojunction/polymer fiber composites and its photocatalytic degradation of methylene blue under Xe lamp irradiation, polym degrad. *Stadtehygiene*, *97*, 125–131.

Marcon, L., Oliveras, J., & Puntes, V. F. (2021). In situ nanoremediation of Soils and Groundwaters from the Nanoparticle's Standpoint: A review. *The Science of the Total Environment*, *791*, 148324. doi:10.1016/j.scitotenv.2021.148324 PMID:34412401

Martin, J. L., Maris, V., & Simberloff, D. S. (2016). The need to respect nature and its limits challenges society and conservation science. *Proceedings of the National Academy of Sciences of the United States of America*. PubMed, *113*(22), 6105–6112. 10.1073/pnas.1525003113

Matos, M. P. S. R., Correia, A. A. S., & Rasteiro, M. G. (2017). Application of carbon nanotubes to immobilize heavy metals in contaminated soils. *Journal of Nanoparticle Research*, *19*(4), 126. doi:10.1007/s11051-017-3830-x

Mehndiratta, P., Jain, A., Srivastava, S., & Gupta, N. (2013). Environmental pollution and nanotechnology. *Environment and Pollution*, *2*(2), 49. doi:10.5539/ep.v2n2p49

Midhat, L., Ouazzani, N., Hejjaj, A., Ouhammou, A., & Mandi, L. (2019). Accumulation of heavy metals in metallophytes from three mining sites (Southern Centre Morocco) and evaluation of their phytoremediation potential. *Ecotoxicology and Environmental Safety*, *169*, 150–160. doi:10.1016/j.ecoenv.2018.11.009 PMID:30445246

Misra, M., & Sachan, G. S. (2021). Nanobioremediation of heavy metals: Perspectives and challenges. *Journal of Basic Microbiology*, 1–16.

Mondal, A., Dubey, B. K., Arora, M., & Mumford, K. (2021). Porous media transport of iron nanoparticles for site remediation application: A review of lab scale column study, transport modelling and field-scale application. *Journal of Hazardous Materials*, *403*, 123443. doi:10.1016/j.jhazmat.2020.123443 PMID:32798796

Mondal, A., Dubey, B. K., Arora, M., & Mumford, K. (2021). Porous media transport of iron nanoparticles for site remediation application: A review of lab scale column study, transport modelling and field-scale application. *Journal of Hazardous Materials, 403*, 123443. doi:10.1016/j.jhazmat.2020.123443

Nair, R., & Chung, Y. T. (2014). Environmental transformation of silver nanoparticles: Effects on aquatic organisms. *Environmental Toxicology and Chemistry*, *33*(2), 462–467.

Navarro, S., Fenoll, J., Vela, N., Ruiz, E., & Navarro, G. (2009). Photocatalytic degradation of eight pesticides in leaching water by use of ZnO under natural sunlight. *Journal of Hazardous Materials*, *172*(2–3), 1303–1310. doi:10.1016/j.jhazmat.2009.07.137 PMID:19729242

Nowack, B., & Mitrano, D. M. (2018). Metal and metal oxide nanoparticles in aquatic systems: An analysis of data and publications. *The Science of the Total Environment*, *628*, 748–768.

Panpatte, D. G., Jhala, Y. K., Shelat, H. N., & Vyas, R. V. (2016). Nanoparticles – The next generation technology for sustainable agriculture. In D. P. Singh, H. B. Singh & R. Prabha (Eds.), Microbial inoculants in sustainable agricultural productivity, Functional applications, 2 (pp. 289–300). Springer.

Parikh, S. J., & James, B. R. (2012). Soil: The foundation of agriculture. *Natl Educ. Knowl.*, *3*, 2.

Pretty, J., & Bharucha, Z. P. (2014). Sustainable intensification in agricultural systems. *Annals of Botany*, *114*(8), 1571–1596. doi:10.1093/aob/mcu205 PMID:25351192

Rajput, V. D., Singh, A., Minkina, T., Rawat, S., Mandzhieva, S., Sushkova, S., Shuvaeva, V., Nazarenko, O., Rajput, P., Komariah, Verma, K. K., Singh, A. K., Rao, M., & Upadhyay, S. K. (2021). Nano-Enabled Products: Challenges and Opportunities for Sustainable Agriculture. *Plants*, *10*(12), 2727. doi:10.3390/plants10122727 PMID:34961197

Raliya, R., Biswas, P., & Tarafdar, J. C. (2015). TiO2 nanoparticle biosynthesis and its physiological effect on mung bean (Vigna radiata L.). *Biotechnology Reports (Amsterdam, Netherlands)*, *13*, 21–29. doi:10.1016/j.btre.2014.10.009 PMID:28626678

Reginatto, C., Cecchin, I., Heineck, K. S., Reddy, K. R., & Thomé, A. (2020). Use of nanoscale Zero-Valent Iron for Remediation of Clayey Soil Contaminated with Hexava lent chromium: Batch and Column Tests. *International Journal of Environmental Research and Public Health*, *17*(3), 1001. doi:10.3390/ijerph17031001 PMID:32033384

Reginatto, C., Cecchin, I., Heineck, K. S., Reddy, K. R., & Thomé, A. (2020). Use of nanoscale zero-valent iron for remediation of clayey soil contaminated with hexavalent chromium: Batch and column tests. *International Journal of Environmental Research and Public Health*, *17*(3), 1001. doi:10.3390/ijerph17031001 PMID:32033384

Sarno, M., Casa, M., Cirillo, C., & Ciambelli, P. (2017). Complete removal of persistent pesticide using reduced graphene oxide-silver nanocomposite. *Chemical Engineering Transactions, 60*, 151–156.

Shahadat, M., Teng, T. T., Rafatullah, M., & Arshad, M. (2015). Titanium-based nanocomposite materials: A review of recent advances and perspectives. *Colloids and Surfaces. B, Biointerfaces*, *126*, 121–137. doi:10.1016/j.colsurfb.2014.11.049 PMID:25543989

Shan, J., Ji, R., Yu, Y., Xie, Z., & Yan, X. (2015). Biochar, activated carbon, and carbon nanotubes have different effects on fate of (14)C-catechol and microbial community in soil. *Scientific Reports*, *5*(1), 16000. doi:10.1038/srep16000 PMID:26515132

Shubair, T., Eljamal, O., Khalil, A. M. E., & Matsunaga, N. (2018). Multilayer system of nanoscale zero valent iron and Nano-Fe/Cu particles for nitrate removal in porous media. *Separation and Purification Technology*, *193*, 242–254. doi:10.1016/j.seppur.2017.10.069

Singh, A., Rajput, V., Varshney, A., Sharma, R., Ghazaryan, K., Minkina, T., . . . El-Ramady, H. (2024b). Revolutionizing Crop Production: Nanoscale Wonders - Current Applications, Advances, and Future Frontiers. *Egyptian Journal of Soil Science, 64*(1). doi:10.21608/ejss.2023.246354.1684

Singh, A., Rajput, V. D., Sharma, R., Ghazaryan, K., & Minkina, T. (2023a). Salinity stress and nanoparticles: Insights into antioxidative enzymatic resistance, signaling, and defense mechanisms. *Environmental Research, 235*, 116585. doi:10.1016/j.envres.2023.116585 PMID:37437867

Singh, A., Rajput, V. D., Tawaha, A. R. M., Al Zoubi, O. M., Habeeb, T., Rawat, S., Ghazaryan, K., & Minkina, T. (2023b). A Review on Crop Responses to Nanofertilizers for Mitigation of Multiple Environmental Stresses. *Ecological Engineering & Environmental Technology, 24*(7), 280–296. doi:10.12912/27197050/169313

Singh, A., Rajput, V. D., Tawaha, A. R. M., Al Zoubi, O. M., Habeeb, T., Rawat, S., Ghazaryan, K., & Minkina, T. (2023c). A Review on Crop Responses to Nanofertilizers for Mitigation of Multiple Environmental Stresses. *Ecological Engineering & Environmental Technology, 24*(7), 280–296. doi:10.12912/27197050/169313

Singh, A., Rajput, V. D., Varshney, A., Ghazaryan, K., & Minkina, T. (2023d). Small Tech, Big Impact: Agri-nanotechnology Journey to Optimize Crop Protection and Production for Sustainable Agriculture. *Plant Stress, 10*, 100253. doi:10.1016/j.stress.2023.100253

Singh, A., Sengar, R. S., Rajput, V. D., Al-Ghzawi, A. L., Shahi, U. P., Ghazaryan, K., Minkina, T., Tawaha, A. R. M. A., Al Zoubi, O. M., & Habeeb, T. (2024a). Impact of Salinity Stress and Zinc Oxide Nanoparticles on Macro and Micronutrient Assimilation: Unraveling the Link between Environmental Factors and Nutrient Uptake. *Journal of Ecological Engineering, 25*(2), 1–9. doi:10.12911/22998993/172947

Singh, A., Sharma, R., Rajput, V. D., Ghazaryan, K., Minkina, T., Al Tawaha, A. R. M., & Varshney, A. (2023e). Green Synthesis of Nanofertilizers and Their Application for Crop Production. *Nanotechnology in the Life Sciences*, 205–231. doi:10.1007/978-3-031-41329-2_8

Singh, A., Sharma, R., Rawat, S., Singh, A. K., Rajput, V. D., Fedorov, Y., . . . Chaplygin, V. (2022). Nanomaterial-plant interaction: Views on the pros and cons. *Toxicity of Nanoparticles in Plants*, 47–68. doi:10.1016/B978-0-323-90774-3.00015-5

Singh, S. P., & Batra, R. D. (2018). Nanotechnology in wastewater treatment: A review. In Novel applications in polymers and waste management (pp. 173–182). Apple Academic Press. Crossref

Struik, P. C., & Kuyper, T. W. (2017). Sustainable intensification in agriculture: The richer shade of green. A review. *Agronomy for Sustainable Development, 37*(5), 39. doi:10.1007/s13593-017-0445-7

Sun, Y., Wang, Y., Deng, X., Liang, H., & Guo, Y. (2020). Nanomaterials for the remediation of soils contaminated by heavy metals and metalloids: A review. *Reviews in Environmental Science and Biotechnology, 19*(4), 769–792.

Tabasum, A., Alghuthaymi, M., Qazi, U. Y., Shahid, I., Abbas, Q., Javaid, R., Nadeem, N., & Zahid, M. (2020). UV-accelerated photocatalytic degradation of pesticide over magnetite and cobalt ferrite decorated graphene oxide composite. *Plants, 10*(1), 6. doi:10.3390/plants10010006 PMID:33374688

Thakur, M., Smita, S., Gupta, P., & Singh, D. (2020). Impact of nanoparticles on the rhizosphere microbiome and its ecological significance. *Archives of Agronomy and Soil Science*, *66*(14), 1977–1994.

Tian, H., Liang, Y., Yang, D., & Sun, Y. (2020). Characteristics of PVP–stabilised NZVI and application to dechlorination of soil–sorbed TCE with ionic surfactant. *Chemosphere*, *239*, 124807. doi:10.1016/j.chemosphere.2019.124807 PMID:31520982

Vardumyan, H., Singh, A., Rajput, V., Minkina, T., El-Ramady, H., & Ghazaryan, K. (2024). Additive-Mediated Phytoextraction of Copper-Contaminated Soils Using Medicago lupulina L. *Egyptian Journal of Soil Science*, *64*(2), 599–618. doi:10.21608/ejss.2024.266169.1714

Veera kumar, P., Sangili, A., Saranya, K., Pandikumar, A., & Lin, K. C. (2021). Palladium and silver nanoparticles embedded on zinc oxide nanostars for photocatalytic degradation of pesticides and herbicides. *Chemical Engineering Journal, 410*, 128434. doi:10.1016/j.cej.2021.128434

Wu, Y., Qi, Y., Song, B., Zhang, M., Jiang, L., & Wang, Y. (2019). Nanoparticles inhibit the microbial activities of As (III)-oxidizing bacteria and the transformation of As. In soil. *Environmental Pollution. III*, *255*, 113134.

Xie, Y., He, Y., Irwin, P. L., Jin, T., & Shi, X. (2011). Antibacterial activity and mechanism of action of zinc oxide nanoparticles against Campylobacter jejuni. *Applied and Environmental Microbiology*, *77*(7), 2325–2331. doi:10.1128/AEM.02149-10 PMID:21296935

Xue, W., Huang, D., Zeng, G., Wan, J., Zhang, C., Xu, R., Cheng, M., & Deng, R. (2018). Nanoscale zero-valent iron coated with rhamnolipid as an effective stabilizer for immobilization of Cd and Pb in river sediments. *Journal of Hazardous Materials*, *341*, 381–389. doi:10.1016/j.jhazmat.2017.06.028 PMID:28806558

You, Y., Zhang, L., Zhang, J., Zheng, J., Zhang, Y., & Chen, Y. (2017). Stimulatory effects of multi-walled carbon nanotubes on plant growth: Relationship with nanotube functionalization. *Plant Physiology and Biochemistry*, *110*, 97–106.

Younis, U., & Malik, A. H. (2021). Nanotechnology and plant growth-promoting rhizobacteria for sustainable agriculture. In *Advances in soil microbiology* (pp. 169–186). Springer.

Zeng, R., Wang, J., Cui, J., Hu, L., & Mu, K. (2010). Photocatalytic degradation of pesticide residues with Re3+-doped Nano-TiO2. *Journal of Rare Earths*, *28*, 353–356. doi:10.1016/S1002-0721(10)60329-8

Zhang, Y., Yang, J., Zhong, L., & Liu, L. (2018). Effect of multi-wall carbon nanotubes on Cr(VI) reduction by citric acid: Implications for their use in soil remediation. *Environmental Science and Pollution Research International*, *25*(24), 23791–23798. doi:10.1007/s11356-018-2438-8 PMID:29876853

Zuverza-Mena, N., Armendariz, R., Peralta-Videa, J. R., & Gardea-Torresdey, J. L. (2016). Effects of silver nanoparticles on radish sprouts: Root growth reduction and modifications in the nutritional value. *Frontiers in Plant Science*, *7*, 90. doi:10.3389/fpls.2016.00090 PMID:26909084

Chapter 12
Nanotechnological Innovations for Soil Pollution Remediation and Environmental Conservation

Omkar Singh
https://orcid.org/0000-0003-1971-185X
Sardar Vallabhbhai Patel University of Agriculture and Technology, India

Shivangi Singh
Sardar Vallabhbhai Patel University of Agriculture and Technology, India

Uday Pratap Shahi
Sardar Vallabhbhai Patel University of Agriculture and Technology, India

Raghavendra Singh
ICAR-Indian Institute of Farming Systems Research, India

Prabhat Kumar Singh
Acharya Narendra Deva University of Agriculture and Technology, India

Karen K. Grigoryan
Yerevan State University, Armenia

Hrant Khachatryan
Armenian National Agrarian University, Armenia

Priyadarshani Rajput
Southern Federal University, Russia

Leonid Perelomov
Leo Tolstoy Tula State Pedagogical University, Russia

Svetlana Kozmenko
Leo Tolstoy Tula State Pedagogical University, Russia

Marina Burachevskaya
Leo Tolstoy Tula State Pedagogical University, Russia

ABSTRACT

Globally, soil contamination threatens the ecosystem and human health. Ecological, financial, and efficiency issues plague excavation, chemical, and bioremediation methods. Nanotechnology is a promise for soil remediation, offering advantages over existing approaches. Nanotechnology technologies for soil pollution treatment and environmental conservation are covered in this chapter. It examines heavy metals, organic pollutants, pesticides, and radioactive contamination in soil and the limitations of cur-

DOI: 10.4018/979-8-3693-1890-4.ch012

rent remediation methods. The chapter next discusses nanotechnology in soil remediation, including nanoparticles, nanomaterials for adsorption and degradation, nano-bioremediation, and nanosensors for pollution monitoring. Case studies and research show nanotechnology's success in soil remediation. Future nanotechnology for soil remediation views and difficulties highlight the need for more research and development to address scalability, cost-effectiveness, and environmental safety.

1. INTRODUCTION

1.1 Significance of Soil Pollution and Its Impact on the Environment

Soil pollution is a major worldwide environmental issue that presents considerable risks to ecosystems, human health, and food security (Rodriguez-Eugenio et al., 2018; Singh et al., 2024b). Human activities, such as industrialization, agricultural practices, and poor waste management, have caused the buildup of many contaminants in the soil (Ashraf et al., 2019). Various pollutants such as heavy metals, organic contaminants, pesticides, and radioactive compounds can harm soil quality, biodiversity, and the overall operation of terrestrial ecosystems (Khalid et al., 2017). Soil pollution has widespread repercussions that go beyond the nearby affected areas. Pollutants in the soil can be absorbed by plants and transferred into the food chain, potentially endangering human health if consumed (Tóth et al., 2016). Contaminated soil can degrade water quality by leaching and surface runoff, posing a threat to aquatic ecosystems and the safety of drinking water sources (Salah et al., 2015). Soil pollution leads to the depletion of arable land, lower agricultural output, and food insecurity, especially in emerging nations (Rai et al., 2019).

Soil contamination severity and extent range among regions and are affected by industrial operations, agricultural methods, population density, and environmental restrictions (Rodriguez-Eugenio et al., 2018). Rapid industrialization and inadequate environmental legislation in numerous developing nations have worsened the issue of soil contamination (Ashraf et al., 2019). Inadequate waste management infrastructure and excessive use of agrochemicals have worsened soil quality, as noted by Khalid et al. (2017).

1.2 Need for Innovative Remediation Techniques

Due to the substantial environmental and health consequences of soil contamination, there is a pressing requirement for efficient remediation methods (Khalid et al., 2017). Traditional procedures like excavation, landfilling, chemical treatment, and bioremediation are commonly employed to deal with soil contamination (Ashraf et al., 2019). Nevertheless, these approaches frequently face constraints regarding effectiveness, expenses, and ecological viability (Rai et al., 2019). Excavation and landfilling refer to the process of physically removing contaminated soil and disposing of it in specific landfill sites (Khalid et al., 2017). Although efficient in eliminating pollutants, this method is costly, takes a significant amount of labour, and relies on the presence of appropriate landfill sites (Ashraf et al., 2019). Transporting contaminated soil to landfill sites might result in additional environmental consequences and potential hazards during processing and disposal, as mentioned by Rai et al. (2019).

Chemical treatment approaches, like chemical oxidation and reduction, use chemical agents to transform contaminants into less hazardous or immobile forms (Khalid et al., 2017). Chemical treatments can be successful but are often costly to operate and can produce additional pollutants that need to be treated further (Ashraf et al., 2019). Moreover, the utilisation of chemical agents might negatively impact soil characteristics and microbial populations, resulting in prolonged ecological disruptions (Rai et al., 2019). Bioremediation, utilising microorganisms to break down or change contaminants, is recognised as an environmentally acceptable and economical method (Rodriguez-Eugenio et al., 2018). Bioremediation has constraints on the types of pollutants it can efficiently address and the duration of the remediation process (Khalid et al., 2017). Bioremediation success relies on parameters like pollutant type, soil conditions, and the existence of appropriate microbial communities (Ashraf et al., 2019). Due to the constraints of traditional remediation methods, there is an increasing focus on creating new and sustainable strategies to tackle soil pollution (Rai et al., 2019). Nanotechnology is a promising field that provides new options for environmental remediation, such as cleaning up contaminated soil (Pandey et al., 2019).

1.3 Potential of Nanotechnology in Addressing Soil Pollution

Nanotechnology, manipulating materials at the nanoscale (1-100 nm), has demonstrated significant potential in tackling environmental issues such as soil pollution (Pandey et al., 2019). Nanomaterials have distinctive characteristics like high surface area to volume ratio, increased reactivity, and adjustable surface chemistry, making them appealing for environmental uses (Karn et al., 2009). Nanotechnology has numerous benefits compared to traditional approaches in soil treatment. Nanomaterials can be engineered to selectively target particular pollutants, improve the effectiveness of cleanup procedures, and reduce the ecological footprint (Gomes et al., 2018; Singh et al., 2023). Nanoparticles including iron oxide, titanium dioxide, and zero-valent iron have been well researched for their capacity to adsorb, decompose, or immobilise different contaminants in soil (Rizwan et al., 2018). Nanotechnology in soil remediation has the advantage of delivering remedial materials directly to damaged locations. Nanomaterials can be designed to possess great mobility and stability in the soil matrix, enabling them to access and engage with contaminants in their original location (Rizwan et al., 2018). This focused distribution method decreases the necessity for excavation and lessens the disruption to the nearby environment (Gomes et al., 2018). Nanotechnology can improve cleanup efficiency by utilising catalytic and reactive nanoparticles (Pandey et al., 2019). Nanoparticles with catalytic properties can speed up the breakdown of organic contaminants, whereas reactive nanomaterials can help convert heavy metals into less harmful or immobile forms (Saikanth et al., 2023).

Nanomaterials have a large surface area that offers numerous active sites for adsorbing and degrading pollutants, resulting in quicker and more efficient cleanup as compared to traditional approaches (Gomes et al., 2018; Singh et al., 2023; Singh O. 2023). In addition, nanotechnology can be combined with other remediation methods, such bioremediation, to produce synergistic outcomes (Rizwan et al., 2018). Nanomaterials can serve as transporters for microbes or enzymes, improving their stability and effectiveness in the soil environment (Pandey et al., 2019; Singh et al., 2023; Singh et al., 2023f). Nanobioremediation, the integration of nanotechnology and bioremediation, has demonstrated encouraging outcomes in breaking down persistent organic pollutants and containing heavy metals (Gomes et al., 2018). Nanotechnology provides potential for monitoring and evaluating soil pollution, in addition to remediation (Ivanova et al., 2018). Nanosensors can be created to identify and measure certain contaminants in the soil, offering immediate and on-location monitoring abilities (Qian et al., 2019). The nanosensors can be incorporated into wireless sensor networks to facilitate remote monitoring and early warning systems for soil contamination (Ivanova et al., 2018). Although nanotechnology shows promise in soil remediation, there are also worries about the environmental impact and toxicity of nanomaterials

(Bundschuh et al., 2018). Nanomaterials released into the environment can result in unforeseen outcomes, including their buildup in food chains and the possible threats to human health and ecosystems (Deng et al., 2017; Sachan et al., 2023). It is essential to carefully evaluate the environmental safety and long-term consequences of employing nanotechnology for soil remediation (Bundschuh et al., 2018).

2. TYPES OF SOIL POLLUTION

Soil contamination can result from various pollutants, each possessing distinct features and environmental consequences. It is essential to comprehend the various forms of soil pollution to create successful cleanup plans. This section will cover the primary types of soil contaminants, such as heavy metals, organic pollutants, pesticides and herbicides, and radioactive substances.

2.1 Heavy Metal Contamination

Heavy metals are prevalent and enduring soil contaminants (Wuana & Okieimen, 2011). Lead, cadmium, mercury, chromium, and arsenic are naturally occurring elements that can build up in the soil because of human activity (Alloway, 2013). Industrial activities, mining, burning fossil fuels, and improper electronic waste disposal are significant contributors to heavy metal pollution in soils (Ashraf et al., 2019). Heavy metals provide substantial dangers to human health and the environment because of their toxicity, inability to degrade naturally, and ability to accumulate in living organisms (Rai et al., 2019; Singh et al., 2024a). High levels of heavy metals can have negative impacts on soil quality, plant growth, and ecosystem functioning (Khalid et al., 2017). Plants cultivated in polluted soils can absorb heavy metals, which can then enter the food chain, posing health hazards to humans and animals (Tóth et al., 2016). Heavy metal toxicity is influenced by factors like the type of metal, its concentration, bioavailability, and soil qualities (Alloway, 2013). Lead is recognised for its neurotoxic effects, especially in youngsters, while cadmium is linked to kidney damage and bone abnormalities (Wuana & Okieimen, 2011). Mercury, namely in its organic form known as methylmercury, is a powerful neurotoxin that has the ability to accumulate in aquatic food chains (Beckers & Rinklebe, 2017).

2.2 Organic Pollutants

Organic pollutants are a significant group of soil contaminants consisting of various manmade and naturally produced substances (Olaniran & Igbinosa, 2011). Pollutants can come from industrial activity, agricultural practices, and poor waste disposal (Gomes et al., 2018). Organic pollutants such as polycyclic aromatic hydrocarbons (PAHs), polychlorinated biphenyls (PCBs), dioxins, and petroleum hydrocarbons are common examples (Pandey et al., 2019; Tarigholizadeh et al., 2023). PAHs are organic chemicals produced when organic resources like fossil fuels and biomass are incompletely burned. They are recognised for their carcinogenic and mutagenic characteristics, as well as their long-lasting presence in the environment (Stogiannidis & Laane, 2015). PCBs are man-made chlorinated chemicals that were often utilised in industrial settings until they were prohibited in the late 1970s (Gomes et al., 2018). PCBs are very stable and have an affinity for fats, which makes them likely to accumulate in the food chain (Desforges et al., 2018). Petroleum hydrocarbons, specifically benzene, toluene, ethylbenzene, and xylenes (BTEX), are frequently found in soil as pollutants due to oil spills and leaks from storage tanks (Gomes et al., 2018). These chemicals can cause immediate and long-term health issues such as neurological abnormalities, respiratory issues, and cancer (Olaniran & Igbinosa, 2011). Organic contaminants in the soil can disturb soil microbial populations and interfere with nitrogen cycle mechanisms (Pandey et al., 2019).

2.3 Pesticides and Herbicides

Pesticides and herbicides are commonly utilised in agriculture to manage pests and weeds, as stated by Carvalho (2017). Although these compounds have been vital in boosting agricultural output, their excessive and incorrect application has resulted in substantial soil pollution (Rai et al., 2019). Pesticides and herbicides can remain in the soil for long periods, which might result in their buildup and their movement into groundwater (Khalid et al., 2017).The toxicity of pesticides and herbicides varies based on their chemical makeup and method of action (Carvalho, 2017). Organochlorines (e.g., DDT) and organophosphates are widely used pesticides renowned for their severe toxicity and long-lasting presence in the environment (Rai et al., 2019). These compounds can harm unintended creatures such as beneficial insects, birds, and aquatic life (Khalid et al., 2017). Pesticide and herbicide residues in the soil can endanger human health by direct contact or by consuming contaminated crops (Carvalho, 2017). Long-term exposure to these substances has been associated with a range of health issues, such as neurological diseases, endocrine disruption, and cancer (Rai et al., 2019). Pesticide and herbicide residues in the soil might cause resistance in target pests and weeds, requiring higher doses or more toxic alternatives to be used (Khalid et al., 2017).

2.4 Radioactive Contamination

Radioactive contamination of soil is an infrequent although potentially serious type of soil pollution (Hu et al., 2010). Radioactive materials can contaminate the soil through different means, including nuclear accidents, incorrect disposal of radioactive waste, and fallout from nuclear weapons testing (Oughton & Salbu, 2012). The primary radionuclides of interest in soil contamination are cesium-137, strontium-90, and plutonium isotopes (Hu et al., 2010).Radioactive contamination of soil presents substantial hazards to human health and the environment because of the ionising radiation released by the radionuclides (Oughton & Salbu, 2012). Ionising radiation exposure can result in DNA damage, which can elevate the likelihood of cancer and genetic alterations (Hu et al., 2010). Plants cultivated on soils contaminated with radioactivity can absorb radionuclides, allowing them to enter the food chain (Oughton & Salbu, 2012). Remediating radioactively contaminated soils is difficult because of the extended half-lives of various radionuclides and their ability to move across the soil profile. Traditional methods of remediation, such excavation and disposal, may not be practical or economical for extensive pollution. Developing new remediation solutions, such as those based on nanotechnology, is essential for dealing with radioactive soil pollution (Hu et al., 2010).

3. CONVENTIONAL SOIL REMEDIATION TECHNIQUES

Traditional methods for soil remediation have been extensively employed to tackle soil contamination and rehabilitate polluted areas. The methods can be roughly classified as physical, chemical, and biological procedures (Khalid et al., 2017). Although successful in some cases, these approaches have limitations that have prompted the exploration of advanced remediation technologies, like nanotechnology-based solutions (Gomes et al., 2018). This section will examine the primary traditional soil remediation methods and their respective constraints.

3.1 Physical Remediation Methods

Physical remediation approaches encompass the physical extraction, confinement, or segregation of polluted soil (Khalid et al., 2017). Excavation and landfilling are prevalent physical remediation methods according to Kuppusamy et al. (2016). The contaminated soil is removed and taken to a specific landfill for disposal according to Khalid et al. (2017). Excavation and landfilling can remove contaminated soil from a site but have limitations such as high costs, greater risk of contamination spread during transportation, and the requirement for sufficient landfill area (Kuppusamy et al., 2016). Soil washing is a physical remediation approach that removes toxins from the soil by utilising water or other solvents (Dermont et al., 2008). The polluted soil is combined with a washing solution to extract pollutants from the soil particles (Khalid et al., 2017). The purified soil can be brought back to the original location, but the solution containing contaminants needs additional treatment or disposal (Dermont et al., 2008). Soil washing is highly efficient in eliminating heavy metals and organic pollutants, although it can be energy-intensive and may necessitate the use of strong chemicals (Kuppusamy et al., 2016). Thermal desorption is a remediation method that involves using heat to vaporise and eliminate toxins from the soil (Khalid et al., 2017). The polluted soil is subjected to temperatures between 100°C to 600°C, which vary according on the contaminant type (Kuppusamy et al., 2016). The evaporated pollutants are subsequently gathered or processed by off-gas treatment systems (Khalid et al., 2017). Thermal desorption is a potent method for cleaning soils polluted with volatile organic compounds (VOCs) and semi-volatile organic compounds (SVOCs), but it requires a lot of energy and can change the soil's physical and chemical characteristics (Kuppusamy et al., 2016).

3.1 Chemical Remediation Methods

Chemical remediation approaches utilise chemical agents to transform pollutants into less hazardous or immobile forms (Khalid et al., 2017). The approaches can be used either on-site or off-site, depending on the particular site characteristics and cleanup objectives (Gomes et al., 2018). Chemical oxidation and reduction are two prevalent methods for chemical cleanup, as stated by Kuppusamy et al. (2016). Chemical oxidation involves the use of potent oxidising agents such hydrogen peroxide, ozone, or persulfates to break down organic pollutants in polluted soil (Gomes et al., 2018). Oxidising chemicals react with pollutants, converting them into less dangerous substances such carbon dioxide and water (Khalid et al., 2017). Chemical oxidation is a successful method for remediation of soils polluted with petroleum hydrocarbons, chlorinated solvents, and pesticides (Kuppusamy et al., 2016). The method can be costly, and the oxidising agents could negatively impact soil quality and microbial communities (Gomes et al., 2018). Chemical reduction utilises reducing agents like zero-valent iron or sodium dithionite to transform pollutants into less harmful or immobile forms (Kuppusamy et al., 2016). This method is very beneficial for treating soils that are polluted with heavy metals and chlorinated chemicals, as stated by Khalid et al. (2017). Reducing agents transfer electrons to pollutants, changing their chemical composition and ability to move (Gomes et al., 2018). Chemical reduction can be expensive, and the durability of the immobilised pollutants over time may be a worry (Kuppusamy et al., 2016).

3.3 Biological Remediation Methods

Biological remediation approaches, or bioremediation, involve utilising microorganisms or plants to break down, change, or store pollutants in the soil (Gomes et al., 2018). Bioremediation can be accomplished by many methods such as biostimulation, bioaugmentation, and phytoremediation (Kuppusamy et al., 2016).

Biostimulation is the process of adding nutrients, oxygen, or other substances to the soil to enhance the growth and function of indigenous microbes that can break down pollutants (Khalid et al., 2017). This method is highly efficient for treating soils contaminated with organic contaminants such petroleum hydrocarbons and pesticides (Kuppusamy et al., 2016). The effectiveness of biostimulation relies on the existence of appropriate microbial communities and the accessibility of the pollutants (Gomes et al., 2018).

Bioaugmentation is the process of introducing particular microbial strains or groups into polluted soil to improve the breakdown of specific pollutants (Kuppusamy et al., 2016). The introduced microbes are chosen for their capacity to break down the targeted pollutants (Khalid et al., 2017). Bioaugmentation is effective for remediating soils contaminated with persistent substances including polychlorinated biphenyls (PCBs) and polycyclic aromatic hydrocarbons (PAHs) (Gomes et al., 2018). The presence and function of the introduced microbes may be restricted by environmental conditions such soil pH, temperature, and moisture content (Kuppusamy et al., 2016).

Phytoremediation is a method of bioremediation that involves using plants to eliminate, break down, or secure pollutants in the soil (Khalid et al., 2017). Plants can absorb heavy metals (phytoextraction), break down organic pollutants (phytodegradation), or enhance microbial activity in the root area (rhizodegradation) (Gomes et al., 2018). Phytoremediation is a cost-effective and environmentally beneficial technology that has limitations such as the root zone depth, slow plant growth, and the risk of contamination transfer into the food chain (Kuppusamy et al., 2016).

3.4 Limitations of Conventional Remediation Techniques

Conventional soil restoration approaches have constraints that can impede their effectiveness and sustainability, despite being widely used (Gomes et al., 2018). The constraints consist of elevated expenses, extended treatment durations, ecological disruption, and the creation of additional waste products (Kuppusamy et al., 2016).Traditional remediation techniques such excavation and landfilling, thermal desorption, and chemical oxidation are costly because of the energy demands, transportation expenses, and the necessity for specialised equipment (Khalid et al., 2017). The expensive nature of these methods may hinder their adoption, especially for extensive polluted areas (Gomes et al., 2018).Traditional remediation methods typically necessitate extended treatment durations, varying from months to years, contingent on the nature and scope of the pollution (Kuppusamy et al., 2016). This may result in extended site disturbance and postponed revitalization of the polluted land (Khalid et al., 2017). Additionally, certain methods like phytoremediation are constrained by the sluggish development of plants and the seasonal fluctuations in their effectiveness (Gomes et al., 2018).Conventional remediation procedures, especially physical and chemical treatments, can lead to considerable environmental disruption (Kuppusamy et al., 2016). Excavation and landfilling can disturb soil structure, eliminate habitats, and heighten the potential for contamination dispersion during transportation (Khalid et al., 2017). Chemical oxidation and reduction can change soil characteristics and harm soil microbial populations (Gomes et al., 2018). Conventional remediation procedures have a drawback of producing additional waste streams, as noted by Kuppusamy et al. (2016). Soil washing generates wastewater contaminated with pollutants that needs additional treatment or disposal (Khalid et al., 2017). Thermal desorption and chemical oxidation can produce gases and residues that must be handled properly to prevent further contamination (Gomes et al., 2018). Due to these constraints, there is an increasing focus on creating novel and sustainable remediation technologies that can address the limits of traditional approaches (Kuppusamy et al., 2016). The upcoming sections will explore nanotechnology-based methods that show potential for improving the effectiveness, cost-efficiency, and environmental friendliness of soil remediation, as highlighted by Gomes et al. (2018).

4. NANOTECHNOLOGICAL INNOVATIONS FOR SOIL REMEDIATION

Nanotechnology is a promising field for creating new solutions to combat soil degradation. Nanomaterials, due to their distinctive characteristics and strong reactivity, present novel possibilities for the restoration of polluted soils (Karn et al., 2009). This section will cover different nanotechnological advancements for soil remediation, such as nanoparticles, nanomaterials for adsorption and degradation of contaminants, nano-bioremediation, and nanosensors for monitoring soil contamination.

4.1 Nanoparticles for Soil Remediation

4.1.1 Iron Nanoparticles

Iron nanoparticles, particularly nano-scale zero-valent iron (nZVI), have been extensively studied for their potential in soil remediation (Karn et al., 2009). nZVI particles are highly reactive due to their large surface area and can effectively degrade or immobilize a wide range of contaminants, including chlorinated solvents, heavy metals, and pesticides (Gomes et al., 2018). The remediation mechanism of nZVI involves reduction reactions, where the nanoparticles donate electrons to the contaminants, converting them into less toxic or immobile forms (Karn et al., 2009). nZVI can be applied in situ by direct injection into the contaminated soil, providing a cost-effective and minimally invasive remediation approach (Gomes et al., 2018).

4.1.2 Titanium Dioxide Nanoparticles

Titanium dioxide (TiO2) nanoparticles have been studied for their ability to catalyse the degradation of organic contaminants in soil, as reported by Gomes et al. (2018). TiO_2 nanoparticles, when exposed to ultraviolet (UV) radiation, produce hydroxyl radicals that are extremely reactive and can break down organic pollutants including polycyclic aromatic hydrocarbons (PAHs) and pesticides (Karn et al., 2009). Doping TiO_2 nanoparticles with other elements or combining them with nanomaterials like graphene can increase their photocatalytic activity (Gomes et al., 2018). Titanium dioxide nanoparticles can be used in polluted soils by either mixing them into the soil or injecting them as a slurry, then exposing them to sunshine or artificial UV light sources (Karn et al., 2009).

4.1.3 Zero-Valent Iron Nanoparticles

Zero-valent iron (ZVI) nanoparticles are extensively researched for their capacity to clean up soils polluted with heavy metals and chlorinated chemicals (Gomes et al., 2018). ZVI nanoparticles can capture heavy metals by adsorption, reduction, and precipitation processes (Karn et al., 2009). ZVI nanoparticles can either convert heavy metal ions to their elemental form or create insoluble metal hydroxides upon contact, thereby trapping them in the soil (Gomes et al., 2018). ZVI nanoparticles have the ability to dechlorinate chlorinated chemicals like trichloroethylene (TCE) by undergoing reductive dehalogenation processes (Karn et al., 2009). ZVI nanoparticles can be used either direct soil mixing or injection for in-situ remediation (Gomes et al., 2018).

4.1.4 Other Nanoparticles

Several different types of nanoparticles, such as silver nanoparticles, iron oxide nanoparticles, and bimetallic nanoparticles, have been studied for their possible use in soil remediation (Gomes et al., 2018). Silver nanoparticles have antibacterial capabilities and can be utilised for remediating soils contaminated with harmful microorganisms (Karn et al., 2009). Iron oxide nanoparticles like magnetite and maghemite have the ability to absorb heavy metals and organic contaminants from polluted soils (Gomes et al., 2018). Bimetallic nanoparticles like iron-palladium and iron-nickel have been created to merge the remediation capacities of several metals and boost the overall effectiveness of remediation (Karn et al., 2009).

4.2 Nanomaterials for Adsorption and Degradation of Pollutants

4.2.1 Carbon Nanotubes

Carbon nanotubes (CNTs) have been studied for their capacity to absorb and eliminate contaminants from polluted soils (Gomes et al., 2018). Carbon nanotubes (CNTs) possess a large specific surface area and powerful adsorption capabilities, which make them efficient for eliminating a wide range of pollutants such as heavy metals, organic pollutants, and pesticides (Karn et al., 2009). CNTs adsorption mechanism includes physical adsorption, electrostatic interactions, and chemical bonding as discussed by Gomes et al. (2018). Carbon nanotubes (CNTs) can be used in contaminated soils either by soil mixing or as a permeable reactive barrier to treat contaminated groundwater (Karn et al., 2009).

4.2.2 Graphene-Based Nanomaterials

Graphene and its derivatives, such graphene oxide (GO) and reduced graphene oxide (rGO), have demonstrated potential in cleaning up polluted soils (Gomes et al., 2018). Nanomaterials possess a large surface area and strong adsorption capability, which enables them to efficiently eliminate heavy metals, organic contaminants, and pesticides from soils (Karn et al., 2009). Graphene-based nanomaterials' adsorption method includes surface complexation, electrostatic interactions, and π-π interactions (Gomes et al., 2018). Graphene-based nanomaterials can improve contaminated soils by mixing them with the soil or adding them as an amendment, which enhances the soil's capacity to absorb and immobilise pollutants (Karn et al., 2009).

4.2.3 Dendrimers

Dendrimers are complex three-dimensional polymeric nanoparticles that have been studied for their possible use in soil remediation (Gomes et al., 2018). Dendrimers are structured molecules with a high concentration of functional groups, which enables them to efficiently adsorb and immobilise contaminants (Karn et al., 2009). Dendrimers remediate contaminants by encapsulating them in their internal cavities or attaching them to their surface functional groups. Dendrimers can enhance contaminated soils' capacity to hold and stabilise pollutants when applied by soil mixing or as a soil supplement (Karn et al., 2009).

4.2.4 Nanoclays

Nanoclays such montmorillonite and kaolinite have been studied for their capacity to absorb and trap contaminants in polluted soils (Gomes et al., 2018). Nanoclays possess a high specific surface area and cation exchange capacity, which makes them efficient in eliminating heavy metals and organic contaminants from soils (Karn et al., 2009). Nanoclays' adsorption mechanism includes ion exchange, surface complexation, and hydrophobic interactions (Gomes et al., 2018). Nanoclays can be used in contaminated soils by soil mixing or as a soil amendment to improve the soil's capacity to absorb and stabilise pollutants (Karn et al., 2009).

4.2 Nano-Bioremediation

4.2.1 Nanoparticle-Assisted Bioremediation

Nanoparticle-assisted bioremediation utilises nanoparticles to improve the effectiveness of microorganisms in breaking down soil pollutants (Gomes et al., 2018). Nanoparticles can create a conducive environment for microbial growth and activity by acting as a supportive structure and supplying necessary nutrients (Karn et al., 2009). Iron oxide nanoparticles can enhance the proliferation and function of iron-reducing bacteria, leading to the decomposition of organic pollutants and the containment of heavy metals in the soil (Gomes et al., 2018). Nanoparticles can serve as electron shuttles, aiding in the electron transport between bacteria and pollutants, therefore improving the biodegradation process (Karn et al., 2009).

4.2.2 Genetically Engineered Microorganisms With Nanoparticles

Genetically modified microorganisms (GEMs) can be paired with nanoparticles to develop sophisticated nano-bioremediation systems for soil cleanup (Gomes et al., 2018). GEMs can be engineered to produce specific enzymes or metabolic pathways that focus on certain pollutants in the soil (Karn et al., 2009). Nanoparticles can improve the efficiency of GEMs for cleanup. GEMs can be modified to produce metal-binding proteins on their surface. These proteins can be combined with nanoparticles to enhance the absorption and fixation of heavy metals in the soil (Gomes et al., 2018). GEMs can be modified to produce enzymes that break down organic contaminants, while nanoparticles can serve as a support structure to improve the enzymes' stability in soil (Karn et al., 2009).

4.3 Nanosensors for Monitoring Soil Pollution

4.3.1 Types of Nanosensors

Nanosensors are analytical tools that utilise nanoparticles or nanostructures to identify and measure contaminants in soil (Gomes et al., 2018). Multiple nanosensors have been created to monitor soil contamination, such as optical, electrochemical, and biological nanosensors (Karn et al., 2009). Optical nanosensors utilise nanomaterials' optical characteristics like fluorescence or surface plasmon resonance to identify contaminants (Gomes et al., 2018). Electrochemical nanosensors utilise the electrical or electrochemical characteristics of nanomaterials to detect the levels of contaminants (Karn et al., 2009). Biological nanosensors utilise biological recognition elements, like enzymes or antibodies, in conjunction with nanomaterials to identify certain pollutants in the soil (Gomes et al., 2018).

4.3.2 Applications in Soil Pollution Monitoring

Nanosensors are widely used in monitoring soil contamination by detecting and measuring heavy metals, organic pollutants, and pesticides (Gomes et al., 2018). Nanosensors can offer immediate and on-location monitoring of soil pollution, allowing for quick and precise evaluation of contamination levels (Karn et al., 2009). Gold nanoparticle-based optical nanosensors have been created to detect heavy metals including mercury and lead in soils (Gomes et al., 2018). Carbon nanotube electrochemical nanosensors were utilised to detect organic pollutants, specifically polycyclic aromatic hydrocarbons (PAHs), in polluted soils (Karn et al., 2009). Enzyme-based biological nanosensors have been used to detect pesticides, specifically organophosphates, in agricultural soils (Gomes et al., 2018). Nanosensors in soil pollution monitoring can help direct remediation operations, evaluate the success of remediation procedures, and guarantee the enduring safety and quality of treated soils (Karn et al., 2009).

5. CASE STUDIES AND RESEARCH FINDINGS

5.1 Successful Implementation of Nanotechnology in Soil Remediation

Many case studies and research have shown the effective use of nanotechnology in soil remediation. The experiments demonstrate the efficacy of different nanomaterials and methods in eliminating or containing pollutants in soil in practical settings (Gomes et al., 2018). An example is the application of nano-scale zero-valent iron (nZVI) to clean soils polluted with chlorinated solvents like trichloroethylene (TCE) (Karn et al., 2009).Elliott and Zhang (2001) conducted a field research where nZVI particles were introduced into a TCE-contaminated aquifer at an industrial site. The findings demonstrated a notable decrease in TCE levels, with a clearance effectiveness of up to 99% attained in a short period after treatment. The nZVI particles efficiently dechlorinated TCE, converting it into less harmful derivatives including ethene and ethane (Elliott & Zhang, 2001). The study showed that employing nZVI is both feasible and effective for in situ cleanup of chlorinated solvent-contaminated soils and groundwater.Nanotechnology has been successfully used in soil remediation by employing iron oxide nanoparticles to immobilise heavy metals. Shipley et al. (2011) used maghemite nanoparticles to treat lead-contaminated soil from a battery recycling site in their investigation. The findings indicated that the maghemite nanoparticles efficiently absorbed and immobilised lead, decreasing its bioavailability and potential environmental impact. The soil that was treated showed a notable decrease in the leaching of lead, suggesting the enduring stability of the trapped lead (Shipley et al., 2011).Nano-bioremediation methods have demonstrated potential in cleaning soil contaminated with organic contaminants. Biswas et al. (2015) utilised a nanoparticle-assisted bioremediation approach to clean soil polluted with polycyclic aromatic hydrocarbons (PAHs). Iron oxide nanoparticles were utilised alongside a PAH-degrading bacterial strain to improve the biodegradation of PAHs in the soil. The study demonstrated a notable rise in PAH breakdown rates when iron oxide nanoparticles were present. These nanoparticles served as a support structure for bacterial growth and improved the accessibility of the PAHs.The case studies and research findings offer proof of the beneficial use of nanotechnology in soil remediation, showcasing its ability to remove or contain various contaminants in real-world settings. The effectiveness of nanotechnology-based remediation methods relies on elements like the type and amount of pollutants, soil characteristics, and environmental circumstances (Gomes et al., 2018).

5.2 Ongoing Research and Development

Ongoing research and development in the field of nanotechnology for soil remediation aim to enhance the efficiency, cost-effectiveness, and sustainability of remediation systems (Pandey et al., 2019). Current research focuses on creating multi-functional nanomaterials capable of targeting various pollutants at the same time (Gomes et al., 2018). Researchers are studying the utilisation of bimetallic nanoparticles like iron-nickel and iron-palladium to improve the overall efficiency of remediation by combining the remediation properties of various metals (Karn et al., 2009).Ongoing research is focused on integrating nanotechnology with other remediation methods as phytoremediation and bioremediation (Pandey et al., 2019). Researchers are investigating the application of nanomaterials to improve the absorption and movement of pollutants by plants, and to boost the growth and function of microorganisms that break down pollutants in the soil (Gomes et al., 2018). Research has demonstrated that using iron oxide nanoparticles can enhance the effectiveness of phytoremediation in soils contaminated with heavy metals by making the metals more accessible to plants (Tripathi et al., 2015). Research on developing eco-friendly ways for producing nanomaterials is currently a significant focus (Pandey et al., 2019). Green synthesis utilises eco-friendly and sustainable techniques, such plant extracts or microbial processes, to create nanomaterials (Gomes et al., 2018). The strategies focus on decreasing the environmental impact and expenses of producing nanomaterials while preserving their effectiveness in cleanup (Karn et al., 2009).

Ongoing research is being conducted to investigate the environmental concerns linked to the application of nanotechnology in soil remediation (Pandey et al., 2019). Research is examining the destiny, movement, and harmful effects of nanomaterials in the soil, along with their possible influence on soil microbial communities and ecosystem functioning (Gomes et al., 2018). Creating risk assessment frameworks and monitoring mechanisms is essential for the safe and responsible use of nanotechnology in soil remediation.Field-scale demonstrations and pilot projects are being carried out alongside laboratory investigations to confirm the efficacy and practicality of nanotechnology-based remediation methods in real-world scenarios (Pandey et al., 2019). The field investigations offer vital insights into the scalability, cost-effectiveness, and long-term performance of nanotechnology-based remediation techniques (Gomes et al., 2018).Ongoing research and development in nanotechnology for soil remediation aim to overcome the drawbacks of traditional remediation methods and offer new solutions for managing polluted soils sustainably. Nanotechnology is anticipated to become more significant in addressing soil contamination and restoring contaminated places as research advances.

6. ENVIRONMENTAL CONCERNS AND SAFETY CONSIDERATIONS

6.1 Potential Risks Associated With Nanotechnology in Soil Remediation

Although nanotechnology shows potential for soil remediation, it is important to carefully assess the possible dangers and environmental issues linked to the application of nanomaterials in the environment (Pandey et al., 2019). An important issue is the possible harmful effects of nanomaterials on soil organisms, plants, and other unintended species (Bundschuh et al., 2018). Nanomaterials can have distinct interactions with biological systems because of their small size and strong reactivity, which may result in negative impacts on soil organisms (Karn et al., 2009).Research has demonstrated that certain nanomaterials, such metal oxide nanoparticles, can be hazardous to soil microorganisms, impacting their growth, variety, and function (Gomes et al., 2018). Silver nanoparticles have antibacterial effects and can alter the microbial community structure in soils (McGee et al., 2017). Titanium dioxide nanoparticles have been discovered to impede the growth and function of soil bacteria and fungi (Simonin et al., 2018). The impact on soil microbial populations can affect soil health, nitrogen cycling, and ecosystem functioning (Bundschuh et al., 2018).

A potential issue is the absorption and buildup of nanomaterials by plants, which may result in their integration into the food chain (Pandey et al., 2019). Research has shown that plants are capable of absorbing and moving nanoparticles from the soil into their tissues, which could impact their growth, development, and yield (Ma et al., 2010). Nanomaterials gathering in edible plant parts can endanger human health through food exposure (Karn et al., 2009). It is crucial to examine the bioaccumulation potential of nanomaterials employed in soil remediation and determine their impact on food safety (Gomes et al., 2018). The impact of nanoparticles on soil organisms and plants might differ based on parameters like the type and size of the nanomaterials, their surface qualities, and the features of the soil (Bundschuh et al., 2018). The solubility of metal oxide nanoparticles has been shown to impact their toxicity, with more soluble nanoparticles often showing increased toxicity (Gomes et al., 2018; Kolesnikov et al., 2024). Soil pH, organic matter concentration, and clay mineralogy can influence how nanomaterials behave and their toxicity in the soil environment (Karn et al., 2009).

7. ENVIRONMENTAL FATE AND TRANSPORT OF NANOMATERIALS

It is essential to comprehend how nanoparticles behave and move in the soil to evaluate their possible dangers and lasting effects (Pandey et al., 2019). Nanomaterials discharged into the soil can undergo transformations such aggregation, dissolution, and sorption, impacting their movement, availability, and harmfulness (Gomes et al., 2018).

Nanomaterials can engage with soil elements such organic matter, clay minerals, and metal oxides through surface adsorption and complexation processes (Karn et al., 2009). These interactions can affect the retention and movement of nanomaterials in the soil profile, as well as their likelihood to leak into groundwater (Baalousha et al., 2016). Nanomaterials' movement in soil can be influenced by elements such soil texture, porosity, and water flow (Bundschuh et al., 2018). Nanomaterials can undergo changes in their physicochemical properties and reactivity when exposed to the soil environment (Gomes et al., 2018). Metal oxide nanoparticles can dissolve, releasing metal ions that are more accessible and harmful to soil organisms (Karn et al., 2009). Nanomaterial aggregation can decrease their surface area and reactivity, which may impact their remediation effectiveness and toxicity.Another crucial factor to address is the long-term behaviour of nanomaterials in the soil environment (Pandey et al., 2019). Some nanomaterials can dissolve or alter over time, while others can stay in the soil for lengthy durations, potentially accumulating and causing long-term consequences on soil ecosystems (Gomes et al., 2018). Thus, it is crucial to carry out extended research to evaluate the endurance, accumulation, and environmental effects of nanomaterials tilizin in soil restoration (Karn et al., 2009).

8. REGULATIONS AND GUIDELINES FOR THE USE OF NANOTECHNOLOGY IN SOIL REMEDIATION

Regulations and guidelines are necessary to ensure the safe and responsible use of nanotechnology in soil remediation due to the potential dangers and uncertainties associated with it (Baalousha et al., 2016). The regulations for nanotechnology in soil remediation are still in the early phases of development, and existing regulations may not fully consider the distinct features and dangers associated with nanoparticles (Gomes et al., 2018).Environmental regulations like the Toxic Substances Control Act (TSCA) in the United States and the Registration, Evaluation, Authorization, and Restriction of Chemicals (REACH) in the European Union establish guidelines for evaluating and controlling chemical substances, including nanomaterials (Karn et al., 2009). The regulations may require modification or extension to effectively tackle the issues presented by nanomaterials in soil remediation (Pandey et al., 2019).Various international organisations, including the Organisation for Economic Cooperation and Development (OECD) and the International Organisation for Standardisation (ISO), have created guidelines and standards for evaluating and controlling the safety of nanomaterials (Baalousha et al., 2016). The guidelines offer suggestions for the analysis, evaluation, and risk evaluation of nanomaterials, together with instructions for the safe management and disposal of waste containing nanoparticles (Gomes et al., 2018).

Guidelines and best practices must be established for soil remediation when using nanomaterials. These should include site characterization, nanomaterial selection and application, monitoring and assessment of remediation effectiveness, and long-term site management. Collaboration among regulators, researchers, industry stakeholders, and environmental organisations is essential for developing comprehensive and science-based guidelines (Gomes et al., 2018).

Moreover, public involvement and communication are necessary when it comes to utilizing nanotechnology for soil remediation (Baalousha et al., 2016). Enhancing transparency and including the public in decision-making processes can foster trust and mitigate any apprehensions regarding the environmental and health ramifications of nanomaterials (Karn et al., 2009). Develop effective risk communication tactics to educate the public on the advantages, dangers, and unknowns linked to nanotechnology-based soil remediation methods (Pandey et al., 2019).

9. FUTURE PERSPECTIVES AND CHALLENGES

Nanotechnology has demonstrated significant promise in remedying soil pollution by providing creative alternatives to overcome the constraints of traditional remediation methods. As research in this sector advances, there are several developments and future outlooks to take into account. One potential field involves creating multi-functional nanomaterials capable of targeting numerous pollutants at the same time (Gomes et al., 2018). Advanced nanomaterials can be designed with precise characteristics including high adsorption capacity, catalytic activity, and magnetic responsiveness. This allows them to efficiently eliminate or break down various contaminants (Pandey et al., 2019). One potential future path is combining nanotechnology with other methods of remediation, like phytoremediation and bioremediation (Karn et al., 2009). Integrating nanotechnology with these biological approaches can improve the overall effectiveness and longevity of the cleanup process. Nanomaterials can enhance the bioavailability of pollutants to plants or microorganisms, aiding in their absorption and breakdown (Gomes et al., 2018). Moreover, the advancement of intelligent nanoparticles capable of reacting to environmental factors like pH, temperature, or light can facilitate the precise discharge of remediation substances and specific transportation to polluted locations (Pandey et al., 2019).Yet, certain obstacles must be overcome to achieve the effective integration of nanotechnology in soil remediation. Scalability and cost-effectiveness are significant challenges for nanotechnology-based remediation methods (Karn et al., 2009).

Although laboratory studies have demonstrated promising results, additional study and optimisation are needed to apply these discoveries on a larger scale. Large-scale manufacturing and utilisation of nanomaterials can be costly, prompting the necessity for the creation of economical and effective techniques for nanomaterial synthesis and distribution (Gomes et al., 2018). One such obstacle is the possible environmental hazards linked to the utilisation of nanomaterials in soil remediation. Nanomaterials can be hazardous to soil creatures and plants, and their long-term effects on the environment are not completely known yet, as mentioned in the preceding section (Bundschuh et al., 2018). It is essential to perform thorough risk assessments and establish suitable safety regulations and recommendations for utilising nanotechnology in soil remediation (Pandey et al., 2019).Additionally, there are areas with insufficient information and research requirements that must be resolved to progress the subject of nanotechnology for soil remediation. An important focus is on comprehending the behaviour and destiny of nanomaterials within the intricate soil environment (Gomes et al., 2018). Further research is required to clarify the relationships between nanomaterials and soil elements such organic matter, clay minerals, and microorganisms, and how these relationships affect the effectiveness of remediation and possible ecological consequences (Karn et al., 2009). Another study requirement is the establishment of standardised procedures and guidelines for the analysis, evaluation, and supervision of nanomaterials in soil remediation purposes (Pandey et al., 2019). Establishing uniform and reliable techniques will aid in comparing and validating study findings and in developing regulatory frameworks for the safe use of nanotechnology in soil remediation (Gomes et al., 2018).

10. CONCLUSION

Globally, soil contamination threatens the ecosystem and human health. Ecological, financial, and efficiency issues plague excavation, chemical, and bioremediation methods. Nanotechnology is promise for soil remediation, offering advantages over existing approaches. Nanotechnology technologies for soil pollution treatment and environmental conservation are covered in this chapter. It examines heavy metals, organic pollutants, pesticides, and radioactive contamination in soil and the limitations of current remediation methods. The chapter next discusses nanotechnology in soil remediation, including nanoparticles, nanomaterials for adsorption and degradation, nano-bioremediation, and nanosensors for pollution monitoring. Case studies and research show nanotechnology's success in soil remediation. The chapter also discusses nanoparticles in soil's environmental and safety dangers, fate, transit, and legislation and recommendations. Future nanotechnology for soil remediation views and difficulties highlights the need for more research and development to address scalability, cost-effectiveness, and environmental safety. An interdisciplinary and collaborative strategy is needed to use nanotechnology for sustainable soil remediation and environmental protection, the chapter concludes.

ACKNOWLEDGMENT

LP & MB by the Ministry of Education and Science of the Russian Federation under the program "Priority 2030", agreement no. 073-03-2022-117/7.

REFERENCES

Alloway, B. J. (2013). Sources of heavy metals and metalloids in soils. In B. J. Alloway (Ed.), *Heavy metals in soils: Trace metals and metalloids in soils and their bioavailability* (pp. 11–50). Springer. doi:10.1007/978-94-007-4470-7_2

Ashraf, M. A., Hussain, I., Rasheed, R., Iqbal, M., Riaz, M., & Arif, M. S. (2017). Advances in microbe-assisted reclamation of heavy metal contaminated soils over the last decade: A review. *Journal of Environmental Management*, *198*, 132–143. doi:10.1016/j.jenvman.2017.04.060 PMID:28456029

Baalousha, M., Cornelis, G., Kuhlbusch, T. A., Lynch, I., Nickel, C., Peijnenburg, W., & van den Brink, N. W. (2016). Modeling nanomaterial fate and uptake in the environment: Current knowledge and future trends. *Environmental Science. Nano*, *3*(2), 323–345. doi:10.1039/C5EN00207A

Beckers, F., & Rinklebe, J. (2017). Cycling of mercury in the environment: Sources, fate, and human health implications: A review. *Critical Reviews in Environmental Science and Technology*, *47*(9), 693–794. doi:10.1080/10643389.2017.1326277

Biswas, B., Sarkar, B., Rusmin, R., & Naidu, R. (2015). Bioremediation of PAHs and VOCs: Advances in clay mineral-microbial interaction. *Environment International*, *85*, 168–181. doi:10.1016/j.envint.2015.09.017 PMID:26408945

Bundschuh, M., Filser, J., Lüderwald, S., McKee, M. S., Metreveli, G., Schaumann, G. E., Schulz, R., & Wagner, S. (2018). Nanoparticles in the environment: Where do we come from, where do we go to? *Environmental Sciences Europe*, *30*(1), 1–17. doi:10.1186/s12302-018-0132-6 PMID:29456907

Carvalho, F. P. (2017). Pesticides, environment, and food safety. *Food and Energy Security*, *6*(2), 48–60. doi:10.1002/fes3.108

Deng, Y., Ezyske, C. M., Suliman, W., Carlson, J., & Ali, N. (2017). Nanomaterials for soil remediation: Synthesis, applications and biotoxicity. *Nanotechnology for Environmental Engineering*, *2*(1), 1–18.

Dermont, G., Bergeron, M., Mercier, G., & Richer-Laflèche, M. (2008). Soil washing for metal removal: A review of physical/chemical technologies and field applications. *Journal of Hazardous Materials*, *152*(1), 1–31. doi:10.1016/j.jhazmat.2007.10.043 PMID:18036735

Desforges, J. P., Hall, A., McConnell, B., Rosing-Asvid, A., Barber, J. L., Brownlow, A., De Guise, S., Eulaers, I., Jepson, P. D., Letcher, R. J., Levin, M., Ross, P. S., Samarra, F., Víkingson, G., Sonne, C., & Dietz, R. (2018). Predicting global killer whale population collapse from PCB pollution. *Science*, *361*(6409), 1373–1376. doi:10.1126/science.aat1953 PMID:30262502

Elliott, D. W., & Zhang, W. X. (2001). Field assessment of nanoscale bimetallic particles for groundwater treatment. *Environmental Science & Technology*, *35*(24), 4922–4926. doi:10.1021/es0108584 PMID:11775172

Gomes, H. I., Dias-Ferreira, C., & Ribeiro, A. B. (2018). Overview of in situ and ex situ remediation technologies for PCB-contaminated soils and sediments and obstacles for full-scale application. *The Science of the Total Environment*, *445*, 237–260. PMID:23334318

Hu, Q. H., Weng, J. Q., & Wang, J. S. (2010). Sources of anthropogenic radionuclides in the environment: A review. *Journal of Environmental Radioactivity*, *101*(6), 426–437. doi:10.1016/j.jenvrad.2008.08.004 PMID:18819734

Ivanova, O. S., Zamyatin, A. V., Zamyatina, A. O., & Vorobyev, A. A. (2018). Nanosensors for detection of heavy metal ions in the environment. *Nanotechnology for Environmental Engineering*, *3*(1), 1–11.

Karn, B., Kuiken, T., & Otto, M. (2009). Nanotechnology and in situ remediation: A review of the benefits and potential risks. *Environmental Health Perspectives*, *117*(12), 1813–1831. doi:10.1289/ehp.0900793 PMID:20049198

Khalid, S., Shahid, M., Niazi, N. K., Murtaza, B., Bibi, I., & Dumat, C. (2017). A comparison of technologies for remediation of heavy metal contaminated soils. *Journal of Geochemical Exploration*, *182*, 247–268. doi:10.1016/j.gexplo.2016.11.021

Kolesnikov, S., Timoshenko, A., Varduni, V., Minnikova, T., Tsepina, N., Mandzhieva, S., & Rajput, V. D. (2024). *Comparative estimation of ecotoxicity of nano- and microparticles of CuO, ZnO, NiO, Co3O4 by the integral biological indicator of soil state*. Nanotechnology for Environmental Engineering., doi:10.1007/s41204-024-00358-9

Kuppusamy, S., Palanisami, T., Megharaj, M., Venkateswarlu, K., & Naidu, R. (2016). In-situ remediation approaches for the management of contaminated sites: A comprehensive overview. In P. de Voogt (Ed.), *Reviews of Environmental Contamination and Toxicology* (Vol. 236, pp. 1–115). Springer.

Ma, X., Geiser-Lee, J., Deng, Y., & Kolmakov, A. (2010). Interactions between engineered nanoparticles (ENPs) and plants: Phytotoxicity, uptake and accumulation. *The Science of the Total Environment*, *408*(16), 3053–3061. doi:10.1016/j.scitotenv.2010.03.031 PMID:20435342

McGee, C. F., Storey, S., Clipson, N., & Doyle, E. (2017). Soil microbial community responses to contamination with silver, aluminium oxide and silicon dioxide nanoparticles. *Ecotoxicology (London, England)*, *26*(3), 449–458. doi:10.1007/s10646-017-1776-5 PMID:28197855

Olaniran, A. O., & Igbinosa, E. O. (2011). Chlorophenols and other related derivatives of environmental concern: Properties, distribution and microbial degradation processes. *Chemosphere*, *83*(10), 1297–1306. doi:10.1016/j.chemosphere.2011.04.009 PMID:21531434

Oughton, D. H., & Salbu, B. (2012). Ecological risk assessment. In A. Pieczynska (Ed.), *Uranium: Characteristics, occurrence and human exposure* (pp. 295–310). Nova Science Publishers.

Pandey, V. C., Abhilash, P. C., & Singh, N. (2019). Sustainable phytoremediation and practices in the management of soil pollution. In *Phytomanagement of Polluted Sites* (pp. 1–19). Elsevier. doi:10.1016/B978-0-12-813912-7.00001-6

Qian, T., Zhao, D., Zhang, F., & Zhang, W. (2019). Nanotechnology for soil remediation: Applications vs. implications. *ACS Sustainable Chemistry & Engineering*, *7*(5), 4395–4404.

Rai, P. K., Lee, S. S., Zhang, M., Tsang, Y. F., & Kim, K. H. (2019). Heavy metals in food crops: Health risks, fate, mechanisms, and management. *Environment International*, *125*, 365–385. doi:10.1016/j.envint.2019.01.067 PMID:30743144

Rizwan, M., Ali, S., Rehman, M. Z. U., Adrees, M., Arshad, M., Qayyum, M. F., Ali, L., Hussain, A., Chatha, S. A. S., & Imran, M. (2019). Alleviation of cadmium accumulation in maize (Zea mays L.) by foliar spray of zinc oxide nanoparticles and biochar to contaminated soil. *Environmental Pollution*, *248*, 358–367. doi:10.1016/j.envpol.2019.02.031 PMID:30818115

Rodriguez-Eugenio, N., McLaughlin, M., & Pennock, D. (2018). *Soil pollution: a hidden reality*. Food and Agriculture Organization of the United Nations.

Sachan, D. S., Reddy, K. J., Sulochna, Saini, Y., Rai, A. K., Singh, O., & Laxman, T. (2023). Assessing grain yield and achieving enhanced quality in maize by next generation fertilizer: A review. *International Journal of Environment and Climate Change*, *13*(8), 626–637. doi:10.9734/ijecc/2023/v13i81991

Saikanth, D. R. K., Singh, B. V., Rai, A. K., Kumar, U. S., Surender, Yadav, B., & Singh, O. (2023). Biochar implementation in rice paddies for addressing greenhouse gas emissions and nutrient loss: A review. *International Journal of Plant and Soil Science*, *35*(18), 610–623. doi:10.9734/ijpss/2023/v35i183326

Salah, E. A. M., Zaidan, T. A., & Al-Rawi, A. S. (2015). Assessment of heavy metals pollution in the sediments of Euphrates River, Iraq. *Journal of Water Resource and Protection*, *7*(13), 1009–1023.

Shipley, H. J., Engates, K. E., & Guettner, A. M. (2011). Study of iron oxide nanoparticles in soil for remediation of arsenic. *Journal of Nanoparticle Research*, *13*(6), 2387–2397. doi:10.1007/s11051-010-9999-x

Simonin, M., Richaume, A., Guyonnet, J. P., Dubost, A., Martins, J. M., & Pommier, T. (2018). Titanium dioxide nanoparticles strongly impact soil microbial function by affecting archaeal nitrifiers. *Scientific Reports*, *8*(1), 1–12. PMID:29311619

Singh, A., Margaryan, G., Harutyunyan, A., Movsesyan, H., Khachatryan, H., Rajput, V., . . . Ghazaryan, K. (2024c). Advancing Agricultural Resilience in Ararat Plain, Armenia: Utilizing Biogenic Nanoparticles and Biochar under Saline Environments to Optimize Food Security and Foster European Trade. Egyptian Journal of Soil Science, 64(2). doi:10.21608/ejss.2024.257000.1703

Singh, A., Rajput, V., Varshney, A., Sharma, R., Ghazaryan, K., Minkina, T., . . . El-Ramady, H. (2024b). Revolutionizing Crop Production: Nanoscale Wonders - Current Applications, Advances, and Future Frontiers. Egyptian Journal of Soil Science, 64(1). doi:10.21608/ejss.2023.246354.1684

Singh, A., Rajput, V. D., Sharma, R., Ghazaryan, K., & Minkina, T. (2023f). Salinity stress and nanoparticles: Insights into antioxidative enzymatic resistance, signaling, and defense mechanisms. *Environmental Research*, *235*, 116585. doi:10.1016/j.envres.2023.116585 PMID:37437867

Singh, O. (2023). Nanotechnology for sustainability and food security in agriculture: A nanopriming long story in short. In A. Singh, V. Rajput, K. Ghazaryan, S. Gupta, & T. Minkina (Eds.), *Nanopriming Approach to Sustainable Agriculture* (pp. 315–339). IGI Global. doi:10.4018/978-1-6684-7232-3.ch014

Singh, O., Shahi, U.P., Dhyani, B.P., Kumar, S., Vivek, S., & Sengar, R.S., & Singh, A. (2023). Effect of zinc oxide nanoparticles application on growth and yield of basmati rice (Oryza sativa) in alkaline soil. *Indian Journal of Agronomy*, *68*(3), 1–6.

Singh, S., Singh, O., & Shahi, U. P. (2023). Nanotechnology: A New Innovation in Biofortification and Food Safety. In A. Singh, V. Rajput, K. Ghazaryan, S. Gupta, & T. Minkina (Eds.), *Nanopriming Approach to Sustainable Agriculture* (pp. 340–351). IGI Global. doi:10.4018/978-1-6684-7232-3.ch015

Singh, Y. K., Singh, B. V., Katiyar, D., Saikanth, D. R. K., Kumar, K., Singh, O., Verma, S., & Kumar, P. (2023). Efficacy of nano fertilizers on yield, attributes and economics of wheat. *International Journal of Environment and Climate Change*, *13*(7), 291–297. doi:10.9734/ijecc/2023/v13i71879

Stogiannidis, E., & Laane, R. (2015). Source characterization of polycyclic aromatic hydrocarbons by using their molecular indices: an overview of possibilities. In D. M. Whitacre (Ed.), *Reviews of environmental contamination and toxicology* (Vol. 234, pp. 49–133). Springer. doi:10.1007/978-3-319-10638-0_2

Tarigholizadeh, S., Sushkova, S., Rajput, V. D., Ranjan, A., Arora, J., Dudnikova, T., Barbashev, A., Mandzhieva, S., Minkina, T., & Wong, M. H. (2023). Transfer and Degradation of PAHs in the Soil–Plant System: A Review. *Journal of Agricultural and Food Chemistry*, *72*(1), 46–64. doi:10.1021/acs.jafc.3c05589 PMID:38108272

Tóth, G., Hermann, T., Da Silva, M. R., & Montanarella, L. (2016). Heavy metals in agricultural soils of the European Union with implications for food safety. *Environment International*, *88*, 299–309. doi:10.1016/j.envint.2015.12.017 PMID:26851498

Tripathi, D. K., Singh, S., Singh, S., Chauhan, D. K., & Dubey, N. K. (2015). Silicon nanoparticles more efficiently alleviate arsenate toxicity than silicon in maize cultivar and hybrid differing in arsenate tolerance. *Frontiers in Environmental Science*, *3*, 46.

Wuana, R. A., & Okieimen, F. E. (2011). Heavy metals in contaminated soils: A review of sources, chemistry, risks and best available strategies for remediation. *International Scholarly Research Notices*.

Chapter 13
Exploring the Potential of Medicinal Plants in Phytomedicine:
Integrating NanoOmics and Nanozymes for Sustainable Agriculture

Barsha Baruah
https://orcid.org/0000-0003-3562-3753
Dibrugarh University, Assam, India

Bobita Kumari
https://orcid.org/0000-0002-1930-8430
Dibrugarh University, Assam, India

Afsana Khatoon
https://orcid.org/0000-0003-3007-4185
Dibrugarh University, Assam, India

Dhitri Borah
https://orcid.org/0000-0001-5423-7183
Biswanath College, Assam, India

Neelesh Kumar
https://orcid.org/0000-0003-2867-4755
Rani Lakshmi Bai Central Agricultural University, India

ABSTRACT

Phytomedicine refers to the use of plants to treat, diagnose and prevent diseases for thousands of years. Medicinal plants, often known as the traditional healers, are potentially rich in phytochemicals with promising pharmacological properties and therefore have been incorporated in traditional medicines. Various medicinal plants and trees are well known for their phytomedicinal benefits and are used in various drugs with lesser side effects. Active components like alkaloids, carotenoids, flavonoids, phenolic acids, tannins, proanthocyanins, sesquiterpenes, glycosides, saponins, volatile oils, minerals, etc. are present in medicinal plants. Some common medicinal plants are used widely such as Azadirachta indica (Neem), Murraya koinigii (Curry leaves), Emblica officinalis (Amla), Myrica esculenta (Kaphal, Bayberry), Terminalia bellerica (Bahera), Terminalia chebula (Haritaki), Mimosa pudica (Humble plant), and Phyllanthus niruri. In this chapter, some phytomedicinal plants and their medicinal properties as well as efficacy are reviewed and discussed.

DOI: 10.4018/979-8-3693-1890-4.ch013

INTRODUCTION

A plant used for medicinal purpose is termed as 'Medicinal Plant'. Plants have always been the most used resources for many purposes, especially when it comes to therapeutic use. Many different kinds of plants may be found in India dispersed over its several states. Most of the diverse plant species are distributed in North-East India. In Assam alone, the Ministry of Environment and Forests (MOEF), the Government of India in the 1980s reported 286 species of plants. Most of the herbal plants of India are already having their uses in the medical field and some are still to be undergone through various scientific approaches. This relationship between plants and human beings has grown and many plants gradually came to be used as medicines. Today, though we have much more advanced science and technology, the common people are still dependent on the traditional medicines to cure various ailments. For primary healthcare, an average of 80% of people still depend on medicinal plants in developing countries (Kumar *et al.*, 2019). Various bioactive compounds (flavonoids, tannins, steroids, polyphenols, glycosides, saponins, anthraquinones, and terpenoids) found in different medicinal plants exhibit good results against various gynecological disorders such as infertility, polycystic ovarian syndrome (PCOS), post-menopausal syndrome, pubertal changes etc. There are innumerable examples of medicinal plants and herbs all over India and study of medicinal plants is considered to be very important (Sharma *et al.*, 2019; Kumar *et al.*, 2021a). Medicinal herbs are considered to be safe for the treatment of ailments as they have fewer side effects. Medicinal plants like aloe vera, tulsi, neem, turmeric, ginger etc. are very commonly used for some common ailments. Plants having such medicinal properties having abundant sources of ingredients are used in developing new drugs (Kumar *et al.*, 2022a). These secondary metabolites produced by plants have a therapeutic role in humans and can be used as drugs In *Ginkgo biloba* leaves, terpenes, lactones, and flavonoids (Briskin *et al.*, 2000) are found. The antagonists of platelets activating factor are represented by ginkgolides which is a bio regulator (Briskin *et al.*, 2000). In hepatic and spleen enlargements etc., this herb is often used (Maruthappan *et al.*, 2010). The phytochemical investigation of *Curcuma xanthorrhiza* and *Curcuma aromatic* shows that it contains flavonoids, tannins, terpenoids and saponins as the major constituents. *Curcuma aromatic* is used against snakebite. *C. xanthorrhiza* is traditionally used to treat several ailments. Therefore, phytomedicine, which is the culmination of several different healthcare sectors, has transformed and strengthened the cornerstone of the healthcare system that is currently in place, and it holds a significant stake in the business (Sharma *et al.*, 2022). For this reason, it appears that the future of medicinal medications generated from plants has a significant amount of room for the discovery of a wide variety of new and innovative therapeutic procedures and commodities. Plants like green tea, turmeric, ginger, and pepper are considered important sources of nutrition and are often acknowledged for their therapeutic role. There are many opportunities for basic research on medicinal plants and also of their phytomedicinal study. So, the study of these medicinal plants is considered to be of utmost importance (Semwal *et al.*, 2023).

THERAPEUTIC AND PHYTOMEDICINAL ROLE OF PLANTS

The great majority of the plants found in northeast India are common, but several of them have garnered a lot of interest as medicinal herbs. The aim of this article is to provide an overview of 5 medicinal herbs that are commonly utilized, with a particular emphasis on their phytoconstituents and the mechanisms of action of their plant secondary metabolites.

Myrica esculenta (Kaiphal)

Myrica esculenta widely known as kaiphal or kataphala in India and Nagatenga in Assam is broadly used in Ayurveda. It has derived its common name "Kaiphal" from sanskrit language whereas in English it is commonly named "Box Myrtle". Myrica trees are companions of oak (*Quercus leucotrichophora*) and pine (*Pinus* sp.) trees. They thrive in soils that are deficient in nitrogen and grow well in these conditions. Some Indian names of *M. esculenta* are kaphal, satsarila, kaphal, kaiphal, kirishivani, maruta, sohphi, kaphal, keiphang, kayaphala, kobuli, katphala, kaiphal (Mann *et al.*, 2015). It is found in India, China, Sri Lanka, Pakistan and some Asian countries.

Phytochemistry and Pharmacological Action

A variety of active phytoconstituents, including tannins, phenolic acids, flavonoids, terpenes, glycosides, steroids, volatile oils, and amino acids, were discovered in the leaves, bark, fruits, and fine branches of *M. esculenta.* These phytoconstituents have a wide variety of pharmacological effects. The fruit contains higher values of fat, protein, fiber, and minerals. Sood et al. 2017 reported bioactive compounds in the plant that belong to the class of alkaloids, diarylheptanoids, ionones, saponins, triterpenoids and volatile compounds.

Tannins and Phenolic Acid

Its bark contains gallic acid along with castalagin (hydrolyzable tannin) (Mann *et al.*, 2015).

Flavonoids

The leaves, fruits, and stem bark contain myricetin. Quercetin is a major component found only in *M. esculenta* leaves. Among glycosidic flavonoids, flavone 3′,4′-dihydroxy-6-methoxy-7-*O*-α-L-rhamnopyranoside is commonly found too (Dawang *et al.,* 1991).

Triterpenopids

Several triterpenoids such as lupeol, oleanolic acid, trihydroxytaraxaranoic acid, dihydroxytaraxerane, dihydroxytaraxaranoic acid, tetrahydroxytaraxenoic acid, 3-epi-ursonic acid and arjunolic acid were reported in the leaves and bark of *M. esculenta* (Nguyen *et al.,* 2010).

Volatile Compounds

The following volatile substances were found in leaves: α-pinene, α-selinene, β-caryophyllene, β-selinen, α-caryophyllene, and α-cadinol.

Steroids

Steroids such as β-rosasterol and β-sitosterol-β-D-glucopyranoside were identified in leaves.

Pharmacological Action

Crude extracts of plants and the active compounds have been investigated for various biological activities such as analgesic, anti-asthmatic, anticancer, antioxidant, anti-inflammatory, antidiabetic and wound healing activity by using several *in-vitro* and *in-vivo* animal models, that scientifically validated the traditional application of this plant (Sood *et al.,* 2017). Some of them have been discussed below.

Anti-Asthmatic Activity

The ethanol extract of bark demonstrated anti-asthmatic action, including anti-anaphylactic activity in guinea pigs produced by egg albumin and spasmolytic activity. These findings provide scientific support for the application of *M. esculenta* bark in traditional medicine to treat allergic responses including asthma and bronchitis.

Anti-Cancer Activity

Potent anticancer activities are exhibited by methanol and acetone extracts of the fruit. Moderate anticancer activity is shown by methanol extract leading to inhibition of cancer cell lines in methyl thiazolyl tetrazolium (MTT) assay. Mann *et al.* (2015) reported the inhibitory effect of the extract on the cancer cells increased with an increase in dose due to the presence of bioactive compounds such as ferulic acid and gallic acid.

Anti-Microbial Activity

The volatile oil present in the bark of *M. esculenta* has strong antibacterial action against both gram-positive and gram-negative bacteria. The methanol and chloroform extracts of the bark exhibited higher antibacterial activity than fruit extracts. Additionally, it was discovered that fruit aqueous extract, ethanol, and methanol extract all exhibited strong antifungal properties.

Anti-Oxidant Activity

Many antioxidants are found in the fruits of this plant. Some studies showed that *M. esculenta* fruits contained higher flavonoid contents than other species of the same genus (Bhatia *et al.,* 2021).

Hypertension, Herbitar, and as Mast-Cell Stabilizer

M. esculenta play a key role in the prophylaxis of hypertension. The leaves of this plant contain some secondary metabolites which manage hypertension (Samundewari *et al.,* 2011). It also plays a good role as a herbitar a poly herbal formulation against CCl4-induced hepatotoxicity in rats (Rastogi *et al.,* 1995). Aqueous extract and ethyl acetate at a concentration of 100-200 mg/kg demonstrated slightly good protection of mast cell degranulation than the standard drug prednisolone (65%) (Bhatia *et al.,* 2011).

Emblica officinalis (Amla)

Amla is an important herb in especially in ayurveda. The fruit is the major part and is utilized for the treatment of common cold and act as a good liver tonic. Throughout the Indian subcontinent, Ayurveda, the ancient medical system, is one of the oldest branches of medicine. As a result, the pharmacological and therapeutic properties of several plants have been studied. This plant has been extensively studied in the family Euphorbiaceae (Hasan *et al.*, 2016). The fruit of *E. officinalis* is widely used and believed to increase defense against diseases.

Phytochemistry of Amla

The plantis rich in fiber, carbohydrates, and iron and is reported as the richest source of vitamin C. Several important herbal and useful drugs have been formulated by the phytoconstituents of *E. officinalis*. The active ingredient found in *E. officinalis* is "Phyllembin" (Jain *et al.*, 2016). The fruit has a lot of quercetin, phyllemblic chemicals, citric acid, pectin, 1-O-galloyl-β-D-glucose, 3-O-galloyl-D-glucose, plunigluconin, pedunculagin, gallic acid, ellagic acid, chebulinic acid, chebulagic acid, and emblicanin A and B. It contains many flavonoids such as quercetin, kaempferol 3-O-α-l (6" ethyl) rhamnopyranoside etc. (Jangpangi *et al.*, 2023). Fruits of *E. officinalis* was also analyzed for their alkaloidal content. Chromatography and IR spectral studies revealed the presence of phyllantine and phyllantidine. Pharmacological study on amla shows that it can relieve pain, fight coughs and colds, lower cholesterol, help the body adapt, and protect the heart, intestines, kidneys, brain, and cancer.

Pharmacological Action

It has been observed that amla contains a variety of bioactive chemicals with various pharmacological actions, including antioxidant, antibacterial, and terpenoids, as well as tannins, flavonoids, saponins, and ascorbic acids. Some of the important pharmacological activities of *E. officinalis* are described below.

Anti-cancer and Anti-Oxidant Activity

Vitamin C is found in good amounts in amla. Hydrolysable tannins with low molecular weight contribute to their antioxidant properties. Tannins such as emblicanin-A and emblicanin-B show protection in the haemolysis of rat peripheral blood erythrocytes against oxygen radicals (Kulkarni *et al.,* 2018). Organic acid gallates and hydrolyzable tannins show cytotoxicity to tumor cells (Lanka *et al.,* 2018).

Anti-Inflammatory and Anti-Microbial Activity

E. officinalis water extract demonstrated inhibitory effect on the synthesis and release of inflammatory mediators in rats. In vitro studies using bronchial epithelial cells shows inhibition of pro-inflammatory, cytokine expression. The water fraction of methanol extract of the plant leaves demonstrated anti-inflammatory activity (Bhandari *et al.,* 2012). Chronic inflammation has been scientifically established to generate free radicals, leading to oxidative and nitrosative stress (Hasan *et al.*, 2016; Singh *et al.*, 2023). This stress, either directly or indirectly, plays a role in the development of cancerous cell transformation by causing genomic instability, changes in epigenetic processes, abnormal gene expression, increased growth of mutated cells, resistance to cell death, tumor blood vessel formation, and the spread of cancer to other parts of the body (Lanka *et al.*, 2018).

The anti-microbial activities of amla extract have been investigated through disc diffusion and agar dilution methods in opposition to *Staphylococcus aureus* (Jain *et al.,* 2016). Bacteria, viruses and fungi showed positive results towards some extracts like alcoholic and aqueous. Its activity is stronger against gram-positive bacteria than against fungi (Lanka *et al.*, 2018). They found that *E. Officinalis* reduces the bacterial colonization of lungs when the rice is fed for a long time (Bhandari *et al.,* 2012).

Anti-Diabetic and Hypoglycemic Activity

Tinospora cordifolia, Trigonella foenum and *E. officinalis* extracts were tested for hypoglycemic effects and oral glucose tolerance test (OGTT) in normal and alloxan-induced diabetic rats and significant, marginal and very less decrease in blood glucose level was observed when different herbal combinations were used (Hasan *et al.,* 2016).

Hepatoprotective Activity

E. officinalis protects against hepatotoxicity-induced liver injury. When amla is given before some other chemicals it shows major decrease in some enzymes like lactate dehydrogenase (LDH), lipid peroxidation (LPO) and DNA synthesis etc. (Bhandari *et al.,* 2012). Hepatoprotective properties in experiments in rats have been shown by alcoholic and aqueous extracts of the fruit. Hepatic results include anti-tuberculosis drugs, arsenic, ethanol, thiacetamide etc. (Lanka *et al.,* 2018; Mavi *et al.*, 2023). Apart from these pharmacological activities, *E. officinalis* show many other activities like analgesic, antitussive, antivenom, immune functions, skin protection, anti-metastatic activity etc.

Azadirachta indica (Neem)

Azadirachta indica A. Juss (Meliaceae) commonly known as neem is native of India and is of great medicinal value and distributed widespread in the world. It is known as 'Arishtha' in Sanskrit meaning 'Reliever of sickness'. It is also known as 'Sarbaroganibarini'. The bioactive isolates of the plant include nimbin, nimbolide, azadirachtin, melicicin, gedunin, valarcin, calanin etc. (Maji *et al.*, 2021).

It is mostly found in India and Burma, Southeast Asia and West Africa etc. In India, it occurs naturally in Sivalik hills, dry forests of Karnataka, Andhra Pradesh and Tamil Nadu to an altitude of approximately 700m. It is cultivated in Pakistan, Peninsular Malaysia, Singapore, Philippines, and Australia.

Phytochemistry

The active phytochemicals found in this plant are azadirachtin, melicicin, gedunin, nimbidin, nimbolides, salanin, nimbin, valassin, meliacin etc. The most important classes of phytochemicals include isoprenoids and non-isoprenoids. The phytochemicals found in each part of the neem have antimicrobial, antiviral, anti-inflammatory plasmoidal, antiseptic, antipyretic, anti-diabetic, and anti-ulcer properties (Maji *et al.*, 2021).

Philippines, and Australia. They grow approximately about 12 meters to 18 meters in height &1.8-2.4 meters in perimeter. The tree is usually a flowering plant that produces flowers that stretches 4-7mm in length and broadness is about 6-10mm during 3-5 years of age .The flowers are generally white in color & has a pleasant jasmine-like odour .The leaves are up to 30cm in length & has a dark green color with seeded drupes& three-lobed stigma. Fruits are green and aromatic with a garlic-like odour. Neem is also commonly known as neem, aristha, margosa, Indian lilac etc (Bempah *et al.,* 2011).

Phytochemistry

The active phytochemicals found in this plant are melicicin, azadirachtin, nimbidin, gedunin nimbolides, nimbin salanin, meliacin, valassin, etc. The most important classes of phytochemicals include isoprenoids and non-isoprenoids. The phytochemicals found in each part of the neem have antimicrobial, antiviral, plasmoidal, antiseptic, anti-inflammatory, antipyretic, diabetis preventive and ulcer preventive properties. (Maji *et al.*, 2021).

Phytochemicals Present in Neem Leaves

The leaves contain more therapeutic properties than any other part of the tree (Maji *et al.*, 2021). Neem leaf is a high source of different types of amino acids and fatty acids. Sulphur-containing compounds such as cyclic trisulphide and tetrasulphide, quercetin, methanolic extract etc. are present in neem leaves (Maithani *et al.*, 2011). They are very potent phytochemicals present in various ailments. Phytochemicals present in the bark of neem include condensed tannin containing (+) catechin, (+) gallocatechin, gallic acid, (-) epicatechin etc. (Maithani *et al.*, 2011). The bark of the neem tree also yields a number of phenolic diterpenoids, c-secomeliacins and limonoides (Maji *et al.*, 2021). Phytochemicals present in fruit and seed are phytosterols in fruits of neem exhibit anti-ulcer activity in stress induced gastric lesion. The seeds of this plant contain many phytochemicals viz. nimbidin, nimbin, nimbolide and nimbidic acid etc. (Maithani *et al.*, 2011) (Figure 14.1). Neem oil also contains steroids and triterpenoids like azaradirachtin.

Phytomedicinal Uses

The therapeutic properties of the leaf, fruit, and bark have been specifically outlined. Important medicinal activities of various parts of neem plant are discussed below:

Anti-Bacterial and Antiviral Activity

The methanol extract has exhibited activity against some bacteria like *Bacillus subtilis* and *Salmonella typhietc*. It was found that an extract of petroleum ether of A. indica leaves proved to be operative against *Candida albicans* (Grover *et al.*, 2011). Neem oil extracts exhibits diverse effects against gram-positive and gram-negative agents. It was also found that neem seed oil had antibacterial properties against many kinds of harmful bacteria.

Antiviral activity can be seen in aqueous extracts of leaves. The extracts of Neem bark stops the entry of HSV-1 into the cells. The extract of Neem leaf is antagonistic towards B-4 coxactive virus (Badam *et al.*, 1999).

Figure 1. Different bioactive compounds in neem (Source: Getty Images, www.Instockphoto.com)

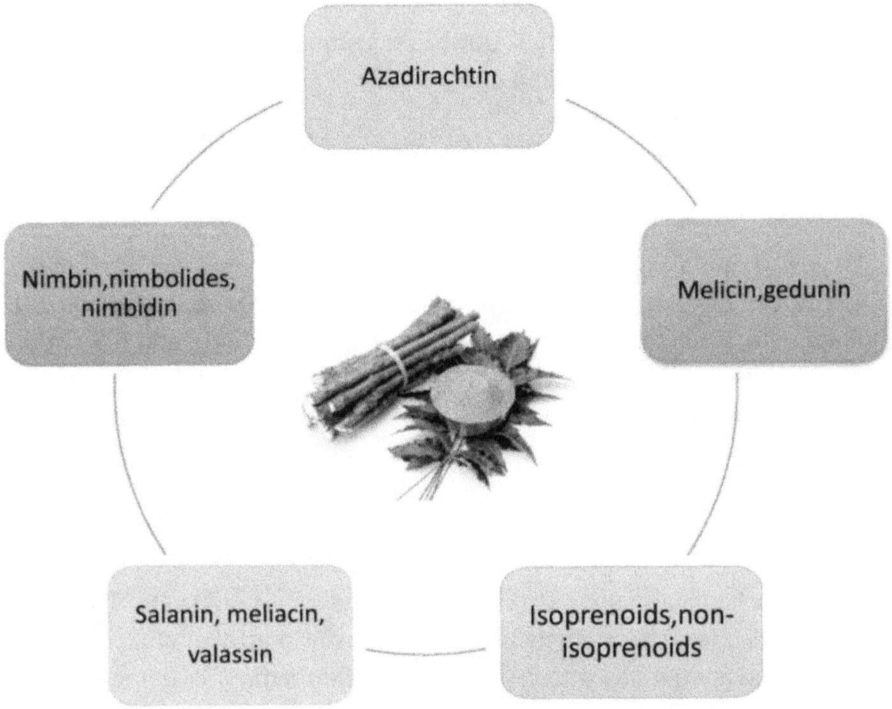

Antioxidant and Anti-Inflammatory Activity

The anti-oxidant power is high in young flowers and leaves. Monoaldehyde (MDA), was reduced by 46.0% and 50.6% for polyherbal vaginal tablet that has proven to be effective in immobilizing sperm.

The neem extract obtained from different parts of Neem Tree plays an efficacious role in reducing inflammation & redness. The dermis of the fruit of *A. indica* and the phytochemicals azadirachtine plays a fantastic role as anti-inflammatory agents. The bark extract immensely proved to be very effective against Stomatitis which is often followed by inflammation in children (Srivastava *et al.*, 2018).

Antidiabetic Activity

The plantroot and extract from the bark helps in maintaining blood-sugar levels as it typically has a promising hypoglycemic action based on pharmacological point of view. Apart from these phytomedical activities of *A. indica* it has various other activities like anti-tumour, anti-larval or larvicidal, anti-pyretic properties etc. (Srivastava *et al.*, 2018).

Nyctanthus arbor-tristis (Night Jasmine)

Nyctanthus arbor-tristis Linn. (Oleaceae), an ornate, sacred tree known for its aromatic natural white flowers across India. It is often known by the name of 'Harsinghar' in Hindi language & 'Night Jasmine' in English as its flowers releases a quite powerful & satisfying aroma during night hours (Srivastava *et al.*, 2018). The flowers usually starts to fall after midnight & hence the name 'arbor-tristis' is given which means 'the sad tree' as by the daybreak the plant loses its beauty & appear toned down during daytime. Therefore, it is often known as Night flowering jasmine and in Hindi it is called Parijat. Its life span is of 5-20 years (Rani *et al.*, 2012; Sahu *et al.*, 2023). It is known with different names like coral jasmine, salikaa, parijata, sepali, sewali etc.

It is found in Southern Asia, Northern-Pakistan, Nepal etc. In India it is usually found in tracts of Jammu and Kashmir, East of Assam, Bengal& Nepal which extends upto the central part of Godavari in South i.e. grows in outer region of Himalayas. The Neem tree grows in outer Himalayas, Nepal to East of Assam, Bengal, Tripura extended through the central region upto Godavari in the South (Kakati *et al.*, 2013). The tree is a tiny native tree and the bark is greyish to greenish in color. They may reach a maximum altitude upto 10 meters. The opposite leaves are 12cm long and 6.5 cm broad with a full edge. The fragrant blooms are clustered in groups in between 2-7 in number. The powder white petals are clinging to dew droplets. The fruit contains a single seed which is heart shaped and brown in color (Jaspal *et al.*, 2021).

Phytochemistry

Contains different phytochemicals such as alkaloids, terpenes, flavonoids etc. The presence of tertiary alkaloids has been revealed in various phytochemical studies primarily of quaternary alkaloids belonging to protoberberines and aporphines with 7-(α-anilino-p-nitrobenzyl)-8- quinolinol and (Milind *et al.*, 2012) (Figure 14.2). Different phytoconstituents which usually exists in various parts of the shrub are given below (Rangari *et al.*, 2008) (Table 14.1).

Pharmacological Action

Nyctanthus arbor-tristis has been found to include saponins, alkaloids, flavoinoids, steroids, tannins, glycosides and phenolic compounds. The encouraging results of *Nyctanthus arbor-tristis* show that the plant works as scavenger of hydrogen peroxide and free radicals effectively. The polyphenolic content as well as its phytochemical compounds attributes to its antioxidant activity (Rani *et al.*, 2012).

Table 1. Different phytoconstituents in N. arbor-tristis.

Plant Part	Phytoconstituents
Bark	Glycosides, alkaloids,
Flower Oil	Anisaldehyde, Phenyl acetaldehyde etc.
Flower	Tanninm, apigenin, carotenoid, anthocyanin, nyctanthin, d-Mannitol, glucose, essential oil, kaemferol, glycosides etc.
Leaves	Carotene, flavanol glycosides-astragaline, d-Mannitol, friedeline, fructose, glucose, Iridoid Gglycosides.
Seeds	3-4 Secotriterpene Acid, Arbortristoside A & B, Lignoceric, Myristic Acids etc.
Stem	β-sitosterol

Figure 2. Some secondary metabolites found in Nyctanthus arbor-tristis (Parijat)

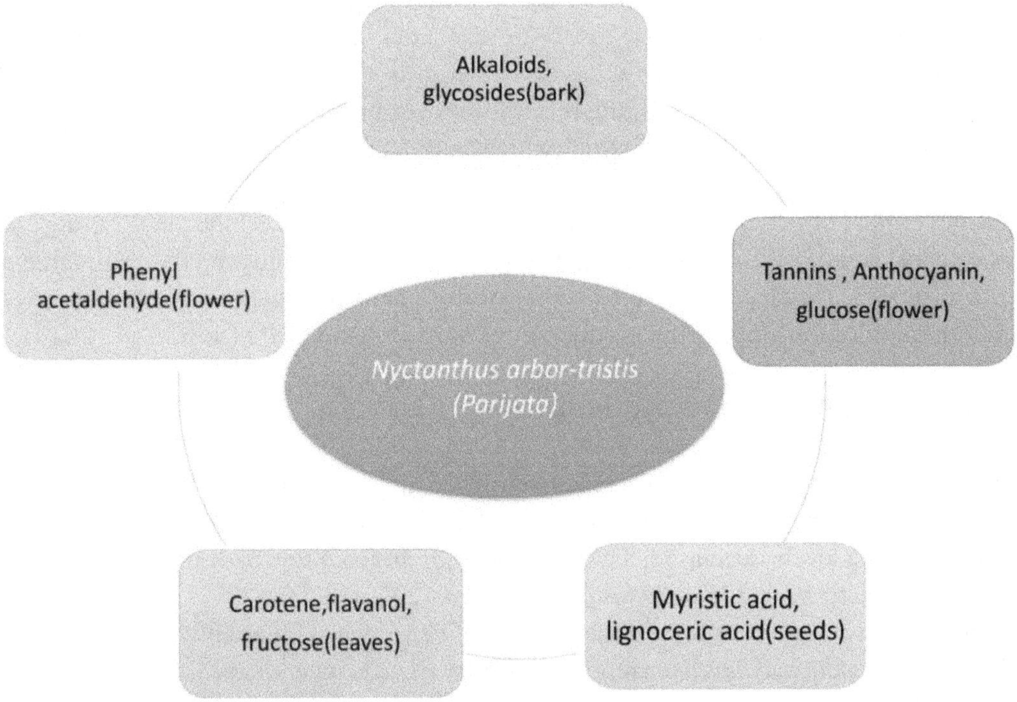

Anti-Fungal and Anti-Microbial Activity

Various components found in *N. arbor-tristis* plant has been analyzed for their antifungal properties which is antagonistic towards the three most common clinically salient pathogenic fungus, namely *Aspergillus flavus, Aspergillus niger* & *Penicillium*, and. The extracts which were obtained from mature as well fresh leaves along with seeds, stems, flowers, bark as well as distilled water, chloroform, and methanol, was assessed using the well diffusion technique. The measurement was based on the "zone of inhibition," which indicates the extent to which fungal growth is inhibited. The findings indicated that just the distilled water extract derived from the had antifungal properties specifically against *A. niger*. Conversely, the chloroform extract obtained from the leaves shown effectiveness alone against *A. flavus*. Methanolic extract of leaves, stem and bark of *N. arbor-tristis*has the highest antifungal activity against both Aspergillus and Penicillium (Rawat *et al.,* 2011).

Bactericidal activities of methanol extracts of leaves have been studied in *Staphylococcus aureus*. The extract was seen active in bacteria *B. subtilis, Streptococcus sp., E. coli,* and *Shigella sp.* were stopped by the extracts of chloroform and ethyl acetate (Rani *et al.*, 2012).

Anti-Histaminic and Anti-Asthmatic Activity

Clonidine-induced catalepsy (catalepsy is a condition characterized by lack of response of limbs to the external stimuli and muscle rigidity and can be induced by neuroleptic drugs) can be inhibited by the petroleum ether extract of *N. arbor-tristis* bark (Rawat *et al.*, 2021; Le Doux *et al.*, 2015). This bark was found to have antihistaminic activity, suggesting its potential usefulness in treating asthma. Asthma is a chronic inflammatory disease characterized by bronchial inhibition, airway inflammation and high reactivity to various stimuli such as allergens, histamine and others. The primary element of the initial asthmatic response when exposed to an allergen is the narrowing of the bronchial tubes. It has been determined that in asthmatic conditions, the reaction of obstructed airways and heightened sensitivity is linked to a lack of nitric oxide production, which is crucial for maintaining proper airflow in the respiratory system.

The ethanol extract of *N. arbor-tristis* leaves was tested for histamine-induced dose-dependent contraction of tracheal smooth muscles which finalized that the extract could stop the contractile responses produced by histamine which confirms the activity of the ethanolic extract of *N. arbor-tristis* leaves on airway smooth muscles thus increasing production of nitrogen oxide, NO (Sopi *et al.*, 2013).

Anti-Diabetic and Anticancer Activity

Ethanololic extract of stem bark of *N. arbor-tristis* shows anti-diabetic properties. In diabetic rats, the triglycerides and serum cholesterol levels were very high but when they were treated with bark of stem of *N. arbor-tristis*, the levels decreased. Thus, ethanol extract from this bark of *N. arbor-tristis* proved to be important in the treatment of diabetes (Suresh *et al.*, 2010).

Significant cytotoxic activities from some extracts of flowers were seen in various studies. Recently, methanolic extract of stem bark has been reported to show significant anticancer activity when compared to 5-fluorouracil in Swiss albino. Apart from these *N. arbor-tristis* shows anticholinesterase activities, anti-allergic activities and antiviral activities because of the presence of a huge range of secondary metabolites (Srivastava *et al.,* 2018).

Terminalia bellirica (Bhibhitaki)

The plant, *Terminalia bellirica (Gaertn.) Roxb.* can be utilized either singly or when combined with other plant-based medicines. It is known as 'Bhibhitaki' in Sanskrit. The fruit of *Terminalia bellirica* is an integral part in 'Triphala' which is often used as a supplement for promoting longevity in alternative medicine systems of ayurveda and siddha (Jagetia *et al.,* 2009). Antioxidant, antidiabetic and many other pharmacological activities are exhibited by the fruit. Various Indian names of *T. bellirica* homorabhomra, bhaira, bahedam, baheda, bahera, thanakkai etc.

The plant is indigenous to Sri Lanka, India, Bangladesh, Bhutan, Thailand, China, Indonesia, Pakistan, Malaysia, Nepal, Cambodia, and Vietnam. In India, it is commonly found in Madhya Pradesh, Uttar Pradesh, Punjab and Maharashtra (Swati *et al.,* 2017). It is found mostly in India with a height of 30-40 m. Flowers are small, greenish yellow with offensive smell. Fruits are grey and velvety the drupes contain single seed. Flowering starts in March-May and fruits ripen in December and February. The bark is brown-grey in colour (Narendra *et al.*, 2018).

Phytochemistry

Phytoconstituents present in various parts of *T. bellerica* are (Narendra *et al.*, 2018). The seed contains cardenolide, cannogenol 3-O-D-galactopyranosyl14-O-L-rhamnopyranoside. The seeds are found to contain phospholipids, tetradecanoic acid, palmitic acid, oleic acid and octadecanoic acid. Fruits contain

Figure 3. Some secondary metabolites available in Terminalia bellirica

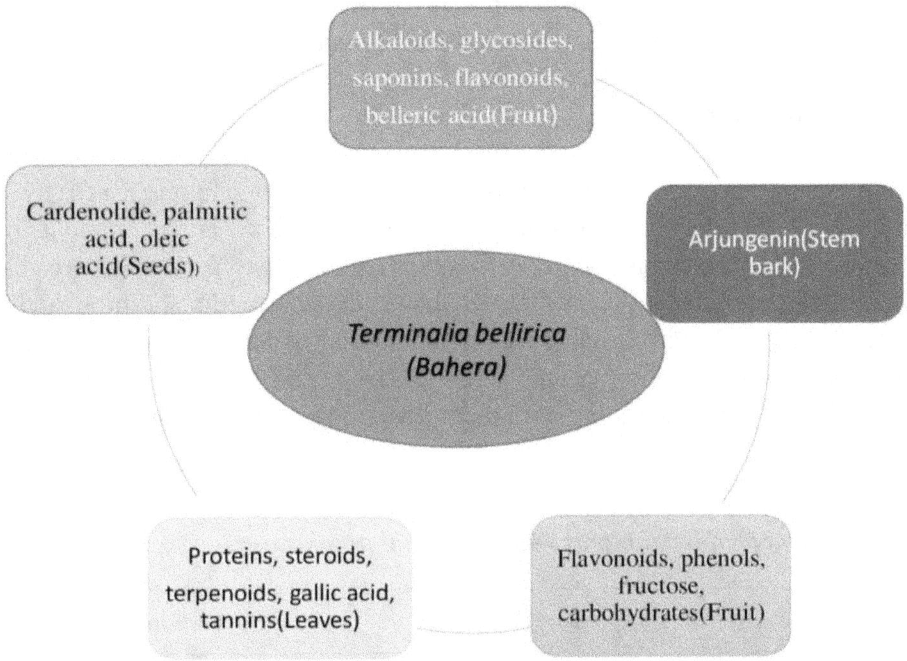

alkaloids, saponins, flavonoids, polysaccharides, arjungenin, bellericosides, ethyl gallates, fructose, phenol, tannins and flavonoids, carbohydrates and proteins etc. The stem bark contains arjungenin and its different components like belleric acid, gallic acid etc. Proteins, steroids, terpenoids, gallic acid, ellagic acid, methyl gallates, quercetine, kaempferol, saponins, tannins, amino acids, proteins, alkaloids, carbohydrates and flavonoids are also prevalent in leaves (Figure 14.3).

Pharmacological Action

Various pharmacological potential of *T. bellirica* has been observed. Some of them are mentioned below.

Analgesic and Antioxidant Activity

The crude extracts of *T. bellerica* shows analgesic, antinociceptive and antisecretory properties hence proving its ethnomedicinal use in diarrhea and pain in unani and ayurveda (Khan *et al.*, 2010). The crude aqueous extract contains enzymatic and non-enzymatic antioxidants (Anindita *et al.*, 2015; Ramesh *et al.*, 2018). Ethanolic fractions has significant potential as a source of antioxidants and are employed in combating illnesses such as cancer, aging, and other conditions associated with oxidative stress.

Anti-Microbial and Anticancer Activity

It was found that some disease-causing organisms were highly sensitive to the extract of methanol and concluded that the dry fruit of the plant possesses potential antimicrobial activity (Elizabeth *et al.*, 2005).

T. bellerica extracts showed inhibitory activity against growth in two cancer cell lines tested. Synergistic effects are shown by doxorubicin or cisplatin in A549 and HepG2 cells. The dose lowering level was not same and specific to each combination and cell line (Pinnai *et al.*, 2008).

Anti-Salmonella Activity

Thealcoholic and aqueous extracts of *T. bellerica* fruit have great anti-salmonella activity (Madani *et al.*, 2008) in their study of the effect of *T. bellerica* against *Salmonella typhi* and *Salmonella typhimurium* found that there was no cell toxicity in in-vitro cells toxicity study. Apart from the above pharmacological activities *T. bellerica shows* analgesic, anti-hypertensive, anti-spasmodic and bronchodilatory properties, antibiofilm activities, β-lactamase inhibitor activities, acute and sub-acute toxicities, antithrombotic and thrombolytic activities and antimutagenic activities (Deb *et al.*, 2015). Various other known plants with immense medicinal properties along with their phytoconstituents are also available (Table 14.2).

Table 2. Some important medicinal plants with medicinal properties

Common Name	Scientific Name	Secondary Metabolites	Medicinal Uses	References
Arjun	*Terminalia arjuna*	Tannins, carbohydrates, alkaloids, terpenoids, steroids, flavonoids and phenols	Cardiotonic in heart failure, cardiomyopathy, Ischemic blood diseases, anaemia	Amalraj *et al.*, 2017)
Tulsi	*Ocimum tenuiflorum*	Tepenes, lactone, xanthenes, eugenol, rosmeric acid apigenic and carnosic acid	Prevent liver, kidney and brain injury, used as hand sanitizer, mouthwash	(Cohen *et al.*, 2014)
Aloe vera	*Aloe barbadensis miller*	Vitmins, lignin, saponins, enzymes, minerals, sugar, salicylic acids and amino acids	Used for acne, burns, weightloss, diabetes, hepatitis and inflammatory	(Surjuse *et al.*, 2008)
Giloy	*Tinospora cordifolia*	Furocoumarins, oxalic acid, saponins, insulin, triterpenoids saponins	Anti-HIV 1 activity, fever, jaundice, asthma snake bite	Kumar *et al.*, 2019
Clove	*Syzygiumaromaticum*	Eugenol, eugenol acetate and galicacid, flavonoids.	Anti-fungal, anticandidal, anti-bacterial activity.	(Rojas *et al.*, 2014)
Ginger	*Zinggiber officinale*	Sesquiterpenoids, monoterpenoids, phenyl, propanoid and diarylheptanoids	Nasea, gastritis, dyspepsia, cold cough	(Kandiannan *et al.*, 1996)
Bryophyllum	*Bryophyllumpinnatum*	Polyphenols, tannins, glycosaponins, flavonoids, steroidal glycosides	Anti-oxidant, antipyretic, anti-inflammatory, antialergic, antiseptic, antidepression	(Latif *et al.*, 2019)
Costus	*Costus speciosus*	Alkaloids, glycosides, steriods, phenolic, flavonoids, tannins	Pharyngitis, tonsillitis in children, pleurisy, snakevenom antidote	(El-far *et al.*, 2018)
Green Chiretta	*Andrographis panicuata*	Diterpenoids like andrographolide, neoandrographolide, dehydroandrographolide	Treatment in cold, fever, laryngitis, malaria, dysentery, diabetes, hypertension	(Mishra *et al.*, 2007)
Ashwagandha	*Withanisomnifera*	Alkaloids, steroidal lactones, saponins	Effective in rheumatoid, osteoarthritis, Alzeimer's diseases, cancer, rejuvenator	(Singh *et al.*, 2011)
Hing	*Ferula asafoetida*	Carbohydrates, resins, gums, phenolic compounds, diterpenes, oleic acid, glucuronic acid	Whooping cough, asthma, bronchitis, ulcer, bronchitis, weak digestion	(Khanal *et al.*, 2021)
Bhringraj	*Eclipta prostrata*	Coumestans, triterpenes, flavonoids, steroids	Anti-hepatotoxic, anti-diabetic, anti-inflammatory, hair rejuvenator	Sarkar *et al.*, 2021

CONCLUSION

Since natural products as well as their derivatives play a great role in the prevention and cure of different ailments with fewer side effects properties their popularity is increasing worldwide. Since ancient times India and its people have been traditionally using various plants as medicines. It has been found from clinical studies that medicinal plants play a great role in prevention of various diseases. Through this review we can certainly conclude that medicinal plants exhibit important secondary metabolites, contributing to the restorative and remedial importance of these plants. These plants are used in the treatment of different diseases. This chapter encourages further research to identify more phytochemicals responsible for biological activities. In the future, further studies could be undertaken to study phytochemical and pharmacological exploration of several other medicinal plants. Although many plants are very effective and safe against many human ailments, it is still to be verified by some clinical trials. We can understand the potential of medicinal plants and their extracts by using the latest scientific and technological tools.

REFERENCES

Alzohairy, M. A. (2016). Therapeutics role of *Azadirachta indica* (Neem) and their active constituents in diseases prevention and treatment. *Evidence-Based Complementary and Alternative Medicine*, *2016*, 7382506. doi:10.1155/2016/7382506 PMID:27034694

Amalraj, A., & Gopi, S. (2017). Medicinal properties of *Terminalia arjuna* (Roxb.) Wight & Arn.: A review. *Journal of Traditional and Complementary Medicine*, *7*(1), 65–78. doi:10.1016/j.jtcme.2016.02.003 PMID:28053890

Badam, L., Joshi, S. P., & Bedekar, S. S. (1999). 'In vitro' antiviral activity of neem (*Azadirachta indica*. A. Juss) leaf extract against group B coxsackieviruses. *The Journal of Communicable Diseases*, *31*(2), 79–90. PMID:10810594

Bamola, A., Semwal, D. K., Semwal, S., & Rawat, U. (2009). Flavonoid glycosides from *Myrica esculenta* leaves. *Journal of the Indian Chemical Society*, *86*, 535–536.

Bempah, K. C., Buah-Kwofie, A., & Asomaning, J. (2011). Morphological studies of neem (*Azadirachta indica* A. Juss.) seed and physicochemical properties of its oil extracts collected in Accra metropolis of Ghana. *Elixir Applied Botany*, *39*, 4951–4953.

Bhandari, P. R., & Kamdod, M. A. (2012). *Emblica officinalis* (Amla): A review of potential therapeutic applications. *International Journal of Green Pharmacy*, *6*(4), 257–269. doi:10.4103/0973-8258.108204

Bhatia, D., & Goyal, P. (2021). A review of pharmacological and pharmacognostic profile of *Myrica nagi*. *Research Journal of Pharmacy and Technology*, *14*(2), 1109–1114. doi:10.5958/0974-360X.2021.00200.6

Biswas, K., Chattopadhyay, I., Banerjee, R. K., & Bandyopadhyay, U. (2002). Biological activities and medicinal properties of neem (*Azadirachta indica*). *Current Science*, *82*(10), 1336–1345.

Briskin, D. P. (2000). Medicinal plants and phytomedicines. Linking plant biochemistry and physiology to human health. *Plant Physiology*, *124*(2), 507–514. doi:10.1104/pp.124.2.507 PMID:11027701

Choudhary, G. P. (2012). Anti-ulcer activity of the ethanolic extract of *Terminalia belerica* Roxb. *Internation Journal of Pharmaceutical Chemical Sciences*, *1*(4), 1293.

Cohen, M. M. (2014). Tulsi-*Ocimum sanctum*: A herb for all reasons. *Journal of Ayurveda and Integrative Medicine*, *5*(4), 251–259. doi:10.4103/0975-9476.146554 PMID:25624701

Cortés-Rojas, D. F., de Souza, C. R. F., & Oliveira, W. P. (2014). Clove (*Syzygium aromaticum*): A precious spice. *Asian Pacific Journal of Tropical Biomedicine*, *4*(2), 90–96. doi:10.1016/S2221-1691(14)60215-X PMID:25182278

Dawang, S., Zuchun, Z., Foo, L. Y., & Wong, H. (1991). Flavonols from *Myrica esculenta* a bark. *Linchan Huaxue Yu Gongye*, *4*, 251–257.

Deb, A., Barua, S., & Das, B. (2016). Pharmacological activities of Baheda (*Terminalia bellerica*): A review. *Journal of Pharmacognosy and Phytochemistry*, *5*(1), 194–197.

El-Far, A., Shaheen, H., Alsenosy, A., El-Sayed, Y., Al Jaouni, S., & Mousa, S. (2018). *Costus speciosus*: Traditional uses, phytochemistry, and therapeutic potentials. *Pharmacognosy Reviews*, *12*(23), 120–127. doi:10.4103/phrev.phrev_29_17

Elizabeth, K. M. (2005). Antimicrobial activity of *Terminalia bellerica*. *Indian Journal of Clinical Biochemistry*, *20*(2), 150–153. doi:10.1007/BF02867416 PMID:23105549

Girish, K., & Shankara, B. S. (2008). Neem-a green treasure. *Electronic Journal of Biology*, *4*(3), 102–111.

Grover, A., Bhandari, B. S., & Rai, N. (2011). Antimicrobial activity of medicinal plants-*Azadirachta indica* A. Juss, *Allium cepa* L. and *Aloe vera* L. *International Journal of Pharmaceutical & Technological. Research*, *3*(2), 1059–1065.

Hasan, M. R., Islam, M. N., & Islam, M. R. (2016). Phytochemistry, pharmacological activities and traditional uses of *Emblica officinalis*: A review. *International Current Pharmaceutical Journal*, *5*(2), 14–21. doi:10.3329/icpj.v5i2.26441

Jagetia, G. C., Baliga, M. S., Malagi, K. J., & Kamath, M. S. (2002). The evaluation of the radioprotective effect of Triphala (an ayurvedic rejuvenating drug) in the mice exposed to γ-radiation. *Phytomedicine*, *9*(2), 99–108. doi:10.1078/0944-7113-00095 PMID:11995956

Jain, P. K., Das, D. E., Pandey, N. A., & Jain, P. R. (2016). Traditional Indian herb *Emblica officinalis* and its medicinal importance. *Innovative Journal of Ayurvedic Science*, *4*(4), 1–15.

Jangpangi, K., Khati, A., Chauhan, R. S., Rajesh, R., & Kumar, N. (2023). Assessment of the haematological and biochemical effects of Himalayan herb *Urtica dioica* leaves diets fed amur carp, *Cyprinus carpio haematopterus*. *Turkish Journal of Fisheries and Aquatic Sciences*, *24*(4), TRJFAS23854. doi:10.4194/TRJFAS23854

Kabra, A., Martins, N., Sharma, R., Kabra, R., & Baghel, U. S. (2019). *Myrica esculenta* Buch.-Ham. ex D. Don: A natural source for health promotion and disease prevention. *Plants*, *8*(6), 149. doi:10.3390/plants8060149 PMID:31159283

Kakoti, B. B., Pradhan, P., Borah, S., Mahato, K., & Kumar, M. (2013). Analgesic and anti-inflammatory activities of the methanolic stem bark extract of *Nyctanthes arbor-tristis* Linn. *BioMed Research International, 2013*, 826295. doi:10.1155/2013/826295 PMID:23984409

Kandiannan, K., Sivaraman, K., Thankamani, C. K., & Peter, K. V. (1996). Agronomy of ginger (*Zingiber officinale* Rosc.). *Journal of Spices and Aromatic Crops*, *5*(1), 1–27.

Khan, A. U., & Gilani, A. H. (2008). Pharmacodynamic evaluation of *Terminalia bellerica* for its anti-hypertensive effect. *Yao Wu Shi Pin Fen Xi*, *16*(3), 6–14.

Khan, K. H. (2009). Roles of *Emblica officinalis* in medicine-A review. *Botany Research International*, *2*(4), 218–228.

Khanal, A., Devkota, H. P., Kaundinnyayana, S., Gyawali, P., Ananda, R., & Adhikari, R. (2021). Culinary herbs and spices in Nepal: A review of their traditional uses, chemical constituents, and pharmacological activities. *Ethnobotany Research and Applications*, *21*, 1–18. doi:10.32859/era.21.40.1-18

Kulkarni, K. V., & Ghurghure, S. M. (2018). Indian gooseberry (*Emblica officinalis*): Complete pharmacognosy review. *International Journal of Chemical Studies*, *2*(2), 5–11.

Kumar, A., Jnanesha, A. C., & Nagaraju, S. (2019). Cultivation and medicinal properties of giloy [*Tinospora cordifolia* (Thunb.) Miers]. *Agrobios*, *18*(1), 57–58.

Kumar, G., Sharma, J. G., Goswami, R. K., Shrivastav, A. K., Kumar, N., Chandra, S., & Chakrabarti, R. (2021b). The study of effect of vitamin C and *Achyranthes aspera* seeds enriched diets on the growth, biochemical composition, digestive enzyme activities and expressions of genes involved in the biosynthesis of fatty acids in snow trout *Schizothorax richardsonii* (Gray, 1832). *Journal of Applied Aquaculture*, *35*(3), 489–509. doi:10.1080/10454438.2021.1985679

Kumar, G., Sharma, J. G., Goswami, R. K., Shrivastav, A. K., Tocher, D. R., Kumar, N., & Chakrabarti, R. (2022b). Freshwater macrophytes: Potential source of minerals and fatty acids for fish, poultry and livestock. *Frontiers in Nutrition*, *9*, 869425. doi:10.3389/fnut.2022.869425 PMID:35479737

Kumar, N., & Khurana, S. M. (2018). Phytochemistry and medicinal potential of the *Terminalia bellirica* Roxb. (Bahera). *Indian Journal of Natural Products and Resources*, *9*(2), 97–107.

Kumar, N., Sharma, J. G., Kumar, G., Shrivastav, A. K., Tiwari, N., Begam, A., & Chakrabarti, R. (2021a). Evaluation of nutritional value of prickly chaff flower (*Achyranthes aspera)* as fish feed ingredient. *The Indian Journal of Animal Sciences*, *91*(3), 239–244. doi:10.56093/ijans.v91i3.114152

Kumar, N., Sharma, J. G., Mittal, P., & Chakrabarti, R. (2022a). Effect of leaves and seeds of *Achyranthes aspera* as feed supplements on the immunological and stress parameters and related-gene expressions of Asian catfish (*Clarias batrachus*). *Veterinary Research Communications*, *47*(1), 99–109. doi:10.1007/s11259-022-09932-5 PMID:35507243

Kumar, N., Sharma, J. G., Singh, S. P., Singh, A., Harikrishna, V., & Chakrabarti, R. (2019). Validation of growth enhancing, immunostimulatory and disease resistance properties of *Achyranthes aspera* in *Labeo rohita* fry in pond conditions. *Heliyon*, *5*(2), e01246. doi:10.1016/j.heliyon.2019.e01246 PMID:30820481

Kumari, S., Krishna, J. M., Joshi, A. B., Gurav, S., Bhandarkar, A. V., Agarwal, A., Deepak, M., & Gururaj, G. M. (2017). A pharmacognostic, phytochemical and pharmacological review of *Terminalia bellerica*. *Journal of Pharmacognosy and Phytochemistry*, *6*(5), 368–376.

Lanka, S. (2018). A review on pharmacological, medicinal and ethnobotanical important plant: *Phyllanthus emblicalinn.* (syn. *Emblica officinalis*). *World Journal of Pharmaceutical Research*, *7*(4), 380–396.

Latif, A., Ashiq, K., Qayyum, M., Ashiq, S., Ali, E., & Anwer, I. (2019). Phytochemical and pharmacological profile of the medicinal herb: *Bryophyllum pinnatum*. *Journal of Animal & Plant Sciences*, *29*(6), 1528–1534.

Le Doux, M. S. (2015). Murine models of caytaxin deficiency. In *Movement Disorders* (pp. 439–452). Academic Press. doi:10.1016/B978-0-12-405195-9.00025-1

Maithani, A., Parcha, V., Pant, G., Dhulia, I., & Kumar, D. (2011). *Azadirachta indica* (neem) leaf: A review. *Journal of Pharmacy Research*, *4*(6), 1824–1827.

Maji, S., & Modak, S. (2021). Neem: Treasure of natural phytochemicals. *Chemical Science Review and Letters*, *10*(39), 396–401.

Mallik, J., Das, P., Karon, B., & Das, S. (2012). A review on phytochemistry and pharmacological activity of *Terminalia belerica*. *International Journal of Drug Formulation and Research*, *3*(6), 1–7.

Mann, S., Satpathy, G., & Gupta, R. K. (2015). *In vitro* evaluation of bio-protective properties of underutilized *Myrica esculenta* Buch.-Ham. ex D. Don fruit of Meghalaya. *Indian Journal of Natural Products and Resources*, *6*(3), 183–188.

Maruthappan, V., & Shree, K. S. (2010). Hypolipidemic activity of haritaki (*Terminalia chebula*) in atherogenic diet induced hyperlipidemic rats. *Journal of Advanced Pharmaceutical Technology & Research*, *1*(2), 229–235. PMID:22247850

Mavi, A. K., Gaur, S., Gaur, G., Babita, Jindal, P., Kumar, N., & Kumar, U. (2023). CAR T-cell therapy: Reprogramming patient's immune cell to treat cancer. *Cellular Signalling*, *105*, 110638. doi:10.1016/j.cellsig.2023.110638 PMID:36822565

Meshram, M. M., Rangari, S. B., Kshirsagar, S. B., Gajbhiye, S., Trivedi, M. R., & Sahane, R. S. (2012). *Nyctanthes arbor-tristis*aherbal panacea. *International Journal of Pharmaceutical Sciences and Research*, *3*(8), 2432–2440.

Mishra, S. K., Sangwan, N. S., & Sangwan, R. S. (2007). Plant review *Andrographis paniculata* (Kalmegh): A review. *Pharmacognosy Reviews*, *1*(2), 283–298.

Mukharjee, D., Lokwani, S., Dave, A., Gonsalves, C., & Sarkar, C. (2021). Bhringraj: A pharmaceutical treasure trove. *Indian Journal of Natural Sciences*, *12*(68), 35231–35239.

Nhiem, N. X., Van Kiem, P., Van Minh, C., Tai, B. H., Cuong, N. X., Thu, V. K., Anh, H. L. T., Jo, S. H., Jang, H. D., Kwon, Y. I., & Kim, Y. H. (2010). A new monoterpenoid glycoside from *Myrica esculenta* and the inhibition of angiotensin I-converting enzyme. *Chemical & Pharmaceutical Bulletin*, *58*(10), 1408–1410. doi:10.1248/cpb.58.1408 PMID:20930414

Pinmai, K., Chunlaratthanabhorn, S., Ngamkitidechakul, C., Soonthornchareon, N., & Hahnvajanawong, C. (2008). Synergistic growth inhibitory effects of *Phyllanthus emblica* and *Terminalia bellerica* extracts with conventional cytotoxic agents: Doxorubicin and cisplatin against human hepatocellular carcinoma and lung cancer cells. *World Journal of Gastroenterology*, *14*(10), 1491–1497. doi:10.3748/wjg.14.1491 PMID:18330936

Rangari, V. D. (2009). Pharmacognosy & phytochemistry. Career publications.

Rani, C., Chawla, S., Mangal, M., Mangal, A. K., Kajla, S., & Dhawan, A. K. (2012). *Nyctanthes arbor-tristis* Linn. (Night Jasmine): A sacred ornamental plant with immense medicinal potentials. *Indian Journal of Traditional Knowledge*, *11*(3), 427–435.

Rastogi, R. P., & Mehrotra, B. N. (1995). *Compendium of Indian medicinal trees*. Central Drug Research Institute.

Rawat, H., Verma, Y., Ayesha, N. S., Negi, N., Pant, H. C., Mishra, A., & Gaurav, N. (2021). *Nyctanthes arbor-tristis*: A traditional herbal plant with miraculous potential in medicine. *International Journal of Botany Studies*, *6*, 427–440.

Reema, S., Deepali, T., Gauri, S., & Pankaj, S. (2018). Nyctanthes arbor-Tristis: A Wonder Indian herbal drug needs healthcare attention. *Biomedical Journal of Scientific & Technical Research*, *5*(3), 1199.

Sahu, A., Kumar, N., Singh, C. P., & Singh, M. (2023). Environmental DNA (eDNA): Powerful technique for biodiversity conservation. *Journal for Nature Conservation*, *71*, 126325. doi:10.1016/j.jnc.2022.126325

Saklani, S. A., Chandra, S. U., Mishra, A. P., & Badoni, P. P. (2012). Nutritional evaluation, antimicrobial activity and phytochemical screening of wild edible fruit of *Myrica nagi* pulp. *International Journal of Pharmacy and Pharmaceutical Sciences*, *4*(3), 407–411.

Semwal, A., Kumar, A., & Kumar, N. (2023). A review on pathogenicity of *Aeromonas hydrophila* and their mitigation through medicinal herbs in aquaculture. *Heliyon*, *9*(3), e14088. doi:10.1016/j.heliyon.2023.e14088 PMID:36938468

Shan, B., Cai, Y. Z., Brooks, J. D., & Corke, H. (2007). The in vitro antibacterial activity of dietary spice and medicinal herb extracts. *International Journal of Food Microbiology*, *117*(1), 112–119. doi:10.1016/j.ijfoodmicro.2007.03.003 PMID:17449125

Sharma, J. G., Kumar, N., Mittal, P., & Chakrabarti, R. (2022). Evaluation of UV-B protective properties of leaves and seeds of *Achyranthes aspera* in Asian catfish *Clarias batrachus* (Linn.). *Photochemical & Photobiological Sciences*, *21*(8), 1341–1356. doi:10.1007/s43630-022-00222-2 PMID:35403962

Sharma, J. G., Kumar, N., Singh, S. P., Singh, A., Harikrishna, V., & Chakrabarti, R. (2019). Evaluation of immunostimulatory properties of prickly chaff flower *Achyranthes aspera* in rohu *Labeo rohita* fry in pond conditions. *Aquaculture (Amsterdam, Netherlands)*, *505*, 183–189. doi:10.1016/j.aquaculture.2019.02.065

Singh, J., Singh, A. P., & Singh, A. P. (2021). *Nyctanthes arbor-tristis*: A comprehensive review. *World Journal of Current Medical and Pharmaceutical Research*, *3*(4), 74–78. doi:10.37022/wjcmpr.v3i4.181

Singh, N., Bhalla, M., de Jager, P., & Gilca, M. (2011). An overview on Ashwagandha: A Rasayana (rejuvenator) of Ayurveda. *African Journal of Traditional, Complementary, and Alternative Medicines*, *8*(5S), 208–213. doi:10.4314/ajtcam.v8i5S.9 PMID:22754076

Singh, R., Chauhan, R. S., Semwal, A., & Kumar, N. (2023). Efficacy of 17α-methyl testosterone on growth and survival of fry of swordtail, *Xiphophorus helleri. Journal of Krishi Vigyan*, *11*(2), 65–69. doi:10.5958/2349-4433.2023.00017.X

Sood, P., & Shri, R. (2018). A review on ethnomedicinal, phytochemical and pharmacological aspects of *Myrica esculenta. Indian Journal of Pharmaceutical Sciences*, *80*(1), 2–13.

Sopi, R. B., & Khan, M. F. H. (2013). Bronchodilatory effect of ethanolic extract of the leaves of *Nyctanthes arbor-tristis. Pharmacognosy Research*, *5*(3), 169–172. doi:10.4103/0974-8490.112422 PMID:23901212

Suresh, V., Jaikumar, S., & Arunachalam, G. (2010). Anti diabetic activity of ethanol extract of stem bark of *Nyctanthes arbor-tristis* linn. *Research Journal of Pharmaceutical, Biological and Chemical Sciences*, *1*(4), 311–317.

Surjushe, A., Vasani, R., & Saple, D. G. (2008). *Aloe vera*: A short review. *Indian Journal of Dermatology*, *54*(3), 163–166. doi:10.4103/0019-5154.44785 PMID:19882025

Yanthan, M., & Misra, A. K. (2013). Molecular approach to the classification of medicinally important actinorhizal genus Myrica. *Indian Journal of Biotechnology*, *12*, 133–136.

Zhang, Y. J., Nagao, T., Tanaka, T., Yang, C. R., Okabe, H., & Kouno, I. (2004). Antiproliferative activity of the main constituents from *Phyllanthus emblica. Biological & Pharmaceutical Bulletin*, *27*(2), 251–255. doi:10.1248/bpb.27.251 PMID:14758047

Zhang, Y. J., Tanaka, T., Iwamoto, Y., Yang, C. R., & Kouno, I. (2001). Novel sesquiterpenoids from the roots of *Phyllanthus Emblica. Journal of Natural Products*, *64*(7), 870–873. doi:10.1021/np010059z PMID:11473414

Chapter 14
Nanoagriculture:
Exploring Environmental Impacts and Sustainable Advantages

Omkar Singh
https://orcid.org/0000-0003-1971-185X
Sardar Vallabhbhai Patel University of Agriculture and Technology, India

Shivangi Singh
Sardar Vallabhbhai Patel University of Agriculture and Technology, India

Prabhat Kumar Singh
Acharya Narendra Deva University of Agriculture and Technology, India

Vijay Kant Singh
Punjab Agriculture University, Ludhiyana, India

Anand Singh
Acharya Narendra Deva University of Agriculture and Technology, India

Sakshi Singh
Tilak Dhari P.G. College, India

Hasmik S. Movsesyan
Yerevan State University, Armenia

Tatiana Minkina
Southern Federal University, Russia

Olga Nazarenko
Independent Researcher, Russia

Savr Mandzhiev
Southern Federal University, Russia

Elizaveta Latsynnik
Southern Federal University, Russia

ABSTRACT

Nanoagriculture can solve food production and agricultural sustainability problems. Nanoagriculture is examined in this chapter, including its uses, environmental impacts, sustainable benefits, and future. The chapter introduces nanoagriculture and its role in modern agriculture. Nanotechnology in agriculture includes nanofertilizers, nanopesticides, nanosensors, and other innovative technologies. Nanoagriculture has pros and cons, including reduced pesticide use and improved soil quality and nanoparticle toxicity and bioaccumulation. An example shows how nanoagriculture affects crop yield and the ecosystem. Nanoagriculture boosts resource efficiency, environmental protection, and economic growth. Nanoagriculture needs more research, clear laws, public involvement, and stakeholder coopera-

DOI: 10.4018/979-8-3693-1890-4.ch014

tion, as the chapter discusses. Nanoagriculture has the potential to revolutionize agriculture and promote sustainable food production, the chapter concludes. It also underlines the importance of responsible creation and execution for safe and effective use. This chapter provides a comprehensive overview of nanoagriculture, including its current condition, future prospects, and the necessary steps to employ it for sustainable agriculture.

1. INTRODUCTION

1.1 Definition of Nanoagriculture

Nanoagriculture is a developing area that combines nanotechnology with agricultural to boost crop yield, optimize resource efficiency, and reduce environmental harm (Prasad et al., 2017). Nanotechnology is used in agriculture for crop protection, nutrient management, and precision farming by applying nanomaterials, nanodevices, and nanotools (Fraceto et al., 2016). Nanoagriculture utilizes the distinct characteristics of materials at the nanoscale, usually between 1 to 100 nanometers, to create new and sustainable agricultural methods (Mukhopadhyay, 2014).

1.2 Significance of Nanoagriculture in Modern Farming Practices

Nanoagriculture has the ability to revolutionize modern farming by tackling the issues in the agricultural sector. Due to the rising global population and higher food requirements, there is an urgent necessity for sustainable intensification of agriculture (FAO, 2017; singh et. al., 2023). Nanoagriculture provides potential solutions to increase crop yield, minimize the environmental impact of agriculture, and guarantee food security (Dasgupta et al., 2015; Kolesnikov et al., 2024). Nanoagriculture can maximize resource efficiency, minimize waste, and reduce dependency on conventional agrochemicals by delivering nutrients precisely, controlling pests effectively, and monitoring crop health in real-time (Kah et al., 2018). Nanoagriculture can aid in the advancement of climate-resilient farming systems by offering instruments to adjust to evolving environmental conditions (Dwivedi et al., 2019; Rajput et al., 2021).

1.3 Purpose and Scope of the Chapter

This chapter aims to offer a thorough examination of nanoagriculture, focusing on its environmental effects and sustainable benefits. The chapter seeks to connect scientific progress in nanotechnology with its practical use in agriculture. The article will explore the diverse uses of nanotechnology in agriculture, such as nanofertilizers, nanopesticides, nanosensors, and other novel methods. The chapter will explore the potential environmental effects of nanoagriculture, examining both the benefits and drawbacks. The chapter will present real-world examples of nanoagriculture application through a case study, emphasizing the results, lessons gained, and recommendations for future adoption. The chapter will investigate the sustainable benefits of nanoagriculture, including enhanced resource efficiency, less environmental impact, and socioeconomic gains. The obstacles and future prospects of nanoagriculture will be examined, focusing on regulatory issues, public opinion, research requirements, and partnerships among stakeholders.

2. APPLICATIONS OF NANOTECHNOLOGY IN AGRICULTURE

2.1 Nanofertilizers

Nanofertilizers are a new type of fertilizers that use nanotechnology to improve nutrient delivery and increase crop productivity. The fertilizers mentioned contain nutrients enclosed in nanomaterials or nanoparticles, enabling precise and regulated delivery of nutrients to the plants (DeRosa et al., 2010; singh O. 2023: Singh et al., 2023).

2.1.1 Enhanced Nutrient Delivery

Nanofertilizers enhance nutrient delivery to crops more effectively than traditional fertilizers. Nutrients are protected from degradation and loss in the soil through nanoencapsulation, which allows for a continuous release over a prolonged period (Subramanian et al., 2015; Rajput et al., 2024). This precise delivery method minimizes nitrogen loss and boosts nutrient absorption by plants, resulting in increased efficiency of fertilizer utilization (Chhipa, 2017; Singh et al., 2023). Nano-nitrogen fertilizers have demonstrated higher nitrogen utilization efficiency and lower nitrate leaching in comparison to traditional fertilizers (Rai et al., 2020; Singh et al., 2022).

2.1.2 Increased Crop Yield

Nanofertilizers have shown great promise in increasing agricultural productivity. Delivering nutrients to plants in a precise and controlled manner enhances growth and development, leading to increased productivity (Manjunatha et al., 2016). Research has shown that using several nanofertilizers, like zinc oxide nanoparticles, chitosan nanoparticles, and nano-phosphorus fertilizers, might lead to higher crop yields (Dimkpa et al., 2019; Malerba & Cerana, 2018; Rameshaiah et al., 2015). The enhancements in yield can help meet the growing food demand and provide food security.

2.2 Nanopesticides

Nanopesticides are pesticide mixtures that utilize nanotechnology to enhance pest control effectiveness and minimize environmental harm (Kah et al., 2018). Nanoencapsulated active substances, nanoemulsions, or nanoparticles having pesticidal characteristics can be part of these formulations (Kookana et al., 2014).

2.2.1 Targeted Pest Control

Nanopesticides provide precise pest control by transporting the active components precisely to the specific pests or afflicted plant areas. This focused distribution method reduces the likelihood of unwanted ecological effects by limiting the exposure of non-target organisms to pesticides (Nuruzzaman et al., 2016). Nanoencapsulation methods enhance the stability and controlled release of active chemicals, hence extending the efficacy of pesticides (Ragaei & Sabry, 2014). Neem oil-based nanopesticides have demonstrated increased insecticidal effectiveness against different crop pests (Forim et al., 2013).

2.2.2 Reduced Environmental Impact

Utilizing nanopesticides can greatly diminish the environmental repercussions of pest management methods. Precise distribution and regulated release of active chemicals reduce the required amount of pesticides, therefore lowering the overall pesticide burden in the environment (Iavicoli et al., 2017). Additionally, enclosing pesticides in biodegradable nanoparticles can aid in their decomposition and decrease their longevity in the ecosystem (Grillo et al., 2016). Nanopesticides have demonstrated the ability to substitute certain extremely toxic traditional pesticides, hence decreasing the environmental hazards linked to insect control.

2.3 Nanosensors

Nanosensors are small devices that use nanomaterials or nanostructures to detect and monitor several factors in the agricultural setting (Sagadevan et al., 2019). The sensors provide great sensitivity, selectivity, and real-time monitoring, allowing for precise implementation of precision agriculture techniques (Prasad et al., 2017).

2.3.1 Monitoring Soil Conditions

Nanosensors can be used to detect various soil variables such moisture content, nutrient levels, pH, and temperature. The sensors offer immediate information on soil condition, enabling farmers to make well-informed choices about irrigation, fertilization, and other agricultural techniques (Mukhopadhyay, 2014). Carbon nanotube-based sensors have been created to monitor soil moisture and nutrient levels, allowing for accurate irrigation and fertilizer delivery (Giraldo et al., 2019).

2.3.2 Detecting Plant Diseases

Nanosensors can be utilized for early identification of plant illnesses, allowing for prompt intervention and disease control. The sensors are capable of identifying particular biomarkers or volatile organic chemicals emitted by sick plants, enabling quick and precise diagnosis (Fang & Ramasamy, 2015). Nanosensors utilizing gold nanoparticles, carbon nanotubes, and quantum dots have demonstrated potential in identifying different plant diseases (Khater et al., 2017). Nanosensors can aid in early disease identification, hence mitigating crop losses and decreasing reliance on extensive pesticide use.

2.4 Nanoenzymes

Nanoenzymes are nanomaterials that demonstrate catalytic capabilities similar to enzymes. Nanomaterials have attracted considerable interest in multiple sectors, including agriculture, because of their distinctive characteristics, such as high stability, cost-effectiveness, and ease of manufacturing in comparison to natural enzymes (Wei & Wang, 2013). Here are some uses of nanoenzymes in agriculture: Nanoenzymes are nanomaterials that demonstrate catalytic capabilities similar to enzymes. Nanomaterials have attracted considerable interest in multiple sectors, including agriculture, because of their distinctive characteristics, such as high stability, cost-effectiveness, and ease of manufacturing in comparison to natural enzymes (Wei & Wang, 2013). Here are some uses of nanoenzymes in agriculture:

2.4.1 Pesticide Degradation

Nanoenzymes demonstrate promise in breaking down pesticide residues in soil and water. Fe3O4 magnetic nanoparticles with peroxidase-like activity were utilized to break down organophosphate insecticides (Pang et al., 2020). Gold nanoparticles having oxidase-like capabilities have been used to break down pesticides such as chlorpyrifos and methyl parathion (Liu et al., 2019). Utilizing nanoenzymes for pesticide breakdown can decrease the ecological footprint of these substances and enhance the quality of soil and water.

2.4.2 Plant growth Promotion

Nanoenzymes can boost plant development by improving nutrient absorption and reducing abiotic stressors. Cerium oxide nanoparticles with catalase-like activity have been demonstrated to enhance plant development and alleviate oxidative stress during drought circumstances (Wu et al., 2017). Iron oxide nanoparticles with peroxidase-like activity have been utilized to improve seed germination and seedling growth, as reported by Iyer et al. (2020). Nanoenzymes used to promote plant growth can enhance agricultural output and improve resistance to environmental challenges.

2.4.3 Disease Management

Nanoenzymes can be employed for the management of plant diseases by inhibiting pathogen growth and enhancing plant defense mechanisms. Silver nanoparticles with peroxidase-like activity have demonstrated antifungal properties against plant pathogens like Fusarium oxysporum and Aspergillus niger (Mishra et al., 2017). Furthermore, copper oxide nanoparticles exhibiting oxidase-like activity have been used to control bacterial leaf blight in rice (Kannan & Jasim, 201Nanoenzymes can be used to control plant diseases by stopping pathogen growth and boosting plant defense mechanisms. Silver nanoparticles with peroxidase-like activity have shown antifungal effects against plant diseases such as Fusarium oxysporum and Aspergillus niger (Mishra et al., 2017). Copper oxide nanoparticles with oxidase-like activity have been utilized to manage bacterial leaf blight in rice (Kannan & Jasim, 2019). Nanoenzymes offer a potential alternative to conventional chemical fungicides and bactericides in disease management, thereby decreasing the environmental impact of disease control methods.9). The use of nanoenzymes in disease management can provide an alternative to traditional chemical fungicides and bactericides, reducing the environmental footprint of disease control practices.

2.4.4 Biosensing and Diagnostic Applications

Nanoenzymes can be used to create biosensors and diagnostic instruments for identifying plant diseases, environmental contaminants, and other significant substances in agriculture. Gold nanoparticles possessing peroxidase-like characteristics were utilized in creating a colorimetric biosensor for identifying the plant disease Xanthomonas oryzae pv. oryzae (Jiang et al., 2018). Platinum nanoparticles with oxidase-like properties have been used to create an electrochemical biosensor for detecting the herbicide methyl parathion (Shi et al., 2016). Nanoenzymes can be used in biosensing and diagnostics to identify plant illnesses and environmental toxins early, which can help in making timely decisions and interventions in agriculture.

2.4.5 Soil Remediation

Nanoenzymes can be used to clean up polluted soils by breaking down contaminants and enhancing soil quality. Manganese oxide nanoparticles having oxidase-like properties were utilized to break down polycyclic aromatic hydrocarbons (PAHs) in soil (Li et al., 2018). Iron oxide nanoparticles having peroxidase-like capabilities have been used to clean up soils contaminated with heavy metals such as chromium and lead (Shi et al., 2017). Nanoenzymes can be utilized in soil remediation to enhance the quality of degraded lands and improve soil conditions for agricultural use.

2.5 Other Applications

Nanotechnology is utilized in different areas of agriculture, such as nanoencapsulation and incorporating nanomaterials into packaging.

Nanoencapsulation is the process of encapsulating active chemicals like nutrients, insecticides, or plant growth regulators into nanocarriers (Rani & Yadav, 2020). This method improves the stability, bioavailability, and controlled release of the enclosed compounds, enhancing their efficacy and decreasing the necessary amounts (Campos et al., 2014). Nanoencapsulation has been utilized for creating slow-release fertilizers, precise pesticide delivery systems, and plant growth enhancers (Nair et al., 2010).

Nanomaterials are being researched for their potential application in agricultural packaging to prolong the shelf life of goods and guarantee food safety (Chaudhry et al., 2008). Nanocomposites and nanocoatings offer protection against moisture, gasses, and UV light, which helps in preventing rotting and maintaining the quality of agricultural products (Duncan, 2011). Antimicrobial nanoparticles like silver and zinc oxide can be added to packing materials to prevent the growth of foodborne bacteria, thus improving food safety (Pradhan et al., 2015).

3. ENVIRONMENTAL IMPACTS OF NANOAGRICULTURE

3.1 Positive Impacts

3.1.1 Reduced Use of Conventional Agrochemicals

Nanoagriculture has the ability to decrease the reliance on traditional agrochemicals like pesticides and fertilizers by creating precise and effective delivery methods. Nanoencapsulated insecticides and nanofertilizers offer controlled release of active components, decreasing the required chemical quantity and lessening their environmental footprint (Kah et al., 2018). Chitosan nanoparticles containing pesticides have demonstrated higher effectiveness and lower environmental harm when compared to traditional formulations (Grillo et al., 2014). Nanofertilizers can improve plant nutrient absorption, leading to decreased fertilizer usage and less nutrient runoff into the environment (DeRosa et al., 2010).

3.1.2 Improved Soil Health

Nanoagriculture can enhance soil health by decreasing the buildup of toxic substances and fostering beneficial soil microbes. Nanoparticles can repair contaminated soils by adsorbing or degrading pollutants such heavy metals and organic contaminants (Rodrigues et al., 2016). Iron oxide nanoparticles have been demonstrated to efficiently eliminate arsenic from polluted soils (Shipley et al., 2010). Moreover, certain nanoparticles like silver nanoparticles have been shown to enhance the proliferation of advantageous soil bacteria, leading to enhanced soil fertility and plant development (Mehmood et al., 2017).

3.1.3 Decreased Water Pollution

Nanoagriculture can reduce water pollution by decreasing the leaching of agrochemicals into groundwater and surface water bodies. Nanoencapsulated agrochemicals with controlled release qualities can reduce the likelihood of chemical runoff and leaching, therefore safeguarding water resources (Kookana et al., 2014). Nanosensors can be used to monitor water quality and identify toxins, allowing for early warning systems and aiding in remediation operations (Parisi et al., 2015).

3.2 Potential Negative Impacts

3.2.1 Nanoparticle Toxicity to Non-Target Organisms

The primary issue with the environmental impact of nanoagriculture is the possible toxicity of nanoparticles to organisms not intended as targets. Nanoparticles have been demonstrated to have harmful impacts on a range of creatures, such as plants, animals, and microbes (Bundschuh et al., 2018). Silver nanoparticles, utilized as antibacterial agents, have been found to induce toxicity in aquatic species and soil microorganisms (Anjum et al., 2013). Titanium dioxide nanoparticles have been shown to cause oxidative stress and genotoxicity in plants (Choi & Park, 2014). Predicting the ecological implications of nanoparticles is problematic due to characteristics like size, shape, surface qualities, and ambient circumstances affecting their toxicity (Holden et al., 2014).

3.2.2 Bioaccumulation and Biomagnification in the Food Chain

Nanoagriculture poses a risk of nanoparticles accumulating and magnifying in the food chain. Nanoparticles can infiltrate the food chain through different pathways, including direct absorption by plants, consumption by animals, or transmission between trophic levels (Cornelis et al., 2014). Nanoparticles can build up in organisms' tissues and increase in concentration as they go up the food chain (Gardea-Torresdey et al., 2014). This may result in heightened exposure and significant health hazards for apex predators, such as humans. Research has demonstrated the buildup of gold nanoparticles in plant tissues and their subsequent movement to herbivorous insects (Judy et al., 2011).

3.2.3 Uncertainties Regarding Long-Term Effects

Although nanoagriculture offers potential advantages, there remain doubts about the lasting environmental impacts of nanoparticles. Nanoparticles' fate and behavior in intricate environmental settings as soil and water remain incompletely comprehended (Baalousha et al., 2016). Nanoparticles can experience several changes, including aggregation, dissolution, and chemical interactions, which can modify their characteristics and environmental effects as time passes (Lowry et al., 2012). The long-term impacts of low-dose exposure to nanoparticles on ecosystems and human health are not well understood (Iavicoli et al., 2017; Singh et al., 2023c). Long-term research is necessary to evaluate the ecological and health hazards linked to the extensive utilization of nanoparticles in agriculture.

4. CASE STUDY 1: WHEAT (TRITICUM AESTIVUM L.)

4.1 Background and Context

Wheat is an essential cereal crop worldwide, playing a large role in meeting the daily caloric and protein requirements of the global population (Shewry & Hey, 2015). Wheat cultivation faces challenges such nutrient deficiencies, environmental stresses, and increasing demand. Pandey et al. (2018) suggest that nanotechnology could offer solutions to enhance wheat productivity and quality.

4.2 Implementation of Nanoagriculture Techniques

Zinc oxide (ZnO) and iron oxide (Fe_3O_4) nanoparticles were used in a case study to treat wheat crops by spraying them on the leaves and applying them to the soil. The nanoparticles were produced by green synthesis techniques to guarantee their environmental friendliness and compatibility with living organisms (Siddiqi & Husen, 2017). The treatment rates and frequency were fine-tuned according to the wheat growth phases and the crop's individual needs.

4.3 Results and Findings

4.3.1 Environmental Impacts

ZnO and Fe3O4 nanoparticles decreased the need for chemical fertilizers by improving the wheat plants' nutrient absorption and usage efficiency (Qureshi et al., 2018). The decrease in chemical fertilizer consumption reduced the likelihood of nutrient leaching and environmental contamination.

4.3.2 Economic and Social Implications

Nanoparticles used in wheat production led to higher grain yield and enhanced grain quality, especially in relation to micronutrient levels (Rameshaiah & Jpallavi, 2015). Increased wheat grain output and quality can lead to higher profits for farmers and enhanced nutritional security for consumers.

5. CASE STUDY 2: MAIZE (ZEA MAYS L.)

5.1 Background and Context

Maize is a significant cereal crop grown globally, acting as a primary food source for millions of individuals and a crucial element in animal feed and industrial uses (Nuss & Tanumihardjo, 2010; Sachan et al., 2023). Maize productivity is frequently constrained by nutritional deficits, drought stress, and pest infestations. Nanofertilizers and nanoparticles have been studied as possible solutions to improve maize productivity (Anusuya & Banu, 2016).

5.2 Implementing Nanoagriculture Techniques

Chitosan-based nanofertilizers were used on maize crops by coating the seeds and spraying the leaves in this case study. The nanofertilizers were designed to offer a gradual and regulated discharge of vital nutrients, including nitrogen, phosphorus, and potassium (NPK) (Abdel-Aziz et al., 2016). Silver nanoparticles (AgNPs) were used on maize plants to evaluate their effectiveness in insect control.

5.3 Results and Discoveries

5.3.1 Ecological Consequences

Chitosan-based nanofertilizers decreased the need for chemical fertilizers in maize production by enhancing nutrient use efficiency and reducing nutrient losses (Yue et al., 2020). Reducing the use of chemical fertilizers can help lessen the environmental effects caused by excessive nutrient release, like eutrophication and greenhouse gas emissions.

5.3.2 Economic and Societal Consequences

Nanofertilizers and AgNPs treatment resulted in higher maize grain yield and enhanced grain quality (Aghdam et al., 2016). Improved maize grain yields and quality can lead to higher profits for farmers and improved food security for communities who depend on maize as a primary crop.

6. CASE STUDY 3: RICE (ORYZA SATIVA L.)

6.1 Background and Context

Rice is the predominant staple meal for a significant section of the global population, especially in Asia (Khush, 2005). Rice production is frequently limited by issues like nutrient deficits, water limitations, and disease outbreaks. Nanotechnology provides new opportunities to address these difficulties and enhance rice productivity and sustainability (Salim et al., 2018).

6.2 Implementation of Nanoagriculture Techniques

SiO_2 NPs and ZnO NPs were used on rice crops by dipping the roots and spraying the leaves in this study. SiO_2 nanoparticles were utilized to increase plants' tolerance to non-living environmental factors, whilst ZnO nanoparticles were used to enhance zinc absorption and fight against fungal infections (Wang et al., 2018).

6.3 Results and Findings

6.3.1 Environmental Impacts

Using SiO2 NPs and ZnO NPs enhanced how efficiently rice plants utilized nutrients and water, decreasing the requirement for excessive fertilizer and water. Optimizing resource utilization can reduce the environmental impact of rice production by decreasing water pollution and saving water resources.

6.3.2 Economic and Social Implications

Nanoparticles used in rice production led to higher grain yield and enhanced grain nutritional quality, namely in zinc concentration (Rico et al., 2019). Increased rice productivity and nutritional content can lead to higher economic profits for farmers and better public health results, particularly in areas where rice is a primary food source.

7. CASE STUDY 4: SORGHUM (SORGHUM BICOLOR L.)

7.1 Background and Context

Sorghum is a resilient cereal crop that is vital for ensuring food security and supporting livelihoods in dry locations around the world (Dahlberg et al., 2011). Sorghum production frequently encounters obstacles such nutritional deficits, pest infestations, and climate variability. Nanotechnology offers possibilities to tackle these challenges and improve sorghum productivity and resilience (Munyi et al., 2020).

7.2 Implementation of Nanoagriculture Techniques

Iron oxide nanoparticles (Fe_3O_4 NPs) and chitosan-based nanopesticides were used on sorghum crops by applying them by seed priming and foliar spraying in this case study. Fe_3O_4 nanoparticles were utilized to enhance iron absorption and mitigate iron shortage in sorghum plants (Deshpande et al., 2017), whilst chitosan-based nanopesticides were applied to manage sorghum pests, including the sorghum shoot fly (Atherigona soccata) (Kumaraswamy et al., 2018).

7.3 Results and Findings

7.3.1 Environmental Impacts

Fe_3O_4 NPs treatment decreased the requirement for chemical iron fertilizers by improving iron absorption and utilization in sorghum plants (Gupta et al., 2019). Reducing the use of chemical fertilizers can help lessen the environmental effects linked to iron fertilization, like soil acidification and water pollution.

7.3.2 Economic and Social Implications

Fe_3O_4 NPs and chitosan-based nanopesticides resulted in higher sorghum grain yield and enhanced grain quality according to Nkurunziza et al. (2019). Increased sorghum yield can improve food and economic security for smallholder farmers in semi-arid and desert locations. Eco-friendly nanopesticides can decrease the need on synthetic pesticides, encouraging safer and more sustainable pest control methods.

8. SUSTAINABLE ADVANTAGES OF NANOAGRICULTURE

8.1 Improved Resource Efficiency

Nanoagriculture has the potential to enhance resource efficiency in agricultural production systems (Prasad et al., 2017). Utilizing the distinct characteristics of nanomaterials and nanodevices might enhance the efficiency of vital resources like water and nutrients (Prasad et al., 2017).

8.1.1 Water Conservation

- Nanotechnology can aid in water conservation in agriculture by advancing smart irrigation systems and water-saving devices (Rai et al., 2016).
- Nanomaterials like carbon nanotubes and silica nanoparticles can be added to soil to improve water retention and decrease evaporation losses (Rai et al., 2016).
- Nanosensors can be used to continuously monitor soil moisture levels, allowing for precise irrigation and reducing water waste (Mura et al., 2016).
- Jha et al. (2019) found that employing nanosensors in irrigation scheduling resulted in a 30% decrease in water usage compared to traditional approaches, while maintaining agricultural yields.

8.1.2 Nutrient Management

- Nanofertilizers and nano-based delivery systems have the potential to enhance nutrient management in agriculture, as stated by Subramanian et al. (2015).
- These technologies enable the precise delivery of nutrients to plants, ensuring they receive the correct amount at the appropriate time, while reducing nutrient loss through leaching or volatilization (Subramanian et al., 2015).
- Nanoencapsulation methods can improve the stability and bioavailability of nutrients, therefore decreasing the requirement for excessive fertilizer use (Qureshi et al., 2018).
- Wu et al. (2016) demonstrated that chitosan-based nanofertilizers improved nitrogen use efficiency by 20-30% in rice production when compared to traditional urea fertilizers.

8.2 Reduced Environmental Footprint

Nanoagriculture can decrease the environmental impact of agricultural practices by limiting the use of agrochemicals, lowering greenhouse gas emissions, and averting land degradation (Iavicoli et al., 2017).

1. Lower greenhouse gas emissions
   - Nanotechnology in agriculture can reduce greenhouse gas emissions from traditional farming methods (Chhipa, 2017).
   - Nanofertilizers and nano-based crop protection solutions have the potential to decrease the reliance on synthetic fertilizers and pesticides, which are significant contributors to nitrous oxide and carbon dioxide emissions (Chhipa, 2017).
   - Nanomaterials can help reduce atmospheric greenhouse gases through soil remediation and carbon sequestration (Singh & Singh, 2019).
   - Xu et al. (2020) showed that using biochar-based nanocomposites in agricultural soils decreased nitrous oxide emissions and improved soil carbon storage (Saikanth et al., 2023).
2. Minimized land degradation
   - Nanoagriculture can mitigate land degradation through sustainable land management practices and by decreasing the environmental effects of intensive farming.
   - Nanoparticles can repair contaminated soils by immobilizing heavy metals and organic contaminants (Rizwan et al., 2017).
   - Nanomaterials can enhance soil structure and fertility by improving water retention, nutrient availability, and microbial activity (Khati et al., 2019).
   - Adisa et al. (2019) found that using zinc oxide nanoparticles on damaged soils enhanced soil quality and boosted agricultural yields, highlighting the promise of nanotechnology in addressing land degradation.

8.3 Socioeconomic Benefits

Nanoagriculture approaches can bring about substantial socioeconomic advantages, such as enhanced food security and economic prospects for farmers (Singh et al., 2018; Sastry et al., 2017).

1. Increased food security
   - Nanoagriculture can boost food security by improving crop yield, minimizing crop losses, and raising the nutritional value of agricultural products (Singh et al., 2018).
   - Nanotechnology applications including nanofertilizers, nanopesticides, and nano-based crop protection products might assist farmers in increasing yields and improving crop quality, especially in difficult environmental conditions (Kashyap et al., 2015).
   - Nanosensors and precision agricultural technology can identify crop stressors and pathogens early, allowing for timely treatments to prevent crop failures (Parisi et al., 2015).
2. Economic opportunities for farmers
   - Nanoagriculture can generate fresh economic prospects for farmers through cost reduction in production, enhancement of crop value, and the exploration of new markets for nano-enhanced agricultural products (Sastry et al., 2017).
   - Nanofertilizers and nanopesticides can decrease farmers' dependence on costly agrochemicals, hence reducing input expenses and enhancing profitability (Fraceto et al., 2016).
   - Nano-based crop protection technologies and smart packaging solutions can prolong the shelf life of agricultural food, hence decreasing post-harvest losses and enhancing the market value of crops (Shukla et al., 2019).

- Nanotechnology-based precision agriculture approaches can help farmers enhance resource utilization, minimize waste, and boost the efficiency of their farming operations (Kah et al., 2018).

9. CHALLENGES AND FUTURE PERSPECTIVES

Nanotechnology in agriculture has the potential to change farming processes, increase crop output, and support sustainable resource management. Yet, the acceptance and extensive use of nanoagriculture encounter several obstacles that must be resolved to guarantee its safe and conscientious advancement.

A major concern is the little knowledge about the long-term impacts of nanomaterials on the environment and human health. Several studies have shown the advantages of nanoparticles and nanofertilizers in enhancing crop growth and nutrient absorption. However, there are worries about their possible toxicity and environmental effects (Iavicoli et al., 2017). Nanoparticles can gather in the soil, water, and food chain, posing potential dangers to non-target creatures and human consumers (Tripathi et al., 2017). It is essential to perform thorough risk assessments and establish standardized protocols to evaluate the safety of nanomaterials in agricultural systems (Kah et al., 2018).

Regulatory and policy concerns of nanoagriculture present another challenge. The existing regulations for agricultural inputs and procedures may not adequately cover the distinct characteristics and potential hazards of nanomaterials (Amenta et al., 2015). Specific rules and guidelines are required to oversee the development, characterisation, and application of nanomaterials in agriculture (Hofmann et al., 2020). The laws should rely on scientific evidence and stakeholder engagement to guarantee the safe and appropriate application of nanotechnology in agricultural activities.

Public opinion and acceptance of nanoagriculture pose a significant barrier. Consumers' apprehensions regarding the safety and environmental effects of nano-enhanced agricultural goods can impact their readiness to embrace and endorse these technologies (Parisi et al., 2015). It is crucial to have effective communication and involvement with the public in order to address these issues and encourage informed decision-making. Implement educational campaigns and outreach programs to increase understanding of the advantages and potential drawbacks of nanoagriculture, promoting trust and transparency among researchers, farmers, and customers (Ranjan et al., 2019).

Further research and development activities are required to fully use the potential of nanoagriculture. Although advancements have been achieved in creating and studying nanomaterials for agricultural use, there are still gaps in understanding their behavior and interactions within intricate agricultural systems (Prasad et al., 2017). Future research should prioritize the development of innovative nanomaterials with improved efficiency, stability, and biocompatibility, while also examining their long-term behavior and impact on the environment (Singh et al., 2021). The combination of nanotechnology with other new technologies like biotechnology, precision agriculture, and smart farming systems can create new and sustainable solutions for agricultural problems.

Partnerships among stakeholders such as researchers, farmers, policymakers, and industry partners are crucial for the effective adoption of nanoagriculture. Interdisciplinary research teams should collaborate to tackle the intricate issues linked to the production and utilization of nanomaterials in agriculture (Lowry et al., 2019). Collaborations can aid in exchanging knowledge, transferring technology, and establishing best practices for the ethical application of nanotechnology in agricultural systems. Collaborations between academia and industry can speed up the process of bringing nano-enhanced agricultural products and technology to market, ensuring global access for farmers (Sastry et al., 2017). Ultimately, nanoagriculture has great potential for tackling the issues of sustainable food production and resource management, but there are substantial obstacles that must be addressed. These factors encompass the requirement for thorough risk evaluations, the establishment of precise regulations and procedures, the encouragement of public approval and involvement, and the progression of research and development endeavors. Nanoagriculture can be advanced as an effective instrument for promoting sustainable and resilient agricultural systems by encouraging partnerships among stakeholders and implementing a responsible, evidence-based strategy to address increasing global concerns.

10. CONCLUSION

Nanoagriculture, the utilization of nanotechnology in agriculture, has arisen as a hopeful method to tackle the urgent issues encountered by contemporary farming systems. This chapter has examined different facets of nanoagriculture, such as its uses, environmental effects, sustainable benefits, and future outlook. Nanoparticles and nanomaterials have shown promise in boosting agricultural productivity, optimizing nutrient utilization, and decreasing the environmental impact of agriculture, as demonstrated in several research and real-life situations. Nanofertilizers and nanopesticides have demonstrated the capacity to enhance crop yields, decrease nutrient losses, and diminish the reliance on toxic agrochemicals. Nanosensors and precision agricultural technologies allow for immediate monitoring of soil conditions and crop health, which helps in making timely interventions and optimizing resources.

The chapter has highlighted the importance of a careful and accountable method to the advancement and execution of nanoagriculture. The possible dangers of using nanomaterials in agriculture, including their toxicity to unintended creatures and the risk of accumulating in the food chain, should not be ignored. Thorough risk evaluations, defined processes, and particular laws are crucial to guarantee the safe and sustainable use of nanotechnology in agricultural systems. The long-lasting benefits of nanoagriculture, such as enhanced resource efficiency, minimized environmental impact, and socioeconomic gains, have been thoroughly examined. Nanoagriculture can help conserve water, reduce greenhouse gas emissions, and minimize soil degradation. It also improves food security and provides economic opportunities for farmers. The chapter has outlined various problems and potential prospects for the progression of nanoagriculture. The requirements consist of conducting additional research to fill knowledge gaps, creating precise regulations and procedures, encouraging public engagement and acceptance, and cultivating collaborations among stakeholders. Nanoagriculture shows great potential for transforming agricultural methods and attaining sustainable food production. We can create inventive solutions for the intricate issues in modern agriculture by combining nanoparticles' distinct features with other advanced technology. Responsible development and implementation of nanoagriculture necessitate a collaborative effort from academics, policymakers, farmers, and the public to guarantee its safe and successful use. Nanoagriculture, through ongoing research, collaboration, and responsible management, can significantly contribute to creating a more sustainable, resilient, and productive agricultural future.

ACKNOWLEDGMENT

TM acknowledge the support by the Strategic Academic Leadership Program of the Southern Federal University ("Priority 2030").

REFERENCES

Abdel-Aziz, H. M. M., Hasaneen, M. N. A., & Omer, A. M. (2016). Nano chitosan-NPK fertilizer enhances the growth and productivity of wheat plants grown in sandy soil. *Spanish Journal of Agricultural Research*, *14*(1), e0902. doi:10.5424/sjar/2016141-8205

Adisa, I. O., Reddy Pullagurala, V. L., Peralta-Videa, J. R., Dimkpa, C. O., Elmer, W. H., Gardea-Torresdey, J. L., & White, J. C. (2019). Recent advances in nano-enabled fertilizers and pesticides: A critical review of mechanisms of action. *Environmental Science. Nano*, *6*(7), 2002–2030. doi:10.1039/C9EN00265K

Aghdam, M. T. B., Mohammadi, H., & Ghorbanpour, M. (2016). Effects of nanoparticulate anatase titanium dioxide on physiological and biochemical performance of Linumusitatissimum (Linaceae) under well-watered and drought stress conditions. *Revista Brasileira de Botânica. Brazilian Journal of Botany*, *39*(1), 139–146. doi:10.1007/s40415-015-0227-x

Amenta, V., Aschberger, K., Arena, M., Bouwmeester, H., Botelho Moniz, F., Brandhoff, P., Gottardo, S., Marvin, H. J. P., Mech, A., Quiros Pesudo, L., Rauscher, H., Schoonjans, R., Vettori, M. V., Weigel, S., & Peters, R. J. (2015). Regulatory aspects of nanotechnology in the agri/feed/food sector in EU and non-EU countries. *Regulatory Toxicology and Pharmacology*, *73*(1), 463–476. doi:10.1016/j.yrtph.2015.06.016 PMID:26169479

Anjum, N. A., Gill, S. S., Duarte, A. C., Pereira, E., & Ahmad, I. (2013). Silver nanoparticles in soil-plant systems. *Journal of Nanoparticle Research*, *15*(9), 1–26. doi:10.1007/s11051-013-1896-7

Anusuya, S., & Banu, K. N. (2016). Silver-chitosan nanoparticles induced biochemical variations of chickpea (Cicer arietinum L.). *Biocatalysis and Agricultural Biotechnology*, *8*, 39–44. doi:10.1016/j.bcab.2016.08.005

Baalousha, M., Cornelis, G., Kuhlbusch, T. A., Lynch, I., Nickel, C., Peijnenburg, W., & Van Den Brink, N. W. (2016). Modeling nanomaterial fate and uptake in the environment: Current knowledge and future trends. *Environmental Science. Nano*, *3*(2), 323–345. doi:10.1039/C5EN00207A

Bundschuh, M., Filser, J., Lüderwald, S., McKee, M. S., Metreveli, G., Schaumann, G. E., Schulz, R., & Wagner, S. (2018). Nanoparticles in the environment: Where do we come from, where do we go to? *Environmental Sciences Europe*, *30*(1), 1–17. doi:10.1186/s12302-018-0132-6 PMID:29456907

Cai, L., Cai, H., Deng, Y., Liu, J., Cui, H., Yu, X., & Huo, Z. (2019). Foliar exposure of Fe3O4 nanoparticles on Oryza sativa L. and its impact on nutrient accumulation, photosynthetic performance and internal molecular response. *Ecotoxicology and Environmental Safety*, *183*, 109587.

Campos, E. V. R., de Oliveira, J. L., Fraceto, L. F., & Singh, B. (2014). Polysaccharides as safer release systems for agrochemicals. *Agronomy for Sustainable Development*, *35*(1), 47–66. doi:10.1007/s13593-014-0263-0

Chaudhry, Q., Scotter, M., Blackburn, J., Ross, B., Boxall, A., Castle, L., Aitken, R., & Watkins, R. (2008). Applications and implications of nanotechnologies for the food sector. *Food Additives and Contaminants*, *25*(3), 241–258. doi:10.1080/02652030701744538 PMID:18311618

Chhipa, H. (2017). Nanofertilizers and nanopesticides for agriculture. *Environmental Chemistry Letters*, *15*(1), 15–22. doi:10.1007/s10311-016-0600-4

Choi, M., & Park, Y. (2014). Phytotoxicity of zinc oxide nanoparticles on the early growth of radish (Raphanus sativus L.) and the detection of nanoparticles in plant tissues. *Journal of Environmental Science and Health. Part C, Environmental Carcinogenesis & Ecotoxicology Reviews*, *32*(4), 375–387.

Cornelis, G., Hund-Rinke, K., Kuhlbusch, T., Van den Brink, N., & Nickel, C. (2014). Fate and bioavailability of engineered nanoparticles in soils: A review. *Critical Reviews in Environmental Science and Technology*, *44*(24), 2720–2764. doi:10.1080/10643389.2013.829767

Dahlberg, J., Berenji, J., Sikora, V., & Latkovic, D. (2011). Assessing sorghum [Sorghum bicolor (L) Moench] germplasm for new traits: food, fuels & unique uses. Maydica, 56(1750), 85-92.

Dasgupta, N., Ranjan, S., Mundekkad, D., Ramalingam, C., Shanker, R., & Kumar, A. (2015). Nanotechnology in agro-food: From field to plate. *Food Research International*, *69*, 381–400. doi:10.1016/j.foodres.2015.01.005

DeRosa, M. C., Monreal, C., Schnitzer, M., Walsh, R., & Sultan, Y. (2010). Nanotechnology in fertilizers. *Nature Nanotechnology*, *5*(2), 91–91. doi:10.1038/nnano.2010.2 PMID:20130583

Deshpande, P., Dapkekar, A., Oak, M. D., Paknikar, K. M., & Rajwade, J. M. (2017). Zinc complexed chitosan/TPP nanoparticles: A promising micronutrient nanocarrier suited for foliar application. *Carbohydrate Polymers*, *165*, 394–401. doi:10.1016/j.carbpol.2017.02.061 PMID:28363565

Dimkpa, C. O., Andrews, J., Fugice, J., Singh, U., Bindraban, P. S., Elmer, W. H., ... White, J. C. (2019). Facile coating of urea with low-dose ZnO nanoparticles promotes wheat performance and enhances Zn uptake under drought stress. *Frontiers in Plant Science*, *10*, 1645. PMID:32174943

Duncan, T. V. (2011). Applications of nanotechnology in food packaging and food safety: Barrier materials, antimicrobials and sensors. *Journal of Colloid and Interface Science*, *363*(1), 1–24. doi:10.1016/j.jcis.2011.07.017 PMID:21824625

Dwivedi, S., Saquib, Q., Al-Khedhairy, A. A., & Musarrat, J. (2019). Nanoagriculture: A new frontier for sustainable agriculture. In Nanoecotoxicology (pp. 1-22). Elsevier.

Fang, Y., & Ramasamy, R. P. (2015). Current and prospective methods for plant disease detection. *Biosensors (Basel)*, *5*(3), 537–561. doi:10.3390/bios5030537 PMID:26287253

FAO. (2017). *The future of food and agriculture – Trends and challenges*. Food and Agriculture Organization of the United Nations.

Forim, M. R., Costa, E. S., da Silva, M. F. G. F., Fernandes, J. B., Mondego, J. M., & Boiça, A. L. Junior. (2013). Development of a new method to prepare nano-/microparticles loaded with extracts of Azadirachta indica, their characterization and use in controlling Plutella xylostella. *Journal of Agricultural and Food Chemistry*, *61*(38), 9131–9139. doi:10.1021/jf403187y PMID:23991702

Fraceto, L. F., Grillo, R., de Medeiros, G. A., Scognamiglio, V., Rea, G., & Bartolucci, C. (2016). Nanotechnology in agriculture: Which innovation potential does it have? *Frontiers in Environmental Science*, *4*, 20. doi:10.3389/fenvs.2016.00020

Gardea-Torresdey, J. L., Rico, C. M., & White, J. C. (2014). Trophic transfer, transformation, and impact of engineered nanomaterials in terrestrial environments. *Environmental Science & Technology*, *48*(5), 2526–2540. doi:10.1021/es4050665 PMID:24499408

Giraldo, J. P., Wu, H., Newkirk, G. M., & Kruss, S. (2019). Nanobiotechnology approaches for engineering smart plant sensors. *Nature Nanotechnology*, *14*(6), 541–553. doi:10.1038/s41565-019-0470-6 PMID:31168083

Grillo, R., Pereira, A. E., Nishisaka, C. S., de Lima, R., Oehlke, K., Greiner, R., & Fraceto, L. F. (2014). Chitosan/tripolyphosphate nanoparticles loaded with paraquat herbicide: An environmentally safer alternative for weed control. *Journal of Hazardous Materials*, *278*, 163–171. doi:10.1016/j.jhazmat.2014.05.079 PMID:24968252

Gupta, R., Anand, G., Gaur, R., & Yadav, D. (2019). Plant-microbiome interactions for sustainable agriculture: A review. *Archives of Microbiology*, *201*(10), 1361–1375. PMID:31332473

Hofmann, T., Lowry, G. V., Lynch, I., Ghosh, S., Owen, R., Boyes, W. K., ... Unrine, J. (2020). Harmonizing across environmental nanomaterial testing media for increased comparability of nanomaterial datasets. *Environmental Science. Nano*, *7*(11), 3491–3499.

Holden, P. A., Klaessig, F., Turco, R. F., Priester, J. H., Rico, C. M., Avila-Arias, H., Mortimer, M., Pacpaco, K., & Gardea-Torresdey, J. L. (2014). Evaluation of exposure concentrations used in assessing manufactured nanomaterial environmental hazards: Are they relevant? *Environmental Science & Technology*, *48*(18), 10541–10551. doi:10.1021/es502440s PMID:25158225

Iavicoli, I., Leso, V., Beezhold, D. H., & Shvedova, A. A. (2017). Nanotechnology in agriculture: Opportunities, toxicological implications, and occupational risks. *Toxicology and Applied Pharmacology*, *329*, 96–111. doi:10.1016/j.taap.2017.05.025 PMID:28554660

Iyer, J. K., Dickey, A., Rouhani, P., Kaul, A., Govindaraju, N., Singh, R. N., & Kaul, R. (2020). Nanoparticles of pure and doped cerium oxide as promising catalysts for the remediation of water pollution: A review. *Environmental Chemistry Letters*, *18*(6), 1917–1948.

Jha, S., Ghosal, M., & Gupta, S. K. (2019). Nanosensors for precision agriculture. In *Nanotechnology for agriculture* (pp. 159–183). Springer.

Jiang, X., Sun, D., Zhang, G., He, N., Gao, Q., Yue, T., ... Li, J. (2018). A colorimetric biosensor based on single-enzyme nanoparticles for detecting the plant pathogen Xanthomonas oryzae pv. oryzae. *Sensors and Actuators. B, Chemical*, *273*, 1077–1084.

Judy, J. D., Unrine, J. M., & Bertsch, P. M. (2011). Evidence for biomagnification of gold nanoparticles within a terrestrial food chain. *Environmental Science & Technology*, *45*(2), 776–781. doi:10.1021/es103031a PMID:21128683

Kah, M., Kookana, R. S., Gogos, A., & Bucheli, T. D. (2018). A critical evaluation of nanopesticides and nanofertilizers against their conventional analogues. *Nature Nanotechnology*, *13*(8), 677–684. doi:10.1038/s41565-018-0131-1 PMID:29736032

Kah, M., Tufenkji, N., & White, J. C. (2019). Nano-enabled strategies to enhance crop nutrition and protection. *Nature Nanotechnology*, *14*(6), 532–540. doi:10.1038/s41565-019-0439-5 PMID:31168071

Kannan, S., & Jasim, B. (2019). Copper oxide nanoparticles as a novel tool to control bacterial leaf blight of rice caused by Xanthomonas oryzae pv. oryzae. *Pesticide Biochemistry and Physiology*, *156*, 1–5.

Kashyap, P. L., Xiang, X., & Heiden, P. (2015). Chitosan nanoparticle based delivery systems for sustainable agriculture. *International Journal of Biological Macromolecules*, *77*, 36–51. doi:10.1016/j.ijbiomac.2015.02.039 PMID:25748851

Khater, M., de la Escosura-Muñiz, A., & Merkoçi, A. (2017). Biosensors for plant pathogen detection. *Biosensors & Bioelectronics*, *93*, 72–86. doi:10.1016/j.bios.2016.09.091 PMID:27818053

Khati, P., Chaudhary, P., & Gangola, S. (2019). Nanomaterials in soil: Environmental applications and ecological implications. In *Nanomaterials in plants, algae, and microorganisms* (pp. 83–105). Academic Press.

Khush, G. S. (2005). What it will take to feed 5.0 billion rice consumers in 2030. *Plant Molecular Biology*, *59*(1), 1–6. doi:10.1007/s11103-005-2159-5 PMID:16217597

Kookana, R. S., Boxall, A. B., Reeves, P. T., Ashauer, R., Beulke, S., Chaudhry, Q., Cornelis, G., Fernandes, T. F., Gan, J., Kah, M., Lynch, I., Ranville, J., Sinclair, C., Spurgeon, D., Tiede, K., & van den Brink, P. J. (2014). Nanopesticides: Guiding principles for regulatory evaluation of environmental risks. *Journal of Agricultural and Food Chemistry*, *62*(19), 4227–4240. doi:10.1021/jf500232f PMID:24754346

Kumar, S., Nehra, M., Dilbaghi, N., Marrazza, G., Hassan, A. A., & Kim, K. H. (2019). Nano-based smart pesticide formulations: Emerging opportunities for agriculture. *Journal of Controlled Release*, *294*, 131–153. doi:10.1016/j.jconrel.2018.12.012 PMID:30552953

Kumaraswamy, R. V., Kumari, S., Choudhary, R. C., Pal, A., Raliya, R., Biswas, P., & Saharan, V. (2018). Engineered chitosan based nanomaterials: Bioactivities, mechanisms and perspectives in plant protection and growth. *International Journal of Biological Macromolecules*, *113*, 494–506. doi:10.1016/j.ijbiomac.2018.02.130 PMID:29481952

Li, Y., Zhao, J., Gao, Y., Huang, Y., & Zhao, X. (2018). MnO2 nanozymes for the degradation of polycyclic aromatic hydrocarbons (PAHs) in contaminated soil. *Environmental Science. Nano*, *5*(10), 2369–2378.

Liu, Y., Wu, H., Li, M., Yin, J. J., & Nie, Z. (2019). Oxidase-like activity of gold nanoparticles and their gold staining enhanced ELISA application. *Journal of Materials Chemistry. B*, *7*(41), 6393–6399.

Lowry, G. V., Avellan, A., & Gilbertson, L. M. (2019). Opportunities and challenges for nanotechnology in the agri-tech revolution. *Nature Nanotechnology*, *14*(6), 517–522. doi:10.1038/s41565-019-0461-7 PMID:31168073

Lowry, G. V., Gregory, K. B., Apte, S. C., & Lead, J. R. (2012). Transformations of nanomaterials in the environment. *Environmental Science & Technology*, *46*(13), 6893–6899. doi:10.1021/es300839e PMID:22582927

Malerba, M., & Cerana, R. (2018). Recent advances of chitosan applications in plants. *Polymers*, *10*(2), 118. doi:10.3390/polym10020118 PMID:30966154

Manjunatha, S. B., Biradar, D. P., & Aladakatti, Y. R. (2016). Nanotechnology and its applications in agriculture: A review. *Journal of Farm Sciences*, *29*(1), 1–13.

Mehmood, A., Murtaza, G., Bhatti, T. M., & Kausar, R. (2017). Phytoextracts synthesis of silver nanoparticles and their impact on microbial and plant growth. *IET Nanobiotechnology / IET*, *11*(7), 830–835.

Mishra, S., Singh, B. R., Singh, A., Keswani, C., Naqvi, A. H., & Singh, H. B. (2017). Biofabricated silver nanoparticles act as a strong fungicide against Bipolaris sorokiniana causing spot blotch disease in wheat. *PLoS One*, *12*(5), e0175750. PMID:28542577

Mukhopadhyay, S. S. (2014). Nanotechnology in agriculture: Prospects and constraints. *Nanotechnology, Science and Applications*, *7*, 63. doi:10.2147/NSA.S39409 PMID:25187699

Munyi, M., Wainaina, I., & Njage, E. (2020). Nanotechnology applications in crop production and food systems: A review. *Journal of Applied Science & Environmental Management*, *24*(4), 655–667.

Mura, S., Greppi, G., Irudayaraj, J., & Seetharaman, K. (2016). Nanosensors for food and agriculture. In *Nanoscience in Food and Agriculture 1* (pp. 259–284). Springer.

Nair, R., Varghese, S. H., Nair, B. G., Maekawa, T., Yoshida, Y., & Kumar, D. S. (2010). Nanoparticulate material delivery to plants. *Plant Science*, *179*(3), 154–163. doi:10.1016/j.plantsci.2010.04.012

Nkurunziza, D., Niyigena, C., & Mugenzi, I. (2019). The effect of nanofertilizers on growth and yield of soybean (Glycine max L. Merr). *Journal of Nanomaterials*.

Nuruzzaman, M., Rahman, M. M., Liu, Y., & Naidu, R. (2016). Nanoencapsulation, nano-guard for pesticides: A new window for safe application. *Journal of Agricultural and Food Chemistry*, *64*(7), 1447–1483. doi:10.1021/acs.jafc.5b05214 PMID:26730488

Nuss, E. T., & Tanumihardjo, S. A. (2010). Maize: A paramount staple crop in the context of global nutrition. *Comprehensive Reviews in Food Science and Food Safety*, *9*(4), 417–436. doi:10.1111/j.1541-4337.2010.00117.x PMID:33467836

Pandey, A. C., Sanjay, S. S., & Yadav, R. S. (2018). Fe3O4 nanoparticles-based magnetic solid-phase extraction coupled with high-performance liquid chromatography for organophosphorus pesticides monitoring in agricultural and environmental water samples. *Environmental Science and Pollution Research International*, *27*(29), 36679–36691.

Parisi, C., Vigani, M., & Rodríguez-Cerezo, E. (2015). Agricultural nanotechnologies: What are the current possibilities? *Nano Today*, *10*(2), 124–127. doi:10.1016/j.nantod.2014.09.009

Pradhan, N., Singh, S., Ojha, N., Shrivastava, A., Barla, A., Rai, V., & Bose, S. (2015). Facets of nanotechnology as seen in food processing, packaging, and preservation industry. *BioMed Research International*, *2015*, 2015. doi:10.1155/2015/365672 PMID:26613082

Prasad, R., Bhattacharyya, A., & Nguyen, Q. D. (2017). Nanotechnology in sustainable agriculture: Recent developments, challenges, and perspectives. *Frontiers in Microbiology*, *8*, 1014. doi:10.3389/fmicb.2017.01014 PMID:28676790

Qureshi, A., Singh, D. K., & Dwivedi, S. (2018). Nano-fertilizers: A novel way for enhancing nutrient use efficiency and crop productivity. *International Journal of Current Microbiology and Applied Sciences*, *7*(2), 3325–3335. doi:10.20546/ijcmas.2018.702.398

Ragaei, M., & Sabry, A. K. H. (2014). Nanotechnology for insect pest control. *International Journal of Science, Environment and Technology*, *3*(2), 528–545.

Rai, V., Acharya, S., & Dey, N. (2016). Implications of nanobiosensors in agriculture. *Journal of Biomaterials and Nanobiotechnology*, *3*(2), 315–324. doi:10.4236/jbnb.2012.322039

Rai, V., Acharya, S., & Dey, N. (2020). Implications of nanobiosensors in agriculture. *Journal of Biomaterials and Nanobiotechnology*, *11*(01), 1.

Rajput, P., Kumar, P., Priya, A., Kumari, S., Shiade, S. R. G., Rajput, V. D., Fathi, A., Pradhan, A., Sarfraz, R., Sushkova, S., Mandzhieva, S., Minkina, T., Soldatov, A., Wong, M. H., & Rensing, C. (2024). Nanomaterials and biochar mediated remediation of emerging contaminants. *The Science of the Total Environment*, *916*, 170064. doi:10.1016/j.scitotenv.2024.170064 PMID:38242481

Rajput, V. D., Singh, A., Minkina, T., Rawat, S., Mandzhieva, S., Sushkova, S., Shuvaeva, V., Nazarenko, O., Rajput, P., Komariah, Verma, K. K., Singh, A. K., Rao, M., & Upadhyay, S. K. (2021). Nano-Enabled Products: Challenges and Opportunities for Sustainable Agriculture. *Plants*, *10*(12), 2727. doi:10.3390/plants10122727 PMID:34961197

Rameshaiah, G. N., Pallavi, J., & Shabnam, S. (2015). Nano fertilizers and nano sensors–an attempt for developing smart agriculture. *International Journal of Engineering Research and General Science*, *3*(1), 314–320.

Rani, K., & Yadav, S. (2020). Nano-encapsulation in agriculture: Emerging opportunities and challenges. *Journal of Plant Biochemistry and Biotechnology*, *29*(4), 606–621.

Ranjan, S., Dasgupta, N., & Lichtfouse, E. (2019). *Nanoscience in food and agriculture*. Springer International Publishing.

Rizwan, M., Ali, S., Qayyum, M. F., Ok, Y. S., Adrees, M., Ibrahim, M., Zia-ur-Rehman, M., Farid, M., & Abbas, F. (2017). Effect of metal and metal oxide nanoparticles on growth and physiology of globally important food crops: A critical review. *Journal of Hazardous Materials*, *322*, 2–16. doi:10.1016/j.jhazmat.2016.05.061 PMID:27267650

Rodrigues, S. M., Trindade, T., Duarte, A. C., Pereira, E., Koopmans, G. F., & Römkens, P. F. (2016). A framework to measure the availability of engineered nanoparticles in soils: Trends in soil tests and analytical tools. *Trends in Analytical Chemistry*, *75*, 129–140. doi:10.1016/j.trac.2015.07.003

Sachan, D. S., Reddy, K. J., Sulochna, Saini, Y., Rai, A. K., Singh, O., & Laxman, T. (2023). Assessing grain yield and achieving enhanced quality in maize by next generation fertilizer: A review. *International Journal of Environment and Climate Change*, *13*(8), 626–637. doi:10.9734/ijecc/2023/v13i81991

Sagadevan, S., Periasamy, M., Ling, T. C., & Dhanalekshmi, K. I. (2019). Application of nanosensors in smart agriculture. In *Nanosensors for smart manufacturing* (pp. 247–268). Elsevier.

Saikanth, D. R. K., Singh, B. V., Rai, A. K., Kumar, U. S., Surender, Yadav, B., & Singh, O. (2023). Biochar implementation in rice paddies for addressing greenhouse gas emissions and nutrient loss: A review. *International Journal of Plant and Soil Science*, *35*(18), 610–623. doi:10.9734/ijpss/2023/v35i183326

Sastry, R. K., Rao, N. H., & Cahoon, R. (2017). Can nanotechnology provide the innovations for a second green revolution in Indian agriculture? In *Nanotechnology in agriculture: Prospects and constraints* (pp. 3–21). Springer.

Shi, C., Zhu, N., Cao, Y., & Wu, P. (2017). Biosynthesis of gold nanoparticles assisted by the intracellular protein extract of Pycnoporus sanguineus and its catalysis in degradation of 4-nitroaniline. *Nanoscale Research Letters*, *12*(1), 1–9. PMID:28050875

Shi, J., Tian, X., Liu, F., Luo, Y., & Lan, S. (2016). An electrochemical sensor based on Pt nanoparticles and molecularly imprinted polymer for sensitive detection of methyl parathion. *Sensors and Actuators. B, Chemical*, *227*, 583–590. doi:10.1016/j.snb.2015.12.092

Shipley, H. J., Engates, K. E., & Grover, V. A. (2010). Removal of Pb (II), Cd (II), Cu (II), and Zn (II) by hematite nanoparticles: Effect of sorbent concentration, pH, temperature, and exhaustion. *Environmental Science and Pollution Research International*, *20*(3), 1727–1736. doi:10.1007/s11356-012-0984-z PMID:22645012

Shukla, P. K., Misra, P., & Kole, C. (2019). *Nanotechnology for agriculture*. Elsevier.

Singh, A., Rajput, V. D., Pandey, D., Sharma, R., Ghazaryan, K., & Minkina, T. (2023c). Nano Zinc-Enabled Strategies in Crops for Combatting Zinc Malnutrition in Human Health. *Frontiers in Bioscience (Landmark Edition)*, *28*(8), 158. doi:10.31083/j.fbl2808158 PMID:37664935

Singh, A., Sharma, R., Rawat, S., Singh, A. K., Rajput, V. D., Fedorov, Y., . . . Chaplygin, V. (2022). Nanomaterial-plant interaction: Views on the pros and cons. *Toxicity of Nanoparticles in Plants*, 47–68. https://doi.org/ doi:10.1016/B978-0-323-90774-3.00015-5

Singh, A., & Singh, N. B. (2019). Application of nanotechnology in agriculture. In *Nanomaterials in biofuels research* (pp. 151–187). Springer.

Singh, J., Kumar, S., Alok, A., Upadhyay, S. K., Rawat, M., Tsang, D. C. W., ... Bolan, N. (2018). The potential of green synthesized zinc oxide nanoparticles as a nutrient source for plant growth. *Journal of Cleaner Production*, *214*, 1061–1070. doi:10.1016/j.jclepro.2019.01.018

Singh, O. (2023). Nanotechnology for sustainability and food security in agriculture: A nanopriming long story in short. In A. Singh, V. Rajput, K. Ghazaryan, S. Gupta, & T. Minkina (Eds.), *Nanopriming Approach to Sustainable Agriculture* (pp. 315–339). IGI Global. doi:10.4018/978-1-6684-7232-3.ch014

Singh, O., Shahi, U.P., Dhyani, B.P., Kumar, S., Vivek, S., & Sengar, R.S., Shivangi, & Singh, A. (2023). Effect of zinc oxide nanoparticles application on growth and yield of basmati rice (Oryza sativa) in alkaline soil. *Indian Journal of Agronomy*, *68*(3), 1–6.

Singh, R. P., Handa, R., & Manchanda, G. (2021). Nanoparticles in sustainable agriculture: An emerging opportunity. *Journal of Controlled Release*, *329*, 1234–1248. doi:10.1016/j.jconrel.2020.10.051 PMID:33122001

Singh, S., Singh, O., & Shahi, U. P. (2023). Nanotechnology: A New Innovation in Biofortification and Food Safety. In A. Singh, V. Rajput, K. Ghazaryan, S. Gupta, & T. Minkina (Eds.), *Nanopriming Approach to Sustainable Agriculture* (pp. 340–351). IGI Global. doi:10.4018/978-1-6684-7232-3.ch015

Singh, Y. K., Singh, B. V., Katiyar, D., Saikanth, D. R. K., Kumar, K., Singh, O., Verma, S., & Kumar, P. (2023). Efficacy of nano fertilizers on yield, attributes and economics of wheat. *International Journal of Environment and Climate Change*, *13*(7), 291–297. doi:10.9734/ijecc/2023/v13i71879

Subramanian, K. S., Manikandan, A., Thirunavukkarasu, M., & Rahale, C. S. (2015). Nano-fertilizers for balanced crop nutrition. In *Nanotechnologies in food and agriculture* (pp. 69–80). Springer. doi:10.1007/978-3-319-14024-7_3

Tarigholizadeh, S., Sushkova, S., Rajput, V. D., Ranjan, A., Arora, J., Dudnikova, T., Barbashev, A., Mandzhieva, S., Minkina, T., & Wong, M. H. (2023). Transfer and Degradation of PAHs in the Soil–Plant System: A Review. *Journal of Agricultural and Food Chemistry*, *72*(1), 46–64. doi:10.1021/acs.jafc.3c05589 PMID:38108272

Tripathi, D. K., Shweta, Singh, S., Singh, S., Pandey, R., Singh, V. P., Sharma, N. C., Prasad, S. M., Dubey, N. K., & Chauhan, D. K. (2017). An overview on manufactured nanoparticles in plants: Uptake, translocation, accumulation and phytotoxicity. *Plant Physiology and Biochemistry*, *110*, 2–12. doi:10.1016/j.plaphy.2016.07.030 PMID:27601425

Wei, H., & Wang, E. (2013). Nanomaterials with enzyme-like characteristics (nanozymes): Next-generation artificial enzymes. *Chemical Society Reviews*, *42*(14), 6060–6093. doi:10.1039/c3cs35486e PMID:23740388

Wu, H., Shabala, L., Shabala, S., & Giraldo, J. P. (2017). Enzyme nanoparticles enhance plant growth by delivering nutrients and protecting against abiotic stresses. *ACS Nano*, *11*(10), 9650–9661. PMID:28872828

Wu, L., Liu, M., & Liang, R. (2016). Preparation and properties of a double-coated slow-release NPK compound fertilizer with superabsorbent and water-retention. *Bioresource Technology*, *99*(3), 547–554. doi:10.1016/j.biortech.2006.12.027 PMID:17320380

Xu, L., Xing, H., Huang, Y., Dai, X., Zhang, W., & Chen, Y. (2020). Biochar-based fertilizer: A green method for enhancing nitrogen retention and reducing nitrogen leaching in vegetable soils. *Chemosphere*, *255*, 126978.

Chapter 15
Plant Secondary Metabolites:
Their Impact on Human Health Within the Context of Harnessing NanoOmics and Nanozymes

Sudeshna Mandal
https://orcid.org/0009-0000-4958-6696
Visva-Bharati, West Bengal, India

Dhitri Borah
https://orcid.org/0000-0001-5423-7183
Biswanath College, Assam, India

Dipjyoti Boruah
Dibrugarh University, Assam, India

Neelesh Kumar
https://orcid.org/0000-0003-2867-4755
Rani Lakshmi Bai Central Agricultural University, India

Avanish Kumar Shrivastav
Maitreyi College, University of Delhi, India

Amarjeet Singh
https://orcid.org/0000-0003-2814-6039
Sri Venkateswara College, University of Delhi, India

ABSTRACT

Secondary metabolites produced by plants are particular to their species or genus and do not contribute to the basic processes of growth, development, and reproduction. These compounds play a key role as feed deterrents against herbivores, protectants from microbial infection, abiotic stresses, UV-protectants, pollinators attractants, antioxidants, and signalling molecules. Based on their chemical structures and production routes, secondary metabolites may be further classified across several classes. Alkaloids, terpenes, and phenolic chemicals make up the largest and most common group. The importance of secondary metabolites found in plants for human nutrition has grown to be an important interesting research area. Secondary metabolites are used by people as medicine to cure illnesses and infections, flavourings, and pesticides in the recent past. Therefore, an overview of plants is provided in this chapter including secondary metabolites and how these compounds are relevant to human health.

DOI: 10.4018/979-8-3693-1890-4.ch015

INTRODUCTION

Plant produces an immense group of organic compounds referred to as plant secondary metabolites (PSM). Plants have to deal with a number of constraints while interacting with environment including dispersion of seed, physical and chemical defences, attack of pathogens, herbivores induced injuries, pollination, food deprivation, and protection against UV radiation. As a result, plants have developed secondary metabolic pathways that help generate various compounds. These secondary metabolites also called phytochemicals possess a variety of structural attributes and are dispersed to restricted species or genera within the plant kingdom (Harborne, 1993). Secondary metabolites' involvement in therapeutics as prospective natural medication sources and also as a dietary precautions component has recently become a crucial component of studies on human nutrition. Earlier secondary metabolites were ignored in aspects of human nutrition as anti-nutritional compounds. However, the discovery of health-promoting characters of secondary metabolites that reached from the inducement of antioxidant process, condensed risk of cancers, diabetes (type-2) and heart diseases (Crozier *et al.*, 2008; Kumar *et al.*, 2019). Secondary metabolites are also gaining importance because of their advantageous uses as oils, fibers, waxes, artificial flavours, dyes, perfumes, drugs, sources of antibiotics, new natural medicines and pesticides (Croteau *et al.*, 2000; Kumar *et al.*, 2022a).

Polyphenols, which include flavonoids, anthocyanins, hydoxycinnamates and the essential antioxidants chorogenate and coumarate are one of the biggest categories of plant secondary metabolites with over 8000 different structures and varied functionalities (Tohge and Fernie, 2017). Flavonoids are known to possess numerous biological properties and can behave as potent antioxidants to protect the human body against oxidative stress. Phenolic compounds and carotenoids are the UV-absorbing photoprotective secondary metabolites. The buildup of these compounds results in modifications to the content of plant-based foods and an increase in secondary plant metabolites that improve health (Kumar *et al.*, 2021a; Sharma *et al.*, 2022).

PLANT SECONDARY METABOLITES

Plant secondary metabolites (PSM) in plants are the naturally occurring chemical compounds derived as products of primary metabolic pathways like photosynthesis, the krebs or citric acid cycle and glycolysis etc. PSM also known as phytochemicals, are clinically evident, naturally produced chemical substances found in plants that are beneficial to human health (Hasler and Blumberg, 1999). These chemical compounds tend to protect the plants from pathogens, insects and resistance to microbial actions and also prevent serious leaf damage from light thus conferring survival in stressful environmental conditions. The scientific foundation for using plants as traditional remedies comes from these biological effects of secondary metabolites. Methods of analysis like chromatography in recent times permit the restoration of many molecules (Hussein and Anssary, 2018). More than 50,000 secondary metabolites have been identified in the plant kingdom and current drugs rely on PSM for actions (Teoh, 2015). These are used as therapeutics in human health as a treatment for many diseases. The generalized classes of plant secondary metabolites of therapeutic importance include saponins, alkaloids, phenolics, terpenes, lipids and carbohydrates. However, many PSMs are still not known, around 150 PSMs have been reported in detail (Morrison and Hark, 1996). PSM are present in various parts of the plants such as stems, roots seeds, leaves and fruits (Costa *et al.*, 1999). The outer layers of plant tissues also acquire some PSM, most especially the pigment molecules. Depending on the growing condition, processing and cooking, the quantity of PSM varies from plant to plant (King and Young, 1999). PSM are not necessary nutrients to humans for survival but these substances are known to possess biological characteristics like antimicrobial, immunostimulatory, antioxidant, anticancer, detoxification enzymes modulation, hormone metabolism and reduction in platelet aggregation, which prevent humans from various diseases (Sharma *et al.*, 2019).

Classes of Plant Secondary Metabolites

The primary metabolites of plants are the substances that give rise to the primary functions like respiration, photosynthesis and translocation. Secondary metabolites are the products obtained from the biosynthetic modifications like hydroxylation, methylation and glycosylation and are derived from primary metabolites. Compared to primary metabolites, PSM have more complex side chains and a more complicated makeup structurally (Ehrlich and Holdren, 1971; Figueiredo *et al.*, 2008; Semwal et al., 2023). Based on biosynthetic pathways, PSM are divided into three main categories: nitrogen-containing chemicals, phenolic compounds, terpenes, and steroids (Figure 1).

Figure 1. Plant secondary metabolite types (Source: Twaij and Hasan, 2022)

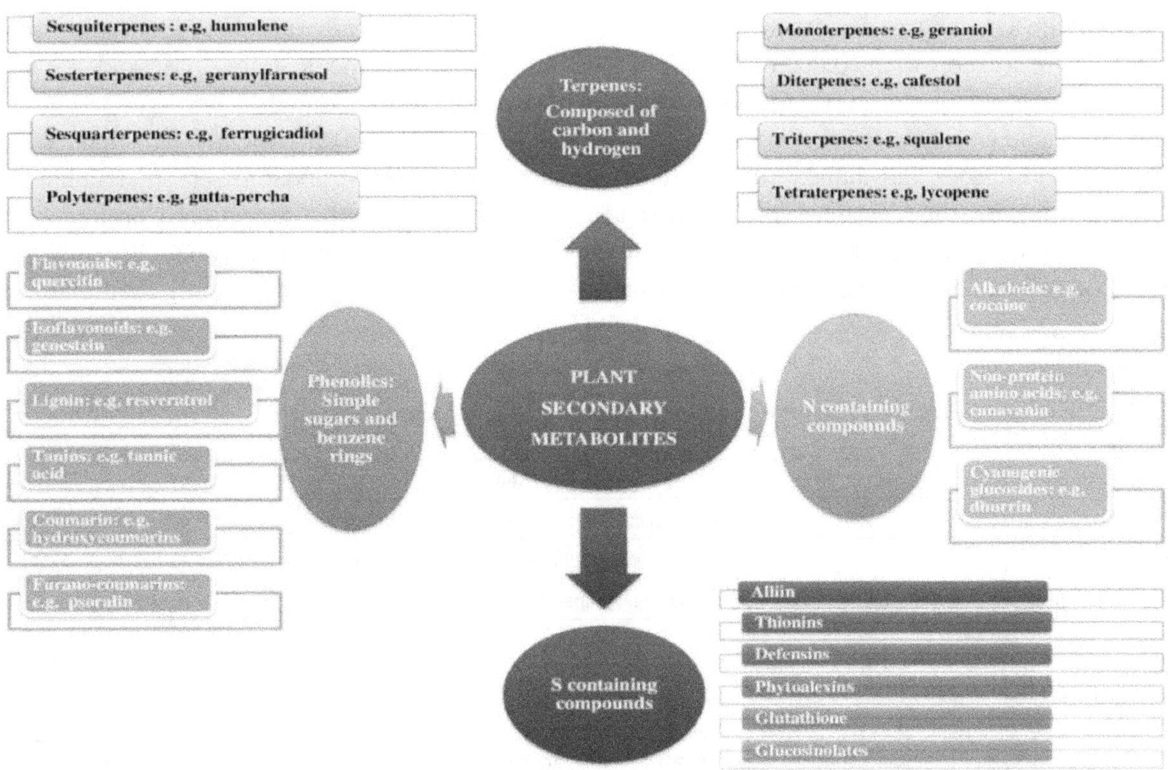

Phenolics

Phenolics are the largest group of plant secondary metabolites containing one or more phenol groups and are frequently found conjugated to organic acids and carbohydrates. Pharmacologically phenolics are valued for anti-inflammatory and antihepatotoxic properties and were the first antiseptic used in surgery. They can be single aromatic-ringed or complex tannins and derived polyphenols. These compounds are categorized as either non-flavonoids or flavonoids.

Flavonoids

Flavonoids are polyphenolswith15-carbon-containing compounds and comprised of two aromatic rings interlinked by 3-carbonresulting in the formation of a third ring. The main subclasses of flavonoids are flavones, flavonol, flavonone, anthocyanidin, Isoflavonone, flavanol, chalcone and flavone. A large concentration of flavonoids is found in fruits and the epidermis of leaves. Such compounds have an essential role in disease resistance, activation of nitrogen-fixing nodules, pigmentation and UV protection (Koes *et al.*, 1994). Human health benefits of flavonoids include anti-allergic, cardioprotective, antidiabetic, anti-inflammatory, anti-oxidative activity, free radical scavenging capacity and anti-cancer activity (Egertand Rimbach, 2011). Flavonoids present in different sources of food proved to be preventive against cardiovascular diseases (Kozłowska and Wegierek, 2014) (Table 16.1). Because flavonoids have strong antioxidant properties and shield the body from free radical oxidation, they slow down the aging process.

Nitrogen-Containing Compounds

Alkaloids are a major group of organic compounds having heterocyclic ring structure of nitrogen atom or atoms. There are more than 20,000 molecules of alkaloids as stimulant medications, insecticides, narcotics, medicines and also in several metabolic reactions. A small dose of alkaloids is effective biologically on humans and animals. Various biological activities of alkaloids on humans include anti-malarial, antioxidant, anti-cancer, anti-diabetic, anti-inflammatory, anti-hypertensive andanti-microbial. In humans, alkaloids stimulate the central nervous system and act on the permeability of membranes, proteins, RNA and DNA. Based on ring structure alkaloids are classified into indole-containing alkaloids (reserpine, vincristine), pyridine and piperidine alkaloids (anabasine, lobeline, canine), pyrrole and pyrrolidine (coca

Table 1. Various sources of flavonoids (Source: Shen et al., 2022).

Flavonoids	Sources
Flavonol	Peaches, pecan nuts, apricot, plum, apple, red wine, cocoa
Flavonone	Orange, citrus fruits, grapefruit juice
Anthocyanidin	purple cabbage, grapes, berries, cherries, red-wine, eggplant
Isoflavonone	Legumes, chickpeas, pueraria, tofu, soy
Flavanol	Apples, cocoa, pecan nuts, cherry, plum, apricot, red wine, tea, peaches
Chalcone	Soybean, citrus, apple, mulberry, ginger
Flavone	Vegetable oils, sweet pepper, wheat sprout, grain mixtures, parsley, carrot, pimiento, celery, chamomile

species, hygrine), isoquinoline alkaloids (nicotine, morphine), quinoline alkaloids (quinidine, quinine) and pyrrolizidine (echimidine) (Badri, 2019; Kumar *et al.*, 2022b). Based on the origin and presence of nitrogen, alkaloids are true alkaloids (morphine and nicotine), proto alkaloids (adrenaline, ephedrine), and pseudo alkaloids (theobromine, caffeine) (Table 16.2).

Sulfur Containing Compounds

Secondary metabolites containing sulfur are a limited category of compounds derived from plants. It includes glucosinolates, glutathione, defensins, and alliin that are endowed with health-promoting activities and act as modulators in several oxidative stress-related diseases in human health releasing reactive oxygen species and turning on detoxifying enzymes in inflammation and also inducing apoptosis. There are different sources of sulfur-containing compounds (Table 16.3). These compounds are helpful for immunological progressions, tumors, heart diseases (breast, esophageal, lung, prostate, colorectal, gastric and skin cancers) and heart diseases. The glucosinolates attained more interest due to their anti-diabetic properties and anti-carcinogenic features. It also lowers the possibility of UV-induced skin lesions. Alliin is an organosulfur compound with the potential of scavenging free radicals, and antioxidant properties and elevates immunity (Quintero *et al.*, 2013).

Terpenes

Terpenes are the most abundant and diversified category of plant secondary metabolites. Terpenes are the compounds with 5 carbon containing isoprene units which organized in several forms. Hemiterpenes are characterized by one isoprene unit, monoterpenes by two, sesquiterpenes by three, diterpenes by four, sesterterpenes by five, triterpenes by six, and tetraterpenes by eight units (Table 16.4.). Terpenoids

Table 2. Various sources of alkaloids (Source: Rajput et al., 2021).

Classes of Alkaloids	Alkaloids	Sources
Indole containing	Reserpine, vincristine	Banana, nuts, milk, chicken, avocado
Piperidine	Anabasine, lobeline, canine	Spices includes white pepper, black pepper
Pyrrolidine	Seneciphylline, echimidine, retronecine	Eggs, soup (borage), milk, honey, salad
Purine alkaloids	Caffeine, theobromine, theophylline	Tea, coffee
Beta carboline alkaloids	Tetrahydrobeta carboline, norharman, harman alkaloid	Tomatoes, kiwi, banana, pineapple and their juices
Steroids alkaloids	Tomatine, solasonine, solanines, chacotriose	Egg, plant, capsicum and potatoes

Table 3. Sources of sulfur-containing compounds

Sulfur Containing Compounds	Sources
Glucosinolates	Brassica crops, cruciferous plants
Glutathione	Garlic, onions, avocados, watermelon
Defensins,	Wheat, barley
Alliin	*Allium sativum*, leaves and bulbs of these alliaceous vegetables

protect from and treat chronic disease like various tumors, heart disease and cancer. Lactone, a sesquiterpene synthesized in shoots and roots of the plant *Artemisia annua*, is beneficial in treating malaria as a drug (Chadwick *et al.*, 2013).

BIOLOGICAL ACTIVITY OF PLANT SECONDARY METABOLITES

The efficiency and mode of action of PSM in preventing illness and advancing health requires extensive investigation. Such studies embodied identifying and separating the constituent chemicals and the potential of biological activity both in vivo as well as in vitro investigations in experimental organisms through clinical and epidemiological studies in humans. Some studies reported that phytochemicals normalize blood pressure, enhance arterial elasticity, lower the cholesterol synthesis or absorption and inhibit low-density lipoprotein cholesterol oxidation (Mathai, 2000; Jangpangi *et al.*, 2023). Several drugs available today in medicine are of plant origin and have beneficial and therapeutic impacts on health of people as a result of many compounds found in many plants including phenolics, flavonoids, terpenoids, alkaloids and many others. It was estimated that more than 80% of the population globally depends on medicines of plant origin to look over the primary healthcare requirements (WHO, 2013). However, the biological properties of PSM remained unfamiliar due to their low concentrations in plants; also, they were more known as metabolic waste products. Earlier PSM also studied more as anti-nutritional factor to animals, but in recent times PSM has emerged more for its ecological and beneficial role to humans and other organisms. The biological activities of different plants and their PSM are discussed below.

Antioxidant Activity

According to several reports, antioxidants guard against the oxidative harm produced by free radicals. Every organism has defense mechanisms to scavenge and stabilize free radicals, but oxidative stress arises when the rate at which free radicals are generated exceeds the capacity of these defense systems (Singh *et al.*, 2023). This harmful process may harm cell components including proteins, lipids and DNA. The defense and management of disorders involving oxidative stress makes use of several drug formulations based on antioxidants. As a consequence of limits on synthetic antioxidants due to their carcinogenicity, the search for naturally occurring antioxidants to replace manufactured ones in food, cosmetics, and medicinal materials has become more and more common. Many medicinal plants are thought to have created natural antioxidants that may combat free radicals in the

Table 4. Various sources of terpenes (Source: Ludwiczuk et al., 2017)

Terpenes	Sources
Hemiterpenes	The plant *Hamamelis japonica* and the leaves of several trees, such as oaks, conifers, poplars, and willows
Monoterpenes	Berries such as grapes, apricots, caraway, spearmint-dill, and citrus fruits
Sesquiterpenes	Lettuce, chicory
Diterpenes	Coffee beans
Tetrapenes	Spinach, kale, green beans, tomatoes, sweet potatoes, maize, orange, peaches, papaya, squash

body, according to certain theories. Due to their inherent therapeutic properties to scavenge free radicals, natural foods have lately attracted a lot of attention. This has led to a great deal of interest in biological systems for scavenging free radicals and the addition of adding inorganic antioxidants to the diet. They could serve as a secure substitute for dangerous synthetic antioxidants. Different plants secondary metabolites have been discovered to exhibit antioxidant activity (Vijayvargia and Vijayvergia, 2016) (Figure 16.2).

Antimicrobial Activity

Antibiotics and chemotherapeutic drugs are used to eradicate infectious illness and also to treat microbial infections. Since the development of antimicrobial medications, infectious disease control has been astonishingly successful. To many of the initial successful medicines that were discovered, certain infections quickly develop resistance. These drug-resistant bacteria are becoming more prevalent, which increases morbidity, death, and health care costs. There is a great deal of variation not only between pathogens that cause distinct clinical diseases in various geographical areas but also sometimes in particular locations. The Asia-Pacific area exhibited the highest rates of antibiotic resistance out of the five global regions examined. Throughout human history, herbal treatments have been used to cure a variety of infectious disorders. In the whole globe, they are the main cause of death (Kumar *et al.*, 2021b). Concern about antibiotic resistance is now widespread. The rise of multidrug-resistant bacteria poses a danger to the therapeutic effectiveness of several currently available medications. Plant extracts and phytochemicals have a major role in treatment, both of which have well-known antibacterial effects. Using a variety of solvent extracts, researchers have investigated the antibacterial

Figure 2. Antioxidant activities of different plants

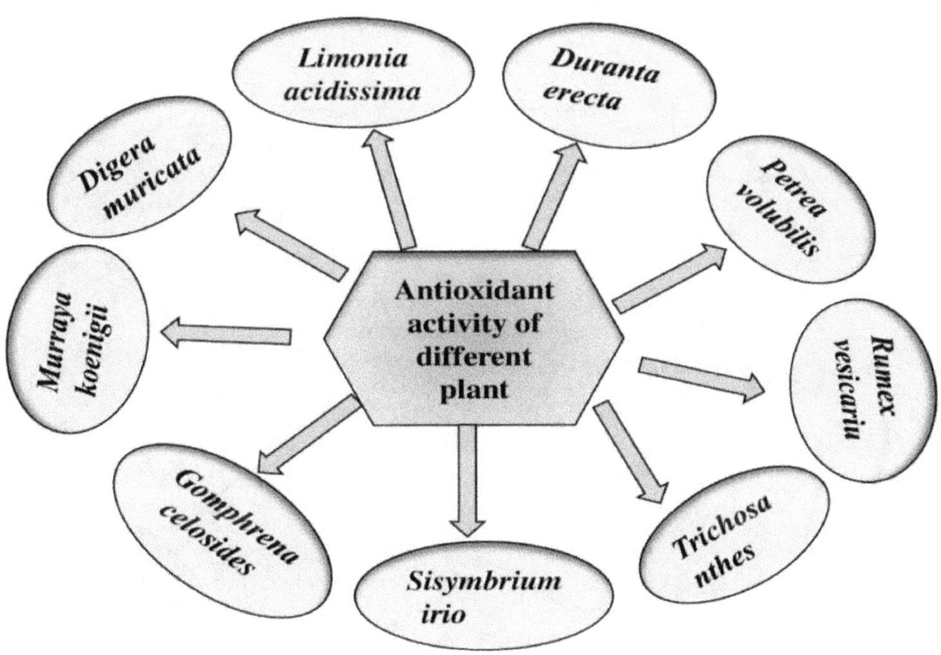

activity of a number of herbs and spices in plant leaves, flowers, stems, roots, and fruits. (Vijayvargia and Vijayvergia, 2014) (Figure 16.3).

Anti-Diabetic Activity

Chronic hyperglycemia or elevated blood glucose levels are hallmarks of diabetes mellitus (DM) is a metabolic disease characterized by a whole or partial lack of insulin production. By 2025, there will probably be 300 million population globally who suffer from this condition, which is on the increase. The greatest number of diabetes patients is anticipated in India. type Iinsulin-dependent and type II insulin-independent are the two types of the disease. Inadequate insulin synthesis causes type I diabetes, which is caused by an absence of beta cells. Individuals suffering from type I diabetes are entirely dependent on outside insulin providers; those with type II diabetes, on the other hand, are not responsive to insulin and can only be managed with medicine, dietary modifications, and physical activity (Mavi *et al.*, 2023). The most prevalent kind of diabetes mellitus worldwide and in non-insulin-dependent diabetes mellitus in India is linked to higher levels of postprandial hyperglycemia.

Alpha-amylase, an essential digestive enzyme located in the pancreas, initiates the process of starch degradation by converting it to maltose and, ultimately, glucose. The rapid dissolution of this dietary carbohydrate causes postprandial hyperglycemia. A correlation has been established between the small intestine's human pancreatic amylase activity with higher postprandial glucose levels; therefore, controlling this enzyme is a crucial element in the treatment of diabetes. Thus, inhibiting enzymes that facilitate the metabolism of carbohydrate, such as alpha-amylase, would have a substantial impact on the management of diabetes. The treatment of postprandial hyperglycemia mostly relies on alpha-amylase inhibitors. It reduces the amount of starch that is hydrolyzed into maltose by inhibiting the activity of the alpha-amylase enzyme, which in turn decreases postprandial hyperglycemia. A broad variety of glycosidases

Figure 3. Antimicrobial activities of different plants

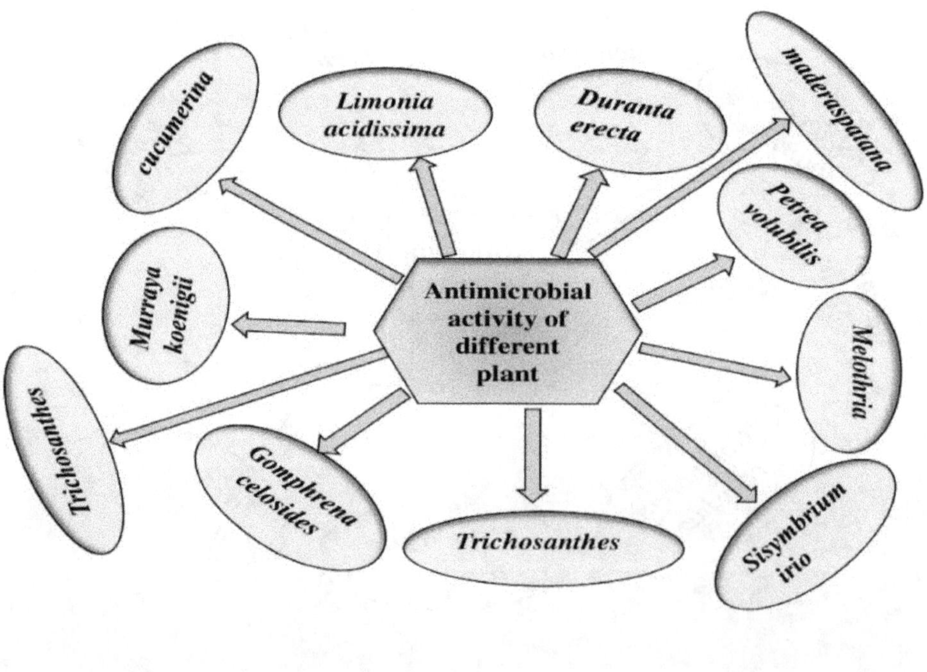

including beta glycosidase and alpha-amylase, are known to be inhibited by the inhibitors, now used in clinical settings (acarbose, miglitol, and voglibose). Biguanides, sulphonylureas, and thiazolidinediones are further treatments for type II diabetes, although they have a variety of unfavorable side effects. The creation of new anti-diabetic medications may be greatly aided by natural extracts from traditional medicinal plants, therefore the hunt for more accurate, reliable, and secure hypoglycemic agents has remain a critical topic of study (Figure 16.4) (Vijayvargia and Vijayvergia, 2014).

MAJOR PLANT SECONDARY METABOLITES AND HUMAN HEALTH

Phytosterols

Since humans are unable to produce them, they must be obtained through food. Several studies show that they have reduced blood cholesterol levels. Plant sterols are crucial chemical building blocks for the production of steroid chemicals, pesticides, antioxidants, and anticancer medications. A list of some of the sterols that have been extracted from different plants has been published. The sterols collected from higher plants most often are sitosterol, campesterol, stigmasterol and lanosterol. Other researchers made a similar discovery, citing -sitosterol, campesterol, and stigmasterol as the most often extracted chemicals from higher plants. Evidence shows that phytosterols have anti-cancer, antioxidant, anti-inflammatory, anti-ovarian, anti-lung, anti-stomach and anti-estrogen-dependent human breast cancer properties (Ju *et al.*, 2004).

Figure 4. Anti-diabetic activity of different plants

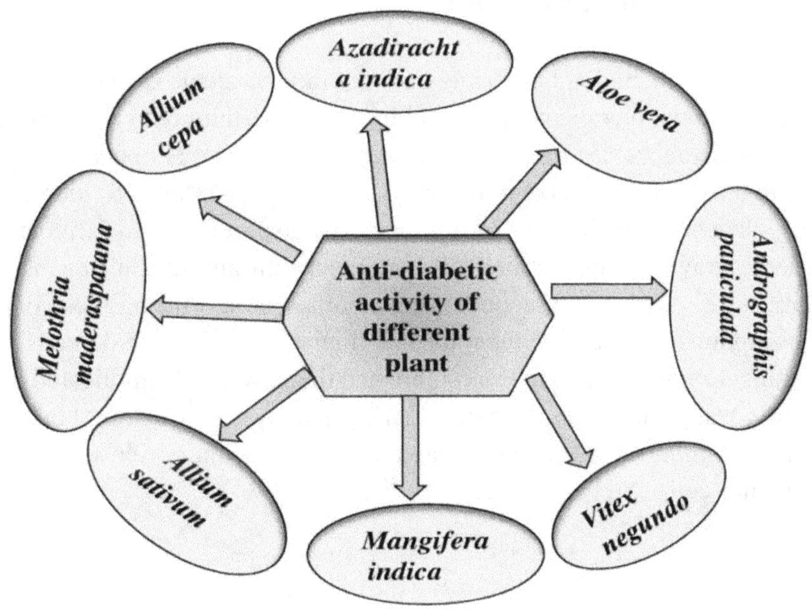

Alkaloids

A group of naturally occurring organic plant chemicals known as alkaloids exhibit neutral and mildly acidic characteristics. Numerous species, including bacteria, fungi, plants, and mammals, generate alkaloids. Quinine, ephedrine, homoharringtonine, vincamine, quinidine, and morphine are some common alkaloids having analgesics, antibacterial (chelerythrine) and antihyperglycemic properties are only a few of the many pharmacological properties (e.g. piperine). Many compounds have many qualities that are used in traditional or contemporary medicine, the identification of these features has served as the foundation for the development of new drugs.

Treatment of Diabetes

When injected into healthy rats, several *Catharanthus roseus* (periwinkle) alkaloids, such as leurosine, catharanthineand vindoline, have been demonstrated to produce a modest hypoglycaemic impact after 2-5 hours. However, it was eventually shown that this was cytotoxic to people with diabetes. The therapy of *Tecoma stans* leaves on non-insulin-dependent diabetes mellitus patients promoted the development of two additional hypoglycaemic alkaloids, such as tecomine and tecostanine (Verpoorte and Memelink, 2017). Administered intravenously to rabbits with alloxan-induced diabetes that were healthy, these alkaloids have been shown to have a fast hypoglycaemic impact; however, it was ineffective in pancreatectomized rudents. However, clinical testing on these alkaloids revealed poor stability. Trigonelline is a different plant alkaloid that is isolated from *Trigonella foenum-graecum* and is chemically known as N-methylnicotinic acid. It is well documented to have therapeutic effects on hypoglycemia and hypolipidemia. The transcriptional activities have been significantly impacted by this alkaloid. Before trigonelline could be used therapeutically to treat diabetes and its repercussions on illnesses of the central nervous system, further study was required, especially to understand the mechanism of action (Zhou *et al.*, 2012).

Treatment of Neurodegenerative Diseases

The coptidisrhizoma extract included berberine, an isoquinoline alkaloid that was also used in regenerative medicine. It has shown therapeutic potential for issues pertaining to the central nervous system, including schizophrenia, cerebral ischemia, and Alzheimer's disease. Hippocampal precursor cells are supported by these in terms of survival and differentiation. This also helps to increase the thickness of remyelinated axons when provided in cases of injured nerve disorders like inflammatory bowel disease, lupus, and myasthenia gravis, demonstrating neurine extension and axonal regeneration (Han *et al.*, 2012). One of the most well-researched secondary metabolites, songorine has been discovered to inhibit the effects of bulbocapnine, a dopamine receptor antagonist, as well as to demonstrate dopaminergic cure and synaptic transmission in mice. It was established that it was also implicated in the carrageenan-induced inflammatory test's anti-inflammatory efficacy. The process of wound healing was examined in order to determine the mouse's regeneration capabilities. Particularly relevant is the regeneration of mesenchymal progenitor cells (Yu *et al.*, 2015).

Treatment of Cancer

An alkaloid called vincristine was discovered in the stem, or bark and leaves of *Catharanthus roseus*. This is an oncolytic alkaloid used in the cure of several cancer types. Along with other antineoplastic medications, it has also been used to treat uterine cervix cancer, breast cancer, different tumors types, sarcomas of specific body parts and Hodgkin's disease (Tyagi *et al.*, 2010).

Terpenes

Terpenes are organic chemicals that are found in plants as secondary metabolites that come from a variety of organismal kingdoms, including bacteria, fungi, animals, and plants. These make up the vast majority of the category of naturally occurring chemicals. Terpenes produced in plants for pigmentation such as plant growth hormones, chlorophyll, gibberellin like abscisic acids and auxin. They are also reported as a wide range of medicinal characteristics and are protective against parasites, fungi, viruses, and cancer.

Treatment of Malaria

One of the crucial secondary terpene metabolites isolated from the bushes of *Artemisia annua*, artemisinin reported as most potential secondary metabolites treating malaria. They lessen the risk by introducing *Plasmodium falciparum* resistance to many drugs that contain a variety of antimalarials. Artemisinin has shown a rapid and successful treatment of malaria symptoms when used in conjunction with other drugs. It has been demonstrated that this significantly impacts *Plasmodium falciparum*, *Plasmodium vivax* and *Plasmodium malaria*, the three vector-borne illness. It was found that this derivative prevented haemoglobin from performing its functions by generating the carbon center molecules needed to remove the insect parasite and alkylate the protein membrane. The best option for treating malarial infections is artemisinin, which has a significant effect (De Ridder *et al.*, 2008).

Treatment of Body Endurance

Among the terpenes obtained from *Cannabis sativa* is tetrahydrocannabinol. The impacts of cannabinoids on both animals and people have been extensively researched. These have a wide range of uses, including painkillers, hunger stimulants, sedatives, lethargy treatments, and sleep aids. It has been made available that MC (Medical cannabis) use is a superior choice for the individual who is looking to stop using tobacco, alcohol, and drugs excessively.

Carotenoids

Carotenoids which are one of the abundant naturally occurring colorful pigments in plants is a tetraterpene that exhibits the role of vitamin A. Epidemiological and interference studies show potential agents for the chemical prevention of carcinogenesis include carotenoids, and cardiovascular disease and have an optimistic impact on the immune system and disease prevention. It induces natural killer cell and antigen-presenting monocyte activity thus enhancing the immunological responses mediated by cells, particularly in older adults. Carotenoids are also found to be beneficial for maintaining skin health by preventing UV-induced erythema (Stahl *et al.*, 2001).

Another carotenoid named lycopene, a lipid-soluble antioxidant synthesized by numerous plants and microbes has also been focused on disease prevention and biological activity. It was reported that above 85% of all sources of lycopene consist of tomatoes and tomato-based foods (Table 16.5). Studies on the anticarcinogenic effect of lycopene show that a higher consumption of lycopene increases the serum levels of lycopene which then significantly reduces the risk of several cancers including prostate, breast, cervical, ovarian and liver. Studies also show the presence of significantly lower lycopene levels in the serum of cancer patients with a higher level of oxidative stress (Rao *et al.*, 1999). Both β-carotene and lycopene were also found to reduce osteoporosis among postmenopausal women. Oxidative stress also shows a positive correlation with hypertension which is found to be decreased significantly after lycopene supplementation. Hypertensive patients were also found to have reduced plasma lycopene as compared to normal individuals. Male infertility, which is one of the common reproductive disorders, is found to be associated with a significantly higher level of ROS and reduced lycopene levels in semen compared to fertile controls (Rao and Rao, 2007).

Flavonoids

Since ancient times, their usage has persisted despite being recognized to possess therapeutic characteristics and play a significant part in effective medical treatments. They have potent anti-cancer properties and are efficient free radical scavengers and water-soluble antioxidants that prevent oxidative cell damage. They serve as anti-inflammatory, antispasmodic, anti-allergic, and antibacterial drugs in addition to being utilized to improve aquaresis (Mills and Bone, 2000). Certain findings suggest that flavonoids have the potential to increase blood supply and reduce blood pressure. Furthermore, the pharmacological effect of flavonoids inhibits the activity of several enzymes, like lipoxygenase, phosphodiesterase, xanthine oxidase, aldose reductase, and cyclooxygenase. They also control other hormones, such as androgens, estrogens and thyroid hormone. It has been shown that their anti-inflammatory actions may affect both the inflammatory proliferative and exudative stages.

Treatment of Diabetes and Retinopathy of Diabetes

Testing on human leukaemia HL-60 cells revealed that sweet potato cyanidin had antioxidative, antimutagenic, and antiproliferative effects (Konczak *et al.*, 2003). Anthocyanin-rich pulps possess antioxidant

Table 5. Various sources of carotenoids (Source: Eggersdorfer and Wyss, 2018)

Carotenoids	Health Benefits	References
Lycopene	Heart health, prevention of cancer	Cheng *et al.*, 2017
α-carotene	Pro-vitamin, vision	Food and Nutrition Board, Institute of Medicine, 2000
β-carotene	Cognitive function of brain, skin-UV protection, fertility, vision	Food and Nutrition Board, Institute of Medicine, 2000; Walk *et al.*, 2017
B-cryptoxanthin	Vision, pro-vitamin	Food and Nutrition Board, Institute of Medicine, 2000
Lutein	Maternal, infant nutrition, eye health, brain-cognitive function, fertility	Food and Nutrition Board, Institute of Medicine, 2000
Zeaxanthin	Eye health	Walk *et al.*, 2017

properties that aid in the mitigation of diabetes-related free radicals. Tart cherries are a good source of anthocyanin. Studies tested different cherries, such as sweet cherries, yellow cherries and sour cherries, to demonstrate the reducing benefits of blood sugar in diabetes (Lachin and Reza, 2012). Researchers discovered that an extract taken from sour cherry seeds had anti-diabetic properties. Zucker diabetic fatty rats were given sour cherry seed extract after, Varga and his colleagues examined the blood sugar levels in the animals. They found that the extract dramatically reduced the amount of blood sugar (Singh *et al.*, 2023). Varga and his team investigated how sour cherry seed extract might affect diabetic eyes and discovered that it lessened the severity of hyper-glycemia and shielded the retina from damage-damage to perfusion (Varga, 2017).

Treatment of Cardiovascular Disease

Proper absorption of fruits high in anthocyanins, which include plant bioactive components, is essential for improving blood pressure, endothelial function, and insulin sensitivity (Cassidy *et al.*, 2015). Flavonoids, therefore, are important fruit components that reduce the risk of cardiovascular disease. Foods rich in anthocyanins change vascular stiffness and can reduce both systolic and diastolic blood pressure (Cassidy *et al.*, 2015). Patients with hypercholesterolemia benefit from anthocyanins' anti-inflammatory properties. The relationship between women's cardiovascular disease and flavonoids has been the subject of earlier research. The risk of myocardial infarction is reduced in younger women who increase their anthocyanin intake. Eating fruits high in anthocyanin and flavonoids reduced men's risk of myocardial infarction and ischemic stroke (Cassidy *et al.*, 2015).

Treatment of Atherosclerosis

Microinflammation, the development of plaques high in lipoprotein, and prothrombogenic state are characteristics of atherosclerosis. Anthocyanins and other potent substances may be found in cornelian cherry extracts. The polyphenolic chemicals found in these fruit extracts have a broad variety of pharmacological actions, including antiparasitic, anticancer, anti-atherosclerotic, neuroprotective, and renal protective characteristics (Hosseinpour *et al.*, 2017).

Treatment of Bone Disorders

Research has been done on integrating the autogenous bone graft in the rabbit's skull because naringin possesses osteoinductive capabilities. Naringin improves antioxidant activity while preserving the amount of calcium present in bones, improving the bone's overall quality. Naringin also possesses antiosteoporotic properties, which is why it inhibits the production of osteoclasts, according to animal studies. As a possible therapy for osteoporosis, naringin inhibits the formation of osteoclasts and releases a cytokine receptor called osteoprotegerins (Xu *et al.*, 2016).

Treatment of Coronary Artery Disease

One important element in the preservation of cardiovascular function is dietary flavanols. Flavanol is mostly found in wine, chocolate-based products, tea, fruits, and vegetables. Flavanols are highly concentrated in cocoa. It has been shown that consuming cocoa flavanols improves endothelial function in healthy individuals as well as those with diabetes and coronary artery disease. Through the mobilisation of circulating angiogenic cells, cocoa flavanols also enhance the ability of endothelium regeneration in individuals with coronary artery disease (Heiss *et al.*, 2010). By reducing the amount of endothelium microparticles in the blood, this cocoa flavanol enhances endothelial function in patients with coronary heart disease. Thus, flavanols could improve endothelium integrity (Horn *et al.*, 2014).

Treatment in Cancer

Chalcone has a wide range of pharmacological actions and is present in both naturally occurring and synthesized substances. A wide spectrum of pharmacological properties, such as anti-inflammatory, antioxidant, anti-tumor and antibacterial, effects, are also present in chalcones and their derivatives. Chalcones have anticancer action via either polymerizing or depolymerizing tubulin, according to recent investigations. Naphthalene moiety is said to be an anticancer agent that inhibits tubulin polymerization (Wanga *et al.*, 2019). Cardamonin is a compound that is mostly present in cardamom spice and other plants. In several malignancies, gastrointestinal, and hematological, including breast and colorectal cancers, it has been shown to act as a chemo-preventive agent. After testing several compounds, researchers discovered that cardamonin limits tumour development by eradicating the population of breast cancer stem cells, which had been amplified after treatment of breast cancer cells both in vitro and in vivo. A tiny chemical called cardamonin is crucial in the inhibition of cancer stem cells that have been enhanced by chemotherapy medicines (Sharma *et al.*, 2016; Sahu *et al.*, 2023). Cardamonin inhibits the enrichment of cancer stem cells when administered in conjunction with chemotherapeutic agents. In breast cancer cells, cardamonin may also lessen the expression of the gene that indicates the stem cell.

Additional Plant Secondary Metabolites With Health Benefits

Some secondary metabolites have been identified with their extraordinary health benefits and other uses (Table 16.6). Various secondary metabolites participate as immunosuppressants, cholesterol-lowering agents, antitumor agents, artemisinin (anti-malarial), shikonin (pigment for cosmetics), sophoro-lipids (surfactants) and betulinic acid and derivatives. Numerous additional natural compounds have secondary functions that might be used in medicine. For instance, Anthracyclines include Doxorubicin, an anticancer drug from the anthracycline class, which is used to treat late-stage cancers. Bleomycin, daunorubicin, valrubicine, epirubicin, and dactinomycin are further members. Taxolisone of the most effective anticancer medications used today; paclitaxel (taxol) was first discovered in the Pacific yew tree *Taxus brevifolia* and is highly effective against breast cancer (Cisse and Mucke, 2009). The epothilones, which have the same mechanism of action against breast cancers as taxol but are efficacious against taxol-resistant tumors, also exhibit a combination of anticancer and antifungal activity (Goodin, 2008).

Table 6. Pharmacological activity of plant secondary metabolites

Secondary Metabolites	Pharmacological Activity of Plant	References
Withanolides, steroidal lactone	Treatment of Parkinson's and Alzheimer's diseases, improves memory, and has cancer-preventing, chemo-preventive effects and immunomodulatory	Rathinamoorthy *et al.*, 2014
Aconitine	Activates Na+ channels, analgesic	Gupta *et al.*, 2014
Atropine	Anti-cholinergic, anti-myopia effects	Gu *et al.*, 2011
Codeine	Antidepressant, analgesic, anti-diarrheal, sedative and hypnotic	Smith *et al.*, 2006
Nicotine	Insecticide, stimulant, anti-inflammatory	Rungsung *et al.*, 2015
Quinine	Antimalarial, analgesic, antipyretic	Rungsung *et al.*, 2015
Solanine	Antifungal, sedative, anticonvulsant, anticarcinogenic, anti-inflammatory.	Takshak, 2018
Tomatine	Antitumor, antifungal	Kúdelová *et al.*, 2013
Berberine	Anti-inflammatory, antibacterial/viral, antidiabetic	Takshak, 2018
Quercetin	Artherosclerosis plaque prevention and mitigation of coronary heart disease	JuŸwiak *et al.*, 2005
Resveratrol	Prevents the aggregation of platelets by inhibitingcyclooxygenase1 activity	Crescente *et al.*, 2009
Theaflavins and thearubigins	Anti-carcinogenic and anti-clastogenic	Bhattacharya *et al.*, 2011
Ferulic acid	Anti-diabetic properties	Nankar *et al.*, 2017
Di and tri terpenoids, limonoids	Blood purifier that protects skin illness, anti-diabetic, inhibiting colon cancer, and anti-allergic	Gupta *et al.*, 2014
Dehydro-pipernonaline, piperidine	Removes cough, cleans lungs, aids in weight reduction when combined with turmeric, hyperlipidemia and fights cancer	Gupta *et al.*, 2014
Tinosporin, isoquinoline alkaloids	Chemopreventive, immunostimulant, diabetic, cardioprotective nourishes the skin and hair.	Mittal *et al.*, 2014
ß-sitosterol, compesterol, emodin and aloin	Anti-viral, anti-tumoranti-diabetic, has healing properties and shows antiseptic effects	Mittal *et al.*, 2014
Emblicanin B	Good for skin, eyes and hairs,	Paarakh *et al.*, 2010
sawertiamarine and mangeferin	Anti-inflammatory, cardioprotective	Joshi *et al.*, 2005
Allicin	Anti-arthritis	Joshi *et al.*, 2005

CONCLUSION

PSMs are very specific to plants and show toxicity at a high concentration. PSM has a huge potential for biological effects in the context of parasites, bacteria, fungi, and human cells. Secondary metabolites have numerous biological properties as preventive roles from neurodegenerative diseases like Parkinson's or Alzheimer's, and cardiovascular diseases to antimicrobial and anticancer in humans. However, the study of PSM needs constant efforts and time to understand their significance in human health. Bioinformatics and omic workflow a recent area to discover the secondary metabolites in various plant kingdoms, also in different environments like aquatic environments, soils and microbial communities. Plants possess a mixture of numerous bioactive compounds but the benefit of pure isolated components will allow the efficiencies of its effect against the disease. Therefore, in-depth knowledge is of utmost necessary in different outlooks of PSM, their synthetic pathways and effects.

REFERENCES

Badri, S., Basu, V. R., Chandra, K., & Anasuya, D. (2019). A review on pharmacological activities of alkaloids. *World Journal of Current Medical and Pharmaceutical Research*, *1*(6), 230–234. doi:10.37022/WJCMPR.2019.01068

Bhattacharya, U., Mukhopadhyay, S., & Giri, A. K. (2011). Comparative antimutagenic and anticancer activity of three fractions of black tea polyphenols the arubigins. *Nutrition and Cancer*, *63*(7), 1122–1132. doi:10.1080/01635581.2011.605985 PMID:21919645

Cassidy, A., Rogers, G., Peterson, J. J., Dwyer, J. T., Lin, H., & Jacques, P. F. (2015). Higher dietary anthocyanin and flavonol intakes are associated with anti-inflammatory effects in a population of US adults. *The American Journal of Clinical Nutrition*, *102*(1), 172–181. doi:10.3945/ajcn.115.108555 PMID:26016863

Chadwick, M., Trewin, H., Gawthrop, F., & Wagstaff, C. (2013). Sesquiterpenoids lactones: Benefits to plants and people. *International Journal of Molecular Sciences*, *14*(6), 12780–12805. doi:10.3390/ijms140612780 PMID:23783276

Cheng, H., Wang, J., Wang, M., & Zhong, S. (2017). Toward privacy-preserving JPEG image retrieval. *Journal of Electronic Imaging*, *26*(4), 043022. doi:10.1117/1.JEI.26.4.043022

Cisse, M., & Mucke, L. (2009). A prion protein connection. *Nature*, *457*(7233), 1090–1091. doi:10.1038/4571090a PMID:19242462

Costa, M. A., Xia, Z. Q., Davin, L. B., & Lewis, N. G. (1999). Toward engineering the metabolic pathways of cancer-preventing lignans in cereal grains and other crops. In *Phytochemicals in Human Health Protection, Nutrition and Plant Defense* (pp. 67–87). Springer. doi:10.1007/978-1-4615-4689-4_4

Crescente, M., Jessen, G., Momi, S., Höltje, H. D., Gresele, P., Cerletti, C., & De Gaetano, G. (2009). Interactions of gallic acid, resveratrol, quercetin and aspirin at the platelet cyclooxygenase-1 level functional and modelling studies. *Thrombosis and Haemostasis*, *102*(08), 336–346. doi:10.1160/TH09-01-0057 PMID:19652885

Croteau, R., Kutchan, T. M., & Lewis, N. G. (2000). Natural products (secondary metabolites). *Biochemistry and Molecular Biology of Plants*, *24*, 1250–1319.

Crozier, A., Clifford, M. N., & Ashihara, H. (Eds.). (2008). *Plant secondary metabolites: occurrence, structure and role in the human diet*. John Wiley & Sons.

De Ridder, S., Van der Kooy, F., & Verpoorte, R. (2008). Artemisia annua as a self-reliant treatment for malaria in developing countries. *Journal of Ethnopharmacology*, *120*(3), 302–314. doi:10.1016/j.jep.2008.09.017 PMID:18977424

Egert, S., & Rimbach, G. (2011). Which sources of flavonoids: Complex diets or dietary supplements? *Advances in Nutrition*, *2*(1), 8–14. doi:10.3945/an.110.000026 PMID:22211185

Eggersdorfer, M., & Wyss, A. (2018). Carotenoids in human nutrition and health. *Archives of Biochemistry and Biophysics*, *652*, 18–26. doi:10.1016/j.abb.2018.06.001 PMID:29885291

Ehrlich, P. R., & Holdren, J. P. (1971). Impact of Population Growth: Complacency concerning this component of man's predicament is unjustified and counterproductive. *Science*, *171*(3977), 1212–1217. doi:10.1126/science.171.3977.1212 PMID:5545198

Figueiredo, A. C., Barroso, J. G., Pedro, L. G., & Scheffer, J. J. (2008). Factors affecting secondary metabolite production in plants: Volatile components and essential oils. *Flavour and Fragrance Journal*, *23*(4), 213–226. doi:10.1002/ffj.1875

Food and Nutrition Board Staff, Panel on Dietary Antioxidants, & Institute of Medicine Staff. (2000). Dietary reference intakes for vitamin C, vitamin E, selenium and carotenoids: a report of the panel on dietary antioxidants and related compounds, subcommittees on upper reference levels of nutrients and of interpretation and use of dietary reference intakes and the standing committee on the scientific evaluation of dietary reference intakes, Food and Nutrition Board, Institute of Medicine, National Academies Press, Washington, D.C. USA.

Goodin, S. (2008). Novel cytotoxic agents: Epothilones. *American Journal of Health-System Pharmacy*, *65*(10, Supplement_3), S10–S15. doi:10.2146/ajhp080089 PMID:18463327

Gu, L., Li, N., Gong, J., Li, Q., Zhu, W., & Li, J. (2011). Berberine ameliorates intestinal epithelial tight-junction damage and down-regulates myosin light chain kinase pathways in a mouse model of endotoxinemia. *The Journal of Infectious Diseases*, *203*(11), 1602–1612. doi:10.1093/infdis/jir147 PMID:21592990

Han, A. M., Heo, H., & Kwon, Y. K. (2012). Berberine promotes axonal regeneration in injured nerves of the peripheral nervous system. *Journal of Medicinal Food*, *15*(4), 413–417. doi:10.1089/jmf.2011.2029 PMID:22316297

Harborne, J. B. (2014). *Introduction to ecological biochemistry*. Academic press.

Hasler, C. M., & Blumberg, J. B. (1999). Symposium on phytochemicals: Biochemistry and physiology. *The Journal of Nutrition*, *129*, 756S–757S. doi:10.1093/jn/129.3.756S PMID:10082785

Heiss, C., Keen, C. L., & Kelm, M. (2010). Flavanols and cardiovascular disease prevention. *European Heart Journal*, *31*(21), 2583–2592. doi:10.1093/eurheartj/ehq332 PMID:20852295

Horn, P., Amabile, N., Angeli, F. S., Sansone, R., Stegemann, B., Kelm, M., Springer, M. L., Yeghiazarians, Y., Schroeter, H., & Heiss, C. (2014). Dietary flavanol intervention lowers the levels of endothelial microparticles in coronary artery disease patients. *British Journal of Nutrition*, *111*(7), 1245–1252. doi:10.1017/S0007114513003693 PMID:24286443

Hosseinpour-Jaghdani, F., Shomali, T., Gholipour-Shahraki, S., Rahimi-Madiseh, M., & Rafieian-Kopaei, M. (2017). Cornus mas: A review on traditional uses and pharmacological properties. *Journal of Complementary & Integrative Medicine*, *14*(3), 20160137. doi:10.1515/jcim-2016-0137 PMID:28782352

Hussein, R. A., & El-Anssary, A. A. (2019). Plants secondary metabolites: The key drivers of the pharmacological actions of medicinal plants. *Herbal Medicine*, *1*(3), 11–30. doi:10.5772/intechopen.76139

Jangpangi, K., Khati, A., Chauhan, R. S., Rajesh, R., & Kumar, N. (2023). Assessment of the haematological and biochemical effects of Himalayan herb *Urtica dioica* leaves diets fed amur carp, *Cyprinus carpio haematopterus. Turkish Journal of Fisheries and Aquatic Sciences*, *24*(4), TRJFAS23854. doi:10.4194/TRJFAS23854

Joshi, P., & Dhawan, V. (2005). *Swertia chirayita*–an overview. *Current Science*, *89*(4), 635–640.

Ju, Y. H., Clausen, L. M., Allred, K. F., Almada, A. L., & Helferich, W. G. (2004). β-sitosterol, β-sitosterol glucoside, and a mixture of β-sitosterol and β-sitosterol glucoside modulate the growth of estrogen-responsive breast cancer cells in vitro and in ovariectomized athymic mice. *The Journal of Nutrition*, *134*(5), 1145–1151. doi:10.1093/jn/134.5.1145 PMID:15113961

JuŸwiak, S., Wójcicki, J., Mokrzycki, K., Marchlewicz, M., Białecka, M., Wenda-Różewicka, L., Gawronska-Szklarz, B., & Droździk, M. (2005). Effect of quercetin on experimental hyperlipidemia and atherosclerosis in rabbits. *Pharmacological Reports*, *57*(57), 604–609. PMID:16227643

King, A. M. Y., & Young, G. (1999). Characteristics and occurrence of phenolic phytochemicals. *Journal of the American Dietetic Association*, *99*(2), 213–218. doi:10.1016/S0002-8223(99)00051-6 PMID:9972191

Koes, R. E., Quattrocchio, F., & Mol, J. N. (1994). The flavonoid biosynthetic pathway in plants: Function and evolution. *BioEssays*, *16*(2), 123–132. doi:10.1002/bies.950160209

Konczak-Islam, I., Yoshimoto, M., Hou, D. X., Terahara, N., & Yamakawa, O. (2003). Potential chemopreventive properties of anthocyanin-rich aqueous extracts from in vitro produced tissue of sweet potato (*Ipomoea batatas* L.). *Journal of Agricultural and Food Chemistry*, *51*(20), 5916–5922. doi:10.1021/jf030066o PMID:13129295

Kozlowska, A., & Szostak-Wegierek, D. (2014). Flavonoids-food sources and health benefits. *Roczniki Panstwowego Zakladu Higieny*, *65*(2), 79–85. PMID:25272572

Kúdelová, J., Seifrtová, M., Suchá, L., Tomšík, P., Havelek, R., & Řezáčová, M. (2013). Alpha-tomatine activates cell cycle checkpoints in the absence of DNA damage in human leukemic MOLT-4 cells. *Journal of Applied Biomedicine*, *11*(2), 93–103. doi:10.2478/v10136-012-0033-8

Kumar, G., Sharma, J. G., Goswami, R. K., Shrivastav, A. K., Kumar, N., Chandra, S., & Chakrabarti, R. (2021a). The study of effect of vitamin C and *Achyranthes aspera* seeds enriched diets on the growth, biochemical composition, digestive enzyme activities and expressions of genes involved in the biosynthesis of fatty acids in snow trout *Schizothorax richardsonii* (Gray, 1832). *Journal of Applied Aquaculture*, *35*(3), 489–509. doi:10.1080/10454438.2021.1985679

Kumar, G., Sharma, J. G., Goswami, R. K., Shrivastav, A. K., Tocher, D. R., Kumar, N., & Chakrabarti, R. (2022a). Freshwater macrophytes: Potential source of minerals and fatty acids for fish, poultry and livestock. *Frontiers in Nutrition*, *9*, 869425. doi:10.3389/fnut.2022.869425 PMID:35479737

Kumar, N., Sharma, J. G., Kumar, G., Shrivastav, A. K., Tiwari, N., Begam, A., & Chakrabarti, R. (2021b). Evaluation of nutritional value of prickly chaff flower (*Achyranthes aspera*) as fish feed ingredient. *The Indian Journal of Animal Sciences*, *91*(3), 239–244. doi:10.56093/ijans.v91i3.114152

Kumar, N., Sharma, J. G., Mittal, P., & Chakrabarti, R. (2022b). Effect of leaves and seeds of *Achyranthes aspera* as feed supplements on the immunological and stress parameters and related-gene expressions of Asian catfish (*Clarias batrachus*). *Veterinary Research Communications*, *47*(1), 99–109. doi:10.1007/s11259-022-09932-5 PMID:35507243

Kumar, N., Sharma, J. G., Singh, S. P., Singh, A., Harikrishna, V., & Chakrabarti, R. (2019). Validation of growth enhancing, immunostimulatory and disease resistance properties of *Achyranthes aspera* in *Labeo rohita* fry in pond conditions. *Heliyon*, *5*(2), e01246. doi:10.1016/j.heliyon.2019.e01246 PMID:30820481

Kumar, S. G., & Sharma, A. (2014). Medicinal properties of *Zingiber officinale* Roscoe-A review. *Journal of Pharmaceutical and Biological Sciences*, *9*, 124–129.

Lachin, T., & Reza, H. (2012). Anti diabetic effect of cherries in alloxan induced diabetic rats. *Recent Patents on Endocrine, Metabolic & Immune Drug Discovery*, *6*(1), 67–72. doi:10.2174/187221412799015308 PMID:22280223

Ludwiczuk, A., Skalicka-Woźniak, K., & Georgiev, M. I. (2017). Terpenoids. In *Pharmacognosy* (pp. 233–266). Academic Press. doi:10.1016/B978-0-12-802104-0.00011-1

Mathai, K. (2000). Nutrition in the adult years. Krause's Food, Nutrition, and Diet Therapy, 271, 274-275.

Mavi, A. K., Gaur, S., Gaur, G., & Kumar, U. (2023). CAR T-cell therapy: Reprogramming patient's immune cell to treat cancer. *Cellular Signalling*, *105*, 110638. doi:10.1016/j.cellsig.2023.110638 PMID:36822565

Mills, S., & Bone, K. (2000). *Principles and practice of phytotherapy: Modern herbal medicine*. Churchill Livingstone.

Mittal, J., Sharma, M. M., & Batra, A. (2014). *Tinospora cordifolia*: A multipurpose medicinal plant-A. *Faslnamah-i Giyahan-i Daruyi*, *2*(2), 32–47.

Morresion, G., & Hark, L. (1999). *Medical Nutrition and Disease*. BlackWell Science.

Nankar, R., Prabhakar, P. K., & Doble, M. (2017). Hybrid drug combination: Combination of ferulic acid and metformin as anti-diabetic therapy. *Phytomedicine*, *37*, 10–13. doi:10.1016/j.phymed.2017.10.015 PMID:29126698

Paarakh, P. M. (2010). *Terminalia arjuna* (Roxb.) Wt. and Arn.: A review. *International Journal of Pharmacology*, *6*(5), 515–534. doi:10.3923/ijp.2010.515.534

Quintero-Fabián, S., Ortuño-Sahagún, D., Vázquez-Carrera, M., & López-Roa, R. I. (2013). Alliin, a garlic (*Allium sativum*) compound, prevents LPS-induced inflammation in 3T3-L1 adipocytes. *Mediators of Inflammation*, *2013*, 381815. doi:10.1155/2013/381815 PMID:24453416

Rajput, A., Sharma, R., & Bharti, R. (2021). Pharmacological activities and toxicities of alkaloids on human health. *Materials Today: Proceedings*, *48*(5), 1407–1415.

Rao, A. V., Fleshner, N., & Agarwal, S. (1999). Serum and tissue lycopene and biomarkers of oxidation in prostate cancer patients: A case-control study. *Nutrition and Cancer*, *33*(2), 159–164. doi:10.1207/S15327914NC330207 PMID:10368811

Rao, A. V., & Rao, L. G. (2007). Carotenoids and human health. *Pharmacological Research, 55*(3), 207–216. doi:10.1016/j.phrs.2007.01.012 PMID:17349800

Rathinamoorthy, R., & Thilagavathi, G. (2014). Terminalia chebula-review on pharmacological and biochemical studies. *International Journal of Pharm Tech Research, 6*, 97–116.

Rungsung, W., Ratha, K. K., Dutta, S., Dixit, A. K., & Hazra, J. (2015). Secondary metabolites of plants in drugs discovery. *World Journal of Pharmaceutical Research, 4*(7), 604–613.

Sahu, A., Kumar, N., Singh, C. P., & Singh, M. (2023). Environmental DNA (eDNA): Powerful technique for biodiversity conservation. *Journal for Nature Conservation, 71*, 126325. doi:10.1016/j.jnc.2022.126325

Semwal, A., Kumar, A., & Kumar, N. (2023). A review on pathogenicity of *Aeromonas hydrophila* and their mitigation through medicinal herbs in aquaculture. *Heliyon, 9*(3), e14088. doi:10.1016/j.heliyon.2023.e14088 PMID:36938468

Sharma, J. G., Kumar, N., Mittal, P., & Chakrabarti, R. (2022). Evaluation of UV-B protective properties of leaves and seeds of *Achyranthes aspera* in Asian catfish *Clarias batrachus* (Linn.). *Photochemical & Photobiological Sciences, 21*(8), 1341–1356. doi:10.1007/s43630-022-00222-2 PMID:35403962

Sharma, J. G., Kumar, N., Singh, S. P., Singh, A., Harikrishna, V., & Chakrabarti, R. (2019). Evaluation of immunostimulatory properties of prickly chaff flower *Achyranthes aspera* in rohu *Labeo rohita* fry in pond conditions. *Aquaculture (Amsterdam, Netherlands), 505*, 183–189. doi:10.1016/j.aquaculture.2019.02.065

Sharma, R., Kumar, R., Kodwani, R., Kapoor, S., Khare, A., Bansal, R., Khurana, S., Singh, S., Thomas, J., Roy, B., Phartyal, R., Saluja, S., & Kumar, S. (2016). A review on mechanisms of anti-tumor activity of chalcones. *Anti-cancer Agents in Medicinal Chemistry, 16*(2), 200–211. doi:10.2174/1871520615666150518093144 PMID:25980813

Shen, N., Wang, T., Gan, Q., Liu, S., Wang, L., & Jin, B. (2022). Plant flavonoids: Classification, distribution, biosynthesis, and antioxidant activity. *Food Chemistry, 383*, 132531. doi:10.1016/j.foodchem.2022.132531 PMID:35413752

Singh, A., Rajput, V. D., Sharma, R., Ghazaryan, K., & Minkina, T. (2023). Salinity stress and nanoparticles: Insights into antioxidative enzymatic resistance, signaling, and defense mechanisms. *Environmental Research, 235*, 116585. doi:10.1016/j.envres.2023.116585 PMID:37437867

Singh, R., Chauhan, R. S., Semwal, A., & Kumar, N. (2023). Efficacy of 17α-methyl testosterone on growth and survival of fry of swordtail, *Xiphophorus helleri. Journal of Krishi Vigyan, 11*(2), 65–69. doi:10.5958/2349-4433.2023.00017.X

Smith, J., Owen, E., Earis, J., & Woodcock, A. (2006). Effect of codeine on objective measurement of cough in chronic obstructive pulmonary disease. *The Journal of Allergy and Clinical Immunology, 117*(4), 831–835. doi:10.1016/j.jaci.2005.09.055 PMID:16630941

Stahl, W., Heinrich, U., Wiseman, S., Eichler, O., Sies, H., & Tronnier, H. (2001). Dietary tomato paste protects against ultraviolet light-induced erythema in humans. *The Journal of Nutrition, 131*(5), 1449–1451. doi:10.1093/jn/131.5.1449 PMID:11340098

Takshak, S. (2018). Bioactive compounds in medicinal plants: A condensed review. *Journal of Pharmacy and Natural Medicines*, *1*(1), 1–35.

Tohge, T., & Fernie, R. (2017). An overview of compounds derived from the shikimate and phenylpropanoid pathways and their medicinal importance. *Mini-Reviews in Medicinal Chemistry*, *17*(12), 1013–1027. doi:10.2174/1389557516666160624123425 PMID:27342231

Twaij, B. M., & Hasan, M. N. (2022). Bioactive secondary metabolites from plant sources: Types, synthesis, and their therapeutic uses. *International Journal of Plant Biology*, *13*(1), 4–14. doi:10.3390/ijpb13010003

Tyagi, G., Jangir, D. K., Singh, P., & Mehrotra, R. (2010). DNA interaction studies of an anticancer plant alkaloid, vincristine, using fourier transform infrared spectroscopy. *DNA and Cell Biology*, *29*(11), 693–699. doi:10.1089/dna.2010.1035 PMID:20662555

Varga, B., Priksz, D., Lampé, N., Bombicz, M., Kurucz, A., Szabó, A. M., Pósa, A., Szabó, R., Kemény-Beke, Á., Remenyik, J., Gesztelyi, R., & Juhász, B. (2017). Protective effect of *Prunus cerasus* (sour cherry) seed extract on the recovery of ischemia/reperfusion-induced retinal damage in zucker diabetic fatty rat. *Molecules (Basel, Switzerland)*, *22*(10), 1782. doi:10.3390/molecules22101782 PMID:29065463

Verpoorte, R., & Memelink, J. (2002). Engineering secondary metabolite production in plants. *Current Opinion in Biotechnology*, *13*(2), 181–187. doi:10.1016/S0958-1669(02)00308-7 PMID:11950573

Vijayvargia, P., & Vijayvergia, R. (2014). A review on *Limonia acidissima* L.: Multipotential medicinal plant. *International Journal of Pharmaceutical Sciences and Research*, *28*(1), 191–195.

Vijayvargia, P., & Vijayvergia, R. (2016). Assessment of phytochemicals and antioxidant activity of *Murraya koenigii* Linn. *International Journal of Pharmaceutical Sciences and Research*, *7*(5), 2163.

Walk, A. M., Khan, N. A., Barnett, S. M., Raine, L. B., Kramer, A. F., Cohen, N. J., Moulton, C. J., Renzi-Hammond, L. M., Hammond, B. R., & Hillman, C. H. (2017). From neuro-pigments to neural efficiency: The relationship between retinal carotenoids and behavioral and neuroelectric indices of cognitive control in childhood. *International Journal of Psychophysiology*, *118*, 1–8. doi:10.1016/j.ijpsycho.2017.05.005 PMID:28528704

Wang, G., Peng, Z., & Li, Y. (2019). Synthesis, anticancer activity and molecular modeling studies of novel chalcone derivatives containing indole and naphthalene moieties as tubulin polymerization inhibitors. *Chemical & Pharmaceutical Bulletin*, *67*(7), 725–728. doi:10.1248/cpb.c19-00217 PMID:30982797

World Health Organization. (2013). *WHO Traditional Medicine Strategy: 2014-2023*. World Health Organization.

Xu, T., Wang, L., Tao, Y., Ji, Y., Deng, F., & Wu, X. H. (2016). The function of naringin in inducing secretion of osteoprotegerin and inhibiting formation of osteoclasts. *Evidence-Based Complementary and Alternative Medicine*, *2016*, 8981650. doi:10.1155/2016/8981650 PMID:26884798

Yu, V. N., Povet'eva, T. N., Suslov, N. I., Shults, E. E., Ziuz'kov, G. N., Aksinenko, S. G., Krapivin, A. V., & Kharina, T. G. (2015). Anxiolytic activity of diterpene alkaloid songorine pharmacology and toxicology. *Bulletin of Experimental Biology and Medicine*, *159*(5), 620–622. doi:10.1007/s10517-015-3029-z PMID:26468026

Zhou, J., Chan, L., & Zhou, S. (2012). Trigonelline: A plant alkaloid with therapeutic potential for diabetes and central nervous system disease. *Current Medicinal Chemistry*, *19*(21), 3523–3531. doi:10.2174/092986712801323171 PMID:22680628

Section 5

Nano–Omics–Based Abiotic Stress Management

Chapter 16
Addressing Abiotic Stress in Agriculture Through Nanotechnology Solutions

Shivangi Singh
Sardar Vallabhbhai Patel University of Agriculture and Technology, India

Omkar Singh
https://orcid.org/0000-0003-1971-185X
Sardar Vallabhbhai Patel University of Agriculture and Technology, India

Uday Pratap Shahi
Sardar Vallabhbhai Patel University of Agriculture and Technology, India

Raghavendra Singh
ICAR-Indian Institute of Farming Systems Research, India

Prabhat Kumar Singh
Acharya Narendra Deva University of Agriculture and Technology, India

Sakshi Singh
Tilak Dhari P.G. College, India

Aleksandr Yesayan
https://orcid.org/0009-0007-2407-3609
Yerevan State University, Armenia

Tatiana Minkina
Southern Federal University, Russia

Ilya Alliluev
Southern Federal University, Russia

Dmitry Bren
Southern Federal University, Russia

Astghik Melkumyan
Southern Federal University, Russia

ABSTRACT

Drought, salt, temperature extremes, nutritional inadequacies, and heavy metal toxicity hinder worldwide agricultural output and food security. Using nanostructures' unique properties, nanotechnology offers various creative solutions to these issues. This chapter details how nanotechnology can reduce drought, salt, temperature, nutrient deficits, and heavy metal toxicity in agriculture. The chapter covers nanotechnology applications including water retention materials, sensors, fertilisers, plant growth regulators, seed coatings, adsorbents, and chelating agents. Case studies and research demonstrate how these methods boost crop stress tolerance and output. The chapter examines risks, toxicity, regulatory

DOI: 10.4018/979-8-3693-1890-4.ch016

challenges, and commercialization limits of creating and exploiting nanotechnology in agriculture. Interdisciplinary collaboration and stakeholder engagement are crucial to nanotechnology's agricultural potential, according to the chapter. Further research and development to solve knowledge gaps and promote responsible innovation are also supported. This chapter provides a comprehensive and current overview of nanotechnology applications for abiotic stress management in agriculture to motivate academics, practitioners, and policymakers. To solve global issues, it develops sustainable and resilient agriculture systems.

1. INTRODUCTION

Definition of Abiotic Stress in Agriculture

Abiotic stress in agriculture refers to the adverse effects on crop growth, development, and productivity caused by non-living environmental factors (Cramer et al., 2011). These factors include drought, salinity, extreme temperatures, nutrient deficiency, and heavy metal toxicity (Fahad et al., 2017). Abiotic stresses pose significant challenges to agricultural production worldwide, leading to reduced crop yields and quality (Gupta et al., 2020). As climate change intensifies, the frequency and severity of abiotic stresses are expected to increase, further exacerbating the impact on global food security (IPCC, 2014).

Impact of Abiotic Stress on Crop Productivity

Abiotic stresses significantly reduce crop productivity, leading to substantial yield losses and economic repercussions for farmers and the agricultural sector (Mantri et al., 2012). Drought stress can cause wilting, leaf senescence, and decreased photosynthetic efficiency, leading to inhibited growth and reduced agricultural production (Farooq et al., 2009). Salinity stress can impair plant water uptake, produce ionic toxicity, and interfere with nutrient absorption, ultimately resulting in decreased growth and productivity (Munns & Tester, 2008). Extreme temperatures, whether hot or cold, can harm crop tissues, disturb physiological functions, and have adverse effects on crop quality and yield (Hasanuzzaman et al., 2013). Nutrient deficiency stress in crops arises from the absence of vital elements like nitrogen, phosphorus, or potassium, resulting in stunted growth, decreased photosynthesis, and diminished yields (Marschner, 2012). Heavy metal toxicity stress, resulting from the buildup of metals like cadmium, lead, or arsenic in the soil, can disrupt plant metabolism, induce oxidative harm, and diminish agricultural yield (Rizwan et al., 2016). The combined impact of these non-living stresses can lead to significant reductions in agricultural production, anywhere from 10% to 90%. The extent of the losses depends on the kind of crop, the intensity and duration of the stress, and the growth stage of the crop (Mantri et al., 2012; Fahad et al., 2017). Nutrient deficiency stress occurs when crops lack essential nutrients such as nitrogen, phosphorus, or potassium, leading to impaired growth, reduced photosynthesis, and lower yields (Marschner, 2012). Heavy metal toxicity stress, caused by the accumulation of metals such as cadmium, lead, or arsenic in the soil, can interfere with plant metabolism, cause oxidative damage, and reduce crop productivity (Rizwan et al., 2016). The cumulative effect of these abiotic stresses can result in substantial yield losses, ranging from 10% to 90% depending on the crop species, the severity and duration of the stress, and the stage of crop development (Mantri et al., 2012; Fahad et al., 2017).

Limitations of Traditional Approaches to Mitigate Abiotic Stress

Historically, efforts to reduce abiotic stress in agriculture have concentrated on developing crop types that can withstand stress, improving irrigation and fertilisation techniques, and applying effective crop management methods (Vurukonda et al., 2016). Although these methods have helped enhance crop resilience to abiotic stresses, they have constraints and may not be adequate to deal with the rising occurrence and severity of these stresses due to climate change (Gupta et al., 2020). Developing stress-tolerant crop varieties using traditional methods is a lengthy and laborious process that may require several years to complete and introduce new varieties (Ashraf, 2010). Additionally, the genetic diversity for breeding purposes may be restricted, and the intricate characteristics of abiotic stress tolerance, which frequently require multiple genes and pathways, present a challenge in achieving substantial enhancements in stress resilience solely through conventional breeding methods (Mickelbart et al., 2015).

Improving irrigation and fertilisation methods can reduce the impact of drought and nutrient deficiencies. However, these methods can be resource-intensive and may not be sustainable in the long run, especially in areas with scarce water resources or low soil fertility (Gupta et al., 2020; singh O. 2023). Furthermore, overuse or incorrect use of irrigation and fertilisers can result in environmental issues such water depletion, soil deterioration, and nutrient runoff (Gupta et al., 2020). Implementing crop management measures including altering planting dates, utilising mulches, or adopting intercropping systems might somewhat reduce the effects of abiotic stresses (Fahad et al., 2017). Nevertheless, these approaches may not be efficient in situations of intense or extended stress and would necessitate substantial alterations in agricultural methods and infrastructure (Vurukonda et al., 2016).

Potential of Nanotechnology in Addressing Abiotic Stress

Nanotechnology manipulates matter at the nanoscale (1-100 nm) and provides innovative solutions for abiotic stress in agriculture (Fraceto et al., 2016; singh et al., 2023). Nanomaterials have distinct physical, chemical, and biological characteristics that can be utilised to create new methods for improving crop resilience to stress and increasing yield (Prasad et al., 2017). Nanotechnology offers the advantage of creating nanomaterials with customised functions to combat certain abiotic stressors. Nanomaterials can be tailored to enhance water retention and soil moisture control, helping crops to endure drought conditions more effectively (Jampílek & Králová, 2017). Nano-sensors can be created to continuously monitor soil moisture levels, enabling farmers to enhance irrigation timing and preserve water resources (Xu et al., 2014).

Nanotechnology can be used to create precise nutrient delivery systems, like nano-fertilizers, that can release nutrients in a controlled and consistent way. This enhances nutrient utilisation and decreases nutrient wastage (DeRosa et al., 2010). Nano-encapsulated plant growth regulators and antioxidants can improve crop tolerance to salinity, heat, and cold stress when applied (Prasad et al., 2017). In addition, nanotechnology can help in treating soil pollution and decreasing the harmful effects of heavy metals. Nano-adsorbents efficiently eliminate heavy metals from the soil, as demonstrated by Anjum et al. (2016). Additionally, nano-encapsulated chelating agents aid in metal detoxification in plants, as shown by Shukla et al. (2019). Nanotechnology has shown promise in dealing with abiotic stress in agriculture, as evidenced by several research papers and case cases (Fraceto et al., 2016; Khan et al., 2017). Despite being in the early stages, the progress and implementation of nanotechnology solutions in agriculture face challenges such as potential risks, toxicity concerns, regulatory frameworks, and commercialization issues.

The growing need for sustainable and resilient agricultural systems, along with the rapid progress in nanotechnology research, offers significant opportunities for nanotechnology to transform how we tackle abiotic stress in agriculture (Prasad et al., 2017; Rajput et al., 2021). By combining the distinctive characteristics of nanomaterials with conventional methods, we can create new and efficient techniques to boost crop resilience, increase productivity, and safeguard global food security amidst climate change challenges (Fraceto et al., 2016).

2. TYPES OF ABIOTIC STRESS IN AGRICULTURE

Abiotic stresses are environmental variables that are not living and have a negative impact on the growth, development, and production of crops (Cramer et al., 2011). Natural or human-induced stresses are anticipated to intensify in intensity and frequency as a result of climate change (IPCC, 2014). This section will cover five primary types of abiotic stress that present considerable obstacles to agricultural production: drought stress, salinity stress, temperature stress (heat and cold), nutrient deficiency stress, and heavy metal toxicity stress.

Drought Stress

Plants undergo drought stress when they face an extended period of water shortage, resulting in hindered growth and decreased yield (Farooq et al., 2009). Drought can result from multiple factors, including less rainfall, increased evaporation rates, and inadequate soil water retention capacity (Fahad et al., 2017). Drought stress impacts plant physiology by decreasing leaf water potential, stomatal conductance, and photosynthetic rate, leading to reduced biomass accumulation and yield (Farooq et al., 2009). Drought stress can cause oxidative damage, membrane instability, and hinder nutrient absorption (Anjum et al., 2011). Plants have developed different strategies to deal with drought stress, including osmotic adjustment, antioxidant defence, and alterations in root structure (Fang & Xiong, 2015). Nevertheless, these systems might not be adequate to uphold crop productivity during intense or extended drought situations (Fahad et al., 2017). Methods to reduce drought impact in agriculture including developing drought-resistant crop varieties, improving irrigation techniques, and adopting water-conserving technologies (Vurukonda et al., 2016).

Salinity Stress

Salinity stress happens when plants are subjected to elevated levels of soluble salts in the soil, leading to disturbances in water intake, ionic toxicity, and hindrance in nutrient absorption (Munns & Tester, 2008). Salinity stress can result from variables including irrigation with saline water, inadequate drainage, and excessive fertilisation (Shrivastava & Kumar, 2015). Salinity stress impacts plant growth and development by diminishing leaf expansion, photosynthetic rate, and biomass accumulation, resulting in decreased yields (Munns & Tester, 2008). Plants have evolved different strategies to deal with salinity stress, including ion exclusion, compartmentalization of harmful ions, and production of suitable solutes (Deinlein et al., 2014). Yet, these systems might not be adequate to sustain agricultural output in conditions of elevated saline levels (Shrivastava & Kumar, 2015). Methods to alleviate salinity stress in agriculture involve developing salt-tolerant crop varieties, enhancing irrigation and drainage practices, and using soil amendments to decrease salt buildup (Machado & Serralheiro, 2017).

Temperature Stress (Heat and Cold)

Temperature stress arises when crops are subjected to temperatures that are not within their ideal range, leading to disturbances in physiological functions and decreased productivity (Hasanuzzaman et al., 2013). Heat stress in crops occurs when temperatures exceed their ideal range, resulting in decreased photosynthetic efficiency, heightened respiration, and hindered reproductive growth (Fahad et al., 2017). Cold stress in crops results from exposure to low temperatures, leading to chilling injury, membrane damage, and decreased enzyme activity (Sanghera et al., 2011). Plants have developed different strategies to deal with temperature stress, including the synthesis of heat shock proteins, antioxidant defence, and alterations in membrane lipid composition (Hasanuzzaman et al., 2013). Nevertheless, these systems might not be adequate to sustain agricultural productivity in severe temperature circumstances (Fahad et al., 2017). To address temperature stress in agriculture, strategies involve developing crop varieties that are resilient to temperature, selecting optimal planting dates, and utilising protective structures like shade nets and greenhouses (Teixeira et al., 2013).

Nutrient Deficiency Stress

Nutrient deficiency stress in crops arises from a lack of important nutrients like nitrogen, phosphorus, or potassium, resulting in stunted growth, decreased photosynthesis, and reduced yields (Marschner, 2012). Nutrient insufficiency can result from causes like poor soil fertility, insufficient fertilisation, and imbalances in soil pH (Fageria et al., 2008). Nutrient deficit stress impacts plant physiology by decreasing leaf chlorophyll content, enzyme activity, and root growth, resulting in reduced yields (Marschner, 2012). Plants have evolved different strategies to deal with nutrient deficiency stress, including boosting root development, improving nutrient absorption efficiency, and transferring nutrients from ageing tissues (Hawkesford et al., 2012). Nevertheless, these systems might not be adequate to sustain agricultural yield in cases of severe or persistent nutrient deficiencies (Fageria et al., 2008). To address nutrient deficiency stress in agriculture, strategies such as improving fertilisation methods, utilising slow-release or controlled-release fertilisers, and implementing precision agricultural techniques are recommended (Bindraban et al., 2015).

Heavy Metal Toxicity Stress

Heavy metal toxicity stress in crops results from the accumulation of large levels of toxic metals like cadmium, lead, or arsenic in plant tissues, causing hindered growth, decreased photosynthesis, and oxidative harm (Rizwan et al., 2016). Heavy metal poisoning can result from industrial pollution, mining operations, and the utilisation of contaminated irrigation water or fertilisers (Sarwar et al., 2017). Heavy metal toxicity stress impacts plant physiology by interfering with enzyme activity, changing gene expression, and triggering oxidative stress (Emamverdian et al., 2015). Plants have developed different strategies to deal with the stress caused by heavy metal toxicity, including metal exclusion, sequestration in vacuoles, and the production of metal-binding proteins (Singh et al., 2016). Nevertheless, these systems

may not be adequate to sustain crop output in the presence of significant levels of metal contamination (Sarwar et al., 2017). Methods to limit the harmful effects of heavy metal toxicity in agriculture including phytoremediation, soil amendments to decrease metal availability, and the cultivation of metal-tolerant crop types (Ali et al., 2013).

3. NANOTECHNOLOGY SOLUTIONS FOR DROUGHT STRESS

Drought stress is a major abiotic stress that significantly impacts agricultural production globally, resulting in decreased crop yields and quality (Fahad et al., 2017). Nanotechnology provides innovative solutions for mitigating drought stress through the creation of materials and technologies designed to boost water retention, improve plant water absorption, and monitor soil moisture levels (Jampílek & Králová, 2017). This section will cover four nanotechnology-based methods to alleviate drought stress in agriculture: nano-based water retention materials, nano-sensors for monitoring soil moisture, nano-fertilizers for enhanced nutrient absorption during drought, and case studies and research illustrations (Singh et al., 2023 e).

Nano-Based Water Retention Materials

Nano-based water retention materials are designed to enhance soil water retention and decrease water loss via evaporation (Calabi-Floody et al., 2018). The materials mentioned are nano-clay composites, nano-hydrogels, and nano-porous materials (Sekhon, 2014). Nano-clay composites like bentonite and zeolite have a large surface area and cation exchange capacity, allowing them to absorb and retain water in the soil (Jiang et al., 2017; Singh at al., 2024). Nano-hydrogels are polymer networks that are cross-linked and have the ability to absorb and release water based on environmental cues like temperature and pH (Saha et al., 2020). Nano-porous materials like silica nanoparticles and carbon nanotubes have a high porosity and surface area, enabling them to retain and gradually release water to plants (Calabi-Floody et al., 2018). Nano-based water retention materials have demonstrated positive outcomes in enhancing soil water retention and plant growth in agriculture during drought circumstances. The incorporation of nano-clay composites into sandy soils enhanced water retention by up to 30% and enhanced the development and productivity of maize plants in times of water scarcity (Jiang et al., 2017; Sachan et al., 2023). Nano-hydrogels applied to soil improved water retention by up to 400% and boosted the development and output of wheat plants during drought conditions (Saha et al., 2020).

Nano-Sensors for Monitoring Soil Moisture

Nano-sensors are small devices capable of detecting and measuring physical, chemical, or biological factors at a very small scale, as described by Fraceto et al. (2016). Nano-sensors can be utilised in drought stress situations to monitor soil moisture levels in real-time, enabling farmers to enhance irrigation scheduling and preserve water resources (Xu et al., 2014). Nano-sensors for monitoring soil moisture can utilise many technologies including capacitive, resistive, or optical sensing (Preethi et al., 2021). Capacitive nano-sensors detect variations in the dielectric constant of the soil, which correlates with the soil moisture content (Wang & Qu, 2009). The sensors are composed of two electrodes divided by a dielectric substance, and the capacitance between the electrodes varies based on the soil moisture level (Wang & Qu, 2009). Resistive nano-sensors detect variations in electrical resistance in the soil, which diminishes as soil moisture content rises (Preethi et al., 2021). The sensors are comprised of two electrodes enclosed in a permeable matrix, and the electrical resistance between the electrodes varies based on the moisture content of the soil (Preethi et al., 2021). Optical nano-sensors detect variations in the optical characteristics of soil, including reflectance or absorbance, that are linked to the soil's moisture level (Kim et al., 2013). The sensors are comprised of a light source and a detector, and the optical signal varies based on the soil moisture level (Kim et al., 2013). Nano-sensors used for monitoring soil moisture in agriculture have demonstrated positive outcomes in enhancing irrigation efficiency and decreasing water usage. Capacitive nano-sensors in a vineyard decreased water usage by 30% without compromising crop output or quality (Brase et al., 2017). Optical nano-sensors in a maize field enhanced irrigation scheduling and boosted water use efficiency by 25% (Kim et al., 2013).

Nano-Fertilizers for Improved Nutrient Uptake Under Drought

Nano-fertilizers are designed materials that can provide nutrients to plants in a precise and regulated way, enhancing nutrient utilisation efficiency and minimising environmental consequences (DeRosa et al., 2010). Nano-fertilizers can improve plant nutrient uptake and boost drought tolerance in the presence of drought stress by supplying necessary nutrients in a bioavailable form (Rane et al., 2015). Nano-fertilizers can be developed using a range of materials, including polymers, clay minerals, and metal oxides (Rane et al., 2015). Polymer-based nano-fertilizers are composed of a polymer matrix that encloses nutrient molecules. These nutrients are gradually released in reaction to environmental stimuli like pH or temperature, as explained by DeRosa et al. (2010). Clay mineral-based nano-fertilizers like montmorillonite and kaolinite have a large surface area and cation exchange capacity, allowing them to absorb and release nutrients in a regulated way (Rane et al., 2015). Metal oxide nano-fertilizers like zinc oxide and iron oxide can supply micronutrients to plants in a readily usable form, enhancing plant development and stress resistance (Dimkpa & Bindraban, 2016). Nano-fertilizers have demonstrated positive outcomes in enhancing plant nutrient absorption and resilience to drought when used in agriculture. Polymer-based nano-fertilizers enhanced nutrient utilisation by up to 30% and boosted the growth and production of soybean plants in drought conditions (Rane et al., 2015). Zinc oxide nano-fertilizers improved the development and drought tolerance of wheat plants by enhancing antioxidant defence and lowering oxidative damage, as demonstrated by Dimkpa et al. (2017).

Case Studies and Research Examples

Numerous case studies and research examples illustrate the capability of nanotechnology in combating drought stress in agriculture. Suriyaprabha et al. (2014) studied how silica nanoparticles impact the growth and drought resistance of maize plants. The study shown that utilising silica nanoparticles enhanced plant growth, biomass accumulation, and water usage efficiency in conditions of drought stress. The authors credited these outcomes to the improved antioxidant defence and decreased oxidative damage in the plants treated with silica nanoparticles.

Abdel-Aziz et al. (2016) assessed the capability of chitosan nanoparticles to act as a water retention agent in sandy soils. The study demonstrated that incorporating chitosan nanoparticles into the soil enhanced water retention by up to 50% and boosted the growth and productivity of wheat plants in drought circumstances. The scientists proposed that chitosan nanoparticles created a protective layer on the soil, which decreased water evaporation and enhanced water accessibility for the plants.

Kottegoda et al. (2017) created a nano-fertilizer using hydroxyapatite nanoparticles that contain urea molecules to slowly release nitrogen to plants. The findings indicated that the nano-fertilizer enhanced nitrogen utilisation efficiency by 50% and boosted the growth and production of rice plants in drought conditions. The authors linked these effects to the regulated nitrogen release from the nano-fertilizer, aligning with the plant's nutrient requirements and decreasing nitrogen losses through leaching and volatilization.

4. NANOTECHNOLOGY SOLUTIONS FOR SALINITY STRESS

Salinity stress significantly impacts agricultural growth and yield in various places globally, especially in arid and semi-arid settings (Shrivastava & Kumar, 2015). Excessive soil salinity can impede plant water uptake, induce ionic toxicity, and hinder nutrient absorption, resulting in diminished growth and yield (Munns & Tester, 2008). Nanotechnology provides novel methods for combating salinity stress in agriculture through the creation of materials and devices that can decrease salt buildup in the soil, boost plant resilience to salt, and enhance crop development in saline environments (Nair et al., 2020). This section will cover four nanotechnology-based methods to alleviate salinity stress in agriculture: nano-adsorbents for soil salt removal, nano-encapsulated plant growth regulators, nano-based seed coatings for improved germination in saline conditions, and case studies and research illustrations.

Nano-Adsorbents for Salt Removal From Soil

Nano-adsorbents are specialised materials designed with a large surface area and a particular attraction to salt ions. They are utilised to eliminate surplus salts from the soil and alleviate salt-induced stress on plants (Alcívar et al., 2018). The materials mentioned are nano-zeolites, nano-clays, and nano-carbon-based adsorbents (Li et al., 2019). Nano-zeolites are porous aluminosilicate minerals with a high capacity to exchange cations and a preference for sodium and chloride ions, which are the primary elements found in salty soils (Alcívar et al., 2018). Nano-clays like montmorillonite and bentonite have a large surface area and charge density that allow them to adsorb and immobilise salt ions in the soil (Li et al., 2019). Nano-carbon-based adsorbents, such graphene oxide and carbon nanotubes, possess distinctive electrical and structural characteristics that enable them to engage with salt ions and eliminate them from the soil solution (Elmer & White, 2018). Nano-adsorbents used in agriculture have demonstrated positive outcomes in decreasing soil salinity and enhancing crop development in saline environments. Nano-zeolites added to saline soils decreased electrical conductivity by 50% and enhanced the growth and yield of tomato plants (Bina et al., 2018). Graphene oxide nano-adsorbents reduced sodium and chloride ion levels by 70% in soil and improved the germination and growth of wheat seedlings under salty conditions (Elmer & White, 2018).

Nano-Encapsulated Plant Growth Regulators

Plant growth regulators (PGRs) are natural or manufactured chemicals that can influence plant growth and development, and improve plant resistance to abiotic challenges such as salinity (Rademacher, 2015).

PGRs in agriculture are generally limited in effectiveness due to their poor stability, quick degradation, and low bioavailability (Grillo et al., 2016). Nanotechnology can address these constraints by enclosing PGRs in nanocarriers like polymeric nanoparticles, liposomes, and nanoemulsions. This method safeguards the PGRs from degradation, enhances their solubility and bioavailability, and facilitates their controlled release in plants (Grillo et al., 2016). Nano-encapsulated plant growth regulators (PGRs) have demonstrated potential in increasing plant resistance to salt and facilitating crop development in salty environments. An instance is the use of salicylic acid enclosed in chitosan nanoparticles, which boosted the salt tolerance of wheat plants by strengthening antioxidant defence, decreasing oxidative harm, and raising photosynthetic efficiency (Behboudi et al., 2018). The application of gibberellic acid enclosed in liposomes through foliar spray improved the development and productivity of soybean plants when exposed to salt stress. This was achieved by boosting the function of antioxidant enzymes and decreasing the build-up of reactive oxygen species (Asgari et al., 2018).

Nano-Based Seed Coatings for Enhanced Germination Under Salinity

Seed germination is a crucial phase in plant growth and is greatly affected by salinity stress (Ibrahim, 2016). Excessive salt levels in the soil might hinder seed germination by decreasing water absorption, interfering with metabolic functions, and leading to oxidative harm to the embryo (Ibrahim, 2016). Nanotechnology can tackle these difficulties by creating nanostructured seed coats that can shield the seeds from salt stress, improve water absorption, and stimulate germination (Hussein et al., 2020). Nano-based seed coatings consist of polymers, clays, and metal oxides that provide a protective layer over seeds, enhancing their ability to withstand salt (Hussein et al., 2020). Tomato seeds coated with chitosan nanoparticles showed improved germination and seedling growth under salt stress by enhancing water uptake, minimising ion leakage, and increasing antioxidant enzyme activity (Behboudi et al., 2019). The use of zinc oxide nanoparticles on maize seeds improved their germination and seedling vigour in saline circumstances by promoting the use of seed reserves and decreasing the presence of reactive oxygen species.

Case Studies and Research Examples

Numerous case studies and research examples illustrate the capability of nanotechnology in mitigating salinity stress in agriculture. Khaled et al. (2018) studied how nano-silicon dioxide impacted the development and salt tolerance of faba bean plants. The study demonstrated that applying nano-silicon dioxide to the leaves enhanced plant growth, levels of photosynthetic pigments, and the activity of antioxidant enzymes when plants were exposed to salt stress. The authors credited these effects to silicon's role in decreasing sodium absorption, enhancing potassium retention, and improving osmotic adjustment in the plants.

Shalaby et al. (2018) assessed the capability of nano-selenium as a seed primer to improve the salt tolerance of sunflower plants. The study found that treating sunflower seeds with nano-selenium improved germination rate, seedling growth, and antioxidant protection in salty circumstances. The scientists proposed that nano-selenium functioned as an antioxidant and growth stimulant, mitigating the negative impacts of salt stress on the plants.

Zhao et al. (2020) created a nano-fertilizer using chitosan nanoparticles containing potassium nitrate to improve the salt tolerance of cotton plants. The study shown that using nano-fertilizer led to higher plant height, biomass, and yield under saline circumstances compared to traditional fertiliser. The authors credited these outcomes to the gradual release of potassium from the nanoparticles, enhancing potassium nutrition and decreasing sodium accumulation in the plants.

5. NANOTECHNOLOGY SOLUTIONS FOR TEMPERATURE STRESS

Temperature stress, encompassing heat and cold stress, is a significant abiotic factor that negatively impacts crop growth, development, and productivity on a global scale (Hasanuzzaman et al., 2013). Heat stress in crops results from exposure to temperatures exceeding their optimal range, causing decreased photosynthesis, heightened respiration, and hindered reproductive growth (Fahad et al., 2017). Cold stress in crops is caused by low temperatures and can result in chilling injury, membrane damage, and decreased enzyme activity (Sanghera et al., 2011). Nanotechnology provides advanced solutions to address temperature stress in agriculture through the creation of materials and devices designed to shield crops from extreme temperatures, monitor crop temperature, and improve plant resilience to temperature stress (Calabi-Floody et al., 2018). This section will cover three nanotechnology-based strategies for combating temperature stress in agriculture: nano-based thermal insulation materials for crop protection, nano-sensors for monitoring crop canopy temperature, and nano-encapsulated antioxidants for heat stress tolerance. It will also include case studies and research examples.

Nano-Based Thermal Insulation Materials for Crop Protection

Nano-based thermal insulation materials are designed to create a shield against severe temperatures and minimise heat exchange between the surroundings and the crop (Mondal & Mishra, 2020). The materials can be used as coatings on greenhouse covers, mulches, or shade nets to help maintain ideal temperatures for crop growth (Mondal & Mishra, 2020). Nano-based thermal insulation materials consist of nano-silica aerogels, nano-clay composites, and nano-phase change materials (Xu et al., 2019; Singh et al., 2023a). Nano-silica aerogels are porous materials with low thermal conductivity and good insulation qualities, suitable for producing lightweight and transparent thermal insulation coatings for greenhouses. Nano-clay composites, including montmorillonite-polymer nanocomposites, have excellent thermal stability and are suitable for creating heat-resistant mulches and shade nets (Pavlidou & Papaspyrides, 2008). Nano-phase change materials, such paraffin wax nanoparticles, have the ability to absorb and release latent heat during phase transitions. This property can be utilised to control the temperature within greenhouses (Xu et al., 2019). Nano-based thermal insulation materials have demonstrated promising outcomes in agriculture by safeguarding crops from temperature stress and enhancing crop development and production. Nano-silica aerogel coatings on greenhouse covers decreased heat transfer by up to 50%, maintaining ideal temperatures for tomato development and leading to a 20% boost in output (Nasirian & Nouri, 2019). Nano-clay composite mulches used in a strawberry field decreased soil temperature by 5°C during the day and boosted yield by 15% compared to the control group (Xing et al., 2019).

Nano-Sensors for Monitoring Crop Canopy Temperature

Nano-sensors are miniature devices that can detect and quantify physical, chemical, or biological factors at the nanoscale (Fraceto et al., 2016). Nano-sensors can be utilised to monitor crop canopy temperature in real-time under temperature stress conditions. This enables farmers to enhance irrigation scheduling, mitigate heat damage, and enhance crop water usage efficiency (Soliman et al., 2020). Nano-sensors used for monitoring crop canopy temperature can utilise many technologies, including infrared thermometry, thermochromic materials, and wireless sensor networks (Soliman et al., 2020).

Infrared thermometry nano-sensors utilise infrared light from the crop canopy to remotely assess its temperature (Zhang et al., 2020). The sensors can be attached to drones or satellites to analyse the regional differences in crop temperature and pinpoint regions experiencing heat stress (Zhang et al., 2020). Thermochromic nano-sensors are materials that alter their colour based on temperature variations, serving as visual indicators of crop temperature (Mohajerani et al., 2019). The sensors can be used as coatings on plants or fruits to monitor their temperature and prevent heat damage (Mohajerani et al., 2019). Wireless nano-sensor networks comprise numerous nano-sensors positioned across the crop canopy. These sensors can wirelessly send temperature data to a central monitoring system (Soliman et al., 2020). These networks offer detailed spatial and temporal information on crop temperature, which can be used for precise watering and managing heat stress (Soliman et al., 2020). Nano-sensors used to monitor crop canopy temperature have demonstrated positive outcomes in enhancing crop water use efficiency and mitigating heat damage. In a maize field, the application of infrared thermometry nano-sensors pinpointed areas experiencing heat stress, allowing for precise watering. This approach led to a 25% decrease in water consumption and a 10% rise in crop output (Zhang et al., 2020). Thermochromic nano-sensors were used on apple fruits to detect early signs of sunburn damage, allowing for appropriate interventions like shading or cooling to reduce fruit loss (Mohajerani et al., 2019).

Nano-Encapsulated Antioxidants for Heat Stress Tolerance

Antioxidants are substances that can eliminate reactive oxygen species (ROS) and shield plants from oxidative harm induced by heat stress (Hasanuzzaman et al., 2013). Antioxidants are typically not very efficient in reducing heat stress due to their unstable nature, quick breakdown, and limited ability to be absorbed by the body (Quintero-Quiroz et al., 2019). Nanotechnology can address these constraints by enclosing antioxidants in nanocarriers like polymeric nanoparticles, liposomes, and cyclodextrins. This method safeguards the antioxidants from deterioration, enhances their solubility and bioavailability, and facilitates their regulated release in plants (Quintero-Quiroz et al., 2019).

Nano-encapsulated antioxidants have demonstrated potential in boosting plant resilience to heat stress and benefiting crop development and productivity in high-temperature environments. Qi et al. (2018) demonstrated that applying ascorbic acid encapsulated in chitosan nanoparticles on wheat plants boosted heat tolerance by boosting antioxidant enzyme activities, decreasing membrane damage, and promoting photosynthetic efficiency. The application of α-tocopherol encapsulated in solid lipid nanoparticles improved the heat tolerance of tomato plants by boosting osmoprotectant accumulation, lowering oxidative stress, and maintaining membrane stability (Alvi et al., 2019).

Case Studies and Research Examples

Numerous case studies and research examples illustrate the capability of nanotechnology in mitigating temperature stress in agriculture. Poursani et al. (2020) studied how zinc oxide nanoparticles impact the heat tolerance of potato plants. The study demonstrated that applying zinc oxide nanoparticles to the leaves enhanced plant growth, levels of photosynthetic pigments, and antioxidant enzyme activity in response to heat stress. The authors credited these results to zinc's role in strengthening membrane integrity, boosting osmotic adjustment, and lowering oxidative damage in the plants.

Fahad et al. (2020) assessed the use of nano-silicon as a thermal insulation material to reduce heat stress in rice plants. The study demonstrated that applying nano-silicon to the leaves decreased leaf temperature by 3°C and boosted grain yield by 12% in the presence of heat stress. The scientists proposed that nano-silicon created a protective layer on the leaf surface, decreasing the absorption of infrared radiation and enhancing the reflection of surplus heat.

Kaushal et al. (2019) created a nano-sensor using carbon nanotubes and graphene to monitor the canopy temperature of soybean plants. The nano-sensor demonstrated precise measurement of canopy temperature with a resolution of 0.1°C and a reaction time of under 1 second. The authors showed that the nano-sensor may be combined with a wireless network to monitor crop temperature in real-time and detect heat stress early.

6. NANOTECHNOLOGY SOLUTIONS FOR NUTRIENT DEFICIENCY STRESS

Nutrient deficiency stress significantly hinders crop development and yield on a large scale across millions of hectares of agricultural land globally (Bindraban et al., 2015). Nutrient insufficiency arises from either the absence of critical elements like nitrogen (N), phosphorus (P), and potassium (K) in the soil, or when these nutrients are present but inaccessible to plants due to variables like pH imbalance, soil texture, or water constraint (Dimkpa & Bindraban, 2016). Nutrient deficiency stress can result in inhibited growth, chlorosis, decreased photosynthesis, and reduced yields (Mengel & Kirkby, 2012). Nanotechnology provides advanced solutions for combating nutrient deficiency stress in agriculture through the creation of nano-fertilizers for precise nutrient delivery, nano-sensors for immediate nutrient tracking, and nano-based soil amendments for enhanced nutrient accessibility (Lowry et al., 2019). This section will cover three nanotechnology-based strategies to alleviate nutrient shortage stress in agriculture, including case studies and research illustrations.

Nano-Fertilizers for Targeted Nutrient Delivery

Nano-fertilizers are designed materials that can provide nutrients to plants in a precise and regulated way, enhancing nutrient utilisation efficiency and minimising environmental consequences (DeRosa et al., 2010; Singh et al., 2023c). Nano-fertilizers can be engineered to gradually release nutrients, aligning with crop nutrient requirements and reducing nutrient losses from leaching or volatilization (Dimkpa & Bindraban, 2016). Nano-fertilizers can be modified with certain ligands or coatings to target specific plant tissues or growth stages (Raliya et al., 2018).

Numerous nano-fertilizers have been created, such as nano-encapsulated fertilisers, nano-sized nutrient particles, and nano-clay-based fertilisers (Lowry et al., 2019). Nano-encapsulated fertilisers are made up of nutrient molecules enclosed in biodegradable polymers or other materials. This encapsulation helps shield the nutrients from degradation and allows for controlled release. Nano-sized nutrient particles like nano-hydroxyapatite or nano-ferrihydrite offer nutrients in a highly accessible form because of their extensive surface area and reactivity (Dimkpa & Bindraban, 2016). Nano-clay-based fertilisers like zeolite or montmorillonite can absorb and release nutrients according to the plant's needs and the soil's cation exchange capacity (Subramanian et al., 2015).

Nano-fertilizers have demonstrated encouraging outcomes in enhancing crop growth, production, and nutrient utilisation in agriculture (Singh et al., 2023). Nano-hydroxyapatite used as a phosphorus fertiliser boosted soybean plant growth by 33% and yield by 20% compared to traditional phosphorus fertilisers (Liu & Lal, 2014). Nano-clay-based fertilisers applied to wheat boosted nitrogen utilisation efficiency by 30% and grain production by 10% compared to conventional fertilisers (Subramanian et al., 2015).

Nano-Sensors for Real-Time Nutrient Monitoring

Nano-sensors are small devices capable of real-time detection and measurement of particular nutrients in soil or plant tissues (Fraceto et al., 2016). Nano-sensors offer detailed spatial and temporal information on nutrient levels and absorption, allowing for precise fertilisation and minimising nutrient wastage (Dubey & Mailapalli, 2016). Nano-sensors for nutrient monitoring can utilise many technologies, including electrochemical, optical, or biological sensing (Kim et al., 2019).

Electrochemical nano-sensors detect alterations in electrical characteristics, like conductivity or potential, caused by the interaction of the target nutrient with the sensor surface (Kim et al., 2019; Singh et al., 2022). Nanomaterials like carbon nanotubes or graphene can be used to create these sensors due to their large surface area and sensitivity (Mao et al., 2019). Optical nano-sensors identify alterations in optical characteristics, such fluorescence or absorbance, caused by the binding of a target nutrient to a particular recognition element, such as an enzyme or an antibody (Kim et al., 2019). Sensors can utilise quantum dots, plasmonic nanoparticles, or other nanomaterials with distinct optical characteristics (Petryayeva & Krull, 2011). Biological nano-sensors utilise living organisms like bacteria or plants as the sensing component to generate a detectable signal, like fluorescence or luminescence, when exposed to nutrient stress (Samanta et al., 2016).

Nano-sensors used for nutrient monitoring in agriculture have demonstrated positive outcomes in enhancing fertilisation efficiency and decreasing environmental consequences. Carbon nanotube-based electrochemical sensors were utilised to detect nitrates in soil, allowing for real-time monitoring of nitrogen levels and precise fertilisation. This led to a 15% decrease in nitrogen fertiliser usage and a 10% boost in maize production (Mao et al., 2019). Quantum dot-based optical sensors were used to detect phosphate in plant tissues, enabling early diagnosis of phosphorus deficiency and guiding foliar fertilisation. This led to a 20% increase in tomato yield (Choi et al., 2016).

Nano-Based Soil Amendments for Improved Nutrient Availability

Nano-based soil supplements are designed materials that can enhance soil characteristics and increase nutrient accessibility to plants (Subramanian et al., 2015; Singh et al., 2023; Singh et al., 2024a). These additions can be tailored to enhance soil water retention, decrease nutrient runoff, or stimulate soil microbial growth (Lowry et al., 2019). Nano-based soil amendments can consist of different components including nano-clays, nano-silica, or nano-organic matter (Dimkpa & Bindraban, 2016).

Nano-clays like zeolite or montmorillonite has a large surface area and cation exchange capacity, allowing them to absorb and hold nutrients in the soil and gradually release them to plants (Subramanian et al., 2015). Nano-silica, like silica nanoparticles or nanotubes, can enhance soil water retention and nutrient availability by forming a network of nanopores and channels in the soil (Haghighi et al., 2019). Nano-organic matter like nano-biochar or nano-compost can improve soil fertility and microbial activity by supplying carbon and nutrients (Mukhopadhyay et al., 2020; Singh et al., 2023).

Nano-based soil supplements have demonstrated positive outcomes in enhancing soil quality, nutrient availability, and crop output in agriculture. Nano-zeolite used as a soil supplement in maize enhanced the soil's cation exchange capacity by 20%, nitrogen utilisation efficiency by 25%, and grain yield by 15% compared to the control (Subramanian et al., 2015). Nano-biochar treatment in sandy soil led to a 30% increase in soil water holding capacity, a 40% increase in accessible phosphorus, and a 20% increase in wheat yield compared to the control (Mukhopadhyay et al., 2020).

Case Studies and Research Examples

Numerous case studies and research examples illustrate the potential of nanotechnology in alleviating nutrient deficit stress in agriculture. Raliya et al. (2018) created a nano-fertilizer using zinc oxide nanoparticles coated with urea for applying on mung bean leaves. The study shown that the nano-fertilizer enhanced zinc uptake by 30%, photosynthetic rate by 20%, and seed output by 25% in comparison to the standard zinc sulphate fertiliser. The authors linked these impacts to the increased bioavailability and movement of zinc from the leaves to the seeds.

Choi et al. (2016) created a nano-sensor using quantum dots to monitor phosphate levels in plant tissues in real-time. The sensor was made of cadmium telluride quantum dots that were modified with a phosphate-binding protein. This modification caused the sensor to emit a fluorescent signal that varied in intensity based on the concentration of phosphate present. The scientists showed that the sensor could identify phosphate deficiency in tomato leaves prior to observable symptoms and direct foliar phosphate fertilisation, leading to a 20% rise in fruit yield.

Haghighi et al. (2019) assessed the impact of nano-silica on soil characteristics and nutrient accessibility in a calcareous soil. The study shown that adding 1% nano-silica to the soil boosted water retention by 25%, phosphorus availability by 30%, and tomato yield by 15% compared to the untreated soil. The authors proposed that nano-silica enhanced soil structure and nutrient retention through the formation of a network of nanopores and the adsorption of nutrients on its surface.

7. NANOTECHNOLOGY SOLUTIONS FOR HEAVY METAL TOXICITY STRESS

Heavy metal toxicity stress is an increasing environmental issue in agriculture due to the buildup of toxic metals including cadmium (Cd), lead (Pb), and arsenic (As) in the soil (Rizwan et al., 2016). Heavy metal contamination can result from human activities such mining, smelting, industrial waste disposal, and the overuse of metal-containing fertilisers and pesticides (Awasthi et al., 2017). Plants can absorb heavy metals, leading to phytotoxicity and decreased crop growth and output (Yadav, 2010). Furthermore, heavy metals have the potential to infiltrate the food chain, leading to significant health hazards for both people and animals (Ali et al., 2013). Nanotechnology provides advanced solutions for dealing with heavy metal toxicity in agriculture through the creation of nano-adsorbents to remove heavy metals from soil, nano-encapsulated chelating agents for metal detoxification, and nano-sensors for detecting heavy metal contamination (Ashfaq et al., 2016). This section will cover three nanotechnology-based ways to reduce heavy metal toxicity stress in agriculture, including case studies and research examples.

Nano-Adsorbents for Heavy Metal Removal From Soil

Nano-adsorbents are specially designed materials having a large surface area and a particular attraction to heavy metals. They are utilised for extracting harmful metals from polluted soils (Anjum et al., 2016). The materials consist of nano-sized metal oxides, carbon nanotubes, graphene oxide, and nano-clay minerals (Kah et al., 2018). Nano-sized metal oxides including iron oxide (Fe_3O_4), titanium dioxide (TiO_2), and manganese oxide (MnO_2) may efficiently adsorb heavy metals because of their extensive surface area and reactive functional groups (Gong et al., 2019). Carbon nanotubes and graphene oxide possess distinctive electrical and structural characteristics that allow them to engage with heavy metals via electrostatic attraction, complexation, and ion exchange mechanisms (Jing et al., 2019). Nano-clay minerals like montmorillonite and bentonite have a layered structure and high cation exchange capacity, enabling them to absorb and immobilise heavy metals in the soil (Uddin, 2017).

Using nano-adsorbents to remove heavy metals from soil has demonstrated positive outcomes in decreasing the availability of metals and their absorption by plants. An experiment conducted by Gong et al. in 2019 showed that adding iron oxide nanoparticles to soil contaminated with Cadmium reduced the uptake of Cadmium by wheat plants by 50% and increased plant biomass by 30% compared to the control group. Graphene oxide application in Pb-contaminated soil reduced Pb buildup in maize shoots by 70% and improved plant growth and photosynthetic performance (Jing et al., 2019).

Nano-Encapsulated Chelating Agents for Metal Detoxification

Chelating agents are organic compounds that can form stable complexes with metal ions, reducing their bioavailability and toxicity in the soil (Shahid et al., 2017). However, the effectiveness of chelating agents in metal detoxification is generally restricted by their poor stability, quick degradation, and probable leakage to groundwater (Sarwar et al., 2017). Nanotechnology can circumvent these constraints by encapsulating chelating chemicals in nanocarriers, such as polymeric nanoparticles, liposomes, and dendrimers, which can preserve the agents from degradation, manage their release, and enhance their targeting to contaminated locations (Li et al., 2020).

Nano-encapsulated chelating compounds have demonstrated promising results in boosting metal detoxification and improving plant development in contaminated soils. For example, the application of ethylenediaminetetraacetic acid (EDTA) encapsulated in chitosan nanoparticles enhanced the Pb removal from a contaminated soil by 40% and reduced the Pb buildup in vetiver grass by 50% compared to free EDTA (Li et al., 2020). Similarly, the use of nitrilotriacetic acid (NTA) encapsulated in dendrimers increased the Cd phytoextraction by Indian mustard by 60% and reduced the Cd toxicity symptoms in the plants (Alvarez-Fernandez et al., 2016).

Nano-Sensors for Detecting Heavy Metal Contamination

Nano-sensors are small devices capable of detecting and measuring heavy metals in soil or water with great sensitivity and accuracy. Nano-sensors offer immediate and on-location monitoring of heavy metal pollution, allowing for early detection and prompt remedial measures (Rathoure, 2020). Nano-sensors for heavy metal detection can utilise many methods, including electrochemical, optical, or biological sensing (Kumar & Fouletier, 2019).

Electrochemical nano-sensors detect alterations in electrical characteristics, like current or potential, caused by the interaction of the target metal with the sensor surface (Kumar & Fouletier, 2019). Nanomaterials including gold nanoparticles, carbon nanotubes, and conductive polymers with high conductivity and catalytic activity can be used to create these sensors (Shtenberg et al., 2018). Optical nano-sensors can identify alterations in optical characteristics like fluorescence or colorimetry when a particular metal attaches to a certain recognition component like a molecular probe or a nanoparticle (Kumar & Fouletier, 2019). Sensors can utilise quantum dots, plasmonic nanoparticles, or upconversion nanoparticles with distinct optical properties to provide a detectable signal when metals bind to them (Zhang et al., 2015). Biological nano-sensors utilise live organisms like bacteria or plants as the sensing component, which can generate a detectable reaction like bioluminescence or colour alteration when in contact with heavy metals (Verma & Bhardwaj, 2015).

Nano-sensors have been effectively used to detect heavy metals in agriculture, aiding in monitoring soil and water pollution and directing repair efforts. An electrochemical sensor utilising gold nanoparticles allowed the detection of arsenic in groundwater with a detection limit of 0.1 ppb, far lower than the WHO drinking water guideline (Shtenberg et al., 2018). A quantum dot-based optical sensor was used to visually detect Cd in rice grains at concentrations as low as 0.1 μM, which is below the safety limit for human ingestion (Zhang et al., 2015).

Case Studies and Research Examples

Numerous case studies and research examples illustrate how nanotechnology can help alleviate heavy metal toxicity stress in agriculture. Gong et al. (2019) studied how iron oxide nanoparticles impacted the cleanup of a soil contaminated with Cd and the growth of wheat plants. The study demonstrated that using iron oxide nanoparticles at a concentration of 5 g/kg soil decreased Cd bioavailability by 50%, boosted plant biomass by 30%, and improved the plants' antioxidant defence system compared to the control group. The authors credited the effects to the iron oxide nanoparticles' high adsorption capacity and stability, which decreased Cd uptake and toxicity in the plants.

Li et al. (2020) created a nano-encapsulated chelating agent using EDTA-loaded chitosan nanoparticles to clean up soil contaminated with lead. The study found that using nano-encapsulated EDTA enhanced Pb elimination from the soil by 40%, decreased Pb buildup in vetiver grass by 50%, and enhanced plant growth and photosynthetic performance compared to free EDTA. The authors proposed that nano-encapsulation shielded EDTA from degradation, regulated its release, and improved its targeting to the polluted soil, leading to a more effective and eco-friendly remediation process.

Zhang and colleagues (2015) created a nano-sensor using quantum dots to visually detect the presence of Cd in rice grains. The sensor was made out of CdTe quantum dots that were modified with a Cd-binding aptamer, causing a fluorescent response when Cd was present. The authors showed that the sensor could detect Cadmium in rice grains at concentrations as low as 0.1 μM, which is below the safe threshold for human consumption. They also presented a straightforward and quick approach for on-site screening of Cadmium contamination in food samples.

8. CHALLENGES AND FUTURE PROSPECTS

Nanotechnology in agriculture may help manage abiotic challenges such drought, salt, temperature extremes, nutritional inadequacy, and heavy metal toxicity. To ensure safety, efficacy, and sustainability, nanotechnology-based agricultural solutions must overcome various obstacles (Faceto et al., 2016).

Potential Risks and Toxicity Concerns of Nanotechnology in Agriculture

Nanomaterial hazards and toxicity make nanotechnology in agriculture difficult (Tripathi et al., 2017). Nanomaterials have unique features and benefits, but they can also behave and respond differently from bulk materials, which might have unforeseen repercussions on the environment and human health (Madkour, 2020). Some nanomaterials can enter the ecosystem, accumulate in the food chain, and poison beneficial soil bacteria, pollinators, and aquatic animals (Simonin et al., 2020). Nanomaterials' durability, bioaccumulation, and transgenerational consequences in the agroecosystem are unknown (Rai et al., 2019). To address these issues, extensive risk assessment studies and standardised methodologies for nanomaterial safety and toxicity in agriculture are needed (Kah et al., 2018). A multidisciplinary strategy combining nanotechnology, toxicity, ecology, and agricultural sciences is needed (Prasad et al., 2017). Safer-by-design strategies like biodegradable and biocompatible nanoparticles can reduce risks and improve the sustainability of nanotechnology-based agricultural solutions (Iavicoli et al., 2017).

Policy and Regulation

Lack of clear and harmonised regulatory frameworks and policy guidelines hinders nanotechnology in agriculture (Amenta et al., 2015). Nanomaterials in agriculture are unregulated, which can hamper product development and commercialization (Kah et al., 2019). The regulatory criteria and licencing processes for nano-based agrochemicals and fertilisers might vary widely, which can hinder commerce and limit worldwide adoption (Hofmann et al., 2021). This requires a coordinated and adaptive regulatory framework to keep up with nanotechnology's rapid advances and ensure the safety and efficacy of nano-based agricultural products (Amenta et al., 2015). Policymakers, regulators, industry stakeholders, and scientists must work together to define, standardise, and approve nano-based goods (Kah et al., 2019). Public and stakeholder participation in decision-making can also boost agricultural nanotechnology trust and adoption (Hofmann et al., 2021).

Commercialization and Scaling Nanotechnology Solutions

The high cost of production, lack of infrastructure and supply chains, and little awareness and acceptance among farmers and consumers make nanotechnology solutions in agriculture difficult to commercialise and scale (Hossain et al., 2021). Specialised equipment, trained labour, and tight quality control can raise the cost and complexity of nanomaterial manufacture (Singh & Gupta, 2016). The complex soil and plant environment makes it difficult to design stable, dispersed, and targeted delivery methods for nano-fertilizers and nano-pesticides (Fraceto et al., 2016). To solve these issues, cost-effective and scalable nanomaterial and nano-based product production methods must be developed, such as green synthesis and manufacturing process optimisation (Singh & Gupta, 2016). Industry, academic, and government collaborations can harness resources, skills, and infrastructure to speed agro nanotechnology commercialization and acceptance (Hossain et al., 2021). Effective communication and education techniques can also help farmers and consumers accept nanotechnology and use it responsibly and sustainably (Prasad et al., 2017).

Future Research Prospects

Despite the challenges, nanotechnology in agriculture offers several potential and research paths. Future research should focus on multi-functional and smart nanomaterials that can handle several abiotic stressors and supply nutrients and agrochemicals selectively (Hossain et al., 2021). Nano-sensors and nano-delivery systems can monitor and control crop stress variables like drought, salt, and nutrient shortage in real time, optimising water and agrochemical use (Fraceto et al., 2016). The potential synergies and interconnections between nanotechnology and other emerging technologies like biotechnology, precision agriculture, and digital agriculture are another attractive study field (Hofmann et al., 2021). Nanotechnology and genome editing methods like CRISPR-Cas can create novel agricultural features and stress-resilient types (Abdallah et al., 2020). Nanotechnology combined with remote sensing and data analytics can create smart and precision agriculture systems that optimise crop management and resource utilisation (Prasad et al., 2017).

9. CONCLUSION

Nanotechnology in agriculture is a promising method to tackle difficulties including drought, salinity, severe temperatures, nutrient deficiencies, and heavy metal toxicity caused by abiotic pressures. Nanotechnology solutions such as nano-based materials for water retention, sensors, fertilisers, plant growth regulators, seed coatings, adsorbents, and chelating agents have shown great promise in increasing crop resilience to stress, enhancing nutrient utilisation, and supporting sustainable agriculture. Nanotechnology provides creative methods to tackle the intricate issues presented by abiotic stressors in agriculture. Researchers and practitioners can exploit the distinctive characteristics and capabilities of nanomaterials to create precise and efficient methods for increasing crop stress resistance, optimising nutrient utilisation, and encouraging sustainable farming techniques. Successful integration of nanotechnology in agriculture necessitates interdisciplinary collaboration, stakeholder involvement, and responsible innovation to guarantee its safety, effectiveness, and fair influence. Nanotechnology, via ongoing research and development, has the potential to transform how we tackle abiotic stress in agriculture and support the attainment of global food security and sustainability objectives.

ACKNOWLEDGMENT

TM acknowledge the support by the Ministry of Science and Higher Education of the Russian Federation (no. FENW-2023-0008) and the Strategic Academic Leadership Program of the Southern Federal University ("Priority 2030").

REFERENCES

Abdallah, N. A., Prakash, C. S., & McHughen, A. G. (2020). Nanotechnology and CRISPR-Cas9: Tools for a bio-based future. *Trends in Biotechnology*, *38*(11), 1177–1180.

Abdel-Aziz, H. M. M., Hasaneen, M. N. A., & Omer, A. M. (2016). Nano chitosan-NPK fertilizer enhances the growth and productivity of wheat plants grown in sandy soil. *Spanish Journal of Agricultural Research*, *14*(1), e0902. doi:10.5424/sjar/2016141-8205

Alcívar, M., Zurita-Silva, A., Sandoval, M., Muñoz, C., & Schoebitz, M. (2018). Reclamation of saline soils in Santa Elena province (Ecuador) using two types of zeolites in the cultivation of Zea mays. *Communications in Soil Science and Plant Analysis*, *49*(22), 2844–2858.

Ali, H., Khan, E., & Sajad, M. A. (2013). Phytoremediation of heavy metals—Concepts and applications. *Chemosphere*, *91*(7), 869–881. doi:10.1016/j.chemosphere.2013.01.075 PMID:23466085

Alvarez-Fernandez, M., Zornoza, P., Diaz-Ravina, M., & Núñez, A. (2016). Application of nanotechnology for the remediation of soils contaminated with heavy metals. In *Soil Remediation and Plants* (pp. 335–354). Academic Press.

Alvi, M. A., Akram, M. S., & Khan, S. (2019). α-Tocopherol-loaded solid lipid nanoparticles improve the heat stress tolerance in tomato (Solanum lycopersicum L.). *Journal of Plant Growth Regulation*, *38*(4), 1445–1455.

Amenta, V., Aschberger, K., Arena, M., Bouwmeester, H., Moniz, F. B., Brandhoff, P., ... Peters, R. J. (2015). Regulatory aspects of nanotechnology in the agri/feed/food sector in EU and non-EU countries. *Regulatory Toxicology and Pharmacology*, *73*(1), 463–476. doi:10.1016/j.yrtph.2015.06.016 PMID:26169479

Anjum, N. A., Gill, S. S., Gill, R., Hasanuzzaman, M., Duarte, A. C., Pereira, E., Ahmad, I., Tuteja, R., & Tuteja, N. (2016). Metal/metalloid stress tolerance in plants: Role of ascorbate, its redox couple, and associated enzymes. *Protoplasma*, *253*(5), 1265–1283. PMID:24682425

Asgari, F., Majd, A., Jonoubi, P., & Najafi, F. (2018). Effects of gibberellic acid and salinity on germination, seed reserve utilization and initial growth of soybean (Glycine max). *Iranian Journal of Plant Physiology*, *8*(2), 2433–2444.

Ashfaq, M., Singh, S., Sharma, A., & Verma, N. (2016). Cytotoxic evaluation of the hierarchal web of carbon micro-nanofibers. *Industrial Crops and Products*, *95*, 334–340.

Ashraf, M. (2010). Inducing drought tolerance in plants: Recent advances. *Biotechnology Advances*, *28*(1), 169–183. doi:10.1016/j.biotechadv.2009.11.005 PMID:19914371

Awasthi, A., Singh, P., Shukla, A. K., & Dubey, N. K. (2017). Nanotechnology and plant–pathogen interactions: An emerging tool for disease management. In *Nanotechnology in Plant Growth Promotion and Protection* (pp. 211–226). John Wiley & Sons, Ltd.

Behboudi, F., Mohamad Sarraf Mamoori, S., & Moghaddam, M. (2018). Nano-chitosan positively affected the performance of Triticum aestivum under salinity stress. *Journal of Plant Molecular Breeding*, *6*(1), 32–42.

Behboudi, F., Tahmasebi Sarvestani, Z., Kassaee, M. Z., Sanavi, S. A. M. M., Sorooshzadeh, A., & Ahmadi, S. B. (2019). The influence of chitosan nanoparticles on seed germination and seedling growth of wheat (Triticum aestivum) under salinity stress. *Agroecology and Sustainable Food Systems*, *43*(6), 696–712.

Bina, F., Zahedi, H., & Zare, M. (2018). The effect of zeolite and nano-zeolite on reducing salinity stress and increasing the yield and biomass growth of tomato. *Journal of Water and Soil Science*, *22*(1), 163–177.

Bindraban, P. S., Dimkpa, C., Nagarajan, L., Roy, A., & Rabbinge, R. (2015). Revisiting fertilisers and fertilisation strategies for improved nutrient uptake by plants. *Biology and Fertility of Soils*, *51*(8), 897–911. doi:10.1007/s00374-015-1039-7

Brase, T., Gries, P., & Sauerborn, K. (2017). Precision farming in vineyards: the use of sensor technology to optimize production. *Proceedings of the 20th GiESCO International Meeting*, Mendoza, Argentina, 415-420.

Calabi-Floody, M., Medina, J., Rumpel, C., Condron, L. M., Hernandez, M., Dumont, M., & de la Luz Mora, M. (2018). Smart fertilizers as a strategy for sustainable agriculture. [). Academic Press.]. *Advances in Agronomy*, *147*, 119–157. doi:10.1016/bs.agron.2017.10.003

Calabi-Floody, M., Medina, J., Suazo, J., Ordiqueo, M., Aponte, H., Mora, M. D. L. L., & Rumpel, C. (2018). Optimization of wheat biochar as a soil amendment to improve soil water retention capacity. *Journal of Soil Science and Plant Nutrition*, *18*(1), 81–93.

Choi, J. H., Kim, G. B., & Cha, H. J. (2016). Quantum dot-based nano-sensor for detection of phosphate in plant tissues. *Biosensors & Bioelectronics*, *77*, 1042–1047.

Cramer, G. R., Urano, K., Delrot, S., Pezzotti, M., & Shinozaki, K. (2011). Effects of abiotic stress on plants: A systems biology perspective. *BMC Plant Biology*, *11*(1), 163. doi:10.1186/1471-2229-11-163 PMID:22094046

Cuce, E., Cuce, P. M., Wood, C. J., & Riffat, S. B. (2014). Toward aerogel based thermal superinsulation in buildings: A comprehensive review. *Renewable & Sustainable Energy Reviews*, *34*, 273–299. doi:10.1016/j.rser.2014.03.017

DeRosa, M. C., Monreal, C., Schnitzer, M., Walsh, R., & Sultan, Y. (2010). Nanotechnology in fertilizers. *Nature Nanotechnology*, *5*(2), 91–91. doi:10.1038/nnano.2010.2 PMID:20130583

Dimkpa, C. O., & Bindraban, P. S. (2016). Fortification of micronutrients for efficient agronomic production: A review. *Agronomy for Sustainable Development*, *36*(1), 7. doi:10.1007/s13593-015-0346-6

Dimkpa, C. O., Singh, U., Bindraban, P. S., Elmer, W. H., Gardea-Torresdey, J. L., & White, J. C. (2017). Zinc oxide nanoparticles alleviate drought-induced alterations in sorghum performance, nutrient acquisition, and grain fortification. *The Science of the Total Environment*, *688*, 926–934. doi:10.1016/j.scitotenv.2019.06.392 PMID:31726574

Dubey, A., & Mailapalli, D. R. (2016). Nanofertilisers, nanopesticides, nanosensors of pest and nanotoxicity in agriculture. In *Sustainable Agriculture Reviews* (pp. 307–330). Springer. doi:10.1007/978-3-319-26777-7_7

Elmer, W. H., & White, J. C. (2018). The use of nanoparticles to enhance the activity of fungicides and herbicides against plant pathogens and weeds. In *Nanoscience in Food and Agriculture 5* (pp. 81–103). Springer.

Fahad, S., Adnan, M., Noor, M., Arif, M., Alam, M., Khan, I. A., Ullah, H., Wahid, F., Mian, I. A., Jamal, Y., Basir, A., Hassan, S., Saud, S., & Amanullah. (2020). Major constraints for global rice production. In Advances in Rice Research for Abiotic Stress Tolerance (pp. 1-22). Woodhead Publishing.

Farooq, M., Wahid, A., Kobayashi, N., Fujita, D., & Basra, S. M. A. (2009). Plant drought stress: Effects, mechanisms and management. *Agronomy for Sustainable Development, 29*(1), 185–212. doi:10.1051/agro:2008021

Fraceto, L. F., Grillo, R., de Medeiros, G. A., Scognamiglio, V., Rea, G., & Bartolucci, C. (2016). Nanotechnology in agriculture: Which innovation potential does it have? *Frontiers in Environmental Science, 4*, 20. doi:10.3389/fenvs.2016.00020

Gong, X., Huang, D., Liu, Y., Zeng, G., Wang, R., Wan, J., Zhang, C., Cheng, M., Qin, X., & Xue, W. (2019). Remediation of contaminated soils by biotechnology with nanomaterials: Bio-behavior, applications, and perspectives. *Critical Reviews in Biotechnology, 39*(8), 1034–1048. PMID:28903604

Grillo, R., Pereira, A. E., Nishisaka, C. S., de Lima, R., Oehlke, K., Greiner, R., & Fraceto, L. F. (2014). Chitosan/tripolyphosphate nanoparticles loaded with paraquat herbicide: An environmentally safer alternative for weed control. *Journal of Hazardous Materials, 278*, 163–171. doi:10.1016/j.jhazmat.2014.05.079 PMID:24968252

Gupta, A., Rico-Medina, A., & Caño-Delgado, A. I. (2020). The physiology of plant responses to drought. *Science, 368*(6488), 266–269. doi:10.1126/science.aaz7614 PMID:32299946

Haghighi, M., Afifipour, Z., & Mozafarian, M. (2019). Effect of silicon and nano-silicon on growth and physiological traits of basil (Ocimum basilicum L.) under vegetative and generative growth stages. *Journal of Plant Nutrition, 42*(10), 1057–1068.

Hasanuzzaman, M., Nahar, K., Alam, M. M., Roychowdhury, R., & Fujita, M. (2013). Physiological, biochemical, and molecular mechanisms of heat stress tolerance in plants. *International Journal of Molecular Sciences, 14*(5), 9643–9684. doi:10.3390/ijms14059643 PMID:23644891

Hofmann, T., Lowry, G. V., Ghoshal, S., Tufenkji, N., Brambilla, D., Dutcher, J. R., ... Wilkinson, K. J. (2021). Technology readiness and overcoming barriers to sustainably implement nanotechnology-enabled plant agriculture. *Nature Food, 2*(6), 416–425.

Hossain, M., Siddique, M. N. A., Bari, M. S., Hasan, M. M., Abedin, M. A., & Shahid, M. (2021). Nanotechnology applications in plant breeding, yield and crop management. In *Nanomaterials for Agriculture and Forestry Applications* (pp. 35–72). Elsevier.

Hussein, M. M., Abou-Baker, N. H., & Tawfik, M. M. (2020). The prospective roles of nano-materials and nano-technology in germination and seedling growth of some crop species. In *Plant Nanobionics* (pp. 177–209). Springer.

Iavicoli, I., Leso, V., Beezhold, D. H., & Shvedova, A. A. (2017). Nanotechnology in agriculture: Opportunities, toxicological implications, and occupational risks. *Toxicology and Applied Pharmacology*, *329*, 96–111. doi:10.1016/j.taap.2017.05.025 PMID:28554660

Ibrahim, E. A. (2016). Seed priming to alleviate salinity stress in germinating seeds. *Journal of Plant Physiology*, *192*, 38–46. doi:10.1016/j.jplph.2015.12.011 PMID:26812088

IPCC. (2014). Climate Change 2014: Synthesis Report. Contribution of Working Groups I, II and III to the Fifth Assessment Report of the Intergovernmental Panel on Climate Change [Core Writing Team, R.K. Pachauri and L.A. Meyer (eds.)]. IPCC, Geneva, Switzerland.

Jampílek, J., & Králová, K. (2017). Nanomaterials for delivery of nutrients and growth-promoting compounds to plants. In *Nanotechnology: An Agricultural Paradigm* (pp. 177–226). Springer. doi:10.1007/978-981-10-4573-8_9

Jiang, C., Zu, C., Lu, D., Zheng, Q., Shen, J., Wang, H., & Li, D. (2017). Effect of exogenous selenium supply on photosynthesis, Na+ accumulation and antioxidative capacity of maize (Zea mays L.) under salinity stress. *Scientific Reports*, *7*(1), 1–14. doi:10.1038/srep42039 PMID:28169318

Jing, X., Jiang, M., Li, Y., Wang, Y., Zhang, X., Tang, L., Zhang, Y., & Chen, Y. (2019). Immobilization of lead contaminated soil using a new class of iron phosphate nanoparticles with reference to their reduction of bioavailability and leachability. *Chemosphere*, *227*, 178–186.

Kah, M., Kookana, R. S., Gogos, A., & Bucheli, T. D. (2018). A critical evaluation of nanopesticides and nanofertilizers against their conventional analogues. *Nature Nanotechnology*, *13*(8), 677–684. doi:10.1038/s41565-018-0131-1 PMID:29736032

Kah, M., Navarro, D., Kookana, R. S., Kirby, J. K., Santra, S., Ozcan, A., & Kabiri, S. (2018). Fate of engineered nanoparticles in natural and engineered soils. Engineered Nanoparticles and the Environment: Biophysicochemical Processes and Toxicity, 30-59.

Kah, M., Tufenkji, N., & White, J. C. (2019). Nano-enabled strategies to enhance crop nutrition and protection. *Nature Nanotechnology*, *14*(6), 532–540. doi:10.1038/s41565-019-0439-5 PMID:31168071

Kaushal, M., Kumar, A., & Kaushal, R. (2019). Carbon nanotube and graphene-based nanomaterials for agriculture. In *Nanomaterials for Agriculture and Forestry Applications* (pp. 111–142). Elsevier.

Khaled, A. Y., Aziz, S. A., Bejo, S. K., Nawi, N. M., & Ismail, I. (2018). Foliar application of nano-silica positively influenced the growth and productivity of faba bean (Vicia faba) grown under different concentrations of NaCl. *Environmental Science and Pollution Research International*, *25*(33), 33137–33146.

Khan, M. N., Mobin, M., Abbas, Z. K., & Siddiqui, M. H. (2017). Nitric oxide-induced tolerance to salt stress in plants: Present and future prospects. In *Nitric Oxide Action in Abiotic Stress Responses in Plants* (pp. 111–127). Springer.

Kim, H., Sudduth, K. A., & Hummel, J. W. (2013). Soil macronutrient sensing for precision agriculture. *Journal of Environmental Monitoring*, *11*(10), 1810–1824. doi:10.1039/b906634a PMID:19809703

Kim, K. Y., Park, S. A., & Kim, J. E. (2019). Nanomaterial-based optical sensors for determining nutrients in the soil and crop. *Biosystems Engineering*, *185*, 141–159.

Kottegoda, N., Sandaruwan, C., Priyadarshana, G., Siriwardhana, A., Rathnayake, U. A., Arachchige, D. M. B., Kumarasinghe, A. R., Dahanayake, D., Karunaratne, V., & Amaratunga, G. A. (2017). Urea-hydroxyapatite nanohybrids for slow release of nitrogen. *ACS Nano*, *11*(2), 1214–1221. doi:10.1021/acsnano.6b07781 PMID:28121129

Kumar, P., & Fouletier, J. (2019). Electrochemical sensors for heavy metals: Principles, operating modes, and recent developments. In *Trends in Environmental Analytical Chemistry* (Vol. 23, p. e00063). Elsevier.

Lahiani, M. H., Souli, H., & Cabot, P. (2021). Nanotechnology for sustainable agriculture and food security. In *Nanomaterials for Agriculture and Forestry Applications* (pp. 1–32). Elsevier.

Li, J., Zhang, M., Sun, B., Xing, G., Song, Y., Gui, R., & Huang, D. (2020). Carbon quantum dots as fluorescence sensors for sensitive and selective detection of heavy metal ions: A review. *Journal of Biomedical Nanotechnology*, *16*(8), 1089–1102. PMID:33397551

Li, Y., Zhang, Q., Zhang, J., Wu, L., Qi, Y., & Zhou, J. M. (2019). Identification of microRNAs involved in pathogen-associated molecular pattern-triggered plant innate immunity. *Plant Physiology*, *152*(4), 2222–2231. doi:10.1104/pp.109.151803 PMID:20164210

Liu, R., & Lal, R. (2014). Synthetic apatite nanoparticles as a phosphorus fertilizer for soybean (Glycine max). *Scientific Reports*, *4*(1), 1–6. doi:10.1038/srep05686 PMID:25023201

Lowry, G. V., Avellan, A., & Gilbertson, L. M. (2019). Opportunities and challenges for nanotechnology in the agri-tech revolution. *Nature Nanotechnology*, *14*(6), 517–522. doi:10.1038/s41565-019-0461-7 PMID:31168073

Madkour, L. H. (2020). Advanced nanomaterials for agro-food application. In *Advanced Nanomaterials for Membrane Synthesis and Its Applications* (pp. 365–381). Elsevier.

Mantri, N., Patade, V., Penna, S., Ford, R., & Pang, E. (2012). Abiotic stress responses in plants: present and future. In *Abiotic Stress Responses in Plants* (pp. 1–19). Springer. doi:10.1007/978-1-4614-0634-1_1

Mao, X., Liu, X., Cheng, J., Liu, X., & Huang, Y. (2019). A portable soil nitrogen detector based on NIRS and a smartphone. *Sensors (Basel)*, *19*(20), 4451. PMID:31615081

Marschner, P. (2012). *Marschner's mineral nutrition of higher plants*. Academic press.

Mengel, K., & Kirkby, E. A. (2012). *Principles of plant nutrition*. Springer Science & Business Media.

Mickelbart, M. V., Hasegawa, P. M., & Bailey-Serres, J. (2015). Genetic mechanisms of abiotic stress tolerance that translate to crop yield stability. *Nature Reviews. Genetics*, *16*(4), 237–251. doi:10.1038/nrg3901 PMID:25752530

Mohajerani, S., Fazel, M. A., Madani, H., & Shahrjerdi, A. (2019). Potential of thermochromic materials for maintaining the quality of horticultural crops in the post-harvest stage. *Food Packaging and Shelf Life*, *22*, 100394.

Mondal, S., & Mishra, S. (2020). Nano-based thermal insulation of buildings. In *Emerging Nanotechnologies for Green Building* (pp. 1–28). Elsevier.

Mukhopadhyay, S. S., Gurav, R. V., & Sabarinathan, N. (2020). Nano amendments for improving soil fertility and crop productivity. In *Nanomaterials for Agriculture and Forestry Applications* (pp. 111–141). Elsevier.

Munns, R., & Tester, M. (2008). Mechanisms of salinity tolerance. *Annual Review of Plant Biology, 59*(1), 651–681. doi:10.1146/annurev.arplant.59.032607.092911 PMID:18444910

Nair, R., Varghese, S. H., Nair, B. G., Maekawa, T., Yoshida, Y., & Kumar, D. S. (2020). Nanoparticulate material delivery to plants. *Plant Science, 179*(3), 154–163. doi:10.1016/j.plantsci.2010.04.012

Nasirian, R., & Nouri, S. (2019). Improving tomato growth and yield by using of nano-silica coating on the greenhouse cover. *Acta Scientiarum Polonorum. Hortorum Cultus, 18*(1), 3–13.

Parisi, C., Vigani, M., & Rodríguez-Cerezo, E. (2015). Agricultural nanotechnologies: What are the current possibilities? *Nano Today, 10*(2), 124–127. doi:10.1016/j.nantod.2014.09.009

Pavlidou, S., & Papaspyrides, C. D. (2008). A review on polymer–layered silicate nanocomposites. *Progress in Polymer Science, 33*(12), 1119–1198. doi:10.1016/j.progpolymsci.2008.07.008

Petryayeva, E., & Krull, U. J. (2011). Localized surface plasmon resonance: Nanostructures, bioassays and biosensing—A review. *Analytica Chimica Acta, 706*(1), 8–24. doi:10.1016/j.aca.2011.08.020 PMID:21995909

Poursani, A. V., Gerami, M., & Nezhad, M. N. (2020). Zinc oxide nanoparticles improve heat tolerance in potato (Solanum tuberosum L.) plants. *Journal of Plant Physiology, 255*, 153293.

Prasad, R., Bhattacharyya, A., & Nguyen, Q. D. (2017). Nanotechnology in sustainable agriculture: Recent developments, challenges, and perspectives. *Frontiers in Microbiology, 8*, 1014. doi:10.3389/fmicb.2017.01014 PMID:28676790

Preethi, B., Sivasubramanian, V., & Muthuraman, P. (2021). Nano-sensors for measuring soil moisture: A review. *Nanotechnology for Environmental Engineering, 6*(1), 1–14.

Qi, M., Liu, Y., & Li, T. (2018). Chitosan nanoparticles enhance the antioxidant activity of ascorbic acid in wheat seedlings under heat stress. *Molecules (Basel, Switzerland), 23*(11), 2908. PMID:30413008

Quintero-Quiroz, C., Botero-Botero, L. R., & Zárate-Triviño, D. G. (2019). Nanotechnology: An alternative for the development of potent antioxidant systems in plants under abiotic stress conditions. *Journal of the Science of Food and Agriculture, 99*(14), 6141–6155.

Rademacher, W. (2015). Plant growth regulators: Backgrounds and uses in plant production. *Journal of Plant Growth Regulation, 34*(4), 845–872. doi:10.1007/s00344-015-9541-6

Rai, V., Das, G., & Saxena, S. (2019). Nanomaterials toward the detection of abiotic stress in plants. In P. K. Singh, A. Kumar, & M. W. Siddiqui (Eds.), *Controlled Release Fertilizers for Sustainable Agriculture* (pp. 213–235). Academic Press.

Rajput, V. D., Singh, A., Minkina, T., Rawat, S., Mandzhieva, S., Sushkova, S., Shuvaeva, V., Nazarenko, O., Rajput, P., Komariah, Verma, K. K., Singh, A. K., Rao, M., & Upadhyay, S. K. (2021). Nano-Enabled Products: Challenges and Opportunities for Sustainable Agriculture. *Plants, 10*(12), 2727. doi:10.3390/plants10122727 PMID:34961197

Raliya, R., Saharan, V., Dimkpa, C., & Biswas, P. (2018). Nanofertilizer for precision and sustainable agriculture: Current state and future perspectives. *Journal of Agricultural and Food Chemistry*, *66*(26), 6487–6503. doi:10.1021/acs.jafc.7b02178 PMID:28835103

Rane, J., Deshpande, M., Jagadish, S. V. K., Nagarajan, M., & Gambhir, P. (2015). Improving water use efficiency by using nano-fertilizers. *Indian Journal of Plant Physiology / Official Publication of the Indian Society for Plant Physiology*, *20*(4), 357–362.

Rathoure, A. K. (2020). Nanomaterials for wastewater treatment and water purification. In *Nanomaterials for Sustainable Energy and Environmental Remediation* (pp. 1–23). Elsevier.

Rizwan, M., Ali, S., Adrees, M., Ibrahim, M., & Rehman, M. Z. u., Farid, M., Qayyum, M. F., Nawaz, R., Hussain, A., Chatha, S. A. S., & Abbas, F. (2016). Mechanisms of silicon-mediated alleviation of drought and salt stress in plants: A review. *Environmental Science and Pollution Research International*, *23*(5), 4416–4431. PMID:26335528

Rizwan, M., Ali, S., Rizvi, H., Rinklebe, J., Tsang, D. C., Meers, E., Ok, Y. S., & Ishaque, W. (2016). Phytomanagement of heavy metals in contaminated soils using sunflower: A review. *Critical Reviews in Environmental Science and Technology*, *46*(18), 1498–1528. doi:10.1080/10643389.2016.1248199

Rizwan, M., Ali, S., Zia-ur-Rehman, M., Adrees, M., Arshad, M., Qayyum, M. F., Ali, L., Hussain, A., Chatha, S. A. S., & Imran, M. (2019). Alleviation of cadmium accumulation in maize (Zea mays L.) by foliar spray of zinc oxide nanoparticles and biochar to contaminated soil. *Environmental Pollution*, *248*, 358–367. doi:10.1016/j.envpol.2019.02.031 PMID:30818115

Sachan, D. S., Reddy, K. J., Sulochna, Saini, Y., Rai, A. K., Singh, O., & Laxman, T. (2023). Assessing grain yield and achieving enhanced quality in maize by next generation fertilizer: A review. *International Journal of Environment and Climate Change*, *13*(8), 626–637. doi:10.9734/ijecc/2023/v13i81991

Saha, A., Sekharan, S., & Manna, U. (2020). Superabsorbent hydrogel (SAH) as a soil amendment for drought management: A review. *Soil & Tillage Research*, *204*, 104736. doi:10.1016/j.still.2020.104736

Samanta, A., Deng, Z., & Liu, Y. (2016). Nanomaterials for microbial biosensors. In *NanoBiosensors* (pp. 135–166). Academic Press.

Sanghera, G. S., Wani, S. H., Hussain, W., & Singh, N. B. (2011). Engineering cold stress tolerance in crop plants. *Current Genomics*, *12*(1), 30–43. doi:10.2174/138920211794520178 PMID:21886453

Sarwar, N., Imran, M., Shaheen, M. R., Ishaque, W., Kamran, M. A., Matloob, A., Rehim, A., & Hussain, S. (2017). Phytoremediation strategies for soils contaminated with heavy metals: Modifications and future perspectives. *Chemosphere*, *171*, 710–721. doi:10.1016/j.chemosphere.2016.12.116 PMID:28061428

Sekhon, B. S. (2014). Nanotechnology in agri-food production: An overview. *Nanotechnology, Science and Applications*, *7*, 31. doi:10.2147/NSA.S39406 PMID:24966671

Servin, A. D., & White, J. C. (2016). Nanotechnology in agriculture: Next steps for understanding engineered nanoparticle exposure and risk. *NanoImpact*, *1*, 9–12. doi:10.1016/j.impact.2015.12.002

Shahid, M., Shamshad, S., Rafiq, M., Khalid, S., Bibi, I., Niazi, N. K., Dumat, C., & Rashid, M. I. (2017). Chromium speciation, bioavailability, uptake, toxicity and detoxification in soil-plant system: A review. *Chemosphere, 178*, 513–533. doi:10.1016/j.chemosphere.2017.03.074 PMID:28347915

Shalaby, T. A., Bayoumi, Y., Alshaal, T., Elhawat, N., Sztrik, A., & El-Ramady, H. (2018). Selenium fortification induces growth, antioxidant activity, yield and nutritional quality of lettuce in salt-affected soil using foliar and soil applications. *Plant and Soil, 423*(1), 319–332.

Shrivastava, P., & Kumar, R. (2015). Soil salinity: A serious environmental issue and plant growth promoting bacteria as one of the tools for its alleviation. *Saudi Journal of Biological Sciences, 22*(2), 123–131. doi:10.1016/j.sjbs.2014.12.001 PMID:25737642

Shtenberg, G., Massad-Ivanir, N., & Segal, E. (2018). Optical and electrochemical sensing platforms based on nanoporous alumina: A review. *Nanomaterials (Basel, Switzerland), 8*(11), 918. PMID:30405041

Shukla, P., Chaurasia, P., Younis, K., Qadri, O. S., Faridi, S. A., & Srivastava, G. (2019). Nanotechnology in sustainable agriculture: Studies from seed priming to post-harvest management. *Nanotechnology for Environmental Engineering, 4*(1), 1–11. doi:10.1007/s41204-019-0058-2

Simonin, M., Cantarel, A. A., Crouzet, A., Gervaix, J., Martins, J. M., & Richaume, A. (2020). Negative effects of copper oxide nanoparticles on carbon and nitrogen cycle microbial activities in contrasting agricultural soils and in presence of plants. *Frontiers in Microbiology, 11*, 1769. PMID:30619181

Singh, A., Rajput, V., Varshney, A., Sharma, R., Ghazaryan, K., Minkina, T., . . . El-Ramady, H. (2024b). Revolutionizing Crop Production: Nanoscale Wonders - Current Applications, Advances, and Future Frontiers. Egyptian Journal of Soil Science, 64(1). doi:10.21608/ejss.2023.246354.1684

Singh, A., Rajput, V. D., Pandey, D., Sharma, R., Ghazaryan, K., & Minkina, T. (2023c). Nano Zinc-Enabled Strategies in Crops for Combatting Zinc Malnutrition in Human Health. *Frontiers in Bioscience (Landmark Edition), 28*(8), 158. doi:10.31083/j.fbl2808158 PMID:37664935

Singh, A., Rajput, V. D., Varshney, A., Ghazaryan, K., & Minkina, T. (2023a). Small Tech, Big Impact: Agri-nanotechnology Journey to Optimize Crop Protection and Production for Sustainable Agriculture. *Plant Stress, 10*, 100253. doi:10.1016/j.stress.2023.100253

Singh, A., Sharma, R., Rajput, V. D., Ghazaryan, K., Minkina, T., Al Tawaha, A. R. M., & Varshney, A. (2023e). Green Synthesis of Nanofertilizers and Their Application for Crop Production. Nanotechnology in the Life Sciences, 205–231. doi:10.1007/978-3-031-41329-2_8

Singh, A., Sharma, R., Rawat, S., Singh, A. K., Rajput, V. D., Fedorov, Y., . . . Chaplygin, V. (2022). Nanomaterial-plant interaction: Views on the pros and cons. Toxicity of Nanoparticles in Plants, 47–68. doi:10.1016/B978-0-323-90774-3.00015-5

Singh, H., & Gupta, R. (2016). Synthesis and characterization of carbon nanotubes for advanced research and development. In *Nanocarbon and Its Composites* (pp. 79–94). Elsevier.

Singh, O. (2023). Nanotechnology for sustainability and food security in agriculture: A nanopriming long story in short. In A. Singh, V. Rajput, K. Ghazaryan, S. Gupta, & T. Minkina (Eds.), *Nanopriming Approach to Sustainable Agriculture* (pp. 315–339). IGI Global. doi:10.4018/978-1-6684-7232-3.ch014

Singh, O., Shahi, U.P., Dhyani, B.P., Kumar, S., Vivek, S., & Sengar, R.S., Shivangi, and Singh, A. (2023). Effect of zinc oxide nanoparticles application on growth and yield of basmati rice (Oryza sativa) in alkaline soil. *Indian Journal of Agronomy*, *68*(3), 1–6.

Singh, O., Singh, S., Singh, V. K., & Singh, A. (2023). Biochar: An organic amendment for sustainable soil health. In *Encyclopedia of Green Materials* (pp. 265–271). Springer Nature Singapore Pte Ltd. doi:10.1007/978-981-16-4921-9_265-1

Singh, S., Singh, O., & Shahi, U. P. (2023). Nanotechnology: A New Innovation in Biofortification and Food Safety. In A. Singh, V. Rajput, K. Ghazaryan, S. Gupta, & T. Minkina (Eds.), *Nanopriming Approach to Sustainable Agriculture* (pp. 340–351). IGI Global. doi:10.4018/978-1-6684-7232-3.ch015

Singh, V. K., Gill, A. A. S., Singh, O., Singh, S., & Shahi, U. P. (2024). Zeolite: A Natural Mineral for Sustainable Agriculture. In *Encyclopedia of Green Materials* (pp. 1–10). Springer Nature Singapore Pte Ltd. doi:10.1007/978-981-16-4921-9_263-1

Singh, Y. K., Singh, B. V., Katiyar, D., Saikanth, D. R. K., Kumar, K., Singh, O., Verma, S., & Kumar, P. (2023). Efficacy of nano fertilizers on yield, attributes and economics of wheat. *International Journal of Environment and Climate Change*, *13*(7), 291–297. doi:10.9734/ijecc/2023/v13i71879

Soliman, A. S., Hassan, M., Abou-Elella, F., & Ahmed, A. H. H. (2020). Application of nanomaterials for precision agriculture and food quality assessment: A review. In *Nanotechnology Applications in Environmental Engineering* (pp. 240–263). IGI Global.

Subramanian, K. S., Manikandan, A., Thirunavukkarasu, M., & Rahale, C. S. (2015). Nano-fertilizers for balanced crop nutrition. In *Nanotechnologies in Food and Agriculture* (pp. 69–80). Springer. doi:10.1007/978-3-319-14024-7_3

Suriyaprabha, R., Karunakaran, G., Yuvakkumar, R., Rajendran, V., & Kannan, N. (2014). Silica nanoparticles for increased silica availability in maize (Zea mays L.) seeds under hydroponic conditions. *Current Nanoscience*, *8*(6), 902–908. doi:10.2174/157341312803989033

Tripathi, D. K., Shweta, Singh, S., Singh, S., Pandey, R., Singh, V. P., Sharma, N. C., Prasad, S. M., Dubey, N. K., & Chauhan, D. K. (2017). An overview on manufactured nanoparticles in plants: Uptake, translocation, accumulation and phytotoxicity. *Plant Physiology and Biochemistry*, *110*, 2–12. doi:10.1016/j.plaphy.2016.07.030 PMID:27601425

Uddin, M. K. (2017). A review on the adsorption of heavy metals by clay minerals, with special focus on the past decade. *Chemical Engineering Journal*, *308*, 438–462. doi:10.1016/j.cej.2016.09.029

Verma, N., & Bhardwaj, A. (2015). Biosensor technology for pesticides—A review. *Applied Biochemistry and Biotechnology*, *175*(6), 3093–3119. doi:10.1007/s12010-015-1489-2 PMID:25595494

Vurukonda, S. S. K. P., Vardharajula, S., Shrivastava, M., & SkZ, A. (2016). Enhancement of drought stress tolerance in crops by plant growth promoting rhizobacteria. *Microbiological Research*, *184*, 13–24. doi:10.1016/j.micres.2015.12.003 PMID:26856449

Wang, Y., & Qu, J. (2009). Research progress of soil moisture sensor based on capacitance. *Sensors & Transducers*, *108*(9), 73–81.

Xing, Y., Yan, W., Zhang, Y., Bian, G., Zhang, Y., Yang, G., & Wang, Q. (2019). Effects of nano-clay composite coatings on the quality of strawberries under ambient temperature. *Coatings*, *9*(6), 360.

Xu, J., Li, M., Niu, W., Wu, J., & Yang, W. (2019). Preparation and application of phase change microcapsules in thermal energy storage and temperature control of greenhouses. *Journal of Cleaner Production*, *247*, 119111.

Xu, L., Zhang, W., & Ding, Y. (2014). Review of nano sensors for monitoring the soil moisture. *Applied Mechanics and Materials*, *651*, 1517–1522. doi:10.4028/www.scientific.net/AMM.494-495.1517

Yadav, S. K. (2010). Heavy metals toxicity in plants: An overview on the role of glutathione and phytochelatins in heavy metal stress tolerance of plants. *South African Journal of Botany*, *76*(2), 167–179. doi:10.1016/j.sajb.2009.10.007

Zhang, C., Wang, H., Hao, H., Li, Y., Jiang, L., & Zhou, J. (2015). Functionalized quantum dots for biosensing applications. In *Nano-Oncologicals* (pp. 73–88). Springer.

Zhang, H., Han, W., De Smet, I., Talboys, P., Loya, R., Hassan, A., Rong, H., Jürgens, G., Paul Knox, J., & Wang, M. H. (2020). Rapid activation of heat stress response pathways by small molecules promotes plant thermotolerance. *Nature Chemical Biology*, *16*(2), 148–152. PMID:31974524

Zhao, Q., Xu, J., Chen, J., Zhang, L., & Zhang, W. (2020). Slow-release of potassium fertilizer encapsulated by starch-based superabsorbent polymer and its effect on growth of cotton. *Carbohydrate Polymers*, *241*, 116255.

Chapter 17
Nano-Omics-Based Abiotic and Biotic Stresses Management

Priyanka Upadhyay
Lovely Professional University, India

Sonia Navvuru
https://orcid.org/0009-0005-5762-6025
University of Alberta, Edmonton, Canada

Praveen Kumar Yadav
ICAR-Indian Agricultural Research Institute, India

Shivani Lalotra
https://orcid.org/0009-0007-1515-1134
Department of Agronomy, School of Agriculture, Lovely Professional University, India

Abhishek Singh
https://orcid.org/0000-0002-0049-6910
Yerevan State University, Armenia

Vishnu D. Rajput
https://orcid.org/0000-0002-6802-4805
Southern Federal University, Russia

Tatiana Minkina
Southern Federal University, Russia

Karen Ghazaryan
https://orcid.org/0000-0002-8973-2595
Faculty of Biology, Yerevan State University, Armenia

ABSTRACT

In the face of escalating pressures from climate change, environmental degradation, and escalating pest and disease burdens, traditional agricultural practices are proving insufficient to ensure food security and sustainable crop production. Nanoomics, a pioneering interdisciplinary field, offers a promising avenue for addressing these challenges by harnessing the unique properties of nanomaterials and integrating economic principles to devise innovative solutions. By addressing both abiotic and biotic stresses, this chapter delves into the diverse strategies and mechanisms employed to enhance crop resilience and productivity. Through a multidisciplinary approach encompassing nanoscience, plant biology, and agricultural economics, this chapter examines the synthesis, characterization, and deployment of nanomaterials for stress detection, monitoring, and mitigation.

INTRODUCTION

Nanoomics, a convergence of nanotechnology and omics sciences, holds immense promise in addressing the complex challenges of abiotic and biotic stress in agriculture (Usmat et al., 2020; Demirer et al., 2021).

DOI: 10.4018/979-8-3693-1890-4.ch017

Abiotic stressors such as drought, salinity, and extreme temperatures, along with biotic stress factors like pests and pathogens, jeopardize global food security by impeding crop growth and yield (Singh et al., 2024; Khan et al., 2021). Nanoomics harnesses the unique properties of nanomaterials to engineer tailored solutions for stress management in agriculture. By leveraging nanoscale phenomena, such as increased surface area and enhanced reactivity, nanomaterials can effectively target and modulate plant responses to stressors (Singh et al., 2022; Singh et al., 2023a). For instance, nanoparticles can improve water use efficiency, nutrient uptake, and stress tolerance, thereby bolstering crop productivity under adverse environmental conditions (Liu et al., 2021; Singh et al., 2022).

Nanotechnology offers a suite of tools to combat abiotic stressors prevalent in agriculture. Nano-enabled irrigation systems and soil amendments enhance water retention and nutrient availability in arid and saline soils, mitigating the impacts of drought and salinity stress on crops (Shola et al., 2023b). Furthermore, nano-fertilizers and smart delivery systems ensure efficient nutrient uptake by plants, minimizing nutrient deficiencies and optimizing growth even in nutrient-poor environments (Singh et al., 2023c). In the realm of biotic stress management, nanotechnology presents novel approaches for pest and disease control. Nanoparticle-based formulations of pesticides and antimicrobial agents offer targeted delivery and prolonged efficacy, reducing chemical usage and environmental contamination (Singh et al., 2021; 2023d; 2023e; 2024a). Additionally, nanobiosensors enable early detection of pathogens and pests, facilitating timely interventions to prevent disease outbreaks and crop losses. Despite its immense potential, the widespread adoption of nanoomics in agriculture faces several challenges, including safety concerns, regulatory hurdles, and ethical considerations (Ghazaryan et al., 2018; Singh et al., 2022). Addressing these challenges requires interdisciplinary collaborations, robust risk assessment frameworks, and stakeholder engagement to ensure responsible and sustainable deployment of nanotechnology in agriculture (Rajput et al., 2021; Ghazaryan et al., 2023). Looking ahead, continued research and innovation in nanoomics hold the promise of transformative solutions to mitigate abiotic and biotic stress, thereby safeguarding global food security and sustainability (Khan et al., 2021; Shoala et al., 2023).

Nanoomics represents a paradigm shift in agricultural research and practice, offering tailored solutions to mitigate abiotic and biotic stressors threatening global food security. By harnessing the unique properties of nanomaterials, nanoomics promises to revolutionize crop protection strategies, enhance resilience, and promote sustainable agriculture. As we navigate the challenges and opportunities of this burgeoning field, collaboration and responsible stewardship will be key to harnessing the full potential of nanotechnology in addressing the pressing challenges of abiotic and biotic stress in agriculture (Singh et al., 2023f; Vardumyan et al., 2024). This chapter aims to elucidate the potential applications of nanoomics in mitigating abiotic and biotic stress in agriculture. By reviewing recent advancements and emerging trends in nanotechnology-enabled interventions, we endeavor to provide insights into how nanoomics can revolutionize crop protection strategies and contribute to sustainable agricultural development.

TYPES OF NANOMATERIALS COMMONLY EMPLOYED IN AGRICULTURE

The burgeoning global population is placing increasing demands on agriculture to boost production, all while contending with diminishing natural resources. Consequently, the development and utilization of sustainable inputs in farming have become imperative. Nanomaterials, new-age materials possessing unique properties, are gaining popularity as versatile and efficient inputs for revolutionizing crop production. Several instances of employing nanomaterials to boost agricultural production have been recorded and will be discussed in detail in the following sections. Nanomaterials comprise a wide range of nanoparticle-based formulations that provide nutrition (nano fertilizers), inhibit pests and diseases (nano fungicides and nano insecticides), improve water use efficiency (TiO_2), smother weeds (nano herbicides), serve as biosensors for monitoring and diagnosis (nano sensors/ nano biosensors), and deliver targeted genes (smart gene delivery - nanocarriers) (Fig. 1).

These new age materials are low contaminating nano formulation that support sustainable agriculture (Fraceto et al., 2016) and can be applied either foliarly or through soil (Fig. 2) (Singh et al., 2024).

Nanoparticles

A nanoparticle (NP) is any material that is between 1-100 nm in size and can have either one- or three-dimensional nanoscale structure. Nanoparticles can be broadly classified as Natural nanoparticles and synthetic nanoparticles. Natural nano-particles of different properties and sizes can be observed, such as nano clays, cytochrome c, lipoproteins, lycopene carotenoid, exosomes, ferritin, magnetosomes, viruses etc. Synthetic, manufactured, or engineered nanoparticles (ENPs), on the other hand, are artificially synthesized under controlled conditions (Phogat et al. 2016). ENPs are briefly classified in figure1.

Nano Sensors

Nano sensors are another advanced application of nanomaterials for having precise knowledge of the status of plants and their surroundings. Sensitive analytical methods and remarkable advancements in nanofabrication technology have allowed the production of responsive sensors. Nano sensors are not only useful in regulating the nutritional or management inputs for the crop but also in analyzing the harmful

Figure 1. Artificially engineered nanoparticles commonly employed in agriculture

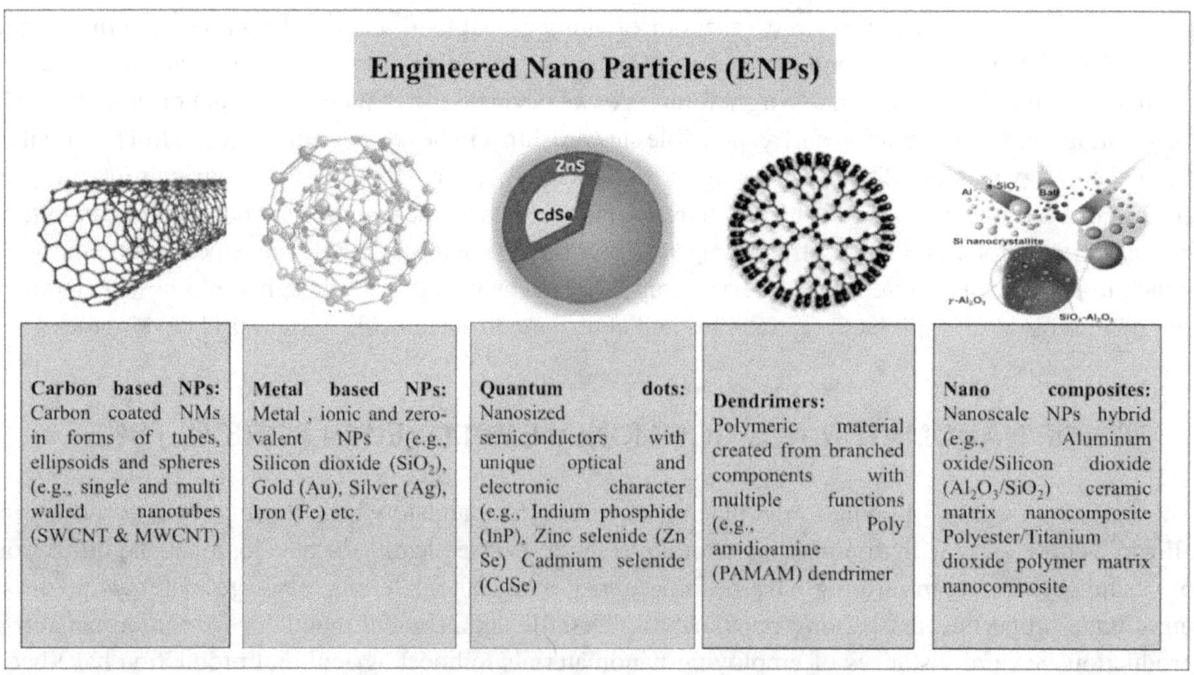

Figure 2. Application of nanoparticles (NPs) via foliar (Figure 1) and soil (Figure 2) methodology. Those administered foliarly (A) are absorbed through stomata, while soil-applied NPs (B) are transported via symplastic and apoplastic pathways, utilizing (C) xylem and phloem to reach various plant parts (Singh et al., 2024c)

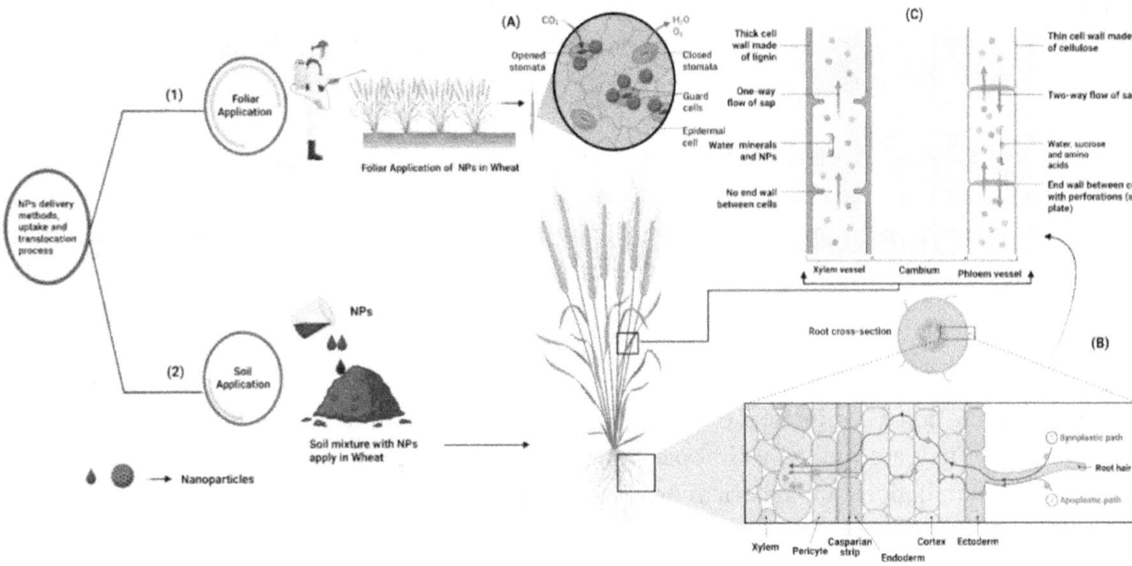

impact, if only on the natural resources. Nano biosensors are small- sized bioreceptors on a transducer that emit signals upon identification of a single or multiple analytes (Arduini et al. 2016). The analytes could be diverse range of proteins, hormones, fertilizers, pesticides, and metabolites. Based on their operating principles, different biosensors can be classified as follows (Alvarado et al. 2017):

Physical: Nanotubes or nanoribbons change their electronic properties when there is a deformity. For instance, a localized deformation of a pipe, for instance, causes a voltage change.

Chemical: The basic idea behind this feature is that different types of molecules can alter the electrical characteristics of nanotubes or nanoribbons, changing the number of electrons that can move around in the system. Consequently, the voltage threshold is altered by the presence of a certain kind of molecule.

Mechanical: Distance, movement, position, and level are the scale of mechanical measurement

Biological: Comprising Biological recognition system (bioreceptor) such as enzymes, proteins, antibodies, enzymes, and transducers such as magnetic detectors, electrochemical detectors, and optical transducers

NMs as sensors aid in faster detection with high sensitivity and low sample requirement. NMs such as metal NPs, CNTs, graphene and QDs are efficient in detecting heavy metals, pesticides and other pollutants in soil and water (Zhang and Fang, 2010). A Combining of fluoro immunoassay-based sensing method and NMs such as cadmium telluride (CdTe) QDs were employed to detect 2,4-dichlorophenoxyactic acid (Vinayaka et al., 2009). Further, employing QDs high size- tunable emission perperty, photostability and fluorescent property of QDs several harmful pesticides were detected (Vinayaka and Thakur, 2010), for instance methiocarb (Hua et al., 2010), chlorpyrifos (Chen et al., 2010), pyrethroid (Kranthi et al., 2009) and atrazine (Cummins et al., 2006).

NMs are also efficient in acting as rapid diagnostic biomarkers for detecting biotic and abiotic stresses. In wheat, AuNP- based immunosensors were used for detecting *Tilletia indica*, a fungal causal organism for karnal bunt (Singh et al., 2010). Yao et al. (2009) used a combination of fluorescent silica nanoparticles (FSNP) with antibody molecules and detected *Xanthomonas axonopodis* pv. *Vesicastoria*, causal organism of bacterial spot disease in Solanaceae. Plants release volatile compounds in response to a pathogen a attack. The plant volatile signature detection enables faster detection and management before much damage is caused. Through a electrochemical sensing system Fang et al. (2014) employed TiO2and SnO2 to detect p-ethylguaiacol, a volatile released upon infection with fungi *Phytophthora cactorum*. Nano sensors are efficient tools for real time crops and environmental interaction monitoring. They enable evaluation of crop under different environmental stress by monitoring the biochemical changes in the plant through nanosensors and meteorological monitoring through hyperspectral imaging cameras, drones, agricultural devices embedded with smartphone etc.

Nano Emulsions

Nano emulsions are non- equilibrated, complex, kinetically stable colloidal systems synthesized through nanotechnology. They are optically isotopic, ranging in size ranging from 20 to 200 nm (Saritha et al., 2022). Nano emulations are spherical droplets in aqueous homogenous suspension of oil or water encapsulating active and functional compounds (Srilatha, 2011, Vega-V´asquez et al 2020). Based on their mode of manufacture and composition their properties and functions differ. In order to improve their stability while storage they are dehydrated by removing water (Srilatha, 2011). Nano emulsions are efficient in maintaining their morphological stability under Extreme circumstances including high pressure, salinity, and temperature which is attributed to their ability to remain unaffected from their change in oil volume ratio (Sun et al 2017). Through their research, Wang et al. (2007) found increased efficiency and stability of water- insoluble pesticide, β- cypermethrin in the form of nano emulsion. Similarly, neem oil nano emulsion had increased larvicidal effect, despite having smaller droplets (Anjali et al., 2012; Narayanan et al., 2013).

APPLICATIONS AND IMPLICATIONS OF NANOMATERIALS IN AGRICULTURE

Safeguarding Crops Against Biotic and Abiotic Stressors

As human civilization has progressed, our interventions in systematic crop production have led to an exponential increase in the cultivation of homogeneous crops. This trend towards uniformity in crop genetics has resulted in a narrowing of their adaptability and resistance to environmental adversaries. Consequently, several pests and diseases that were once minor issues for crops with a wide genetic base are now highly prevalent. Biotic factors such as insects, pathogenic diseases, bacterial diseases, and viral diseases, as well as abiotic factors such as drought, waterlogging, and heat stress, contribute to massive crop losses.

Nanoparticles in Combatting Biotic Stressors

To manage biotic challenges, chemical management has been popular among farmers for many years. However, in recent years, it has been realized that pesticides and fungicides, besides their superficial benefits, are toxic to the environment, degrade soil fertility, and pose hazards to human and animal health (Meena et al., 2020). Therefore, it is imperative to design safer chemicals that are efficient and less harmful. The use of nanotechnology to produce nano pesticides, nano nematodes, and other plant protective nanomaterials has emerged as a novel and advanced solution that minimizes waste, enhances food quality, and reduces global crop losses (Jampílek et al., 2017).

Nanostructured alumina (NSA) is a nanoengineered material synthesized through the oxidation of metal with a fixed electric charge via the triboelectrification process. The negatively charged NSA particles interact with the positive charges of insects through strong electrical bonding, leading to dehydration, detachment of the cuticle layer, and ultimately the death of the insect (Stadler et al., 2018). Furthermore, nanoparticles have been shown to enhance the efficiency of conventional transgenically produced Bt-cotton in insect resistance. Zhao et al. (2020) found the potential of CuNPs in upregulating the exogenous protein expression of Bt toxin from Bacillus thuringiensis, thereby enhancing bollworm resistance. The photocatalytic activity and antibacterial quality of TiO2 and ZnO have been found to deter the spot disease pathogen X. perforans in tomato crops (Paret et al., 2013, Singh et al., 2024b). More examples of NMs and NPS efficient in mitigating biotic stress are presented in Table 1.

Nanoparticles in Combatting Abiotic Stressors

In the quest to boost production, crops have been spread from their natural habitats to extreme conditions, exposing them to high and low temperatures, drought, salt stress, metal toxicity, UV radiation, waterlogging, and other challenges. These unnatural conditions induce stress in the crops, affecting their molecular mechanisms, triggering the expression of stress-responsive genes and prompting the production of antioxidant enzymes vital for scavenging harmful substances in these harsh environments (Rejeb et al., 2014). The most important plant metabolic process, photosynthesis, is highly susceptible to oxidative and osmotic stress. The main components of the photosynthetic machinery, Rubisco, photosystem II (PS II) and ATP are majorly affected by stress (eg., salinity, waterlogging, drought, heat, nutrient deficiency). Therefore, reducing these stresses is essential for the normal functioning of a plant. Organic osmolytes such as trehalose, polyols, and amino acids like glycine and proline play a significant protective role in facing these environmental adversaries (Burg and Ferraris, 2008). Various studies have demonstrated the involvement of NPs in enhancing these protective mechanisms (Table 2).

The main signature of the negative influence of an abiotic stress is expressed by the ROS species released in plant organelles. Although the plants are equipped with enzymatic machinery to combat certain level of oxidative stress, but there many instances where these stresses exceed the defenses, and we lose the crop. In such instances, NMs facilitate the mitigation of stress by accumulating osmolytes, delivering nutrients and aminoacids and activating specific genes. SiO2 NPs has been reported to enhance plant transpiration rate, total chlorophyll content, water use efficiency and carbonic anhydrase activity when exposed to salt stress in *Cucurbita pepo* (Siddiqui et al., 2014). Another study by Mingyu et al. (2008) in spinach found similar benefits of TiO_2 NPs (anatase) in inhibiting linolenic acid in electron transport chain, modifying photoreduction and reducing the rate of Oxygen production in chloroplasts. NPs have the ability to replicate the actions of antioxidant enzymes in the form of nano-enzymes that can scavenge from oxidative stress (Sharifi et al., 2020).

Table 1. Nanomaterial-based detection and mitigation of biotic stress in agriculture

Nanomaterial (NMs)	Nanoparticle (NP) Component	Stress/Causal Organism	Role	Reference
NPs	$Cu(OH)_2$ NPs (Kocide 3000)	*Escherichia coli, Bacillus subtilis* and fungal pathogens; *F. oxysporum, C. lunata, A. alternata,* and *P. Destructive*	Antibacterial efficiency superior to eh conventional fungicide; bavistin (Devistin, carbendazim 50% WP)	Yoon et al., 2007
	Cu & K_2SiO_3 NPs	Clavibacter *michiganensis*	Combination reduces infection in hydroponically grown tomato plants by stimulating enzymatic and non-enzymatic metabolites essential for defense	Cumplido-Nájera et al., 2019
	CuO & Cu NPs	*S. littoralis*	Insecticidal against Bt-transgenic and conventional cotton by enhancing exogenous gene expression of Bt toxin	Ayoub et al., 2018
	CeO_2 NPs	Tomato Fusarium wilt	Reduces disease severity by 53 to 57%	Adisa et al., 2018
	AgNPs	*Meloidogyne spp.*	Kills or Significantly reduces nematode population when applied in appropriate concentration	Cromwell et al., 2014
	Ti O_2 NPs	Spot disease in tomato caused by *X. perforans*	Significantly lower bacterial spot severity due to strong photocatalytic activity and antibacterial potential	Paret et al., 2013
	MgO NPs	*Ralstonia solanacearum* infection in Tomato	Root pretreatment significantly reduces disease incidence, mediated through upregulation of the salicylic acid-inducible PR1, jasmonic acid-inducible LoxA, ethylene-inducible Osm, and GluA associated to systemic resistance	Imada et al., 2016
	MgO NPs	Tobacco bacterial wilt induced by *R. solanacearum*	Nematicide included via physical injury of cell membranes and ROS accumulation	Imada et al., 2016
	Si NPs	*Aspergillus spp* infection in maize	causes phenolic chemicals to be regulated, which increases resistance	Suriyaprabha et al., 2012
	DNA directed Ag NPs grown on graphene oxide (GO) AgNPs	Bacterial spot of tomato caused by *Xanthomonas perforans*	Significantly declines X. perforans activity	Ocsoy et al., 2013
	Artemisia absinthium extracted biosynthesized AgNPs	Phytophthora	Suppresses infection and enhances plant survivability	Ali et al., 2015
Nanotubes	SWCNTs, multiwalled carbon nanotubes (MWCNTs), graphene oxide (GO), reduced graphene oxide (rGO), fullerene (C60)	*Fusarium graminearum* and *Fusarium poae*	Strong anti fungal activity by inhibiting water uptake and inducing plasmolysis	Wang et al., 2014
Nanosheets	$Cu_3(PO_4)_2{\cdot}3H_2O$	*Fusarium oxysporum* f.sp. *niveum*	Suppress root fungal disease progression in water melon (*Citrullus lanatus*)	Borgatta et al., 2018
Nanoformulations of conventional pesticides Cellulose/silica nanocomposites	Tebuconazole	Rust and molds	Doubles release rate efficiency	Mattos and Magalhães, 2016

continued on following page

Table 1. Continued

Nanomaterial (NMs)	Nanoparticle (NP) Component	Stress/Causal Organism	Role	Reference
Nanoclusters	Ag core-DHPAC shell nanocluster	*Phytophthora capsici, Phytophthora colocasiae,* and *Phytophthora nicotianae*	Reduces infection by upto 80%	Ho et al., 2015
Porous hollow silica	2,4-dichlorophenoxyacetate	Weeds	Enhanced duration of herbicide activity	Bin Hussein et al., 2005

Table 2. Nanomaterials in managing of abiotic stress in agriculture

NPs	Type of the Abiotic Stress	Crop	Role	Reference
TiO_2	Drought	Wheat (*Triticum aestivum*)	Increases seedling dry weight, total chlorophyll, carotenoids, relative water content, antioxidant enzymes and improves stomatal conductance and transpiration rate	Faraji and Sepehri (2020)
		Wheat (T. *aestivuml. c. v* " Pisshtaz)	Enhances growth and yield, Increases starch and gluten content	Jaberzadeh et al., 2013
		Flax (*Linum usitatissimum*)	Decreases MDA activity and H2O2 levels. Elevates carotenoid and chlorophyll content	Aghdam et al., 2016
		Chickpea (*Cicer arietinum* L.)	Reduces electrolyte leakage index and MDA levels	Mohammadi et al., 2013
ZnO	Drought	Soybean (*Glycine max*)	Improves percentage and rate of germination but decreases dry weight	Sedghi et al., 2013
	Salinity stress	Okra (*Abelmoschus esculentus* L. *Moench*)	Increases photosynthetic pigments, elevates SOD and CAT activity, decreases proline and total soluble sugar contents	Alabdallah and Alzahrani, 2020
ZnO and Si	Salinity stress	Mango (*Mangifera indica* L.)	Enhances nutrient uptake, carbon assimilation and plant growth	Elsheery et al., 2020
Yttrium doped Fe_2O_3	Drought	Canola (*Brassica napus*)	Increases leaf growth rate and chlorophyll content, decreases H2O2a and MDA levels	Palmqvist et al., 2017
Fe_2O_3	Salinity	Pipermint (*Mentha piperita* L.)	Suppresses antioxidant activity, proline and MDA levels	Askary et al., 2017
$FeSO_4$	Salinity	Sunflower (*Helianthus annuus* L. cvs. Alstar, Olsion, Yourflor, Hysun36, and Hysun33)	Reduced sodium (Na) content, increased leaf area, shoot dry weight, sub-stomatal CO2 concentration (Ci), net carbon dioxide (CO2) assimilation rate (A), chlorophyll content, Fv/Fm, and iron (Fe) content	Torabian et al., 2017
FeNPs	Cadmium toxicity and drought	Wheat (*Triticum aestivum*)	Increased photosynthesis, decreased concentration of cadmium, increased growth and physiology, and Fe concentration	Adrees et al., 2020
SiO_2	Water deficit and Salinity stress	Banana (*Musa acuminata*, "Grand Nain")	Increases shoot growth, photosynthesis, and chlorophyll content. Stabilizes K+ and Na+ balance and decreases cell wall damage	Mahmoud et al., 2020
	Salinity	Tomato (*Solanum lycopersicum* L.)	Elevates and downregulates salt stress related gene expression to promote tolerance	Almutairi, 2016
AgNPs	Salinity stress	Wheat (*Triticum aestivum*)	Reduced activity of antioxidant enzymes, elevated activity of POD, proline, and sugar content	Mohamed et al., 2017
	Heat stress	Wheat (*Triticum aestivum*)	Enhances morphological growth of the plant, including increased leaf area and number, improved root and shoot length, and increased root number	Iqbal et al., 2019
	Salinity stress	Fenugreek (*Trigonella foenum-graecum*)	Improves germination rate, freshness, and dry mass	Hojjat and Kamyab (2017)

continued on following page

Table 2. Continued

NPs	Type of the Abiotic Stress	Crop	Role	Reference
Al_2O_3	Flooding stress	Soybean (*Glycine max* L. cv. Enrei)	Lengthened hypocotyl, enhanced protein synthesis linked to glycolysis, and increased protein in the mitochondrial membrane	Mustafa and Komatsu (2016)
Cu NPs	Drought	Maize (*Zea mays*)	Increased overall seed quantity, grain yield, anthocyanin, chlorophyll, and carotenoid levels, as well as higher leaf water content and plant biomass	Van Ha et al., 2020
SeNPs	High and low temperature stress	*Lycopersicum esculentum* Mill. cv. "Halil"	Increased root volume, fresh and dried weight, and diameter as well as improved shoot freshness	Haghighi et al., 2014
SeNPs	Heat stress	Sorghum (*Sorghum bicolor* L.)	Improved antioxidant system and thylakoid membrane composite and integrity	Djanaguiraman et al., 2018
SiNPs	UV-B stress	Wheat (*Triticum aestivum*)	Boosted antioxidants to guard against oxidative damage caused by UV- B	Tripathi et al., 2017
Chitosan-PVA and CuNPs	Salinity	Tomato (*Solanum lycopersicum* L.)	Increases chlorophyll a, b and carotenoid, Vit C, lycopene and SOD content and improved growth	Hernández-Hernández et al., 2018
CNTs and graphene	Drought	Cotton (*Gossypium hirustum*)	Improves fiber biomass, root and shoot growth of young seedlings	Pandey et al., 2019
Poly(acrylic) CeO NPs	Multiple stress	*Arabidopsis thaliana*	Enhanced rates of photosystem II quantum yield, carbon absorption, and rubisco carboxylation to scavenge ROS and photosynthetic processes	Wu et al., 2017

Enhancing Nutrient Use Efficiency and Fertigation with Nanoparticles

Major and minor nutrients in appropriate amounts are essential for the growth and metabolism of a plant. Deficiency of nutrients makes the plants more susceptible to pests and diseases. The extraneous application of crop nutrients, such as ammonium salts, phosphate compounds, nitrates, and urea, has undeniably led to drastic increases in crop yields. However, these progressive yields have come at a cost: environmental pollution from their industrial production, degradation of inert soil microflora, toxic salt accumulation in the soil, and nutrient leaching into water bodies, resulting in eutrophication. Therefore, nanomaterials, as nutrient delivery agents, are effective in targeted and rationed nutrient release. Nanoparticles have become an important source for the major and micronutrients required for many biological processes in a plant, including photosynthesis, seedling vigor, root and shoot growth. Due to their higher surface tension compared to typical surfaces, nanoparticles hold materials from plants more firmly than traditional surfaces, making them potentially useful in slowing the release of fertilizers. Furthermore, nano coatings shield the surfaces of larger particles (Brady, 1999; Santoso et al., 1995).

The mode of nutrient application is a major factor in determining its bioavailability. Foliar application of nanoparticles facilitates higher nutrient use efficiency (NUE) and promotes rapid plant growth (Mahil et al., 2019). Once entering the root through symplastic and apoplastic pathways, they move to the xylem tissue, reaching the aerial parts of the plants (leaves and stems) (Sun et al., 2014). Further, upon entering the cell, these particles proceed either apoplastically or symplastically, passing through the plasmodesmata of different cells to arrive at the cytoplasm. They start to spread to distinct cytoplasmic organelles in the cytoplasm and take part in a variety of cellular metabolic processes. Therefore, their ability to pass through the narrow plasmodesmata (50-60 nm) between the cells enables higher NUE with increased productivity of about 6-17%, in addition to low nutrient losses (Iqbal, 2019). The various reported nanomaterials and nanoparticles role in nutrient delivery and fertilization of crops are presented in Table 3 and Fig. 3.

Nanoparticles for Soil Remediation and Regeneration

Today, soil pollution is one of the major concerns that need quick attention (Galdames et al. 2020). The United Nations Environment Programme (UNEP) and the Food and Agriculture Organization (FAO) have declared the next ten years as critical for ecosystem restoration. Recognizing this imperative, the 73rd session of the UN General Assembly has formally adopted a resolution to celebrate 2021–2030 as the UN Decade on Ecosystem Restoration (UN-DER), aimed at preventing, halting, and reversing the degradation of ecosystems worldwide (source:_www.unenvironment.org).

The two primary causes of soil pollution resulting in deterioration are injudicious industrial disposals and the unregulated use of fertilizers and toxic pesticides in agriculture. These agricultural inputs are often laden with heavy metals (HMs), which pose not only toxicity concerns but also present challenges in disposal due to their rapid accumulation and non-biodegradability (Naikoo et al., 2020). HMs negatively impact crop physiology by disrupting cellular function, damaging cell membranes, modifying enzyme specificity, and disrupting DNA conformation (Naikoo et al 2019; Raghib et al 2020). Although

Table 3. Nano fertilizer for efficient nutrient delivery and fertilization of crops

Nano Fertilizer	Crop/Plant	Role/Advantage	Reference
SiO_2 NPs	Tomato (*L. esculentum* Mill. cv. super strain B)	Improves seed germination, mean germination time, seed germination index, and seedling fresh weight	Siddiqui and Al-Whaibi, 2014
Mesoporous silica nanoparticles (MSN)	Wheat (*T. aestivum* L.) and Lupin (*Lupinus angustifolius*)	Improves photosynthetic activities, seed germination, total protein, plant biomass,and chlorophyll content.	Sun et al 2016
Mn NPs and $MnSO_4$	Mung bean (*Vigna radiata* L.)	Increases fresh dry biomass, shoot length and photophosphorylation	Pradhan et al 2013
MgO NPs	Tobacco (*Nicotiana tabacum*)	Increases seedling growth, morpho-physiological activities, and Mg uptake	Cai et al 2018
Cu-chitosan NPs	Maize (*Zea mays*)	Increased starch content and α-amylase activity to improve seedling growth	Saharan et al 2016
MoS_2 NPs	Rice (*Oryza sativa* L.)	Increases leaf chlorophyll content, plant biomass and upregulates aquaporin gene	Li Y. et al 2018
Fe_2O_3NPs	Soybean (*Glycine max* L., Oxley)	Increases chlorophyll content of hydroponically cultivated soybean	Ghafariyan et al 2013
α-Fe_2O_3NPs	Legumes	Improves root length by 88–366%	Palchoudhury et al., 2018
γ-Fe_2O_3	Maize (*Zea mays*)	Enhanced the root length and germination rate by 11.5% and 27.2%, respectively	Li et al., 2016
	Peanut (*Arachis hypogaea*)	Improves rate of photosynthesis and plant biomass	Alidoust and Isoda, 2013
ZnONPs	Mung bean (*Vigna radiata* L.)	Increases overall plant biomass	Dhoke et al 2013
Quantum dots (CQDs)	Mung bean (*Vigna radiata* L.)	Increases rate of photosynthesis and electron transport by promoting acyclic photophosphorylation, ATP synthesis, and oxygen release	Chandra et al 2014
SWCNT-Based NPs	Spinach (*Spinacia oleracea*)	Increases photosynthetic activity by 31%, by increasing electron transfer rate	Giraldo et al. (2014)

Figure 3. A schematic representation of Nano fertilizer for efficient nutrient delivery and fertilization of crops (Singh et al., 2024c)

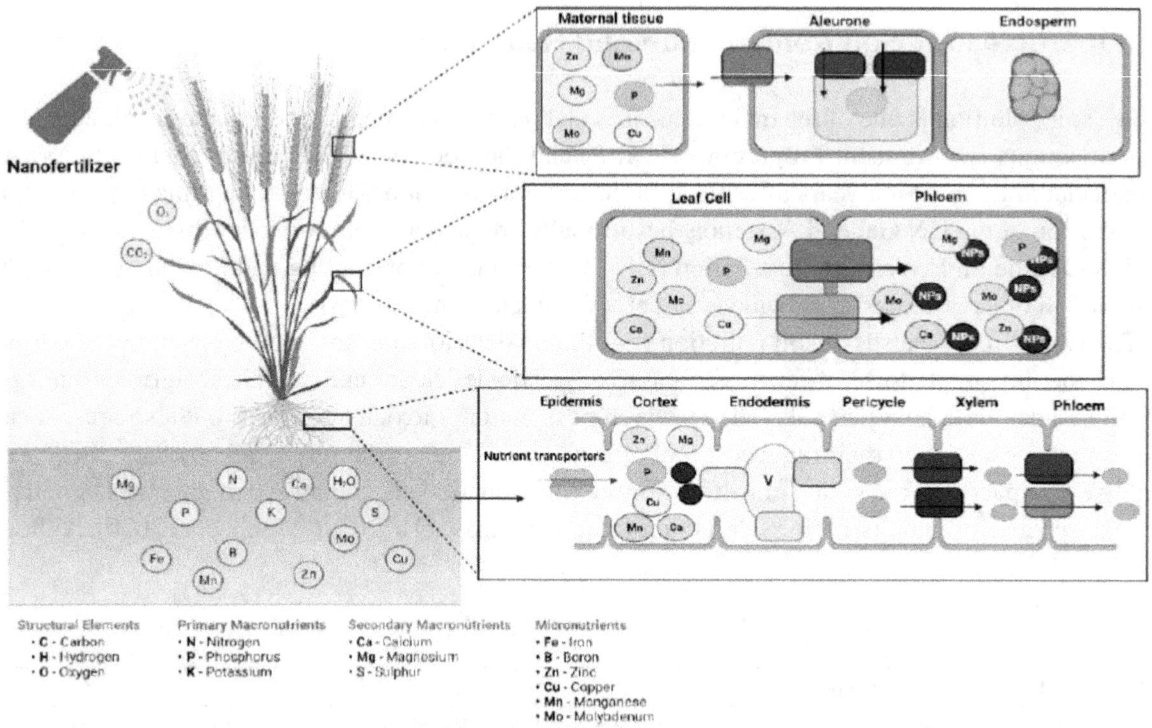

some crops are well-known for their phytoremediation properties in scavenging these HMs, this strategy is costly and time-consuming. Therefore, the use of nanoparticles for remediation is in demand due to their high reactivity, cost-effectiveness, operability, and in-situ management (Xue et al. 2018; Fajardo et al. 2019). ENMs, such as nanoscale calcium peroxide, nanoscale zero-valent iron(nZVI) particles, and nanoscale metal oxides, have been popularly used in the destruction of organics (e.g., gasoline), soil remediation for the degradation of halogenated organic compounds, and adsorption of metals, respectively (Mueller and Nowack. 2010).

The application of nZVI particles has shown promising results in the remediation of contaminated sites, particularly for subsurface contamination of deterring halogenated aliphatic hydrocarbons (Arnold and Roberts 2000). Chemical analysis conducted by Carmen et al. (2019) revealed its significant role in stabilizing elevated concentrations of zinc and lead. Following nZVI treatment, there is a considerable reduction in the bioavailability and toxicity of heavy metals (HMs), facilitating the implementation of bioremediation to expedite the detoxification process and achieve faster reclamation of the contaminated sites. Hematite (Fe2O3) and Magnetite (Fe3O4) nanoparticles have been reported to be effective in removing arsenic (As3+), (As5+), lead (Pb), cadmium (Cd),chromium (Cr), nickel (Ni), selenium (Se), cobalt (Co), molybdenum (Mo), vanadium (V), thorium (Th), antimony (Sb), thallium (Tl), and uranium (U) to negligible concentrations, suggesting their budding potential in remediation of HM-contaminated soils (Shipley et al. 2010). Further, remediating property of Iron nanoparticles (FeNP) in soil contaminated with arsenic (As) (Shipley et al. 2010), metals and chlorinated organic compounds, (Lowry and Johnson 2004; Zhang 2003) have also been found.

Precision Farming With Nanoparticles

The industrial and green revolutions are the two major accelerators that completely changed the way agriculture had been done sustainably until then. Uncontrolled use of chemicals and fertilizers deteriorated and exploited natural resources (Soil, water bodies, and atmosphere). The Concept of Precision farming emerged during the 1980s when it was first theorized by Dr. Pierre Robert who is now revered as the father of modern precision farming. Precision farming aims total or management in a coherent and holistic manner (Lowenberg-DeBoer 2015), especially exploiting high spatial and temporal variability of crop and environmental traits (Zhang et al. 2002, Lowenberg-DeBoer 2015). It allows farmers to customize their agricultural management decisions based on specific spatial and temporal considerations. By employing site-specific sensing, sampling, and management techniques, farmers can address the individualized needs of different areas within a field, treating it as a diverse and dynamic entity.

Nanotechnology application has a significant role from the beginning of sowing by using seeds primed with NPs, providing assured nutrition and protection during the initial establishment period. With the aid of nanotechnology in the form of NMs, nanosensors, and artificial intelligence (drones and robots), precision farmers can make smart decisions regarding the quality of inputs, sowing, management of pests and diseases, harvesting, and facilitate real-time soil and plant health monitoring. NMs in the form of nano fertilizers provide targeted nutrient delivery. Additionally, quantum dots and nanobiosensors enable stress monitoring so that management practices can be implemented quickly. Therefore, nanotechnology plays a major role in obtaining higher yields from precise input application with limited resources at different stages of crop cultivation in precision agriculture, as shown in Fig 4.

Figure 4. Efficient incorporation of nano technology in precision farming

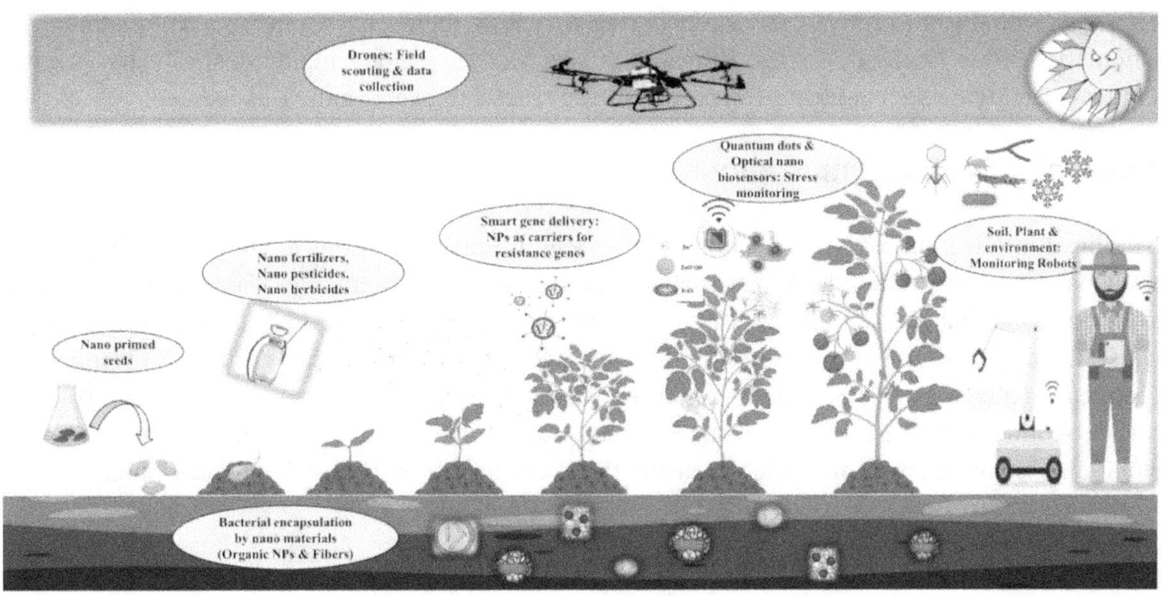

CHALLENGES AND LIMITATIONS

While nanoomics holds immense promise in addressing abiotic and biotic stress in agriculture, several challenges and limitations must be addressed to realize its full potential.

Safety and Environmental Concerns

The use of nanomaterials in agriculture raises concerns about their potential adverse effects on human health and the environment (Singh et al., 2021). Nanoparticles may exhibit unexpected interactions with living organisms, leading to unintended consequences such as bioaccumulation and toxicity (Khot et al., 2020; Mishra and Singh, 2021). Moreover, the release of nanoparticles into the environment could pose risks to ecosystems and biodiversity. Rigorous risk assessment protocols and regulatory frameworks are needed to ensure the safe and responsible use of nanomaterials in agriculture.

Regulatory Hurdles

The regulatory landscape for nanotechnology in agriculture is complex and often fragmented, posing challenges for the commercialization and widespread adoption of nanoomics-based solutions. Regulatory agencies may lack standardized protocols for assessing the safety and efficacy of nanoproducts, leading to delays in approval processes and market entry (Prasad et al., 2014). Streamlining regulatory procedures and harmonizing international standards are essential to facilitate the development and deployment of nanoomics technologies.

Scalability and Cost-Effectiveness

The scalability of nanoomics-based solutions remains a significant challenge, particularly concerning manufacturing processes and production costs. Nanomaterial synthesis methods often require specialized equipment and expertise, making large-scale production expensive and impractical (Singh et al., 2023). Additionally, the high cost of nanomaterials may limit their accessibility to smallholder farmers, particularly in developing countries. Research efforts focused on developing cost-effective synthesis techniques and scalable manufacturing processes are crucial to overcoming this barrier.

Ethical and Societal Implications

The ethical implications of nanotechnology in agriculture raise questions about equity, social justice, and technological dependency. Concerns about unequal access to nanoomics-based solutions, particularly among marginalized communities, underscore the importance of addressing issues of equity and inclusivity in agricultural innovation (Rajput et al., 2021). Furthermore, the potential for technological dependency on nanomaterials may exacerbate existing disparities in resource allocation and exacerbate socio-economic inequalities. Ethical considerations must be integrated into the development and implementation of nanoomics technologies to ensure their responsible and equitable deployment.

Knowledge Gaps and Uncertainties

Despite significant advances in nanotechnology, many fundamental questions regarding the interactions between nanomaterials and biological systems remain unanswered. The complex and dynamic nature of nanobiointeractions presents challenges for predictive modeling and risk assessment, leading to uncertainties about the long-term effects of nanoomics-based interventions on ecosystems and human health (Singh et al., 2022). Interdisciplinary research efforts aimed at elucidating the mechanisms underlying nanobiointeractions and addressing knowledge gaps are essential for advancing the field of nanoomics and ensuring its safe and sustainable application in agriculture.

FUTURE PERSPECTIVES AND CONCLUSION

Rapid global population growth, coupled with limitations in conventional agricultural technologies, has spurred a critical food production challenge. Nanocompounds have emerged as pivotal contributors to enhancing plant growth and productivity. Due to their small size, these nanomaterials can easily penetrate the soil, fostering beneficial microbial populations responsible for nutrient provision through mineralization of various micro- and macronutrients (Chaudhary et al., 2023). Study the impact of nanocompounds on different crops at various concentrations is imperative to ensure their safe and effective application in agriculture. The urgent need for sustainable agricultural progress demands interdisciplinary strategies to optimize resources and meet escalating demands, with nanotechnology emerging as a promising tool to fortify crop resilience against biotic and abiotic stress (Singh et al., 2022; 2024). Despite being in a nascent stage, the integration of data from engineered nanomaterial (ENM) and plant interaction studies requires heightened attention. Establishing a knowledge base from multiomic tools not only aids in comprehending these interactions but also facilitates the design of ENM-based sensors capable of detecting stress signaling molecules and enabling controlled agrochemical release (Arturo and Keller, 2020). Multiomics, with their high-throughput capabilities, explore ENM-plant interactions across growth stages and environmental conditions, determining optimal application. In the realm of omic studies, a predominant reliance on untargeted approaches to identify biomarkers underscores the increasing need for precise reporting of environmental conditions, plant age, ENM characteristics, stability in exposure media, exposure routes, instrumental conditions, and analytical challenges, emphasizing the importance of standardized protocols in the research community. The interaction between plants and nanomaterials in the food and agro-industry has spurred research on nanoparticle (NP) toxicity in microorganisms, with limited exploration of plant genotoxicity caused by NPs. Emphasizing the pivotal role of transcriptomics and proteomics, recent studies have uncovered intricate plant responses to nanomaterial stress, revealing differential protein expressions in detoxification, oxidation reduction, signal transduction, and hormonal signaling pathways (Munir et al., 2023). Notably, proteomic and metabolomic studies of Arabidopsis thaliana leaves during tripartite interactions involving plants, pathogens, and nanoparticles have highlighted the differential expression of proteins and metabolites, influencing bioenergy, metabolism, and plant defense. The findings suggest that nanomaterials not only possess pathogen-killing capabilities but also enhance the antimicrobial properties of plants, inducing primary and secondary immune responses by modulating the proteome and metabolome (Kumari et al., 2020). Recent research into silver nanoparticles underscore the impact of particle size on induced phytotoxicity, offering insights into nuanced plant reactions to NP stress and contributing to a comprehensive understanding of plant-NP interactions. In the quest for sustainable agriculture, the exploration of nanomaterial applications demands continued interdisciplinary efforts, rigorous reporting standards, and a deeper integration of multi-omics tools to unlock the full potential of nanotechnology in addressing global food security challenges. Addressing the challenges and limitations of nano-omics in mitigating abiotic and biotic stress in agriculture requires concerted efforts from researchers, policymakers, and stakeholders. By prioritizing safety, sustainability, and ethical considerations, we can harness the transformative potential of nanotechnology to develop innovative solutions that enhance crop resilience, promote environmental stewardship, and safeguard global food security for future generations.

ACKNOWLEDGMENT

YSU internal grant is supported by KG. AS is supported by the 23PostDoc-4D007 grant provided by the Science Committee of the Republic of Armenia. VDR and TM are supported by the Strategic Academic Leadership Program of Southern Federal University, known as "Priority 2030," and the Ministry of Science and Higher Education of the Russian Federation (grant number: FENW-2023-0008).

REFERENCES

Adisa, I. O., Reddy Pullagurala, V. L., Rawat, S., Hernandez-Viezcas, J. A., Dimkpa, C. O., Elmer, W. H., et al. (2018). Role of cerium compounds in Fusarium wilt suppression and growth enhancement in tomato (Solanum lycopersicum). Journal of Agricultural and Food Chemistry, 66, 5959–5970. doi: . 8b01345 doi:10.1021/acs.jafc

Adrees, M., Khan, Z. S., Ali, S., Hafeez, M., Khalid, S., Ur Rehman, M. Z., Hussain, A., Hussain, K., Shahid Chatha, S. A., & Rizwan, M. (2020). Simultaneous mitigation of cadmium and drought stress in wheat by soil application of iron nanoparticles. *Chemosphere*, *238*, 124681. doi:10.1016/j.chemosphere.2019.124681 PMID:31524618

Aghdam, M. T., Mohammadi, H., & Ghorbanpour, M. (2016). Effects of nanoparticulate anatase titanium dioxide on physiological and biochemical performance of Linum usitatissimum (Linaceae) under well-watered and drought stress conditions. *Revista Brasileira de Botânica. Brazilian Journal of Botany*, *39*(1), 139–146. doi:10.1007/s40415-015-0227-x

Alabdallah, N. M., & Alzahrani, H. S. (2020). The potential mitigation effect of ZnO nanoparticles on [Abelmoschus esculentus L. Moench] metabolism under salt stress conditions. *Saudi Journal of Biological Sciences*, *27*(11), 3132–3137. doi:10.1016/j.sjbs.2020.08.005 PMID:33100874

Ali, M., Kim, B., Belfield, K. D., Norman, D., Brennan, M., & Ali, G. S. (2015). Inhibition of Phytophthora parasitica and P. capsici by silver nanoparticles synthesized using aqueous extract of Artemisia absinthium. *Phytopathology*, *105*(9), 1183–1190. doi:10.1094/PHYTO-01-15-0006-R PMID:25871856

Alidoust, D., & Isoda, A. (2013). Effect of γFe_2O_3 nanoparticles on photosynthetic characteristic of soybean (Glycine max (L.) Merr.): Foliar spray versus soil amendment. *Acta Physiologiae Plantarum*, *35*(12), 3365–3375. doi:10.1007/s11738-013-1369-8

Almutairi, Z. M. (2016). Effect ofnano-silicon application on the expression ofsalt tolerance genes in germinating tomato ('Solanum lycopersicum' L.) seedlings under salt stress. *Plant Omics*, *9*, 106.

Alvarado, M.A., Guzmán, N.M., Solís, N.M. and Vega Baudrit, J., 2017. Recycling and Elimination of Wastes obtained from Agriculture by using Nanotechnology: Nanosensors.

Anjali, C. H., Sharma, Y., Mukherjee, A., & Chandrasekaran, N. (2012). Neem oil (Azadirachta indica) nanoemulsion—A potent larvicidal agent against Culex quinquefasciatus. *Pest Management Science*, *68*(2), 158–163. doi:10.1002/ps.2233 PMID:21726037

Arduini, F., Cinti, S., Scognamiglio, V., & Moscone, D. (2016). Nanomaterials in electrochemical biosensors for pesticide detection: Advances and challenges in food analysis. *Mikrochimica Acta*, *183*(7), 2063–2083. doi:10.1007/s00604-016-1858-8

Arnold, W. A., & Roberts, A. L. (2000). Pathways and kinetics of chlorinated ethylene and chlorinated acetylene reaction with Fe (0) particles. *Environmental Science & Technology*, *34*(9), 1794–1805. doi:10.1021/es990884q

Askary, M., Talebi, S. M., Amini, F., & Bangan, A. D. (2017). Effects of iron nanoparticles on Mentha piperita L. under salinity stress. *Biologija (Vilnius, Lithuania)*, *63*(1), 65–75. doi:10.6001/biologija.v63i1.3476

Ayoub, H. A., Khairy, M., Elsaid, S., Rashwan, F. A., & Abdel-Hafez, H. F. (2018). Pesticidal activity of nanostructured metal oxides for generation of alternative pesticide formulations. *Journal of Agricultural and Food Chemistry*, *66*(22), 5491–5498. doi:10.1021/acs.jafc.8b01600 PMID:29757642

Barrett, B., Nitze, I., Green, S., & Cawkwell, F. (2014). Assessment of multi-temporal, multi-sensor radar and ancillary spatial data for grasslands monitoring in Ireland using machine learning approaches. *Remote Sensing of Environment*, *152*(2), 109–124. doi:10.1016/j.rse.2014.05.018

Bin Hussein, M. Z., Yahaya, A. H., Zainal, Z., & Kian, L. H. (2005). Nanocomposite-based controlled release formulation of an herbicide, 2, 4-dichlorophenoxyacetate incapsulated in zinc–aluminiumlayered double hydroxide. *Science and Technology of Advanced Materials*, *6*(8), 956–962. doi:10.1016/j.stam.2005.09.004

Borgatta, J., Ma, C., Hudson-Smith, N., Elmer, W., Plaza Pérez, C. D., De La Torre-Roche, R., Zuverza-Mena, N., Haynes, C. L., White, J. C., & Hamers, R. J. (2018). Copper based nanomaterials suppress root fungal disease in watermelon (Citrullus lanatus): Role of particle morphology, composition and dissolution behavior. *ACS Sustainable Chemistry & Engineering*, *61*(11), 4847–14856. doi:10.1021/acssuschemeng.8b03379

Burg, M. B., & Ferraris, J. D. (2008). Intracellular organic osmolytes: Function and regulation. *The Journal of Biological Chemistry*, *283*(12), 7309–7313. doi:10.1074/jbc.R700042200 PMID:18256030

Cai, L., Liu, M., Liu, Z., Yang, H., Sun, X., Chen, J., Xiang, S., & Ding, W. (2018). MgONPs can boost plant growth: Evidence from increased seedling growth, morpho-physiological activities, andMg uptake in tobacco (Nicotiana tabacumL.). *Molecules (Basel, Switzerland)*, *23*(12), 3375. doi:10.3390/molecules23123375 PMID:30572666

Chandra, S., Pradhan, S., Mitra, S., Patra, P., Bhattacharya, A., Pramanik, P., & Goswami, A. (2014). High throughput electron transfer from carbon dots to chloroplast: A rationale of enhanced photosynthesis. *Nanoscale*, *6*(7), 3647–3655. doi:10.1039/C3NR06079A PMID:24562190

Chaudhary, P., Chaudhary, A., Khati, P., Kumar, G., Tyagi, J., & Behera, M. (2023). Nanotechnology and Omics Approach in Agrobiotechnology. In *Agricultural and Environmental Nanotechnology: Novel Technologies and their Ecological Impact* (pp. 341–352). Springer Nature Singapore. doi:10.1007/978-981-19-5454-2_12

Chen, Y., Ren, H. L., Liu, N., Sai, N., Liu, X., Liu, Z., Gao, Z., & Ning, B. (2010). A fluoroimmunoassay based on quantum dot–streptavidin conjugate for the detection of chlorpyrifos. *Journal of Agricultural and Food Chemistry*, *58*(16), 8895–8903. doi:10.1021/jf101778t PMID:23654227

Cromwell, W. A., Yang, J., Starr, J. L., & Jo, Y. K. (2014). Nematicidal effects of silver nanoparticles on root-knot nematode in bermudagrass. *Journal of Nematology*, *46*, 261–266. PMID:25275999

Cummins, C. M., Koivunen, M. E., Stephanian, A., Gee, S. J., Hammock, B. D., & Kennedy, I. M. (2006). Application of europium (III) chelate-dyed nanoparticle labels in a competitive atrazine fluoroimmunoassay on an ITO waveguide. *Biosensors & Bioelectronics*, *21*(7), 1077–1085. doi:10.1016/j.bios.2005.04.003 PMID:16368482

Demirer, G. S., Silva, T. N., Jackson, C. T., Thomas, J. B., Ehrhardt, D. W., Rhee, S. Y., Mortimer, J. C., & Landry, M. P. (2021). Nano-technology to advance CRISPR–Cas genetic engineering of plants. *Nature Nanotechnology*, *16*, 243–250. doi:10.1038/s41565-021-00854-y PMID:33712738

Dhoke, S. K., Mahajan, P., Kamble, R., and Khanna, A. (2013). Effect of nanoparticles suspension on the growth of mung (Vigna radiata) seedlings by foliar spray method. Nanotechnology development, 3:e1. doi:10.4081/nd.2013.e

Djanaguiraman, M., Belliraj, N., Bossmann, S. H., & Prasad, P. V. (2018). High temperature stress alleviation by selenium nanoparticle treatment in grain sorghum. *ACS Omega*, *3*(3), 2479–2491. doi:10.1021/acsomega.7b01934 PMID:31458542

Elsheery, N. I., Helaly, M. N., El-Hoseiny, H. M., & Alam-Eldein, S. M. (2020). Zinc oxide and silicone nanoparticles to improve the resistance mechanism and annual productivity of salt-stressed mango trees. *Agronomy (Basel)*, *10*(4), 558. doi:10.3390/agronomy10040558

Fajardo, C., Costa, G., Nande, M., Martín, C., Martín, M., & Sánchez-Fortún, S. (2019). Heavy metals immobilization capability of two iron-based nanoparticles (nZVI and Fe3O4): Soil and freshwater bioassays to assess ecotoxicological impact. *The Science of the Total Environment*, *656*, 421–432. doi:10.1016/j.scitotenv.2018.11.323 PMID:30513432

Fajardo, C., Sánchez-Fortún, S., Costa, G., Nande, M., Botías, P., García-Cantalejo, J., Mengs, G., & Martín, M. (2020). Evaluation of nanoremediation strategy in a Pb, Zn and Cd contaminated soil. *The Science of the Total Environment*, *706*, 136041. doi:10.1016/j.scitotenv.2019.136041 PMID:31855644

Fang, Y., Umasankar, Y., & Ramasamy, R. P. (2014). Electrochemical detection of p-ethylguaiacol, a fungi infected fruit volatile using metal oxide nanoparticles. *Analyst*, *139*(15), 3804–3810. doi:10.1039/C4AN00384E PMID:24895939

Faraji, J., & Sepehri, A. (2020). Exogenous nitric oxide improves the protective effects oftio 2 nanoparticles on growth, antioxidant system, and photosynthetic performance ofwheat seedlings under drought stress. *Journal of Soil Science and Plant Nutrition, 20*(2), 703–714. doi:10.1007/s42729-019-00158-0

Fraceto, L. F., Grillo, R., de Medeiros, G. A., Scognamiglio, V., Rea, G., & Bartolucci, C. (2016). Nanotechnology in agriculture: Which innovation potential does it have? *Frontiers in Environmental Science, 4*, 20. doi:10.3389/fenvs.2016.00020

Galdames, A., Ruiz-Rubio, L., Orueta, M., Sánchez-Arzalluz, M., & Vilas-Vilela, J. L. (2020). Zero-valent iron nanoparticles for soil and groundwater remediation. *International Journal of Environmental Research and Public Health, 17*(16), 5817. doi:10.3390/ijerph17165817 PMID:32796749

Ghafariyan, M. H., Malakouti, M. J., Dadpour, M. R., Stroeve, P., & Mahmoudi, M. (2013). Effects of-magnetite nanoparticles on soybean chlorophyll. *Environmental Science & Technology, 47*, 10645–10652. doi:10.1021/es402249b PMID:23951999

Ghazaryan, K. A., Harutyunyan, A. S., Khachatryan, H. E., Singh, A., Minkina, T. M., Rajput, V. D., & Movsesyan, H. S. (2023). Study of tolerance and phytodesalination potential of wheat, oat, emmer, and barley for sustainable saline agriculture. *Applied Ecology and Environmental Research, 21*(5), 4853–4882. doi:10.15666/aeer/2105_48534882

Ghazaryan, K. A., Movsesyan, H. S., Khachatryan, H. E., & Ghazaryan, N. P. (2018). Geochemistry of Potentially Toxic Trace Elements in Soils of Mining Area: A Case Study from Zangezur Copper and Molybdenum Combine, Armenia. *Bulletin of Environmental Contamination and Toxicology, 101*(6), 732–737. doi:10.1007/s00128-018-2443-0 PMID:30229278

Giraldo, J. P., Landry, M. P., Faltermeier, S. M., McNicholas, T. P., Iverson, N. M., Boghossian, A. A., Reuel, N. F., Hilmer, A. J., Sen, F., Brew, J. A., & Strano, M. S. (2014). Plant nanobionics approach to augment photosynthesis and biochemical sensing. *Nature Materials, 13*(4), 400–408. doi:10.1038/nmat3890 PMID:24633343

Haghighi, M., Abolghasemi, R., & da Silva, J. A. T. (2014). Low and high temperature stress affect the growth characteristics of tomato in hydroponic culture with Se and nano-Se amendment. *Scientia Horticulturae, 178*, 231–240. doi:10.1016/j.scienta.2014.09.006

Hernández-Hernández, H., González-Morales, S., Benavides-Mendoza, A., Ortega-Ortiz, H., Cadenas-Pliego, G., & Juárez-Maldonado, A. (2018). Effects of chitosan–PVA and Cu nanoparticles on the growth and antioxidant capacity of tomato under saline stress. *Molecules (Basel, Switzerland), 23*(1), 178. doi:10.3390/molecules23010178 PMID:29337864

Ho, V. A., Le, P. T., Nguyen, T. P., Nguyen, C. K., Nguyen, V. T., & Tran, N. Q. (2015). Silver core-shell nanoclusters exhibiting strong growth inhibition of plant-pathogenic fungi. *Journal of Nanomaterials, 2015*, 241614. doi:10.1155/2015/241614

Hojjat, S. S., & Kamyab, M. (2017). The effect of silver nanoparticle on Fenugreek seed germination under salinity levels. *Russian Agricultural Sciences, 43*(1), 61–65. doi:10.3103/S1068367417010189

Hua, X., Qian, G., Yang, J., Hu, B., Fan, J., Qin, N., Li, G., Wang, Y., & Liu, F. (2010). Development of an immunochromatographic assay for the rapid detection of chlorpyrifos-methyl in water samples. *Biosensors & Bioelectronics*, *26*(1), 189–194. doi:10.1016/j.bios.2010.06.005 PMID:20598518

Imada, K., Sakai, S., Kajihara, H., Tanaka, S., & Ito, S. (2016). Magnesium oxide nanoparticles induce systemic resistance in tomato against bacterial wilt disease. *Plant Pathology*, *65*(4), 551–560. doi:10.1111/ppa.12443

Iqbal, M., Raja, N. I., Hussain, M., Ejaz, M., & Yasmeen, F. (2019). Effect of silver nanoparticles on growth ofwheat under heat stress. *Iranian Journal of Science and Technology. Transaction A, Science*, *43*(2), 387–395. doi:10.1007/s40995-017-0417-4

Iqbal, M. A. (2019). *"Nano-fertilizers for sustainable crop production under changing climate: a global perspective," in Sustainable Crop Production*. IntechOpen.

Jaberzadeh, A., Moaveni, P., Moghadam, H. R., & Zahedi, H. (2013). Influence of bulk and nanoparticles titanium foliar application on some agronomic traits, seed gluten and starch contents of wheat subjected to water deficit stress. *Notulae Botanicae Horti Agrobotanici Cluj-Napoca*, *41*(1), 201–207. doi:10.15835/nbha4119093

Jampílek, J., & Kráľová, K. 2017. Nanopesticides: preparation, targeting, and controlled release. In New Pesticides and Soil Sensors (pp. 81-127). Academic Press. doi:10.1016/B978-0-12-804299-1.00004-7

Khan, M., Khan, A. U., Hasan, M. A., Yadav, K. K., Pinto, M. M., Malik, N., Yadav, V. K., Khan, A. H., Islam, S., & Sharma, G. K. (2021). Agro-nanotechnology as an emerging field: A novel sustainable approach for improving plant growth by reducing biotic stress. *Applied Sciences (Basel, Switzerland)*, *11*(5), 2282. doi:10.3390/app11052282

Khot, L. R., Sankaran, S., Maja, J. M., Ehsani, R., & Schuster, E. W. (2020). Applications of nanomaterials in agricultural production and crop protection: A review. *Crop Protection (Guildford, Surrey)*, *133*, 105221.

Kranthi, K. R., Davis, M., Mayee, C. D., Russell, D. A., Shukla, R. M., Satija, U., Kshirsagar, M., Shiware, D., & Kranthi, S. (2009). Development of a colloidal-gold based lateralflow immunoassay kit for 'quality-control' assessment of pyrethroid and endosulfan formulations in a novel single strip format. *Crop Protection (Guildford, Surrey)*, *28*(5), 428–434. doi:10.1016/j.cropro.2009.01.003

Kumari, M., Pandey, S., Mishra, S. K., Giri, V. P., Agarwal, L., Dwivedi, S., Pandey, A. K., Nautiyal, C. S., & Mishra, A. (2020, April 17). Omics-Based Mechanistic Insight Into the Role of Bioengineered Nanoparticles for Biotic Stress Amelioration by Modulating Plant Metabolic Pathways. *Frontiers in Bioengineering and Biotechnology*, *8*, 242. doi:10.3389/fbioe.2020.00242 PMID:32363178

Li, J., Hu, J., Ma, C., Wang, Y., Wu, C., Huang, J., & Xing, B. (2016). Uptake, translocation and physiological effects of magnetic iron oxide (γ- Fe2O3) nanoparticles in corn (Zea mays L.). *Chemosphere*, *159*, 326–334. doi:10.1016/j.chemosphere.2016.05.083 PMID:27314633

Li, Y., Jin, Q., Yang, D., & Cui, J. (2018). Molybdenum sulfide induce growth enhancement effect of rice (Oryza sativa L.) through regulating the synthesis of chlorophyll and the expression of aquaporin gene. *Journal of Agricultural and Food Chemistry*, *66*(16), 4013–4021. doi:10.1021/acs.jafc.7b05940 PMID:29630363

Liu, C., Zhou, H., & Zhou, J. (2021). The applications of nanotechnology in crop production. *Molecules (Basel, Switzerland)*, *26*(23), 7070. doi:10.3390/molecules26237070 PMID:34885650

Lowenberg-DeBoer, J. (2015). *The precision agriculture revolution: Making the modern farmer*. Foreign Affairs, May/June.

Mahil, T., Baburai, N., & Aravinda, K. (2019). Foliar application of nanofertilizers in agricultural crops-A review. *Journal of Farm Sciences*, *32*, 239–249.

Majumdar, S., & Keller, A. A. (2020). Omics to address the opportunities and challenges of nanotechnology in agriculture. *Critical Reviews in Environmental Science and Technology*. Advance online publication. doi:10.1080/10643389.2020.1785264

Mattos, B. D., & Magalhães, W. L. (2016). Biogenic nanosilica blended by nanofibrillated cellulose as support for slow-release of tebuconazole. *Journal of Nanoparticle Research*, *18*(9), 274. doi:10.1007/s11051-016-3586-8

Meena, R. S., Kumar, S., Datta, R., Lal, R., Vijayakumar, V., Brtnicky, M., Sharma, M. P., Yadav, G. S., Jhariya, M. K., Jangir, C. K., Pathan, S. I., Dokulilova, T., Pecina, V., & Marfo, T. (2020). Impact of agrochemicals on soil microbiota and management: A review. *Land (Basel)*, *9*(2), 34. doi:10.3390/land9020034

Mishra, A., & Singh, H. B. (2021). *Nanoparticles in agro-products: Benefits, applications and concerns," in Nanomaterials for agriculture and forestry applications*. Springer.

Mohamed, A. K. S., Qayyum, M. F., Abdel-Hadi, A. M., Rehman, R. A., Ali, S., & Rizwan, M. (2017). Interactive effect of salinity and silver nanoparticles on photosynthetic and biochemical parameters of wheat. *Archives of Agronomy and Soil Science*, *63*(12), 1736–1747. doi:10.1080/03650340.2017.1300256

Mohammadi, R., Maali-Amiri, R., & Abbasi, A. (2013). Effect of TiO2 nanoparticles on chickpea response to cold stress. *Biological Trace Element Research*, *152*(3), 403–410. doi:10.1007/s12011-013-9631-x PMID:23456351

Munir, N., Gulzar, W., Abideen, Z., Hasanuzzaman, M., El-Keblawy, A., & Zhao, F. (2023). Plant–Nanoparticle Interactions: Transcriptomic and Proteomic Insights. *Agronomy (Basel)*, *13*(8), 2112. doi:10.3390/agronomy13082112

Mustafa, G., & Komatsu, S. (2016). Insights into the response of soybean mitochondrial proteins to various sizes of aluminum oxide nanoparticles under flooding stress. *Journal of Proteome Research*, *15*(12), 4464–4475. doi:10.1021/acs.jproteome.6b00572 PMID:27780359

Naikoo, M. I., Dar, M. I., Khan, F. A., Raghib, F., & Rajakaruna, N. (2019). Trophic transfer and bioaccumulation of lead along soil–plant–aphid–ladybird food chain. *Environmental Science and Pollution Research International*, *26*(23), 23460–23470. doi:10.1007/s11356-019-05624-x PMID:31201701

Naikoo, M. I., Kafeel, U., Naushin, F., & Khan, F. A. (2020). Halophytes in India and Their Role in Phytoremediation. In M.-N. Grigore (Ed.), *Handbook of Halophytes: From Molecules to Ecosystems towards Biosaline Agriculture* (pp. 1–21). Springer International Publishing. doi:10.1007/978-3-030-17854-3_115-1

Narayanan, A., Sharma, P., & Moudgil, B. M. (2013). Applications of engineered particulate systems in agriculture and food industry. *Kona Powder and Particle Journal, 30*(0), 221–235. doi:10.14356/kona.2013021

Ocsoy, I., Paret, M. L., Ocsoy, M. A., Kunwar, S., Chen, T., You, M., & Tan, W. (2013). Nanotechnology in plant disease management: DNA-directed silver nanoparticles on graphene oxide as an antibacterial against Xanthomonas perforans. *ACS Nano, 7*(10), 8972–8980. doi:10.1021/nn4034794 PMID:24016217

Palchoudhury, S., Jungjohann, K. L., Weerasena, L., Arabshahi, A., Gharge, U., Albattah, A., Miller, J., Patel, K., & Holler, R. A. (2018). Enhanced legume root growth with presoaking in α-Fe2O3 nanoparticle fertilizer. *RSC Advances, 8*(43), 24075–24083. doi:10.1039/C8RA04680H PMID:35539206

Palmqvist, N. M., Seisenbaeva, G. A., Svedlindh, P., & Kessler, V. G. (2017). Maghemite nanoparticles acts as nanozymes, improving growth and abiotic stress tolerance in Brassica napus. *Nanoscale Research Letters, 12*(1), 631. doi:10.1186/s11671-017-2404-2 PMID:29260423

Paret, M. L., Vallad, G. E., Averett, D. R., Jones, J. B., & Olson, S. M. (2013). Photocatalysis: Effect of light-activated nanoscale formulations of TiO2 on Xanthomonas perforans and control ofbacterial spot of tomato. *Phytopathology, 103*(3), 228–236. doi:10.1094/PHYTO-08-12-0183-R PMID:23190116

Phogat, N., Ali Khan, S., Shankar, S., & Ansary, A., & Uddin, I. (2016). Fate of inorganic nanoparticles in agriculture. *Advanced Materials Letters, 7*(1), 3–12. doi:10.5185/amlett.2016.6048

Pradhan, S., Patra, P., Das, S., Chandra, S., Mitra, S., Dey, K. K., Akbar, S., Palit, P., & Goswami, A. (2013). Photochemical modulation of biosafe manganese nanoparticles on Vigna radiata: A detailed molecular, biochemical, and biophysical study. *Environmental Science & Technology, 47*(22), 13122–13131. doi:10.1021/es402659t PMID:24144189

Prasad, R., Kumar, V., & Prasad, K. S. (2014). Nanotechnology in sustainable agriculture: Present concerns and future aspects. *African Journal of Biotechnology, 13*(6), 705–713. doi:10.5897/AJBX2013.13554

Raghib, F., Naikoo, M. I., Khan, F. A., Alyemeni, M. N., & Ahmad, P. (2020). Interaction of ZnO nanoparticle and AM fungi mitigates Pb toxicity in wheat by upregulating antioxidants and restricted uptake of Pb. *Journal of Biotechnology, 323*, 254–263. doi:10.1016/j.jbiotec.2020.09.003 PMID:32905797

Rajput, V. D., Singh, A., Minkina, T., Rawat, S., Mandzhieva, S., Sushkova, S., Shuvaeva, V., Nazarenko, O., Rajput, P., Komariah, Verma, K. K., Singh, A. K., Rao, M., & Upadhyay, S. K. (2021). Nano-Enabled Products: Challenges and Opportunities for Sustainable Agriculture. *Plants, 10*(12), 2727. doi:10.3390/plants10122727 PMID:34961197

Saharan, V., Kumaraswamy, R. V., Choudhary, R. C., Kumari, S., Pal, A., Raliya, R., & Biswas, P. (2016). Cu-chitosan nanoparticle mediated sustainable approach to enhance seedling growth in maize by mobilizing reserved food. *Journal of Agricultural and Food Chemistry, 64*(31), 6148–6155. doi:10.1021/acs.jafc.6b02239 PMID:27460439

Santoso, D., Lefroy, R. D., & Blair, G. J. (1995). Sulfur and phosphorus dynamics in an acid soil/crop system. *Soil Research (Collingwood, Vic.)*, *33*(1), 113–124. doi:10.1071/SR9950113

Saritha, G. N. G., Anju, T., & Kumar, A. (2022). Nanotechnology-Big impact: How nanotechnology is changing the future of agriculture? *Journal of Agriculture and Food Research*, *10*, 100457. doi:10.1016/j.jafr.2022.100457

Sedghi, M., Hadi, M., & Toluie, S. G. (2013). Effect of nano zinc oxide on the germination parameters of soybean seeds under drought stress. Annals of West University of Timisoara, ser. *Biology (Basel)*, *16*, 73–78.

Sharifi, M., Faryabi, K., Talaei, A. J., Shekha, M. S., Ale-Ebrahim, M., Salihi, A., Nanakali, N. M. Q., Aziz, F. M., Rasti, B., Hasan, A., & Falahati, M. (2020). Antioxidant properties of gold nanozyme: A review. *Journal of Molecular Liquids*, *297*, 112004. doi:10.1016/j.molliq.2019.112004

Shoala, T. (2023). The dual role of nanotechnology in the management of biotic and abiotic stresses in plants. In *Nanotechnology in Agriculture and Agroecosystems* (pp. 51–70). Elsevier. doi:10.1016/B978-0-323-99446-0.00011-8

Singh, A., Agrawal, S., Rajput, V. D., Ghazaryan, K., Movsesyan, H. S., Minkina, T., . . . Gupta, S. K. (2023a). Seed Nanopriming. Nanopriming Approach to Sustainable Agriculture, 290–313. doi:10.4018/978-1-6684-7232-3.ch013

Singh, A., Margaryan, G., Harutyunyan, A., S. Movsesyan, H., Khachatryan, H., Rajput, V., . . . Ghazaryan, K. (2024a). Advancing Agricultural Resilience in Ararat Plain, Armenia: Utilizing Biogenic Nanoparticles and Biochar under Saline Environments to Optimize Food Security and Foster European Trade. Egyptian Journal of Soil Science, 64(2). doi:10.21608/ejss.2024.257000.1703

Singh, A., Rajput, V. D., Pandey, D., Sharma, R., Ghazaryan, K., & Minkina, T. (2023b). Nano Zinc-Enabled Strategies in Crops for Combatting Zinc Malnutrition in Human Health. *Frontiers in Bioscience (Landmark Edition)*, *28*(8), 158. doi:10.31083/j.fbl2808158 PMID:37664935

Singh, A., Rajput, V. D., Sharma, R., Ghazaryan, K., & Minkina, T. (2023c). Salinity stress and nanoparticles: Insights into antioxidative enzymatic resistance, signaling, and defense mechanisms. *Environmental Research*, *235*, 116585. doi:10.1016/j.envres.2023.116585 PMID:37437867

Singh, A., Rajput, V. D., Tawaha, A. R. M., Al Zoubi, O. M., Habeeb, T., Rawat, S., Ghazaryan, K., & Minkina, T. (2023d). A Review on Crop Responses to Nanofertilizers for Mitigation of Multiple Environmental Stresses. *Ecological Engineering & Environmental Technology*, *24*(7), 280–296. doi:10.12912/27197050/169313

Singh, A., Rajput, V. D., Varshney, A., Ghazaryan, K., & Minkina, T. (2023e). Small Tech, Big Impact: Agri-nanotechnology Journey to Optimize Crop Protection and Production for Sustainable Agriculture. *Plant Stress*, *10*, 100253. doi:10.1016/j.stress.2023.100253

Singh, A., Rajput, V. D., Varshney, A., Sharma, R., Ghazaryan, K., Minkina, T., Alexiou, A., & El-Ramady, H. (2024c). Revolutionizing Crop Production: Nanoscale Wonders-Current Applications, Advances, and Future Frontiers. *Egyptian Journal of Soil Science*, *64*(1), 0. doi:10.21608/ejss.2023.246354.1684

Singh, A., Rawat, S., Rajput, V. D., Ghazaryan, K., Minkina, T., Al Tawaha, A. R. M., . . . Sandhu, K. S. (2023f). Nanoparticle-Based Priming-Enhanced Agricultural Practices. Nanopriming Approach to Sustainable Agriculture, 142–157. doi:10.4018/978-1-6684-7232-3.ch006

Singh, A., Rawat, S., Rajput, V. D., Ghazaryan, K., Minkina, T., Al Tawaha, A. R. M., & Varshney, A. (2023g). Impact of Nanofertilizers for the Mitigation of Multiple Environmental Stresses. Nanotechnology in the Life Sciences, 431–454. doi:10.1007/978-3-031-41329-2_16

Singh, A., Sengar, R. S., Rajput, V. D., Al-Ghzawi, A. L., Shahi, U. P., Ghazaryan, K., Minkina, T., Tawaha, A. R. M. A., Al Zoubi, O. M., & Habeeb, T. (2024b). Impact of Salinity Stress and Zinc Oxide Nanoparticles on Macro and Micronutrient Assimilation: Unraveling the Link between Environmental Factors and Nutrient Uptake. *Journal of Ecological Engineering*, *25*(2), 1–9. doi:10.12911/22998993/172947

Singh, A., Sharma, R., Rawat, S., Singh, A. K., Rajput, V. D., Fedorov, Y., . . . Chaplygin, V. (2022). Nanomaterial-plant interaction: Views on the pros and cons. Toxicity of Nanoparticles in Plants, 47–68. doi:10.1016/B978-0-323-90774-3.00015-5

Singh, S., Singh, M., Agrawal, V. V., & Kumar, A. (2010). An attempt to develop surface plasmon resonance based immunosensor for Karnal bunt (Tilletia indica) diagnosis based on the experience of nano-gold based lateral flow immuno-dipstick test. *Thin Solid Films*, *519*(3), 1156–1159. doi:10.1016/j.tsf.2010.08.061

Srilatha, B. (2011). Nanotechnology in agriculture. *Journal of Nanomedicine & Nanotechnology*, *2*(7).

Stadler, T., Buteler, M., Valdez, S. R., & Gitto, J. G. (2018). Particulate nanoinsecticides: A new concept in insect pest management. *Insectic. Agricul. Toxicol.*, *83*, 72448. doi:10.5772/intechopen.72448

Sun, D., Hussain, H. I., Yi, Z., Rookes, J. E., Kong, L., & Cahill, D. M. (2016). Mesoporous silica nanoparticles enhance seedling growth and photosynthesis in wheat and lupin. *Chemosphere*, *152*, 81–91. doi:10.1016/j.chemosphere.2016.02.096 PMID:26963239

Sun, X., Zhang, Y., Chen, G., & Gai, Z. (2017). Application of nanoparticles in enhanced oil recovery: A critical review of recent progress. *Energies*, *10*(3), 345. doi:10.3390/en10030345

Torabian, S., Zahedi, M., & Khoshgoftar, A. H. (2017). Effects of foliar spray of nano-particles of FeSO4 on the growth and ion content of sunflower under saline condition. *Journal of Plant Nutrition*, *40*(5), 615–623. doi:10.1080/01904167.2016.1240187

Usman, M., Farooq, M., Wakeel, A., Nawaz, A., Cheema, S. A., Rehman, H., Ashraf, I., & Sanaullah, M. (2020). Nanotechnology in agriculture: Current status, challenges, and future opportunities. *The Science of the Total Environment*, *721*, 137778. doi:10.1016/j.scitotenv.2020.137778 PMID:32179352

Van HaC.Van NguyenD.NguyenH. M.LeN. T.NguyenK. H.LeH. M.. (2020). Copper nanoparticle application enhances plant growth and grain yield in maize under drought stress conditions. bioRxiv [Preprint]. doi:10.1101/2020.02.24.963132

Vardumyan, H., Singh, A., Rajput, V., Minkina, T., El-Ramady, H., & Ghazaryan, K. (2024). Additive-Mediated Phytoextraction of Copper-Contaminated Soils Using Medicago lupulina L. *Egyptian Journal of Soil Science*, *64*(2), 599–618. doi:10.21608/ejss.2024.266169.1714

Vega-Vásquez, P., Mosier, N. S., & Irudayaraj, J. (2020). Nanoscale drug delivery systems: From medicine to agriculture. *Frontiers in Bioengineering and Biotechnology*, *8*, 79. doi:10.3389/fbioe.2020.00079 PMID:32133353

Vinayaka, A. C., Basheer, S., & Thakur, M. S. (2009). Bioconjugation of CdTe quantum dot for the detection of 2, 4-dichlorophenoxyacetic acid by competitive fluoroimmunoassay based biosensor. *Biosensors & Bioelectronics*, *24*(6), 1615–1620. doi:10.1016/j.bios.2008.08.042 PMID:18930650

Vinayaka, A. C., & Thakur, M. S. (2010). Focus on quantum dots as potential fluorescent probes for monitoring food toxicants and foodborne pathogens. *Analytical and Bioanalytical Chemistry*, *397*(4), 1445–1455. doi:10.1007/s00216-010-3683-y PMID:20401471

Wang, L., Li, X., Zhang, G., Dong, J., & Eastoe, J. (2007). Oil-in-water nanoemulsions for pesticide formulations. *Journal of Colloid and Interface Science*, *314*(1), 230–235. doi:10.1016/j.jcis.2007.04.079 PMID:17612555

Wang, X., Liu, X., Chen, J., Han, H., & Yuan, Z. (2014). Evaluation and mechanism of antifungal effects of carbon nanomaterials in controlling plant fungal pathogen. *Carbon*, *68*, 798–806. doi:10.1016/j.carbon.2013.11.072

Wu, H., Santana, I., Dansie, J., & Giraldo, J. P. (2017). In vivo delivery of nanoparticles into plant leaves. *Current Protocols in Chemical Biology*, *9*(4), 269–284. doi:10.1002/cpch.29 PMID:29241293

Xue, W., Huang, D., Zeng, G., Wan, J., Cheng, M., Zhang, C., Hu, C., & Li, J. (2018). Performance and toxicity assessment of nanoscale zero valent iron particles in the remediation of contaminated soil: A review. *Chemosphere*, *210*, 1145–1156. doi:10.1016/j.chemosphere.2018.07.118 PMID:30208540

Yao, K. S., Li, S. J., Tzeng, K. C., Cheng, T. C., Chang, C. Y., Chiu, C. Y., Liao, C. Y., Hsu, J. J., & Lin, Z. P. (2009). Fluorescence silica nano probe as a biomarker for rapid detection of plant pathogens. *Advanced Materials Research*, *79*, 513–516. doi:10.4028/www.scientific.net/AMR.79-82.513

Yoon, K. Y., Byeon, J. H., Park, J. H., & Hwang, J. (2007). Susceptibility constants of Escherichia coli and Bacillus subtilis to silver and copper nanoparticles. *The Science of the Total Environment*, *373*(2-3), 572–575. doi:10.1016/j.scitotenv.2006.11.007 PMID:17173953

Zhang, L., & Fang, M. (2010). Nanomaterials in pollution trace detection and environmental improvement. *Nano Today*, *5*(2), 128–142. doi:10.1016/j.nantod.2010.03.002

Zhang, N., Wang, M., & Wang, N. (2002). Precision agriculture: A worldwide overview. *Computers and Electronics in Agriculture*, *36*(2-3), 113–132. doi:10.1016/S0168-1699(02)00096-0

Zhao, L., Lu, L., Wang, A., Zhang, H., Huang, M., Wu, H., Xing, B., Wang, Z., & Ji, R. (2020). Nanobiotechnology in agriculture: Use of nanomaterials to promote plant growth and stress tolerance. *Journal of Agricultural and Food Chemistry*, *68*(7), 1935–1947. doi:10.1021/acs.jafc.9b06615 PMID:32003987

Section 6

Future Role of Nanopriming-Based Approaches for Food Security and Sustainable Agriculture

Chapter 18
Nanoparticles in Food: Sources, Application, Health Risks, and Regulations

Kingsley Erhons Enerijiofi
https://orcid.org/0000-0001-7513-4052
Glorious Vision University, Nigeria

Ndukwe Maduka
https://orcid.org/0000-0003-1375-8665
Federal University, Otuoke, Nigeria

Ruth Chiamaka Osaro-Matthew
Michael Okpara University of Agriculture, Umudike, Nigeria

Cyril Ohikhatemen Ahonsi
Nigeria Natural Medicine Development Agency, Federal Ministry of Innovation Science and Technology, Nigeria

ABSTRACT

The advent of nanotechnology in the twentieth century and its application in various aspects of food production, processing, and packaging offers a lot of benefits to the increasing world population facing food shortages, malnutrition, hunger, and foodborne illnesses, especially in developing countries. However, public health concerns, ethical considerations, and arguments over labeling of food products containing nanoparticles could influence the level of acceptability of the product by consumers. There are many commercial nanofoods of high quality manufactured by leading food companies in the world. The various types of nanomaterials used in the food industry, research findings from recent studies on nanomodified foods, and public health risk associated with ingestion of nanoparticles in foods are discussed. In addition, the permissible limit of some nanoparticles in food, existing laws and regulations on nanofoods, and the challenges limiting wide availability of nanofoods in developing countries are looked into in detail.

DOI: 10.4018/979-8-3693-1890-4.ch018

INTRODUCTION

The world population of about 6 billion is grappling with the challenge of malnutrition, food security, safety and shortages (Ameta et al., 2020). By 2050, the population is expected to reach 9.08 billion and quantity of food required to feed people living on planet earth by then will be 100% of today's output (Hannon et al., 2015; Cerqueira and Pastrana, 2019; Islam, 2019). It is estimated that 2 billion people across the world are experiencing micronutrient deficiencies which is largely attributed to nutrient loss during conventional food processing and storage. The application of nanotechnology is a highly promising strategy to tackle these problems (Enerijiofi et al., 2024; Otchere et al., 2023). Massive production of food aided by technology and its sustainability will drastically reduce food shortages and hunger (Food Safety Authority of Ireland; 2008; McClemens and Xiao, 2017). The need to meet food demand by the growing world population especially in developing countries has led to the application of nanotechnology which offers the people a whole lot of benefits (Suthar et al., 2021; Enerijiofi et al., 2023). It involves the use of resources at the nanoscale level to biofabricate and/or manufacture new materials (Ozimek et al., 2010; Chellaram et al., 2019; Ameta et al., 2020). Nestle, H. J. Heinz, Unilever and Altria are some of the leading food manufacturers in the world investing heavily in nanotechnology research and development (Ozimek et al., 2010; Rajput et al., 2021). In 2030, it is projected that food containing nanomaterials in Japanese markets will reach 250 billion yen (Higashisaka et al., 2015). It is interesting that some developing countries notably Brazil, South Africa, Thailand, Chile, Philippines, Argentina and Mexico have started taking advantage of nanotechnology to address numerous challenges facing them (Salamanca-Buentello et al., 2005). Figure 1 shows the application of nanotechnology in different fields.

Materials within the nanoscale range which are materials with at least one internal or external dimension measuring between 1 and 100 nm are regarded as nanomaterials (Jeevanandam et al., 2018; Sharma et al., 2017). In recent years, nanozymes derived from nanomaterials is receiving increasing attention of researchers (Wu et al., 2021; Songca, 2022). Nanozymes which function as biosensors is useful in monitoring food safety and quality (Arshad et al., 2022). It is useful in detecting foodborne pathogens, toxins, heavy metals, pesticides, antibiotic and drug residues in foods (Payal et al., 2021; Lang et al., 2022). Furthermore, the application of nanozymes help to extend the shelf life of food, ensure that nutritive properties of food products are maintained, develop nanovitamins and improve packaging materials which possess spoilage indicators (Baker, 2020). Nanoparticles (NPs) are regarded as smallest objects with a large surface area and mass transfer than big sized particle (Vickers, 2017; Singh et al., 2022, 2023a). The widespread use of nanoparticles in food industry facilitate the exposure and transfer to humans from the environment to food. Nanoparticles could enter the food chain through the use of pesticides and during food processing (Enerijiofi and Obade, 2018; Naseer *et al.,* 2018). Since nanoparticles have the ability to penetrate bio-permeable barriers, the potential safety hazard of NPs on human is drawing attention of many researchers (Chen et al., 2023). The use of 'omics' techniques which deals with genomics, proteomics, transcriptomics and metabolomics are useful in toxicity studies on nanoparticles (Fröhlich, 2017). A recent concept known as 'foodomics' which involve the application of 'omics' technology in food analysis has made it possible for scientists to research on food nutrients and other constituents in detail (Balkir et al., 2021; Valdés et al., 2022; Taslim et al., 2023).

Nanoparticles are used in the food industry as preservatives, wrapping materials, antimicrobials, flavouring agents and packaging of food. It has been well established that the use of nanomaterials in food enhance its quality, appeal and shelf life. Nanomaterials possess lots of appealing biological and chemical activities and reactivities (Bajpei et al., 2012). Nanomaterials are used in packaging food to enhance

Figure 1. Possible application of nanotechnology in food processing, safety, packaging and nutrition of food science, and other fields
Source: Biswas et al., (2022).

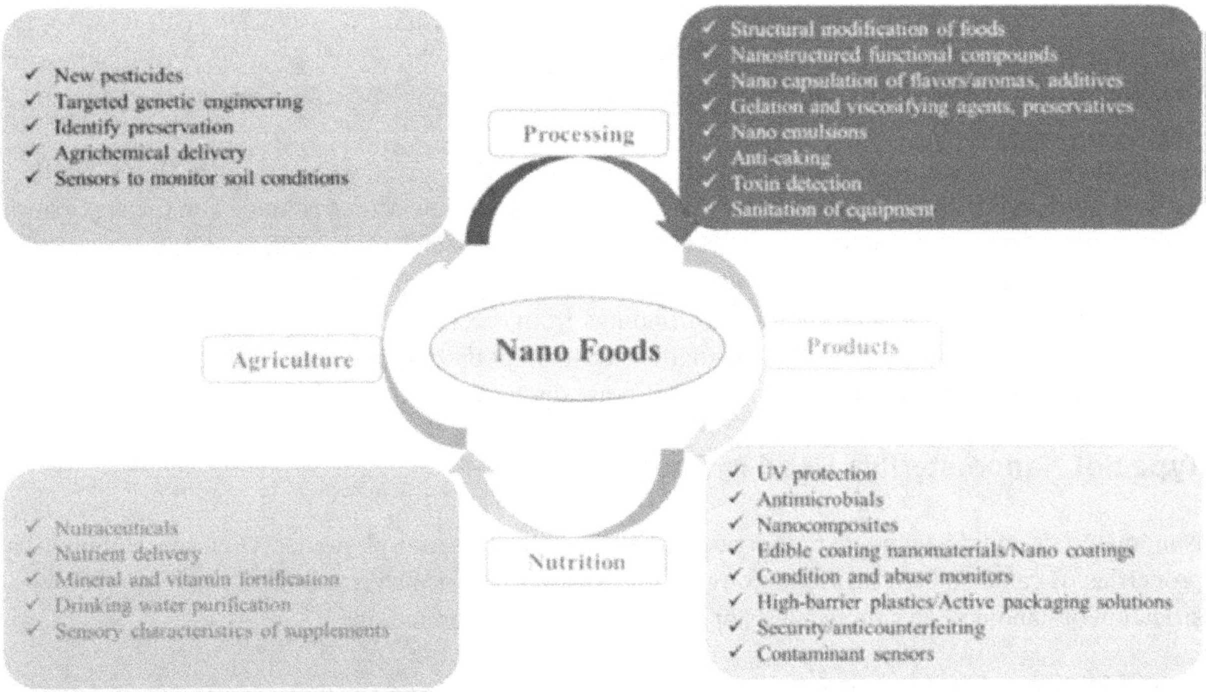

mechanical barriers; biomarkers in food safety and quality; for easy detection of microbial pathogens and enhancement of nutrient bioavailability (Mohammed et al., 2022). One of the ways nanoparticles enter the body system is by ingestion of food and water (Maharramov et al., 2019). In developed countries, it is estimated that daily consumption of nanoparticles by adults is above 10^{12} particles per day (Martirosyan and Schneider, 2014). Nanotechnology makes it possible to produce foods that are friendly to the environment which also improve public health. Consumption of such foods could prevent obesity and other diseases associated with unhealthy foods such as diabetes and hypertension (Chaudhry et al., 2008).

Some of the nanoparticles in food are carbon-based NPs, nanoselenium NPs, rare earth oxide. Based on the mode of synthesis, nanoparticles could be physical, chemical and or biological (Singh et al., 2023b). Nanoparticles can also be categorized as biodegradable or non-biodegradable. Studies have shown that organic NPs are generally non-toxic and do not accumulate in the body due to their biodegradable nature (Frenzel et al., 2015). The toxicity of NPs are mainly linked with the chemical reactivity, e.g. some inorganic NPs release ions that lead to undesirable chemical reactions such as silver nanoparticles (AgNPs) (McClemens and Xiao, 2017). Also, positively charged NPs that are insoluble are readily absorbed in human intestinal barriers making it more bioavailable. Nanoparticles with smaller diameters are easily absorbed (Amela et al., 2020). However, the physical and chemical behaviour of NPs in the human system is not fully understood and therefore, require further investigation (Li and Cummmins, 2020).

Due to limited information in the level of exposure of humans to NPs and hazard characterization, a unified risk assessment method to identify NPs in food will advance knowledge. It will form the basis to regulate nanostructured foods (Shah et al., 2017; Mellia et al., 2022). Research and development of nanostructured food starting from the laboratory stage up to industrial scale is faced with the challenge of sustainability. In this regard, important lessons should be drawn from conventional foods (Hessel *et al.*, 2023). The sustainable development of nanotechnology need in-depth understanding of synthesis, deposition, exposure, environmental release and potential health hazard associated with it. Also, the vast area of nanotechnology has not been fully investigated of which over–regulation of this nascent technology may hamper its development whereas under-regulation may expose man to the adverse effects (Yin et al., 2017).

Globally, the number of published articles on nanotechnology in the web of science (WoS) between the year 2000 to 2019 is 2,718,619 (Lateef et al., 2021). This review article is hinged on the application of nanotechnology in food products. The various types of nanomaterials used in food industry are considered, some commercial nanofood products are listed and permissible limits of nanomaterials in food are highlighted. Of importance are research findings from recent studies on nanostructured foods and its public health implications. Existing regulations on nanofoods and challenges in developing countries limiting the availability of nanofood in the market are discussed in this book chapter.

Types of Nanomaterials Used in Food Industry

Nanomaterials could be organic or inorganic which vary in sizes, morphologies, chemical and physical properties. It can also be classified into two groups based on the material and shapes. However, classification of nanomaterials on the basis of material is more encompassing.

Carbon-Based Nanomaterial

Basically, nanomaterials in this group contain carbon. The ability of carbon atoms to covalently bond with one another indefinitely has resulted in various allotropic forms of carbon nanomaterials (CNMs) (Agustín et al., 2018; Nasir et al., 2018). Carbon nanomaterials can exist in various morphologies such as spheres, hollow tubes or pellisoids. Carbon-based nanomaterials are mainly synthesized through green nanotechnology. They are regarded as safe and environmentally friendly, hence its multifaceted application in the food industry. Carbon-based nanomaterials can be grouped based on their spatial dimensions into zero-dimensional (fullerenes), one- dimensional (carbon nanotubes), two-dimensional (graphene) and multidimensional (pristine graphene coil) and other grapheme derivatives (Nasir et al., 2018; Qu et al., 2018; Mohammad et al., 2019)

Carbon Dots

Carbon dots are distinct particles spherical in nature, zero-dimensional semiconductors with diameter of less 10. The chemical and physical properties of carbon dots are of high-quality. The properties include good tensile strength, water soluble, photoluminescence, low toxicity, highly photostable, and biocompatible (Qu et al., 2018; Salimi et al., 2021; Deepika et al., 2023). Carbon dots on the surface carry active functional groups like amino, hydroxyl and carboxyl groups which enable cross-linking in a chemical reaction (Kousheh et al., 2020). The primary material of the carbon dot determine its chemical and physical properties. Carbon dots which have wide applications in food industry are mainly of ecofriendly origin such as plant and animal (Qu et al., 2018; Singh et al., 2018; Salimi et al., 2021).

Carbon Nanotubes

Carbon nanotubes (CNTs) are hollow tubular form and one-dimensional carbon based nanomaterials. It was first reported by Iijima and Ichihashi (1993). Carbon nanotubes are classified as single-walled carbon nanotubes (SWCNTs) with one graphene layer, and multiple walled carbon nanotubes (MWCNTs) with a central hollow core surrounded by more than one cylindrical shells (Mashkour et al., 2018; Pattanshetti et al., 2020). The physical, mechanical and chemical and electrical properties of carbon nanotubes are good. This could be partly attributed to its sp2 and *pi* (π) bonds (Mohammad et al., 2019). Carbon nanotubes are highly resistant to water; possess a high thermal and electrical conductivity, high surface to volume ratio with a good antimicrobial activity. All these properties account for great importance of CNTs in the food industry especially in food packaging.

Carbon Nanofibers

Carbon nanofibers have properties that are rare which include multiple active sites readily available for physical and chemical interactions with other molecules. Carbon nanofibers have a large surface area that is chemically active and can be easily functionalized (Ezati et al., 2022). The multiple application of carbon nanofibers have been attributed to the possibility of controlling their surface acid-base properties. Mechanically it has been established that carbon nanofibers have excellent mechanical properties as it does not disintegrate in a liquid phase reaction involving vigorous agitation (Mohamed, 2019). The size and dimensions of carbon nanofiber makes it suitable for filtration system, as it can easily be easily separated from reactants and products in the liquid phase (Cao et al., 2021). Carbon nanofibers have good electrical conductivity and resistivity which suggest that they can be used for the construction of electric devices at the nanoscale range (Ezati et al., 2022).

Nano Carbon Black

Carbon black which is a near pure elemental carbon consists of about 98 – 99.7% of carbon. Going by the ISO 2012 criteria, carbon black is considered a nanostructured material since it is within the nanoscale. Carbon black is a quasi-graphite produced at a high temperature from incomplete combustion of aromatic hydrocarbons (Arduini *et al.*, 2020). The good electrical conductivity, easy and versatile functionalization properties of carbon black makes it a good nanofiller (Li et al., 2019). Carbon black is widely used in the production of packaging material as well as biosensors in the food industry (Oliveira et al., 2020).

Carbon Nano-Onions

Carbon nano-onions also known as multi-shell fullerenes are extensively used in the food industry due to their excellent properties which include low toxicity, biocompatibility, low densities, large surface area, multilayer morphology, and ease to functionalization (Grand Tovar et al., 2020; Wang et al., 2021). Carbon nano-onions are commonly produced by thermal annealing of nanodiamond under inert atmosphere or vacuum. Carbon nano onions have versatile application in the food industry. It is also used for bioimaging (Bartelmes and Giordani, 2014; Grand Tovar et al., 2020), as food grade lubricants (Wang et al., 2021), biosensors and polymer compositing (Grand Tovar et al., 2020).

Graphenes

Graphenes are allotropic form of carbon with single layer of atom all over the surface. Each atom on the surface is available for chemical reaction (Reina et al., 2017; Aunkor et al., 2020). Graphenes are semi-metal with outstanding electronic characteristics. They are excellent thermal conductors, with high tensile strength making it a suitable material for food packaging (Aunkor et al., 2020). The usefulness of graphenes in food industry cannot be overemphasized. A graphene base sorbents are considered superior owing to its sorption capacity (Szunerits et al., 2018). Performance enhanced food packaging materials are produced with graphene (de Carvalho and Junior, 2020). Graphenes are also used in detecting toxins such as pesticides and mycotoxins in food (Pan et al., 2019). Antibacterial activity of graphenes against some food borne pathogens have been reported (Aunkor et al., 2020).

Metal-Based Nanomaterial

Metal based nanomaterial consist of metal and metal oxide nanoparticles such as gold (Au), silver (Ag), copper oxide (CuO), zinc Oxides (ZnO) titanium dioxide (TiO_2), silicon oxide (SiO) and magnesium Oxide (MgO) (Meena et al., 2015; Singh et al., 2015; Mirzaei and Darroudi, 2020; Dash et al., 2022; Singh et al., 2023c,d). These nanomaterials are extensively used in the construction of nanocomposite material for food packaging, additives and food preservatives. Metal based nanomaterials generally have broad spectrum antimicrobial activities, good mechanical properties, biocompatibility and barrier capability (Mirzaei and Darroudi, 2020). Au and Ag oxides which are the most prominent in food industry are chemically inert. They have excellent photothermal performance and ease of functionalization (Talarska et al., 2021). CuO based nanomaterials are important in the food packaging system due to their thermal and chemical stability properties. CuO is relatively cheap and also photocatalytic. TiO_2 are non-toxic, thermostable, cheap and photostable. Due to these properties, TiO_2 provides UV protection for food packaging materials. ZnO based nanomaterials are widely used as food additive. ZnO is photocatalytic, have antioxidant effect, antimicrobial property, chemical sensing ability and also regarded as safe (FDA, 2011; Zanet et al., 2019). MgO based nanomaterials are thermostable (Wang et al., 2020), gas impermeable, easily recyclable, have high thermal conductivity, hence, their prominence in the food packaging system. SiO has a low toxicity. It has excellent adsorption properties, highly stable and can be functionalized. SiO based nanomaterials are commonly used nanomaterials in the food industry especially as anticaking agent (Singh et al., 2017).

Organic-Based Nanomaterials

Organic based nanomaterials are formed from organic matter exclusive of carbon. Some examples include liposomes, polymers, micelles, dendrimers, chitosan, and nanogels (Suh et al., 2016). Organic based nanomaterials are readily degradable due to the non-covalent (weak) interactions between atoms, non-toxic, highly biocompatible, however, they are not thermal and photo stable. These nanomaterials are widely applied in the nanoencapsulation of food ingredients, as biopreservatives (Tayel, 2016; Chang et al., 2020) and also to boost the organoleptic and nutritional properties of food products (Singh et al., 2017).

Composite-Based Nanomaterials

Composite based nanomaterials are hybrid composite material made up of two or more materials on the nanoscale dimension (Cui et al., 2015; Cui et al., 2016; Grand Tovar et al., 2020). Carbon-based nanomaterials, organic based nanomaterials, and metal-based nanomaterials can be blended with any form of polymer bulk materials, ceramics or metal to achieve desired properties for a specific purpose (Jeevanandam et al., 2018; Vasile, 2018). Compositing nanomaterials help in addressing the shortcomings of individual nanomaterial and significantly enhance the intrinsic performance of nanomaterials. Various polymer composites with improved barrier properties have been reported (Cui et al., 2015; Cui et al., 2016; Vasile, 2018). It is of great value to food packaging system. Chen et al. (2020) reported that nanocomposite packaging material of poly (vinyl alcohol) and silica aerogel showed improved thermal stability and steam resistance. Figure 2 shows nanosystems in different forms and its food applications.

Current Studies on Nanomaterials in Foods

The last two decades has witnessed a lot of advanced studies on nanostructured foods especially the involvement of nanoparticles in food packaging (Remya *et al.*, 2019; Ameta *et al.*, 2020; Madhusha *et*

Figure 2. Nanosystems in different forms and its food applications
Source: Lugani et al. (2021)

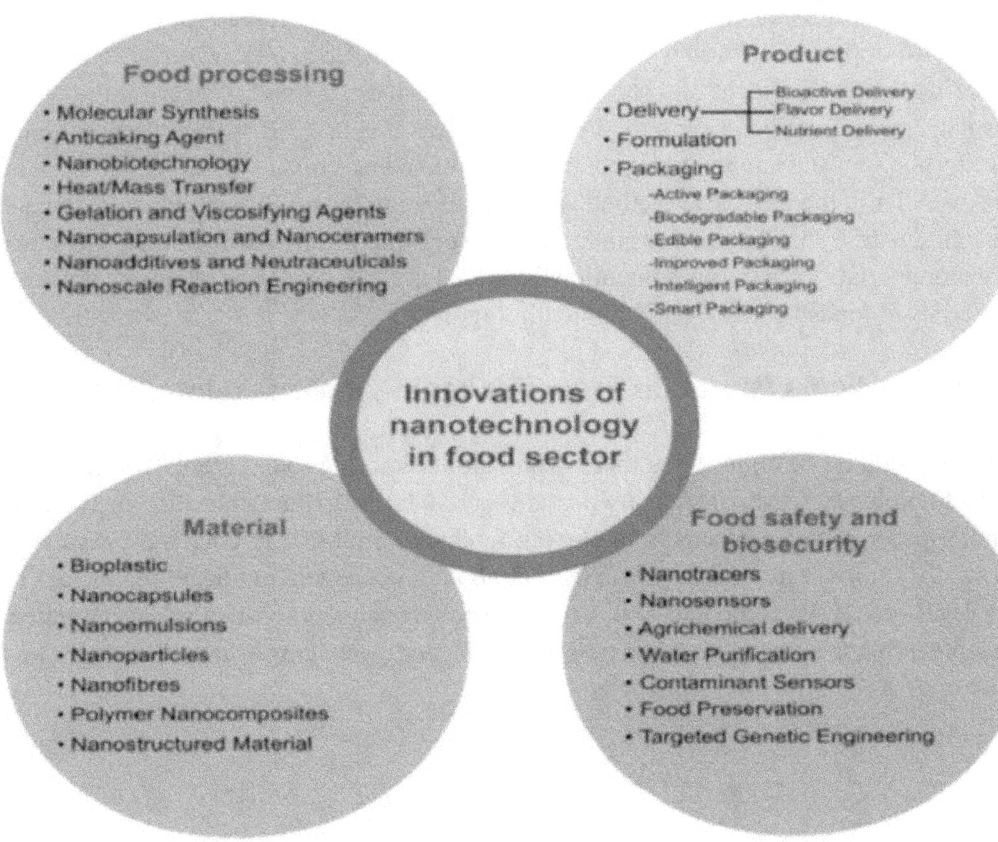

al., 2021). Mazumder and Sivashankar (2020) compiled a list of some commercially available products containing nanomaterials which include milk bottles produced with nanosilver. The product is used for storage. Milk can be poured inside the bottles and used to feed toddlers. The manufacturer of the milk bottle promoted the product for people to buy. According to Chung *et al.* (2017), advanced research on nanofood packaging is aimed at developing excellently packaged food products. The freshness and shelf life of such packaged foods are superior to conventional products.

Recent studies on functionality of nanomaterials in processing and preservation of food were reported by Foujdar *et al.* (2022). In one of the studies, it was observed that during storage of bread fortified with microencapsulated garlic oil, the cohesiveness and springiness of the product did not change. Further, the garlic oil was protected from colour change and oxidation. Interestingly, the sensory attributes of the product was better than commercially available garlic bread. Another study showed that physicochemical properties of ice-cream was enhanced by adding curcumin nanoemulsion as a milk protein emulsifier. The sensory properties of the product was not different from the control.

According to recent reports, nanomaterials used in food packaging could migrate into food (Paidari et al. 2021; Chadha *et al.*, 2022; Barage et al., 2022). A decade ago, Bott *et al.* (2014) reported that migration of TiN nanoparticles from low density polyethylene (LDPE) could not be measured at detection limit (0.09-0.11 $\mu g\ kg^{-1}$) in food simulants (95% ethanol and isooctane; 0.24 $\mu g\ kg^{-1}$ for 3% acetic acid). Paidari *et al.* (2021) highlighted several migration studies on nanoparticles which include Ag, ZnO, clay, Ti, Cu, Cr and Fe using different matrixes and detection methods. A related study showed that 0.46 mg/kg of Ag and 0.82 mg/kg of Cu migrated from the surface into food simulants. ICP-MS and prior acid digestion methods were used in the study.

Three methods used to detect and measure the concentration of nanoparticles in foods include microscopic, spectroscopy and quantitative analysis (Paidari *et al.*, 2021). However, there are some challenges hindering the development of a suitable method for detecting and measuring the concentration of nanoparticles at different points in a food life cycle. Szakal *et al.* (2014) listed the challenges which include (i) Morphology and nanomaterial composition are heterogeneous (ii) Complex food matrix (iii) Limitations peculiar to the analytical method and (iv) Diversity of the alimentary tract. In carrying out research that has to do with nanomaterials in foods, it is very important to differentiate between naturally occurring nanomaterials from the ones intentionally added to food. The areas of current research interest on nanomaterials in foods have been highlighted in Figure 3.

Risk of Nanoparticles Migration From Packaging Materials Into Food

Istiqola and Syafiuddin (2020) reported that silver nanoparticles (AgNPs) are useful in advanced food packaging technology. However, its toxicity becomes a risk factor to human health when the concentration of AgNPs that migrate into foods is above recommended limit (Mazumder and Sivashankar, 2020; Chadha *et al.*, 2022). Research findings carried out about a decade ago indicate that acidic meals had the highest level of AgNPs compared with various types of nanocomposites which migrated into the food. A recent study reported that metal oxides used for food packaging which include SiO_2, TiO_2 and ZnO did not migrate into the food (Barage *et al.*, 2022).

Figure 3. Areas of current research on nanomaterials in foods
Source: Primary

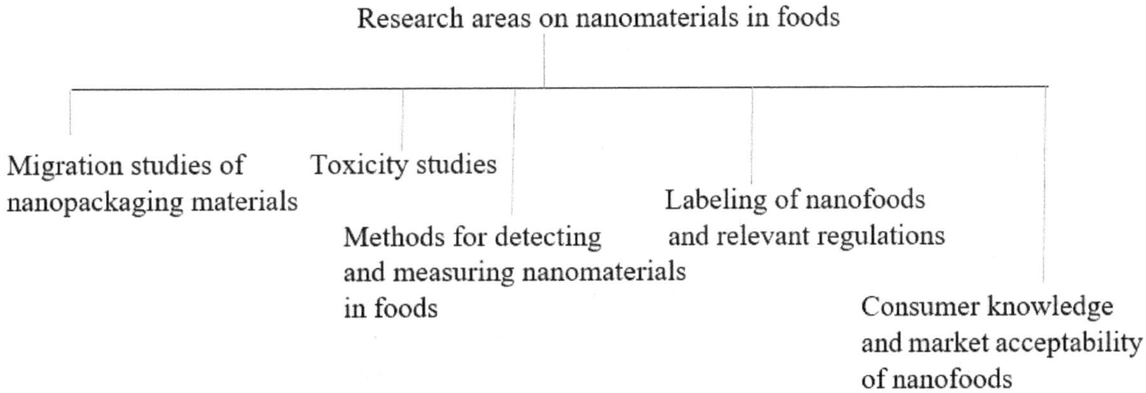

Toxicity Concerns of Nanomaterials and Antimicrobial Properties

Despite the numerous benefits associated with nanotechnology applications in foods, a lot of concern about the safety of nanofoods have been raging over the years (Ameta *et al.*, 2020). Presently, there is limited information on the potential toxicity of nanomaterials directly or indirectly associated with foods. Brand *et al.* (2019) adopted a pragmatic approach to identify, assess and prioritize the potential risks associated with nanomaterials in foods. Toxicity testing of titanium dioxide, silica and silver nanoparticles commonly used in the food industry have been carried out. Higashisaka et al. (2015) reported that 30 mg/kg bw/day of silver nanoparticles, 3.5 g silica particles/kg body and 500 mg/kg bw/day of titanium dioxide nanoparticles measuring 140 nm in diameter had no adverse effects in rats. On the contrary, Naseer et al. (2018) reported that nanoparticles used in foods have toxic effects which include oxidative stress, damage to DNA, decrease in cell viability, cytotoxicity, increase in reactive oxygen species (ROS), among others. According to He *et al.* (2019), a team of researchers working in French Agricultural Research Institute raised alarm in 2017 that titanium dioxide (E171) earlier approved by the United States Food Drug and Administration (USFDA) as a safe nanomaterial in foods is carcinogenic. The report prompted some food manufactures in France to place a ban on TiO_2. The controversy surrounding the ban on food additive TiO_2 (E171) necessitated another study that involved the use of concepts applicable in sociology of controversy and organizational routines better known as innovation studies (Boutillier *et al.*, 2022). Findings from the study suggest that nanoparticles detected in the food additive E171 instigated the controversy because of the potential health effect in humans. According to Ozcakir (2023), harmful effect of TiO_2 nanoparticles on human cells is influenced by size of the nanoparticles. A team of researchers that made this observation, therefore, recommended the use of TiO_2 nanoparticles within the range of 110-300 nm because it is less harmful to human cells.

Chadha *et al.* (2022) reviewed the synthesis and properties of various nanomaterials used in food packaging and highlighted their advantages and disadvantages. They expressed concern about the toxicity of nanomaterials in foods. In a related study, Barage *et al.* (2022) raised societal concerns in the use of various nanomaterials in food packaging. Agriopoulou *et al.* (2020) reported that the use of edible coatings and films to package highly perishable fruits with the intention of extending the shelf life of the product could be harmful to human health when it is consumed. Jagtiani (2021) reported that sputtering powder used to produce coatings (ploylactic acid-functionalized aluminium-doped zinc oxide) possess inhibitory properties against *Escherichia coli*. According to Madhusha *et al.* (2021), advanced studies that involves incorporating silver nanoparticle and chistosan could have a synergistic effect that will enhance its antimicrobial properties which is very useful parameter in modern food packaging.

Despite the fact that toxicity concerns with regards to nanoparticles in food has not been well addressed, the number of commercially available nanofoods in the markets especially in developed countries is increasing. Ethical issues could influence sales and consumer acceptability of nanostructured foods (Thompson, 2015; Kushwaha and Pant, 2023). A pilot study was conducted by Hasim *et al.* (2019) using a survey approach to determine the influence of four factors namely: trust, knowledge, psychological and perceived benefits on consumer acceptability that drives purchase intentions towards a typical nanofood called P3 sweetener. The study involved sixty-five respondents in Malaysia and result obtained showed that Cronback's Alpha Coefficient (a) is > 0.70; meaning that all the factors showed significant reliability test.

Labeling of Nanofoods

Labeling of nanofoods commercially available in the market is still a controversial topic among experts. Only few research articles have been published in this regard. According to Inobeme *et al.* (2022), nano-modified foods available in the market are not labeled to show that it contains nanomaterials. Therefore, the general public are unaware they are consuming nanofoods. Hasmin *et al.* (2021) used doctrinal approach and content analysis to evaluate the World Trade Organization (WTO) agreements with regards to mandatory labeling of nanofoods. Findings from the study suggests that mandatory labeling of nanofoods will amount to restrictions on international trade. However, the researchers recommended that specific guidelines on labelling of nanonfoods should be drafted by Codex Alimentarius Commission (CAC) and Codex Committee on Food Labelling (CCFL) and implemented in a similar manner with biotechnology foods. A related study earlier carried out involved the use of analytical framework of consumers and suppliers that were heterogeneous and imperfectly competitive, respectively. According to the report, labeling of nanofoods increased the cost of production, reduced the uncertainty of consumers towards the product and possibly affected consumer perception which they considered to be harmful (Bratovcic, 2020).

Public Health Implications of Nanoparticles in Food

The application of nanotechnology in food production, processing and packaging has numerous benefits, but the general public, especially those consuming the product and experts in nanotechnology have expressed health concern about the accumulation of nanoparticles from food in human body (Mazumder and Sivashankar, 2020; Jagtiani, 2021; Barage et al., 2022; Singh et al., 2024a,b). Consumption of nanomeals could impair the normal functioning of the human body (Sharara, 2021). Nanosilver used in manufacturing food packaging material could migrate into the body and become toxic to human fibroblast and carcinoma cells. Consequently, the DNA and mitochondria could be damaged (Mazumder and Sivashankar, 2020). Very tiny sizes of nanoparticles enhances its migration into the cells, reaching parts of the body such as brain, liver, heart, kidneys, spleen, bone marrow and nervous system (Jain et al., 2017). Vital organs in the body such as kidney and liver could suffer from adverse effects due to long term exposure to nanoparticles (Naseer et al., 2018; Biswas et al., 2022). There is possibility that single oral intake of ZnO nanoparticles will damage the kidney and lung as well as cause hepatic injury (Sahoo et al., 2022). Although nanoparticles could be toxic to cells, there are indications that nanoparticles could demonstrate some level of anticancer activity. Additional studies in this regard is required (Oberdorster *et al.,* 2005; Foujdar et al., 2022). According to Onyeka *et al.* (2022), nanoparticles with poor solubility could cause cancer because it has a greater surface area to volume ratio that increases biological and chemical reactivities. According to Naseer et al. (2018), nanoparticles could cause inflammation of the gastrointestinal tract, Alzheimer's disease and Parkinson's syndrome. Ingestion of nanoparticles in food increases the risk of Crohn's disease and colon cancer in humans. In the course of preparing nanoparticles meant for food industries, it could be released into the air or get in contact with the skin which endangers the health of workers (Martirosyan and Schneider, 2014; Ameta et al., 2020; Rezaei *et al.* (2021). Despite public health concern associated with nanoparticles in food, its long term effect on human health is yet to be fully substantiated due to lack of data and other limitations (Sari and Piyal, 2019; Chaudhary et al., 2020; Barage et al., 2022).

Permissible Limit of Nanoparticles in Food

Based on a series of toxicity testing of nanoparticles in food, advisory boards on the safety of food for public consumption in different countries have recommended permissible limits of nanoparticles which can be ingested through nanofoods. The Scientific Committee on Food of the European Food Safety Authority (EFSA) recommend that silicon intake per day derived from nanosilica (E551) which is a food additive registered in European Union (EU) for an individual weighing 60 kg is between 20-50 mg per day (Higashisaka et al. 2015). The EFSA also stated that quantity of silver permitted to migrate from AgNPs in food packaging to water and food is 0.05 mg/L and 0.05 mg/kg, respectively (Istiqola and Syafiuddin, 2020). The quantity of TiO_2 to be used as a food colour additive approved by the United States Food and Drug Administration (USFDA) is not permitted to exceed 1% w/w (Martirosyan and Schneider, 2014).

Some Food Products Commercially Available Containing Nanomaterials

According to statistics released after a survey was carried out by the U.S. Food and Drug Administration (USFDA) in 2008, the number of food and drink items containing nanomaterials across the world which were sold to consumers were above 80 (Higashisaka et al., 2015). The number of nanofood products categorized as 'food and beverage' is 204 according to information obtained from March 2013 Project on Emerging Nanotechnologies (Martirosyan and Schneider, 2014). Food manufacturers in Australia were using nanocapsules of tuna fish oil to supply omega-3 fatty acids to white bread particularly the popular brands such as Tip top bread after it was mixed with nanocapsules (Ozimek et al., 2010; Ameta et al., 2020). Advertisement of nano selenium was presented to consumers as a tea additive. The USFDA has approved the sale of SiO_2 (E551), MgO (E530) and TiO_2 (E1717) as anti-caking agent, flavour carriers and colouring additives, respectively. The addition of TiO_2 (E171) in white sauces, cake icings, candies, gums, and puddings pie is permitted by the authorities (Jagtiani, 2021; Sahoo et al., 2022). The product known as NovaSOL (AQUANOVA) was developed to add omega-3 fatty acid, vitamins, Co Q10 and phytoextracts into food and beverages. Nutritional supplements referred as nano-ceutricals (RBC Life Sciences, Inc.) formulated as cocoa nanoclusters has an increased surface area which enhances its chocolate flavour. The product is ideal for consumers that want to shed weight (Sponton et al., 2019). Supplemented functional drink, food supplement and so many commercially available nanofood products manufactured by food companies using different product names in the USA, Israel, Australia, Korea, and Germany include Kimchi, nutra leaseanola active oil, fortified fruit juice, tip top bread, nanoslim beverage, oat nutritional drink, oat chocolate and oat, vanilla nutritional drink etc. (Singh et al., 2023). Canola active oil, a brand of canola cooking oil is manufactured by Shemen Industries, Tel Aviv, Israel. Qinhuangdao Taiji Ring Nano-Products Co., Ltd, Hebei, People's Republic of China is a producer of a tea branded Nanotea. Nanoceuticals Slim Shake Chocolate is produced by RBC Life Sciences Inc., Irving, Texas, USA (Paul and Bhojane, 2021).

Existing Laws and Regulations on Nanofoods

According to recent reports, there is no official regulation specifically for nanomaterials vis-à-vis nanofoods in any country in the world (Amenta et al., 2015; Onyeaka et al., 2022). It is surprising that countries such as Japan and China which are large scale producers of nanomaterials lack a proper regulation specifically for nanotechnology (Nile et al., 2020). The absence of specific laws and statutory regulations with regards to nanotechnologies and its applications in foods in Malaysia has been reported (Ismail et al., 2018).

Guidelines and regulations in the application of nanotechnology in foods are drawn from relevant provisions in food regulations. The Food Standards Australia New Zealand Act 1991 (Cth) (FSANZ Act) handle all aspects of food quality, labelling and packaging in Australia and New Zealand. Therefore, nanofoods could be included as part of the category of foods covered by FSANZ Act and Food Code. It is debatable whether substances already in existence in large scale is different when it is reformulated at the nanoscale. FSANZ Act 1991 and Food Standards Australia New Zealand Food Standards Code which covers different categories of food does not specifically regulate nanofoods and products related to it (Ludlow, 2009).

Today, it is worthy to note that Switzerland and member countries in the European Union (EU) have relevant provisions specifically for nano-based products in terms of risk assessment. Generally, the legal provisions by the EU with regards to nano-based products is considered to be more stringent when it is compared with the US. Due to the leading role of EU member countries in regulating nano-based products, it is considered as a good example to non-EU countries. African countries seem to be trailing behind non-EU countries in establishing relevant laws and regulations on nano-based products. In 2002, the South African Nanotechnology Initiative (SANi) was launched (Amenta et al., 2015). According to Bumbudsanpharoke and Ko (2015), no country in Africa has any information on the government website regarding safety assessment or specific regulation on nanomaterials and its various applications which include nanofoods.

The need for specific laws, regulations or simply guidelines on production, handling and the use of nanofoods should not be given little attention (Couto and Almeida, 2022). Otherwise, there is possibility that nanotoxicity catastrophe will befall mankind in future (Onyeaka et al., 2022). Particle size, physicochemical properties, food safety risks and changes that occurred as a result of processing are the various parameters taken into consideration while drafting regulations with regards to nanofoods (Ramkumar et al., 2019). Provision of food information to consumers - Regulation (EU) No. 1169/20119) and Plastic Food Contact Materials and Articles (10/2011) are relevant in regulating the use of nanomaterials in foods. In the European Union (EU), Reg. 1333/2008, Reg. 1332/2008, Reg. 450/2009 and Regulation (EU) 2015/2283 1 of 25 November 2015 are aimed at regulating the use of nanomaterials as food additives, food enzymes, active and intelligent materials and articles designed to make contact with food and novel foods, respectively (Amenta et al., 2015; Sodano et al., 2018). A review of thirteen European legislations which are considered to be relevant in regulating the use of metallic nanoparticles (MNPs) in the food sector was carried out by Couto and Almeida (2022).

When approval to use food additives and ingredients have been given or packaging of these items in its bulk form, subjecting any of them to risk assessment because its nanoform is intended to be used is not a regulatory requirement neither is it compulsory to label the product as a nanofood (Hansen and Baun, 2012). The need to disclose to the general public that a particular food contain nanoparticles has been stressed by stakeholders. For that reason, a list of nanomaterials permitted in food was published by the Food Safety Authority (FSA) in the UK (Amenta et al., 2015). According to the European Union (EU) Regulation 1169/2011, it is compulsory for manufacturers to declare that a product is a nanofood. In the US, such a requirement is not applicable because the concerns raised about the safety of nanofoods are yet to be resolved. Going ahead to label such products as nanofoods could mislead customers. According to the US Food and Drug Administration (FDA), labeling of nanofoods is optional. Labeling or not labeling nanofoods is still a controversial topic yet to be resolved (Hasmin et al., 2020).

Before food enzymes, additives and flavourings manufactured using nanomaterials are marketed, it has to be approved in accordance with Regulation (EC) 1331/2008 (Amenta et al., 2015). In the US, food manufacturers who make use of nanomaterials as indirect additives are required to obtain a pre-market clearance. According to the Food Contact Notification (FCN) system, the FDA are mandated to reply food manufacturers within 120 days of receiving notification to use nanomaterials. If they fail in their responsibility, the manufacturers are permitted to push their products to the market without obtaining clearance. A situation whereby a producer make use of the same chemical, he is not required to get approval from the FDA (Onyeaka et al., 2022). The FDA request manufacturers of foods and additives outside the category of generally regarded as food 'GRAS' to submit information regarding its physical characteristics, physical and chemical effect including the analytical method used to determine the quantity of the substance added to the food and the safety of its intended use. It is now left for FDA to set limits depending on the scientific information regarding the risk associated with the substance. The stand of FDA regarding the use of food supplements is that manufacturers are solely responsible for the safety of their product. They do not need to register the product before pushing it into the market. What is required of them is to make sure that information contained in the product label is truthful and not misleading. The FDA will step in as a regulatory agency if the product is not safe. The position of FDA has received knocks from environmental non-governmental organizations (NGOs) and civil society institutions for lacking an effective oversight in the use of nanomaterials in foods (IRGC, 2008).

What is considered the most important system to regulate nanotechnologies across member states in the European Union is known as REACH (Registration, Evaluation, Authorization and Restriction of Chemicals). The functions of REACH is similar with Regulation 258/97 on novel foods and novel food ingredients which makes it compulsory to obtain a pre-market authorization in respect to nanomaterials (Grobe et al., 2008). Scientific Committee on Emerging and Newly Identified Health Risks (SCENIHR) is saddled with the responsibility of carrying out risk assessment of nanotechnology in the European Union (EU) before the product is approved for human use (Nile et al., 2020). Adoption of the recommendation by the Organization for Economic Cooperation and Development (OECD) Council on the Safety Testing and Assessment of Manufactured Nanomaterials took place on 19th September, 2013 (Couto and Almeida, 2022).

The United States Food and Drug Administration (USFDA) is charged with the responsibility of regulating nanofoods and food packaging in the US. A similar body in Australia is called Food Standards

Australia and New Zealand (FSANZ) while the European Union Novel Foods Regulation (EC 258-97) takes full control of regulating nanofoods and food ingredients (Berekaa, 2015; Paul and Bhojane, 2021). According to EU Novel Foods Regulation (EC 258-97), pre-market approval must be obtained for every new ingredient and product. Overtime, the existing nanofoods regulations put in place have be revised (Hansen and Baun, 2012). The European Commission (EC) statutory guideline EC 1935/2004 is designed to regulate the use of nanoparticles to package food in Europe (Amenta et al., 2015). The status of regulations and legislations on nanomaterials in food packaging in different countries was summarized by Bumbudsanpharoke and Ko (2015). According to the report, the Brazilian Congress in 2013 rejected the first bill for labeling of food and drugs that contains nanostructure. A similar thing happened in China where regulatory authorities rejected the request to use nano-materials or food ingredients. Countries mentioned where there is no specific regulation or organization on nanomaterial for food packaging include Australia/New Zealand, Indonesia, Malaysia, China and South Africa. Sodano (2018) highlighted neoliberalism sponsored by government in different countries in the world as one of the major reasons for poor regulation of nanofoods. Limited knowledge about the risks associated with nanofoods is hindering the implementation of laws and regulations in production, handling, and the use of nanofoods.

Challenges Limiting the Wide Availability of Nanofoods in Developing Countries

Over the years, the application of nanotechnology in food which has translated to ready availability of nanofoods in the market is dominated by developed countries such as the US, China, Japan, European Union (EU) member countries; leaving developing countries, most of them struggling with technical know-how, production of nanofoods, quality control and handling of safety concerns (Chaudhry and Castle, 2011; Kasa, 2016; Lugani et al., 2021).

In Nigeria, abysmally low research output in the field of nanotechnology between 2010-2020 compare with other developing countries in Africa such as Egypt, Tunisia, Morocco, South Africa and Algeria was reported by Lateef et al. (2021). Poor funding was identified as one of the greatest challenge limiting research development in nanotechnology. Deficiency in the curricular used in higher institutions in Nigeria and lack of experts in the field of nanotechnology are other challenges (Elegbede and Lateef, 2019). According to Lateef et al. (2021), Nigeria is occupying the fourth position in Africa in terms of research publications in nanotechnology.

Naseri and Davoodi (2011) listed some challenges developing countries are facing with regards to commercialization of nanotechnology. They include infrastructural, managerial, cultural and economic issues. Nearly a decade after the research was published, Isam (2019) reported that lack of infrastructure and experts in the use of nanomaterials in foods is a very big challenge in developing countries. Very weak regulations specifically for nanotechnology in most developing countries is also a limiting factor towards availability of nanofood in the market (Cruz-Lopes et al., 2021). Existing food regulatory bodies in these countries are finding it very difficult to cope with the demands of nanotech applications which is continuously changing. Insufficient toxicology data to ascertain the long term health effect of nanofood has made it difficult to assess the safety of nanofood. Due to weak economy, international trade barriers and relatively high price of nanofood could hamper the availability of nanofood in developing countries (Chaudhry et al., 2011). Intellectual property on commercialized nanofood acquired by very big food companies operating in developed countries is a challenge to smaller companies doing business in developing countries to invest in research and massive production of nanofood (Sponton et al., 2019; Lawal et al., 2023).

CONCLUSION

The benefits of applying nanotechnology in food production, processing and packaging is greater than public health concerns expressed towards consumption of nanostructured foods. As food companies especially the ones in developed countries are releasing new nanomaterial-based products into the market to make huge profits, scientists should go a step further to determine the long term effect in human populations across different age groups and races. A periodic review of existing international and national regulations on nanostructured foods is in the overall interest of consumers. In line with current research efforts, the safety and quality of nanostructured food will achieve a reliable and faster results by deploying nanozymes and 'omics' techniques. Developing countries should increase funding for nanotechnology research to boost food production, improve food safety and socioeconomic wellbeing of her teeming population.

CONFLICT OF INTERESTS

The authors declare that no conflict of interests exist.

REFERENCES

Agriopoulou, S., Stamatelopoulou, E., Skiada, V., Tsarouhas, P., & Varzakas, T. (2020). Emerging nanomaterials applications for food packaging and preservation. Safety issues and risk assessment. *Proceedings*, *4*(x), 1–8.

Agustín, G. C., Escarpa, A., & Carlos, D. G. (2018). Carbon-based nanomaterials in analytical chemistry. In *Detection Science* (1st ed., pp. 1–45). Royal Society of Chemistry.

Alfei, S., Marengo, B., & Zuccari, G. (2017). "Nanotechnology application in food packaging: A plethora of opportunities versus pending risks assessment and public concerns," Food Research Application of Nanotechnology in Food Science: Perception and Overview. *Frontiers in Microbiology*, *8*, 1501. doi:10.3389/fmicb.2017.01501

Amenta, V., Aschberger, K., Arena, M., Bouwmeester, H., Moniz, F. B., Brandhoff, P., Gottardo, S., Marvin, H. J. P., Mech, A., Pesudo, L. Q., Rauscher, H., Schoonjans, R., Vettori, M. V., Weigel, S., & Peters, R. J. (2015). Regulatory aspects of nanotechnology in the agri/feed/food sector in EU and non-EU countries. *Regulatory Toxicology and Pharmacology*, *73*(1), 463–476. doi:10.1016/j.yrtph.2015.06.016 PMID:26169479

Ameta, S. K., Rai, A. K., Hiran, D., Ameta, R., & Ameta, S. C. (2020). Use of nanomaterials in food science. In *Biogenic nano-particles and their use in agro-ecosystems* (Vol. 457). Springer. doi:10.1007/978-981-15-2985-6_24

Arduini, F., Cinti, S., Mazzaracchio, V., Scognamiglio, V., Amine, A., & Moscone, D. (2020). Carbon black as an outstanding and affordable nanomaterial for electrochemical (bio) sensor design. *Biosensors & Bioelectronics*, *156*(112033), 112033. Advance online publication. doi:10.1016/j.bios.2020.112033 PMID:32174547

Arshad, F., Mohd-Naim, N. F., Chandrawati, R., Cozzolino, D., & Ahmed, M. U. (2022). Nanozyme-based sensors for detection of food biomarkers: A review. *RSC Advances*, *12*(40), 26160–26175. doi:10.1039/D2RA04444G PMID:36275095

Aunkor, M. T. H., Raihan, T., Prodhan, S. H., Metselaar, H. S. C., Malik, S. U. F., & Azad, A. K. (2020). Antibacterial activity of graphene oxide nanosheet against multidrug resistant superbugs isolated from infected patients. *Royal Society Open Science*, *7*(7), 200640. doi:10.1098/rsos.200640 PMID:32874659

Bajpei, S. K., Chand, N., & Chaurasia, V. (2012). Nano zinc oxide loaded calcium alignate film with potential antibacterial properties. *Food and Bioprocess Technology*, *5*(5), 1871–1881. doi:10.1007/s11947-011-0587-6

Baker, S. (2020). Nanozymes: emerging prospect in food science and technology. SSRN Electronic Journal, 1-7. DOI: doi:10.2139/ssrn.3619997

Balkir, P., Kemahlioglu, K., & Yucel, U. (2021). Foodomics: A new approach in food quality and safety. *Trends in Food Science & Technology*, *108*, 49–57. doi:10.1016/j.tifs.2020.11.028

Barage, S., Lakkakula, J., Sharma, A., Roy, A., Alghamdi, S., Almehmadi, M., Hossain, M. J., Allahyani, M., & Abdulaziz, O. (2022). Nanomaterial in food packaging: A comprehensive review. *Journal of Nanomaterials*, *2022*, 1–12. doi:10.1155/2022/6053922

Bartelmess, J., & Giordan, S. (2014). Carbon nano-onions (multilayer fullerenes): Chemistry and applications. *Beilstein Journal of Nanotechnology*, *5*, 1980–1998. doi:10.3762/bjnano.5.207 PMID:25383308

Berekaa, M. M. (2015). Nanotechnoogy in food industry; advances in food processing, packaging and food safety. *International Journal of Current Microbiology and Applied Sciences*, *4*(5), 345–357.

Biswas, R., Alam, M., Sarkar, A., Haque, M. I., Hasan, M. M., & Hoque, M. (2022). Application of nanotechnology in food: Processing, preservation, packaging and safety assessment. *Heliyon*, *8*(11), 1–16. doi:10.1016/j.heliyon.2022.e11795 PMID:36444247

Bott, J., Störmer, A., & Franz, R. (2014). A model study into the migration potential of nanoparticles from plastics nanocomposites for food contact. *Food Packaging and Shelf Life*, *2*(2), 73–40. doi:10.1016/j.fpsl.2014.08.001

Boutillier, S.; Fourmentin, S. & Laperche, B. (2020). Food additives and the future of health: An analysis of the ongoing controversy on titanium dioxide. Futures, 122. doi:10.1016/j.futures.2020.102598. hal-03490868

Brand, W., van Kesteren, P. C. E., & Oomen, A. G. (2019). Potential health risks of nanomaterials in food: a methodology to identify signals and prioritise risks. RIVM Letter Report 2019-0191. National Institute for Public Health and the Environment, The Netherlands. doi:10.21945/RIVM-2019-0191

Bratovcic, A. (2020). Nanomaterials in food processing and packaging, its toxicity and food labeling. ACTA Scientific Nutritional Health, 4(9), 7-13.

Bumbudsanpharoke, N., & Ko, S. (2015). Nano-food packaging: An overview of market migration research, and safety regulations. *Journal of Food Science*, *80*(5), 910–923. doi:10.1111/1750-3841.12861 PMID:25881665

Cao, Y., Chen, G., Wan, Y., & Luo, J. (2021). Nanofiltration membrane for bio-separation: Process-oriented materials innovation. *Engineering in Life Sciences*, *21*(6), 405–416. doi:10.1002/elsc.202000100 PMID:34140851

Cerqueira, M. A., & Pastrana, L. M. (2019). Does the future of food pass by using nanotechnologies. *Frontiers in Sustainable Food Systems*, *3*, 1–5. doi:10.3389/fsufs.2019.00016

Chadha, U., Bhardwaj, P., Selvaraj, S. K., Arasu, K., Praveena, S., Pavan, A., Khanna, M., Singh, P., Singh, S., Chakravorty, A., Badoni, B., Banavoth, M., Sonar, P., & Paramasivam, V. (2022). Current trends and future perspectives of nanomaterials in food packaging application. *Journal of Nanomaterials*, *2022*, 1–32. doi:10.1155/2022/2745416

Chang, S. H.; Chen, C. H. & Tsai, G. J. (2020). Effects of chitosan on *Clostridium perfringens* and application in the preservation of pork sausage. Marine Drugs, 18(2), 1-15. doi:10.3390/md18020070

Chaudhary, P., Fatima, F., & Kumar, A. (2020). Relevance of nanomaterials in food packaging and its advanced future prospects. *Journal of Inorganic and Organometallic Polymers and Materials*, *30*(12), 5180–5192. doi:10.1007/s10904-020-01674-8 PMID:32837459

Chaudhry, O., Scotter, M., Blackburn, J., Ross, B., Boxall, A., Castle, L., Aitken, R., & Watkins, R. (2008). Applications and implications of nanotechnologies for food sector. *Food Additives and Contaminants*, *24*(3), 241–258. doi:10.1080/02652030701744538 PMID:18311618

Chaudhry, Q., & Castle, L. (2011). Food applications of nanotechnologies: An overview of opportunities and challenges for developing countries. *Trends in Food Science & Technology*, *22*(11), 595–603. doi:10.1016/j.tifs.2011.01.001

Chellaram, C., Murugaboopathi, G., John, A. A., Siakumar, R., Ganesan, S., Krithika, S., & Priya, G. (2014).. . *Significance of Nanotechnology in Food*, *8*, 109–113.

Chen, C., Ding, R., Yang, S., Wang, J., Chen, W., Zong, L., & Xie, J. (2020). Development of thermal insulation packaging film based on poly (vinyl alcohol) incorporated with silica aerogel for food packaging application. *Lebensmittel-Wissenschaft + Technologie*, *129*, 109568. doi:10.1016/j.lwt.2020.109568

Chen, J., Guo, Y., Zhang, X., Liu, J., Gong, P., Su, Z., Fan, L., & Li, G. (2023). Emerging nanoparticles in food: Sources, application and safety. *Journal of Agricultural and Food Chemistry*, *71*(8), 3564–3582. doi:10.1021/acs.jafc.2c06740 PMID:36791411

Chung, I. M., Rajakumar, G., Gomathi, T., Park, S. K., Kim, S. H., & Thiruvengadam, M. (2017). Nanotechnology for human food: Advances and perspectives. *Frontiers in Life Science*, *10*(1), 63–72. doi:10.1080/21553769.2017.1365775

Conidi, C., Castro-Muñoz, R., & Cassano, A. (2020). Nanofiltration in beverage industry. In *Nanotechnology in the Beverage Industry* (pp. 525–548). Elsevier. doi:10.1016/B978-0-12-819941-1.00018-3

Couto, C., & Almeida, A. (2022). Metallic nanoparticles in the food sector: A mini-review. *Foods*, *11*(3), 1–22. doi:10.3390/foods11030402 PMID:35159552

Cui, Y., Kumar, S., Kona, B. R., & van Houcke, D. (2015). Gas barrier properties of polymer/clay nanocomposites. *RSC Advances*, *5*(78), 63669–63690. doi:10.1039/C5RA10333A

Cui, Y., Kundalwal, S., & Kumar, S. (2016). Gas barrier performance of graphene/polymer nanocomposites. *Carbon*, *98*, 131–333. doi:10.1016/j.carbon.2015.11.018

Dash, K. K., Deka, P., Bangar, S. P., Chaudhary, V., Trif, M., & Rusu, A. (2022). Applications of Inorganic Nanoparticles in Food Packaging: A Comprehensive Review. *Polymers*, *14*(3), 521. doi:10.3390/polym14030521 PMID:35160510

de Carvalho, A. P. A., & Junior, C. A. C. (2020). Green strategies for active food packagings: A systematic review on active properties of graphene-based nanomaterials and biodegradable polymers. *Trends in Food Science & Technology*, *103*, 130–143. doi:10.1016/j.tifs.2020.07.012

Deepika, G., Kumar, L., & Gaikwad, K. K. (2023). Carbon dots for food packaging applications Sustainable. *Food Technology*, *1*(2), 185–199. doi:10.1039/D2FB00020B

Elegbede, J. A., & Lateef, A. (2019). Green nanotechnology in Nigeria: The research landscape, challenges and prospects. *Annals of Science and Technology*, *4*(2), 6–38. doi:10.2478/ast-2019-0008

Enerijiofi, K. E., Musa, S. I., Igiebor, F. A., Lawani, M., Chuka, N. E., Odozi, E. B., & Ikhajiagbe, B. (2024). *Nanotechnological approaches for enhanced microbial activities, sustainable agriculture and the environment. In sustainable agriculture nanotechnology, biotechnology and management* (V. D. Rajput, A. Singh, K. Ghazaryan, & T. M. Minkina, Eds.). Walter de Gruyter.

Enerijiofi, K. E., & Obade, E. (2018). Assessment of the impact of herbicide contaminated top soils on its physiochemical, microbiological and enzymatic properties. *FUW Trends in Science and Technology Journal*, *3*(2), 802–806.

Ezati, P., Rhim, J. W., Molaei, R., Priyadarshi, R., & Han, S. (2022). Cellulose nanofibre-based coating integrated with nitrogen-functionalized carbon dots for active packaging applications of fresh fruit. *Postharvest Biology and Technology*, *186*, 111845. Advance online publication. doi:10.1016/j.postharvbio.2022.111845

FDA. (2011). *Part 182—Substances Generally Recognized as Safe*. Food and Drug Administration.

Food Safety Authority of Ireland (2008). The relevance of food safety of applications of nanotechnology in food and feed industries. Author.

Foujdar, R., Kumar, S., Chauhan, A. K., & Bera, M. B. (2022). Role of nanoparticles for food processing and preservation In: Nanotechnology horizons in food process engineering, vol. 1. Goyal, M. R.; Malik, J. A.; Kumar, S. and Watharkar, R. B. (Eds.). Apple Academic Press, Inc. Co-published with CRC Press (Taylor and Francis). doi:10.1201/9781003305378_4

Frenzel, M., Krolak, E., & Wanger, A. E. (2015). Physicochemical properties of WPL coated liposomes serving as stable transporters in a real food matrix. *Lebensmittel-Wissenschaft + Technologie*, *63*(1), 527–534. doi:10.1016/j.lwt.2015.03.055

Fröhlich, E. (2017). Role of omics techniques in the toxicity testing of nanoparticles. *Journal of Nanobiotechnology*, *15*(1), 1–22. doi:10.1186/s12951-017-0320-3 PMID:29157261

Grande Tovar, C. D., Castro, J. I., Valencia, C. H., Navia Porras, D. P., Herminsul, M., Hernandez, J., Valencia Zapata, M. E., & Chaur, M. N. (2020). Nanocomposite films of chitosan-grafted carbon nano-onions for biomedical applications. *Molecules (Basel, Switzerland)*, *25*(5), 1203. doi:10.3390/molecules25051203 PMID:32155970

Grobe, A., Renn, O., & Dialogik, A. J. (2018). *Risk governance of nanotechnology applications in food and cosmetics. International Risk Governance Council (IRGC).*

Guo, S., Luo, J., Yang, Q., Qiang, X., Feng, S., & Wan, Y. (2019). Decoloration of molasses by ultrafiltration and nanofiltration: Unraveling the mechanisms of high sucrose retention. *Food and Bioprocess Technology*, *12*(1), 39–53. doi:10.1007/s11947-018-2189-z

Hannon, J. C., Kerry, J., Cruz-Romero, M., Morris, M., & Cummins, E. (2015). Advances and challenges for the use of engineered nanoparticles in food contact materials. *Trends in Food Science & Technology*, *43*(1), 43–62. doi:10.1016/j.tifs.2015.01.008

Hansen, S. F., & Baun, A. (2012). European regulation affecting nanomaterials – review of limitations and future recommendations. *Dose-Response*, *10*(3), 364–383. doi:10.2203/dose-response.10-029. Hansen PMID:22942870

Hasim, M. A.; Jabar, J. & Murad, M. A. (2019). A preliminary research on consumer acceptance in nanofood towards purchase intention: a pilot research. International Journal of Recent Technology and Engineering, 8(2S3), 352-356.

Hasmin, N. A., Zainol, Z. A., Ismail, R., & Mahmood, A. (2020). Regulatory challenges of nanofood labelling. *Journal of Politics and Law*, *13*(2), 241–249. doi:10.5539/jpl.v13n2p241

Hasmin, N. A., Zainol, Z. A., Ismail, R., & Matmin, J. (2021). Disclosure of nanomaterials under nanotechnology product inventory, voluntary certification, and voluntary labelling. *Pertanika Journal of Social Science & Humanities*, *29*(1), 157–174. doi:10.47836/pjssh.29.1.09

He, X., Deng, H., & Hwang, H. (2019). The current application of nanotechnology in food and agriculture. *Yao Wu Shi Pin Fen Xi*, *27*, 1–21. doi:10.1016/j.jfda.201812.002 PMID:30648562

Hessel, V., Escribà-Gelonch, M., Schmidt, S., Tran, N. N., Davey, K., Al-Ani, L. A., Julkapli, N. M., Wahab, Y. A., Khalil, I., Woo, M. W., & Gras, S. (2023). Nanofood process technology: Insights on how sustainability informs process design. *ACS Sustainable Chemistry & Engineering*, *11*(31), 11437–11458. doi:10.1021/acssuschemeng.3c01223 PMID:37564955

Higashisaka, K., Yoshioka, Y. & Tsutsumi, Y. (2015). Applications and safety of nanomaterials used in the food industry. Food Safety Commission, Cabinet Office, Government of Japan, 39-47. doi:10.14252/foodsafetyfscj.2015005

Iijima, S., & Ichihashi, T. (1993). Single-shell carbon nanotubes of 1-nm diameter. *Nature*, *363*(6430), 603–605. doi:10.1038/363603a0

Inobeme, A., Mathew, J. T., Adetunji, C. O., Maliki, M., Onyeachu, B. I., Adekoya, M. A., Eziukwu, C. A., Kelani, T., Inobeme, J., Ajai, A. I., Jacob, J. O., Olori, E., & Okonkwo, S. (2022). Advances, prospects and concerns in nanomaterials for food processing. *Nanotechnology and Advanced Material Science*, *5*(3), 1–2.

Islam, M. T. (2019). Applications of nanomaterials for future food security: Challenges and prospects. *Malaysian Journal of Halal Research Journal*, *2*(1), 6–9. doi:10.2478/mjhr-2019-0002

Ismail, S., Budin, S., & Ali, S. A. M. (2019). The nanotechnology application and workforce health and safety – a study of the Malaysia laws, statutory regulations and guidelines on nanotechnology. *Journal of Physics: Conference Series*, 1–9. doi:10.1088/1742-6596/1349/1/012031

ISO 4656. Rubber Compounding Ingredients - Carbon Black - Determination of Oil Absorption Number (OAN) and Oil Absorption Number of Compressed Sample (COAN). ACS Omega 2012, https://pubs.acs.org/journal/acsodf Article

Istiqola, A., & Syafiuddin, A. (2020). A review of silver nanoparticles in food packaging technologies: Regulation, methods, properties, migration, and future challenges. *Journal of the Chinese Chemical Society (Taipei)*, *67*(11), 1–15. doi:10.1002/jccs.202000179

Jagtiani, E. (2021). Advancements in nanotechnology for food science and industry. *Food Frontiers*, *3*(1), 56–82. doi:10.1002/fft2.104

Jain, A., Ranjan, S., Dasgupta, N., & Ramalingam, C. (2017). Nanomaterials in food and agriculture: An overview on their safety concerns and regulatory issues. *Critical Reviews in Food Science and Nutrition*, *0*(0), 1–21. doi:10.1080/10408398.2016.1160363 PMID:27052385

Jeevanandam, J., Barhoum, A., Chan, Y. S., Dufresne, A., & Danquah, M. K. (2018). Review on nanoparticles and nanostructured materials: History, sources, toxicity and regulations. *Beilstein Journal of Nanotechnology*, *9*, 1050–1074. doi:10.3762/bjnano.9.98 PMID:29719757

Kasa, Y. (2013). The application of nanotechnology and its risk in developing countries, 4(7): 249-259. *International Journal of Advanced Research*, *4*(2), 269–259.

Khali, I., Yehye, W. A., Etxenerria, A. E., Alhadi, A. A., Dezfooli, S. M., Julkapli, N. B. M., Bsirun, W. J., & Seyfoddin, A. (2019). Nanoantioxidants: Recent trends in antioxidant delivery applications. *Antioxidants*, *9*(1), 1–30. doi:10.3390/antiox9010024 PMID:31888023

Kousheh, S. A., Moradi, M., Tajik, H., & Molaei, R. (2020). Preparation of antimicrobial/ultraviolet protective bacterial nanocellulose film with carbon dots synthesized from lactic acid bacteria. *International Journal of Biological Macromolecules*, *155*, 216–225. doi:10.1016/j.ijbiomac.2020.03.230 PMID:32240732

Kumar, S., Mudai, A., Roy, B., Basumatary, I. B., Mukherjee, A., & Dutta, J. (2020). Biodegradable hybrid nanocomposite of chitosan/gelatin and green synthesized zinc oxide nanoparticles for food packaging. *Foods*, *9*(9), 1143. doi:10.3390/foods9091143 PMID:32825205

Kushwaha, S., & Pant, H. (2023). Role of nanoparticles in maintaining food safety and tackling malnutrition. *International Journal of Clinical Biochemistry and Research, 10*(1), 19–27. doi:10.18231/j.ijcbr.2023.004

Lang, Y., Zhnag, B., Cai, D., Tu, W., Zhnag, J., Shentu, X., Ye, Z., & Yu, X. (2022). Determination methods of the risk factors in food based on nanozymes: A review. *Biosensors (Basel), 13*(1), 1–19. doi:10.3390/bios13010069 PMID:36671904

Lateef, A., Azeez, M. A., Suaibu, O. B., & Adigun, G. O. (2021). A decade of nanotechnology research in Nigeria (2010-2020): A scientometric analysis. *Journal of Nanoparticle Research, 23*(9), 1–27. doi:10.1007/s11051-021-05322-1

Lawal, M., Payne, J., Onyeaka, H., Alao, A. M., & Okoampah, E. (2023). Boosting food safety in Ghana: Exploring the future of nanotechnology. *Nano Select*, 1–13. doi:10.1002/nano.202300078

Li, Y., & Cummmins, E. (2020). Hazard characterization of silver nanoparticles for human exposure routes. *Journal of Environmental Science and Health. Part A, Toxic/Hazardous Substances & Environmental Engineering, 55*(6), 704–725. doi:10.1080/10934529.2020.1735852 PMID:32167009

Li, Y., Huang, X., Zeng, L., Li, R., Tian, H., Fu, X., Wang, Y., & Zhong, W. H. (2019). A review of the electrical and mechanical properties of carbon nanofiller-reinforced polymer composites. *Journal of Materials Science, 54*(2), 1036–1076. doi:10.1007/s10853-018-3006-9

Ludlow, K. (2009). The readiness of Australian food regulation for the use of nanotechnology in food and food packaging. *University of Tasmania Law Review, 26*(2), 177–203.

Lugani, Y.; Sooch, B. S.; Singh, P. & Kumar, S. (2021). Nanobiotechnology applications in food sector and future innovations. Microbial Biotechnology in Food and Health, 197-225. doi:10.1016/B978-0-12-819813-1.00008-6

Madhusha, C., Munaweera, I., & Kottegoda, N. (2021). Functional nanomaterials as smart food packaging: A brief review. *African Journal of Agriculture and Food Science, 4*(1), 58–78.

Maharramov, A. M., Hasanova, U. A., Suleymanova, I. A., Osmanova, G. E., & Hajiyeva, N. E. (2019). The engineered nanoparticles in food chain: potential toxicity and effects. Springer Nature Applied Sciences. doi:10.1007/s42452-019-1412-5

Martirosyan, A., & Schneider, Y. J. (2014). Engineered nanomaterials in food: Implications for food safety and consumer health. *International Journal of Environmental Research and Public Health, 11*(6), 5720–5750. doi:10.3390/ijerph110605720 PMID:24879486

Mashkour, M., Sharifinia, M., Yousefi, H., & Afra, E. (2018). MWCNT-coated cellulose nanopapers: Droplet-coating, process factors, and electrical conductivity performance. *Carbohydrate Polymers, 202*, 504–512. doi:10.1016/j.carbpol.2018.09.006 PMID:30287029

Mazumder, N., & Sivashankar, S. (2020). Nanopackaging of food: An overview. *Research Journal of Pharmaceutical, Biological and Chemical Sciences, 11*(4), 1–12. doi:10.33887/rjpbcs/2020.11.4.1

McClemens, D. T., & Xiao, H. (2017). Is nano safe in food? Establishing the factors impacting the gastro intestinal fate and toxicity of organic and inorganic food-grade nanoparticle. *NPJ Science of Food*, *1*(1), 6. doi:10.1038/s41538-017-0005-1 PMID:31304248

Meena Kumari, M., Jacob, J., & Philip, D. (2015). Green synthesis and applications of Au–Ag bimetallic nanoparticles. *Spectrochimica Acta. Part A: Molecular and Biomolecular Spectroscopy*, *137*, 185–192. doi:10.1016/j.saa.2014.08.079 PMID:25218228

Mellia, J. O., Galea, R., Nag, R., Cummins, E., Gatt, R., & Valdramidis, V. (2022). Nanoparticle food applications and their toxicity: Current trends and needs in risk assessment strategies. *Journal of Food Protection*, *85*(2), 355–372. doi:10.4315/JFP-21-184 PMID:34614149

Mirzaei, H., & Darroudi, M. (2017). Zinc oxide nanoparticles: Biological synthesis and biomedical applications. *Ceramics International*, *43*(1), 907–914. doi:10.1016/j.ceramint.2016.10.051

Mohamed, A. (2019). Chapter eight- synthesis, characterization, and applications carbon nanofibers: In carbon based nanofillers and their rubber nanocomposites. Elsevier.

Mohammad, M. R., Ahmed, D. S., & Mohammed, M. K. (2019). Synthesis of Ag-doped TiO_2 nanoparticles coated with carbon nanotubes by the sol−gel method and their antibacterial activities. *Journal of Sol-Gel Science and Technology*, *90*(3), 498–509. doi:10.1007/s10971-019-04973-w

Mohammed, Z. G., Ahmad, F., Ibrahim, S. A., & Zaidi, S. (2022). Application of nanotechnology in different aspects of the food industry. *Discover Food*, *2*(1), 12. doi:10.1007/s44187-022-00013-9

Naseer, B., Srivastava, G., Qadri, O. S., Faridi, S. A., Islam, R. V. I., & Younis, K. (2018). Importance and health hazard of nanoparticles used in food industry. *Nanotechnology Reviews*, *7*(6), 623–641. Doi.org/10.1515/intrev-2018-0076. doi:10.1515/ntrev-2018-0076

Naseri, R., & Davoodi, R. (2011). Commercialization of nanotechnology in developing countries. *3rd International Conference on Information and Financial Engineering*, 385-389.

Nasir, S., Hussein, M. Z., Zainal, Z., & Yusof, N. A. (2018). Carbon-based nanomaterials/allotropes: A glimpse of their synthesis, properties and some applications. *Materials (Basel)*, *11*(2), 295. doi:10.3390/ma11020295 PMID:29438327

Nile, S. H., Baskar, V., Selvaraj, D., Nile, A., Xiao, J., & Kai, G. (2020). Nanotechnologies in food science: Applications, recent trends, and future perspectives. *Nano-Micro Letters*, *12*(45), 1–34. doi:10.1007/s40820-020-0383-9 PMID:34138283

Oberdorster, G., Oberdorster, E., & Oberdorster, J. (2005). Nanotoxicity: An emerging discipline evolving from studies of ultrafine particle. *Environmental Health Perspectives*, *113*(7), 823–839. doi:10.1289/ehp.7339 PMID:16002369

Oliveira, T. M. B. F., Ribeiro, F. W. P., Sousa, C. P., Salazar-Banda, G. R., de Lima-Neto, P., Correia, A. N., & Morais, S. (2020). Current overview and perspectives on carbon-based (bio) sensors for carbamate pesticides electroanalysis. *Trends in Analytical Chemistry*, *124*(115779), 115779. doi:10.1016/j.trac.2019.115779

Onyeka, H., Passaretti, P., & Miri, T. (2022). The safety of nanomaterials in food production and packaging. *Current Research in Food Science*, *5*, 763–774. doi:10.1016/j.crfs.2022.04.005 PMID:35520272

Onyeka, H., Passaretti, P., Miri, T., & Al-Sharify, Z. T. (2022). The safety of nanomaterials in food production and packaging. *Current Research in Food Science*, *5*, 763–774. doi:10.1016/j.crfs.2022.04.005 PMID:35520272

Otchere, E., McKay, B. M., English, M. M., & Aryee, A. N. A. (2023). Current trends in nano-delivery systems for functional foods: A systematic review. *PeerJ*, *11*, e14980. doi:10.7717/peerj.14980 PMID:36949757

Ozcakir, G. (2023). Applications of nanomaterials in food industry: A review. *Materials Proceedings*, *5*, x. doi:10.3390/xxxxx

Ozimek, L., Pospiech, E., & Narine, S. S. (2010). Nanotechnology in food and meat processing. *Acta Scientiarum Polonorum. Technologia Alimentaria*, *9*(4), 401–412.

Paidari, S., Tahergorabi, R., Anari, E. S., Nafchi, A. M., Zamindar, N., & Goli, M. (2021). Migration of various nanoparticles into food samples: A review. *Foods*, *10*(9), 214. doi:10.3390/foods10092114 PMID:34574224

Pan, M., Yin, Z., Liu, K., Du, X., Liu, H., & Wang, S. (2019). Carbon-based nanomaterials in sensors for food safety. *Nanomaterials (Basel, Switzerland)*, *9*(9), 1–23. doi:10.3390/nano9091330 PMID:31533228

Pattanshetti, A., Pradeep, N., Chaitra, V., & Uma, V. (2020). Synthesis of multi-walled carbon nanotubes (MWCNTs) from plastic waste & analysis of garlic coated gelatin/MWCNTs nanocomposite films as food packaging material. *SN Applied Sciences*, *2*(4), 730. doi:10.1007/s42452-020-2442-8

Paul, N., & Bhojane, P. (2021). Role of nanoparticles in agriculture and food industry. *Journal of Emerging Technologies and Innovative Research*, *8*(11), 221–234.

Payal, A., Krishnamoorthy, S., Elumalai, A., Moses, J. A., & Anandharamakrishnan, C. (2021). A review on recent developments and applications of nanozymes in food safety and quality analysis. *Food Analytical Methods*, *14*(8), 1–23. doi:10.1007/s12161-021-01983-9

Petkoska, A. T., Daniloski, D., D'Cunha, N. M., Naumovski, N., & Broach, A. T. (2021). Edible packaging: Sustainable solutions and novel trends in food packaging. *Food Research International*, *140*, 109981. doi:10.1016/j.foodres.2020.109981 PMID:33648216

Pruksasri, S., Lanner, B., & Novalin, S. (2020). Nanofiltration as a potential process for the reduction of sugar in apple juices on an industrial scale. LWT- Food. *Sciences et Techniques (Paris)*, *133*, 110118.

Qu, J. H., Wei, Q., & Sun, D. W. (2018). Carbon dots: Principles and their applications in food quality and safety detection. *Critical Reviews in Food Science and Nutrition*, *58*(14), 2466–2475. doi:10.1080/10408398.2018.1437712 PMID:29420064

Rajput, V. D., Singh, A., Minkina, T., Rawat, S., Mandzhieva, S., Sushkova, S., Shuvaeva, V., Nazarenko, O., Rajput, P., Komariah, Verma, K. K., Singh, A. K., Rao, M., & Upadhyay, S. K. (2021). Nano-Enabled Products: Challenges and Opportunities for Sustainable Agriculture. *Plants*, *10*(12), 2727. doi:10.3390/plants10122727 PMID:34961197

Ramkumar, C., Vishwanatha, A., & Saini, R. (2019). Regulatory aspects of nanotechnology for food industry. Nanotechnology applications in dairy science. doi:10.1201/9780429425370-7

Reina, G., Gonzalez-Dominguez, J. M., Criado, A., Vazquez, E., Bianco, A., & Prato, M. (2017). Promises, facts and challenges for graphene in biomedical applications. *Chemical Society Reviews*, *46*(15), 4400–4416. doi:10.1039/C7CS00363C PMID:28722038

Remya, S., Mohan, C. O., Sreelakshmi, K. R., & Bindu, J. (2019). Packaging applications of nanomaterials. ITEC training programme on 'Protocols for the production of high value secondary products from industrial fish and shellfish processing from 25/11/2019 to 21/12/2019 at ICAR-CIFT, Kochi. 215-221pp.

Rezaei, A., Daeihamed, M., Capanoglu, E., Tomas, M., Akbari-Alavijeh, S., Shaddel, R., Khoshnoudi-Nia, S., Boostani, S., Rostamabadi, H., Falsafi, S. R., & Jafari, S. M. (2021). Possible health risks associated with nanostructures in food. In *Safety and regulatory issues of nanoencapsulated food ingredients*. Elsevier Inc. doi:10.1016/B978-0-12-815725-1.00002-1

Sahoo, M., Panigrahi, C., Vishwakarma, S., & Kumar, J. (2022). A review on nanotechnology: Applications in food industry, future opportunities, challenges and potential risks. *Journal of Nanotechnology and Nanomaterials*, *3*(1), 28–33.

Salamanca-Buentello, F.; Persad, D. L.; Court, E. B.; Martin, D. K.; Daar, A. S. & Singer, P. A. (2005). Nanotechnology and the developing world. PLos Medicine, 2(5), 383-386.

Salgado, C. M., Palacio, L., Prádanos, P., Hernández, A., González-Huerta, C., & Pérez-Magari, S. (2015). Comparative study of red grape must nanofiltration: Laboratory and pilot plant scales. *Food and Bioproducts Processing*, *94*, 610–620. doi:10.1016/j.fbp.2014.08.007

Salimi, F., Moradi, M., Tajik, H., & Molaei, R. (2021). Optimization and characterization of eco-friendly antimicrobial nanocellulose sheet prepared using carbon dots of white mulberry (*Morus Alba* L.). *Journal of the Science of Food and Agriculture*, *101*(8), 3439–3447. doi:10.1002/jsfa.10974 PMID:33289129

Sari, G., & Piyal, B. (2019). Nanomaterials and their effects on health. *Istanbul Medical Journal*, *20*(6), 497–501. doi:10.4274/imj.galenos.2018.71598

Shah, M., Fawcell, D., Sharma, S., Tripathy, S. K., & Poinern, G. E. J. (2015). Green synthesis of metallic nanoparticles via biological entities. *Materials (Basel)*, *8*(11), 7278–7308. doi:10.3390/ma8115377 PMID:28793638

Sharara, M. S. (2021). Application of nanotechnology in food industry and some of its hazard effect on human body: An overview. *Alexandria Journal of Food Science and Technology*, *18*(2), 23–32. doi:10.21608/ajfs.2022.217792

Singh, A., Agrawal, S., Rajput, V. D., Ghazaryan, K., Movsesyan, H. S., Minkina, T., . . . Gupta, S. K. (2023d). Seed Nanopriming. Nanopriming Approach to Sustainable Agriculture, 290–313. doi:10.4018/978-1-6684-7232-3.ch013

Singh, A.; Margaryan, G.; Harutyunyan, A. S.; Movsesyan, H.; Khachatryan, H.; Rajput, V., ... Ghazaryan, K. (2024c). Advancing agricultural resilience in Ararat plain, Armenia: Utilizing biogenic nanoparticles and biochar under saline environments to optimize food security and foster European trade. Egyptian Journal of Soil Science, 64(2). doi:10.21608/ejss.2024.257000.1703

Singh, A.; Rajput, V.; Varshney, A.; Sharma, R.; Ghazaryan, K.; Minkina, T., ... El-Ramady, H. (2024b). Revolutionizing crop production: Nanoscale wonders - Current applications, advances, and future frontiers. Egyptian Journal of Soil Science, 64(1). doi:10.21608/ejss.2023.246354.1684

Singh, A., Rajput, V. D., Pandey, D., Sharma, R., Ghazaryan, K., & Minkina, T. (2023c). Nano zinc-enabled strategies in crops for combatting zinc malnutrition in human health. *Frontiers in Bioscience (Landmark Edition)*, *28*(8), 158. doi:10.31083/j.fbl2808158 PMID:37664935

Singh, A., Rajput, V. D., Varshney, A., Ghazaryan, K., & Minkina, T. (2023a). Small tech, big impact: Agri-nanotechnology journey to optimize crop protection and production for sustainable agriculture. *Plant Stress*, *10*, 100253. doi:10.1016/j.stress.2023.100253

Singh, A., Rawat, S., Rajput, V. D., Ghazaryan, K., Minkina, T., Al Tawaha, A. R. M., . . . Sandhu, K. S. (2023b). Nanoparticle-based priming-enhanced agricultural practices. Nanopriming Approach to Sustainable Agriculture, 142–157. doi:10.4018/978-1-6684-7232-3.ch006

Singh, A., Sharma, R., Rajput, V. D., Ghazaryan, K., Minkina, T., Al Tawaha, A. R. M., & Varshney, A. (2023e). Green synthesis of nanofertilizers and their application for crop production. Nanotechnology in the Life Sciences, 205–231. doi:10.1007/978-3-031-41329-2_8

Singh, A., Sharma, R., Rawat, S., Singh, A. K., Rajput, V. D., Fedorov, Y., . . . Chaplygin, V. (2022). Nanomaterial-plant interaction: Views on the pros and cons. Toxicity of Nanoparticles in Plants, 47–68. doi:10.1016/B978-0-323-90774-3.00015-5

Singh, P., Kim, Y. J., Wang, C., Mathiyalagan, R., El-Agamy Farh, M., & Yang, D. C. (2015). Biogenic silver and gold nanoparticles synthesized using red ginseng root extract, and their applications. *Artificial Cells, Nanomedicine, and Biotechnology*, *44*, 811–816. doi:10.3109/21691401.2015.1008514 PMID:25706249

Singh, R., Dutt, S., Sharma, P., Sundramoorthy, A. K., Dubey, A., Singh, A., & Arya, S. (2023). Future of nanotechnology in food industry: Challenges in processing, packaging, and food safety. *Global Challenges (Hoboken, NJ)*, *7*(4), 1–28. doi:10.1002/gch2.202200209 PMID:37020624

Singh, T., Shukla, S., Kumar, P., Wahla, V., Bajpai, V. K., & Rather, I. A. (2017). Application of nanotechnology in food science: Perception and overview. *Frontiers in Microbiology*, *8*, 1501. doi:10.3389/fmicb.2017.01501 PMID:28824605

Singh, T., Shukla, S., Kumar, P., Wahla, V., Bajpai, V. K., Rather, I. A., Song, Y., Lu, F., Li, H., Wang, H., Zhang, M., Liu, Y., & Kang, Z. (2018). Degradable Carbon Dots from Cigarette Smoking with Broad- Spectrum Antimicrobial Activities against Drug-Resistant Bacteria. *ACS Applied Bio Materials*, *1*(6), 1871–1879. doi:10.1021/acsabm.8b00421 PMID:34996288

Sodano, V. (2018). Nano-food regulatory issues in the European union. *AIP Conference Proceedings*, 1–8. doi:10.1063/1.5047772

Songca, S. P. (2022). Applications of nanozymology in the detection and identification of viral, bacterial and fungal pathogens. *International Journal of Molecular Sciences*, *23*(9), 1–25. doi:10.3390/ijms23094638 PMID:35563029

Sponton, O. E., Perez, A. A., Grabois, M., & Santiago, L. G. (2019). Nanosupplements: improvements in nutritional functionality of food products. Food Applications of Nanotechnology, 99-119. DOI: doi:10.1201/9780429297038-5

Suh, S., Meng, X., & Ko, S. (2016). Proof of concept study for different-sized chitosan nanoparticles as carbon dioxide (CO_2) indicators in food quality monitoring. *Talanta*, *1*(161), 265–270. doi:10.1016/j.talanta.2016.08.055 PMID:27769405

Suthar, T. R., Gavane, A. B., Shah, A. Y., & Devkatte, A. N. (2021). Nanotechnoogy, a revolutionary technique in the food industry: A systematic review. Sustainability, Agriculture. *Food and Environmental Research*, *9*(1), 127–141. doi:10.7770/safer-VONO-art2256

Szakal, C., Roberts, S. M., Westerhoff, P., Bartholomaeus, A., Buck, N., Illuminato, I., Canady, R., & Rogers, M. (2014). Measurement of nanomaterials in foods: Integrative consideration of challenges and future prospects. *ACS Nano*, *8*(4), 3128–3135. doi:10.1021/nn501108g PMID:24673283

Szunerits, S., & Boukherroub, R. (2018). Graphene-based nanomaterials in innovative electrochemistry. *Current Opinion in Electrochemistry*, *10*, 24–30. doi:10.1016/j.coelec.2018.03.016

Talarska, P., Boruczkowski, M., & Żurawski, J. (2021). Current knowledge of silver and gold nanoparticles in laboratory research—Application, toxicity, cellular uptake. *Nanomaterials (Basel, Switzerland)*, *11*(9), 2454. doi:10.3390/nano11092454 PMID:34578770

Taslim, N. A., Nurkolis, F., Hardinsyah, H., Yusuf, V. M., Gunawan, W. B., Samtiya, M., Mayulu, N., Assa, Y. A., & Tallei, T. E. (2023). Foodomics approaches to facilitate the verification of the authenticity of foods: A possible strategy to screen, validate, and standardize food matrices. *Nutrición Clínica, Dietética Hospitalaria*, *43*(1), 73–80. doi:10.12873/431hardinsyah

Tayel, A. A. (2016). Microbial chitosan as a biopreservative for fish sausages. *International Journal of Biological Macromolecules*, *93*, 41–46. doi:10.1016/j.ijbiomac.2016.08.061 PMID:27565293

Thompson, P. B. (2015). Risks and ethics in the context of food nanotechnology and the delivery of bioactive ingredients. In Nanotechnology and functional foods: Effective delivery of bioactive ingredients. John Wiley and Sons, Ltd. doi:10.1002/9781118462157.ch19

Valdés, A., Álvarez-Rivera, G., Socas-Rodríguez, B., Herrero, M., Ibáñez, E., & Cifuentes, A. (2022). Foodomics: Analytical opportunities and challenges. *Analytical Chemistry*, *94*(1), 366–381. doi:10.1021/acs.analchem.1c04678 PMID:34813295

Vasile, C. (2018). Polymeric nanocomposites and nanocoatings for food packaging: A review. Materials, 11, 1834. doi:10.3390/ma11110183

Vickers, N. J. (2017). Animal communication: When I'm caching you will answer food? *Current Biology*, *27*(14), 713–715. doi:10.1016/j.cub.2017.05.064 PMID:28743020

Wang, W., Qian, S., Gong, L., Ni, Z., & Ren, H. (2021). Effects of carbon nanoonions on the tribological performance of food grade calcium sulfonate complex grease. *Lubrication Science*, *33*(8), 460–470. doi:10.1002/ls.1567

Wang, Y., Cen, C., Chen, J., & Fu, L. (2020). MgO/carboxymethyl chitosan nanocomposite improves thermal stability, waterproof and antibacterial performance for food packaging. *Carbohydrate Polymers*, *236*, 116078. doi:10.1016/j.carbpol.2020.116078 PMID:32172891

Wu, L., Zhou, S., Wang, G., Yun, Y., Liu, G., & Zhnag, W. (2021). Nanozyme appications: a glimpse of insight in food safety. Frontiers in Bioengineering and Biotechnology, 9, 1-15. doi:10.3389/fbioe.2021.727886

Yin, C., Zgao, W., Liu, R., Liu, R., Wang, Z., Zhu, L., Chen, W., & Liu, S. (2017). TiO_2 particles in seafood and surimi products; attention should be paid to their exposure and uptake through food. *Chemosphere*, *188*, 541–547. doi:10.1016/j.chemosphere.2017.08.168 PMID:28910729

Zanet, V., Vidic, J., Auger, S., Vizzini, P., Lippe, G., Iacumin, L., Comi, G., & Manzano, M. (2019). Activity evaluation of pure and doped zinc oxide nanoparticles against bacterial pathogens and *Saccharomyces cerevisiae*. Journal of Applied Microbiology, 127(5), 1391–1402. doi:10.1111/jam.14407

Chapter 19
Nanozymes:
New Frontiers in Food Safety and Security

Annika Arvind Gupta
https://orcid.org/0000-0001-9956-476X
B.K. Birla College, India

Sonali Zankar Patil
B.K. Birla College, India

ABSTRACT

Food is complex, and food hygiene and security are critical components. With increased consumer demand for safe and healthy food items, a plethora of new, multidisciplinary, and multifaceted concepts have developed in recent decades. Among standard approaches for food safety evaluation, enzyme-based biological sensing technologies play an important role in food safety monitoring. However, long-term problems, such as high cost and limited stability, impede enzyme-based biosensing devices. Aside from their medicinal efficacies, nanozymes are gaining popularity in the agrifood sector for safety and quality monitoring. These enzyme-mimetic nanoparticles have been demonstrated to identify and monitor food quality in real time while maintaining great selectivity, stability, dependability, and recyclability. Food-based applications, nanozymes derived from nanomaterials such as carbon, metals and their oxides, metal-organic frameworks, and others have significant advantages over natural enzymes. This study summarises current improvements in food safety assessment based on nanozymes.

INTRODUCTION

Nanomaterials have gotten a great deal of interest in areas including science, science, climate, aviation, medication and wellbeing, assets, and energy (Wu et al., 2019a; Huang et al., 2019d; Rajput et al., 2021). A wide range of nanomaterials have been made and explored as far as their distinct, electrical, mechanical, attractive, optical, and synergist capacities. Omics encompasses many sub-disciplines, such as transcriptomics, genomes, proteomics, and metabolomics, to give a comprehensive understanding of the molecular landscape (Worheide et al., 2021). The purpose of omics is to get a full understanding

DOI: 10.4018/979-8-3693-1890-4.ch019

of intricate biological systems at the molecular level, providing insights that are beyond the limits of traditional techniques. This multidisciplinary approach investigates the composition, structure, function, and interactions of individual molecules, resulting in a more comprehensive understanding of biological systems. omics uses high-throughput technology to effectively collect and evaluate molecular data. The combination of nanotechnology with omics has resulted in the formation of a new field of study known as nanoomics, which entails employing nanotechnology to separate and analyse chemicals from biological fluids for further omics research and analysis (Gardner et al., 2022). Nanoomics has enabled dramatic advances in disease research. The complex insights gained by nanoomics analysis open the door for personalised therapeutic methods, in which therapies are matched to unique patient profiles. Precision medicine represents a significant shift in healthcare, resulting in more effective and minimally intrusive procedures. Researchers are becoming increasingly interested in nanozyme, a particular kind of nanomaterial that mimics enzymes. Nanozymes have shown many purposes in vitro recognition and living frameworks (Liang and Yan 2019). They can possibly work as immediate substitutes for regular chemicals in analytic techniques, especially immunoassays (Wu et al., 2019b).

Nanozymes are more effectively controlled and sanitized than regular catalysts. Moreover, the morphology, size (e.g., nanowires, nanosheets, nanospheres, nanorods), and nanozyme surface groups may all add to their enzyme-similar activity (Gao et al., 2020; Wang et al., 2019). In this sense, nanozymes may be taken advantage of as a flexible medium in the improvement of scientific methods. Nanozymes, as a sort of nanomaterial, are frequently combined with neutralizer or DNA successions to give a sign enhancement technique (Tao et al., 2020; Zhu et al., 2017). Scientific procedures with decreased recognition cutoff points can be concocted when incorporated with standard standards of optical, electrochemical, or colorimetric examines. Food foreign substances, then again, are poisonous mixtures that are hindering to people and have ascended in intricacy because of new general wellbeing challenges, creative security crises, and changing purchaser assumptions (Huang et al., 2019a). Analytical techniques face significant difficulties because of the complexity of food matrices and contaminants. Alternaria, for instance, can deliver various poisonous auxiliary metabolites. For example, alternariol monomethyl ether, alternariol, tentoxin, altenuene, and tenuazonic acid, which are usually found in sorghum, sunflower seeds, oats, tomatoes, wine, brews, squeezed apple, and refreshments (Pinto and Patriarca, 2017). A few scientific techniques have been created to follow their presence in food sources or the food store network. Obviously, wet-science based scientific techniques have progressively been superseded areas of strength for by that give huge enhancements in exactness, accuracy, and discovery limits. New innovation can beat the constraints of conventional methodologies, like tedious examination, drawn-out techniques, and expensive expenses. The presentation of imaginative "fast" discovery strategies has significantly decreased location time, possibly settling the essential issues of ordinary insightful systems. Thus, other applications of nanozymes in food contaminations identification gives some understanding into this idea. This article gives an outline of new nanozyme-based advances, with an accentuation on their varieties, surface adjustments, and applications in food poisons examination.

CLASSIFICATION

As a rule, nanozymes are classified into four sorts considering their parts: carbon-based, metal oxide-based, metallic-based, and supplementary nanozymes (Figure 1). Metal nanoparticles, for example, platinum nanoparticles (Pt NP), gold nanoparticles (Au NP), and palladium nanoparticles (Pd NP), are utilized in metallic-based nanozymes. Some bimetallic nanozymes that have been created to further develop their catalase action incorporate Au+Pt, Pt+Pd, and Au+Ag (Nandhakumar et al., 2020). Ternary metallic nanozymes, for example, Ag+Au+Pt NP and Pt+Pd+Au NP were additionally explored. Regardless of their benefits, such as their ease of preparation and modification, metallic-based nanozymes may encounter issues such as metal toxicity and spontaneous aggregation. Metal oxide-based nanozymes, instead of metal-based ones, have an overflow of gatherings with different functionalities (Singh et al., 2023a,b,c). Fe_3O_4 nanozymes, for instance, can work as catalase, oxidase-like proteins, peroxidase, or, while CeO_2 nanozymes can work as catalase, superoxide dismutase, and oxidase-like compounds (Nicolini et al., 2015). There is likewise various extra metal oxide-based nanozymes, like MnO_2 NP, CuO NP, and V_2O_5 NP (Yang F. et al., 2018). Nanozymes based on metal oxide-materials are steadier and simpler to change with the utilization of surface practical gatherings.

Carbon nanodots, graphene oxide (GO), and carbon nanotubes (CNT) are the vital parts of nanozymes made of carbon (Garg and Bisht 2016). Cui et al., (2011) for instance, has conceived a joined aqueous/hydrogen decrease approach for mass union of winding carbon nanotubes by acetylene pyrolysis. Wang et al. (2009) proposed an electrochemiluminescence (ECL) sensor in light of Discs nanocrystals delivered on the outer layer of multiwalled carbon nanotubes (Cds/MWCNT). Compact discs/MWCNT might respond with H_2O_2 to make solid and industrious ECL signals because of its peroxidase-like action. Such occasions might be tracked down in a wide range of fields, including natural designing and manufactured science (Sun et al., 2018).

Recently, it was discovered that Prussian blue (PB), covalent organic frameworks (COFs) and metal-organic frameworks (MOFs) can mimic the activity of enzymes (Zhou et al., 2020). MOFs have been broadly considered and executed in an assortment of basic detecting, energy, and synergist spaces, including poisons discovery, gas capacity and partition, and electrocatalysis, because of their porosity and enormous explicit surface region, as well as the variety of structures and functionalities (Lustig et al., 2017). Zhou et al. (2020) for instance, presented the first chiral nanozyme in view of copying a characteristic chemical and the unrivalled design of COFs, which exhibited more action than the local protein. Researchers have made an original sort of microelectrode for in vivo H_2O_2 observing by electrodepositing PB onto carbon fibre microelectrodes worked from CNTs. Each of the techniques depend on the remarkable abilities of newly found nanozymes.

FUNCTIONS AND PERFORMANCES

Nanozymes, which are a type of artificial enzyme (II), are useful for more than just catalysis. Nanozymes can be made from a variety of sources, including metal-to-metal oxides and carbon later. This makes them widely available. By imitating a wide variety of natural enzymes, many nanozymes carry out multienzyme tasks. For instance, contingent upon the pH, Au NP and CeO2 NP can show peroxidase, superoxide dismutase, and catalase exercises (Wu et al., 2019c). Michaelis Menten energy tests are ordinarily performed to contrast and regular proteins. This permits the models for substrate particularity (Km), synergist rate steady (kcat), and reactant effectiveness (kcat/Km) to be joined. Nanozymes, as a rule, emulate oxidase, hydrolase, superoxide dismutase, and catalase capability. Reactant cycles and energy have been broadly analyzed comparable to factors like pH, temperature, and, surprisingly, disintegrated oxygen. However, as previously stated, a single type of nanomaterial may possess a variety

Figure 1. Classification of nanozymes

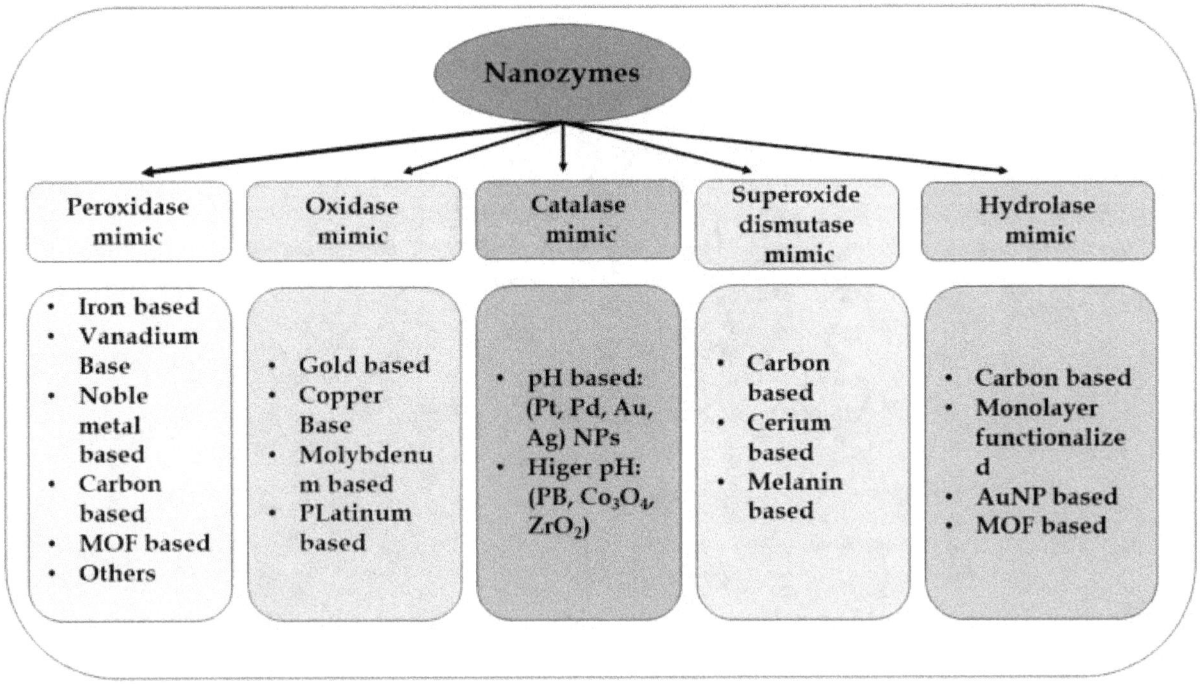

of pH-dependent or structurally related functionalities (size, shape, surface groups, defects, and so on). An aide is expected to make and develop nanozymes because of the muddled collaboration between physicochemical characteristics and reactant highlights.

ENGINEERING AND MODIFICATION

Various examinations have shown that underlying qualities like size, shape, and surface gatherings assume a huge part in the reactant movement of nanozymes (Jiao et al., 2019; Liu and Liu, 2017). The particular surface area of nanoparticles every now and again climbs as their size reduces, achieving an outrageous shortfall of surface particle coordination. Accordingly, the surface dynamic destinations have developed, and the reactant effectiveness has moved along. To put it another way, the greater the catalytic activity of the nanozymes, the smaller they are. In this viewpoint, the size of the compound might be controlled to adjust its movement. Besides, the morphology and point of interaction construction of nanozymes may change during the response cycle because of differed response conditions, influencing reactant action (Vernekar et al., 2016). The reactant movement of nanoparticles might be incredibly upgraded by specifically uncovering the gem surface to high action or specific energy (Kuang et al., 2014). The variable cross section game plan of iotas with unmistakable appearance structures prompts changed surface movement and synergist exhibitions in morphology-subordinate nanozymes. Moreover, nanozyme functionalities can be transformed utilizing various surface change procedures (charges, coatings, functionalization, and loadings) (Figure 2). Such successful modification examples include small molecules (Chang et al., 2016), ions (Huang et al., 2019c), amino acids

Figure 2. Engineering and modification carried out on nanozymes

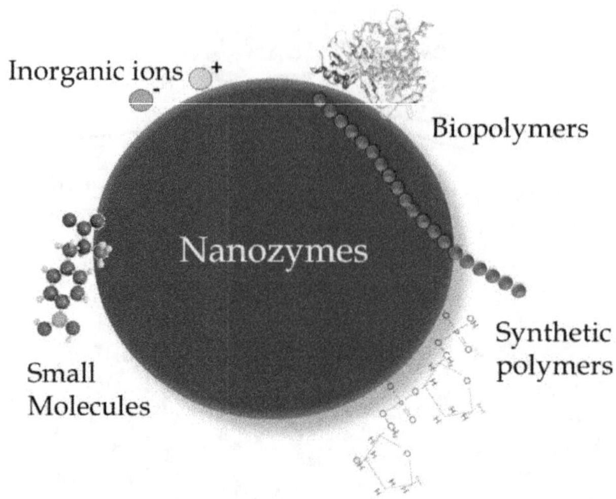

and peptides (Fan et al., 2017), nucleotides and nucleic acids (Huang et al., 2018a), polymers (Wu Y. et al., 2019), and proteins (Su et al., 2019). The system can work on the functionalities of nanozymes and widen their functions. Nanozymes with unmistakable surfaces can give touchy and exact analyte distinguishing proof and location.

APPLICATIONS IN FOOD INDUSTRY

1. Pesticide Detection

According to Bajwa and Sandhu (2014), one of the main inputs utilized in agriculture for safeguarding seeds and crops both after and before harvesting is pesticides. Pesticide remains left in food products can have a adverse effect on human health even though they have greatly benefited society economically (Jiang et al., 2008). Furthermore, as pesticide residues are easily absorbed by both initial and secondary agricultural products, the widespread use of pesticides has caused serious food safety issues. Therefore, several nations and organizations worldwide have set maximum residue limits (Minimum Recommended Levels) for pesticides in foods to guarantee food safety for consumers (Jallow et al., 2017).

Organophosphorus pesticides (OPPs) are considered the most harmful pesticides and have been identified as the main contaminants in fruit, water, and medicinal plants (Yang Q. et al., 2018). For methyl-paraoxon (MP) analysis, for instance, Wei et al. developed a dual-mode approach employing nanoceria as nanozyme (Wei et al., 2019). MP could hydrolyse into para-nitrophenol (p-NP), which has a distinctive absorption peak and brilliant yellow hue and can be readily evaluated using colorimetric and spectroscopic techniques, due to the enzyme-like activity of nanoceria. The LODs for both techniques were 0.42 μM.

2. Pathogen Detection

Foodborne microbes that are pathogenic, that may enable microbes within food to grow and create a large number of dangerous metabolites, have sparked significant interest in the subject of food safety. Accidents involving poisoning from eating contaminated food happen regularly. The drawbacks of traditional approaches included their high complexity, long processing times, and low sensitivity. Pathogen detection has been facilitated by the widespread development of emerging analytical techniques based on nanozymes.

Viruses are a type of foodborne pathogenic microorganism that can infect food, in addition to bacteria. It is noted that there is a dearth of information regarding viruses and food safety concerns. Currently, rotavirus, astrovirus, enterovirus, norovirus, and the hepatitis A and E viruses are the most frequent foodborne viruses. The human norovirus (NoV) is one of the most common viruses that causes outbreaks of foodborne diseases across the world (Patel et al., 2009).

3. Toxin Detection

According to Harms et al. (2018), a toxin is anything that an organism produces that obstructs the function of other living things and poisons the human body. Toxins in trace quantities in the body can be fatal or cause serious harm. For instance, several fungi (mostly Aspergillus, Penicillium, and Fusarium) create secondary metabolites called mycotoxins during their growth, which can readily result in physiological problems in both people and animals (Zain 2011). Mycotoxin-infected feed can cause grains to get contaminated or animal foods (such as meat, milk, and eggs) from animals can become contaminated with mycotoxin.

Additionally, the identification of bacterial toxins such as enterotoxins, botulinum neurotoxins, and Escherichia coli toxin is covered. Ching et al. (2012), for example, described the utilisation of Au nanoparticles in just one lateral flow detector for detecting botulinum neurotoxins A and B. If the toxin is present, it binds to the gold-conjugated mab and travels to the test's capturing line at the same time, resulting in a resolved red line. This is a representation of a typical lateral flow assay that may be adapted to use nanozyme-based analytical techniques.

4. Veterinary Drug Residue

Veterinary feed additives and are frequently used to control physiological functions in animals or to prevent, diagnose, and cure diseases in animals (Rocca et al., 2017; Stolker and Brinkman 2005). Any veterinarian medication used on food animals may leave behind remains in the form of dairy products, eggs, and meat that may contain parent chemicals, conjugates, or metabolites. These deposits can then enter the person food chain and have direct harmful consequences. Veterinary medications can be categorized into many groups according to their purposes, including sedatives, β-adrenergic receptor blockers, growth promoters, antibiotics, anthelmintics, antiprotozoal agents, and trypanosomiasis drugs (Winckler and Grafe 2001).

Zhang et al. (2002) created a TMB/H_2O_2 colorimetric detection technique to detect antibiotics made from tetracycline (TCs) that use TCs-specific aptamers (Apt) depending on the peroxidase-like capabilities of gold nanoclusters. The sensor accurately detects tetracycline in dairy and pharmaceuticals at levels ranging from 1 to 16 μM, with a limit of detection of 46 nM. Tian et al. (2020) have created a ratio electrochemical sensor that utilises the signals amplifying features of planar VS_2/AuNPs nanocomposites and $CoFe_2O_4$ nanozyme for the quantitative determination of kanamycin.

5. Food Additive

Food additives are a type of substance that is supplementary to food or during food manufacturing with the intention of enhancing food's taste, visual appeal, and edibility. Numerous issues with food safety will arise from the overuse of additives or the illicit use of non-food additives. Excessive amounts of food additives may cause cancer growth and other detrimental long-term effects in animals (Dolatabadi and Kashanian 2010). It is crucial to examine these inevitable additions in food samples. This section discusses the detection of additives to food using nanozyme-based techniques.

6. Heavy Metal and Radioactive Elements Detection

Heavy metal ions (Pb^{2+}, Cd^{2+} and Hg^{2+}) in food have been a growing issue due to their possible damage to public health (Wu et al., 2021). Even in tiny amounts found in food, prolonged exposure to these heavy metal ions can result in serious illnesses such renal failure, neurological abnormalities, cognitive impairments, and cardiovascular problems (Zhang et al., 2019). Furthermore, at the trace level, Cu^{2+} is a necessary element in the human body. For instance, it can act as a catalyst for the creation of heme, although consuming high amounts might be harmful. Thus, it is imperative to keep an eye out for heavy metals in food and drinking water. Currently, one of the frontiers in the detection of harmful heavy metal ions is the analytical approach based on nanozymes.

For the removal and detection of organic mercury (MeHg), Liu et al., for instance, created an Au/Ni-Fe LDH/rGO nanocomposite that serves as a surface-enhanced Raman scattering (SERS) substrate and an enzyme mimic (Liu et al., 2021). MeHg can be detected with a limit of quantification (LOQ) of 10 nM and can be eliminated based on the substance of the nanozyme, which is noteworthy considering the various uses of nanozymes.

Figure 3. Applications of nanozymes in food industry

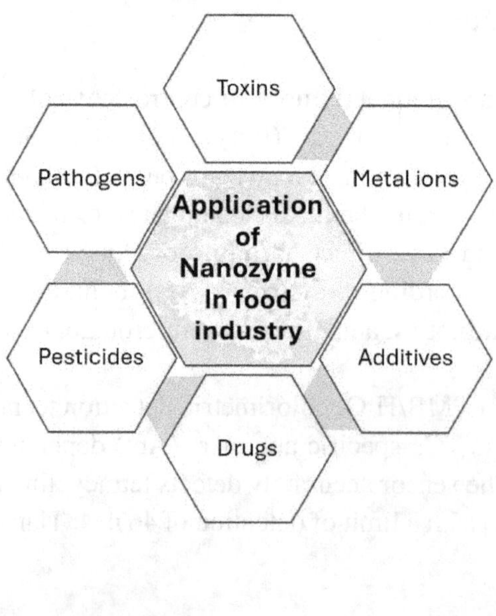

Perspectives and Challenges

Nanozymes-based scientific methodologies have filled in ubiquity because of their exceptional elements. A new path that addresses the limitations of existing analytical methods must be opened in order to propel the advancement of nanozyme research in food safety. Luckily, most of nanozymes are utilized in the advancement of speedy identification techniques like fluorescence, colorimetry, electrochemistry, and biosensors, giving a few forthcoming opportunities to fulfil the requirements of logical exploration. We summed up the scientific methodologies based on nanozymes for the speedy and delicate recognition of food toxins in this view. In spite of the fact that nanozymes can work on logical execution, they are novel fake compounds with a few issues that should be tended to.

- We examine the ideas and systems of nanozymes. Despite the fact that there have been numerous distributions distributed on nanozymes, hardly any trial research has focused on hypothetical review and system explanation. Investigation of the basic standards and cycles of nanozymes is basic since it might assist with unveiling the standard of construction movement association and guide the specific plan of nanozymes with expected efforts.
- We make a predictable framework and set of rules. Nanozymes are based on the idea of enzymes, but their properties are significantly different from those of regular enzymes. Thus, laid out strategies for portraying nanozyme execution are hazardous. The Michaelis-Menten component, for instance, is in many cases utilized while examining regular catalysts, but it is obvious that normal chemicals catalyse a response in a similar medium, rather than nanozymes, which happen in a heterogeneous cycle on the outer layer of nanomaterials. Thus, a reliable methodology and set of principles ought to be created to more readily portray nanozyme execution.
- We create regulated, functional nanozymes. Nanozyme activities and functionality are influenced by shape, dimension, and surface groups, thus it is desirable to generate high-performance nanozymes. The ability to controllably construct nanozymes and boost their functions via altering their surfaces is an essential area.
- We put superior execution nanozymes under serious scrutiny. Different nanozymes for signal age and enhancement have been depicted in the improvement of upgraded logical systems. Be that as it may, when applied to genuine applications, nanozymes' reactant action remains rather unobtrusive. Nanozymes have less sorts than typical chemicals, and they can catalyze one exceptional substrate. Thus, developing nanozymes with strong synergist activity, diverse enzymatic movement, and robust substrate selectivity is critical.
- We utilize different procedures. It is energizing that nanozyme-based discovery approaches are shutting the hole to commonsense food testing systems. In any case, it is extremely difficult to accomplish every one of the increases in a solitary discovery approach. Subsequently, creating multimode nanozymes-based approaches for the fast, exact, delicate, and particular discovery of food contaminations is a choice. Combining it with atomic etching, for example, can considerably increase the specificity and specificity of nanozymes. Overall, nanozymes are currently in the early stages of creating second-generation counterfeit substances. Nanozymes are renowned for their powerful activities, which range between in vitro identification to in vivo monitoring. We believe they are going to soon have a lot of promise for analysing food pollutants. The issues described above will serve as the foundation for future nanozyme research.

CONCLUSION

The combination of nanotechnology with omics exemplifies cross-disciplinary collaboration. This integration enables better knowledge and manipulation of biological processes. As we navigate the complex world of molecules within living beings, nanoomics has the potential to revolutionise sectors as diverse as healthcare, environmental studies, and the agricultural sector, paving the path for fresh solutions and transformational discoveries. Nanozymes, or enzyme mimetics based on nanomaterials, have garnered a lot of attention lately. They have distinct advantages over natural enzymes, including superior stability, robustness, and ease of scaling up at a low cost of production—a crucial substitute. These nanozymes have been used for many types of food and food product detection because of their natural enzyme-like activity. Nanozymes are only beginning to grow into second generation artificial enzymes. Because nanozymes are so excellent at in vitro identification and in vivo evaluation, we believe the area of food contamination assessment will considerably benefit from their application soon.

REFERENCES

Bajwa, U., & Sandhu, K. S. (2014). Effect of Handling and Processing on Pesticide Residues in Food-a Review. *Journal of Food Science and Technology*, *51*(2), 201–220. doi:10.1007/s13197-011-0499-5 PMID:24493878

Chang, Y.-C., Lin, Y.-S., Xiao, G.-T., Chiu, T.-C., & Hu, C.-C. (2016). A Highly Selective and Sensitive Nanosensor for the Detection of Glyphosate. *Talanta*, *161*, 94–98. doi:10.1016/j.talanta.2016.08.029 PMID:27769503

Ching, K. H., Lin, A., McGarvey, J. A., Stanker, L. H., & Hnasko, R. (2012). Rapid and Selective Detection of Botulinum Neurotoxin Serotype-A and -B with a Single Immunochromatographic Test Strip. *Journal of Immunological Methods*, *380*(1-2), 23–29. doi:10.1016/j.jim.2012.03.008 PMID:22504369

Cui, R., Han, Z., & Zhu, J.-J. (2011). Helical Carbon Nanotubes: Intrinsic Peroxidase Catalytic Activity and its Application for Biocatalysis and Biosensing. *Chemistry (Weinheim an der Bergstrasse, Germany)*, *17*(34), 9377–9384. doi:10.1002/chem.201100478 PMID:21769953

Dolatabadi, J. E. N., & Kashanian, S. (2010). A Review on DNA Interaction with Synthetic Phenolic Food Additives. *Food Research International*, *43*(5), 1223–1230. doi:10.1016/j.foodres.2010.03.026

Fan, K., Wang, H., Xi, J., Liu, Q., Meng, X., Duan, D., Gao, L., & Yan, X. (2017). Optimization of Fe3O4 Nanozyme Activity via Single Amino Acid Modification Mimicking an Enzyme Active Site. *Chemical Communications (Cambridge)*, *53*(2), 424–427. doi:10.1039/C6CC08542C PMID:27959363

Gao, L., Fan, K., & Yan, X. (2020). Iron Oxide Nanozyme: A Multifunctional Enzyme Mimetics for Biomedical Application. *Nanozymology*, *7*, 105–140. doi:10.1007/978-981-15-1490-6_5

Gardner, L., Kostarelos, K., Mallick, P., Dive, C., & Hadjidemetriou, M. (2022, August). Nano-omics: Nanotechnology-based multidimensional harvesting of the blood-circulating cancerome. *Nature Reviews. Clinical Oncology*, *19*(8), 551–561. doi:10.1038/s41571-022-00645-x PMID:35739399

Garg, B., & Bisht, T. (2016). Carbon Nanodots as Peroxidase Nanozymes for Biosensing. *Molecules (Basel, Switzerland), 21*(12), 1653. doi:10.3390/molecules21121653 PMID:27918453

Harms, A., Brodersen, D. E., Mitarai, N., & Gerdes, K. (2018). Toxins, Targets, and Triggers: An Overview of Toxin-Antitoxin Biology. *Molecular Cell, 70*(5), 768–784. doi:10.1016/j.molcel.2018.01.003 PMID:29398446

Huang, L., Chen, K., Zhang, W., Zhu, W., Liu, X., Wang, J., Wang, R., Hu, N., Suo, Y., & Wang, J. (2018a). ssDNATailorable Oxidase-Mimicking Activity of Spinel MnCo2O4 for Sensitive Biomolecular Detection in Food Sample. *Sensors and Actuators. B, Chemical, 269*, 87–95. doi:10.1016/j.snb.2018.04.150

Huang, L., Sun, D.-W., Pu, H., Wei, Q., Luo, L., & Wang, J. (2019b). A Colorimetric Paper Sensor Based on the Domino Reaction of Acetylcholinesterase and Degradable γ-MnOOH Nanozyme for Sensitive Detection of Organophosphorus Pesticides. *Sensors and Actuators. B, Chemical, 290*, 573–580. doi:10.1016/j.snb.2019.04.020

Huang, L., Zhu, Q., Zhu, J., Luo, L., Pu, S., Zhang, W., Zhu, W., Sun, J., & Wang, J. (2019c). Portable Colorimetric Detection of Mercury(II) Based on a Non-Noble Metal Nanozyme with Tunable Activity. *Inorganic Chemistry, 58*(2), 1638–1646. doi:10.1021/acs.inorgchem.8b03193 PMID:30604958

Huang, Y., Ren, J., Ren, J., & Qu, X. (2019d). Nanozymes: Classification, Catalytic Mechanisms, Activity Regulation, and Applications. *Chemical Reviews, 119*(6), 357–4412. doi:10.1021/acs.chemrev.8b00672 PMID:30801188

Jallow, M., Awadh, D., Albaho, M., Devi, V., & Ahmad, N. (2017). Monitoring of Pesticide Residues in Commonly Used Fruits and Vegetables in Kuwait. *International Journal of Environmental Research and Public Health, 14*(8), 833. doi:10.3390/ijerph14080833 PMID:28757570

Kuang, Q., Wang, X., Jiang, Z., Xie, Z., & Zheng, L. (2014). High-energy-surface Engineered Metal Oxide Micro- and Nanocrystallites and Their Applications. *Accounts of Chemical Research, 47*(2), 308–318. doi:10.1021/ar400092x PMID:24341353

Liang, M., & Yan, X. (2019). Nanozymes: From New Concepts, Mechanisms, and Standards to Applications. *Accounts of Chemical Research, 52*(8), 2190–2200. doi:10.1021/acs.accounts.9b00140 PMID:31276379

Liu, B., & Liu, J. (2017). Surface Modification of Nanozymes. *Nano Research, 10*(4), 1125–1148. doi:10.1007/s12274-017-1426-5

Liu, B., & Liu, J. (2017). Surface Modification of Nanozymes. Nano Res., 10, 1125–1148. doi:10.1007/s12274-017-1426

Lustig, W. P., Mukherjee, S., Rudd, N. D., Desai, A. V., Li, J., & Ghosh, S. K. (2017). Metal-organic Frameworks: Functional Luminescent and Photonic Materials for Sensing Applications. *Chemical Society Reviews, 46*(11), 3242–3285. doi:10.1039/C6CS00930A PMID:28462954

Nandhakumar, P., Kim, G., Park, S., Kim, S., Kim, S., Park, J. K., Lee, N.-S., Yoon, Y. H., & Yang, H. (2020). Metal Nanozyme with Ester Hydrolysis Activity in the Presence of Ammonia-Borane and its Use in a Sensitive Immunosensor. *Angewandte Chemie International Edition*, *59*(50), 22419–22422. doi:10.1002/anie.202009737 PMID:32875647

Nicolini, V., Gambuzzi, E., Malavasi, G., Menabue, L., Menziani, M. C., Lusvardi, G., Pedone, A., Benedetti, F., Luches, P., D'Addato, S., & Valeri, S. (2015). Evidence of Catalase Mimetic Activity in Ce3+/Ce4+ Doped Bioactive Glasses. *The Journal of Physical Chemistry B*, *119*(10), 4009–4019. doi:10.1021/jp511737b PMID:25710332

Patel, M. M., Hall, A. J., Vinjé, J., & Parashar, U. D. (2009). Noroviruses: A Comprehensive Review. *Journal of Clinical Virology*, *44*(1), 1–8. doi:10.1016/j.jcv.2008.10.009 PMID:19084472

Pinto, V. E. F., & Patriarca, A. (2017). Alternaria Species and Their Associated Mycotoxins. *Methods in Molecular Biology (Clifton, N.J.)*, *1542*, 13–32. doi:10.1007/978-1-4939-6707-0_2 PMID:27924529

Rajput, V. D., Singh, A., Minkina, T., Rawat, S., Mandzhieva, S., Sushkova, S., Shuvaeva, V., Nazarenko, O., Rajput, P., Komariah, Verma, K. K., Singh, A. K., Rao, M., & Upadhyay, S. K. (2021). Nano-Enabled Products: Challenges and Opportunities for Sustainable Agriculture. *Plants*, *10*(12), 2727. doi:10.3390/plants10122727 PMID:34961197

Rocca, L. M., Gentili, A., PerezFernandez, V., & Tomai, P. (2017). Veterinary Drugs Residues: a Review of the Latest Analytical Research on Sample Preparation and LC-MS Based Methods. Food Additives Contaminants A-Chemistry Anal. Control. Expo. Risk Assess., 34, 766–784.

Singh, A., Rajput, V. D., Varshney, A., Ghazaryan, K., & Minkina, T. (2023b). Small Tech, Big Impact: Agri-nanotechnology Journey to Optimize Crop Protection and Production for Sustainable Agriculture. *Plant Stress*, *10*, 100253. doi:10.1016/j.stress.2023.100253

Singh, A., Rawat, S., Rajput, V. D., Ghazaryan, K., Minkina, T., Al Tawaha, A. R. M., . . . Sandhu, K. S. (2023c). Nanoparticle-Based Priming-Enhanced Agricultural Practices. *Nanopriming Approach to Sustainable Agriculture*, 142–157. doi:10.4018/978-1-6684-7232-3.ch006

Singh, A., Sengar, R. S., Rajput, V. D., Al-Ghzawi, A. L., Shahi, U. P., Ghazaryan, K., Minkina, T., Tawaha, A. R. M. A., Al Zoubi, O. M., & Habeeb, T. (2024a). Impact of Salinity Stress and Zinc Oxide Nanoparticles on Macro and Micronutrient Assimilation: Unraveling the Link between Environmental Factors and Nutrient Uptake. *Journal of Ecological Engineering*, *25*(2), 1–9. doi:10.12911/22998993/172947

Stolker, A. A. M., & Brinkman, U. A. T. (2005). Analytical Strategies for Residue Analysis of Veterinary Drugs and Growth-Promoting Agents in Food-Producing Animals-A Review. *Journal of Chromatography. A*, *1067*(1-2), 15–53. doi:10.1016/j.chroma.2005.02.037 PMID:15844509

Su, Y., Wu, D., Chen, J., Chen, G., Hu, N., Wang, H., Wang, P., Han, H., Li, G., & Wu, Y. (2019). Ratiometric Surface Enhanced Raman Scattering Immunosorbent Assay of Allergenic Proteins via Covalent Organic Framework Composite Material Based Nanozyme Tag Triggered Raman Signal "Turn-On" and Amplification. *Analytical Chemistry*, *91*(18), 11687–11695. doi:10.1021/acs.analchem.9b02233 PMID:31418273

Sun, H., Zhou, Y., Ren, J., & Qu, X. (2018). Carbon Nanozymes: Enzymatic Properties, Catalytic Mechanism, and Applications. *Angewandte Chemie International Edition*, *57*(30), 9224–9237. doi:10.1002/anie.201712469 PMID:29504678

Tao, X., Wang, X., Liu, B., & Liu, J. (2020). Conjugation of Antibodies and Aptamers on Nanozymes for Developing Biosensors. *Biosensors & Bioelectronics*, *168*, 112537. doi:10.1016/j.bios.2020.112537 PMID:32882473

Tian, L., Zhang, Y., Wang, L., Geng, Q., Liu, D., Duan, L., Wang, Y., & Cui, J. (2020). Ratiometric Dual Signal-Enhancing-Based Electrochemical Biosensor for Ultrasensitive Kanamycin Detection. *ACS Applied Materials & Interfaces*, *12*(47), 52713–52720. doi:10.1021/acsami.0c15898 PMID:33170623

Wang, H., Wan, K., & Shi, X. (2019). Recent Advances in Nanozyme Research. *Advanced Materials*, *31*(45), 1805368. doi:10.1002/adma.201805368 PMID:30589120

Wang, X.-F., Zhou, Y., Xu, J.-J., & Chen, H.-Y. (2009). Signal-on Electrochemiluminescence Biosensors Based on CdS-Carbon Nanotube Nanocomposite for the Sensitive Detection of Choline and Acetylcholine. *Advanced Functional Materials*, *19*(9), 1444–1450. doi:10.1002/adfm.200801313

Wei, J., Yang, L., Luo, M., Wang, Y., & Li, P. (2019). Nanozyme-assisted Technique for Dual Mode Detection of Organophosphorus Pesticide. *Ecotoxicology and Environmental Safety*, *179*, 17–23. doi:10.1016/j.ecoenv.2019.04.041 PMID:31022651

Winckler, C., & Grafe, A. (2001). Use of Veterinary Drugs in Intensive Animal Production. *Journal of Soils and Sediments*, *1*(2), 66–70. doi:10.1007/BF02987711

Worheide, M. A., Krumsiek, J., Kastenmuller, G., & Arnold, M. (2021). Multi-omics integration in biomedical research - A metabolomics-centric review. *Analytica Chimica Acta*, *2*(1141), 144–162. doi:10.1016/j.aca.2020.10.038 PMID:33248648

Wu, J., Wang, X., Wang, Q., Lou, Z., Li, S., Zhu, Y., Qin, L., & Wei, H. (2019). Nanomaterials with Enzyme-like Characteristics (Nanozymes): Next-Generation Artificial Enzymes (II). *Chemical Society Reviews*, *48*(4), 1004–1076. doi:10.1039/C8CS00457A PMID:30534770

Wu, J., Yang, Q., Li, Q., Li, H., & Li, F. (2021a). Two-Dimensional MnO2 Nanozyme-Mediated Homogeneous Electrochemical Detection of Organophosphate Pesticides without the Interference of H2O2 and Color. *Analytical Chemistry*, *93*(8), 4084–4091. doi:10.1021/acs.analchem.0c05257 PMID:33588528

Wu, L., Li, G., Xu, X., Zhu, L., Huang, R., & Chen, X. (2019b). Application of Nano-ELISA in Food Analysis: Recent Advances and Challenges. *Trends in Analytical Chemistry*, *113*, 140–156. doi:10.1016/j.trac.2019.02.002

Wu, L., Yan, H., Wang, J., Liu, G., & Xie, W. (2019c). Tyrosinase Incorporated with Au-Pt@SiO2 Nanospheres for Electrochemical Detection of Bisphenol A. *Journal of the Electrochemical Society*, *166*(8), B562–B568. doi:10.1149/2.0141908jes

Yang, F., Jiang, X., Zhong, X., Wei, S., & Yuan, R. (2018). Highly Sensitive Electrochemiluminescence Detection of Mucin1 Based on V2O5 Nanospheres as Peroxidase Mimetics to Catalyze H2O2 for Signal Amplification. *Sensors and Actuators. B, Chemical*, *265*, 126–133. doi:10.1016/j.snb.2018.03.031

Zain, M. E. (2011). Impact of Mycotoxins on Humans and Animals. *Journal of Saudi Chemical Society*, *15*(2), 129–144. doi:10.1016/j.jscs.2010.06.006

Zhang, Z., Tian, Y., Huang, P., & Wu, F. Y. (2020). Using Target-specific Aptamers to Enhance the Peroxidase-like Activity of Gold Nanoclusters for Colorimetric Detection of Tetracycline Antibiotics. *Talanta*, *208*, 120342. doi:10.1016/j.talanta.2019.120342 PMID:31816712

Zhou, Y., Wei, Y., Ren, J., & Qu, X. (2020). A Chiral Covalent Organic Framework (COF) Nanozyme with Ultrahigh Enzymatic Activity. *Materials Horizons*, *7*(12), 3291–3297. doi:10.1039/D0MH01535K

Zhu, X., Mao, X., Wang, Z., Feng, C., Chen, G., & Li, G. (2017). Fabrication of Nanozyme@DNA Hydrogel and its Application in Biomedical Analysis. *Nano Research*, *10*(3), 959–970. doi:10.1007/s12274-016-1354-9

Chapter 20
Nanotechnology Applications in Livestock Farming and Animal Health

Mukul Machhindra Barwant
https://orcid.org/0000-0002-5639-4251
Sanjivani Arts Commerce and Science College, India

Arshad Rasool
Minhaj University, Lahore, Pakistan

Vishnu Kiran Mannam
Indian Broiler Group, India

Abrar Hussain
University of Karachi, Pakistan

Muhammad Abdullah
The Islamia University of Bahawalpur, Pakistan

Zaibun nisa Memon
Shah Abdul Latif University, Pakistan

Muhammad Irfan Ahamad
https://orcid.org/0000-0001-6992-2036
Henan University, China

Madeeha Shirani
University of Gujrat, Pakistan

Konain Zahra Maqbool
University of Baltistan, Pakistan

Muhammad Majeed
https://orcid.org/0000-0001-7663-2563
University of Gujrat, Pakistan

Muhammad Asif
Punjab University, Pakistan

Usama Ahmad Khan
https://orcid.org/0009-0003-9577-290X
Ghazi University, Pakistan

ABSTRACT

As the global demand for high-quality animal products continues to rise, the livestock industry faces unprecedented challenges in ensuring the health, productivity, and sustainability of livestock farming. Nanotechnology, with its ability to manipulate materials at the nanoscale, presents a promising frontier for innovative solutions in animal health and husbandry. These applications of nanotechnology in livestock farming, focusing on disease prevention, nutritional enhancement, and environmental sustainability. The first section of the research delves into nanomaterials as novel tools for disease prevention

DOI: 10.4018/979-8-3693-1890-4.ch020

and control in livestock. The second segment highlights the use of nanotechnology to address nutritional challenges in livestock farming. Nanoscale nutrient delivery systems are investigated for their capacity to enhance the bioavailability of essential nutrients, such as vitamins and minerals, thereby improving animal growth, reproduction, and overall health. Nanosensors are used for real-time monitoring of nutritional parameters, feeding strategies for livestock nutrition.

1. INTRODUCTION

The Greek word "Nano," which means dwarf, was utilised when the phrase "nanotechnology" was originally introduced in 1970 to refer to industrial technology at ultrafine dimensions. The International Organisation for Standardisation has released a statement that defines nanotechnology as a scientific application of knowledge for controlling and manipulating matter at the nanoscale (Haris et al., 2023). In order to produce and manipulate tools, materials, and functional structures with nanometer-scale dimensions, we may work at the molecular level, frequently atom by atom, thanks to the fascinating and quickly developing field of nanotechnology (Rajput et al., 2021). For millions of years, nature has been executing "nanotechnological feats." Biological systems merge wet chemistry and electro-chemistry into a single living system through the arrangement of atoms and molecules. It is utilised by the body's cells for disease diagnosis and therapy. It could significantly affect how animals are diagnosed and treated. The distinct size-dependent characteristics of nanoparticles have a wide range of diagnostic uses, including imaging, biosensors, and nanoprobes for contrast agents in magnetic resonance imaging (Chandra et al., 2023; Singh et al., 2023a). The word nano means "dwarf" in Latin, and according to the Nanotechnology Initiative, the term was initially used by Nobel laureate in physics Richard Fenman for research and technological development at the atomic level in southern California. With the help of this fascinating and quickly developing field of study, we may make, manipulate, and use tools, materials, and functional structures with nanometer-scale dimensions by working at the atomic, molecular, and macromolecular levels.

The third and final section examines nanotechnology's role in promoting environmental sustainability within the livestock industry. Nanomaterial-based solutions for waste management, water purification, and efficient nutrient utilization are discussed, offering potential pathways to mitigate the environmental impact of intensive farming practices. Furthermore, the research explores nanomaterials as alternatives to traditional pesticides and fertilizers, aiming to reduce the ecological footprint of livestock operations. By synthesizing the latest research findings and technological developments, this research abstract provides valuable insights into the transformative potential of nanotechnology in addressing critical challenges faced by the livestock industry. The convergence of nanotechnology and animal sciences opens new avenues for sustainable, efficient, and ethically responsible livestock farming practices, ultimately contributing to global food security and the well-being of both animals and humans. For the first time, the concept of nanotechnology was developed in 1959 (Tundisi et al., 2023). In the 1980s, however, a lot more research was done on the fundamental concepts of popularisation and definition. By creating new, small-scale instruments and materials that are advantageous to living things, biotechnology has revolutionised nearly every field in veterinary and animal sciences, while nanotechnology has created new opportunities for applications in molecular biology (Singh et al., 2023b). Metallic nanoparticles are among the many types of nanomaterials that are utilised in medicine delivery, animal nutrition, animal breeding, animal reproduction, and the value addition of animal goods (Chehelgerdi et al., 2023). Nanotechnology can be used to observe, measure, manipulate, and manufacture objects at the nanometer (nm) scale (Singh et al., 2023c,d). Although it has not been as developed in agronomy as it has been in other disciplines, this is because there has been less investment in this field. A nanometer is a length unit of 109, or one billionth of a metre, according to the French acronym SI (Système International d'Unitès). Nanotechnologies are the study and manipulation of matter at the nanoscale, or at dimensions of roughly 1–100 nm, where special phenomena like enhanced physical, chemical, and biological properties may allow for novel applications (Ansari, 2023). Particles with a minimum dimension of less than one micron and the ability to reach atomic and molecular length scales of around 0.2 nanometers are known as nanoparticles.

The surfaces of nanoparticles can carry liquid droplets or gases in an amorphous or crystalline form. (Azeem et al., 2023). The creation and application of materials with nanoscale dimensions in various facets of life is the subject of nanotechnology. Because of their nanoscale size, nanoparticles exhibit highly specialised properties due to their high surface-to-volume ratios. Nanotechnology is a relatively new and developing field of technology. Its primary uses include the creation of novel techniques for producing new goods, creating novel chemicals and materials, and replacing the current generation of equipment with higher-performing models that use less energy and resources and cause less environmental damage. Additionally, nanotechnology can be used to remediate damaged environmental areas (Ansari, 2023).

The use of nanotechnology and nanoparticles in food production, animal breeding, and animal productivity including the production of milk and meat is expanding quickly. At the molecular level, it was used to produce materials, alter their structure, and improve the flavour and texture of food. Production, processing, shipping, storage, traceability, safety, and security are all significantly impacted by this technology (Manimegalai et al., 2023; Singh et al., 2024a). In research and development contexts, numerous uses of nanotechnology for food and agricultural production are being researched. Animal husbandry raises a number of global issues, such as food security, disease prevention, human health, and environmental sustainability (Lin et al., 2023). According to the European Union Commission's proposal on a code of conduct for ethical research in the fields of nanoscience and nanotechnology, "nano-object" refers to any and all nanomaterials, nanostructured materials, nanoparticles and the way they aggregate at the nanoscale, nanosystems, and nanoproducts (Fadiji et al., 2023a; Singh et al., 2024b). For producers raising dogs, dairy products, and pigs, breeding management is a costly and time-consuming issue. Nano-enabled drug delivery systems and diagnostic platforms are explored for their potential to enhance the efficacy of vaccines, antimicrobials, and diagnostics. The ability of nanoparticles to target specific pathogens while minimizing adverse effects holds significant promise for revolutionizing disease management in various livestock species.

Nanotechnology studies individual nanoparticles and their unique application for the meat industry, ranging from meat design, achieving food security, meat safety, overcoming food allergies, eliminating pesticide use, meat packaging, restoring meat damage, and sensory evaluation to processes such as filtration, separation, encapsulation, etc. One solution that is currently being studied is a nanotube implanted under the skin to provide real-time measurement of changes in the level of estradiol in the blood (Biswas et al., 2023; S. Li et al., 2023). One of the biggest risks associated with breeding cattle and poultry is disease. The foundation of a thriving cattle and poultry population is a clean and healthy breeding facility, which is also critical for worker safety and health as well as for the enhancement of the surrounding water and air quality. One of the most important trace minerals for animals is selenium (Se), which is used extensively in animal husbandry. A small amount of selenium added to feed can boost an animal's immune system and antioxidant capacity, preserve the animal's overall health, and encourage the growth of chickens and cattle. Because nano trace elements may directly into an animal's body, their utilisation rate is significantly higher than that of regular inorganic trace elements (Chandra et al., 2023). Look it up In the fishing business, nanotechnologies are widely used for things like water purification, fishpond sterilisation, fish feed, and disease control in aquatic environments. Fish development may be enhanced by dietary supplements containing nanoparticles of minerals such as iron, selenium, and other supplements. (Khalil et al., 2023). The method can be used to lower the cost of water treatment in commercial fish ponds and aquariums. Scientists think that fishponds protected from pollution and disease could be possible with nanotechnologies. Applying nanotechnologies to various conservation and packaging methods to ensure seafood safety by postponing mildew and microbiological deterioration is another potential use case (R. Kumar et al., 2023).

Numerous enigmas surrounding animal health, reproduction, and appropriate hygienic practises throughout the upbringing and maintenance of food animals could be resolved by nanotechnology. It is possible to use the technology application, particularly with livestock. With the passage of time, nanotechnology has evolved from a theory or notion into a new enabling technology with the potential to completely transform livestock and agricultural development globally (Kumari et al., 2023). Because of its unique characteristics (big surface area and quantum size effects), nanoparticulate matter should be regarded as a separate form of matter in addition to the solid, liquid, gaseous, and plasma states. Fullerenes and carbon nanotubes are examples of materials in crystalline nanoparticle form, whereas graphite and diamond are examples of typical crystalline solid forms (S. Li et al., 2023).With dwindling resources and a growing world population, agriculture's role as a source of food is becoming more and more crucial (Kumari et al., 2023). As one of the main engines of the economy, agri-food production is extremely important. It may also provide pathways to crops with extra value. Because of the enormous demand that climate change, energy and resource shortages, and a rapidly expanding global population are putting on food and water supplies, agricultural practises are frequently in the public view (Liu et al., 2023).Nanotechnology holds great promise for transforming agriculture and related industries, such as fisheries and aquaculture. Target farming, which uses nanosized particles with special qualities to increase crop and livestock productivity, is the current emphasis of nano-agriculture.

The application of nanotechnology in the food and agriculture industries is relatively new when compared to its usage in pharmaceuticals and drug delivery. Plant protection, plant growth monitoring, animal and plant disease detection, increased global food production, improved food quality, and waste reduction are all possible with nanotechnology (Lo Presti et al., 2023). Recent years have seen a surge in interest in nanotechnology among researchers, manufacturers, and even the general public. The field is generally viewed as new and rapidly developing, involving the creation, processing, and application of structures, devices, and systems by manipulating shape and size at the nanoscale scale. Nanoparticles, or NPs, are defined as discrete entities with three dimensions of less than 100 nm. The combination of NPs' small size, chemical composition, and surface .One application of nanoparticles in insects is the potential for managing insect pests, as reported by researchers.There have been reports on the use of nanotechnology to control the polyphagous pest Helicoverpa armigera (García et al,.2010).The head louse Pediculus humanus and the fourth instar larvae of Anopheles subpictus and Culex-quinque fasciatus were the targets of maximal mortality in the pediculocidal and larvicidal activity of synthesised silver nanoparticles, which were prepared using an aqueous leaf extract of Tinospora cordifolia. The produced silver nanoparticles shown remarkable larvicidal properties against lice and mosquitoes(Banerjee et al., 2024).

Opportunities in nanobiotechnology include uses in food, agriculture, and energy. Some of the most popular applications of nanobiotechnology in the food processing sectors include bacterium detection and nanosensors/ nanobiosensors for food product quality monitoring. In storage rooms, the grain bulk may be reliably inspected for the presence of fungi or insects using the nanosensors. Industry now has new tools thanks to nanobiotechnology that allow it to change genes and even create entirely new species. This is because it makes it possible for foreign DNA and drugs that alter genes to be carried via nanoparticles, nanofibers, and nanocapsules. Furthermore, synthetic biology a recent field that combines genetic engineering, nanotechnology, and informatics may be used to create novel plant kinds (Arif et al., 2021).

2. NANOSCALE SENSORS FOR MONITORING ANIMAL HEALTH

Sensor is an instrument that is used to measure the physiological and behavioral changes in a subject. In the last few decades the researchers has developed different new application of sensors. Besides its primary focus on robotics, military, entertainment, etc. they are currently used in an important global issues like natural disaster, health monitoring, and disease control (Helwatkar et al., 2014). The use of sensors in animal health is although limited, but they got the attention of the researchers to develop such sensors that have this potential. Currently, different types of technologies are used in animal health monitoring to measure their behavioral changes, disease status, and the treatment methods. Majorly two types of sensor i.e. attached or on-cow sensors and non-attached or off cow sensors. The on-cow sensors are fitted inside or outside of the cow's body while the no attached sensors are outside the cow's body and perform its functions (Helwatkar et al., 2014). Numerous developments were performed in the applications of sensors. These developmental processes include the increase measurement of cow's activity, fabricate data about cow's status, advice to promote a product, and the understanding of farmers with the sensors (Helwatkar et al., 2014).

2.1. Types of Sensors in Animal's Health

The development in sensors system arise a question about its usage and familiarity with the farmers in term of its economic values and the disease management system. To cope the problem sensors were developed that measure the physiological status automatically. Two types of sensor-based approaches are in used.

Non-invasive; which may be immobile sensors or mobile sensors. The immobile sensor is fixed at the barn's point from which the cows pass regularly. This remains fixed as it is known that a cow has a repetitive routine from which they pass. Examples include temperature measurement of the udder and measurement of the breath composition. Surveillance cameras are also considered an immobile sensor which gives complete information about the cow's activities in the herd. The benefits of these immobile sensors are the use of one or few sensors in the whole herd, while the disadvantage is the limit detection of cow's activity (Majeed, Hayat Bhatti, et al., 2020).

The mobile sensors are those sensors which are attached with the cow's body (neck, ankle, etc.) and can monitor the overall spectrum activity of the cow. The known example include accelerometers, vibration sensors, thermometer, humidity meter, and pedometers, etc. pedometers are simple and cost-effective instrument that can measure the all spectrum activity (lying, oestrus time etc.) of the animals and give a correct results. Recently, GPS-based positioning sensors are attached to cow' collar and used for the monitoring of cow's activities like easting, lying, walking, standing, etc. both pros and cons are associated with this system but it is found effective. The latest and widely used sensor is MESM-based accelerometers, which give high data and consumed less power (Majeed, Bhatti, & Amjad, 2021).

Invasive; these sensors are implemented inside the cow i.e. fitted in the rumen. These internal sensors are used to measure the various physiological processes with high accuracy. Internal temperature, vaginal pressure during birth, and pH of rumen's fluid etc. are measured with these sensors. As these sensors are not affected by external conditions so they give very correct result. Different hurdles like implementation inside the body, energy requirements, and reuse capacity can limit their potential (Zain et al., 2024).

2.2. Nonosensors for Disease Detection

The field of biosensors is growing fast and besides other application they are also used for disease detection. Any analytical devices that are used to convert biochemical information into a signal are called biosensors (Basha, 2019). These signals are recorded in the form of optical, magnetic, thermal, or electrical signal. Normally biosensors have two parts i.e. the biological and transduction. The former recognize interaction with the target and the latter convert the signal into a detectable signal. Easy to operate, rapid action, and less cost make the availability of various types of biosensors. The use of nanomaterials can aid to their selectivity, sensitivity, and speed and thus increase their application (Al-Sahli et al., 2024). The size of nanomaterials is ranging from 1-100 nm which give them distinguished properties and make them feasible to functionalize with vast biological properties. Additional properties of nanomaterials also help them to enhance their functions, i.e. the high extinction of gold nanoparticle than routine dyes can enhance their detection. All these properties make the nanosensors a promising agent for disease detection especially infectious diseases (Deng et al., 2021).

Nanosensors that syndicate nanomaterials with biosensors can enhance their diagnostic potential. For example the infectious disease diagnostic potential is enhanced by the easy operation, sensitivity, selectivity, and speed of nanosensors (Nastasijevic et al., 2021). The uses of biosensors in animals provide the information about the animal health, reproductive cycle detection, and welfare monitoring. With the

Table 1. Different types of sensors and their description used in animal's health

Sr.#	Type of Sensors	Description	References
1.	Pressure sensors	They are specifically used in pastures and stable. They record the jaw movement which is shown as peak.	(Nastasijevic et al., 2021; Pramanik et al., 2020)
2.	Acoustic sensors	They are used to analyse the jaw's movement and grazing	(Nastasijevic et al., 2021)
3.	Acceleration sensors	Mostly used to monitor the feeding behaviour and jaw movement.	(Nastasijevic et al., 2021)
4.	Breath sensors	They help in disease diagnosis by detection and identification of volatile organic compounds.	(Nastasijevic et al., 2021)
5.	Perspiration biosensors	Actually developed for human health monitoring but can also use to measure the animals physical stress.	(Nastasijevic et al., 2021)
6.	Tear fluid biosensors	Help to identify metabolite in blood while analysing tear sample	(Chaupoo et al., 2023; Nastasijevic et al., 2021)
7.	Progesterone biosensors	They are developed by integrating an aptamer specific with the progesterone binding receptors.	(Nastasijevic et al., 2021)

advancement of technology the monitoring sensing is developed that help the farmers to increase the animal's productivity (S. Kumar, 2010). In animals the disease status is checked by targeting the disease markers with biosensors. Complexity of the animal's environment becomes a challenge for biosensors to detect the animal's pathogens (Nastasijevic et al., 2021). The glory application of biosensors in different areas is their cost effectiveness.

Table 2. Various animal diseases and their respective sensors in animal's health

Sr.#	Sensor	Disease	Descriptions	References
1.	point-of-care (PoC) biosensor	Bovine Respiratory Disease (BRD)	These are sensitive and specific to anti-IgE in cattle sample	(Nastasijevic et al., 2021)
2.	Bovine Viral Diarrhoea (BVD) sensor	Bovine Viral Diarrhoea (BVD)	These biosensors are used for the detection of BVD in cattle's serum	(Nastasijevic et al., 2021)
3.	Avian Influenza virus (AIV) sensors	Avian Influenza virus (AIV)	This sensor can detect the immobilized H7N1 antibodies	(Nastasijevic et al., 2021)
4.	Foot and Mouth Diseases (FMD) sensor	Foot and Mouth Diseases (FMD)	The sensor can detect the antibodies against FMD proteins	(Muktar et al., 2015)
5.	Temperature sensor	Fever	High and low temperature effects occur	(Al-Sahli et al., 2024)
6.	Accelerometer	Discomfort	The animal's activity become low	(Al-Sahli et al., 2024)
7.	Accelerometer, Pedometers	Lameness	Changes occur in the walking and sitting of animals	(Payne et al., 2023)
8.	Accelerometer	Oestrus	Changes in hormonal level occur	(Zhou et al., 2023)
9.	Pressure Sensor	Ovarian Cysts	More or less grazing by animals, yield difference	(Zhou et al., 2023)
10.	Accelerometer (Pedometer)	Ketosis	Grazing pattern is affected	(Grillo et al., 2021)
11.	Accelerometer	Milk Fever	The movement and motion is affected	(Sangeeta Dash et al., 2023a)
12.	Temp Sensor	Heifer Diarrhea	High fever and high temperature occur	(Al-Sahli et al., 2024; Victoria et al., 2024)

Table 3. Biosensors used for the detection of pathogenic microbes in animal related environment

Sr.#	Biosensors	Pathogens	Reference
1.	aptamer-based biosensors	*Salmonella Enteritidis Escherichia coli* O157:H7	(Nastasijevic et al., 2021)
2.	DNA-based sensor	Campylobacter	(Nastasijevic et al., 2021)
3.	Cell-based sensors	*Clostridium perfringens*	(Nastasijevic et al., 2021)
4.	Antibody-based biosensors	*Escherichia coli*	(Nastasijevic et al., 2021)
5.	conductometric-based biosensors	*E. coli*	(Nastasijevic et al., 2021)
6.	semi-specific biosensor	*Aeromonas hydrophila*	(Nastasijevic et al., 2021)
7.	Colorimetric biosensors	*S. typhimurium*, *L. monocytogenes*, E. coli O157:H7	(Deng et al., 2021)
8.	Fluorescent biosensors	*S. aureus*, *V. parahemolyticus*, *S. typhimurium*	(Deng et al., 2021)
9.	Electrochemical biosensors	*E. coli*	(Deng et al., 2021)
10.	Silicon microring resonator-based biosensors are	*S. pneumonia*	(Deng et al., 2021)

2.3. Biosensors in Animal Husbandry and Feeds

The use of biosensors are considering helpful in the development of new approaches that can help to enhance the fees supplements methods and approaches. They can better shows the animal's behavior and response to various changes and thus enables the researcher to develop new methods for understanding of animal's responses (Pandey et al., 2023). Biosensors can also help to characterize the animal's feeds and their products like milk, meat, and eggs, etc. the important aspect of using biosensors in veterinary filed is they help in the decision to select a suitable feeds for animals (Singh et al., 2023)

2.4. Nanoscale Wearable for Tracking Animal Behaviour

The advancement in technology attracts the researchers to develop wearable health monitoring devices that helps to monitor the individual health status. Different wearable electronic devices are developed which mostly based on the micro pressure sensors (Hu et al., 2023). Currently, the wearable pressure sensors have been combined with artificial intelligence and new materials, constantly developing in the direction of intelligence, miniaturization and flexibility. Furthermore, they are more comfortable to wear and can monitor the physiological signals of the body for a long time. These advantages of flexible wearable pressure sensors may facilitate solving the needs of personalized health care (Hu et al., 2023).

2.5. Nano Encapsulation for Controlled Nutrient Release

Encapsulation is the enclosing of one substance in the wall of another substance. The phenomena of encapsulation happened naturally in many filed pharmaceutical, cosmetics, agriculture, food, printing, textile and defense(Liu et al., 2023). Microencapsulation involves the formation of micron-sized capsules that contain active ingredients like food ingredients, enzymes, drugs, etc. that are intended to be released under specific conditions. When the size of the materials is in nano scale they are called nanoencapsulation. Encapsulation both micro and nano-encapsulation are important as they entrap a substance within their shell and protect them from the external harsh conditions, thus helping to extend their stability and life. They also provide the release of materials under specified conditions. Micro and nano-encapsulation is important in the release of drug at specific position and specific time. For instance, if a drug is required to be released in the intestine it is mandatory that the drug must be encapsulated to prevent them being dissolved in the stomach hence encapsulated in a shell that show resistance to low pH of the stomach and can released when the pH goes up as in the intestine (Kumari et al., 2023).

The controlled release of nutrient at specific time and specific location is necessary for different functions. The controlled released is affected by the ratio and preparation of the coating layer. For example, the hydroxyl and carbonyl groups of lignin can generate insoluble chelates and complexes with Fe and Zn metal ions. The slow release mechanism is puzzling to visualize (Dutta et al., 2022). Initially the released were measured by the rate curve of multiple processes, but recently various nanoscale nutrients were found that act through encapsulation, emulsion, entrapment, and surface adsorption. Generally, micronutrient release from a carrier occurs via water penetration, nutrient dissolution, and nutrient release through porous channels (Dutta et al., 2022).

3. NANOTECHNOLOGY IN ANIMAL NUTRITION AND HEALTH

Nanotechnology and Nanobiotechnology is currently used in various applications in biological processes. The term nanotechnology was coined by Richard P. Feynman in late 1950s. due to the unique properties like small size, more active, soluble, stable structure, and less oxidative inactivation leads to their different areas of applications (Khider et al., 2023). Nanotechnology helps in the treatment of animal diseases in therapeutics, diagnostics, tissue engineering, and vaccine production. Nanomaterials and nanotechnology have already been used in animal health and production, animal breeding and reproduction, and animal nutrition. In animals post-weaning diarrhea is observed so the farmer worried and try to (Späti et al., 2021).

The advantages of using nano materials in animal science include low cost, less additives, growth promoting, and immunomodulatory effect. Other benefits include pathogens inhibition, regulation of rumen fermentation, and shows reproductive diseases. Nanomaterials also used to treat the animals disease. Nanozinc has been found to boost milk yield and lower the number of somatic cells in dairy cows with recessive mastitis (Wang et al., 2022). It is documented that proper animal health more depends on their nutrition and nanotechnology has the potential to effects the animal's performance as they are used as supplements. The positive effects of nanomaterials on animal's performance are well established and known. The enhance production are also associated with potent antibacterial, antifungal, antiprotozoal, and even antiviral activity of silver nanoparticles (Ullah et al., 2022). Zinc is also used in animals in the form of nanosulphide, nano zinc-sulphate and nano zinc-methionine and found to improve the growth performance of animals . Nanotechnology can improve food evaluation and act as a new too(Hassan, Haq, Majeed, et al., 2022) for delivering nutrients to target tissues and a tool for explaining the metabolism and physiology of nutrients(Haq et al., 2022a).

The main advantage of nanomaterials treatment is their self-control ability. When it comes to the use of veterinary pharmaceuticals in livestock farms, nanomaterials can transport drugs directly to target cells, reduce drug dosage, drug leftovers and withdrawal time in pasture animals. Liquid vitamins prepared by nanotechnology can be used in poultry feed. This nanoscale vitamin is meant to transfer vitamins and other nutrients straight into the bloodstream via the gastrointestinal tract, boosting bioavailability. NPs can also reduce the need for preservatives and eliminate feed odors that are irritating to animals (Hassan, Haq, Ahmad, et al., 2022). They can also increase nutrient dispersibility and feed durability. Feed ingredients can be microencapsulated to protect them from light and oxidation, to prevent them from being degraded by proteases and other digestive enzymes, and to keep them stable at various pH levels and temperatures. They have superior dispersibility in use, allowing for better mixing of fat-soluble additives in the feed, and extending their service life during storage. Nanotechnology has promise in animal medicine and production, but it needs to be tested further in industrial practice before it can be fully developed and utilized (Wang et al., 2022).

The nanomaterials are also helping the livestock sustainability in the environments and their impacts on various factors. The required criteria that help in the environment's different aspects include the large scale production, automation, and cost effectiveness (Aziz et al., 2023). One of the importance of nanotechnology is the early detection of animal's disease status. Initially if an animal were caught infection it takes several days or even month to appear the symptoms which leads to the wide spreading of the disease effects. At this stage the nanotechnology play a crucial role and help in the early detection of disease pathogens. This is possible by putting a small input device in the animal mouth which can sense the physiological changes in the animals and thus help in the early detection of any animal abnormalities

(Rehman et al., 2023). Nanotechnology is also used in disease treatment in animals, tool for molecular and cellular biology, pathogens detection, and virus detection.

3.1. The Physiological Role of Nanoparticles in Animal's Health

Feed additives are substances which are added to the animal are feed for different purposes. These additives are diverse in nature and comprised of but not limited to probiotics, enzymes, organic substances, and recently the nanomaterials. The special properties of the nanoparticles make them promising substances in different application of veterinary sciences. The antimicrobial properties of nanoelements like zinc, copper, and chromium nanoparticle additives. Figure 2 summarized the key physiological role of nanoparticle in the animal's health.

4. NANOTECHNOLOGY IN DISEASE PREVENTION AND TREATMENT

Nanotechnology and nanoparticles emphasize crucial role in the disease prevention and treatment (Figure 1). Some of the applications of nanotechnology in the prophylaxis and treatment of diseases would include drug delivery target to the desired site, use of imaging and diagnostics techniques, theranostics, antigens and adjuvants delivery to immune cells by immunotherapy, nanosensor use in the biomarker's detection, use of nanoparticle in the treatment of infectious bacterial and fungal diseases, tailored treatments to genetic disorders and repair and treatment of damaged cells. By designing nanoparticles, it becomes possible to deliver drugs to the area of a disease minimizing any negative effects, on healthy tissues. Modifying the surface of nanoparticles allows them to specifically target cells or tissues enhancing the efficiency of drug delivery (Muhammad, Basit, et al., 2024). Utilizing nanoparticles can enhance the sensitivity of imaging techniques like MRI, CT scans and ultrasound resulting in more disease diagnosis. For molecular and cellular level imaging, quantum dots and other nanomaterials can serve as contrast agents. Nanotechnology has made it possible to combine therapy with diagnostics in what's known as theranostics. This allows for real time monitoring of treatment effectiveness. Theranostic nanoparticles not carry agents but also provide imaging signals that help track their distribution and effects(Aziz et al., 2023) . Employing nanoparticles can significantly improve the effectiveness of immunotherapy by delivering antigens, adjuvants or immunomodulatory agents to immune cells. Furthermore, they contribute to enhancing the activation of the system, against cancer cells or pathogens (Hassan, Haq, Ahmad, et al., 2022).

Designing nanosensors enables us to detect biomarkers associated with diseases effectively. This facilitates diagnosis and intervention when necessary. Additionally, these nanosensors can continuously monitor biomarker levels in time. Researchers have been exploring the use of nanoparticles to combat viral infections. These tiny particles can be designed to target and eliminate pathogens while also reducing the risk of resistance (Liu et al., 2023). Nanotechnology offers the possibility of medicine, where treatments can be tailored based on an individual's genetic profile and the specific characteristics of their disease (Loiseau et al., 2023). Nanotechnology has opened up new possibilities, in regenerative medicine enabling innovative approaches to repair and restore damaged tissues and organs

4.1. Nanoparticles for Vaccine Delivery

Nanoparticles have demonstrated great potential in the field of vaccine delivery, providing benefits such as better stability, controlled release, improved immune response, and targeting specific cells or tissues (Hussain et al., 2023a; Kumari et al., 2023). Some highlights about using nanoparticles for vaccine delivery include:

4.2. Better Stability

The nanoparticles protect the antigens of vaccines against degradation and hence enhances their stability when stored or transported. This provides a sustained and prolonged immune response compared to conventional vaccines through nanoparticles that enable controlled release of vaccine antigens (Bashir et al., 2023a). These can also act as adjuvants thereby enhancing immune response towards vaccine antigens. Thus, they can activate both innate and adaptive immunity leading to stronger and long-lasting immunity. Surface modification allows nanoparticles to deliver vaccine antigens in antigen-presenting cells (APCs) or lymph nodes among other target cells or tissues. By mimicking the natural infection process, nanoparticles co-deliver both antigens and adjuvants thus promoting a stronger immune response . Nanoparticles may contain multiple antigens or epitopes allowing the use of multi-valent vaccines that are more effective than single valent ones. Nanoparticles are useful in mucosal vaccine delivery to provide protection at mucosal surfaces where many pathogens enter the body. This is very crucial in the prevention of infections at respiratory, gastrointestinal, or genitourinary sites (Arshad et al., 2022).

4.3. Types of Nanoparticles

Commonly used nanocarriers for vaccine delivery include lipid-based nanoparticles, polymeric nanoparticles, virus-like particles (VLPs), and other nanocarriers. The effectiveness and safety of lipid-based vesicles such as lipid nanoparticles have been widely investigated. Nanoparticles have also been employed in delivering DNA and RNA vaccines. Lipid nanoparticle formulations have been considered for encapsulation and protection of mRNA or DNA to enable their cellular entry and expression of the encoded antigens (Haq et al., 2022b). Pfizer-BioNTech and Moderna's COVID-19 vaccines use lipid nanoparticles to deliver mRNA that contains SARS-CoV-2 spike protein.

4.4. Nanomaterials for Antimicrobial Applications

The applications of nanomaterials for antimicrobial applications include the use of various nanoparticles such as Silver Nanoparticles, Metal Oxide Nanoparticles, Carbon-Based Nanomaterials and Polymeric Nanomaterials (Khoja et al., 2022). These days, silver nanoparticles are among the most of nanomaterials that have been widely studied as antimicrobials. They are particularly effective against a wide range of bacteria, fungi and viruses. By interfering with microbial cell membranes, disrupting cellular processes or causing oxidative stress, they result in the death of microorganisms. Metal oxide nanoparticles including zinc oxide (ZnO) and titanium dioxide (TiO2) exhibit antimicrobial properties. When illuminated, such particles produce reactive oxygen species (ROS), which damage microbial cells through peroxidation. In various instances, coatings and dressings employ metal oxide nanoparticles for their antimicrobial effects (Majeed, Bhatti, et al., 2020). Antimicrobial activity has been found in materials consisting of carbon nanotubes (CNTs) or graphene-based materials. Physical disruption of microbial membranes followed by interactions with components within cells lead to microbial cell death by them. Such carbon-based nanomaterials can be utilized in medical devices, water purification and wound healing. Nanoparticles made from biocompatible polymers like chitosan and poly (lactic-co-glycolic acid) (PLGA) can be antimicrobial substances. Thereby these materials can be applied for drug delivery, wound healing. Textiles can be incorporated with antimicrobial nanoparticles to produce fabrics with enhanced properties. In antimicrobial textiles, silver nanoparticles are usually used to prevent growth of bacteria and fungi (Khan et al., 2022). Healthcare settings, food packaging, and public spaces use nanomaterials to form coatings with antibacterial properties. Preventing the spread of bacteria can be done through these coatings, thus reducing infections (Khoja et al., 2022). Wound healing and topical applications are some areas where hydrogels incorporating antimicrobial nanoparticles are useful.

Additionally, these nanohydrogels can be programmed to release antimicrobial agents on demand, accelerating healing while preventing infection.

5. NANOSCALE IMAGING AND DIAGNOSTICS IN DIAGNOSTICS IN VETERINARY MEDICINE

The field of nanoscale imaging and diagnostics holds significant promise for veterinary medicine, offering advanced methodologies that can revolutionize disease detection, monitoring, and research. It is an area characterized by rapid growth and development, with numerous avenues being explored in this burgeoning field (Haq et al., 2022a). One prominent application lies in the use of nanoparticle imaging agents. These nanoparticles function as contrast agents across various imaging techniques such as magnetic resonance imaging (MRI), computed tomography (CT), and ultrasound. By leveraging these nanostructures, veterinarians can diagnose a wide range of diseases afflicting animals including cancer, cardiovascular disorders, and neurological conditions (Majeed et al., 2023). Another intriguing area involves the utilization of nanoparticle-based biosensors. Scientists are actively working on developing innovative biosensor systems employing nanomaterials to detect specific biomarkers associated with various diseases in animals. By enabling early disease detection along with continuous monitoring of physiological parameters within animal subjects themselves utilizing these sophisticated sensor platforms.

Nanotechnology has also made its mark on veterinary pathology through the introduction of nanoparticles into diagnostic protocols thereby improving resolution quality while increasing sensitivity levels significantly compared to conventional methods alone This enhanced visualization capability empowers pathologists to better discern tissue structures cellular changes or pathological abnormalities during their examination providing valuable insights ultimately leading towards accurate diagnoses .Moreover Nanoparticles have proved invaluable when it comes to conducting non-invasive studies aimed at understanding biological processes occurring at both molecular cellular levels Many researchers utilize tailored designed particles perform real-time tracking experiments visualizing events otherwise impossible observe Furthermore they serve useful tools studying disease progression evaluating treatment efficacy investigating fundamental physiological mechanisms play role animal health overall. Lastly, researchers have successfully harnessed power nanoparticles facilitate drug delivery targeted therapy approaches Animal patients benefit from highly efficient precisely controlled administration therapeutic substances capable addressing illnesses more effectively minimizing side effects commonly encountered traditional medication regimes (Majeed et al., 2023).

5.1. Diagnostic Imaging Using Diagnostic Imaging Using Nanotechnology

Nanotechnology has revolutionized the field of diagnostic imaging by incorporating nanomaterials and advanced techniques to enhance the precision, sensitivity, and specificity of medical imaging methods. One notable application is molecular imaging with nanoparticles, which involves designing nanoparticles that can specifically target biomolecules or cells for precise molecular visualization. This breakthrough technology assists in early disease diagnosis by detecting and visualizing intricate molecular processes (Majeed, Lu, et al., 2022). Another significant role played by nanoparticles in diagnostic imaging is their util zation as contrast agents. These tiny particles are engineered to improve tissue visibility across diverse imaging modalities such as magnetic resonance imaging (MRI) and computed tomography (CT) scans. By serving as contrast agents, nanoparticles contribute to superior image quality while aiding in the identification of abnormalities within anatomical structures In MRI applications, nanoparticle-based contrasts significantly boost spatial resolution and sensitivity levels compared to conventional approaches. As a result, these enhanced capabilities facilitate accurate detection and characterization of tumors along with effective vascular imagery (Majeed, Khan, et al., 2022). Likewise, when utilized during CT scans, nanoparticles act as efficient contrasting materials capable of producing high-resolution images. The targeted use enables improved visualization not only for soft tissues but also blood vessels while enabling specific organ-focused screenings. Moreover, ultrasound imaging can benefit tremendously from nanoparticle integration through their ability to optimize acoustic properties within tissues. They impart enhanced clarity on ultrasound images pertaining to tumor observation alongside facilitating comprehensive examination concerning blood flow patterns at an unprecedented level. Through continuous advancements in nanotechnology-driven diagnostic procedures encompassing various forms of medical imagining modalities like MRI's heightened resolution abilities facilitated via nanoparticles acting as qualified contrasting components it becomes abundantly clear that this cutting-edge methodology holds immense potential for improving healthcare outcomes overall.

5.2. Nanoscale Imaging for Productive Health

Nanoscale imaging methods are getting more and more important in reproductive health. They're giving us new insights into the cellular and molecular processes tied to having babies, development, and reproductive issues (Sinha et al., 2023a). Nanoscale imaging lets us see inside reproductive cells—their structures, little organs and molecular processes. And helps us in understanding how the cells work, form gametes, and fuse during fertilization. Nanoparticles could be used for imaging and getting drugs to conditions like endometriosis, ovarian cancer and other gynaecological problems and finds its applications in targeted imaging and treatment in the female reproductive system. Nanoscale Imaging for Male Reproductive Health helps us in analyzing the super tiny structure of sperm cells and the male reproductive system thus figuring out sperm movement, shape, and interactions with reproductive tissues. Nanoscale Imaging of Ovulation and Fertilization studies the ovulation and fertilization happen at the nanoscale thereby investigating the molecular events that lead to successful baby-making. Improving imaging used in ART, including IVF and ICSI using nanotechnology helps women to get pregnant by enhancing the techniques (Majeed, Bhatti, Pieroni, et al., 2021).

6. NANOTECHNOLOGY IN LIVESTOCK WASTE MANAGEMENT

6.1. Nanomaterials for Waste Treatment and Odor Control

Maintaining hygienic environment for the livestock is the most crucial field of application of nanotechnology. As there is drastic increase in harmful diseases in animals in recent times, maintaining a hygienic environment for the livestock by managing their waste production is crucial. Livestock is known to be one of the main sources of emission of greenhouse gases (GHG) especially methane and Hydrogen sulfide and volatile organic compounds (VOCs) accounting to 26% of the emission from livestock manure. Globally, research is undergoing to find various treatment methods like manipulation in feed content, application of catalytic processes addition of microbial additives (Sinha et al., 2023b), anaerobic digestion (Waheed et al., 2022), application of probiotics, acetogens, bacteriocins, and organic acids (Boadi et al., 2004) etc., for mitigating GHG and VOCs from livestock farming but found not so effective. Recently, a number of nanoparticles (NPs) or nanomaterials found to effective in the mitigation of gaseous emissions from animal wastes. NPs are used in the industrial sector for the removal of trace amount of pollutants (Tassadduq et al., 2022). Similarly, NP application may help in GHG mitigation which is not found to be controlled by conventional methods (Hussain et al., 2023b).

Among different nanoparticles, zinc oxide nanoparticles (nZnO) and copper oxide nanoparticles (nCuO) have high inhibitory effect on methane production (Luna-delRisco et al., 2011, Mu et al., 2011). About 19%–77% of methane reduction is reported from waste activated sludge in comparison with a control (Majeed, Tariq, et al., 2022). When nZnO is mixed with swine manure slurry, a reduction of the concentration of methane and hydrogen sulphide by 54% and 98% respectively is recorded (Majeed, Tariq, et al., 2021). nZnO compared to zirconium oxide NP (nZrO2) at application rates of 100, 250, 500 mg/L and 3 g/L in swine manure or dairy manure shown that nZnO is more effective than nZrO2 in combating methane and hydrogen sulphide compared to a control (Muhammad, Waheed, et al., 2024).NPs are effective in reducing gaseous emissions either by directly absorbing gases, destroying gas-producing microorganisms or changing the dangerous pollutants into more acceptable chemicals by different chemical reactions. NPs are applied in different forms like liquid, gel, or powder depending on the substrate to be treated (Ali et al., 2023). The effectiveness of nanotechnology in the management of livestock waste depends on the selection of nanomaterials suitable for the given contaminants or pollutants and the existing environmental conditions (Waheed et al., 2023).

Even though various nanomaterials have been produced in laboratory-scale conditions so far, only few of them have been finally manufactured at industrial scale. Metal (mostly iron) and carbon-based nanomaterials are presently among the most commonly used, while zero-valent iron (nZVI) is the most extensively used nanomaterial in the environmental protection industry.

Several studies have shown that nZVI is a very effective material in the reductive degradation of organic contaminants (e.g. chlorinated hydrocarbons). In addition, owing to its very high surface area, nZVI can act as an adsorbent for heavy metals (e.g. hexavalent chromium), which are transformed to non-toxic forms via redox reactions occurred on nZVI surface (Sinha et al., 2023c). Nano-sized titanium dioxide is another common metal-based nanomaterial. These nanoparticles are mainly used as photocatalysts owing to their high photosensitivity, but also as adsorbents for heavy metals. When this nanomaterial is exposed to ultraviolet (UV) light (sunlight) it produces hydroxyl radicals, which are highly reactive and can oxidise the contaminants. Owing to the production of hydroxyl radicals they exhibit strong antimicrobial activity. The low toxicity and the low production cost are two more advantages of titanium dioxide nanoparticles.Among carbon-based nanomaterials, the carbon nanotubes (CNTs) are the most common. CNTs are allotropes of carbon with a cylindrical nanostructure and, depending on their manufacturing process, CNTs can be categorised as single- or multi-wall CNTs. CNTs exhibit high specific surface area and are characterized by highly assessable adsorption sites. Their adjustable surface chemistry (mainly owing to their tubular structure) and the capability for surface functionalisation enhance their sorption capacity. As a result, CNTs act mainly as adsorbents for heavy metals, polar and non-polar organic compounds and oils. The regeneration and the reuse of CNTs is their main advantage compared with other nanomaterials and obviously towards other conventional carbonaceous products like activated carbon (Yuvaraj et al., 2023).

Nanomaterials can be also used in ex-situ waste management technologies. These technologies are commonly based on processes such as adsorption, membrane filtration and separation and photocatalysis. Owing to their size, nanoparticles exhibit high surface area enhancing, thus, the adsorption of pollutants on their surface. As with regards to membrane filtration and separation, Membranes act as barriers depending on their pore size and the contaminant molecular size. Nanomaterials can be applied for creating nanocomposite membranes either by their embedment in the membrane matrix or by the deposition on the surface. The addition of nanomaterials, such CNTs, in polymeric membranes results in improving the mechanical properties of the membrane and increases the water permeability and the resistance to fouling (Karnwal et al., 2023).

A greater portion of meat is wasted while a very small percentage consumed which lead to major solid waste deposition in environment. It is estimated that approximately 50–54% of each cow, 52% of each sheep or goat, 60–62% of each pig, 68–72% of each chicken, and 78% of each turkey is consumed by humans while the remaining portion goes as waste (Nguyen et al., 2023). It is cumbersome process to dispose the animal remains after evisceration (the removal of the body's interior organs such as the heart, lungs, intestines and kidneys). Although the litter collected in slaughterhouses is biodegradable, but due to irresponsible disposition lead it to fester, producing foul smell and land pollution. It is a big environmental issue globally and the solution lies in recycling such waste into something useful. Nanotechnology is also helpful in such situation. Useful nanoparticles like zinc oxide can be synthesized from such wastes (Panda et al., 2023).

7. ETHICAL AND REGULATORY CONSIDERATIONS

7.1. Ethical Implication of Nanotechnology in Livestock Farming

Livestock farming and nanotechnology is likely to spark some of the ethical and social concerns that genetic engineering and biotechnology have for animal production (Y. Li et al., 2023). Animal welfare, safety of animal-derived products, risks to the environment and human health, and industry consolidation are among the many concerns that are likely to extend from biotechnology to nanotechnology. Furthermore, for many of the newest nanotechnology products proposed for livestock production, biotechnology and nanotechnology are inseparable, and the two converge in a particular products or applications. It is a very young field, and the impact of nanomaterials on human bodies and the environment, especially in the long term, it is not always known. Nanomaterials may exhibit enhanced biological activity as they easily penetrate biological membranes. Nanoparticles may therefore exhibit toxic properties, posing a threat to humans and animals. Currently, the largest risk associated with nanotechnology is connected to the lack of control over it, there are no technical means to monitor for example the environment for the presence and impact of nanoparticles and nanomaterials. A new concept called 'Nano-ethics', which is defined as moral reflection on the development and application of nanotechnology, or the manipulation of matter at the molecular level. In this regard, dilemmas regarding the potential harmfulness of nanoparticles are beginning to appear. In the case of nanotechnology, questions are raised about their availability and impact on the welfare of mankind. There arises a justified fear that nanotechnology will become another element strengthening the division of the world into developed countries and the ones lagging behind or these in control and the controlled ones (Periakaruppan et al., 2023).The ethical questions related to the process of research and testing nanotechnology include probably the most pressing issue – the toxicity of some nanoparticles. For example, nanoparticles can penetrate blood-brain barrier. Some experts, however, point out the fact that, as a result, the nanoparticles may accumulate in the bodies of living organisms and cause complications (Malik et al., 2023). On the other hand, nanocolloids do not tend to settle, and because of that their concentration in the air may not change in nature, what increases the probability of contamination. The ethical issues also include the problem of the intellectual property of a nanoscientist, who encounters problems with patenting. Particularly in countries, where a possibility to register a patent 6-12 months after the publication of results does not exist, the chances to get a patent disappear due to competitive interests of the researcher to publish (Fujita et al., 2023). During the production of the conveniences of nanotechnology, a question arises of how to ensure that they are not being reduced to an object of enrichment for companies that might not pay attention to the moral dimension of the nanotechnology application. At present, we see products containing nanoparticles appear on the market without having to comply with control criteria or without containing warnings. While individual companies embrace the dimension of responsibility, we need an engagement of the governments and other public stakeholders, but there is a question of who is responsible for the decisions in this area. Other ethical problem related to the sale of nanotechnologies, is a privilege of the rich, who can afford better healthcare and products of higher quality. In this case, the key moral category is justice and it is necessary to define criteria for the redistribution of the conveniences of nanotechnology (Lamichhane et al., 2024).

In the context of nanoproducts disposal, we may talk about moral issues connected with responsible removal of the dangerous substances. It is important to establish an effective control by individual governments and it is also appropriate to emphasize prevention (Bhatia et al., 2024).

7.2. Regulatory Frameworks and Challenges

At the international level, the European Union has been consistently working on regulation nanotechnology. The current regulations, as regards the application of nanomaterials, consist mainly of two regulations, the so-called REACH (Registration, Evaluation and Authorisation of Chemicals) and CLP (Classification, Labelling and Packaging) (Sangeeta Dash et al., 2023b). The provisions on nanomaterials can also be found in sectoral regulations. The regulations concerned the use of biocides, cosmetic products, transmission of information to consumers about food and food additives. Since nanotechnologies are also used in medicine, a Directive on the Community Code relating to medicinal products for human use appeared in 2001. The procedures for the authorization of medicinal products have also been defined. To ensure safety it is advisable to compile a register of nanomaterials and products containing nanomaterials, such a register will facilitate the supervision of entities introducing nanomaterials to the market and ensure the transparency of data on products for buyers. At the moment such an obligation is relevant for cosmetic products (Goyal et al., 2023).

In the United States of America in 2008, the US National Research Council issued a report calling for greater regulation of nanotechnology. Australia and Canada are also rather active on nanoregulation. Both have important programmes on Environmental Health and Safety (EHS) research and have published in-depth reviews of their regulations to assess eventual limits when dealing with nanotechnology. Even though no specific laws have been set up, the adoption of a precautionary approach principle, when dealing with nanotechnology application, is envisaged in both countries (Rimpika et al., 2023). In Japan, China, Korea and Taiwan, which are deeply involved in nanotechnology, there are also, at different levels, important research initiatives dealing with EHS issues such as risk assessment and risk management of nanomaterials and nano-related products. They do participate in the worldwide debate on nanoregulation but no specific initiatives on the matter have been taken so far in these countries. Currently, the actions to regulate nanotechnology focus on the national level, and at the international level initiatives of the nature of joint research programs are observed (Bashir et al., 2023b)

Regulatory bodies such as the United States Environmental Protection Agency and the Food and Drug Administration in the U.S. or the Health and Consumer Protection Directorate of the European Commission have started dealing with the potential risks posed by nanoparticles. So far, neither engineered nanoparticles nor the products and materials that contain them are subject to any special regulatsion regarding production, handling or labelling.There has been a different approach to the matter of regulation of nanotechnology. Some authors see the possibility of influencing the strengthening of the safety of use of nanomaterials, through non-judicial means (Mubarok et al., 2024). Due to the dynamics of change in this area, some people see an urgent need for the use of soft law (Liao et al., 2023). Soft law includes resolutions, guidelines, explanations, messages, programs, plans, mostly issued by the authorities involved in the legislative processes, the result of which is a generally applicable legislative act. Construction of soft law instruments is intended to constitute a foundation for the future hard legal regulations. Also important are the voluntary programs aimed at gathering information useful for the design of legal instruments (Balasubramaniam et al., 2023).

The biggest challenge at the moment is primarily related to the establishment of the validated methods and instruments for the detection, characterization and analysis of nanomaterials, development of complete information on the risks associated with nanomaterials and the development of methods to assess the exposure to nanomaterials. Despite the studies conducted so far, we are still not able to quantitatively assess the risk involved. The toxicity of individual nanoparticles varies greatly, making it difficult to establish a common criterion. Evaluation of the toxicity of nanoparticles is carried out with respect to individual organisms: mammals, protozoa, shellfish, algae and plants. Toxicity to mammals is tested on rodents. There were few attempts to assess the health status of people occupationally exposed. Sparse studies on human cell lines revealed distinct DNA damage (Swain et al., 2023). Due to the lack of full knowledge of the risks posed by nanomaterials, it should be covered by multi-faceted and diverse legislation. The new provisions must be constructed on the basis of the precautionary principle and the principle of producer responsibility in order to ensure safe production, use and disposal of nanomaterials before they are placed on the market. The precautionary principle enables quick reaction in the face of possible risks to the health of humans, animals or plants, or for the protection of the environment. According to the Commission, the precautionary principle can be relied on if the phenomenon, product, or activity poses potential security risks, identified within the framework of scientific and objective evaluation, if such an assessment allows the identification with sufficient certainty of such a threat.A difficult issue is the ability to regulate nanotechnology at the international level, under the Future Framework Agreement (Fadiji et al., 2023b). Some authors are unanimous about the fact that the regulation of nanotechnology at the international level is a big challenge because of the different ways of the use of nanomaterials. It seems, however, that in the future comprehensive regulations in the field of nanotechnology will be created. It may be significantly affected by the creation of the best practices for dealing with nanomaterials, which will be applied at the international level (Ahmad et al., 2023).

7.3 Public Perception and Awareness

Although nanotechnology is viewed as one of the key technologies of the twenty-first century and has major potential benefits, it has to be embraced with a precautionary measure, given that not much is known about its unintended effects on account of being new. Despite there being a lot of applications for nanotechnology in many fields and increasingly more applications are being employed in the field of agriculture, the general global population seems to know little about nanotechnology. Interestingly, however, as reported by, a large proportion of the US and the European public have equally very limited knowledge about nanotechnology. These researchers concluded that despite the US public possessing little knowledge, a majority believed that benefits of nanotechnology outweigh the risks as compared with the majority of the European public who viewed nanotechnology with less optimism.It should be quite obvious that if public knowledge of nanotechnology in such highly advanced societies where literacy levels are relatively high is limited, then the situation is worse in other parts of world that are also grappling with high illiteracy levels. As expected, the level of knowledge of nanotechnology is much higher among the highly educated than those with least education.When introducing nanotechnologies into the practice, we may encounter moral issues arising from the need to inform the public. Considering some of the unpredictable consequences of nanotechnologies, it is complicated to provide users with the relevant data, and it is not possible to impose a moratorium until we understand all the dangers, because a lot of countries have already invested a lot of financial resources in the nanotechology industry (Dehghanian et al., 2023). Therefore, it is arguable to what extent the unresearched innovation should be used and how to inform the consumers. The problem is more visible in the areas, where the informed consent for the use of nanotechnology is needed, for example in some spheres of nanomedicine. In this context, it is necessary to mention another ethical dimension of nanotechnology, since its users can influence their health condition or capabilities. Liberty is directly proportional to the level of responsibility and it is necessary to define a framework to determine the admissible activities in this area.

When people know little about a technology, their perception and acceptability will to a large extent depend on how social and ethical issues concerning the technology are handled by industry and researchers.

As stressed by (Dorjee et al., 2023), when knowledge is missing, people can use heuristics, such as trust, to assess the risks and benefits of a new technology. Thus, people are more likely to accept assurances about the safety of nanotechnologies, including nanotechnology in agriculture when they have higher levels of trust in the institutions, researchers and industry, emanating particularly legislative history.

Another aspect that affects public perception about nanotechnology is the way the media reports issues on the technology. In less developed countries, the level of coverage of nanotechnology is very low and this is coupled with low levels of research in the technology. For developed countries with high levels of application of nanotechnology, the reporting or coverage of nanotechnology may be modest probably due to the specialized nature of the field and hence requiring specialist journalists who may be fewer (Ahmed et al., 2023).One way to involve the public and key stakeholders is through upstream dialogue. Scholars have called for upstream public engagement (UPE) to involve the public in discussions about emerging technological products well prior to market entry (Wilsdon and Willis, 2004). UPE can be complemented by multiple other upstream endeavors, including real-time technology assessment (RTAA), whereby engineers and scientists consider the social consequences of their work prior to the development of products from it (Liu et al., 2023; Loiseau et al., 2023). In addition, upstream oversight assessment (UOA) has been proposed as a targeted activity that focuses on assessing the technical features of R&D projects, examining potential risks and benefits, exploring data needs for addressing the them, and suggesting societal oversight challenges and opportunities.

Based on the factors that influences the perception and awareness of the nanotechnology in agriculture, there are quite a number of ethical issues that arise. In the face of the potential risks that nanotechnology in agriculture pose to human health and environment, should the industry continue using these nanosystems despite the uncertainty? Should there be a mandatory requirement for labeling of nanoenabled agricultural products, particularly food stuffs? Is it ethical that public/government institutions should continue funding development of nanotechnology products for use in agriculture despite the current levels of uncertainty? How much information should the public be made aware in relation to the nanotechnology in the whole agricultural value chain? Is it ethical for an industry to release nanosystems for use in agriculture to the public who have no idea about the potential negative impact on their health? Should there be regulations set in regulating nanotechnology in agriculture to increase public perception and acceptability? If these issues/questions are not fully addressed while the nanotechnology in agriculture is still in the development stage, the negative perception and hence reluctance of acceptance of this technology, similar to what was witnessed to genetically modified (GM) food stuffs, particularly in the European Union region may be experienced again.

7.4 Future Directions

As nanotechnology continues to develop and garner more attention, its applications in the animal production industry will become more expansive. The regular inclusion of nano-supplements to fortify livestock feed is likely possible in the near future; however, it will take longer for nanoparticles to fully replace antibiotics in feed as many biocidal candidates must still be tested in vivo before undergoing clinical trials and food safety tests in accordance with government regulations. External uses for nanoparticles have already been integrated into some aspects of animal production, i.e. antiseptic wound dressings, and more are to follow (Khalil et al., 2023). For studies interested in nanoparticles with anti-cancer properties, it is important to investigate nanoparticle cytotoxicity in both cancer cell lines and normal, healthy cell lines. Only using cancer cells and claiming the nanoparticle under investigation has anti-cancer properties may be misleading, as the nanoparticle may be cytotoxic to all cell types. In vivo studies are needed for verification of nanoparticle functions seen in in vitro research.

While the process of introducing a new drug to market is costly and complicated, veterinary drugs and medicine must be cost-effective and affordable, human medicine is subsidized by the national health department in most countries and animal owners must pay the full price of prescription drugs for the treatment of their animals. Furthermore, the market for veterinary animals is less than that for human health medicine. In addition to that species-specific drug formulation and delivery methods are required for livestock. In the future, nanotechnology will reform veterinary science and technology and increase livestock production to fill knowledge gaps for the welfare of livestock. Despite tremendous applications, nanotechnology also has some limitations in drug delivery systems, diagnosis, bioimaging and food safety concerns because it requires highly sensitive, degradable, affordable, cost-effective devices and less toxic nanoparticles. However, more research is needed to confirm the toxicity of nanoparticles.

8. CONCLUSION

One of the most cutting-edge technologies nowadays is nanotechnology. This technology has a plethora of uses. The field of nanotechnology, also known as nano science, holds so much promise that it has been dubbed "the next industrial revolution." It will have a significant impact on global politics as well as society, the environment, and animals. Future developments in nanotechnology could lead to the development of therapeutic nanoparticles, rapid, accurate, and less expensive drug delivery systems, as well as diagnostic equipment made of nanomaterials and nanomedicine. These nanoparticles are essential for monitoring, imaging, and earlier and more rapid illness detection. Those nanomaterials can be used to identify diseases both at the molecular and single-cell levels. It offers efficient medication administration and therapeutic results by Numerous new antibacterial goods have been developed thanks to the wide range of nanoparticle engineering. Outstanding efficacy outcomes; prompt and focused response, as well as high Levels of biodegradability and bioavailability. It impacts advancements. on improved animal health conditions and decreased production/economic losses. Additionally, nanotechnology plays a significant role in animal products like milk and meat. In the future, it will offer new goods and procedures with the aim of improving the product's performance, extending its freshness and shelf life, and raising the standard and safety of food derived from animals. To boost the production of low-fat and low-cholesterol animal products, ongoing research on nanotechnology is necessary, especially on nano biotechnology in veterinary medicine and animal science. Further investigation into nanoparticles, nanomaterials, and nanomedicine is necessary to enhance the efficiency of animal disease diagnosis and therapy. It is imperative that governments prioritise the study of nanotechnology and provide support to scientists and researchers working in this field. Although they are still relatively new in veterinary medicine and animal production, nanoparticles have long been employed as therapeutic and diagnostic tools in the human medical area. Because of the growing worry over microbial antibiotic resistance, the cattle industry has recently faced increased production demands that revolve around the use of antibiotics as growth promoters. Laws and regulations are being changed to stop the use of antibiotics in animal feed, as several nations are reporting an increase in the frequency of antibiotic-resistant bacteria. This establishes the need that appropriate substitutes be developed and added to feed. Numerous studies have demonstrated indications that nanoparticles could be promising candidates for antimicrobials and for promoting animal growth. The state of nanotechnological uses in animal production today and their future directions will. The need for livestock production has grown dramatically on a global scale in recent years, mostly as a result of rising population and market demand. In the end, farm automation managing livestock with a minimal labour and expense approach requires novel technologies. The preservation of animal health, which is closely correlated with yield, is a crucial aspect of farm management. The topic of animal monitoring has seen the emergence of several sensor techniques, one of which being nanotechnology.

REFERENCES

Ahmad, N., Naeem, M., Ali, H., Alabbosh, K. F., Hussain, H., Khan, I., Siddiqui, S. A., Khan, A. A., & Iqbal, B. (2023). From challenges to solutions: The impact of melatonin on abiotic stress synergies in horticultural plants via redox regulation and epigenetic signaling. *Scientia Horticulturae*, *321*, 112369. doi:10.1016/j.scienta.2023.112369

Ahmed, T., Luo, J., Noman, M., Ijaz, M., Wang, X., Masood, H. A., Manzoor, N., Wang, Y., & Li, B. (2023). Microbe-mediated nanoparticle intervention for the management of plant diseases. *Crop Health*, *1*(1), 3. Advance online publication. doi:10.1007/s44297-023-00006-9

Al-Sahli, S. A., Al-Otibi, F., Alharbi, R. I., Amina, M., & Al Musayeib, N. M. (2024). Silver nanoparticles improve the fungicidal properties of Rhazya stricta decne aqueous extract against plant pathogens. *Scientific Reports*, *14*(1), 1297. doi:10.1038/s41598-024-51855-5 PMID:38221517

Ali, H., Muhammad, Z., Majeed, M., Aziz, R., Khan, A., Mangrio, W. M., Abdo, H. G., Almohamad, H., & Al Dughairi, A. A. (2023). Vegetation diversity pattern during spring season in relation to topographic and edaphic variables in sub-tropical zone. *Botanical Studies*, *64*(1), 25. Advance online publication. doi:10.1186/s40529-023-00398-5 PMID:37716923

Ansari, M. A. (2023). Nanotechnology in Food and Plant Science: Challenges and Future Prospects. In Plants (Vol. 12, Issue 13). Multidisciplinary Digital Publishing Institute (MDPI). doi:10.3390/plants12132565

Arif, U., Bhatti, K. H., Ajaib, M., Wagay, N. A., Majeed, M., Zeb, J., Hameed, A., & Kiani, J. (2021). Ethnobotanical indigenous knowledge of Tehsil Charhoi, District Kotli, Azad Jammu and Kashmir, Pakistan. *Ethnobotany Research and Applications*, *22*. Advance online publication. doi:10.32859/era.22.50.1-24

Arshad, F., Waheed, M., Harun, N., Fatima, K., Khan, B. A., Fatima, K., Abbas, Z., Jabeen, S., & Majeed, M. (2022). Indigenous farmer's perception about fodder and foraging species of Semi-arid lowlands of Pakistan: A case study of District Kasur, Pakistan. *Taiwania*, *67*(4), 510–523. doi:10.6165/tai.2022.67.510

Azeem, M., Pirjan, K., Qasim, M., Mahmood, A., Javed, T., Muhammad, H., Yang, S., Dong, R., Ali, B., & Rahimi, M. (2023). Salinity stress improves antioxidant potential by modulating physio-biochemical responses in Moringa oleifera Lam. *Scientific Reports*, *13*(1), 2895. Advance online publication. doi:10.1038/s41598-023-29954-6 PMID:36807545

Aziz, A., Anwar, M. M., Majeed, M., Fatima, S., Mehdi, S. S., Mangrio, W. M., Elbouzidi, A., Abdullah, M., Shaukat, S., Zahid, N., Mahmoud, E. A., Casini, R., Yessoufou, K., & Elansary, H. O. (2023). Quantifying Landscape and Social Amenities as Ecosystem Services in Rapidly Changing Peri-Urban Landscape. *Land (Basel)*, *12*(2), 477. Advance online publication. doi:10.3390/land12020477

Balasubramaniam, T., Shen, G., Esmaeili, N., & Zhang, H. (2023). Plants' Response Mechanisms to Salinity Stress. In Plants (Vol. 12, Issue 12). MDPI. doi:10.3390/plants12122253

Banerjee, S., Mondal, S., Islam, J., Sarkar, R., Saha, B., & Sen, A. (2024). Rhizospheric nano-remediation salvages arsenic genotoxicity: Zinc-oxide nanoparticles articulate better oxidative stress management, reduce arsenic uptake, and increase yield in Pisum sativum (L.). *The Science of the Total Environment*, *913*, 169493. doi:10.1016/j.scitotenv.2023.169493 PMID:38151134

Basha, S. (2019). Future Prospects of Nanotechnology Innovations in Animal Production. *Egyptian Journal of Animal Production*, *57*(1), 67–71. doi:10.21608/ejap.2019.98191

Bashir, S. M., Altaf, M., Hussain, T., Umair, M., Majeed, M., Mangrio, W. M., Khan, A. M., Gulshan, A. B., Hamed, M. H., Ashraf, S., Amjad, M. S., Bussmann, R. W., Abbasi, A. M., Casini, R., Alataway, A., Dewidar, A. Z., Al-Yafrsi, M., Amin, M. H., & Elansary, H. O. (2023a). Vernacular Taxonomy, Cultural and Ethnopharmacological Applications of Avian and Mammalian Species in the Vicinity of Ayubia National Park, Himalayan Region. *Biology (Basel)*, *12*(4), 609. Advance online publication. doi:10.3390/biology12040609 PMID:37106809

Bashir, S. M., Altaf, M., Hussain, T., Umair, M., Majeed, M., Mangrio, W. M., Khan, A. M., Gulshan, A. B., Hamed, M. H., Ashraf, S., Amjad, M. S., Bussmann, R. W., Abbasi, A. M., Casini, R., Alataway, A., Dewidar, A. Z., Al-Yafrsi, M., Amin, M. H., & Elansary, H. O. (2023b). Vernacular Taxonomy, Cultural and Ethnopharmacological Applications of Avian and Mammalian Species in the Vicinity of Ayubia National Park, Himalayan Region. *Biology (Basel)*, *12*(4), 609. Advance online publication. doi:10.3390/biology12040609 PMID:37106809

Bhatia, D., Paul, S., Acharjee, T., & Ramachairy, S. S. (2024). Biosensors and their widespread impact on human health. *Sensors International*, *5*, 100257. Advance online publication. doi:10.1016/j.sintl.2023.100257

Bhattacharya, R., Bhattacharya, R., Majeed, M., Bhandari, S., Aziz, R., Sinha, D., & Niyogi, S. S. (2023). Environmental Rehabilitation of Industrial Waste Dumping Site. In *Biohydrometallurgical Processes* (pp. 44–89). CRC Press., doi:10.1201/9781003451457-4

Biswas, P., Polash, S. A., Dey, D., Kaium, M. A., Mahmud, A. R., Yasmin, F., Baral, S. K., Islam, M. A., Rahaman, T. I., Abdullah, A., Ema, T. I., Khan, D. A., Bibi, S., Chopra, H., Kamel, M., Najda, A., Fouda, M. M. A., Rehan, U. S. M., Mheidat, M., ... Hasan, M. N. (2023). Advanced implications of nanotechnology in disease control and environmental perspectives. *Biomedicine and Pharmacotherapy*, *158*, 114172. doi:10.1016/j.biopha.2022.114172 PMID:36916399

Chandra, J., Hasan, N., Nasir, N., Wahab, S., Thanikachalama, P. V., Sahebkar, A., Ahmad, F. J., & Kesharwani, P. (2023). Nanotechnology-empowered strategies in treatment of skin cancer. *Environmental Research*, *235*, 116649. doi:10.1016/j.envres.2023.116649 PMID:37451568

Chaupoo, A. S., Dhakad, A., Beese, S., Kumar, A., Kumar, P., Pandey, S. K., & Kurdekar, A. K. (2023). Transforming Horticulture: The Influence of Nanotechnology on Crop Enhancement. *International Journal of Environment and Climate Change*, *13*(11), 3381–3400. doi:10.9734/ijecc/2023/v13i113512

Chehelgerdi, M., Chehelgerdi, M., Allela, O. Q. B., Pecho, R. D. C., Jayasankar, N., Rao, D. P., Thamaraikani, T., Vasanthan, M., Viktor, P., Lakshmaiya, N., Saadh, M. J., Amajd, A., Abo-Zaid, M. A., Castillo-Acobo, R. Y., Ismail, A. H., Amin, A. H., & Akhavan-Sigari, R. (2023). Progressing nanotechnology to improve targeted cancer treatment: overcoming hurdles in its clinical implementation. In Molecular Cancer (Vol. 22, Issue 1). BioMed Central Ltd. doi:10.1186/s12943-023-01865-0

Dehghanian, Z., Asgari Lajayer, B., Biglari Quchan Atigh, Z., Nayeri, S., Ahmadabadi, M., Taghipour, L., Senapathi, V., Astatkie, T., & Price, G. W. (2023). Micro (nano) plastics uptake, toxicity and detoxification in plants: Challenges and prospects. *Ecotoxicology and Environmental Safety*, *268*, 115676. doi:10.1016/j.ecoenv.2023.115676 PMID:37979355

Deng, J., Zhao, S., Liu, Y., Liu, C., & Sun, J. (2021). Nanosensors for Diagnosis of Infectious Diseases. *ACS Applied Bio Materials*, *4*(5), 3863–3879. doi:10.1021/acsabm.0c01247 PMID:35006812

Dorjee, L., Gogoi, R., Kamil, D., Kumar, R., Mondal, T. K., Pattanayak, S., & Gurung, B. (2023). Essential oil-grafted copper nanoparticles as a potential next-generation fungicide for holistic disease management in maize. *Frontiers in Microbiology*, *14*, 1204512. Advance online publication. doi:10.3389/fmicb.2023.1204512 PMID:37485521

Dutta, S., Pal, S., Panwar, P., Sharma, R. K., & Bhutia, P. L. (2022). Biopolymeric Nanocarriers for Nutrient Delivery and Crop Biofortification. *ACS Omega*, *7*(30), 25909–25920. doi:10.1021/acsomega.2c02494 PMID:35936412

Fadiji, A. E., Yadav, A. N., Santoyo, G., & Babalola, O. O. (2023). Understanding the plant-microbe interactions in environments exposed to abiotic stresses: An overview. *Microbiological Research*, *271*, 127368. doi:10.1016/j.micres.2023.127368 PMID:36965460

Fujita, H., Ohta, S., Nakamura, N., Somiya, M., & Horie, M. (2023). Progress of Endogenous and Exogenous Nanoparticles for Cancer Therapy and Diagnostics. In Genes (Vol. 14, Issue 2). MDPI. doi:10.3390/genes14020259

Goyal, V., Rani, D., Mehrotra, S., Deng, C., & Wang, Y. (2023). Unlocking the Potential of Nano-Enabled Precision Agriculture for Efficient and Sustainable Farming. In Plants (Vol. 12, Issue 21). Multidisciplinary Digital Publishing Institute (MDPI). doi:10.3390/plants12213744

Grillo, R., Fraceto, L. F., Amorim, M. J. B., Scott-Fordsmand, J. J., Schoonjans, R., & Chaudhry, Q. (2021). Ecotoxicological and regulatory aspects of environmental sustainability of nanopesticides. *Journal of Hazardous Materials*, *404*, 124148. doi:10.1016/j.jhazmat.2020.124148 PMID:33059255

Haq, S. M., Tariq, A., Li, Q., Yaqoob, U., Majeed, M., Hassan, M., Fatima, S., Kumar, M., Bussmann, R. W., Moazzam, M. F. U., & Aslam, M. (2022). Influence of Edaphic Properties in Determining Forest Community Patterns of the Zabarwan Mountain Range in the Kashmir Himalayas. *Forests*, *13*(8), 1214. Advance online publication. doi:10.3390/f13081214

Haris, M., Hussain, T., Mohamed, H. I., Khan, A., Ansari, M. S., Tauseef, A., Khan, A. A., & Akhtar, N. (2023). Nanotechnology – A new frontier of nano-farming in agricultural and food production and its development. *The Science of the Total Environment*, *857*, 159639. doi:10.1016/j.scitotenv.2022.159639 PMID:36283520

Hassan, M., Haq, S. M., Ahmad, R., Majeed, M., Sahito, H. A., Shirani, M., Mubeen, I., Aziz, M. A., Pieroni, A., Bussmann, R. W., Alataway, A., Dewidar, A. Z., Al-Yafrsi, M., Elansary, H. O., & Yessoufou, K. (2022). Traditional Use of Wild and Domestic Fauna among Different Ethnic Groups in the Western Himalayas—A Cross Cultural Analysis. *Animals (Basel)*, *12*(17), 2276. Advance online publication. doi:10.3390/ani12172276 PMID:36077997

Hassan, M., Haq, S. M., Majeed, M., Umair, M., Sahito, H. A., Shirani, M., Waheed, M., Aziz, R., Ahmad, R., Bussmann, R. W., Alataway, A., Dewidar, A. Z., El-Abedin, T. K. Z., Al-Yafrsi, M., Elansary, H. O., & Yessoufou, K. (2022). Traditional Food and Medicine: Ethno-Traditional Usage of Fish Fauna across the Valley of Kashmir: A Western Himalayan Region. *Diversity*, *14*(6), 455. Advance online publication. doi:10.3390/d14060455

Helwatkar, A., Riordan, D., & Walsh, J. (2014). Sensor technology for animal health monitoring. *International Journal on Smart Sensing and Intelligent Systems*, *7*(5), 1–6. doi:10.21307/ijssis-2019-057

Hu, J., Dun, G., Geng, X., Chen, J., Wu, X., & Ren, T. L. (2023). Recent progress in flexible micropressure sensors for wearable health monitoring. *Nanoscale Advances*, *5*(12), 3131–3145. doi:10.1039/D2NA00866A PMID:37325539

Hussain, S., Raza, A., Abdo, H. G., Mubeen, M., Tariq, A., Nasim, W., Majeed, M., Almohamad, H., & Al Dughairi, A. A. (2023). Relation of land surface temperature with different vegetation indices using multi-temporal remote sensing data in Sahiwal region, Pakistan. *Geoscience Letters*, *10*(1), 33. Advance online publication. doi:10.1186/s40562-023-00287-6

Jamil, M. D., Waheed, M., Akhtar, S., Bangash, N., Chaudhari, S. K., Majeed, M., Hussain, M., Ali, K., & Jones, D. A. (2022). Invasive Plants Diversity, Ecological Status, and Distribution Pattern in Relation to Edaphic Factors in Different Habitat Types of District Mandi Bahauddin, Punjab, Pakistan. *Sustainability (Basel)*, *14*(20), 13312. Advance online publication. doi:10.3390/su142013312

Karnwal, A., Dohroo, A., & Malik, T. (2023). Unveiling the Potential of Bioinoculants and Nanoparticles in Sustainable Agriculture for Enhanced Plant Growth and Food Security. In *BioMed Research International* (Vol. 2023). Hindawi Limited. doi:10.1155/2023/6911851

Khalil, H. S., Maulu, S., Verdegem, M., & Abdel-Tawwab, M. (2023). Embracing nanotechnology for selenium application in aquafeeds. In Reviews in Aquaculture (Vol. 15, Issue 1, pp. 112–129). John Wiley and Sons Inc. doi:10.1111/raq.12705

Khan, A. M., Li, Q., Saqib, Z., Khan, N., Habib, T., Khalid, N., Majeed, M., & Tariq, A. (2022). MaxEnt Modelling and Impact of Climate Change on Habitat Suitability Variations of Economically Important Chilgoza Pine (Pinus gerardiana Wall.) in South Asia. *Forests*, *13*(5), 715. Advance online publication. doi:10.3390/f13050715

Khoja, A. A., Haq, S. M., Majeed, M., Hassan, M., Waheed, M., Yaqoob, U., Bussmann, R. W., Alataway, A., Dewidar, A. Z., Al-Yafrsi, M., Elansary, H. O., Yessoufou, K., & Zaman, W. (2022). Diversity, Ecological and Traditional Knowledge of Pteridophytes in the Western Himalayas. *Diversity*, *14*(8), 628. Advance online publication. doi:10.3390/d14080628

Kumar, R., Kumar, M., & Luthra, G. (2023). Fundamental approaches and applications of nanotechnology: A mini review. *Materials Today: Proceedings*. Advance online publication. doi:10.1016/j.matpr.2022.12.172

Kumar, S. (2010). Nanotechnology and animal health. *Veterinary World*, *3*(12), 567–569. doi:10.5455/vetworld.2010.567-569

Kumari, R., Suman, K., Karmakar, S., Mishra, V., Lakra, S. G., Saurav, G. K., & Mahto, B. K. (2023). Regulation and safety measures for nanotechnology-based agri-products. In Frontiers in Genome Editing (Vol. 5). Frontiers Media SA. doi:10.3389/fgeed.2023.1200987

Lamichhane, J. R., Barbetti, M. J., Chilvers, M. I., Pandey, A. K., & Steinberg, C. (2024). Exploiting root exudates to manage soil-borne disease complexes in a changing climate. In Trends in Microbiology (Vol. 32, Issue 1, pp. 27–37). Elsevier Ltd. doi:10.1016/j.tim.2023.07.011

Li, S., Chen, L., & Fu, Y. (2023). Nanotechnology-based ocular drug delivery systems: recent advances and future prospects. In Journal of Nanobiotechnology (Vol. 21, Issue 1). BioMed Central Ltd. doi:10.1186/s12951-023-01992-2

Li, Y., Zhang, P., Li, M., Shakoor, N., Adeel, M., Zhou, P., Guo, M., Jiang, Y., Zhao, W., Lou, B. Z., & Rui, Y. (2023). Application and mechanisms of metal-based nanoparticles in the control of bacterial and fungal crop diseases. In Pest Management Science (Vol. 79, Issue 1, pp. 21–36). John Wiley and Sons Ltd. doi:10.1002/ps.7218

Liao, Z., Fan, J., Lai, Z., Bai, Z., Wang, H., Cheng, M., Zhang, F., & Li, Z. (2023). Response network and regulatory measures of plant-soil-rhizosphere environment to drought stress. In *Advances in Agronomy* (Vol. 180, pp. 93–196). Academic Press Inc. doi:10.1016/bs.agron.2023.03.002

Lin, H., Jie, B., Ye, J., Zhai, Y., Luo, Z., Shao, G., Chen, R., Zhang, X., & Yang, Y. (2023). Recent advance of macroscopic metal-organic frameworks for water treatment: A review. *Surfaces and Interfaces*, *36*, 102564. doi:10.1016/j.surfin.2022.102564

Liu, D., Dai, X., Ye, L., Wang, H., Qian, H., Cheng, H., & Wang, X. (2023). Nanotechnology meets glioblastoma multiforme: Emerging therapeutic strategies. In Wiley Interdisciplinary Reviews: Nanomedicine and Nanobiotechnology (Vol. 15, Issue 1). John Wiley and Sons Inc. doi:10.1002/wnan.1838

Lo Presti, D., Di Tocco, J., Massaroni, C., Cimini, S., De Gara, L., Singh, S., Raucci, A., Manganiello, G., Woo, S. L., Schena, E., & Cinti, S. (2023). Current understanding, challenges and perspective on portable systems applied to plant monitoring and precision agriculture. *Biosensors & Bioelectronics*, *222*, 115005. doi:10.1016/j.bios.2022.115005 PMID:36527829

Loiseau, B., Carrière, S. D., Jougnot, D., Singha, K., Mary, B., Delpierre, N., Guérin, R., & Martin-StPaul, N. K. (2023). The geophysical toolbox applied to forest ecosystems – A review. *The Science of the Total Environment*, *899*, 165503. doi:10.1016/j.scitotenv.2023.165503 PMID:37454861

Majeed, M., Bhatti, K. H., & Amjad, M. S. (2021). Impact of climatic variations on the flowering phenology of plant species in jhelum district, punjab, pakistan. *Applied Ecology and Environmental Research*, *19*(5), 3343–3376. doi:10.15666/aeer/1905_33433376

Majeed, M., Bhatti, K. H., Amjad, M. S., Abbasi, A. M., Bussmann, R. W., Nawaz, F., Rashid, A., Mehmood, A., Mahmood, M., Khan, W. M., & Ahmad, K. S. (2020). Ethno-veterinary uses of poaceae in Punjab, Pakistan. *PLoS ONE, 15*(11). doi:10.1371/journal.pone.0241705

Majeed, M., Bhatti, K. H., Pieroni, A., Sõukand, R., Bussmann, R. W., Khan, A. M., Chaudhari, S. K., Aziz, M. A., & Amjad, M. S. (2021). Gathered wild food plants among diverse religious groups in Jhelum District, Punjab, Pakistan. *Foods*, *10*(3), 594. Advance online publication. doi:10.3390/foods10030594 PMID:33799901

MajeedM.Hayat BhattiK.Shoaib AmjadM.Mehmood AbbasiA.RashidA.Nawaz Muhammad Nawaz ShareefF.MehmoodA.MahmoodM. (2020). *Ethno-veterinary practices of Poaceae taxa in Punjab, Pakistan*. doi:10.21203/rs.3.rs-15532/v1

Majeed, M., Khan, A. M., Habib, T., Anwar, M. M., Sahito, H. A., Khan, N., & Ali, K. (2022). Vegetation analysis and environmental indicators of an arid tropical forest ecosystem of Pakistan. *Ecological Indicators*, *142*, 109291. Advance online publication. doi:10.1016/j.ecolind.2022.109291

Majeed, M., Lu, L., Anwar, M. M., Tariq, A., Qin, S., El-Hefnawy, M. E., El-Sharnouby, M., Li, Q., & Alasmari, A. (2023). Prediction of flash flood susceptibility using integrating analytic hierarchy process (AHP) and frequency ratio (FR) algorithms. *Frontiers in Environmental Science*, *10*, 1037547. Advance online publication. doi:10.3389/fenvs.2022.1037547

Majeed, M., Lu, L., Haq, S. M., Waheed, M., Sahito, H. A., Fatima, S., Aziz, R., Bussmann, R. W., Tariq, A., Ullah, I., & Aslam, M. (2022). Spatiotemporal Distribution Patterns of Climbers along an Abiotic Gradient in Jhelum District, Punjab, Pakistan. *Forests*, *13*(8), 1244. Advance online publication. doi:10.3390/f13081244

Majeed, M., Tariq, A., Anwar, M. M., Khan, A. M., Arshad, F., Mumtaz, F., Farhan, M., Zhang, L., Zafar, A., Aziz, M., Abbasi, S., Rahman, G., Hussain, S., Waheed, M., Fatima, K., & Shaukat, S. (2021). Monitoring of land use–Land cover change and potential causal factors of climate change in Jhelum district, Punjab, Pakistan, through GIS and multi-temporal satellite data. *Land (Basel)*, *10*(10), 1026. Advance online publication. doi:10.3390/land10101026

Majeed, M., Tariq, A., Anwar, M. M., Khan, A. M., Arshad, F., Mumtaz, F., Farhan, M., Zhang, L., Zafar, A., Aziz, M., Abbasi, S., Rahman, G., Hussain, S., Waheed, M., Fatima, K., & Shaukat, S. (2021). Monitoring of Land Use–Land Cover Change and Potential Causal Factors of Climate Change in Jhelum District, Punjab, Pakistan, through GIS and Multi-Temporal Satellite Data; (pp. 129-147). https://www.mdpi.com/books/book/6906-land-use-land-cover-and-natural-hazards-interactions-changes-and-impacts

Majeed, M., Tariq, A., Haq, S. M., Waheed, M., Anwar, M. M., Li, Q., Aslam, M., Abbasi, S., Mousa, B. G., & Jamil, A. (2022). A Detailed Ecological Exploration of the Distribution Patterns of Wild Poaceae from the Jhelum District (Punjab), Pakistan. *Sustainability (Basel)*, *14*(7), 3786. Advance online publication. doi:10.3390/su14073786

Malik, S., Muhammad, K., & Waheed, Y. (2023). Nanotechnology: A Revolution in Modern Industry. In Molecules (Vol. 28, Issue 2). MDPI. doi:10.3390/molecules28020661

Manimegalai, S., Vickram, S., Deena, S. R., Rohini, K., Thanigaivel, S., Manikandan, S., Subbaiya, R., Karmegam, N., Kim, W., & Govarthanan, M. (2023). Carbon-based nanomaterial intervention and efficient removal of various contaminants from effluents – A review. *Chemosphere*, *312*, 137319. doi:10.1016/j.chemosphere.2022.137319 PMID:36410505

Mubarok, S., Nuraini, A., Hamdani, J. S., Suminar, E., Kusumiyati, K., Budiarto, R., Lestari, F. W., Rahmat, B. P. N., & Ezura, H. (2024). Antioxidative response of parthenocarpic tomato, iaa9-3 and iaa9-5, under heat stress condition. *Plant Physiology and Biochemistry*, *207*, 108333. doi:10.1016/j.plaphy.2024.108333 PMID:38181640

Muhammad, M., Basit, A., Majeed, M., Shah, A. A., Ullah, I., Mohamed, H. I., Khan, A., & Ghanaim, A. M. (2024). Bacterial pigments and their applications. In *Bacterial Secondary Metabolites* (pp. 277–298). Elsevier. doi:10.1016/B978-0-323-95251-4.00019-3

Muhammad, M., Waheed, A., Wahab, A., Majeed, M., Nazim, M., Liu, Y. H., Li, L., & Li, W. J. (2024). Soil salinity and drought tolerance: An evaluation of plant growth, productivity, microbial diversity, and amelioration strategies. In *Plant Stress* (Vol. 11). Elsevier B.V. doi:10.1016/j.stress.2023.100319

Muktar, Y., Bikila, T., & Keffale, M. (2015). Application of Nanotechnology for Animal Health and Production Improvement: A Review. *World Applied Sciences Journal*, *33*(10), 1588–1596. doi:10.5829/idosi.wasj.2015.33.10.96253

Nastasijevic, I., Mitrovic, R., & Jankovic, S. (2021). Biosensors for animal health and meat safety monitoring: Farm-to-slaughterhouse continuum. *IOP Conference Series. Earth and Environmental Science*, *854*(1), 012063. Advance online publication. doi:10.1088/1755-1315/854/1/012063

Naveed, Majeed, Jabeen, Hanif, Naveed, Saleem, & Khan. (2024a). Plant Disease Diagnosis with Artificial Intelligence (AI). Microbial Data Intelligence and Computational Techniques for Sustainable Computing (pp 217–234). Springer. doi:10.1007/978-981-99-9621-6_15

Naveed, Zahid, Fatima, Saleem, Majeed, Abid, Javed, Wazir, & Qasim. (2024b). Disease Diagnosis with Artificial Intelligence (AI). In Microbial Data Intelligence and Computational Techniques for Sustainable Computing (pp 235–258). Springer. doi:10.1007/978-981-99-9621-6_16

Naveed, Memon, Abdullah, Makhdoom, Azeem, Mehmood, Salahuddin, Rajpoot, & Majeed. (2024c). Use Cases and Future Aspects of Intelligent Techniques in Microbial Data Analysis. In Microbial Data Intelligence and Computational Techniques for Sustainable Computing (pp 259–280). Springer. doi:10.1007/978-981-99-9621-6_17

Nguyen, N. T. T., Nguyen, L. M., Nguyen, T. T. T., Tran, U. P. N., Nguyen, D. T. C., & Van Tran, T. (2023). A critical review on the bio-mediated green synthesis and multiple applications of magnesium oxide nanoparticles. *Chemosphere*, *312*, 137301. doi:10.1016/j.chemosphere.2022.137301 PMID:36410506

PandaP.ChakrobortyS.BabuS.KrishnaN. (2023). *The Use of Silver Nanoparticles in Environmental Remediation*. doi:10.20944/preprints202301.0330.v1

Pandey, P., Tripathi, A., Dwivedi, S., Lal, K., & Jhang, T. (2023). Deciphering the mechanisms, hormonal signaling, and potential applications of endophytic microbes to mediate stress tolerance in medicinal plants. In Frontiers in Plant Science (Vol. 14). Frontiers Media SA. doi:10.3389/fpls.2023.1250020

Payne, P. R., Finlay-Smits, S., Small, B., Cave, V., & Kean, J. (2023). What's that bug? Community participation in biosecurity in Mount Maunganui, New Zealand. *Biological Invasions*, *25*(2), 593–610. doi:10.1007/s10530-022-02937-w

Periakaruppan, R., Vanathi, P., & Vidhya, D. (2023). Management of soil nutrient deficiency by nano-metal oxides. *Nanometal Oxides in Horticulture and Agronomy*, 291–320. doi:10.1016/B978-0-323-91809-1.00003-2

Pramanik, P. K. D., Solanki, A., Debnath, A., Nayyar, A., El-Sappagh, S., & Kwak, K. S. (2020). Advancing Modern Healthcare with Nanotechnology, Nanobiosensors, and Internet of Nano Things: Taxonomies, Applications, Architecture, and Challenges. *IEEE Access : Practical Innovations, Open Solutions*, *8*, 65230–65266. doi:10.1109/ACCESS.2020.2984269

Rajput, V. D., Singh, A., Minkina, T., Rawat, S., Mandzhieva, S., Sushkova, S., Shuvaeva, V., Nazarenko, O., Rajput, P., Komariah, Verma, K. K., Singh, A. K., Rao, M., & Upadhyay, S. K. (2021). Nano-Enabled Products: Challenges and Opportunities for Sustainable Agriculture. *Plants*, *10*(12), 2727. doi:10.3390/plants10122727 PMID:34961197

Rehman, A., Aziz, A., Anwar, M. M., Majeed, M., Albanai, J. A., Almohamad, H., & Abdo, H. G. (2023). Quantifying the impacts of urbanization on urban green, evidences from Maga City, Lahore Pakistan. *Discover Sustainability*, *4*(1), 48. Advance online publication. doi:10.1007/s43621-023-00169-z

Rimpika, A., Anushi, Manasa, S., Sharma, S., Thakur, A., Shilpa, & Sood, A.Anusha K. N. (2023). An Overview of Precision Farming. *International Journal of Environment and Climate Change*, *13*(12), 441–456. doi:10.9734/ijecc/2023/v13i123701

Sangeeta Dash, E., Yadav, A. L., Bora, R., & Kumar Pandey, A. (2023). *Research Trends in Agriculture Science* (Vol. I). www.bhumipublishing.com

Singh, A., Margaryan, G., Harutyunyan, A., Movsesyan, H., Khachatryan, H., Rajput, V., . . . Ghazaryan, K. (2024b). Advancing Agricultural Resilience in Ararat Plain, Armenia: Utilizing Biogenic Nanoparticles and Biochar under Saline Environments to Optimize Food Security and Foster European Trade. *Egyptian Journal of Soil Science, 64*(2). https://doi.org/ doi:10.21608/ejss.2024.257000.1703

Singh, A., Rajput, V., Varshney, A., Sharma, R., Ghazaryan, K., Minkina, T., . . . El-Ramady, H. (2024a). Revolutionizing Crop Production: Nanoscale Wonders - Current Applications, Advances, and Future Frontiers. *Egyptian Journal of Soil Science, 64*(1). https://doi.org/ doi:10.21608/ejss.2023.246354.1684

Singh, A., Rajput, V. D., Varshney, A., Ghazaryan, K., & Minkina, T. (2023a). Small Tech, Big Impact: Agri-nanotechnology Journey to Optimize Crop Protection and Production for Sustainable Agriculture. *Plant Stress*, *10*, 100253. doi:10.1016/j.stress.2023.100253

Singh, A., Rawat, S., Rajput, V. D., Ghazaryan, K., Minkina, T., Al Tawaha, A. R. M., . . . Sandhu, K. S. (2023d). Nanoparticle-Based Priming-Enhanced Agricultural Practices. *Nanopriming Approach to Sustainable Agriculture*, 142–157. doi:10.4018/978-1-6684-7232-3.ch006

Singh, A., Rawat, S., Rajput, V. D., Ghazaryan, K., Minkina, T., Al Tawaha, A. R. M., & Varshney, A. (2023b). Impact of Nanofertilizers for the Mitigation of Multiple Environmental Stresses. *Nanotechnology in the Life Sciences*, 431–454. doi:10.1007/978-3-031-41329-2_16

Singh, A., Sengar, R. S., Rajput, V. D., Al-Ghzawi, A. L., Shahi, U. P., Ghazaryan, K., Minkina, T., Tawaha, A. R. M. A., Al Zoubi, O. M., & Habeeb, T. (2024c). Impact of Salinity Stress and Zinc Oxide Nanoparticles on Macro and Micronutrient Assimilation: Unraveling the Link between Environmental Factors and Nutrient Uptake. *Journal of Ecological Engineering*, *25*(2), 1–9. doi:10.12911/22998993/172947

Singh, B. K., Delgado-Baquerizo, M., Egidi, E., Guirado, E., Leach, J. E., Liu, H., & Trivedi, P. (2023). Climate change impacts on plant pathogens, food security and paths forward. In Nature Reviews Microbiology (Vol. 21, Issue 10, pp. 640–656). Nature Research. doi:10.1038/s41579-023-00900-7

Sinha, D., Maurya, A. K., Abdi, G., Majeed, M., Agarwal, R., Mukherjee, R., Ganguly, S., Aziz, R., Bhatia, M., Majgaonkar, A., Seal, S., Das, M., Banerjee, S., Chowdhury, S., Adeyemi, S. B., & Chen, J. T. (2023a). Integrated Genomic Selection for Accelerating Breeding Programs of Climate-Smart Cereals. In Genes (Vol. 14, Issue 7). Multidisciplinary Digital Publishing Institute (MDPI). doi:10.3390/genes14071484

Sinha, D., Maurya, A. K., Abdi, G., Majeed, M., Agarwal, R., Mukherjee, R., Ganguly, S., Aziz, R., Bhatia, M., Majgaonkar, A., Seal, S., Das, M., Banerjee, S., Chowdhury, S., Adeyemi, S. B., & Chen, J. T. (2023b). Integrated Genomic Selection for Accelerating Breeding Programs of Climate-Smart Cereals. In Genes (Vol. 14, Issue 7). Multidisciplinary Digital Publishing Institute (MDPI). doi:10.3390/genes14071484

Späti, K., Huber, R., & Finger, R. (2021). Benefits of Increasing Information Accuracy in Variable Rate Technologies. *Ecological Economics*, *185*, 107047. Advance online publication. doi:10.1016/j.ecolecon.2021.107047

Swain, R., Sahoo, S., Behera, M., & Rout, G. R. (2023). Instigating prevalent abiotic stress resilience in crop by exogenous application of phytohormones and nutrient. In *Frontiers in Plant Science* (Vol. 14). Frontiers Media S.A. doi:10.3389/fpls.2023.1104874

Tariq, A., Mumtaz, F., Majeed, M., & Zeng, X. (2023). Spatio-temporal assessment of land use land cover based on trajectories and cellular automata Markov modelling and its impact on land surface temperature of Lahore district Pakistan. *Environmental Monitoring and Assessment*, *195*(1), 114. Advance online publication. doi:10.1007/s10661-022-10738-w PMID:36385403

Tassadduq, S. S., Akhtar, S., Waheed, M., Bangash, N., Nayab, D. E., Majeed, M., Abbasi, S., Muhammad, M., Alataway, A., Dewidar, A. Z., Elansary, H. O., & Yessoufou, K. (2022). Ecological Distribution Patterns of Wild Grasses and Abiotic Factors. *Sustainability (Basel)*, *14*(18), 11117. Advance online publication. doi:10.3390/su141811117

Tundisi, L. L., Ataide, J. A., Costa, J. S. R., Coêlho, D. de F., Liszbinski, R. B., Lopes, A. M., Oliveira-Nascimento, L., de Jesus, M. B., Jozala, A. F., Ehrhardt, C., & Mazzola, P. G. (2023). Nanotechnology as a tool to overcome macromolecules delivery issues. *Colloids and Surfaces. B, Biointerfaces*, *222*, 113043. doi:10.1016/j.colsurfb.2022.113043 PMID:36455361

Ullah, I., Aslam, B., Shah, S. H. I. A., Tariq, A., Qin, S., Majeed, M., & Havenith, H. B. (2022). An Integrated Approach of Machine Learning, Remote Sensing, and GIS Data for the Landslide Susceptibility Mapping. *Land (Basel)*, *11*(8), 1265. Advance online publication. doi:10.3390/land11081265

Victoria, J., Tripathi, S., Prakash, V., Tiwari, K., Mahra, S., Sharma, A., Rana, S., Kandhol, N., Sahi, S., Tripathi, D. K., & Sharma, S. (2024). Encapsulated nanopesticides application in plant protection: Quo vadis? *Plant Physiology and Biochemistry*, *206*, 108225. doi:10.1016/j.plaphy.2023.108225 PMID:38147708

Waheed, M., Arshad, F., Majeed, M., Fatima, S., Mukhtar, N., Aziz, R., Mangrio, W. M., Almohamad, H., Al Dughairi, A. A., Al-Mutiry, M., & Abdo, H. G. (2022). Community Structure and Distribution Pattern of Woody Vegetation in Response to Soil properties in Semi-Arid Lowland District Kasur Punjab, Pakistan. *Land (Basel)*, *11*(12), 2145. Advance online publication. doi:10.3390/land11122145

Waheed, M., Arshad, F., Majeed, M., Haq, S. M., Aziz, R., Bussmann, R. W., Ali, K., Subhan, F., Jones, D. A., & Zaitouny, A. (2023). *Potential distribution of a noxious weed (Solanum viarum Du-nal), current status, and future invasion risk based on MaxEnt modeling*. Geology, Ecology, and Landscapes. doi:10.1080/24749508.2023.2179752

Wang, K., Lu, X., Lu, Y., Wang, J., Lu, Q., Cao, X., Yang, Y., & Yang, Z. (2022). Nanomaterials in Animal Husbandry: Research and Prospects. *Frontiers in Genetics*, *13*(June), 1–11. doi:10.3389/fgene.2022.915911 PMID:35846144

Yuvaraj, M., Sathya Priya, R., Jagathjothi, N., Saranya, M., Suganthi, N., Sharmila, R., Cyriac, J., Anitha, R., & Subramanian, K. S. (2023). Silicon nanoparticles (SiNPs): Challenges and perspectives for sustainable agriculture. *Physiological and Molecular Plant Pathology*, *128*, 102161. doi:10.1016/j.pmpp.2023.102161

Zain, M., Ma, H., Ur Rahman, S., Nuruzzaman, M., Chaudhary, S., Azeem, I., Mehmood, F., Duan, A., & Sun, C. (2024). Nanotechnology in precision agriculture: Advancing towards sustainable crop production. In Plant Physiology and Biochemistry (Vol. 206). Elsevier Masson s.r.l. doi:10.1016/j.plaphy.2023.108244

Zhou, J., Liu, G., Guo, Z., Wang, M., Qi, C., Chen, G., Huang, X., Yan, S., & Xu, D. (2023). Stimuli-responsive pesticide carriers based on porous nanomaterials: A review. *Chemical Engineering Journal*, *455*, 140167. doi:10.1016/j.cej.2022.140167

Compilation of References

. Bhattacharyya, A., Duraisamy, P., Govindarajan, M., Buhroo, A. A., & Prasad, R. (2016). Nano-biofungicides: emerging trend in insect pest control. *Advances and applications through fungal nanobiotechnology*, 307-319.

. Singh, A., Rajput, V. D., Al Tawaha, A. R. M., Al Zoubi, O. M., Habeeb, T., Rawat, S., ... Minkina, T. (2023). A review on crop responses to nanofertilizers for mitigation of multiple environmental stresses. *Ecological Engineering & Environmental Technology, 24*.

Aacharya, R., & Chhipa, H. (2020). Nanocarbon fertilizers: Implications of carbon nanomaterials in sustainable agriculture production. In *Carbon Nanomaterials for Agri-Food and Environmental Applications* (pp. 297–321). Elsevier. doi:10.1016/B978-0-12-819786-8.00015-3

Abbas, S. S., Haneef, M., Lohani, M., Tabassum, H., & Khan, A. F. (2016). Nanomaterials used as a plants growth enhancer: An update. *International Journal of Pharmaceutical Sciences Review and Research*, *5*, 17–23.

Abdallah, N. A., Prakash, C. S., & McHughen, A. G. (2020). Nanotechnology and CRISPR-Cas9: Tools for a bio-based future. *Trends in Biotechnology*, *38*(11), 1177–1180.

Abdel Latef, A. A. H., Zaid, A., Abu Alhmad, M. F., & Abdelfattah, K. E. (2020). The impact of priming with Al_2O_3 nanoparticles on growth, pigments, osmolytes, and antioxidant enzymes of Egyptian Roselle (*Hibiscus sabdariffa* L.) cultivar. *Agronomy (Basel)*, *10*(5), 681. doi:10.3390/agronomy10050681

Abdel-Aziz, H. M. M., Hasaneen, M. N. A., & Omer, A. M. (2016). Nano chitosan-NPK fertilizer enhances the growth and productivity of wheat plants grown in sandy soil. *Spanish Journal of Agricultural Research*, *14*(1), e0902. doi:10.5424/sjar/2016141-8205

Abdennouri, M., Baâlala, M., Galadi, A., El Makhfouk, M., Bensitel, M., Nohair, K., Barka, N., Boussaoud, A., Barka, N., . . . Barka, N. (2016). Photocatalytic degradation of pesticides by titanium dioxide and titanium pillared purified clays. *Arabian Journal of Chemistry*, *9*, S313–S318. https://doi.org/ doi:10.1016/j.arabjc.2011.04.005

Abdi, O., & Kazemi, M. (2015). A review study of biosorption of heavy metals and comparison between different biosorbents. *Journal of Materials & Environmental Sciences*, *6*, 1386–1399.

Adisa, I. O., Reddy Pullagurala, V. L., Rawat, S., Hernandez-Viezcas, J. A., Dimkpa, C. O., Elmer, W. H., et al. (2018). Role of cerium compounds in Fusarium wilt suppression and growth enhancement in tomato (Solanum lycopersicum). Journal of Agricultural and Food Chemistry, 66, 5959–5970. doi: . 8b01345 doi:10.1021/acs.jafc

Adisa, I. O., Reddy Pullagurala, V. L., Peralta-Videa, J. R., Dimkpa, C. O., Elmer, W. H., Gardea-Torresdey, J. L., & White, J. C. (2019). Recent advances in nano-enabled fertilizers and pesticides: A critical review of mechanisms of action. *Environmental Science. Nano*, *6*(7), 2002–2030. doi:10.1039/C9EN00265K

Adrees, M., Khan, Z. S., Ali, S., Hafeez, M., Khalid, S., Ur Rehman, M. Z., Hussain, A., Hussain, K., Shahid Chatha, S. A., & Rizwan, M. (2020). Simultaneous mitigation of cadmium and drought stress in wheat by soil application of iron nanoparticles. *Chemosphere*, *238*, 124681. doi:10.1016/j.chemosphere.2019.124681 PMID:31524618

Afzal, S., Singh, M. P., Chaudhary, N., & Singh, N. K. (2022). Application of nanoparticles in developing resilience against abiotic stress in rice plant (*Oryza sativa* L.). In *Plant Perspectives to Global Climate Changes* (pp. 151–172). Academic Press. doi:10.1016/B978-0-323-85665-2.00014-5

Aghdam, M. T. B., Mohammadi, H., & Ghorbanpour, M. (2016). Effects of nanoparticulate anatase titanium dioxide on physiological and biochemical performance of Linumusitatissimum (Linaceae) under well-watered and drought stress conditions. *Revista Brasileira de Botânica. Brazilian Journal of Botany*, *39*(1), 139–146. doi:10.1007/s40415-015-0227-x

Agriopoulou, S., Stamatelopoulou, E., Skiada, V., Tsarouhas, P., & Varzakas, T. (2020). Emerging nanomaterials applications for food packaging and preservation. Safety issues and risk assessment. *Proceedings*, *4*(x), 1–8.

Agrios, G. (2012). *Plant Pathology*. Elsevier. Retrieved from http://books.google.ie/books?id= xLdSrKhThNEC&printsec=frontcover&dq=0323144039&hl=&cd=1&source=gbs_api

Agustín, G. C., Escarpa, A., & Carlos, D. G. (2018). Carbon-based nanomaterials in analytical chemistry. In *Detection Science* (1st ed., pp. 1–45). Royal Society of Chemistry.

Ahamad, T., Naushad, M., Al-Saeedi, S. I., Almotairi, S., & Alshehri, S. M. (2020). Fabrication of MoS2/ZnS embedded in N/S doped carbon for the photocatalytic degradation of pesticide. *Materials Letters*, *263*, 127271. doi:10.1016/j.matlet.2019.127271

Ahmad, F., & Nadeem, H. (2022). Mass Spectroscopy as an analytical tool to harness the production of secondary plant metabolites: The way forward for drug discovery. In *Gene, Drug, and Tissue Engineering* (pp. 77–103). Springer US.

Ahmad, M. A., Adeel, M., Shakoor, N., Javed, R., Ishfaq, M., Peng, Y., Zain, M., Azeem, I., Ali, I., Usman, M., Wu, Z., Gohari, G., Xu, M., Rui, Y., Zhang, Z., White, J. C., & Deng, X. (2023). Modifying engineered nanomaterials to produce next generation agents for environmental remediation. *The Science of the Total Environment*, *894*, 164861. doi:10.1016/j.scitotenv.2023.164861 PMID:37343875

Ahmad, N., Naeem, M., Ali, H., Alabbosh, K. F., Hussain, H., Khan, I., Siddiqui, S. A., Khan, A. A., & Iqbal, B. (2023). From challenges to solutions: The impact of melatonin on abiotic stress synergies in horticultural plants via redox regulation and epigenetic signaling. *Scientia Horticulturae*, *321*, 112369. doi:10.1016/j.scienta.2023.112369

Ahmed, S., Khan, F. S. A., Mubarak, N. M., Khalid, M., Tan, Y. H., Mazari, S. A., Karri, R. R., & Abdullah, E. C. (2021). Emerging pollutants and their removal using visible-light responsive photocatalysis – A comprehensive review. *Journal of Environmental Chemical Engineering*, *9*(6), 106643. doi:10.1016/j.jece.2021.106643

Ahmed, T., Luo, J., Noman, M., Ijaz, M., Wang, X., Masood, H. A., Manzoor, N., Wang, Y., & Li, B. (2023). Microbe-mediated nanoparticle intervention for the management of plant diseases. *Crop Health*, *1*(1), 3. Advance online publication. doi:10.1007/s44297-023-00006-9

Ahmed, T., Noman, M., Ijaz, M., Ali, S., Rizwan, M., Ijaz, U., Hameed, A., Ahmad, U., Wang, Y., Sun, G., & Li, B. (2021). Current trends and future prospective in nanoremediation of heavy metals contaminated soils: A way forward towards sustainable agriculture. *Ecotoxicology and Environmental Safety*, *227*, 112888. doi:10.1016/j.ecoenv.2021.112888 PMID:34649136

Alabdallah, N. M., & Alzahrani, H. S. (2020). The potential mitigation effect of ZnO nanoparticles on [Abelmoschus esculentus L. Moench] metabolism under salt stress conditions. *Saudi Journal of Biological Sciences*, *27*(11), 3132–3137. doi:10.1016/j.sjbs.2020.08.005 PMID:33100874

Alcívar, M., Zurita-Silva, A., Sandoval, M., Muñoz, C., & Schoebitz, M. (2018). Reclamation of saline soils in Santa Elena province (Ecuador) using two types of zeolites in the cultivation of Zea mays. *Communications in Soil Science and Plant Analysis*, *49*(22), 2844–2858.

Alfei, S., Marengo, B., & Zuccari, G. (2017). "Nanotechnology application in food packaging: A plethora of opportunities versus pending risks assessment and public concerns," Food Research Application of Nanotechnology in Food Science: Perception and Overview. *Frontiers in Microbiology*, *8*, 1501. doi:10.3389/fmicb.2017.01501

Alharbi, O. M. L., Basheer, A. A., Khattab, R. A., & Ali, I. (2018). Health and environmental effects of persistent organic pollutants. *Journal of Molecular Liquids*, *263*, 442–453. doi:10.1016/j.molliq.2018.05.029

Ali, B., Saleem, M. H., Ali, S., Shahid, M., Sagir, M., Tahir, M. B., Qureshi, K. A., Jaremko, M., Selim, S., Hussain, A., Rizwan, M., Ishaq, W., & Rehman, M. (2022). Mitigation of salinity stress in barley genotypes with variable salt tolerance by application of zinc oxide nanoparticles. *Frontiers in Plant Science*, *13*, 973782. doi:10.3389/fpls.2022.973782 PMID:36072329

Alidokht, L., Anastopoulos, I., Ntarlagiannis, D., Soupios, P., Tawabini, B., Kalderis, D., & Khataee, A. (2021). Recent advances in the application of nanomaterials for the remediation of arsenic-contaminated water and soil. *Journal of Environmental Chemical Engineering*, *9*(4), 105533. doi:10.1016/j.jece.2021.105533

Alidoust, D., & Isoda, A. (2013). Effect of γFe2O3 nanoparticles on photosynthetic characteristic of soybean (Glycine max (L.) Merr.): Foliar spray versus soil amendment. *Acta Physiologiae Plantarum*, *35*(12), 3365–3375. doi:10.1007/s11738-013-1369-8

Ali, H., Khan, E., & Sajad, M. A. (2013). Phytoremediation of heavy metals—Concepts and applications. *Chemosphere*, *91*(7), 869–881. doi:10.1016/j.chemosphere.2013.01.075 PMID:23466085

Ali, H., Muhammad, Z., Majeed, M., Aziz, R., Khan, A., Mangrio, W. M., Abdo, H. G., Almohamad, H., & Al Dughairi, A. A. (2023). Vegetation diversity pattern during spring season in relation to topographic and edaphic variables in subtropical zone. *Botanical Studies*, *64*(1), 25. Advance online publication. doi:10.1186/s40529-023-00398-5 PMID:37716923

Ali, M., Kim, B., Belfield, K. D., Norman, D., Brennan, M., & Ali, G. S. (2015). Inhibition of Phytophthora parasitica and P. capsici by silver nanoparticles synthesized using aqueous extract of Artemisia absinthium. *Phytopathology*, *105*(9), 1183–1190. doi:10.1094/PHYTO-01-15-0006-R PMID:25871856

Ali, Q., Zia, M. A., Kamran, M., Shabaan, M., Zulfiqar, U., Ahmad, M., Iqbal, R., & Maqsood, M. F. (2023). Nanoremediation for heavy metal contamination: A review. *Hybrid Advances*, *4*, 100091. doi:10.1016/j.hybadv.2023.100091

Ali, S. S., Al-Tohamy, R., Koutra, E., Moawad, M. S., Kornaros, M., Mustafa, A. M., & Sun, J. (2021). Nanobiotechnological advancements in agriculture and food industry: Applications, nanotoxicity, and future perspectives. *The Science of the Total Environment*, *792*, 148359. doi:10.1016/j.scitotenv.2021.148359 PMID:34147795

Ali, S., Tyagi, A., & Bae, H. (2023). ROS interplay between plant growth and stress biology: Challenges and future perspectives. *Plant Physiology and Biochemistry*, *203*, 108032. doi:10.1016/j.plaphy.2023.108032 PMID:37757722

Alloway, B. J. (2013). Sources of heavy metals and metalloids in soils. In B. J. Alloway (Ed.), *Heavy metals in soils: Trace metals and metalloids in soils and their bioavailability* (pp. 11–50). Springer. doi:10.1007/978-94-007-4470-7_2

Allwood, J. W., Williams, A., Uthe, H., van Dam, N. M., Mur, L. A., Grant, M. R., & Pétriacq, P. (2021). Unravelling plant responses to stress—The importance of targeted and untargeted metabolomics. *Metabolites*, *11*(8), 558. doi:10.3390/metabo11080558 PMID:34436499

Almutairi, Z. M. (2016). Effect ofnano-silicon application on the expression ofsalt tolerance genes in germinating tomato ('Solanum lycopersicum' L.) seedlings under salt stress. *Plant Omics*, *9*, 106.

Alsaeedi, A., Elgarawani, M. M., Alshaal, T., & Elhawat, N. (2022). Silicon-and nanosilicon-mediated drought and waterlogging stress tolerance in plants. In *Silicon and Nano-silicon in Environmental Stress Management and Crop Quality Improvement* (pp. 121–152). Academic Press. doi:10.1016/B978-0-323-91225-9.00005-4

Al-Sahli, S. A., Al-Otibi, F., Alharbi, R. I., Amina, M., & Al Musayeib, N. M. (2024). Silver nanoparticles improve the fungicidal properties of Rhazya stricta decne aqueous extract against plant pathogens. *Scientific Reports*, *14*(1), 1297. doi:10.1038/s41598-024-51855-5 PMID:38221517

Altabbaa, S., Mann, N. A., Chauhan, N., Utkarsh, K., Thakur, N., & Mahmoud, G. A. E. (2023). Era connecting nanotechnology with agricultural sustainability: Issues and challenges. *Nanotechnology for Environmental Engineering*, *8*(2), 481–498. doi:10.1007/s41204-022-00289-3

Altug, H., Oh, S. H., Maier, S. A., & Homola, J. (2022). Advances and applications of nanophotonic biosensors. In Nature Nanotechnology (Vol. 17, Issue 1, pp. 5–16). Nature Research. doi:10.1038/s41565-021-01045-5

Alvarado, M.A., Guzmán, N.M., Solís, N.M. and Vega Baudrit, J., 2017. Recycling and Elimination of Wastes obtained from Agriculture by using Nanotechnology: Nanosensors.

Alvarez-Fernandez, M., Zornoza, P., Diaz-Ravina, M., & Núñez, A. (2016). Application of nanotechnology for the remediation of soils contaminated with heavy metals. In *Soil Remediation and Plants* (pp. 335–354). Academic Press.

Alvi, M. A., Akram, M. S., & Khan, S. (2019). α-Tocopherol-loaded solid lipid nanoparticles improve the heat stress tolerance in tomato (Solanum lycopersicum L.). *Journal of Plant Growth Regulation*, *38*(4), 1445–1455.

Alzohairy, M. A. (2016). Therapeutics role of *Azadirachta indica* (Neem) and their active constituents in diseases prevention and treatment. *Evidence-Based Complementary and Alternative Medicine*, *2016*, 7382506. doi:10.1155/2016/7382506 PMID:27034694

Amalraj, A., & Gopi, S. (2017). Medicinal properties of *Terminalia arjuna* (Roxb.) Wight & Arn.: A review. *Journal of Traditional and Complementary Medicine*, *7*(1), 65–78. doi:10.1016/j.jtcme.2016.02.003 PMID:28053890

Ambaye, T. G., Chebbi, A., Formicola, F., Prasad, S., Gomez, F. H., Franzetti, A., & Vaccari, M. (2022). Remediation of soil polluted with petroleum hydrocarbons and its reuse for agriculture: Recent progress, challenges, and perspectives. *Chemosphere*, *293*, 133572. doi:10.1016/j.chemosphere.2022.133572 PMID:35016966

Amenta, V., Aschberger, K., Arena, M., Bouwmeester, H., Moniz, F. B., Brandhoff, P., ... Peters, R. J. (2015). Regulatory aspects of nanotechnology in the agri/feed/food sector in EU and non-EU countries. *Regulatory Toxicology and Pharmacology*, *73*(1), 463–476. doi:10.1016/j.yrtph.2015.06.016 PMID:26169479

Ameta, S. K., Rai, A. K., Hiran, D., Ameta, R., & Ameta, S. C. (2020). Use of nanomaterials in food science. In *Biogenic nano-particles and their use in agro-ecosystems* (Vol. 457). Springer. doi:10.1007/978-981-15-2985-6_24

Androvitsaneas, P., Young, A. B., Schneider, C., Maier, S., Kamp, M., Höfling, S., Knauer, S., Harbord, E., Hu, C. Y., Rarity, J. G., & Oulton, R. (2016). Charged quantum dot micropillar system for deterministic light-matter interactions. *Physical Review. B*, *93*(24), 241409. doi:10.1103/PhysRevB.93.241409

Anjali, C. H., Sharma, Y., Mukherjee, A., & Chandrasekaran, N. (2012). Neem oil (Azadirachta indica) nanoemulsion—A potent larvicidal agent against Culex quinquefasciatus. *Pest Management Science*, *68*(2), 158–163. doi:10.1002/ps.2233 PMID:21726037

Anjum, N. A., Gill, S. S., Duarte, A. C., Pereira, E., & Ahmad, I. (2013). Silver nanoparticles in soil-plant systems. *Journal of Nanoparticle Research*, *15*(9), 1–26. doi:10.1007/s11051-013-1896-7

Anjum, N. A., Gill, S. S., Gill, R., Hasanuzzaman, M., Duarte, A. C., Pereira, E., Ahmad, I., Tuteja, R., & Tuteja, N. (2016). Metal/metalloid stress tolerance in plants: Role of ascorbate, its redox couple, and associated enzymes. *Protoplasma*, *253*(5), 1265–1283. PMID:24682425

Ansari, M. A. (2023). Nanotechnology in Food and Plant Science: Challenges and Future Prospects. In Plants (Vol. 12, Issue 13). Multidisciplinary Digital Publishing Institute (MDPI). doi:10.3390/plants12132565

Anton, N., & Vandamme, T. F. (2011). Nanoemulsions and micro emulsions: Clarification. *International Journal of Pharmaceutics*, *377*, 142–147. doi:10.1016/j.ijpharm.2009.05.014 PMID:19454306

Anusuya, S., & Banu, K. N. (2016). Silver-chitosan nanoparticles induced biochemical variations of chickpea (Cicer arietinum L.). *Biocatalysis and Agricultural Biotechnology*, *8*, 39–44. doi:10.1016/j.bcab.2016.08.005

Aqeel, U., Aftab, T., Khan, M. M. A., Naeem, M., & Khan, M. N. (2022). A comprehensive review of impacts of diverse nanoparticles on growth, development and physiological adjustments in plants under changing environment. *Chemosphere*, *291*, 132672. doi:10.1016/j.chemosphere.2021.132672 PMID:34756946

Arduini, F., Cinti, S., Mazzaracchio, V., Scognamiglio, V., Amine, A., & Moscone, D. (2020). Carbon black as an outstanding and affordable nanomaterial for electrochemical (bio) sensor design. *Biosensors & Bioelectronics*, *156*(112033), 112033. Advance online publication. doi:10.1016/j.bios.2020.112033 PMID:32174547

Arduini, F., Cinti, S., Scognamiglio, V., & Moscone, D. (2016). Nanomaterials in electrochemical biosensors for pesticide detection: Advances and challenges in food analysis. *Mikrochimica Acta*, *183*(7), 2063–2083. doi:10.1007/s00604-016-1858-8

Arif, U., Bhatti, K. H., Ajaib, M., Wagay, N. A., Majeed, M., Zeb, J., Hameed, A., & Kiani, J. (2021). Ethnobotanical indigenous knowledge of Tehsil Charhoi, District Kotli, Azad Jammu and Kashmir, Pakistan. *Ethnobotany Research and Applications*, *22*. Advance online publication. doi:10.32859/era.22.50.1-24

Arnold, W. A., & Roberts, A. L. (2000). Pathways and kinetics of chlorinated ethylene and chlorinated acetylene reaction with Fe (0) particles. *Environmental Science & Technology*, *34*(9), 1794–1805. doi:10.1021/es990884q

Arora, D., Arora, A., Bala, R., Panghal, V., & Kumar, S. (2023). Enhancement in Phytoremediation Efficiency of Tagetus erecta with the Application of Nano-scale Zero Valent Iron (nZVI) for the Restoration of Lead Contaminated Soil: An Approach Toward Sustainability. *Water, Air, and Soil Pollution*, *234*(8), 535. Advance online publication. doi:10.1007/s11270-023-06540-0

Arsenov, D., Beljin, J., Jović, D., Maletić, S., Borišev, M., & Borišev, I. (2023). Nanomaterials as endorsed environmental remediation tools for the next generation: Eco-safety and sustainability. *Journal of Geochemical Exploration*, *253*, 107283. doi:10.1016/j.gexplo.2023.107283

Arshad, F., Mohd-Naim, N. F., Chandrawati, R., Cozzolino, D., & Ahmed, M. U. (2022). Nanozyme-based sensors for detection of food biomarkers: A review. *RSC Advances*, *12*(40), 26160–26175. doi:10.1039/D2RA04444G PMID:36275095

Arshad, F., Waheed, M., Harun, N., Fatima, K., Khan, B. A., Fatima, K., Abbas, Z., Jabeen, S., & Majeed, M. (2022). Indigenous farmer's perception about fodder and foraging species of Semi-arid lowlands of Pakistan: A case study of District Kasur, Pakistan. *Taiwania*, *67*(4), 510–523. doi:10.6165/tai.2022.67.510

Asgari, F., Majd, A., Jonoubi, P., & Najafi, F. (2018). Effects of gibberellic acid and salinity on germination, seed reserve utilization and initial growth of soybean (Glycine max). *Iranian Journal of Plant Physiology*, *8*(2), 2433–2444.

Ashfaq, M., Singh, S., Sharma, A., & Verma, N. (2016). Cytotoxic evaluation of the hierarchal web of carbon micro-nanofibers. *Industrial Crops and Products*, *95*, 334–340.

Ashraf, M. (2010). Inducing drought tolerance in plants: Recent advances. *Biotechnology Advances*, *28*(1), 169–183. doi:10.1016/j.biotechadv.2009.11.005 PMID:19914371

Ashraf, M. A., Hussain, I., Rasheed, R., Iqbal, M., Riaz, M., & Arif, M. S. (2017). Advances in microbe-assisted reclamation of heavy metal contaminated soils over the last decade: A review. *Journal of Environmental Management*, *198*, 132–143. doi:10.1016/j.jenvman.2017.04.060 PMID:28456029

Ashraf, S., Siddiqa, A., Shahida, S., & Qaisar, S. (2019). Titanium-based nanocomposite materials for arsenic removal from water: A review. *Heliyon*, *5*(5), e01577. doi:10.1016/j.heliyon.2019.e01577 PMID:31193228

Askary, M., Talebi, S. M., Amini, F., & Bangan, A. D. (2017). Effects of iron nanoparticles on Mentha piperita L. under salinity stress. *Biologija (Vilnius, Lithuania)*, *63*(1), 65–75. doi:10.6001/biologija.v63i1.3476

Aswathi, V. P., Meera, S., Maria, C. G. A. & Nidhin, M. (2022). Green synthesis of nanoparticles from biodegradable waste extracts and their applications: a critical review. *Nanotechnology for Environmental Engineering, 8*(2), 377–397. doi:10.1007/s41204-022-00276-8

Atta, A. M., Mohamed, N. H., Hegazy, A. K., Moustafa, Y. M., Mohamed, R. R., Safwat, G. & Diab, A. A. (2020). Green Technology for Remediation of Water Polluted with Petroleum Crude Oil: Using of Eichhornia crassipes (Mart.) Solms Combined with Magnetic Nanoparticles Capped with Myrrh Resources of Saudi Arabia. *Nanomaterials, 10*(2), 262. doi:10.3390/nano10020262

Aunkor, M. T. H., Raihan, T., Prodhan, S. H., Metselaar, H. S. C., Malik, S. U. F., & Azad, A. K. (2020). Antibacterial activity of graphene oxide nanosheet against multidrug resistant superbugs isolated from infected patients. *Royal Society Open Science*, *7*(7), 200640. doi:10.1098/rsos.200640 PMID:32874659

Awasthi, A., Singh, P., Shukla, A. K., & Dubey, N. K. (2017). Nanotechnology and plant–pathogen interactions: An emerging tool for disease management. In *Nanotechnology in Plant Growth Promotion and Protection* (pp. 211–226). John Wiley & Sons, Ltd.

Ayangbenro, A. S., & Babalola, O. O. (2017). A new strategy for heavy metal polluted environments: A review of microbial biosorbents. *International Journal of Environmental Research and Public Health*, *14*(1), 94. doi:10.3390/ijerph14010094 PMID:28106848

Ayoub, H. A., Khairy, M., Elsaid, S., Rashwan, F. A., & Abdel-Hafez, H. F. (2018). Pesticidal activity of nanostructured metal oxides for generation of alternative pesticide formulations. *Journal of Agricultural and Food Chemistry*, *66*(22), 5491–5498. doi:10.1021/acs.jafc.8b01600 PMID:29757642

Azeem, M., Pirjan, K., Qasim, M., Mahmood, A., Javed, T., Muhammad, H., Yang, S., Dong, R., Ali, B., & Rahimi, M. (2023). Salinity stress improves antioxidant potential by modulating physio-biochemical responses in Moringa oleifera Lam. *Scientific Reports*, *13*(1), 2895. Advance online publication. doi:10.1038/s41598-023-29954-6 PMID:36807545

Aziz, A., Anwar, M. M., Majeed, M., Fatima, S., Mehdi, S. S., Mangrio, W. M., Elbouzidi, A., Abdullah, M., Shaukat, S., Zahid, N., Mahmoud, E. A., Casini, R., Yessoufou, K., & Elansary, H. O. (2023). Quantifying Landscape and Social Amenities as Ecosystem Services in Rapidly Changing Peri-Urban Landscape. *Land (Basel)*, *12*(2), 477. Advance online publication. doi:10.3390/land12020477

Aziz, N., Faraz, M., Pandey, R., Shakir, M., Fatma, T., Varma, A., Barman, I., & Prasad, R. (2015). Facile algae-derived route to biogenic silver nanoparticles: Synthesis, antibacterial, and photocatalytic properties. *Langmuir*, *31*(42), 11605–11612. doi:10.1021/acs.langmuir.5b03081 PMID:26447769

Baalousha, M., Cornelis, G., Kuhlbusch, T. A., Lynch, I., Nickel, C., Peijnenburg, W., & van den Brink, N. W. (2016). Modeling nanomaterial fate and uptake in the environment: Current knowledge and future trends. *Environmental Science. Nano*, *3*(2), 323–345. doi:10.1039/C5EN00207A

Baalousha, M., Stolpe, B., Lead, J. R., & Colman, B. P. (2016). Nanoparticle dispersion in environmentally important matrices and implications for exposure: A review. *Reviews of Environmental Contamination and Toxicology*, *236*, 53–97.

Badam, L., Joshi, S. P., & Bedekar, S. S. (1999). 'In vitro' antiviral activity of neem (*Azadirachta indica*. A. Juss) leaf extract against group B coxsackieviruses. *The Journal of Communicable Diseases*, *31*(2), 79–90. PMID:10810594

Badino, P., Gras, A., Johanna Groot, M., Shoaib Amjad, M., & Morales-de la Nuez, A. (n.d.). *Quantitative ethnoveterinary study on plant resource utilization by indigenous communities in high-altitude regions*. Academic Press.

Badri, S., Basu, V. R., Chandra, K., & Anasuya, D. (2019). A review on pharmacological activities of alkaloids. *World Journal of Current Medical and Pharmaceutical Research*, *1*(6), 230–234. doi:10.37022/WJCMPR.2019.01068

Baghalian, K., Hajirezaei, M. R., & Schreiber, F. (2014). Plant metabolic modeling: Achieving new insight into metabolism and metabolic engineering. *The Plant Cell*, *26*(10), 3847–3866. doi:10.1105/tpc.114.130328 PMID:25344492

Bajpei, S. K., Chand, N., & Chaurasia, V. (2012). Nano zinc oxide loaded calcium alignate film with potential antibacterial properties. *Food and Bioprocess Technology*, *5*(5), 1871–1881. doi:10.1007/s11947-011-0587-6

Bajwa, U., & Sandhu, K. S. (2014). Effect of Handling and Processing on Pesticide Residues in Food- a Review. *Journal of Food Science and Technology*, *51*(2), 201–220. doi:10.1007/s13197-011-0499-5 PMID:24493878

Baker, S. (2020). Nanozymes: emerging prospect in food science and technology. SSRN Electronic Journal, 1-7. doi:10.2139/ssrn.3619997

Balasubramaniam, T., Shen, G., Esmaeili, N., & Zhang, H. (2023). Plants' Response Mechanisms to Salinity Stress. In Plants (Vol. 12, Issue 12). MDPI. doi:10.3390/plants12122253

Balasubramanian, K., & Burghard, M. (2006). Biosensors based on carbon nanotubes. *Analytical and Bioanalytical Chemistry*, *385*(3), 452–468. doi:10.1007/s00216-006-0314-8 PMID:16568294

Balaure, P. C., Gudovan, D., & Gudovan, I. (2017). Nanopesticides: a new paradigm in crop protection. In New pesticides and soil sensors (pp. 129-192). Academic Press.

Balkir, P., Kemahlioglu, K., & Yucel, U. (2021). Foodomics: A new approach in food quality and safety. *Trends in Food Science & Technology*, *108*, 49–57. doi:10.1016/j.tifs.2020.11.028

Bamola, A., Semwal, D. K., Semwal, S., & Rawat, U. (2009). Flavonoid glycosides from *Myrica esculenta* leaves. *Journal of the Indian Chemical Society*, *86*, 535–536.

Banerjee, S., Mondal, S., Islam, J., Sarkar, R., Saha, B., & Sen, A. (2024). Rhizospheric nano-remediation salvages arsenic genotoxicity: Zinc-oxide nanoparticles articulate better oxidative stress management, reduce arsenic uptake, and increase yield in Pisum sativum (L.). *The Science of the Total Environment*, *913*, 169493. doi:10.1016/j.scitotenv.2023.169493 PMID:38151134

Bao, W., Wang, J., Wang, Q., O'Hare, D., & Wan, Y. (2016). Layered double hydroxide nanotransporter for molecule delivery to intact plant cells. *Scientific Reports*, *6*(1), 26738. doi:10.1038/srep26738 PMID:27221055

Baragaño, D., Alonso, J., Gallego, J. R., Lobo, M. C., & Gil-Díaz, M. (2020a). Magnetite nanoparticles for the remediation of soils co-contaminated with As and PAHs. *Chemical Engineering Journal*, *399*, 125809. doi:10.1016/j.cej.2020.125809

Baragaño, D., Alonso, J., Gallego, J. R., Lobo, M. C., & Gil-Díaz, M. (2020b). Zero valent iron and goethite nanoparticles as new promising remediation techniques for As-polluted soils. *Chemosphere*, *238*, 124624. doi:10.1016/j.chemosphere.2019.124624 PMID:31472353

Baragaño, D., Forján, R., Welte, L., & Gallego, J. L. R. (2020). Nanoremediation of As and metals polluted soils by means of graphene oxide nanoparticles. *Scientific Reports*, *10*(1), 1896. doi:10.1038/s41598-020-58852-4 PMID:32024880

Baragaño, D., Gallego, J. L. R., María Menéndez-Aguado, J., Marina, M. A., & Sierra, C.Baragaño. (2021). As sorption onto Fe-based nanoparticles and recovery from soils by means of wet high intensity magnetic separation. *Chemical Engineering Journal*, *408*, 127325. doi:10.1016/j.cej.2020.127325

Barage, S., Lakkakula, J., Sharma, A., Roy, A., Alghamdi, S., Almehmadi, M., Hossain, M. J., Allahyani, M., & Abdulaziz, O. (2022). Nanomaterial in food packaging: A comprehensive review. *Journal of Nanomaterials*, *2022*, 1–12. doi:10.1155/2022/6053922

Bardos, P., Merly, C., Kvapil, P., & Koschitzky, H. P. (2018). Status of nanoremediation and its potential for future deployment: Risk-benefit and benchmarking appraisals. *Remediation (New York, N.Y.)*, *28*(3), 43–56. doi:10.1002/rem.21559

Barik, T. K., Sahu, B., & Swain, V. (2008). Nanosilica—From medicine to pest control. *Parasitology Research*, *103*(2), 253–258. doi:10.1007/s00436-008-0975-7 PMID:18438740

Barrett, B., Nitze, I., Green, S., & Cawkwell, F. (2014). Assessment of multi-temporal, multi-sensor radar and ancillary spatial data for grasslands monitoring in Ireland using machine learning approaches. *Remote Sensing of Environment*, *152*(2), 109–124. doi:10.1016/j.rse.2014.05.018

Bartelmess, J., & Giordan, S. (2014). Carbon nano-onions (multilayer fullerenes): Chemistry and applications. *Beilstein Journal of Nanotechnology*, *5*, 1980–1998. doi:10.3762/bjnano.5.207 PMID:25383308

Basak, G., Hazra, C., & Sen, R. (2020). Biofunctionalized nanomaterials for in situ clean-up of hydrocarbon contamination: A quantum jump in global bioremediation research. *Journal of Environmental Management*, *256*, 109913. doi:10.1016/j.jenvman.2019.109913 PMID:31818738

Basha, S. (2019). Future Prospects of Nanotechnology Innovations in Animal Production. *Egyptian Journal of Animal Production*, *57*(1), 67–71. doi:10.21608/ejap.2019.98191

Bashir, S. M., Altaf, M., Hussain, T., Umair, M., Majeed, M., Mangrio, W. M., Khan, A. M., Gulshan, A. B., Hamed, M. H., Ashraf, S., Amjad, M. S., Bussmann, R. W., Abbasi, A. M., Casini, R., Alataway, A., Dewidar, A. Z., Al-Yafrsi, M., Amin, M. H., & Elansary, H. O. (2023b). Vernacular Taxonomy, Cultural and Ethnopharmacological Applications of Avian and Mammalian Species in the Vicinity of Ayubia National Park, Himalayan Region. *Biology (Basel)*, *12*(4), 609. Advance online publication. doi:10.3390/biology12040609 PMID:37106809

Baulcombe, D. (2004). RNA silencing in plants. *Nature*, *431*(7006), 356–363. doi:10.1038/nature02874 PMID:15372043

Beckers, F., & Rinklebe, J. (2017). Cycling of mercury in the environment: Sources, fate, and human health implications: A review. *Critical Reviews in Environmental Science and Technology*, *47*(9), 693–794. doi:10.1080/10643389.2017.1326277

Behboudi, F., Mohamad Sarraf Mamoori, S., & Moghaddam, M. (2018). Nano-chitosan positively affected the performance of Triticum aestivum under salinity stress. *Journal of Plant Molecular Breeding*, *6*(1), 32–42.

Behboudi, F., Tahmasebi Sarvestani, Z., Kassaee, M. Z., Sanavi, S. A. M. M., Sorooshzadeh, A., & Ahmadi, S. B. (2019). The influence of chitosan nanoparticles on seed germination and seedling growth of wheat (Triticum aestivum) under salinity stress. *Agroecology and Sustainable Food Systems*, *43*(6), 696–712.

Bempah, K. C., Buah-Kwofie, A., & Asomaning, J. (2011). Morphological studies of neem (*Azadirachta indica* A. Juss.) seed and physicochemical properties of its oil extracts collected in Accra metropolis of Ghana. *Elixir Applied Botany*, *39*, 4951–4953.

Berekaa, M. M. (2015). Nanotechnology in food industry; advances in food processing, packaging and food safety. *International Journal of Current Microbiology and Applied Sciences*, *4*(5), 345–357.

Berekaa, M. M. (2015). Nanotechnoogy in food industry; advances in food processing, packaging and food safety. *International Journal of Current Microbiology and Applied Sciences*, *4*(5), 345–357.

Bhagat, J., Nishimura, N., & Shimada, Y. (2021). Toxicological interactions of microplastics/nanoplastics and environmental contaminants: Current knowledge and future perspectives. *Journal of Hazardous Materials*, *405*, 123913. doi:10.1016/j.jhazmat.2020.123913 PMID:33127190

Bhalla, G., Bhalla, B., Kumar, V., & Sharma, A. (2022). Bioremediation and phytoremediation of pesticides residues from contaminated water: a novel approach. *Pesticides Remediation Technologies from Water and Wastewater*, 339–363. doi:10.1016/B978-0-323-90893-1.00016-7

Bhandari, P. R., & Kamdod, M. A. (2012). *Emblica officinalis* (Amla): A review of potential therapeutic applications. *International Journal of Green Pharmacy*, *6*(4), 257–269. doi:10.4103/0973-8258.108204

Bhardwaj, A. K., Arya, G., Kumar, R., Hamed, L., Pirasteh-Anosheh, H., Jasrotia, P., Kashyap, P. L., & Singh, G. P. (2022). Switching to nanonutrients for sustaining agroecosystems and environment: The challenges and benefits in moving up from ionic to particle feeding. *Journal of Nanobiotechnology*, *20*(1), 19. doi:10.1186/s12951-021-01177-9 PMID:34983548

Bhaskar, B., Ahammed, S. K., Chaitanya, B. H., Rasheed, V. A., & Prasad, T. N. V. K. V. (2016). Silver nanoparticles: mycogenesis, characterization and its anti-plant pathogenic applications. Int J Res Appl Nat Soc Sci, 4(10), 105-114.

Bhatia, D., & Goyal, P. (2021). A review of pharmacological and pharmacognostic profile of *Myrica nagi. Research Journal of Pharmacy and Technology*, *14*(2), 1109–1114. doi:10.5958/0974-360X.2021.00200.6

Bhatia, D., Paul, S., Acharjee, T., & Ramachairy, S. S. (2024). Biosensors and their widespread impact on human health. *Sensors International*, *5*, 100257. Advance online publication. doi:10.1016/j.sintl.2023.100257

Bhattacharya, B., Roy, P., Bhattacharya, S., Prasad, B., & Mandal, A. K. (2023). Nanotechnology and sustainable development: Overcoming the obstacles by adopting ethical practices for future farming. In Engineered Nanomaterials for Sustainable Agricultural Production, Soil Improvement and Stress Management (pp. 431-445). Academic Press.

Bhattacharya, R., & Bhattacharya, R. **Majeed, M.,** Bhandari, S., Aziz, R., Sinha, D., ... & Niyogi, S. S. Environmental Rehabilitation of Industrial Waste Dumping Site. In *Biohydrometallurgical Processes* (pp. 44–89). CRC Press. doi:10.1201/9781003451457-4

Bhattacharya, U., Mukhopadhyay, S., & Giri, A. K. (2011). Comparative antimutagenic and anticancer activity of three fractions of black tea polyphenols the arubigins. *Nutrition and Cancer*, *63*(7), 1122–1132. doi:10.1080/01635581.2011.605985 PMID:21919645

Bin Hussein, M. Z., Yahaya, A. H., Zainal, Z., & Kian, L. H. (2005). Nanocomposite-based controlled release formulation of an herbicide, 2, 4-dichlorophenoxyacetate incapsulated in zinc–aluminiumlayered double hydroxide. *Science and Technology of Advanced Materials*, *6*(8), 956–962. doi:10.1016/j.stam.2005.09.004

Bina, F., Zahedi, H., & Zare, M. (2018). The effect of zeolite and nano-zeolite on reducing salinity stress and increasing the yield and biomass growth of tomato. *Journal of Water and Soil Science*, *22*(1), 163–177.

Bindraban, P. S., Dimkpa, C., Nagarajan, L., Roy, A., & Rabbinge, R. (2015). Revisiting fertilisers and fertilisation strategies for improved nutrient uptake by plants. *Biology and Fertility of Soils*, *51*(8), 897–911. doi:10.1007/s00374-015-1039-7

Biswas, B., Sarkar, B., Rusmin, R., & Naidu, R. (2015). Bioremediation of PAHs and VOCs: Advances in clay mineral-microbial interaction. *Environment International*, *85*, 168–181. doi:10.1016/j.envint.2015.09.017 PMID:26408945

Biswas, K., Chattopadhyay, I., Banerjee, R. K., & Bandyopadhyay, U. (2002). Biological activities and medicinal properties of neem (*Azadirachta indica*). *Current Science*, *82*(10), 1336–1345.

Biswas, P., Polash, S. A., Dey, D., Kaium, M. A., Mahmud, A. R., Yasmin, F., Baral, S. K., Islam, M. A., Rahaman, T. I., Abdullah, A., Ema, T. I., Khan, D. A., Bibi, S., Chopra, H., Kamel, M., Najda, A., Fouda, M. M. A., Rehan, U. S. M., Mheidat, M., ... Hasan, M. N. (2023). Advanced implications of nanotechnology in disease control and environmental perspectives. *Biomedicine and Pharmacotherapy*, *158*, 114172. doi:10.1016/j.biopha.2022.114172 PMID:36916399

Biswas, R., Alam, M., Sarkar, A., Haque, M. I., Hasan, M. M., & Hoque, M. (2022). Application of nanotechnology in food: Processing, preservation, packaging and safety assessment. *Heliyon*, *8*(11), 1–16. doi:10.1016/j.heliyon.2022.e11795 PMID:36444247

Blanco-Rojo, R., & Vaquero, M. P. (2019). Iron bioavailability from food fortification to precision nutrition. A review. *Innovative Food Science & Emerging Technologies*, *51*, 126–138. doi:10.1016/j.ifset.2018.04.015

Blundell, S. P., & Owens, G. (2021). Evaluation of enhancement techniques for the dechlorination of DDT by nanoscale zero-valent iron. *Chemosphere*, *264*(1), 128324. doi:10.1016/j.chemosphere.2020.128324 PMID:33022509

Bolade, O. P., Akinsiku, A. A., Oluwafemi, O. S., Williams, A. B., & Benson, N. U. (2021). Biogenic iron oxide nanoparticles and activated sodium persulphate for hydrocarbon remediation in contaminated soil. *Environmental Technology & Innovation*, *23*, 101719. doi:10.1016/j.eti.2021.101719

Bonnell, D. A., & Huey, B. D. (2001). Basic principles of scanning probe microscopy. *Scanning Probe Microscopy & Spectroscopy: Theory, Techniques, and Applications*, 7-42.

Borgatta, J., Ma, C., Hudson-Smith, N., Elmer, W., Plaza Pérez, C. D., De La Torre-Roche, R., Zuverza-Mena, N., Haynes, C. L., White, J. C., & Hamers, R. J. (2018). Copper based nanomaterials suppress root fungal disease in watermelon (Citrullus lanatus): Role of particle morphology, composition and dissolution behavior. *ACS Sustainable Chemistry & Engineering*, *61*(11), 4847–14856. doi:10.1021/acssuschemeng.8b03379

Bott, J., Störmer, A., & Franz, R. (2014). A model study into the migration potential of nanoparticles from plastics nanocomposites for food contact. *Food Packaging and Shelf Life*, *2*(2), 73–40. doi:10.1016/j.fpsl.2014.08.001

Boulkhessaim, S., Gacem, A., Khan, S. H., Amari, A., Yadav, V. K., Harharah, H. N., Elkhaleefa, A. M., Yadav, K. K., Rather, S. U., Ahn, H. J., & Jeon, B. H. (2022). Emerging trends in the remediation of persistent organic pollutants using nanomaterials and related processes: A review. *Nanomaterials (Basel, Switzerland)*, *12*(13), 2148. doi:10.3390/nano12132148 PMID:35807983

Boutillier, S.; Fourmentin, S. & Laperche, B. (2020). Food additives and the future of health: An analysis of the ongoing controversy on titanium dioxide. Futures, 122. doi:10.1016/j.futures.2020.102598.hal-03490868

Brand, W., van Kesteren, P. C. E., & Oomen, A. G. (2019). Potential health risks of nanomaterials in food: a methodology to identify signals and prioritise risks. RIVM Letter Report 2019-0191. National Institute for Public Health and the Environment, The Netherlands. doi:10.21945/RIVM-2019-0191

Brase, T., Gries, P., & Sauerborn, K. (2017). Precision farming in vineyards: the use of sensor technology to optimize production. *Proceedings of the 20th GiESCO International Meeting*, Mendoza, Argentina, 415-420.

Bratovcic, A. (2020). Nanomaterials in food processing and packaging, its toxicity and food labeling. ACTA Scientific Nutritional Health, 4(9), 7-13.

Briffa, J., Sinagra, E., & Blundell, R. (2020). Heavy metal pollution in the environment and their toxicological effects on humans. *Heliyon*, *6*(9), e04691. doi:10.1016/j.heliyon.2020.e04691 PMID:32964150

Briskin, D. P. (2000). Medicinal plants and phytomedicines. Linking plant biochemistry and physiology to human health. *Plant Physiology*, *124*(2), 507–514. doi:10.1104/pp.124.2.507 PMID:11027701

Bulovic, V., Mandell, A., & Perlman, A. (2004). Molecular memory device. *US 20050116256 A*, *1*, 388.

Bulovic, V., Mandell, A., & Perlman, A. (2004). *U.S. Patent No. 6,781,868*. Washington, DC: U.S. Patent and Trademark Office.

Bumbudsanpharoke, N., & Ko, S. (2015). Nano-food packaging: An overview of market, migration research, and safety regulations. *Journal of Food Science*, *80*(5), R910–R923. doi:10.1111/1750-3841.12861 PMID:25881665

Bundschuh, M., Filser, J., Lüderwald, S., McKee, M. S., Metreveli, G., Schaumann, G. E., Schulz, R., & Wagner, S. (2018). Nanoparticles in the environment: Where do we come from, where do we go to? *Environmental Sciences Europe*, *30*(1), 1–17. doi:10.1186/s12302-018-0132-6 PMID:29456907

Burg, M. B., & Ferraris, J. D. (2008). Intracellular organic osmolytes: Function and regulation. *The Journal of Biological Chemistry*, *283*(12), 7309–7313. doi:10.1074/jbc.R700042200 PMID:18256030

Cai, L., Cai, H., Deng, Y., Liu, J., Cui, H., Yu, X., & Huo, Z. (2019). Foliar exposure of Fe3O4 nanoparticles on Oryza sativa L. and its impact on nutrient accumulation, photosynthetic performance and internal molecular response. *Ecotoxicology and Environmental Safety*, *183*, 109587.

Cai, L., Liu, M., Liu, Z., Yang, H., Sun, X., Chen, J., Xiang, S., & Ding, W. (2018). MgONPs can boost plant growth: Evidence from increased seedling growth, morpho-physiological activities, andMg uptake in tobacco (Nicotiana tabacumL.). *Molecules (Basel, Switzerland)*, *23*(12), 3375. doi:10.3390/molecules23123375 PMID:30572666

Calabi-Floody, M., Medina, J., Rumpel, C., Condron, L. M., Hernandez, M., Dumont, M., & de la Luz Mora, M. (2018). Smart fertilizers as a strategy for sustainable agriculture. [). Academic Press.]. *Advances in Agronomy*, *147*, 119–157. doi:10.1016/bs.agron.2017.10.003

Calabi-Floody, M., Medina, J., Suazo, J., Ordiqueo, M., Aponte, H., Mora, M. D. L. L., & Rumpel, C. (2018). Optimization of wheat biochar as a soil amendment to improve soil water retention capacity. *Journal of Soil Science and Plant Nutrition*, *18*(1), 81–93.

Camilli, L., Pisani, C., Gautron, E., Scarselli, M., Castrucci, P., D'Orazio, F., Passacantando, M., Moscone, D., & De Crescenzi, M. (2014). A three-dimensional carbon nanotube network for water treatment. *Nanotechnology*, *25*(6), 065701. doi:10.1088/0957-4484/25/6/065701 PMID:24434944

Campbell, B. M., Thornton, P., Zougmoré, R., Van Asten, P., & Lipper, L. (2014). Sustainable intensification: What is its role in climate smart agriculture? *Current Opinion in Environmental Sustainability*, *8*, 39–43. doi:10.1016/j.cosust.2014.07.002

Campos, E. V. R., de Oliveira, J. L., Fraceto, L. F., & Singh, B. (2014). Polysaccharides as safer release systems for agrochemicals. *Agronomy for Sustainable Development*, *35*(1), 47–66. doi:10.1007/s13593-014-0263-0

Campos, E. V. R., Oliveira, J. L. D., da Silva, C. M. G., Pascoli, M., Pasquoto, T., Lima, R., Abhilash, P. C., & Fernandes Fraceto, L. (2015). Polymeric and solid lipid nanoparticles for sustained release of carbendazim and tebuconazole in agricultural applications. *Scientific Reports*, *5*(1), 13809. doi:10.1038/srep13809 PMID:26346969

Campos-Garcia, J., Martinez, D. S. T., Alves, O. L., Leonardo, A. F. G., & Barbieri, E. (2015). Ecotoxicological effects of carbofuran and oxidised multiwalled carbon nanotubes on the freshwater fish Nile tilapia: Nanotubes enhance pesticide ecotoxicity. *Ecotoxicology and Environmental Safety*, *111*, 131–137. doi:10.1016/j.ecoenv.2014.10.005 PMID:25450925

Cao, Y., Chen, G., Wan, Y., & Luo, J. (2021). Nanofiltration membrane for bio-separation: Process-oriented materials innovation. *Engineering in Life Sciences*, *21*(6), 405–416. doi:10.1002/elsc.202000100 PMID:34140851

Carsanba, E., Favaretto, C., Moraes, F. C., Cucinelli Neto, R. P., Borin, E., & Bordin, J. R. (2021). A graphene-based biosensor for early detection of Xylella fastidiosa in citrus. *Scientific Reports*, *11*(1), 1–10. doi:10.1038/s41598-021-87496-1 PMID:33414495

Carvalho, F. P. (2017). Pesticides, environment, and food safety. *Food and Energy Security*, *6*(2), 48–60. doi:10.1002/fes3.108

Cassidy, A., Rogers, G., Peterson, J. J., Dwyer, J. T., Lin, H., & Jacques, P. F. (2015). Higher dietary anthocyanin and flavonol intakes are associated with anti-inflammatory effects in a population of US adults. *The American Journal of Clinical Nutrition*, *102*(1), 172–181. doi:10.3945/ajcn.115.108555 PMID:26016863

Cembrowska-Lech, D., Krzemińska, A., Miller, T., Nowakowska, A., Adamski, C., Radaczyńska, M., Mikiciuk, M., & Mikiciuk, M. (2023). An integrated multi-omics and artificial intelligence framework for advance plant phenotyping in horticulture. *Biology (Basel)*, *12*(10), 1298. doi:10.3390/biology12101298 PMID:37887008

Cerqueira, M. A., & Pastrana, L. M. (2019). Does the future of food pass by using nanotechnologies. *Frontiers in Sustainable Food Systems*, *3*, 1–5. doi:10.3389/fsufs.2019.00016

Chadha, U., Bhardwaj, P., Selvaraj, S. K., Arasu, K., Praveena, S., Pavan, A., Khanna, M., Singh, P., Singh, S., Chakravorty, A., Badoni, B., Banavoth, M., Sonar, P., & Paramasivam, V. (2022). Current trends and future perspectives of nanomaterials in food packaging application. *Journal of Nanomaterials*, *2022*, 1–32. doi:10.1155/2022/2745416

Chadwick, M., Trewin, H., Gawthrop, F., & Wagstaff, C. (2013). Sesquiterpenoids lactones: Benefits to plants and people. *International Journal of Molecular Sciences*, *14*(6), 12780–12805. doi:10.3390/ijms140612780 PMID:23783276

Chahine, N. O., Collette, N. M., Thomas, C. B., Genetos, D. C., & Loots, G. G. (2014). Nanocomposite scaffold for chondrocyte growth and cartilage tissue engineering: Effects of carbon nanotube surface functionalization. *Tissue Engineering. Part A*, *20*(17-18), 2305–2315. doi:10.1089/ten.tea.2013.0328 PMID:24593020

Chandra, J., Hasan, N., Nasir, N., Wahab, S., Thanikachalama, P. V., Sahebkar, A., Ahmad, F. J., & Kesharwani, P. (2023). Nanotechnology-empowered strategies in treatment of skin cancer. *Environmental Research*, *235*, 116649. doi:10.1016/j.envres.2023.116649 PMID:37451568

Chandra, R., Kumar, V., Tripathi, S., & Sharma, P. (2017). Phytoremediation of industrial pollutants and life cycle assessment. In *Phytoremediation of environmental pollutants* (pp. 441–470). CRC Press. doi:10.4324/9781315161549-18

Chandra, S., Pradhan, S., Mitra, S., Patra, P., Bhattacharya, A., Pramanik, P., & Goswami, A. (2014). High throughput electron transfer from carbon dots to chloroplast: A rationale of enhanced photosynthesis. *Nanoscale*, *6*(7), 3647–3655. doi:10.1039/C3NR06079A PMID:24562190

Chang, S. H.; Chen, C. H. & Tsai, G. J. (2020). Effects of chitosan on *Clostridium perfringens* and application in the preservation of pork sausage. Marine Drugs, 18(2), 1-15. doi:10.3390/md18020070

Chang, M. C., & Kang, H. Y. (2009). Remediation of pyrene-contaminated soil by synthesized nanoscale zero-valent iron particles. *Journal of Environmental Science and Health. Part A, Toxic/Hazardous Substances & Environmental Engineering*, *44*(6), 576–582. doi:10.1080/10934520902784609 PMID:19337920

Chang, Y.-C., Lin, Y.-S., Xiao, G.-T., Chiu, T.-C., & Hu, C.-C. (2016). A Highly Selective and Sensitive Nanosensor for the Detection of Glyphosate. *Talanta*, *161*, 94–98. doi:10.1016/j.talanta.2016.08.029 PMID:27769503

Chaudhary, P., Chaudhary, A., Khati, P., Kumar, G., Tyagi, J., & Behera, M. (2023). Nanotechnology and Omics Approach in Agrobiotechnology. In *Agricultural and Environmental Nanotechnology: Novel Technologies and their Ecological Impact* (pp. 341–352). Springer Nature Singapore. doi:10.1007/978-981-19-5454-2_12

Chaudhary, P., Fatima, F., & Kumar, A. (2020). Relevance of nanomaterials in food packaging and its advanced future prospects. *Journal of Inorganic and Organometallic Polymers and Materials*, *30*(12), 5180–5192. doi:10.1007/s10904-020-01674-8 PMID:32837459

Chaudhary, S., Dheri, G. S., & Brar, B. S. (2021). Nanomics: An emerging tool for sustainable agriculture. *Environmental Nanotechnology, Monitoring & Management*, *15*, 100455. doi:10.1016/j.enmm.2021.100455

Chaudhry, Q., & Castle, L. (2011). Food applications of nanotechnologies: An overview of opportunities and challenges for developing countries. *Trends in Food Science & Technology*, *22*(11), 595–603. doi:10.1016/j.tifs.2011.01.001

Chaudhry, Q., Scotter, M., Blackburn, J., Ross, B., Boxall, A., Castle, L., Aitken, R., & Watkins, R. (2008). Applications and implications of nanotechnologies for the food sector. *Food Additives and Contaminants*, *25*(3), 241–258. doi:10.1080/02652030701744538 PMID:18311618

Chauhan, A. K., Kataria, N., & Garg, V. K. (2020). Green fabrication of ZnO nanoparticles using Eucalyptus spp. leaves extract and their application in wastewater remediation. *Chemosphere*, *247*, 125803. doi:10.1016/j.chemosphere.2019.125803 PMID:31972482

Chauhan, R., Yadav, H. O. S., & Sehrawat, N. (2020). Nanobioremediation: A new and a versatile tool for sustainable environmental cleanup-Overview. *Journal of Materials & Environmental Sciences*, *11*, 564–573.

Chaupoo, A. S., Dhakad, A., Beese, S., Kumar, A., Kumar, P., Pandey, S. K., & Kurdekar, A. K. (2023). Transforming Horticulture: The Influence of Nanotechnology on Crop Enhancement. *International Journal of Environment and Climate Change*, *13*(11), 3381–3400. doi:10.9734/ijecc/2023/v13i113512

Chehelgerdi, M., Chehelgerdi, M., Allela, O. Q. B., Pecho, R. D. C., Jayasankar, N., Rao, D. P., Thamaraikani, T., Vasanthan, M., Viktor, P., Lakshmaiya, N., Saadh, M. J., Amajd, A., Abo-Zaid, M. A., Castillo-Acobo, R. Y., Ismail, A. H., Amin, A. H., & Akhavan-Sigari, R. (2023). Progressing nanotechnology to improve targeted cancer treatment: overcoming hurdles in its clinical implementation. In Molecular Cancer (Vol. 22, Issue 1). BioMed Central Ltd. doi:10.1186/s12943-023-01865-0

Chellaram, C., Murugaboopathi, G., John, A. A., Siakumar, R., Ganesan, S., Krithika, S., & Priya, G. (2014)... *Significance of Nanotechnology in Food*, *8*, 109–113.

Chen, C., Ding, R., Yang, S., Wang, J., Chen, W., Zong, L., & Xie, J. (2020). Development of thermal insulation packaging film based on poly (vinyl alcohol) incorporated with silica aerogel for food packaging application. *Lebensmittel-Wissenschaft + Technologie*, *129*, 109568. doi:10.1016/j.lwt.2020.109568

Chen, C., Tsyusko, O. V., McNear, D. H. Jr, Judy, J., Lewis, R. W., & Unrine, J. M. (2017). Effects of biosolids from a wastewater treatment plant receiving manufactured nanomaterials on Medicago truncatula and associated soil microbial communities at low nanomaterial concentrations. *The Science of the Total Environment*, *609*, 799–806. doi:10.1016/j.scitotenv.2017.07.188 PMID:28768212

Cheng, C., Fu, J., Su, H., & Ren, L. (2023). Recent Advancements in Agriculture Robots: Benefits and Challenges. In Machines (Vol. 11, Issue 1). MDPI. doi:10.3390/machines11010048

Cheng, H., Wang, J., Wang, M., & Zhong, S. (2017). Toward privacy-preserving JPEG image retrieval. *Journal of Electronic Imaging*, *26*(4), 043022. doi:10.1117/1.JEI.26.4.043022

Chen, J., Guo, Y., Zhang, X., Liu, J., Gong, P., Su, Z., Fan, L., & Li, G. (2023). Emerging nanoparticles in food: Sources, application and safety. *Journal of Agricultural and Food Chemistry*, *71*(8), 3564–3582. doi:10.1021/acs.jafc.2c06740 PMID:36791411

Chen, Y., Ren, H. L., Liu, N., Sai, N., Liu, X., Liu, Z., Gao, Z., & Ning, B. (2010). A fluoroimmunoassay based on quantum dot–streptavidin conjugate for the detection of chlorpyrifos. *Journal of Agricultural and Food Chemistry*, *58*(16), 8895–8903. doi:10.1021/jf101778t PMID:23654227

Chhipa, H. (2017). Nanofertilizers and nanopesticides for agriculture. *Environmental Chemistry Letters*, *15*(1), 15–22. doi:10.1007/s10311-016-0600-4

Chhipa, H. (2019). Applications of nanotechnology in agriculture. *Angewandte Chemie International Edition*, *58*(13), 4318–4322. doi:10.1002/anie.201814313 PMID:30714282

Chhipa, H. (2019). Applications of nanotechnology in agriculture. In *Nanoscience in Food and Agriculture 5* (pp. 115–142). Springer. doi:10.1007/978-3-030-14024-7_6

Ching, K. H., Lin, A., McGarvey, J. A., Stanker, L. H., & Hnasko, R. (2012). Rapid and Selective Detection of Botulinum Neurotoxin Serotype-A and -B with a Single Immunochromatographic Test Strip. *Journal of Immunological Methods*, *380*(1-2), 23–29. doi:10.1016/j.jim.2012.03.008 PMID:22504369

Choi, J. H., Kim, G. B., & Cha, H. J. (2016). Quantum dot-based nano-sensor for detection of phosphate in plant tissues. *Biosensors & Bioelectronics*, *77*, 1042–1047.

Choi, M., & Park, Y. (2014). Phytotoxicity of zinc oxide nanoparticles on the early growth of radish (Raphanus sativus L.) and the detection of nanoparticles in plant tissues. *Journal of Environmental Science and Health. Part C, Environmental Carcinogenesis & Ecotoxicology Reviews*, *32*(4), 375–387.

Choudhary, G. P. (2012). Anti-ulcer activity of the ethanolic extract of *Terminalia belerica* Roxb. *Internation Journal of Pharmaceutical Chemical Sciences*, *1*(4), 1293.

Chung, I. M., Rajakumar, G., Gomathi, T., Park, S. K., Kim, S. H., & Thiruvengadam, M. (2017). Nanotechnology for human food: Advances and perspectives. *Frontiers in Life Science*, *10*(1), 63–72. doi:10.1080/21553769.2017.1365775

Cicek, S., & Nadaroglu, H. (2015). The use of nanotechnology in the agriculture. *Advances in Nano Research*, *3*(4), 207–223. doi:10.12989/anr.2015.3.4.207

Cisse, M., & Mucke, L. (2009). A prion protein connection. *Nature*, *457*(7233), 1090–1091. doi:10.1038/4571090a PMID:19242462

Cohen, M. M. (2014). Tulsi-*Ocimum sanctum*: A herb for all reasons. *Journal of Ayurveda and Integrative Medicine*, *5*(4), 251–259. doi:10.4103/0975-9476.146554 PMID:25624701

Conidi, C., Castro-Muñoz, R., & Cassano, A. (2020). Nanofiltration in beverage industry. In *Nanotechnology in the Beverage Industry* (pp. 525–548). Elsevier. doi:10.1016/B978-0-12-819941-1.00018-3

Cornelis, G., Hund-Rinke, K., Kuhlbusch, T., Van den Brink, N., & Nickel, C. (2014). Fate and bioavailability of engineered nanoparticles in soils: A review. *Critical Reviews in Environmental Science and Technology*, *44*(24), 2720–2764. doi:10.1080/10643389.2013.829767

Cortés-Rojas, D. F., de Souza, C. R. F., & Oliveira, W. P. (2014). Clove (*Syzygium aromaticum*): A precious spice. *Asian Pacific Journal of Tropical Biomedicine*, *4*(2), 90–96. doi:10.1016/S2221-1691(14)60215-X PMID:25182278

Costa, M. A., Xia, Z. Q., Davin, L. B., & Lewis, N. G. (1999). Toward engineering the metabolic pathways of cancer-preventing lignans in cereal grains and other crops. In *Phytochemicals in Human Health Protection, Nutrition and Plant Defense* (pp. 67–87). Springer. doi:10.1007/978-1-4615-4689-4_4

Cota-Arriola, O., Onofre Cortez-Rocha, M., Burgos-Hernández, A., Marina Ezquerra-Brauer, J., & Plascencia-Jatomea, M. (2013). Controlled release matrices and micro/nanoparticles of chitosan with antimicrobial potential: Development of new strategies for microbial control in agriculture. *Journal of the Science of Food and Agriculture*, *93*(7), 1525–1536. doi:10.1002/jsfa.6060 PMID:23512598

Couto, C., & Almeida, A. (2022). Metallic nanoparticles in the food sector: A mini-review. *Foods*, *11*(3), 1–22. doi:10.3390/foods11030402 PMID:35159552

Cramer, G. R., Urano, K., Delrot, S., Pezzotti, M., & Shinozaki, K. (2011). Effects of abiotic stress on plants: A systems biology perspective. *BMC Plant Biology*, *11*(1), 163. doi:10.1186/1471-2229-11-163 PMID:22094046

Crescente, M., Jessen, G., Momi, S., Höltje, H. D., Gresele, P., Cerletti, C., & De Gaetano, G. (2009). Interactions of gallic acid, resveratrol, quercetin and aspirin at the platelet cyclooxygenase-1 level functional and modelling studies. *Thrombosis and Haemostasis*, *102*(08), 336–346. doi:10.1160/TH09-01-0057 PMID:19652885

Cromwell, W. A., Yang, J., Starr, J. L., & Jo, Y. K. (2014). Nematicidal effects of silver nanoparticles on root-knot nematode in bermudagrass. *Journal of Nematology*, *46*, 261–266. PMID:25275999

Croteau, R., Kutchan, T. M., & Lewis, N. G. (2000). Natural products (secondary metabolites). *Biochemistry and Molecular Biology of Plants*, *24*, 1250–1319.

Crozier, A., Clifford, M. N., & Ashihara, H. (Eds.). (2008). *Plant secondary metabolites: occurrence, structure and role in the human diet*. John Wiley & Sons.

Cuce, E., Cuce, P. M., Wood, C. J., & Riffat, S. B. (2014). Toward aerogel based thermal superinsulation in buildings: A comprehensive review. *Renewable & Sustainable Energy Reviews*, *34*, 273–299. doi:10.1016/j.rser.2014.03.017

Cui, R., Han, Z., & Zhu, J.-J. (2011). Helical Carbon Nanotubes: Intrinsic Peroxidase Catalytic Activity and its Application for Biocatalysis and Biosensing. *Chemistry (Weinheim an der Bergstrasse, Germany)*, *17*(34), 9377–9384. doi:10.1002/chem.201100478 PMID:21769953

Cui, Y., Kumar, S., Kona, B. R., & van Houcke, D. (2015). Gas barrier properties of polymer/clay nanocomposites. *RSC Advances*, *5*(78), 63669–63690. doi:10.1039/C5RA10333A

Cui, Y., Kundalwal, S., & Kumar, S. (2016). Gas barrier performance of graphene/polymer nanocomposites. *Carbon*, *98*, 131–333. doi:10.1016/j.carbon.2015.11.018

Cummins, C. M., Koivunen, M. E., Stephanian, A., Gee, S. J., Hammock, B. D., & Kennedy, I. M. (2006). Application of europium (III) chelate-dyed nanoparticle labels in a competitive atrazine fluoroimmunoassay on an ITO waveguide. *Biosensors & Bioelectronics*, *21*(7), 1077–1085. doi:10.1016/j.bios.2005.04.003 PMID:16368482

Dahlberg, J., Berenji, J., Sikora, V., & Latkovic, D. (2011). Assessing sorghum [Sorghum bicolor (L) Moench] germplasm for new traits: food, fuels & unique uses. Maydica, 56(1750), 85-92.

Daniel, A. I., Keyster, M., & Klein, A. (2023). Biogenic zinc oxide nanoparticles: A viable agricultural tool to control plant pathogenic fungi and its potential effects on soil and plants. *The Science of the Total Environment*, *897*, 165483. doi:10.1016/j.scitotenv.2023.165483 PMID:37442458

Dar, J., & Soytong, K. (2014). Construction and characterization of copolymer nanomaterials loaded with bioactive compounds from Chaetomium species. *Agricultural Technology (Thailand)*, *10*(4), 823–831.

Dasgupta, N., Ranjan, S., Mundekkad, D., Ramalingam, C., Shanker, R., & Kumar, A. (2015). Nanotechnology in agro-food: From field to plate. *Food Research International*, *69*, 381–400. doi:10.1016/j.foodres.2015.01.005

Dash, K. K., Deka, P., Bangar, S. P., Chaudhary, V., Trif, M., & Rusu, A. (2022). Applications of Inorganic Nanoparticles in Food Packaging: A Comprehensive Review. *Polymers*, *14*(3), 521. doi:10.3390/polym14030521 PMID:35160510

Das, S., Wolfson, B. P., Tetard, L., Tharkur, J., Bazata, J., & Santra, S. (2015). Effect of N-acetyl cysteine coated CdS: Mn/ZnS quantum dots on seed germination and seedling growth of snow pea (Pisum sativum L.): imaging and spectroscopic studies. *Environmental Science. Nano*, *2*(2), 203–212. doi:10.1039/C4EN00198B

Davies, S. R., Kearnes, M., & Macnaghten, P. (2019). Nanotechnology and public engagement: a new kind of (social) science? In *Nano meets macro* (pp. 473–500). Jenny Stanford Publishing. doi:10.1201/9780429067150-32

Dawang, S., Zuchun, Z., Foo, L. Y., & Wong, H. (1991). Flavonols from *Myrica esculenta* a bark. *Linchan Huaxue Yu Gongye*, *4*, 251–257.

de Carvalho, A. P. A., & Junior, C. A. C. (2020). Green strategies for active food packagings: A systematic review on active properties of graphene-based nanomaterials and biodegradable polymers. *Trends in Food Science & Technology*, *103*, 130–143. doi:10.1016/j.tifs.2020.07.012

de la Rosa, G., Vázquez-Núñez, E., Molina-Guerrero, C., Serafín-Muñoz, A. H., & Vera-Reyes, I. (2021). Interactions of nanomaterials and plants at the cellular level: Current knowledge and relevant gaps. *Nanotechnology for Environmental Engineering*, *6*(1), 1–19. doi:10.1007/s41204-020-00100-1

De Ridder, S., Van der Kooy, F., & Verpoorte, R. (2008). Artemisia annua as a self-reliant treatment for malaria in developing countries. *Journal of Ethnopharmacology*, *120*(3), 302–314. doi:10.1016/j.jep.2008.09.017 PMID:18977424

Deb, A., Barua, S., & Das, B. (2016). Pharmacological activities of Baheda (*Terminalia bellerica*): A review. *Journal of Pharmacognosy and Phytochemistry*, *5*(1), 194–197.

Debnath, M., Pandey, M., & Bisen, P. S. (2011). An omics approach to understand the plant abiotic stress. *OMICS: A Journal of Integrative Biology*, *15*(11), 739–762. doi:10.1089/omi.2010.0146 PMID:22122668

Debnath, N., Das, S., Seth, D., Chandra, R., Bhattacharya, S. C., & Goswami, A. (2011). Entomotoxic effect of silica nanoparticles against Sitophilus oryzae (L.). *Journal of Pest Science*, *84*(1), 99–105. doi:10.1007/s10340-010-0332-3

Decarolis, D., Odarchenko, Y., Herbert, J. J., Qiu, C., Longo, A., & Beale, A. M. (2020). Identification of the key steps in the self-assembly of homogeneous gold metal nanoparticles produced using inverse micelles. *Physical Chemistry Chemical Physics*, *22*(34), 18824–18834. doi:10.1039/C9CP03473K PMID:31475258

Deepika, G., Kumar, L., & Gaikwad, K. K. (2023). Carbon dots for food packaging applications Sustainable. *Food Technology*, *1*(2), 185–199. doi:10.1039/D2FB00020B

Dehghanian, Z., Asgari Lajayer, B., Biglari Quchan Atigh, Z., Nayeri, S., Ahmadabadi, M., Taghipour, L., Senapathi, V., Astatkie, T., & Price, G. W. (2023). Micro (nano) plastics uptake, toxicity and detoxification in plants: Challenges and prospects. *Ecotoxicology and Environmental Safety*, *268*, 115676. doi:10.1016/j.ecoenv.2023.115676 PMID:37979355

Demirer, G. S., Silva, T. N., Jackson, C. T., Thomas, J. B., Ehrhardt, D. W., Rhee, S. Y., Mortimer, J. C., & Landry, M. P. (2021). Nano-technology to advance CRISPR–Cas genetic engineering of plants. *Nature Nanotechnology*, *16*, 243–250. doi:10.1038/s41565-021-00854-y PMID:33712738

Deng, J., Zhao, S., Liu, Y., Liu, C., & Sun, J. (2021). Nanosensors for Diagnosis of Infectious Diseases. *ACS Applied Bio Materials*, *4*(5), 3863–3879. doi:10.1021/acsabm.0c01247 PMID:35006812

Deng, Y., Ezyske, C. M., Suliman, W., Carlson, J., & Ali, N. (2017). Nanomaterials for soil remediation: Synthesis, applications and biotoxicity. *Nanotechnology for Environmental Engineering*, *2*(1), 1–18.

Dermont, G., Bergeron, M., Mercier, G., & Richer-Laflèche, M. (2008). Soil washing for metal removal: A review of physical/chemical technologies and field applications. *Journal of Hazardous Materials*, *152*(1), 1–31. doi:10.1016/j.jhazmat.2007.10.043 PMID:18036735

DeRosa, M. C., Monreal, C., Schnitzer, M., Walsh, R., & Sultan, Y. (2010). Nanotechnology in fertilizers. *Nature Nanotechnology*, *5*(2), 91–91. doi:10.1038/nnano.2010.2 PMID:20130583

Desforges, J. P., Hall, A., McConnell, B., Rosing-Asvid, A., Barber, J. L., Brownlow, A., De Guise, S., Eulaers, I., Jepson, P. D., Letcher, R. J., Levin, M., Ross, P. S., Samarra, F., Víkingson, G., Sonne, C., & Dietz, R. (2018). Predicting global killer whale population collapse from PCB pollution. *Science*, *361*(6409), 1373–1376. doi:10.1126/science.aat1953 PMID:30262502

Deshpande, P., Dapkekar, A., Oak, M. D., Paknikar, K. M., & Rajwade, J. M. (2017). Zinc complexed chitosan/TPP nanoparticles: A promising micronutrient nanocarrier suited for foliar application. *Carbohydrate Polymers*, *165*, 394–401. doi:10.1016/j.carbpol.2017.02.061 PMID:28363565

Devi, P., Saini, S., & Kim, K. H. (2019). The advanced role of carbon quantum dots in nanomedical applications. *Biosensors & Bioelectronics*, *141*, 111158. doi:10.1016/j.bios.2019.02.059 PMID:31323605

Dhoke, S. K., Mahajan, P., Kamble, R., and Khanna, A. (2013). Effect of nanoparticles suspension on the growth of mung (Vigna radiata) seedlings by foliar spray method. Nanotechnology development, 3:e1. doi:10.4081/nd.2013.e

Dimkpa, C. O., Andrews, J., Fugice, J., Singh, U., Bindraban, P. S., Elmer, W. H., ... White, J. C. (2019). Facile coating of urea with low-dose ZnO nanoparticles promotes wheat performance and enhances Zn uptake under drought stress. *Frontiers in Plant Science*, *10*, 1645. PMID:32174943

Dimkpa, C. O., & Bindraban, P. S. (2016). Fortification of micronutrients for efficient agronomic production: A review. *Agronomy for Sustainable Development*, *36*(1), 7. doi:10.1007/s13593-015-0346-6

Dimkpa, C. O., & Bindraban, P. S. (2018). Nanofertilizers: New products for the industry? *Journal of Agricultural and Food Chemistry*, *66*(26), 6462–6473. doi:10.1021/acs.jafc.7b02150 PMID:28535672

Dimkpa, C. O., McLean, J. E., Martineau, N., & Britt, D. W. (2011). Interaction of silver nanoparticles with an environmentally beneficial bacterium, Pseudomonas chlororaphis. *Journal of Hazardous Materials*, *199*(1-3), 68–76. doi:10.1016/j.jhazmat.2011.01.118 PMID:21339046

Dimkpa, C. O., Singh, U., Bindraban, P. S., Elmer, W. H., Gardea-Torresdey, J. L., & White, J. C. (2017). Zinc oxide nanoparticles alleviate drought-induced alterations in sorghum performance, nutrient acquisition, and grain fortification. *The Science of the Total Environment*, *688*, 926–934. doi:10.1016/j.scitotenv.2019.06.392 PMID:31726574

Ditta, A., Arshad, M., & Ibrahim, M. (2015). Nanoparticles in sustainable agricultural crop production: applications and perspectives. *Nanotechnology and plant sciences: nanoparticles and their impact on plants*, 55-75.

Dixit, R., Wasiullah, X., Malaviya, D., Pandiyan, K., Singh, U. B., Sahu, A., Shukla, R., Singh, B., Rai, J., Sharma, P., Lade, H., & Paul, D. (2015). Bioremediation of heavy metals from soil and aquatic environment: An overview of principles and criteria of fundamental processes. *Sustainability (Basel)*, *7*(2), 2189–2212. doi:10.3390/su7022189

Djanaguiraman, M., Belliraj, N., Bossmann, S. H., & Prasad, P. V. (2018). High temperature stress alleviation by selenium nanoparticle treatment in grain sorghum. *ACS Omega*, *3*(3), 2479–2491. doi:10.1021/acsomega.7b01934 PMID:31458542

Dolatabadi, J. E. N., & Kashanian, S. (2010). A Review on DNA Interaction with Synthetic Phenolic Food Additives. *Food Research International*, *43*(5), 1223–1230. doi:10.1016/j.foodres.2010.03.026

Dong, G., Wang, Y., Gong, L., Wang, M., Wang, H., He, N., Zheng, Y., & Li, Q. (2013). Formation of soluble Cr(III) end-products and nanoparticles during Cr(VI) reduction by bacillus cereus strain XMCr-6. *Biochemical Engineering Journal*, *70*, 166–172. doi:10.1016/j.bej.2012.11.002

Dorjee, L., Gogoi, R., Kamil, D., Kumar, R., Mondal, T. K., Pattanayak, S., & Gurung, B. (2023). Essential oil-grafted copper nanoparticles as a potential next-generation fungicide for holistic disease management in maize. *Frontiers in Microbiology*, *14*, 1204512. Advance online publication. doi:10.3389/fmicb.2023.1204512 PMID:37485521

Drostkar, E., Talebi, R., & Kanouni, H. (2016). Foliar application of Fe, Zn and NPK nano-fertilizers on seed yield and morphological traits in chickpea under rainfed condition. *Journal of Resources and Ecology*, *4*, 221–228.

Dubey, A., & Mailapalli, D. R. (2016). Nanofertilisers, nanopesticides, nanosensors of pest and nanotoxicity in agriculture. In *Sustainable Agriculture Reviews* (pp. 307–330). Springer. doi:10.1007/978-3-319-26777-7_7

Duncan, T. V. (2011). Applications of nanotechnology in food packaging and food safety: Barrier materials, antimicrobials and sensors. *Journal of Colloid and Interface Science*, *363*(1), 1–24. doi:10.1016/j.jcis.2011.07.017 PMID:21824625

Dutta, S., Pal, S., Panwar, P., Sharma, R. K., & Bhutia, P. L. (2022). Biopolymeric Nanocarriers for Nutrient Delivery and Crop Biofortification. *ACS Omega*, *7*(30), 25909–25920. doi:10.1021/acsomega.2c02494 PMID:35936412

Du, W., Tan, W., Peralta-Videa, J. R., Gardea-Torresdey, J. L., Ji, R., Yin, Y., & Guo, H. (2017). Interaction of metal oxide nanoparticles with higher terrestrial plants: Physiological and biochemical aspects. *Plant Physiology and Biochemistry*, *110*, 210–225. doi:10.1016/j.plaphy.2016.04.024 PMID:27137632

Dwivedi, S., Saquib, Q., Al-Khedhairy, A. A., & Musarrat, J. (2019). Nanoagriculture: A new frontier for sustainable agriculture. In Nanoecotoxicology (pp. 1-22). Elsevier.

Dwivedi, S., Saquib, Q., Al-Khedhairy, A. A., & Musarrat, J. (2016). Understanding the role of nanomaterials in agriculture. In D. Singh, H. Singh, & R. Prabha (Eds.), *Microbial Inoculants in Sustainable Agricultural Productivity* (pp. 271–288). Springer. doi:10.1007/978-81-322-2644-4_17

Egert, S., & Rimbach, G. (2011). Which sources of flavonoids: Complex diets or dietary supplements? *Advances in Nutrition*, *2*(1), 8–14. doi:10.3945/an.110.000026 PMID:22211185

Eggersdorfer, M., & Wyss, A. (2018). Carotenoids in human nutrition and health. *Archives of Biochemistry and Biophysics*, *652*, 18–26. doi:10.1016/j.abb.2018.06.001 PMID:29885291

Ehrlich, P. R., & Holdren, J. P. (1971). Impact of Population Growth: Complacency concerning this component of man's predicament is unjustified and counterproductive. *Science*, *171*(3977), 1212–1217. doi:10.1126/science.171.3977.1212 PMID:5545198

Ekambaram, P. (2012). a. Abdul hasan sathali and k. Priyanka; Solid lipid nanoparticles: A review; sci. Revs. *Chemical Communications (Cambridge)*, *2*(1).

Ekrami, E., Pouresmaieli, M., Hashemiyoon, E., Noorbakhsh, N., & Mahmoudifard, M. (2022). Nanotechnology: A sustainable solution for heavy metals remediation. *Environmental Nanotechnology, Monitoring & Management*, *18*, 100718. doi:10.1016/j.enmm.2022.100718

Elamawi, R. M. A., & El-Shafey, R. A. (2013). Inhibition effects of silver nanoparticles against rice blast disease caused by Magnaporthe grisea. *Egyptian Journal of Agricultural Research*, *91*(4), 1271–1283. doi:10.21608/ejar.2013.165104

Elegbede, J. A., & Lateef, A. (2019). Green nanotechnology in Nigeria: The research landscape, challenges and prospects. *Annals of Science and Technology*, *4*(2), 6–38. doi:10.2478/ast-2019-0008

El-Far, A., Shaheen, H., Alsenosy, A., El-Sayed, Y., Al Jaouni, S., & Mousa, S. (2018). *Costus speciosus*: Traditional uses, phytochemistry, and therapeutic potentials. *Pharmacognosy Reviews*, *12*(23), 120–127. doi:10.4103/phrev.phrev_29_17

Elizabeth, K. M. (2005). Antimicrobial activity of *Terminalia bellerica. Indian Journal of Clinical Biochemistry*, *20*(2), 150–153. doi:10.1007/BF02867416 PMID:23105549

Elliott, D. W., & Zhang, W. X. (2001). Field assessment of nanoscale bimetallic particles for groundwater treatment. *Environmental Science & Technology*, *35*(24), 4922–4926. doi:10.1021/es0108584 PMID:11775172

Elmer, W. H., & White, J. C. (2018). The use of nanoparticles to enhance the activity of fungicides and herbicides against plant pathogens and weeds. In *Nanoscience in Food and Agriculture 5* (pp. 81–103). Springer.

El-Shahawi, M. S., Hamza, A., Bashammakh, A. S., & Al-Saggaf, W. T. (2010). An overview on the accumulation, distribution, transformations,toxicity and analytical methods for the monitoring of persistent organic pollutants. *Talanta*, *80*(5), 1587–1597. doi:10.1016/j.talanta.2009.09.055 PMID:20152382

Elshayb, O. M., Nada, A. M., Farroh, K. Y., Al-Huqail, A. A., Aljabri, M., Binothman, N., & Seleiman, M. F. (2022). Utilizing Urea–Chitosan Nanohybrid for Minimizing Synthetic Urea Application and Maximizing Oryza sativa L. Productivity and N Uptake. *Agriculture*, *12*(7), 944. Advance online publication. doi:10.3390/agriculture12070944

Elshayb, O. M., Nada, A. M., Sadek, A. H., Ismail, S. H., Shami, A., Alharbi, B. M., Alhammad, B. A., & Seleiman, M. F. (2022). The Integrative Effects of Biochar and ZnO Nanoparticles for Enhancing Rice Productivity and Water Use Efficiency under Irrigation Deficit Conditions. *Plants*, *11*(11), 1416. Advance online publication. doi:10.3390/plants11111416 PMID:35684189

Elsheery, N. I., Helaly, M. N., El-Hoseiny, H. M., & Alam-Eldein, S. M. (2020). Zinc oxide and silicone nanoparticles to improve the resistance mechanism and annual productivity of salt-stressed mango trees. *Agronomy (Basel)*, *10*(4), 558. doi:10.3390/agronomy10040558

Enerijiofi, K. E., Musa, S. I., Igiebor, F. A., Lawani, M., Chuka, N. E., Odozi, E. B., & Ikhajiagbe, B. (2024). *Nanotechnological approaches for enhanced microbial activities, sustainable agriculture and the environment. In sustainable agriculture nanotechnology, biotechnology and management* (V. D. Rajput, A. Singh, K. Ghazaryan, & T. M. Minkina, Eds.). Walter de Gruyter.

Enerijiofi, K. E., & Obade, E. (2018). Assessment of the impact of herbicide contaminated top soils on its physiochemical, microbiological and enzymatic properties. *FUW Trends in Science and Technology Journal*, *3*(2), 802–806.

Ezati, P., Rhim, J. W., Molaei, R., Priyadarshi, R., & Han, S. (2022). Cellulose nanofibre-based coating integrated with nitrogen-functionalized carbon dots for active packaging applications of fresh fruit. *Postharvest Biology and Technology*, *186*, 111845. Advance online publication. doi:10.1016/j.postharvbio.2022.111845

Fadiji, A. E., Yadav, A. N., Santoyo, G., & Babalola, O. O. (2023). Understanding the plant-microbe interactions in environments exposed to abiotic stresses: An overview. *Microbiological Research*, *271*, 127368. doi:10.1016/j.micres.2023.127368 PMID:36965460

Fahad, S., Adnan, M., Noor, M., Arif, M., Alam, M., Khan, I. A., Ullah, H., Wahid, F., Mian, I. A., Jamal, Y., Basir, A., Hassan, S., Saud, S., & Amanullah. (2020). Major constraints for global rice production. In Advances in Rice Research for Abiotic Stress Tolerance (pp. 1-22). Woodhead Publishing.

Fajardo, C., Costa, G., Nande, M., Martín, C., Martín, M., & Sánchez-Fortún, S. (2019). Heavy metals immobilization capability of two iron-based nanoparticles (nZVI and Fe3O4): Soil and freshwater bioassays to assess ecotoxicological impact. *The Science of the Total Environment*, *656*, 421–432. doi:10.1016/j.scitotenv.2018.11.323 PMID:30513432

Fajardo, C., Sánchez-Fortún, S., Costa, G., Nande, M., Botías, P., García-Cantalejo, J., Mengs, G., & Martín, M. (2020). Evaluation of nanoremediation strategy in a Pb, Zn and Cd contaminated soil. *The Science of the Total Environment*, *706*, 136041. doi:10.1016/j.scitotenv.2019.136041 PMID:31855644

Fang, Y., & Ramasamy, R. P. (2015). Current and prospective methods for plant disease detection. *Biosensors (Basel)*, *5*(3), 537–561. doi:10.3390/bios5030537 PMID:26287253

Fang, Y., Umasankar, Y., & Ramasamy, R. P. (2014). Electrochemical detection of p-ethylguaiacol, a fungi infected fruit volatile using metal oxide nanoparticles. *Analyst*, *139*(15), 3804–3810. doi:10.1039/C4AN00384E PMID:24895939

Fan, K., Wang, H., Xi, J., Liu, Q., Meng, X., Duan, D., Gao, L., & Yan, X. (2017). Optimization of Fe3O4 Nanozyme Activity via Single Amino Acid Modification Mimicking an Enzyme Active Site. *Chemical Communications (Cambridge)*, *53*(2), 424–427. doi:10.1039/C6CC08542C PMID:27959363

FAO. (2017). *The future of food and agriculture – Trends and challenges*. Food and Agriculture Organization of the United Nations.

FAO. (2021). The State of Food and Agriculture 2021: Making agrifood systems more resilient to shocks and stresses. Food and Agriculture Organization of the United Nations. https://www.fao.org/3/cb4476en/cb4476en.pdf

Faraji, J., & Sepehri, A. (2020). Exogenous nitric oxide improves the protective effects oftio 2 nanoparticles on growth, antioxidant system, and photosynthetic performance ofwheat seedlings under drought stress. *Journal of Soil Science and Plant Nutrition*, *20*(2), 703–714. doi:10.1007/s42729-019-00158-0

Farkas, G. L., & Kiraaly, Z. (1962). Role of phenolic compounds in the physiology of plant diseases and disease resistance. *Journal of Phytopathology*, *44*(2), 105–150. doi:10.1111/j.1439-0434.1962.tb02005.x

Farooq, M., Wahid, A., Kobayashi, N., Fujita, D., & Basra, S. M. A. (2009). Plant drought stress: Effects, mechanisms and management. *Agronomy for Sustainable Development*, *29*(1), 185–212. doi:10.1051/agro:2008021

Fatima, F., Hashim, A., & Anees, S. (2021). Efficacy of nanoparticles as nanofertilizer production: A review. *Environmental Science and Pollution Research International*, *28*(2), 1292–1303. doi:10.1007/s11356-020-11218-9 PMID:33070292

Fazeli Sangani, M., Owens, G., & Fotovat, A. (2018). Transport of engineered nanoparticles in soils and aquifers. doi:10.1139/er-2018-0022

FDA. (2011). *Part 182—Substances Generally Recognized as Safe*. Food and Drug Administration.

Fei, L., Bilal, M., Qamar, S. A., Imran, H. M., Riasat, A., Jahangeer, M., Ghafoor, M., Ali, N., & Iqbal, H. M. N. (2022). Nano-remediation technologies for the sustainable mitigation of persistent organic pollutants. *Environmental Research*, *211*, 113060. doi:10.1016/j.envres.2022.113060 PMID:35283076

Feizi, H., Rezvani Moghaddam, P., Shahtahmassebi, N., & Fotovat, A. (2012). Impact of bulk and nanosized titanium dioxide (TiO_2) on wheat seed germination and seedling growth. *Biological Trace Element Research*, *146*(1), 101–106. doi:10.1007/s12011-011-9222-7 PMID:21979242

Fernández, P., & Grimalt, J. O. (2003). On the global distribution of persistent organic pollutants. *Chimia, 57*(9), 514–521. doi:10.2533/000942903777679000

Feynman, R. P. (1996). *No ordinary genius: the illustrated Richard Feynman.* WW Norton & Company.

Figueiredo, A. C., Barroso, J. G., Pedro, L. G., & Scheffer, J. J. (2008). Factors affecting secondary metabolite production in plants: Volatile components and essential oils. *Flavour and Fragrance Journal, 23*(4), 213–226. doi:10.1002/ffj.1875

Fischer, R. A., & Connor, D. J. (2018). Issues for cropping and agricultural science in the next 20 years. *Field Crops Research, 222*, 121–142. doi:10.1016/j.fcr.2018.03.008

Flood, J. (2010). The importance of plant health to food security. *Food Security, 2*(3), 215–231. doi:10.1007/s12571-010-0072-5

Floros, J. D., Newsome, R., Fisher, W., Barbosa-Cánovas, G. V., Chen, H., Dunne, C. P., German, J. B., Hall, R. L., Heldman, D. R., Karwe, M. V., Knabel, S. J., Labuza, T. P., Lund, D. B., Newell-McGloughlin, M., Robinson, J. L., Sebranek, J. G., Shewfelt, R. L., Tracy, W. F., Weaver, C. M., & Ziegler, G. R. (2010). Feeding the world today and tomorrow: the importance of food science and technology: an IFT scientific review. *Comprehensive Reviews in Food Science and Food Safety, 9*(5), 572–599. doi:10.1111/j.1541-4337.2010.00127.x PMID:33467827

Fogel, R., & Limson, J. (2016). Developing biosensors in developing countries: South Africa as a case study. *Biosensors (Basel), 6*(1), 5. doi:10.3390/bios6010005 PMID:26848700

Food and Nutrition Board Staff, Panel on Dietary Antioxidants, & Institute of Medicine Staff. (2000). Dietary reference intakes for vitamin C, vitamin E, selenium and carotenoids: a report of the panel on dietary antioxidants and related compounds, subcommittees on upper reference levels of nutrients and of interpretation and use of dietary reference intakes and the standing committee on the scientific evaluation of dietary reference intakes, Food and Nutrition Board, Institute of Medicine, National Academies Press, Washington, D.C. USA.

Food Safety Authority of Ireland (2008). The relevance of food safety of applications of nanotechnology in food and feed industries. Author.

Forgiarini, A., Esquena, J., Gonzalez, C., & Solans, C. (2001). Formation of nano-emulsions by low-energy emulsification methods at constant temperature. *Langmuir, 17*(7), 2076–2083. doi:10.1021/la001362n

Forim, M. R., Costa, E. S., da Silva, M. F. G. F., Fernandes, J. B., Mondego, J. M., & Boiça, A. L. Junior. (2013). Development of a new method to prepare nano-/microparticles loaded with extracts of Azadirachta indica, their characterization and use in controlling Plutella xylostella. *Journal of Agricultural and Food Chemistry, 61*(38), 9131–9139. doi:10.1021/jf403187y PMID:23991702

Foujdar, R., Kumar, S., Chauhan, A. K., & Bera, M. B. (2022). Role of nanoparticles for food processing and preservation In: Nanotechnology horizons in food process engineering, vol. 1. Goyal, M. R.; Malik, J. A.; Kumar, S. and Watharkar, R. B. (Eds.). Apple Academic Press, Inc. Co-published with CRC Press (Taylor and Francis). doi:10.1201/9781003305378_4

Fraceto, L. F., Grillo, R., de Medeiros, G. A., Scognamiglio, V., Rea, G., & Bartolucci, C. (2016). Nanotechnology in agriculture: Which innovation potential does it have? *Frontiers in Environmental Science, 4*, 20. doi:10.3389/fenvs.2016.00020

Franco-Lagos, C. L., Sanchez, E., Palacio-Marquez, A., Perez-Alvarez, S., Terrazas-Gomez, M., Villalobos-Cano, O., & Ramirez-Estrada, C. A. (2023). Efficacy of the application of boron nanofertilizer on biomass, yield, nitrogen assimilation and photosynthetic activity in green beans. *Notulae Botanicae Horti Agrobotanici Cluj-Napoca, 51*(1), 12795–12795. doi:10.15835/nbha51112795

Frenzel, M., Krolak, E., & Wanger, A. E. (2015). Physicochemical properties of WPL coated liposomes serving as stable transporters in a real food matrix. *Lebensmittel-Wissenschaft + Technologie*, *63*(1), 527–534. doi:10.1016/j.lwt.2015.03.055

Fröhlich, E. (2017). Role of omics techniques in the toxicity testing of nanoparticles. *Journal of Nanobiotechnology*, *15*(1), 1–22. doi:10.1186/s12951-017-0320-3 PMID:29157261

Fujita, H., Ohta, S., Nakamura, N., Somiya, M., & Horie, M. (2023). Progress of Endogenous and Exogenous Nanoparticles for Cancer Therapy and Diagnostics. In Genes (Vol. 14, Issue 2). MDPI. doi:10.3390/genes14020259

Fu, L., Wang, Z., Dhankher, O. P., & Xing, B. (2020). Nanotechnology as a new sustainable approach for controlling crop diseases and increasing agricultural production. *Journal of Experimental Botany*, *71*(2), 507–519. doi:10.1093/jxb/erz314 PMID:31270541

Gade, A., Ingle, P., Nimbalkar, U., Rai, M., Raut, R., Vedpathak, M., Jagtap, P., & Abd-Elsalam, K. A. (2023). Nanofertilizers: The Next Generation of Agrochemicals for Long-Term Impact on Sustainability in Farming Systems. *Agrochemicals*, *2*(2), 257–278. doi:10.3390/agrochemicals2020017

Galdames, A., Mendoza, A., Orueta, M., de Soto García, I. S., Sánchez, M., Virto, I., & Vilas, J. L. (2017). Development of new remediation technologies for contaminated soils based on the application of zero-valent iron nanoparticles and bioremediation with compost. *Resource-Efficient Technologies*, *3*(2), 166–176. doi:10.1016/j.reffit.2017.03.008

Galdames, A., Ruiz-Rubio, L., Orueta, M., Sánchez-Arzalluz, M., & Vilas-Vilela, J. L. (2020). Zero-valent iron nanoparticles for soil and groundwater remediation. *International Journal of Environmental Research and Public Health*, *17*(16), 5817. doi:10.3390/ijerph17165817 PMID:32796749

Ganie, A. S., Bano, S., Khan, N., Sultana, S., Rehman, Z., Rahman, M. M., Sabir, S., Coulon, F., & Khan, M. Z. (2021). Nanoremediation technologies for sustainable remediation of contaminated environments: Recent advances and challenges. *Chemosphere*, *275*, 130065. doi:10.1016/j.chemosphere.2021.130065 PMID:33652279

Gao, L., Fan, K., & Yan, X. (2020). Iron Oxide Nanozyme: A Multifunctional Enzyme Mimetics for Biomedical Application. *Nanozymology*, *7*, 105–140. doi:10.1007/978-981-15-1490-6_5

Gardea-Torresdey, J. L., Rico, C. M., & White, J. C. (2014). Trophic transfer, transformation, and impact of engineered nanomaterials in terrestrial environments. *Environmental Science & Technology*, *48*(5), 2526–2540. doi:10.1021/es4050665 PMID:24499408

Gardner, L., Kostarelos, K., Mallick, P., Dive, C., & Hadjidemetriou, M. (2022, August). Nano-omics: Nanotechnology-based multidimensional harvesting of the blood-circulating cancerome. *Nature Reviews. Clinical Oncology*, *19*(8), 551–561. doi:10.1038/s41571-022-00645-x PMID:35739399

Garg, B., & Bisht, T. (2016). Carbon Nanodots as Peroxidase Nanozymes for Biosensing. *Molecules (Basel, Switzerland)*, *21*(12), 1653. doi:10.3390/molecules21121653 PMID:27918453

Garg, D., Sridhar, K., Stephen Inbaraj, B., Chawla, P., Tripathi, M., & Sharma, M. (2023). Nano-biofertilizer formulations for agriculture: A systematic review on recent advances and prospective applications. *Bioengineering (Basel, Switzerland)*, *10*(9), 1010. doi:10.3390/bioengineering10091010 PMID:37760112

Gauba, A., Hari, S. K., Ramamoorthy, V., Vellasamy, S., Govindan, G., & Arasu, M. V. (2023). The versatility of green synthesized zinc oxide nanoparticles in sustainable agriculture: A review on metal-microbe interaction that rewards agriculture. *Physiological and Molecular Plant Pathology*, *125*, 102023. doi:10.1016/j.pmpp.2023.102023

Ghafariyan, M. H., Malakouti, M. J., Dadpour, M. R., Stroeve, P., & Mahmoudi, M. (2013). Effects ofmagnetite nanoparticles on soybean chlorophyll. *Environmental Science & Technology*, *47*, 10645–10652. doi:10.1021/es402249b PMID:23951999

Ghazaryan, K. A., Movsesyan, H. S., Khachatryan, H. E., & Ghazaryan, N. P. (2018). Geochemistry of Potentially Toxic Trace Elements in Soils of Mining Area: A Case Study from Zangezur Copper and Molybdenum Combine, Armenia. *Bulletin of Environmental Contamination and Toxicology*, *101*(6), 732–737. doi:10.1007/s00128-018-2443-0 PMID:30229278

Ghazaryan, K., Harutyunyan, A., Khachatryan, H., Singh, A., Minkina, T., Rajput, V., & Movsesyan, H. (2023). Study of tolerance and phytodesalination potential of wheat, oat, emmer, and barley for sustainable saline agriculture. *Applied Ecology and Environmental Research*, *21*(5), 4853–4882. doi:10.15666/aeer/2105_48534882

Ghormade, V., Deshpande, M. V., & Paknikar, K. M. (2011). Perspectives for nano-biotechnology enabled protection and nutrition of plants. *Biotechnology Advances*, *29*(6), 792–803. doi:10.1016/j.biotechadv.2011.06.007 PMID:21729746

Ghouri, M. Z., Khan, Z., Khan, S. H., Ismail, M., Aftab, S. O., Sultan, Q., & Ahmad, A. (2020). Nanotechnology: Transformation of agriculture and food security. *Bioscience*, *3*, 19.

Giannousi, K., Avramidis, I., & Dendrnou-Samara, C. (2013). Synthesis, characterization and evaluation of copper based nanoparticles as agrochemicals against Phytopthora infestans. *RSC Advances*, *3*(44), 21743–21752. doi:10.1039/c3ra42118j

Gibney, E., Cami, J., & Hedberg, J. (2015). Buckyballs in space solve 100-year-old riddle. *Nature*, *523*, 296–297.

Giraldo, J. P., Landry, M. P., Faltermeier, S. M., McNicholas, T. P., Iverson, N. M., Boghossian, A. A., Reuel, N. F., Hilmer, A. J., Sen, F., Brew, J. A., & Strano, M. S. (2014). Plant nanobionics approach to augment photosynthesis and biochemical sensing. *Nature Materials*, *13*(4), 400–408. doi:10.1038/nmat3890 PMID:24633343

Giraldo, J. P., Wu, H., Newkirk, G. M., & Kruss, S. (2019). Nanobiotechnology approaches for engineering smart plant sensors. *Nature Nanotechnology*, *14*(6), 541–553. doi:10.1038/s41565-019-0470-6 PMID:31168083

Girish, K., & Shankara, B. S. (2008). Neem-a green treasure. *Electronic Journal of Biology*, *4*(3), 102–111.

Gogos, A., Knauer, K., & Bucheli, T. D. (2012). Nanomaterials in plant protection and fertilization: Current state, foreseen applications, and research priorities. *Journal of Agricultural and Food Chemistry*, *60*(39), 9781–9792. doi:10.1021/jf302154y PMID:22963545

Gomes, H. I., Dias-Ferreira, C., & Ribeiro, A. B. (2018). Overview of in situ and ex situ remediation technologies for PCB-contaminated soils and sediments and obstacles for full-scale application. *The Science of the Total Environment*, *445*, 237–260. PMID:23334318

Gong, J. L., Wang, B., Zeng, G. M., Yang, C. P., Niu, C. G., Niu, Q. Y., Zhou, W. J., & Liang, Y. (2009). Removal of cationic dyes from aqueous solution using magnetic multi-wall carbon nanotube nanocomposite as adsorbent. *Journal of Hazardous Materials*, *164*(2–3), 1517–1522. doi:10.1016/j.jhazmat.2008.09.072 PMID:18977077

Gong, X., Huang, D., Liu, Y., Peng, Z., Zeng, G., Xu, P., Cheng, M., Wang, R., & Wan, J. (2018). Remediation of contaminated soils by biotechnology with nanomaterials: Bio-behavior, applications, and perspectives. *Critical Reviews in Biotechnology*, *38*(3), 455–468. doi:10.1080/07388551.2017.1368446 PMID:28903604

Gong, X., Huang, D., Liu, Y., Zeng, G., Wang, R., Wan, J., Zhang, C., Cheng, M., Qin, X., & Xue, W. (2017). Stabilized Nanoscale Zerovalent Iron Mediated Cadmium Accumulation and Oxidative Damage of Boehmeria nivea (L.) Gaudich Cultivated in Cadmium Contaminated Sediments. *Environmental Science & Technology*, *51*(19), 11308–11316. doi:10.1021/acs.est.7b03164 PMID:28850225

González-Melendi, P., Fernández-Pacheco, R., Coronado, M. J., Corredor, E., Testillano, P. S., Risueño, M. C., & Pérez-de-Luque, A. (2008). Nanoparticles as smart treatment-delivery systems in plants: Assessment of different techniques of microscopy for their visualization in plant tissues. *Annals of Botany*, *101*(1), 187–195. doi:10.1093/aob/mcm283 PMID:17998213

Goodin, S. (2008). Novel cytotoxic agents: Epothilones. *American Journal of Health-System Pharmacy*, *65*(10, Supplement_3), S10–S15. doi:10.2146/ajhp080089 PMID:18463327

Gorgzadeh, A., Amiri, P. A., Yasamineh, S., Naser, B. K., & abdulallah, K. (2024). The potential use of nanozyme in aging and age-related diseases. *Biogerontology*. Advance online publication. doi:10.1007/s10522-024-10095-w PMID:38466515

Goswami, A., Roy, I., Sengupta, S., & Debnath, N. (2010). Novel applications of solid and liquid formulations of nanoparticles against insect pests and pathogens. *Thin Solid Films*, *519*(3), 1252–1257. doi:10.1016/j.tsf.2010.08.079

Goswami, P., Mathur, J., & Srivastava, N. (2022). Silica nanoparticles as novel sustainable approach for plant growth and crop protection. *Heliyon*, *8*(7), e09908. doi:10.1016/j.heliyon.2022.e09908 PMID:35847613

Gottschalk, F., Sonderer, T., Scholz, R. W., & Nowack, B. (2013). Modeled environmental concentrations of engineered nanomaterials (TiO2, ZnO, Ag, CNT, fullerenes) for different regions. *Environmental Science & Technology*, *46*(21), 11584–11592. PMID:20000512

Gouin, S. (2004). Microencapsulation: Industrial appraisal of existing technologies and trends. *Trends in Food Science & Technology*, *15*(7-8), 330–347. doi:10.1016/j.tifs.2003.10.005

Govekar, S., Kumar, R., Suresh, R., & Sridharan, K. (2018). Nanotechnology to sustain a clean environment. In Emerging trends of nanotechnology in environment and sustainability: A review-based approach (pp. 3–11). doi:10.1007/978-3-319-71327-4_1

Goyal, V., Rani, D., Mehrotra, S., Deng, C., & Wang, Y. (2023). Unlocking the Potential of Nano-Enabled Precision Agriculture for Efficient and Sustainable Farming. In Plants (Vol. 12, Issue 21). Multidisciplinary Digital Publishing Institute (MDPI). doi:10.3390/plants12213744

Grande Tovar, C. D., Castro, J. I., Valencia, C. H., Navia Porras, D. P., Herminsul, M., Hernandez, J., Valencia Zapata, M. E., & Chaur, M. N. (2020). Nanocomposite films of chitosan-grafted carbon nano-onions for biomedical applications. *Molecules (Basel, Switzerland)*, *25*(5), 1203. doi:10.3390/molecules25051203 PMID:32155970

Grillo, R., Abhilash, P. C., & Fraceto, L. F. (2016). Nanotechnology applied to bio-encapsulation of pesticides. *Journal of Nanoscience and Nanotechnology*, *16*(1), 1231–1234. doi:10.1166/jnn.2016.12332 PMID:27398594

Grillo, R., Fraceto, L. F., Amorim, M. J. B., Scott-Fordsmand, J. J., Schoonjans, R., & Chaudhry, Q. (2021). Ecotoxicological and regulatory aspects of environmental sustainability of nanopesticides. *Journal of Hazardous Materials*, *404*, 124148. doi:10.1016/j.jhazmat.2020.124148 PMID:33059255

Grillo, R., Fraceto, L. F., & Pereira, A. E. S. (2021). Nanotechnology in agriculture: Challenges and opportunities. In *Nanomaterials for Agriculture and Forestry Applications* (pp. 51–77). Elsevier. doi:10.1016/B978-0-12-822840-4.00003-2

Grillo, R., Pereira, A. E., Nishisaka, C. S., de Lima, R., Oehlke, K., Greiner, R., & Fraceto, L. F. (2014). Chitosan/tripolyphosphate nanoparticles loaded with paraquat herbicide: An environmentally safer alternative for weed control. *Journal of Hazardous Materials*, *278*, 163–171. doi:10.1016/j.jhazmat.2014.05.079 PMID:24968252

Grobe, A., Renn, O., & Dialogik, A. J. (2018). *Risk governance of nanotechnology applications in food and cosmetics. International Risk Governance Council (IRGC).*

Grover, A., Bhandari, B. S., & Rai, N. (2011). Antimicrobial activity of medicinal plants-*Azadirachta indica* A. Juss, *Allium cepa* L. and *Aloe vera* L. *International Journal of Pharmaceutical & Technological. Research*, *3*(2), 1059–1065.

Gruère, G. P. (2012). Implications of nanotechnology growth in food and agriculture in OECD countries. *Food Policy*, *37*(2), 191–198. doi:10.1016/j.foodpol.2012.01.001

Gu, L., Li, N., Gong, J., Li, Q., Zhu, W., & Li, J. (2011). Berberine ameliorates intestinal epithelial tight-junction damage and down-regulates myosin light chain kinase pathways in a mouse model of endotoxinemia. *The Journal of Infectious Diseases*, *203*(11), 1602–1612. doi:10.1093/infdis/jir147 PMID:21592990

Guo, S., Luo, J., Yang, Q., Qiang, X., Feng, S., & Wan, Y. (2019). Decoloration of molasses by ultrafiltration and nanofiltration: Unraveling the mechanisms of high sucrose retention. *Food and Bioprocess Technology*, *12*(1), 39–53. doi:10.1007/s11947-018-2189-z

Gupta, A., Eral, H. B., Hatton, T. A., & Doyle, P. S. (2016). Nanoemulsions: Formation, properties and applications. *Soft Matter*, *12*(11), 2826–2841. doi:10.1039/C5SM02958A PMID:26924445

Gupta, A., Rico-Medina, A., & Caño-Delgado, A. I. (2020). The physiology of plant responses to drought. *Science*, *368*(6488), 266–269. doi:10.1126/science.aaz7614 PMID:32299946

Gupta, R., Anand, G., Gaur, R., & Yadav, D. (2019). Plant-microbiome interactions for sustainable agriculture: A review. *Archives of Microbiology*, *201*(10), 1361–1375. PMID:31332473

Gupta, V. K., Fakhri, A., Agarwal, S., Al-Salem, S., Atarod, M., Karami, C., Wang, C., & Varjani, S. (2019). A critical review on nanocomposites: From chemical and physical methods of synthesis to properties and applications. *Environmental Chemistry Letters*, *17*(2), 629–654.

Haghighi, M., Abolghasemi, R., & da Silva, J. A. T. (2014). Low and high temperature stress affect the growth characteristics of tomato in hydroponic culture with Se and nano-Se amendment. *Scientia Horticulturae*, *178*, 231–240. doi:10.1016/j.scienta.2014.09.006

Haghighi, M., Afifipour, Z., & Mozafarian, M. (2019). Effect of silicon and nano-silicon on growth and physiological traits of basil (Ocimum basilicum L.) under vegetative and generative growth stages. *Journal of Plant Nutrition*, *42*(10), 1057–1068.

Hajirostamlo, B., Mirsaeedghazi, N., Arefnia, M., Ali, M., & Shariati, E. A. F. (2015). The role of research and development in agriculture and its dependent concepts in agriculture. *Science and Engineering*, *4*(1), 79–81.

Han, A. M., Heo, H., & Kwon, Y. K. (2012). Berberine promotes axonal regeneration in injured nerves of the peripheral nervous system. *Journal of Medicinal Food*, *15*(4), 413–417. doi:10.1089/jmf.2011.2029 PMID:22316297

Hannon, J. C., Kerry, J., Cruz-Romero, M., Morris, M., & Cummins, E. (2015). Advances and challenges for the use of engineered nanoparticles in food contact materials. *Trends in Food Science & Technology*, *43*(1), 43–62. doi:10.1016/j.tifs.2015.01.008

Hansen, S. F., & Baun, A. (2012). European regulation affecting nanomaterials – review of limitations and future recommendations. *Dose-Response*, *10*(3), 364–383. doi:10.2203/dose-response.10-029.Hansen PMID:22942870

Haq, S. M., Tariq, A., Li, Q., Yaqoob, U., Majeed, M., Hassan, M., Fatima, S., Kumar, M., Bussmann, R. W., Moazzam, M. F. U., & Aslam, M. (2022a). Influence of Edaphic Properties in Determining Forest Community Patterns of the Zabarwan Mountain Range in the Kashmir Himalayas. *Forests*, *13*(8), 1214. Advance online publication. doi:10.3390/f13081214

Hara, T., Takeda, T. A., Takagishi, T., Fukue, K., Kambe, T., & Fukada, T. (2017). Physiological roles of zinc transporters: Molecular and genetic importance in zinc homeostasis. *The Journal of Physiological Sciences; JPS*, *67*(2), 283–301. doi:10.1007/s12576-017-0521-4 PMID:28130681

Harborne, J. B. (2014). *Introduction to ecological biochemistry*. Academic press.

Haris, M., Hussain, T., Mohamed, H. I., Khan, A., Ansari, M. S., Tauseef, A., Khan, A. A., & Akhtar, N. (2023). Nanotechnology – A new frontier of nano-farming in agricultural and food production and its development. *The Science of the Total Environment*, *857*, 159639. doi:10.1016/j.scitotenv.2022.159639 PMID:36283520

Harms, A., Brodersen, D. E., Mitarai, N., & Gerdes, K. (2018). Toxins, Targets, and Triggers: An Overview of Toxin-Antitoxin Biology. *Molecular Cell*, *70*(5), 768–784. doi:10.1016/j.molcel.2018.01.003 PMID:29398446

Hasan, M. K., Shopan, J., & Ahammed, G. J. (2020). Nanomaterials and soil health for agricultural crop production: current status and future prospects. *Nanomaterials for agriculture and forestry applications*, 289-312.

Hasan, M. R., Islam, M. N., & Islam, M. R. (2016). Phytochemistry, pharmacological activities and traditional uses of *Emblica officinalis*: A review. *International Current Pharmaceutical Journal*, *5*(2), 14–21. doi:10.3329/icpj.v5i2.26441

Hasanuzzaman, M., Nahar, K., Alam, M. M., Roychowdhury, R., & Fujita, M. (2013). Physiological, biochemical, and molecular mechanisms of heat stress tolerance in plants. *International Journal of Molecular Sciences*, *14*(5), 9643–9684. doi:10.3390/ijms14059643 PMID:23644891

Hasim, M. A.; Jabar, J. & Murad, M. A. (2019). A preliminary research on consumer acceptance in nanofood towards purchase intention: a pilot research. International Journal of Recent Technology and Engineering, 8(2S3), 352-356.

Hasler, C. M., & Blumberg, J. B. (1999). Symposium on phytochemicals: Biochemistry and physiology. *The Journal of Nutrition*, *129*, 756S–757S. doi:10.1093/jn/129.3.756S PMID:10082785

Hasmin, N. A., Zainol, Z. A., Ismail, R., & Mahmood, A. (2020). Regulatory challenges of nanofood labelling. *Journal of Politics and Law*, *13*(2), 241–249. doi:10.5539/jpl.v13n2p241

Hasmin, N. A., Zainol, Z. A., Ismail, R., & Matmin, J. (2021). Disclosure of nanomaterials under nanotechnology product inventory, voluntary certification, and voluntary labelling. *Pertanika Journal of Social Science & Humanities*, *29*(1), 157–174. doi:10.47836/pjssh.29.1.09

Hassanisaadi, M., Barani, M., Rahdar, A., Heidary, M., Thysiadou, A., & Kyzas, G. Z. (2022). Role of agrochemical-based nanomaterials in plants: Biotic and abiotic stress with germination improvement of seeds. *Plant Growth Regulation*, *97*(2), 375–418. doi:10.1007/s10725-021-00782-w

Hassan, M., Haq, S. M., Ahmad, R., Majeed, M., Sahito, H. A., Shirani, M., Mubeen, I., Aziz, M. A., Pieroni, A., Bussmann, R. W., Alataway, A., Dewidar, A. Z., Al-Yafrsi, M., Elansary, H. O., & Yessoufou, K. (2022). Traditional Use of Wild and Domestic Fauna among Different Ethnic Groups in the Western Himalayas—A Cross Cultural Analysis. *Animals (Basel)*, *12*(17), 2276. Advance online publication. doi:10.3390/ani12172276 PMID:36077997

Hassan, M., Haq, S. M., Majeed, M., Umair, M., Sahito, H. A., Shirani, M., Waheed, M., Aziz, R., Ahmad, R., Bussmann, R. W., Alataway, A., Dewidar, A. Z., El-Abedin, T. K. Z., Al-Yafrsi, M., Elansary, H. O., & Yessoufou, K. (2022). Traditional Food and Medicine: Ethno-Traditional Usage of Fish Fauna across the Valley of Kashmir: A Western Himalayan Region. *Diversity*, *14*(6), 455. Advance online publication. doi:10.3390/d14060455

Hatfaludi, T., Liska, M., Zellinger, D., Ousman, J. P., Szostak, M., árpád Ambrus, Jalava, K., & Lubitz, W. (2004). Bacterial ghost technology for pesticide delivery. *Journal of Agricultural and Food Chemistry*, *52*(18), 5627–5634. doi:10.1021/jf049489w PMID:15373403

Hayles, J., Johnson, L., Worthley, C., & Losic, D. (2017). Nanopesticides: a review of current research and perspectives. *New pesticides and soil sensors*, 193-225.

Heikal, Y. M., El-Esawi, M. A., El-Ballat, E. M., & Abdel-Aziz, H. M. (2023). Applications of nanoparticles for mitigating salinity and drought stress in plants: An overview on the physiological, biochemical and molecular genetic aspects. *New Zealand Journal of Crop and Horticultural Science*, *51*(3), 297–327. doi:10.1080/01140671.2021.2016870

Heiss, C., Keen, C. L., & Kelm, M. (2010). Flavanols and cardiovascular disease prevention. *European Heart Journal*, *31*(21), 2583–2592. doi:10.1093/eurheartj/ehq332 PMID:20852295

He, J., Li, J., Gao, Y., He, X., & Hao, G. (2023). Nano-based Smart Formulations: A potential solution to the Hazardous Effects of Pesticide on the Environment. *Journal of Hazardous Materials*, *456*, 131599. doi:10.1016/j.jhazmat.2023.131599 PMID:37210783

He, L., Liu, Y., Mustapha, A., & Lin, M. (2011). Antifungal activity of zinc oxide nanoparticles against Botrytis cinerea and Penicillium expansum. *Microbiological Research*, *166*(3), 207–215. doi:10.1016/j.micres.2010.03.003 PMID:20630731

Helwatkar, A., Riordan, D., & Walsh, J. (2014). Sensor technology for animal health monitoring. *International Journal on Smart Sensing and Intelligent Systems*, *7*(5), 1–6. doi:10.21307/ijssis-2019-057

Hernández-Hernández, H., González-Morales, S., Benavides-Mendoza, A., Ortega-Ortiz, H., Cadenas-Pliego, G., & Juárez-Maldonado, A. (2018). Effects of chitosan–PVA and Cu nanoparticles on the growth and antioxidant capacity of tomato under saline stress. *Molecules (Basel, Switzerland)*, *23*(1), 178. doi:10.3390/molecules23010178 PMID:29337864

Hessel, V., Escribà-Gelonch, M., Schmidt, S., Tran, N. N., Davey, K., Al-Ani, L. A., Julkapli, N. M., Wahab, Y. A., Khalil, I., Woo, M. W., & Gras, S. (2023). Nanofood process technology: Insights on how sustainability informs process design. *ACS Sustainable Chemistry & Engineering*, *11*(31), 11437–11458. doi:10.1021/acssuschemeng.3c01223 PMID:37564955

Hew, P. S., Selamat, J., Jambari, N. N., Murugesu, S., Sanny, M., Khatib, A., & Sukor, R. (2024). Quality and safety of food product–current assessment, issues, and metabolomics as a way forward. *Food Chemistry Advances*, *4*, 100632. doi:10.1016/j.focha.2024.100632

He, X., Deng, H., & Hwang, H. (2019). The current application of nanotechnology in food and agriculture. *Yao Wu Shi Pin Fen Xi*, *27*, 1–21. doi:10.1016/j.jfda.201812.002 PMID:30648562

Hidalgo, J., Anza, M., Epelde, L., Becerril, J. M., & Garbisu, C. (2022). Zero-valent iron nanoparticles and organic amendment assisted rhizoremediation of mixed contaminated soil using Brassica napus. *Environmental Technology & Innovation*, *28*, 102621. doi:10.1016/j.eti.2022.102621

Higashisaka, K., Yoshioka, Y. & Tsutsumi, Y. (2015). Applications and safety of nanomaterials used in the food industry. Food Safety Commission, Cabinet Office, Government of Japan, 39-47. doi:10.14252/foodsafetyfscj.2015005

Hoang, C. V., Thoai, D. N., Cam, N. T. D., Phuong, T. T. T., Lieu, N. T., Hien, T. T. T., Nhiem, D. N., Pham, T.-D., Tung, M. H. T., Tran, N. T. T., Mechler, A., & Vo, Q. V. (2022). Large-scale synthesis of Nanosilica from silica sand for plant stimulant applications. *ACS Omega*, *7*(45), 41687–41695. doi:10.1021/acsomega.2c05760 PMID:36406494

Hoffmann, M., Hotze, E. M., & Wiesner, M. R. (2007). Reactive oxygen species generation on nanoparticulate material. Mcgraw-Hill Professional.

Hofmann, T., Lowry, G. V., Ghoshal, S., Tufenkji, N., Brambilla, D., Dutcher, J. R., ... Wilkinson, K. J. (2021). Technology readiness and overcoming barriers to sustainably implement nanotechnology-enabled plant agriculture. *Nature Food*, *2*(6), 416–425.

Hofmann, T., Lowry, G. V., Lynch, I., Ghosh, S., Owen, R., Boyes, W. K., ... Unrine, J. (2020). Harmonizing across environmental nanomaterial testing media for increased comparability of nanomaterial datasets. *Environmental Science. Nano*, *7*(11), 3491–3499.

Hojjat, S. S., & Kamyab, M. (2017). The effect of silver nanoparticle on Fenugreek seed germination under salinity levels. *Russian Agricultural Sciences*, *43*(1), 61–65. doi:10.3103/S1068367417010189

Holden, P. A., Klaessig, F., Turco, R. F., Priester, J. H., Rico, C. M., Avila-Arias, H., Mortimer, M., Pacpaco, K., & Gardea-Torresdey, J. L. (2014). Evaluation of exposure concentrations used in assessing manufactured nanomaterial environmental hazards: Are they relevant? *Environmental Science & Technology*, *48*(18), 10541–10551. doi:10.1021/es502440s PMID:25158225

Hooshmand, S., Kassanos, P., Keshavarz, M., Duru, P., Kayalan, C. I., Kale, İ., & Bayazit, M. K. (2023). Wearable Nano-Based Gas Sensors for Environmental Monitoring and Encountered Challenges in Optimization. In Sensors (Basel, Switzerland) (Vol. 23, Issue 20). doi:10.3390/s23208648

Horn, P., Amabile, N., Angeli, F. S., Sansone, R., Stegemann, B., Kelm, M., Springer, M. L., Yeghiazarians, Y., Schroeter, H., & Heiss, C. (2014). Dietary flavanol intervention lowers the levels of endothelial microparticles in coronary artery disease patients. *British Journal of Nutrition*, *111*(7), 1245–1252. doi:10.1017/S0007114513003693 PMID:24286443

Hossain, M., Siddique, M. N. A., Bari, M. S., Hasan, M. M., Abedin, M. A., & Shahid, M. (2021). Nanotechnology applications in plant breeding, yield and crop management. In *Nanomaterials for Agriculture and Forestry Applications* (pp. 35–72). Elsevier.

Hosseinpour-Jaghdani, F., Shomali, T., Gholipour-Shahraki, S., Rahimi-Madiseh, M., & Rafieian-Kopaei, M. (2017). Cornus mas: A review on traditional uses and pharmacological properties. *Journal of Complementary & Integrative Medicine*, *14*(3), 20160137. doi:10.1515/jcim-2016-0137 PMID:28782352

Hou, J., Hu, C., Li, P., & Lin, D. (2022). Multidimensional bioresponses in nematodes contribute to the antagonistic toxic interaction between pentachlorophenol and TiO2 nanoparticles in soil. *Journal of Hazardous Materials*, *424*, 127587. doi:10.1016/j.jhazmat.2021.127587 PMID:34740505

Ho, V. A., Le, P. T., Nguyen, T. P., Nguyen, C. K., Nguyen, V. T., & Tran, N. Q. (2015). Silver core-shell nanoclusters exhibiting strong growth inhibition of plant-pathogenic fungi. *Journal of Nanomaterials*, *2015*, 241614. doi:10.1155/2015/241614

Huang, L., Chen, K., Zhang, W., Zhu, W., Liu, X., Wang, J., Wang, R., Hu, N., Suo, Y., & Wang, J. (2018a). ssDNATailorable Oxidase-Mimicking Activity of Spinel MnCo2O4 for Sensitive Biomolecular Detection in Food Sample. *Sensors and Actuators. B, Chemical*, *269*, 87–95. doi:10.1016/j.snb.2018.04.150

Huang, L., Sun, D.-W., Pu, H., Wei, Q., Luo, L., & Wang, J. (2019b). A Colorimetric Paper Sensor Based on the Domino Reaction of Acetylcholinesterase and Degradable γ-MnOOH Nanozyme for Sensitive Detection of Organophosphorus Pesticides. *Sensors and Actuators. B, Chemical*, *290*, 573–580. doi:10.1016/j.snb.2019.04.020

Huang, L., Zhu, Q., Zhu, J., Luo, L., Pu, S., Zhang, W., Zhu, W., Sun, J., & Wang, J. (2019c). Portable Colorimetric Detection of Mercury(II) Based on a Non-Noble Metal Nanozyme with Tunable Activity. *Inorganic Chemistry*, *58*(2), 1638–1646. doi:10.1021/acs.inorgchem.8b03193 PMID:30604958

Huang, Y., Ren, J., & Qu, X. (2019). Nanozymes: Classification, catalytic mechanisms, activity regulation, and applications. *Chemical Reviews*, *119*(6), 4357–4412. doi:10.1021/acs.chemrev.8b00672 PMID:30801188

Hua, S., Gong, J. L., Zeng, G. M., Yao, F. B., Guo, M., & Ou, X. M. (2017). Remediation of organochlorine pesticides contaminated lake sediment using activated carbon and carbon nanotubes. *Chemosphere*, *177*, 65–76. doi:10.1016/j.chemosphere.2017.02.133 PMID:28284117

Hua, X., Qian, G., Yang, J., Hu, B., Fan, J., Qin, N., Li, G., Wang, Y., & Liu, F. (2010). Development of an immunochromatographic assay for the rapid detection of chlorpyrifos-methyl in water samples. *Biosensors & Bioelectronics*, *26*(1), 189–194. doi:10.1016/j.bios.2010.06.005 PMID:20598518

Hua, Z., Ma, S., Ouyang, Z., Liu, P., Qiang, H., & Guo, X. (2023). The review of nanoplastics in plants: Detection, analysis, uptake, migration and risk. *Trends in Analytical Chemistry*, *158*, 116889. doi:10.1016/j.trac.2022.116889

Hu, J., Dun, G., Geng, X., Chen, J., Wu, X., & Ren, T. L. (2023). Recent progress in flexible micro-pressure sensors for wearable health monitoring. *Nanoscale Advances*, *5*(12), 3131–3145. doi:10.1039/D2NA00866A PMID:37325539

Hull, M., & Bowman, D. (Eds.). (2018). *Nanotechnology environmental health and safety: risks, regulation, and management*. William Andrew.

Hu, Q. H., Weng, J. Q., & Wang, J. S. (2010). Sources of anthropogenic radionuclides in the environment: A review. *Journal of Environmental Radioactivity*, *101*(6), 426–437. doi:10.1016/j.jenvrad.2008.08.004 PMID:18819734

Hussain, S., Raza, A., Abdo, H. G., Mubeen, M., Tariq, A., Nasim, W., Majeed, M., Almohamad, H., & Al Dughairi, A. A. (2023b). Relation of land surface temperature with different vegetation indices using multi-temporal remote sensing data in Sahiwal region, Pakistan. *Geoscience Letters*, *10*(1), 33. Advance online publication. doi:10.1186/s40562-023-00287-6

Hussein, M. M., Abou-Baker, N. H., & Tawfik, M. M. (2020). The prospective roles of nano-materials and nano-technology in germination and seedling growth of some crop species. In *Plant Nanobionics* (pp. 177–209). Springer.

Hussein, R. A., & El-Anssary, A. A. (2019). Plants secondary metabolites: The key drivers of the pharmacological actions of medicinal plants. *Herbal Medicine*, *1*(3), 11–30. doi:10.5772/intechopen.76139

Iavicoli, I., Leso, V., Beezhold, D. H., & Shvedova, A. A. (2017). Nanotechnology in agriculture: Opportunities, toxicological implications, and occupational risks. *Toxicology and Applied Pharmacology*, *329*, 96–111. doi:10.1016/j.taap.2017.05.025 PMID:28554660

Ibrahim, E. A. (2016). Seed priming to alleviate salinity stress in germinating seeds. *Journal of Plant Physiology*, *192*, 38–46. doi:10.1016/j.jplph.2015.12.011 PMID:26812088

Iijima, S., & Ichihashi, T. (1993). Single-shell carbon nanotubes of 1-nm diameter. *Nature*, *363*(6430), 603–605. doi:10.1038/363603a0

Ilk, S., Saglam, N., & Özgen, M. (2017). Kaempferol loaded lecithin/chitosan nanoparticles: Preparation, characterization, and their potential applications as a sustainable antifungal agent. *Artificial Cells, Nanomedicine, and Biotechnology*, *45*(5), 907–916. doi:10.1080/21691401.2016.1192040 PMID:27265551

Imada, K., Sakai, S., Kajihara, H., Tanaka, S., & Ito, S. (2016). Magnesium oxide nanoparticles induce systemic resistance in tomato against bacterial wilt disease. *Plant Pathology*, *65*(4), 551–560. doi:10.1111/ppa.12443

Inani, H., Jain, R. K., & Jain, P. K. (2021). Development of carbon nanotube-based biosensor for real-time monitoring of soil nitrate levels in maize. *Journal of Hazardous Materials*, *408*, 124957. doi:10.1016/j.jhazmat.2020.124957

Inobeme, A., Mathew, J. T., Adetunji, C. O., Maliki, M., Onyeachu, B. I., Adekoya, M. A., Eziukwu, C. A., Kelani, T., Inobeme, J., Ajai, A. I., Jacob, J. O., Olori, E., & Okonkwo, S. (2022). Advances, prospects and concerns in nanomaterials for food processing. *Nanotechnology and Advanced Material Science*, *5*(3), 1–2.

Ion, A. C., Ion, I., Culetu, A., & Gherase, D. (2010). Carbon-based nanomaterials. Environmental applications. Romania.

IPCC. (2014). Climate Change 2014: Synthesis Report. Contribution of Working Groups I, II and III to the Fifth Assessment Report of the Intergovernmental Panel on Climate Change [Core Writing Team, R.K. Pachauri and L.A. Meyer (eds.)]. IPCC, Geneva, Switzerland.

Iqbal, M., Umar, S., & Mahmooduzzafar. (2019). Nano-fertilization to enhance nutrient use efficiency and productivity of crop plants. *Nanomaterials and plant potential*, 473-505.

Iqbal, A., Haq, A. U., Cerrón-Calle, G. A., Naqvi, S. A. R., Westerhoff, P., & Garcia-Segura, S. (2021). Green synthesis of flower-shaped copper oxide and nickel oxide nanoparticles via Capparis decidua leaf extract for synergic adsorption-photocatalytic degradation of pesticides. *Catalysts*, *11*(7), 806. doi:10.3390/catal11070806

Iqbal, M. A. (2019). *"Nano-fertilizers for sustainable crop production under changing climate: a global perspective," in Sustainable Crop Production.* IntechOpen.

Iqbal, M., Raja, N. I., Hussain, M., Ejaz, M., & Yasmeen, F. (2019). Effect of silver nanoparticles on growth ofwheat under heat stress. *Iranian Journal of Science and Technology. Transaction A, Science*, *43*(2), 387–395. doi:10.1007/s40995-017-0417-4

Islam, M. T. (2019). Applications of nanomaterials for future food security: Challenges and prospects. *Malaysian Journal of Halal Research Journal*, *2*(1), 6–9. doi:10.2478/mjhr-2019-0002

Ismail, M., Prasad, R., Ibrahim, A. I., & Ahmed, A. I. (2017). Modern prospects of nanotechnology in plant pathology. *Nanotechnology: An Agricultural Paradigm*, 305-317.

Ismail, S., Budin, S., & Ali, S. A. M. (2019). The nanotechnology application and workforce health and safety – a study of the Malaysia laws, statutory regulations and guidelines on nanotechnology. *Journal of Physics: Conference Series*, 1–9. doi:10.1088/1742-6596/1349/1/012031

ISO 4656. Rubber Compounding Ingredients - Carbon Black - Determination of Oil Absorption Number (OAN) and Oil Absorption Number of Compressed Sample (COAN). ACS Omega 2012, https://pubs.acs.org/journal/acsodf Article

Istiqola, A., & Syafiuddin, A. (2020). A review of silver nanoparticles in food packaging technologies: Regulation, methods, properties, migration, and future challenges. *Journal of the Chinese Chemical Society (Taipei)*, *67*(11), 1–15. doi:10.1002/jccs.202000179

Ivanova, O. S., Zamyatin, A. V., Zamyatina, A. O., & Vorobyev, A. A. (2018). Nanosensors for detection of heavy metal ions in the environment. *Nanotechnology for Environmental Engineering*, *3*(1), 1–11.

Iyer, J. K., Dickey, A., Rouhani, P., Kaul, A., Govindaraju, N., Singh, R. N., & Kaul, R. (2020). Nanoparticles of pure and doped cerium oxide as promising catalysts for the remediation of water pollution: A review. *Environmental Chemistry Letters*, *18*(6), 1917–1948.

Jaberzadeh, A., Moaveni, P., Moghadam, H. R., & Zahedi, H. (2013). Influence of bulk and nanoparticles titanium foliar application on some agronomic traits, seed gluten and starch contents of wheat subjected to water deficit stress. *Notulae Botanicae Horti Agrobotanici Cluj-Napoca*, *41*(1), 201–207. doi:10.15835/nbha4119093

Jafari, A., & Hatami, M. (2022). Foliar-applied nanoscale zero-valent iron (nZVI) and iron oxide (Fe_3O_4) induce differential responses in growth, physiology, antioxidative defense and biochemical indices in Leonurus cardiaca L. *Environmental Research*, *215*, 114254. doi:10.1016/j.envres.2022.114254 PMID:36096173

Jagetia, G. C., Baliga, M. S., Malagi, K. J., & Kamath, M. S. (2002). The evaluation of the radioprotective effect of Triphala (an ayurvedic rejuvenating drug) in the mice exposed to γ-radiation. *Phytomedicine*, *9*(2), 99–108. doi:10.1078/0944-7113-00095 PMID:11995956

Jagtiani, E. (2021). Advancements in nanotechnology for food science and industry. *Food Frontiers*, *3*(1), 56–82. doi:10.1002/fft2.104

Jain, A., Ranjan, S., Dasgupta, N., & Ramalingam, C. (2018). Nanomaterials in food and agriculture: An overview on their safety concerns and regulatory issues. *Critical Reviews in Food Science and Nutrition*, *58*(2), 297–317. doi:10.1080/10408398.2016.1160363 PMID:27052385

Jain, D., & Kothari, S. L. (2014). Green synthesis of silver nanoparticles and their application in plant virus inhibition. *Journal of Mycology and Plant Pathology*, *44*(1), 21–24.

Jain, P. K. (2021). Nanomics-based agriculture for sustainable crop production. In *Nanomaterials for Agriculture and Forestry Applications* (pp. 1–25). Elsevier. doi:10.1016/B978-0-12-822840-4.00001-9

Jain, P. K., Das, D. E., Pandey, N. A., & Jain, P. R. (2016). Traditional Indian herb *Emblica officinalis* and its medicinal importance. *Innovative Journal of Ayurvedic Science*, *4*(4), 1–15.

Jain, S., Cherukupalli, S. K., Mahmood, A., Gorantla, S., Rapalli, V. K., Dubey, S. K., & Singhvi, G. (2019). Emerging nanoparticulate systems: Preparation techniques and stimuli responsive release characteristics. *Journal of Applied Pharmaceutical Science*, *9*(8), 130–143. doi:10.7324/JAPS.2019.90817

Jallow, M., Awadh, D., Albaho, M., Devi, V., & Ahmad, N. (2017). Monitoring of Pesticide Residues in Commonly Used Fruits and Vegetables in Kuwait. *International Journal of Environmental Research and Public Health*, *14*(8), 833. doi:10.3390/ijerph14080833 PMID:28757570

Jamil, M. D., Waheed, M., Akhtar, S., Bangash, N., Chaudhari, S. K., Majeed, M., Hussain, M., Ali, K., & Jones, D. A. (2022). Invasive Plants Diversity, Ecological Status, and Distribution Pattern in Relation to Edaphic Factors in Different Habitat Types of District Mandi Bahauddin, Punjab, Pakistan. *Sustainability (Switzerland)*, *14*(20). doi:10.3390/su142013312

Jampílek, J., & Kráľová, K. 2017. Nanopesticides: preparation, targeting, and controlled release. In New Pesticides and Soil Sensors (pp. 81-127). Academic Press. doi:10.1016/B978-0-12-804299-1.00004-7

Jampílek, J., & Králová, K. (2017). Nanomaterials for delivery of nutrients and growth-promoting compounds to plants. In *Nanotechnology: An Agricultural Paradigm* (pp. 177–226). Springer. doi:10.1007/978-981-10-4573-8_9

Jampílek, J., Krátký, M., Konefał, R., & Králík, J. (2019). Nanomaterials for soil remediation: A review. *Environmental Chemistry Letters*, *17*(3), 1263–1270.

Janatova, A., Bernardos, A., Smid, J., Frankova, A., Lhotka, M., Kourimská, L., & Kloucek, P. (2015). Long-term antifungal activity of volatile essential oil components released from mesoporous silica materials. *Industrial Crops and Products*, *67*, 216–220. doi:10.1016/j.indcrop.2015.01.019

Jangpangi, K., Khati, A., Chauhan, R. S., Rajesh, R., & Kumar, N. (2023). Assessment of the haematological and biochemical effects of Himalayan herb *Urtica dioica* leaves diets fed amur carp, *Cyprinus carpio haematopterus*. *Turkish Journal of Fisheries and Aquatic Sciences*, *24*(4), TRJFAS23854. doi:10.4194/TRJFAS23854

Javed, T., Singhal, R. K., Shabbir, R., Shah, A. N., Kumar, P., Jinger, D., ... Siuta, D. (2022). Recent advances in agronomic and physio-molecular approaches for improving nitrogen use efficiency in crop plants. *Frontiers in Plant Science*, *13*, 877544. doi:10.3389/fpls.2022.877544 PMID:35574130

Jeevanandam, J., Barhoum, A., Chan, Y. S., Dufresne, A., & Danquah, M. K. (2018). Review on nanoparticles and nanostructured materials: History, sources, toxicity and regulations. *Beilstein Journal of Nanotechnology*, *9*, 1050–1074. doi:10.3762/bjnano.9.98 PMID:29719757

Jenne, M., Kambham, M., Tollamadugu, N. P., Karanam, H. P., Tirupati, M. K., Reddy Balam, R., & Yagireddy, M. (2018). The use of slow releasing nanoparticle encapsulated Azadirachtin formulations for the management of Caryedon serratus O. (groundnut bruchid). *IET Nanobiotechnology / IET*, *12*(7), 963–967. doi:10.1049/iet-nbt.2017.0276 PMID:30247138

Jha, A., Pathania, D., Damathia, B., Raizada, P., Rustagi, S., Singh, P., ... Chaudhary, V. (2023). Panorama of biogenic nano-fertilizers: A road to sustainable agriculture. *Environmental Research*, *235*, 116456. doi:10.1016/j.envres.2023.116456 PMID:37343760

Jha, S., Ghosal, M., & Gupta, S. K. (2019). Nanosensors for precision agriculture. In *Nanotechnology for agriculture* (pp. 159–183). Springer.

Jiang, C., Zu, C., Lu, D., Zheng, Q., Shen, J., Wang, H., & Li, D. (2017). Effect of exogenous selenium supply on photosynthesis, Na+ accumulation and antioxidative capacity of maize (Zea mays L.) under salinity stress. *Scientific Reports*, *7*(1), 1–14. doi:10.1038/srep42039 PMID:28169318

Jiang, X., Sun, D., Zhang, G., He, N., Gao, Q., Yue, T., ... Li, J. (2018). A colorimetric biosensor based on single-enzyme nanoparticles for detecting the plant pathogen Xanthomonas oryzae pv. oryzae. *Sensors and Actuators. B, Chemical*, *273*, 1077–1084.

Jing, X., Jiang, M., Li, Y., Wang, Y., Zhang, X., Tang, L., Zhang, Y., & Chen, Y. (2019). Immobilization of lead contaminated soil using a new class of iron phosphate nanoparticles with reference to their reduction of bioavailability and leachability. *Chemosphere*, *227*, 178–186.

Joshi, P., & Dhawan, V. (2005). *Swertia chirayita*–an overview. *Current Science*, *89*(4), 635–640.

Jo, Y. K., Kim, B. H., & Jung, G. (2009). Antifungal activity of silver ions and nanoparticles on phytopathogenic fungi. *Plant Disease*, *93*(10), 1037–1043. doi:10.1094/PDIS-93-10-1037 PMID:30754381

Judy, J. D., Unrine, J. M., & Bertsch, P. M. (2011). Evidence for biomagnification of gold nanoparticles within a terrestrial food chain. *Environmental Science & Technology*, *45*(2), 776–781. doi:10.1021/es103031a PMID:21128683

Ju, Y. H., Clausen, L. M., Allred, K. F., Almada, A. L., & Helferich, W. G. (2004). β-sitosterol, β-sitosterol glucoside, and a mixture of β-sitosterol and β-sitosterol glucoside modulate the growth of estrogen-responsive breast cancer cells in vitro and in ovariectomized athymic mice. *The Journal of Nutrition*, *134*(5), 1145–1151. doi:10.1093/jn/134.5.1145 PMID:15113961

JuŸwiak, S., Wójcicki, J., Mokrzycki, K., Marchlewicz, M., Białecka, M., Wenda-Różewicka, L., Gawronska-Szklarz, B., & Droździk, M. (2005). Effect of quercetin on experimental hyperlipidemia and atherosclerosis in rabbits. *Pharmacological Reports*, *57*(57), 604–609. PMID:16227643

Kabiri, S., Degryse, F., Tran, D. N., da Silva, R. C., McLaughlin, M. J., & Losic, D. (2017). Graphene oxide: A new carrier for slow release of plant micronutrients. *ACS Applied Materials & Interfaces*, *9*(49), 43325–43335. doi:10.1021/acsami.7b07890 PMID:29160685

Kabra, A., Martins, N., Sharma, R., Kabra, R., & Baghel, U. S. (2019). *Myrica esculenta* Buch.-Ham. ex D. Don: A natural source for health promotion and disease prevention. *Plants*, *8*(6), 149. doi:10.3390/plants8060149 PMID:31159283

Kah, M., Navarro, D., Kookana, R. S., Kirby, J. K., Santra, S., Ozcan, A., & Kabiri, S. (2018). Fate of engineered nanoparticles in natural and engineered soils. Engineered Nanoparticles and the Environment: Biophysicochemical Processes and Toxicity, 30-59.

Kah, M., Beulke, S., Tiede, K., & Hofmann, T. (2013). Nanopesticides: State of knowledge, environmental fate, and exposure modeling. *Critical Reviews in Environmental Science and Technology*, *43*(16), 1823–1867. doi:10.1080/10643389.2012.671750

Kah, M., & Hofmann, T. (2014). Nanopesticide research: Current trends and future priorities. *Environment International*, *63*, 224–235. doi:10.1016/j.envint.2013.11.015 PMID:24333990

Kah, M., Kookana, R. S., Gogos, A., & Bucheli, T. D. (2018). A critical evaluation of nanopesticides and nanofertilizers against their conventional analogues. *Nature Nanotechnology*, *13*(8), 677–684. doi:10.1038/s41565-018-0131-1 PMID:29736032

Kah, M., Navarro, D., Kookana, R. S., Kirby, J. K., Santra, S., Ozcan, A., & Kabiri, S. (2019). Impact of (nano) formulations on the distribution and wash-off of copper pesticides and fertilisers applied on citrus leaves. *Environmental Chemistry*, *16*(6), 401–410. doi:10.1071/EN18279

Kah, M., Tufenkji, N., & White, J. C. (2019). Nano-enabled strategies to enhance crop nutrition and protection. *Nature Nanotechnology*, *14*(6), 532–540. doi:10.1038/s41565-019-0439-5 PMID:31168071

Kakoti, B. B., Pradhan, P., Borah, S., Mahato, K., & Kumar, M. (2013). Analgesic and anti-inflammatory activities of the methanolic stem bark extract of *Nyctanthes arbor-tristis* Linn. *BioMed Research International*, *2013*, 826295. doi:10.1155/2013/826295 PMID:23984409

Kalluri, R., Narayanan, J., & Thangavel, K. (2021). A gold nanoparticle-based colorimetric biosensor for rapid detection of arsenic in soil samples. *Journal of Hazardous Materials*, *403*, 123875. doi:10.1016/j.jhazmat.2020.123875

Kandasamy, S., & Prema, R. S. (2015). Methods of synthesis of nano particles and its applications. Academic Press.

Kandiannan, K., Sivaraman, K., Thankamani, C. K., & Peter, K. V. (1996). Agronomy of ginger (*Zingiber officinale* Rosc.). *Journal of Spices and Aromatic Crops*, *5*(1), 1–27.

Kannan, S., & Jasim, B. (2019). Copper oxide nanoparticles as a novel tool to control bacterial leaf blight of rice caused by Xanthomonas oryzae pv. oryzae. *Pesticide Biochemistry and Physiology*, *156*, 1–5.

Karn, B., Kuiken, T., & Otto, M. (2009). Nanotechnology and in situ remediation: A review of the benefits and potential risks. *Environmental Health Perspectives*, *117*(12), 1813–1831. doi:10.1289/ehp.0900793 PMID:20049198

Karnwal, A., Dohroo, A., & Malik, T. (2023). Unveiling the potential of bioinoculants and nanoparticles in sustainable agriculture for enhanced plant growth and food security. *BioMed Research International*, *2023*, 1–23. Advance online publication. doi:10.1155/2023/6911851 PMID:38075309

Kasa, Y. (2013). The application of nanotechnology and its risk in developing countries, 4(7): 249-259. *International Journal of Advanced Research*, *4*(2), 269–259.

Kashyap, P. L., Prasanna, R., & Kumar, R. (2021). Nanomics for sustainable agriculture. In *Nanomaterials for Agriculture and Forestry Applications* (pp. 27–50). Elsevier. doi:10.1016/B978-0-12-822840-4.00002-0

Kashyap, P. L., Xiang, X., & Heiden, P. (2015). Chitosan nanoparticle based delivery systems for sustainable agriculture. *International Journal of Biological Macromolecules*, *77*, 36–51. doi:10.1016/j.ijbiomac.2015.02.039 PMID:25748851

Kaushal, M., Kumar, A., & Kaushal, R. (2019). Carbon nanotube and graphene-based nanomaterials for agriculture. In *Nanomaterials for Agriculture and Forestry Applications* (pp. 111–142). Elsevier.

Keller, A. A., Adeleye, A. S., Conway, J. R., Garner, K. L., Zhao, L., Cherr, G. N., & Zuverza-Mena, N. (2017). Comparative environmental fate and toxicity of copper nanomaterials. *NanoImpact*, *7*, 28–40. doi:10.1016/j.impact.2017.05.003

Khaled, A. Y., Aziz, S. A., Bejo, S. K., Nawi, N. M., & Ismail, I. (2018). Foliar application of nano-silica positively influenced the growth and productivity of faba bean (Vicia faba) grown under different concentrations of NaCl. *Environmental Science and Pollution Research International*, *25*(33), 33137–33146.

Khalid, S., Shahid, M., Niazi, N. K., Murtaza, B., Bibi, I., & Dumat, C. (2017). A comparison of technologies for remediation of heavy metal contaminated soils. *Journal of Geochemical Exploration*, *182*, 247–268. doi:10.1016/j.gexplo.2016.11.021

Khali, I., Yehye, W. A., Etxenerria, A. E., Alhadi, A. A., Dezfooli, S. M., Julkapli, N. B. M., Bsirun, W. J., & Seyfoddin, A. (2019). Nanoantioxidants: Recent trends in antioxidant delivery applications. *Antioxidants*, *9*(1), 1–30. doi:10.3390/antiox9010024 PMID:31888023

Khalil, H. S., Maulu, S., Verdegem, M., & Abdel-Tawwab, M. (2023). Embracing nanotechnology for selenium application in aquafeeds. In Reviews in Aquaculture (Vol. 15, Issue 1, pp. 112–129). John Wiley and Sons Inc. doi:10.1111/raq.12705

Khamitova, K. K., Kayupov, B. A., Yegemova, S. S., Gabdullin, M. T., Abdullin, K. A., Ismailov, D. V., & Kerimbekov, D. S. (2019). The use of fullerenes as a biologically active molecule. *International Journal of Nanotechnology*, *16*(1-3), 100–108. doi:10.1504/IJNT.2019.102396

Khan, A. A., & Paquiza, L. (2011). Characterization and ion-exchange behavior of thermally stable nanocomposite polyaniline zirconium titanium phosphate: Its analytical application in separation of toxic metals. *Desalination*, *265*(1–3), 242–254. doi:10.1016/j.desal.2010.07.058

Khan, A. A., Paquiza, L., & Khan, A. (2010). An advanced nano-composite cation-exchanger polypyrrolezirconium titanium phosphate as a Th(IV)-selective potentiometric sensor: Preparation, characterization and its analytical application. *Journal of Materials Science*, *45*(13), 3610–3625. doi:10.1007/s10853-010-4407-6

Khan, A. M., Li, Q., Saqib, Z., Khan, N., Habib, T., Khalid, N., Majeed, M., & Tariq, A. (2022). MaxEnt Modelling and Impact of Climate Change on Habitat Suitability Variations of Economically Important Chilgoza Pine (Pinus gerardiana Wall.) in South Asia. *Forests*, *13*(5), 715. Advance online publication. doi:10.3390/f13050715

Khan, A. S. (2014, May). Ethics and nanotechnology. In *2014 IEEE International Symposium on Ethics in Science, Technology and Engineering* (pp. 1-14). IEEE.

Khan, A. U., & Gilani, A. H. (2008). Pharmacodynamic evaluation of *Terminalia bellerica* for its antihypertensive effect. *Yao Wu Shi Pin Fen Xi*, *16*(3), 6–14.

Khan, A. U., Khan, M., Cho, M. H., & Khan, M. M. (2020). Selected nanotechnologies and nanostructures for drug delivery, nanomedicine and cure. *Bioprocess and Biosystems Engineering*, *43*(8), 1339–1357. doi:10.1007/s00449-020-02330-8 PMID:32193755

Khanal, A., Devkota, H. P., Kaundinnyayana, S., Gyawali, P., Ananda, R., & Adhikari, R. (2021). Culinary herbs and spices in Nepal: A review of their traditional uses, chemical constituents, and pharmacological activities. *Ethnobotany Research and Applications*, *21*, 1–18. doi:10.32859/era.21.40.1-18

Khandelwal, N., Barbole, R. S., Banerjee, S. S., Chate, G. P., Biradar, A. V., Khandare, J. J., & Giri, A. P. (2016). Budding trends in integrated pest management using advanced micro-and nano-materials: Challenges and perspectives. *Journal of Environmental Management*, *184*, 157–169. doi:10.1016/j.jenvman.2016.09.071 PMID:27697374

Khan, I., Saeed, K., & Khan, I. (2019). Nanoparticles: Properties, applications and toxicities. *Arabian Journal of Chemistry*, *12*(7), 908–931. doi:10.1016/j.arabjc.2017.05.011

Khan, K. H. (2009). Roles of *Emblica officinalis* in medicine-A review. *Botany Research International*, *2*(4), 218–228.

Khan, M. N., Mobin, M., Abbas, Z. K., & Siddiqui, M. H. (2017). Nitric oxide-induced tolerance to salt stress in plants: Present and future prospects. In *Nitric Oxide Action in Abiotic Stress Responses in Plants* (pp. 111–127). Springer.

Khan, M. Z., Mondal, P. K., & Sabir, S. (2011a). Bioremediation of 2-chlorophenol containing wastewater by aerobic granules-kinetics and toxicity. *Journal of Hazardous Materials*, *190*(1–3), 222–228. doi:10.1016/j.jhazmat.2011.03.029 PMID:21466913

Khan, M., Khan, A. U., Hasan, M. A., Yadav, K. K., Pinto, M. M., Malik, N., Yadav, V. K., Khan, A. H., Islam, S., & Sharma, G. K. (2021). Agro-nanotechnology as an emerging field: A novel sustainable approach for improving plant growth by reducing biotic stress. *Applied Sciences (Basel, Switzerland)*, *11*(5), 2282. doi:10.3390/app11052282

Khan, N. A., Khan, S. U., Ahmed, S., Farooqi, I. H., Dhingra, A., Hussain, A., & Changani, F. (2019). Applications of nanotechnology in water and wastewater treatment: A review. *Asian Journal of Water, Environment and Pollution*, *16*(4), 81–86. doi:10.3233/AJW190051

Khanna, V. K., Kumar, A., Jain, P. K., & Bhalla, V. (2019). Nanomaterials-based biosensors for food contaminant detection. *Environmental Chemistry Letters*, *17*(2), 845–859. doi:10.1007/s10311-018-00833-z

Khan, S. H. (2020). Green nanotechnology for the environment and sustainable development. In *Green materials for wastewater treatment* (pp. 13–46). Springer. doi:10.1007/978-3-030-17724-9_2

Khan, S. H., Pathak, B., & Fulekar, M. H. (2020). A study on the influence of metal (Fe, Bi, and Ag) doping on structural, optical, and antimicrobial activity of ZnO nanostructures. *Advanced Composites and Hybrid Materials*, *3*(4), 551–569. doi:10.1007/s42114-020-00174-0

Khater, M., de la Escosura-Muñiz, A., & Merkoçi, A. (2017). Biosensors for plant pathogen detection. *Biosensors & Bioelectronics*, *93*, 72–86. doi:10.1016/j.bios.2016.09.091 PMID:27818053

Khati, P., Chaudhary, P., & Gangola, S. (2019). Nanomaterials in soil: Environmental applications and ecological implications. In *Nanomaterials in plants, algae, and microorganisms* (pp. 83–105). Academic Press.

Khatri, A., & Bhateria, R. (2023). Efficacy of Nanofertilizers Over Chemical Fertilizers in Boosting Agronomic Production. *Nature Environment and Pollution Technology*, *22*(2), 767–776. doi:10.46488/NEPT.2023.v22i02.019

Khepar, V., Ahuja, R., Sidhu, A., & Samota, M. K. (2023). Nano-sulfides of Fe and Mn efficiently augmented the growth, antioxidant defense system, and metal assimilation in rice seedlings. *ACS Omega*, *8*(33), 30231–30238. doi:10.1021/acsomega.3c03012 PMID:37636944

Khiyami, M. A., Almoammar, H., Awad, Y. M., Alghuthaymi, M. A., & Abd-Elsalam, K. A. (2014). Plant pathogen nanodiagnostic techniques: Forthcoming changes? *Biotechnology, Biotechnological Equipment*, *28*(5), 775–785. doi:10.1080/13102818.2014.960739 PMID:26740775

Khodakovskaya, M. V., De Silva, K., Biris, A. S., Dervishi, E., & Villagarcia, H. (2012). Carbon nanotubes induce growth enhancement of tobacco cells. *ACS Nano*, *6*(3), 2128–2135. doi:10.1021/nn204643g PMID:22360840

Khodakovskaya, M., Dervishi, E., Mahmood, M., Xu, Y., Li, Z., Watanabe, F., & Biris, A. S. (2009). Carbon nanotubes are able to penetrate plant seed coat and dramatically affect seed germination and plant growth. *ACS Nano*, *3*(10), 3221–3227. doi:10.1021/nn900887m PMID:19772305

Khoja, A. A., Haq, S. M., Majeed, M., Hassan, M., Waheed, M., Yaqoob, U., Bussmann, R. W., Alataway, A., Dewidar, A. Z., Al-Yafrsi, M., Elansary, H. O., Yessoufou, K., & Zaman, W. (2022). Diversity, Ecological and Traditional Knowledge of Pteridophytes in the Western Himalayas. *Diversity*, *14*(8), 628. Advance online publication. doi:10.3390/d14080628

Khoshnevisan, B., Shafiei, H., Rajaeifar, M. A., & Tabatabaei, M. (2021). Nano-based smart pesticide formulations for sustainable agricultural development. *Environmental Science and Pollution Research International*, *28*(16), 19839–19856. doi:10.1007/s11356-021-12592-8

Khot, L. R., Sankaran, S., Maja, J. M., Ehsani, R., & Schuster, E. W. (2012). Applications of nanomaterials in agricultural production and crop protection: A review. *Crop Protection (Guildford, Surrey)*, *35*, 64–70. doi:10.1016/j.cropro.2012.01.007

Khush, G. S. (2005). What it will take to feed 5.0 billion rice consumers in 2030. *Plant Molecular Biology*, *59*(1), 1–6. doi:10.1007/s11103-005-2159-5 PMID:16217597

Kim, H., Sudduth, K. A., & Hummel, J. W. (2013). Soil macronutrient sensing for precision agriculture. *Journal of Environmental Monitoring*, *11*(10), 1810–1824. doi:10.1039/b906634a PMID:19809703

Kim, K. Y., Park, S. A., & Kim, J. E. (2019). Nanomaterial-based optical sensors for determining nutrients in the soil and crop. *Biosystems Engineering*, *185*, 141–159.

Kim, S. W., Jung, J. H., Lamsal, K., Kim, Y. S., Min, J. S., & Lee, Y. S. (2012). Antifungal effects of silver nanoparticles (AgNPs) against various plant pathogenic fungi. *Mycobiology*, *40*(1), 53–58. doi:10.5941/MYCO.2012.40.1.053 PMID:22783135

King, A. M. Y., & Young, G. (1999). Characteristics and occurrence of phenolic phytochemicals. *Journal of the American Dietetic Association*, *99*(2), 213–218. doi:10.1016/S0002-8223(99)00051-6 PMID:9972191

Klaine, S. J., Koelmans, A. A., Horne, N., Carley, S., Handy, R. D., Kapustka, L., Nowack, B., & von der Kammer, F. (2012). Paradigms to assess the environmental impact of manufactured nanomaterials. *Environmental Toxicology and Chemistry*, *31*(1), 3–14. doi:10.1002/etc.733 PMID:22162122

Koes, R. E., Quattrocchio, F., & Mol, J. N. (1994). The flavonoid biosynthetic pathway in plants: Function and evolution. *BioEssays*, *16*(2), 123–132. doi:10.1002/bies.950160209

Kolbert, Z., Szőllősi, R., Rónavári, A., & Molnár, Á. (2022). Nanoforms of essential metals: From hormetic phytoeffects to agricultural potential. *Journal of Experimental Botany*, *73*(6), 1825–1840. doi:10.1093/jxb/erab547 PMID:34922354

Kolesnikov, S., Timoshenko, A., Varduni, V., Minnikova, T., Tsepina, N., Mandzhieva, S., & Rajput, V. D. (2024). *Comparative estimation of ecotoxicity of nano- and microparticles of CuO, ZnO, NiO, Co3O4 by the integral biological indicator of soil state*. Nanotechnology for Environmental Engineering. doi:10.1007/s41204-024-00358-9

Konczak-Islam, I., Yoshimoto, M., Hou, D. X., Terahara, N., & Yamakawa, O. (2003). Potential chemopreventive properties of anthocyanin-rich aqueous extracts from in vitro produced tissue of sweet potato (*Ipomoea batatas* L.). *Journal of Agricultural and Food Chemistry*, *51*(20), 5916–5922. doi:10.1021/jf030066o PMID:13129295

Kong, X. P., Zhang, B. H., & Wang, J. (2021). Multiple roles of mesoporous silica in safe pesticide application by nanotechnology: A review. *Journal of Agricultural and Food Chemistry*, *69*(24), 6735–6754. doi:10.1021/acs.jafc.1c01091 PMID:34110151

Konstantatos, G., & Sargent, E. H. (2009). Solution-processed quantum dot photodetectors. *Proceedings of the IEEE*, *97*(10), 1666–1683. doi:10.1109/JPROC.2009.2025612

Kookana, R. S., Boxall, A. B., Reeves, P. T., Ashauer, R., Beulke, S., Chaudhry, Q., Cornelis, G., Fernandes, T. F., Gan, J., Kah, M., Lynch, I., Ranville, J., Sinclair, C., Spurgeon, D., Tiede, K., & van den Brink, P. J. (2014). Nanopesticides: Guiding principles for regulatory evaluation of environmental risks. *Journal of Agricultural and Food Chemistry*, *62*(19), 4227–4240. doi:10.1021/jf500232f PMID:24754346

Kottegoda, N., Sandaruwan, C., Priyadarshana, G., Siriwardhana, A., Rathnayake, U. A., Arachchige, D. M. B., Kumarasinghe, A. R., Dahanayake, D., Karunaratne, V., & Amaratunga, G. A. (2017). Urea-hydroxyapatite nanohybrids for slow release of nitrogen. *ACS Nano*, *11*(2), 1214–1221. doi:10.1021/acsnano.6b07781 PMID:28121129

Kousheh, S. A., Moradi, M., Tajik, H., & Molaei, R. (2020). Preparation of antimicrobial/ultraviolet protective bacterial nanocellulose film with carbon dots synthesized from lactic acid bacteria. *International Journal of Biological Macromolecules*, *155*, 216–225. doi:10.1016/j.ijbiomac.2020.03.230 PMID:32240732

Kozlowska, A., & Szostak-Wegierek, D. (2014). Flavonoids-food sources and health benefits. *Roczniki Panstwowego Zakladu Higieny*, *65*(2), 79–85. PMID:25272572

Kranthi, K. R., Davis, M., Mayee, C. D., Russell, D. A., Shukla, R. M., Satija, U., Kshirsagar, M., Shiware, D., & Kranthi, S. (2009). Development of a colloidal-gold based lateralflow immunoassay kit for 'quality-control' assessment of pyrethroid and endosulfan formulations in a novel single strip format. *Crop Protection (Guildford, Surrey)*, *28*(5), 428–434. doi:10.1016/j.cropro.2009.01.003

Krishnaraj, C., Ramachandran, R., Mohan, K., & Kalaichelvan, P. T. (2012). Optimization for rapid synthesis of silver nanoparticles and its effect on phytopathogenic fungi. *Spectrochimica Acta. Part A: Molecular and Biomolecular Spectroscopy*, *93*, 95–99. doi:10.1016/j.saa.2012.03.002 PMID:22465774

Kuang, Q., Wang, X., Jiang, Z., Xie, Z., & Zheng, L. (2014). High-energy-surface Engineered Metal Oxide Micro- and Nanocrystallites and Their Applications. *Accounts of Chemical Research*, *47*(2), 308–318. doi:10.1021/ar400092x PMID:24341353

Kúdelová, J., Seifrtová, M., Suchá, L., Tomšík, P., Havelek, R., & Řezáčová, M. (2013). Alpha-tomatine activates cell cycle checkpoints in the absence of DNA damage in human leukemic MOLT-4 cells. *Journal of Applied Biomedicine*, *11*(2), 93–103. doi:10.2478/v10136-012-0033-8

Kulkarni, K. V., & Ghurghure, S. M. (2018). Indian gooseberry (*Emblica officinalis*): Complete pharmacognosy review. *International Journal of Chemical Studies*, *2*(2), 5–11.

Kumar Das, P., Mohanty, C., Krishna Purohit, G., Mishra, S., & Palo, S. (2022). Nanoparticle assisted environmental remediation: Applications, toxicological implications and recommendations for a sustainable environment. *Environmental Nanotechnology, Monitoring & Management*, *18*, 100679. doi:10.1016/j.enmm.2022.100679

Kumar, A., Bisht, B. S., Joshi, V. D., & Dhewa, T. (2019). Review on bioremediation of polluted environment: A management tool. *International Journal of Environmental Science and Technology*, *16*(7), 3439–3462.

Kumar, A., Jnanesha, A. C., & Nagaraju, S. (2019). Cultivation and medicinal properties of giloy [*Tinospora cordifolia* (Thunb.) Miers]. *Agrobios*, *18*(1), 57–58.

Kumaraswamy, R. V., Kumari, S., Choudhary, R. C., Pal, A., Raliya, R., Biswas, P., & Saharan, V. (2018). Engineered chitosan based nanomaterials: Bioactivities, mechanisms and perspectives in plant protection and growth. *International Journal of Biological Macromolecules*, *113*, 494–506. doi:10.1016/j.ijbiomac.2018.02.130 PMID:29481952

Kumar, D. R., Kumar, P. S., Gandhi, M. R., Al-Dhabi, N. A., Paulraj, M. G., & Ignacimuthu, S. (2016). Delivery of chitosan/dsRNA nanoparticles for silencing of wing development vestigial (vg) gene in Aedes aegypti mosquitoes. *International Journal of Biological Macromolecules*, *86*, 89–95. doi:10.1016/j.ijbiomac.2016.01.030 PMID:26794313

Kumar, G., Sharma, J. G., Goswami, R. K., Shrivastav, A. K., Kumar, N., Chandra, S., & Chakrabarti, R. (2021b). The study of effect of vitamin C and *Achyranthes aspera* seeds enriched diets on the growth, biochemical composition, digestive enzyme activities and expressions of genes involved in the biosynthesis of fatty acids in snow trout *Schizothorax richardsonii* (Gray, 1832). *Journal of Applied Aquaculture*, *35*(3), 489–509. doi:10.1080/10454438.2021.1985679

Kumar, G., Sharma, J. G., Goswami, R. K., Shrivastav, A. K., Tocher, D. R., Kumar, N., & Chakrabarti, R. (2022b). Freshwater macrophytes: Potential source of minerals and fatty acids for fish, poultry and livestock. *Frontiers in Nutrition, 9*, 869425. doi:10.3389/fnut.2022.869425 PMID:35479737

Kumari, R., Suman, K., Karmakar, S., Mishra, V., Lakra, S. G., Saurav, G. K., & Mahto, B. K. (2023). Regulation and safety measures for nanotechnology-based agri-products. In Frontiers in Genome Editing (Vol. 5). Frontiers Media SA. doi:10.3389/fgeed.2023.1200987

Kumari, A., Singla, R., Guliani, A., & Walia, S. (2020). Nanopesticides: Opportunities and challenges for crop protection. In *Nanopesticides* (pp. 151–171). Springer. doi:10.1007/978-3-030-44873-8_8

Kumari, M., Pandey, S., Mishra, S. K., Giri, V. P., Agarwal, L., Dwivedi, S., Pandey, A. K., Nautiyal, C. S., & Mishra, A. (2020, April 17). Omics-Based Mechanistic Insight Into the Role of Bioengineered Nanoparticles for Biotic Stress Amelioration by Modulating Plant Metabolic Pathways. *Frontiers in Bioengineering and Biotechnology, 8*, 242. doi:10.3389/fbioe.2020.00242 PMID:32363178

Kumari, S., Krishna, J. M., Joshi, A. B., Gurav, S., Bhandarkar, A. V., Agarwal, A., Deepak, M., & Gururaj, G. M. (2017). A pharmacognostic, phytochemical and pharmacological review of *Terminalia bellerica. Journal of Pharmacognosy and Phytochemistry, 6*(5), 368–376.

Kumar, N., & Khurana, S. M. (2018). Phytochemistry and medicinal potential of the *Terminalia bellirica* Roxb. (Bahera). *Indian Journal of Natural Products and Resources, 9*(2), 97–107.

Kumar, N., Samota, S. R., Venkatesh, K., & Tripathi, S. C. (2023). Global trends in use of nano-fertilizers for crop production: Advantages and constraints–A review. *Soil & Tillage Research, 228*, 105645. doi:10.1016/j.still.2023.105645

Kumar, N., Sharma, J. G., Kumar, G., Shrivastav, A. K., Tiwari, N., Begam, A., & Chakrabarti, R. (2021a). Evaluation of nutritional value of prickly chaff flower (*Achyranthes aspera)* as fish feed ingredient. *The Indian Journal of Animal Sciences, 91*(3), 239–244. doi:10.56093/ijans.v91i3.114152

Kumar, N., Sharma, J. G., Mittal, P., & Chakrabarti, R. (2022a). Effect of leaves and seeds of *Achyranthes aspera* as feed supplements on the immunological and stress parameters and related-gene expressions of Asian catfish (*Clarias batrachus*). *Veterinary Research Communications, 47*(1), 99–109. doi:10.1007/s11259-022-09932-5 PMID:35507243

Kumar, N., Sharma, J. G., Singh, S. P., Singh, A., Harikrishna, V., & Chakrabarti, R. (2019). Validation of growth enhancing, immunostimulatory and disease resistance properties of *Achyranthes aspera* in *Labeo rohita* fry in pond conditions. *Heliyon, 5*(2), e01246. doi:10.1016/j.heliyon.2019.e01246 PMID:30820481

Kumar, P., & Dubey, K. K. (2019). Citric acid cycle regulation: Back bone for secondary metabolite production. In *New and Future Developments in Microbial Biotechnology and Bioengineering* (pp. 165–181). Elsevier. doi:10.1016/B978-0-444-63504-4.00013-X

Kumar, P., & Fouletier, J. (2019). Electrochemical sensors for heavy metals: Principles, operating modes, and recent developments. In *Trends in Environmental Analytical Chemistry* (Vol. 23, p. e00063). Elsevier.

Kumar, R., Kumar, M., & Luthra, G. (2023). Fundamental approaches and applications of nanotechnology: A mini review. *Materials Today: Proceedings*. Advance online publication. doi:10.1016/j.matpr.2022.12.172

Kumar, S. (2010). Nanotechnology and animal health. *Veterinary World, 3*(12), 567–569. doi:10.5455/vetworld.2010.567-569

Kumar, S. (2020). Abiotic stresses and their effects on plant growth, yield and nutritional quality of agricultural produce. *International Journal of Food Science and Agriculture, 4*(4), 367–378. doi:10.26855/ijfsa.2020.12.002

Kumar, S. G., & Sharma, A. (2014). Medicinal properties of *Zingiber officinale* Roscoe-A review. *Journal of Pharmaceutical and Biological Sciences*, *9*, 124–129.

Kumar, S., Mudai, A., Roy, B., Basumatary, I. B., Mukherjee, A., & Dutta, J. (2020). Biodegradable hybrid nanocomposite of chitosan/gelatin and green synthesized zinc oxide nanoparticles for food packaging. *Foods*, *9*(9), 1143. doi:10.3390/foods9091143 PMID:32825205

Kumar, S., Nehra, M., Dilbaghi, N., Marrazza, G., Hassan, A. A., & Kim, K. H. (2019). Nano-based smart pesticide formulations: Emerging opportunities for agriculture. *Journal of Controlled Release*, *294*, 131–153. doi:10.1016/j.jconrel.2018.12.012 PMID:30552953

Kunduru, K. R., Nazarkovsky, M., Farah, S., Pawar, R. P., Basu, A., & Domb, A. J. (2017). Nanotechnology for water purification: Applications of nanotechnology methods in wastewater treatment. In A. M. Grumezescu (Ed.), Water purification (pp. 33–74). Academic Press.

Kuppusamy, S., Palanisami, T., Megharaj, M., Venkateswarlu, K., & Naidu, R. (2016). In-situ remediation approaches for the management of contaminated sites: A comprehensive overview. In P. de Voogt (Ed.), *Reviews of Environmental Contamination and Toxicology* (Vol. 236, pp. 1–115). Springer.

Kushwaha, S., & Pant, H. (2023). Role of nanoparticles in maintaining food safety and tackling malnutrition. *International Journal of Clinical Biochemistry and Research*, *10*(1), 19–27. doi:10.18231/j.ijcbr.2023.004

Kuzma, J., & VerHage, P. (2006). *Nanotechnology in agriculture and food production: anticipated applications*. Project on emerging nanotechnologies.

Lachin, T., & Reza, H. (2012). Anti diabetic effect of cherries in alloxan induced diabetic rats. *Recent Patents on Endocrine, Metabolic & Immune Drug Discovery*, *6*(1), 67–72. doi:10.2174/187221412799015308 PMID:22280223

Lahiani, M. H., Souli, H., & Cabot, P. (2021). Nanotechnology for sustainable agriculture and food security. In *Nanomaterials for Agriculture and Forestry Applications* (pp. 1–32). Elsevier.

Lai, F., Wissing, S. A., Müller, R. H., & Fadda, A. M. (2006). Artemisia arborescens L essential oil-loaded solid lipid nanoparticles for potential agricultural application: Preparation and characterization. *AAPS PharmSciTech*, *7*(1), E10–E18. doi:10.1208/pt070102 PMID:28290017

Lamichhane, J. R., Barbetti, M. J., Chilvers, M. I., Pandey, A. K., & Steinberg, C. (2024). Exploiting root exudates to manage soil-borne disease complexes in a changing climate. In Trends in Microbiology (Vol. 32, Issue 1, pp. 27–37). Elsevier Ltd. doi:10.1016/j.tim.2023.07.011

Lamsal, K., Kim, S. W., Jung, J. H., Kim, Y. S., Kim, K. S., & Lee, Y. S. (2011). Application of silver nanoparticles for the control of Colletotrichum species in vitro and pepper anthracnose disease in field. *Mycobiology*, *39*(3), 194–199. doi:10.5941/MYCO.2011.39.3.194 PMID:22783103

Lang, Y., Zhnag, B., Cai, D., Tu, W., Zhnag, J., Shentu, X., Ye, Z., & Yu, X. (2022). Determination methods of the risk factors in food based on nanozymes: A review. *Biosensors (Basel)*, *13*(1), 1–19. doi:10.3390/bios13010069 PMID:36671904

Lanka, S. (2018). A review on pharmacological, medicinal and ethnobotanical important plant: *Phyllanthus emblicalinn.* (syn. *Emblica officinalis*). *World Journal of Pharmaceutical Research*, *7*(4), 380–396.

Lateef, A., Azeez, M. A., Suaibu, O. B., & Adigun, G. O. (2021). A decade of nanotechnology research in Nigeria (2010-2020): A scientometric analysis. *Journal of Nanoparticle Research*, *23*(9), 1–27. doi:10.1007/s11051-021-05322-1

Latif, A., Ashiq, K., Qayyum, M., Ashiq, S., Ali, E., & Anwer, I. (2019). Phytochemical and pharmacological profile of the medicinal herb: *Bryophyllum pinnatum. Journal of Animal & Plant Sciences*, *29*(6), 1528–1534.

Lauterwasser, C. (2005). Small sizes that matter: Opportunities and risks of Nanotechnologies. *Report in co-operation with the OECD International Futures Programme*.

Lawal, M., Payne, J., Onyeaka, H., Alao, A. M., & Okoampah, E. (2023). Boosting food safety in Ghana: Exploring the future of nanotechnology. *Nano Select*, 1–13. doi:10.1002/nano.202300078

Le Doux, M. S. (2015). Murine models of caytaxin deficiency. In *Movement Disorders* (pp. 439–452). Academic Press. doi:10.1016/B978-0-12-405195-9.00025-1

Lee, G. W. J., & Tadros, T. F. (1982). Formation and stability of emulsions produced by dilution of emulsifiable concentrates. Part I. An investigation of the dispersion on dilution of emulsifiable concentrates containing cationic and non-ionic surfactants. *Colloids and Surfaces*, *5*(2), 105–115. doi:10.1016/0166-6622(82)80066-8

Li, S., Chen, L., & Fu, Y. (2023). Nanotechnology-based ocular drug delivery systems: recent advances and future prospects. In Journal of Nanobiotechnology (Vol. 21, Issue 1). BioMed Central Ltd. doi:10.1186/s12951-023-01992-2

Li, Y., Zhang, P., Li, M., Shakoor, N., Adeel, M., Zhou, P., Guo, M., Jiang, Y., Zhao, W., Lou, B. Z., & Rui, Y. (2023). Application and mechanisms of metal-based nanoparticles in the control of bacterial and fungal crop diseases. In Pest Management Science (Vol. 79, Issue 1, pp. 21–36). John Wiley and Sons Ltd. doi:10.1002/ps.7218

Liang, J., Xia, X., Zaman, W. Q., Zhang, W., Lin, K., Hu, S., & Lin, Z. (2017). Bioaccumulation and toxic effects of decabromodiphenyl ether in the presence of nanoscale zero-valent iron in an earthworm–soil system. *Chemosphere*, *169*, 78–88. doi:10.1016/j.chemosphere.2016.11.012 PMID:27863305

Liang, M., & Yan, X. (2019). Nanozymes: From New Concepts, Mechanisms, and Standards to Applications. *Accounts of Chemical Research*, *52*(8), 2190–2200. doi:10.1021/acs.accounts.9b00140 PMID:31276379

Liao, Z., Fan, J., Lai, Z., Bai, Z., Wang, H., Cheng, M., Zhang, F., & Li, Z. (2023). Response network and regulatory measures of plant-soil-rhizosphere environment to drought stress. In *Advances in Agronomy* (Vol. 180, pp. 93–196). Academic Press Inc. doi:10.1016/bs.agron.2023.03.002

Li, H., Yang, Y., Wang, Y., Wang, C., Li, W., & Wu, L. (2011). Self-assembly and ion-trapping properties of inorganic nanocapsule-surfactant hybrid spheres. *Soft Matter*, *7*(6), 2668–2673. doi:10.1039/c0sm01044h

Li, J., Hu, J., Ma, C., Wang, Y., Wu, C., Huang, J., & Xing, B. (2016). Uptake, translocation and physiological effects of magnetic iron oxide (γ- Fe2O3) nanoparticles in corn (Zea mays L.). *Chemosphere*, *159*, 326–334. doi:10.1016/j.chemosphere.2016.05.083 PMID:27314633

Li, J., Zhang, M., Sun, B., Xing, G., Song, Y., Gui, R., & Huang, D. (2020). Carbon quantum dots as fluorescence sensors for sensitive and selective detection of heavy metal ions: A review. *Journal of Biomedical Nanotechnology*, *16*(8), 1089–1102. PMID:33397551

Li, L., Wang, X., Zhou, Y., & Xing, B. (2021). Effectiveness and limitation of A-nZVI for restoration of a highly As-contaminated soil. *Journal of Cleaner Production*, *284*, 124691. doi:10.1016/j.jclepro.2020.124691

Li, M., Huang, Q., & Wu, Y. (2011). A novel chitosan-poly (lactide) copolymer and its submicron particles as imidacloprid carriers. *Pest Management Science*, *67*(7), 831–836. doi:10.1002/ps.2120 PMID:21370387

Lin, H., Jie, B., Ye, J., Zhai, Y., Luo, Z., Shao, G., Chen, R., Zhang, X., & Yang, Y. (2023). Recent advance of macroscopic metal-organic frameworks for water treatment: A review. *Surfaces and Interfaces*, *36*, 102564. doi:10.1016/j.surfin.2022.102564

Li, Q., Chen, X., Zhuang, J., & Chen, X. (2016). Decontaminating soil organic pollutants with manufactured nanoparticles. *Environmental Science and Pollution Research International*, *23*(12), 11533–11548. doi:10.1007/s11356-016-6255-7 PMID:26906002

Liu, B., & Liu, J. (2017). Surface Modification of Nanozymes. Nano Res., 10, 1125–1148. doi:10.1007/s12274-017-1426

Liu, D., Dai, X., Ye, L., Wang, H., Qian, H., Cheng, H., & Wang, X. (2023). Nanotechnology meets glioblastoma multiforme: Emerging therapeutic strategies. In Wiley Interdisciplinary Reviews: Nanomedicine and Nanobiotechnology (Vol. 15, Issue 1). John Wiley and Sons Inc. doi:10.1002/wnan.1838

Liu, B., & Liu, J. (2017). Surface Modification of Nanozymes. *Nano Research*, *10*(4), 1125–1148. doi:10.1007/s12274-017-1426-5

Liu, C., Zhou, H., & Zhou, J. (2021). The applications of nanotechnology in crop production. *Molecules (Basel, Switzerland)*, *26*(23), 7070. doi:10.3390/molecules26237070 PMID:34885650

Liu, J., Li, W., Chen, H., Li, S., Yang, L., Peng, K., Cai, C., & Huang, X. (2023). Applications of functional nanoparticle–stabilized surfactant foam in petroleum-contaminated soil remediation. *Journal of Hazardous Materials*, *443*, 130267. doi:10.1016/j.jhazmat.2022.130267 PMID:36444047

Liu, R., & Lal, R. (2014). Synthetic apatite nanoparticles as a phosphorus fertilizer for soybean (Glycine max). *Scientific Reports*, *4*(1), 1–6. doi:10.1038/srep05686 PMID:25023201

Liu, R., & Lal, R. (2015). Potentials of engineered nanoparticles as fertilizers for increasing agronomic productions. *The Science of the Total Environment*, *514*, 131–139. doi:10.1016/j.scitotenv.2015.01.104 PMID:25659311

Liu, X., He, B., Xu, Z., Yin, M., Yang, W., Zhang, H., & Shen, J. (2015). A functionalized fluorescent dendrimer as a pesticide nanocarrier: Application in pest control. *Nanoscale*, *7*(2), 445–449. doi:10.1039/C4NR05733C PMID:25429649

Liu, X., Wei, Z., Ma, Y., Liu, J., & Liu, F. (2021). Effects of biochar amendment and reduced irrigation on growth, physiology, water-use efficiency and nutrients uptake of tobacco (Nicotiana tabacum L.) on two different soil types. *The Science of the Total Environment*, *770*, 144769. doi:10.1016/j.scitotenv.2020.144769 PMID:33736368

Liu, Y., Laks, P., & Heiden, P. (2002). Controlled release of biocides in solid wood. II. Efficacy against Trametes versicolor and Gloeophyllum trabeum wood decay fungi. *Journal of Applied Polymer Science*, *86*(3), 608–614. doi:10.1002/app.10897

Liu, Y., Wu, H., Li, M., Yin, J. J., & Nie, Z. (2019). Oxidase-like activity of gold nanoparticles and their gold staining enhanced ELISA application. *Journal of Materials Chemistry. B*, *7*(41), 6393–6399.

Li, X., Zhu, Y., Li, Y., Zhang, D., Zhang, H., & Zhang, Z. (2019). Nanomaterials as novel stabilizers for soil stabilization: Mechanism, ecological risk, and environmental implication. *The Science of the Total Environment*, *646*, 1566–1575.

Li, Y., & Cummmins, E. (2020). Hazard characterization of silver nanoparticles for human exposure routes. *Journal of Environmental Science and Health. Part A, Toxic/Hazardous Substances & Environmental Engineering*, *55*(6), 704–725. doi:10.1080/10934529.2020.1735852 PMID:32167009

Li, Y., Huang, X., Zeng, L., Li, R., Tian, H., Fu, X., Wang, Y., & Zhong, W. H. (2019). A review of the electrical and mechanical properties of carbon nanofiller-reinforced polymer composites. *Journal of Materials Science*, *54*(2), 1036–1076. doi:10.1007/s10853-018-3006-9

Li, Y., Jin, Q., Yang, D., & Cui, J. (2018). Molybdenum sulfide induce growth enhancement effect of rice (Oryza sativa L.) through regulating the synthesis of chlorophyll and the expression of aquaporin gene. *Journal of Agricultural and Food Chemistry*, *66*(16), 4013–4021. doi:10.1021/acs.jafc.7b05940 PMID:29630363

Li, Y., Zhang, Q., Zhang, J., Wu, L., Qi, Y., & Zhou, J. M. (2019). Identification of microRNAs involved in pathogen-associated molecular pattern-triggered plant innate immunity. *Plant Physiology*, *152*(4), 2222–2231. doi:10.1104/pp.109.151803 PMID:20164210

Li, Y., Zhao, J., Gao, Y., Huang, Y., & Zhao, X. (2018). MnO2 nanozymes for the degradation of polycyclic aromatic hydrocarbons (PAHs) in contaminated soil. *Environmental Science. Nano*, *5*(10), 2369–2378.

Li, Z., Liu, X., Jiang, M., & Guo, L. (2018). Specific detection of bioavailable phenanthrene by a whole cell bioreporter in the presence of copper. *Environmental Pollution*, *235*, 381–389. doi:10.1016/j.envpol.2017.12.077

Li, Z., Wang, L., Li, Y., Feng, Y., & Feng, W. (2019). Carbon-based functional nanomaterials: Preparation, properties and applications. *Composites Science and Technology*, *179*, 10–40. doi:10.1016/j.compscitech.2019.04.028

Lo Presti, D., Di Tocco, J., Massaroni, C., Cimini, S., De Gara, L., Singh, S., Raucci, A., Manganiello, G., Woo, S. L., Schena, E., & Cinti, S. (2023). Current understanding, challenges and perspective on portable systems applied to plant monitoring and precision agriculture. *Biosensors & Bioelectronics*, *222*, 115005. doi:10.1016/j.bios.2022.115005 PMID:36527829

Loiseau, B., Carrière, S. D., Jougnot, D., Singha, K., Mary, B., Delpierre, N., Guérin, R., & Martin-StPaul, N. K. (2023). The geophysical toolbox applied to forest ecosystems – A review. *The Science of the Total Environment*, *899*, 165503. doi:10.1016/j.scitotenv.2023.165503 PMID:37454861

Lombi, E., Donner, E., Dusinska, M., & Wickson, F. (2019). A One Health approach to managing the applications and implications of nanotechnologies in agriculture. *Nature Nanotechnology*, *14*(6), 523–531. doi:10.1038/s41565-019-0460-8 PMID:31168074

Lowenberg-DeBoer, J. (2015). *The precision agriculture revolution: Making the modern farmer*. Foreign Affairs, May/June.

Lowry, G. V., Avellan, A., & Gilbertson, L. M. (2019). Opportunities and challenges for nanotechnology in the agri-tech revolution. *Nature Nanotechnology*, *14*(6), 517–522. doi:10.1038/s41565-019-0461-7 PMID:31168073

Lowry, G. V., Gregory, K. B., Apte, S. C., & Lead, J. R. (2012). Transformations of nanomaterials in the environment. *Environmental Science & Technology*, *46*(13), 6893–6899. doi:10.1021/es300839e PMID:22582927

Ludlow, K. (2009). The readiness of Australian food regulation for the use of nanotechnology in food and food packaging. *University of Tasmania Law Review*, *26*(2), 177–203.

Ludwiczuk, A., Skalicka-Woźniak, K., & Georgiev, M. I. (2017). Terpenoids. In *Pharmacognosy* (pp. 233–266). Academic Press. doi:10.1016/B978-0-12-802104-0.00011-1

Lugani, Y.; Sooch, B. S.; Singh, P. & Kumar, S. (2021). Nanobiotechnology applications in food sector and future innovations. Microbial Biotechnology in Food and Health, 197-225. doi:10.1016/B978-0-12-819813-1.00008-6

Luna-Sanguino, G., Ruíz-Delgado, A., Tolosana-Moranchel, A., Pascual, L., Malato, S., Bahamonde, A., & Faraldos, M. (2020). Solar photocatalytic degradation of pesticides over TiO2 -rGO nanocomposites at pilot plant scale. *The Science of the Total Environment*, *737*, 140286. doi:10.1016/j.scitotenv.2020.140286 PMID:32783863

Lustig, W. P., Mukherjee, S., Rudd, N. D., Desai, A. V., Li, J., & Ghosh, S. K. (2017). Metal-organic Frameworks: Functional Luminescent and Photonic Materials for Sensing Applications. *Chemical Society Reviews*, *46*(11), 3242–3285. doi:10.1039/C6CS00930A PMID:28462954

Lv, W., Geng, H., Zhou, B., Chen, H., Yuan, R., Ma, C., & Wang, F. (2022). The behavior, transport, and positive regulation mechanism of ZnO nanoparticles in a plant-soil-microbe environment. *Environmental Pollution*, *315*, 120368. doi:10.1016/j.envpol.2022.120368 PMID:36216179

Ma, A., Wei, Y., Zhou, Z., Xu, W., Ren, F., Ma, H., & Wang, J. (2012). Preparation Bi2S3eTiO2 heterojunction/polymer fiber composites and its photocatalytic degradation of methylene blue under Xe lamp irradiation, polym degrad. *Stadtehygiene*, *97*, 125–131.

Macnaghten, P., Kearnes, M. B., & Wynne, B. (2005). Nanotechnology, governance, and public deliberation: What role for the social sciences? *Science Communication*, *27*(2), 268–291. doi:10.1177/1075547005281531

Madhusha, C., Munaweera, I., & Kottegoda, N. (2021). Functional nanomaterials as smart food packaging: A brief review. *African Journal of Agriculture and Food Science*, *4*(1), 58–78.

Madkour, L. H. (2020). Advanced nanomaterials for agro-food application. In *Advanced Nanomaterials for Membrane Synthesis and Its Applications* (pp. 365–381). Elsevier.

Maharramov, A. M., Hasanova, U. A., Suleymanova, I. A., Osmanova, G. E., & Hajiyeva, N. E. (2019). The engineered nanoparticles in food chain: potential toxicity and effects. Springer Nature Applied Sciences. doi:10.1007/s42452-019-1412-5

Mahil, T., Baburai, N., & Aravinda, K. (2019). Foliar application of nanofertilizers in agricultural crops- A review. *Journal of Farm Sciences*, *32*, 239–249.

Maithani, A., Parcha, V., Pant, G., Dhulia, I., & Kumar, D. (2011). *Azadirachta indica* (neem) leaf: A review. *Journal of Pharmacy Research*, *4*(6), 1824–1827.

Majeed, M., Bhatti, K. H., Amjad, M. S., Abbasi, A. M., Bussmann, R. W., Nawaz, F., Rashid, A., Mehmood, A., Mahmood, M., Khan, W. M., & Ahmad, K. S. (2020). Ethno-veterinary uses of poaceae in Punjab, Pakistan. *PLoS ONE*, *15*(11 November). doi:10.1371/journal.pone.0241705

Majeed, M., Muhammad, M., Nawaz, S., Naz, T., Iqbal, M. M., Zahid, N., ... Mehboob, M. (2023b). Nanopriming for Crop Management for Sustainable Agriculture: An Overview. *Nanopriming Approach to Sustainable Agriculture*, 110-141. doi:10.4018/978-1-6684-7232-3.ch005

Majeed, M., Tariq, A., Anwar, M. M., Khan, A. M., Arshad, F., Mumtaz, F., Farhan, M., Zhang, L., Zafar, A., Aziz, M., Abbasi, S., Rahman, G., Hussain, S., Waheed, M., Fatima, K., & Shaukat, S. (2021). Monitoring of Land Use–Land Cover Change and Potential Causal Factors of Climate Change in Jhelum District, Punjab, Pakistan, through GIS and Multi-Temporal Satellite Data (pp. 129-147). https://www.mdpi.com/books/book/6906-land-use-land-cover-and-natural-hazards-interactions-changes-and-impacts

Majeed, M., Tariq, A., Anwar, M. M., Khan, A. M., Arshad, F., Mumtaz, F., Farhan, M., Zhang, L., Zafar, A., Aziz, M., Abbasi, S., Rahman, G., Hussain, S., Waheed, M., Fatima, K., & Shaukat, S. (2021). Monitoring of Land Use–Land Cover Change and Potential Causal Factors of Climate Change in Jhelum District, Punjab, Pakistan, through GIS and Multi-Temporal Satellite Data (pp. 129-147).https://www.mdpi.com/books/book/6906-land-use-land-cover-and-natural-hazards-interactions-changes-and-impacts

Majeed, M., Tariq, A., Anwar, M. M., Khan, A. M., Arshad, F., Mumtaz, F., Farhan, M., Zhang, L., Zafar, A., Aziz, M., Abbasi, S., Rahman, G., Hussain, S., Waheed, M., Fatima, K., & Shaukat, S. (2021). Monitoring of Land Use–Land Cover Change and Potential Causal Factors of Climate Change in Jhelum District, Punjab, Pakistan, through GIS and Multi-Temporal Satellite Data; (pp. 129-147). https://www.mdpi.com/books/book/6906-land-use-land-cover-and-natural-hazards-interactions-changes-and-impacts

Majeed, M., Bhatti, K. H., & Amjad, M. S. (2021). Impact of climatic variations on the flowering phenology of plant species in jhelum district, punjab, pakistan. *Applied Ecology and Environmental Research*, *19*(5), 3343–3376. doi:10.15666/aeer/1905_33433376

Majeed, M., Bhatti, K. H., Pieroni, A., Sõukand, R., Bussmann, R. W., Khan, A. M., Chaudhari, S. K., Aziz, M. A., & Amjad, M. S. (2021). Gathered wild food plants among diverse religious groups in Jhelum District, Punjab, Pakistan. *Foods*, *10*(3), 594. Advance online publication. doi:10.3390/foods10030594 PMID:33799901

MajeedM.Hayat BhattiK.Shoaib AmjadM.Mehmood AbbasiA.RashidA.Nawaz Muhammad Nawaz ShareefF.MehmoodA. MahmoodM. (2020). *Ethno-veterinary practices of Poaceae taxa in Punjab, Pakistan.* doi:10.21203/rs.3.rs-15532/v1

Majeed, M., Khan, A. M., Habib, T., Anwar, M. M., Sahito, H. A., Khan, N., & Ali, K. (2022). Vegetation analysis and environmental indicators of an arid tropical forest ecosystem of Pakistan. *Ecological Indicators*, *142*, 109291. Advance online publication. doi:10.1016/j.ecolind.2022.109291

Majeed, M., Lu, L., Anwar, M. M., Tariq, A., Qin, S., El-Hefnawy, M. E., El-Sharnouby, M., Li, Q., & Alasmari, A. (2023). Prediction of flash flood susceptibility using integrating analytic hierarchy process (AHP) and frequency ratio (FR) algorithms. *Frontiers in Environmental Science*, *10*, 1037547. Advance online publication. doi:10.3389/fenvs.2022.1037547

Majeed, M., Lu, L., Haq, S. M., Waheed, M., Sahito, H. A., Fatima, S., Aziz, R., Bussmann, R. W., Tariq, A., Ullah, I., & Aslam, M. (2022). Spatiotemporal Distribution Patterns of Climbers along an Abiotic Gradient in Jhelum District, Punjab, Pakistan. *Forests*, *13*(8), 1244. Advance online publication. doi:10.3390/f13081244

Majeed, M., Muhammad, M., Hussain, T., Ali, M., Naz, T., Nawaz, S., & Abbas, G. (2023a). Nanopriming-Based Management of Biotic Stresses for Sustainable Agriculture. In *Nanopriming Approach to Sustainable Agriculture* (pp. 263–289). IGI Global. doi:10.4018/978-1-6684-7232-3.ch012

Majeed, M., Tariq, A., Anwar, M. M., Khan, A. M., Arshad, F., Mumtaz, F., Farhan, M., Zhang, L., Zafar, A., Aziz, M., Abbasi, S., Rahman, G., Hussain, S., Waheed, M., Fatima, K., & Shaukat, S. (2021). Monitoring of land use–Land cover change and potential causal factors of climate change in Jhelum district, Punjab, Pakistan, through GIS and multi-temporal satellite data. *Land (Basel)*, *10*(10), 1026. Advance online publication. doi:10.3390/land10101026

Majeed, M., Tariq, A., Haq, S. M., Waheed, M., Anwar, M. M., Li, Q., Aslam, M., Abbasi, S., Mousa, B. G., & Jamil, A. (2022). A Detailed Ecological Exploration of the Distribution Patterns of Wild Poaceae from the Jhelum District (Punjab), Pakistan. *Sustainability (Basel)*, *14*(7), 3786. Advance online publication. doi:10.3390/su14073786

Maji, S., & Modak, S. (2021). Neem: Treasure of natural phytochemicals. *Chemical Science Review and Letters*, *10*(39), 396–401.

Majumdar, S., & Keller, A. A. (2021). Omics to address the opportunities and challenges of nanotechnology in agriculture. *Critical Reviews in Environmental Science and Technology*, *51*(22), 2595–2636. doi:10.1080/10643389.2020.1785264

Mak, A. C., Osterfeld, S. J., Yu, H., Wang, S. X., Davis, R. W., Jejelowo, O. A., & Pourmand, N. (2010). Sensitive giant magnetoresistive-based immunoassay for multiplex mycotoxin detection. *Biosensors & Bioelectronics*, *25*(7), 1635–1639. doi:10.1016/j.bios.2009.11.028 PMID:20047828

Malerba, M., & Cerana, R. (2016). Chitosan effects on plant systems. *International Journal of Molecular Sciences*, *17*(7), 996. doi:10.3390/ijms17070996 PMID:27347928

Malerba, M., & Cerana, R. (2018). Recent advances of chitosan applications in plants. *Polymers*, *10*(2), 118. doi:10.3390/polym10020118 PMID:30966154

Malik, S., Muhammad, K., & Waheed, Y. (2023). Nanotechnology: A Revolution in Modern Industry. In Molecules (Vol. 28, Issue 2). MDPI. doi:10.3390/molecules28020661

Mallik, J., Das, P., Karon, B., & Das, S. (2012). A review on phytochemistry and pharmacological activity of *Terminalia belerica. International Journal of Drug Formulation and Research*, *3*(6), 1–7.

Manimegalai, S., Vickram, S., Deena, S. R., Rohini, K., Thanigaivel, S., Manikandan, S., Subbaiya, R., Karmegam, N., Kim, W., & Govarthanan, M. (2023). Carbon-based nanomaterial intervention and efficient removal of various contaminants from effluents – A review. *Chemosphere*, *312*, 137319. doi:10.1016/j.chemosphere.2022.137319 PMID:36410505

Manjunatha, S. B., Biradar, D. P., & Aladakatti, Y. R. (2016). Nanotechnology and its applications in agriculture: A review. *Journal of Farm Sciences*, *29*(1), 1–13.

Mann, S., Satpathy, G., & Gupta, R. K. (2015). *In vitro* evaluation of bio-protective properties of underutilized *Myrica esculenta* Buch.-Ham. ex D. Don fruit of Meghalaya. *Indian Journal of Natural Products and Resources*, *6*(3), 183–188.

Mantri, N., Patade, V., Penna, S., Ford, R., & Pang, E. (2012). Abiotic stress responses in plants: present and future. In *Abiotic Stress Responses in Plants* (pp. 1–19). Springer. doi:10.1007/978-1-4614-0634-1_1

Manzoor, N., Ahmed, T., Noman, M., Shahid, M., Nazir, M. M., Ali, L., Alnusaire, T. S., Li, B., Schulin, R., & Wang, G. (2021). Iron oxide nanoparticles ameliorated the cadmium and salinity stresses in wheat plants, facilitating photosynthetic pigments and restricting cadmium uptake. *The Science of the Total Environment*, *769*, 145221. doi:10.1016/j.scitotenv.2021.145221 PMID:33736258

Mao, X., Liu, X., Cheng, J., Liu, X., & Huang, Y. (2019). A portable soil nitrogen detector based on NIRS and a smartphone. *Sensors (Basel)*, *19*(20), 4451. PMID:31615081

Marchiol, L., Iafisco, M., Fellet, G., & Adamiano, A. (2020). Nanotechnology support the next agricultural revolution: Perspectives to enhancement of nutrient use efficiency. *Advances in Agronomy*, *161*, 27–116. doi:10.1016/bs.agron.2019.12.001

Marcon, L., Oliveras, J., & Puntes, V. F. (2021). In situ nanoremediation of soils and groundwaters from the nanoparticle's standpoint: A review. *The Science of the Total Environment*, *791*, 148324. doi:10.1016/j.scitotenv.2021.148324 PMID:34412401

Marschner, P. (2012). *Marschner's mineral nutrition of higher plants*. Academic press.

Martin, J. L., Maris, V., & Simberloff, D. S. (2016). The need to respect nature and its limits challenges society and conservation science. *Proceedings of the National Academy of Sciences of the United States of America*. PubMed, *113*(22), 6105–6112. 10.1073/pnas.1525003113

Martinelli, F., Scalenghe, R., Davino, S., Panno, S., Scuderi, G., Ruisi, P., & Dandekar, A. M. (2015). Advanced methods of plant disease detection. A review. *Agronomy for Sustainable Development*, *35*(1), 1–25. doi:10.1007/s13593-014-0246-1

Martirosyan, A., & Schneider, Y. J. (2014). Engineered nanomaterials in food: Implications for food safety and consumer health. *International Journal of Environmental Research and Public Health*, *11*(6), 5720–5750. doi:10.3390/ijerph110605720 PMID:24879486

Maruthappan, V., & Shree, K. S. (2010). Hypolipidemic activity of haritaki (*Terminalia chebula*) in atherogenic diet induced hyperlipidemic rats. *Journal of Advanced Pharmaceutical Technology & Research*, *1*(2), 229–235. PMID:22247850

Mashkour, M., Sharifinia, M., Yousefi, H., & Afra, E. (2018). MWCNT-coated cellulose nanopapers: Droplet-coating, process factors, and electrical conductivity performance. *Carbohydrate Polymers*, *202*, 504–512. doi:10.1016/j.carbpol.2018.09.006 PMID:30287029

Mathai, K. (2000). Nutrition in the adult years. Krause's Food, Nutrition, and Diet Therapy, 271, 274-275.

Mathew, S., Tiwari, D. K., & Tripathi, D. (2021). Interaction of carbon nanotubes with plant system: A review. *Carbon Letters*, *31*(2), 167–176. doi:10.1007/s42823-020-00195-1

Matos, M. P. S. R., Correia, A. A. S., & Rasteiro, M. G. (2017). Application of carbon nanotubes to immobilize heavy metals in contaminated soils. *Journal of Nanoparticle Research*, *19*(4), 1–11. doi:10.1007/s11051-017-3830-x

Mattos, B. D., & Magalhães, W. L. (2016). Biogenic nanosilica blended by nanofibrillated cellulose as support for slow-release of tebuconazole. *Journal of Nanoparticle Research*, *18*(9), 274. doi:10.1007/s11051-016-3586-8

Maurel, C., Boursiac, Y., Luu, D. T., Santoni, V., Shahzad, Z., & Verdoucq, L. (2015). Aquaporins in plants. *Physiological Reviews*, *95*(4), 1321–1358. doi:10.1152/physrev.00008.2015 PMID:26336033

Mavi, A. K., Gaur, S., Gaur, G., Babita, Jindal, P., Kumar, N., & Kumar, U. (2023). CAR T-cell therapy: Reprogramming patient's immune cell to treat cancer. *Cellular Signalling*, *105*, 110638. doi:10.1016/j.cellsig.2023.110638 PMID:36822565

Ma, X., Geiser-Lee, J., Deng, Y., & Kolmakov, A. (2017). Interactions between engineered nanoparticles (ENPs) and plants: Phytotoxicity, uptake and accumulation. *The Science of the Total Environment*, *408*(16), 3053–3061. doi:10.1016/j.scitotenv.2010.03.031 PMID:20435342

Mazumder, N., & Sivashankar, S. (2020). Nanopackaging of food: An overview. *Research Journal of Pharmaceutical, Biological and Chemical Sciences*, *11*(4), 1–12. doi:10.33887/rjpbcs/2020.11.4.1

McCartney, H. A., Foster, S. J., Fraaije, B. A., & Ward, E. (2003). Molecular diagnostics for fungal plant pathogens. *Pest Management Science: Formerly. Pest Management Science*, *59*(2), 129–142. doi:10.1002/ps.575

McClemens, D. T., & Xiao, H. (2017). Is nano safe in food? Establishing the factors impacting the gastro intestinal fate and toxicity of organic and inorganic food-grade nanoparticle. *NPJ Science of Food*, *1*(1), 6. doi:10.1038/s41538-017-0005-1 PMID:31304248

McGee, C. F., Storey, S., Clipson, N., & Doyle, E. (2017). Soil microbial community responses to contamination with silver, aluminium oxide and silicon dioxide nanoparticles. *Ecotoxicology (London, England)*, *26*(3), 449–458. doi:10.1007/s10646-017-1776-5 PMID:28197855

Meena Kumari, M., Jacob, J., & Philip, D. (2015). Green synthesis and applications of Au–Ag bimetallic nanoparticles. *Spectrochimica Acta. Part A: Molecular and Biomolecular Spectroscopy*, *137*, 185–192. doi:10.1016/j.saa.2014.08.079 PMID:25218228

Meena, R. S., Kumar, S., Datta, R., Lal, R., Vijayakumar, V., Brtnicky, M., Sharma, M. P., Yadav, G. S., Jhariya, M. K., Jangir, C. K., Pathan, S. I., Dokulilova, T., Pecina, V., & Marfo, T. (2020). Impact of agrochemicals on soil microbiota and management: A review. *Land (Basel)*, *9*(2), 34. doi:10.3390/land9020034

Mehmood, A., Murtaza, G., Bhatti, T. M., & Kausar, R. (2017). Phytoextracts synthesis of silver nanoparticles and their impact on microbial and plant growth. *IET Nanobiotechnology / IET*, *11*(7), 830–835.

Mehndiratta, P., Jain, A., Srivastava, S., & Gupta, N. (2013). Environmental pollution and nanotechnology. *Environment and Pollution*, *2*(2), 49. doi:10.5539/ep.v2n2p49

Mehta, M., Sharma, M., Pathania, K., Jena, P. K., & Bhushan, I. (2021). Degradation of synthetic dyes using nanoparticles: A mini-review. *Environmental Science and Pollution Research International*, *28*(36), 49434–49446. doi:10.1007/s11356-021-15470-5 PMID:34350572

Mellia, J. O., Galea, R., Nag, R., Cummins, E., Gatt, R., & Valdramidis, V. (2022). Nanoparticle food applications and their toxicity: Current trends and needs in risk assessment strategies. *Journal of Food Protection*, *85*(2), 355–372. doi:10.4315/JFP-21-184 PMID:34614149

Mengel, K., & Kirkby, E. A. (2012). *Principles of plant nutrition*. Springer Science & Business Media.

Meng, Y., Li, Y., Galvani, C. D., Hao, G., Turner, J. N., Burr, T. J., & Hoch, H. C. (2005). Upstream migration of Xylella fastidiosa via pilus-driven twitching motility. *Journal of Bacteriology*, *187*(16), 5560–5567. doi:10.1128/JB.187.16.5560-5567.2005 PMID:16077100

Meshram, M. M., Rangari, S. B., Kshirsagar, S. B., Gajbhiye, S., Trivedi, M. R., & Sahane, R. S. (2012). *Nyctanthes arbor-tristis*aherbal panacea. *International Journal of Pharmaceutical Sciences and Research*, *3*(8), 2432–2440.

Mickelbart, M. V., Hasegawa, P. M., & Bailey-Serres, J. (2015). Genetic mechanisms of abiotic stress tolerance that translate to crop yield stability. *Nature Reviews. Genetics*, *16*(4), 237–251. doi:10.1038/nrg3901 PMID:25752530

Midhat, L., Ouazzani, N., Hejjaj, A., Ouhammou, A., & Mandi, L. (2019). Accumulation of heavy metals in metallophytes from three mining sites (Southern Centre Morocco) and evaluation of their phytoremediation potential. *Ecotoxicology and Environmental Safety*, *169*, 150–160. doi:10.1016/j.ecoenv.2018.11.009 PMID:30445246

Mills, S., & Bone, K. (2000). *Principles and practice of phytotherapy: Modern herbal medicine*. Churchill Livingstone.

Mirzaei, H., & Darroudi, M. (2017). Zinc oxide nanoparticles: Biological synthesis and biomedical applications. *Ceramics International*, *43*(1), 907–914. doi:10.1016/j.ceramint.2016.10.051

Mirzajani, F., Askari, H., Hamzelou, S., Schober, Y., Rompp, A., Ghassempour, A., & Spengler, B. (2014). Proteomics study of silver nanoparticles toxicity on *Oryza sativa* L. *Ecotoxicology and Environmental Safety*, *108*, 335–339. doi:10.1016/j.ecoenv.2014.07.013 PMID:25124680

Mishra, A., & Singh, H. B. (2021). *Nanoparticles in agro-products: Benefits, applications and concerns," in Nanomaterials for agriculture and forestry applications*. Springer.

Mishra, S. K., Sangwan, N. S., & Sangwan, R. S. (2007). Plant review *Andrographis paniculata* (Kalmegh): A review. *Pharmacognosy Reviews*, *1*(2), 283–298.

Mishra, S., Singh, B. R., Singh, A., Keswani, C., Naqvi, A. H., & Singh, H. B. (2017). Biofabricated silver nanoparticles act as a strong fungicide against Bipolaris sorokiniana causing spot blotch disease in wheat. *PLoS One*, *12*(5), e0175750. PMID:28542577

Mishra, S., & Singh, H. B. (2015). Biosynthesized silver nanoparticles as a nanoweapon against phytopathogens: Exploring their scope and potential in agriculture. *Applied Microbiology and Biotechnology*, *99*(3), 1097–1107. doi:10.1007/s00253-014-6296-0 PMID:25547832

Misra, M., & Sachan, G. S. (2021). Nanobioremediation of heavy metals: Perspectives and challenges. *Journal of Basic Microbiology*, 1–16.

Mittal, J., Sharma, M. M., & Batra, A. (2014). *Tinospora cordifolia*: A multipurpose medicinal plant-A. *Faslnamah-i Giyahan-i Daruyi*, *2*(2), 32–47.

Mody, V. V., Cox, A., Shah, S., Singh, A., Bevins, W., & Parihar, H. (2014). Magnetic nanoparticle drug delivery systems for targeting tumor. *Applied Nanoscience*, *4*(4), 385–392. doi:10.1007/s13204-013-0216-y

Mohajerani, S., Fazel, M. A., Madani, H., & Shahrjerdi, A. (2019). Potential of thermochromic materials for maintaining the quality of horticultural crops in the post-harvest stage. *Food Packaging and Shelf Life*, *22*, 100394.

Mohamed, A. (2019). Chapter eight- synthesis, characterization, and applications carbon nanofibers: In carbon based nanofillers and their rubber nanocomposites. Elsevier.

Mohamed, A. K. S., Qayyum, M. F., Abdel-Hadi, A. M., Rehman, R. A., Ali, S., & Rizwan, M. (2017). Interactive effect of salinity and silver nanoparticles on photosynthetic and biochemical parameters of wheat. *Archives of Agronomy and Soil Science*, *63*(12), 1736–1747. doi:10.1080/03650340.2017.1300256

Mohammadi, R., Maali-Amiri, R., & Abbasi, A. (2013). Effect of TiO2 nanoparticles on chickpea response to cold stress. *Biological Trace Element Research*, *152*(3), 403–410. doi:10.1007/s12011-013-9631-x PMID:23456351

Mohammad, M. R., Ahmed, D. S., & Mohammed, M. K. (2019). Synthesis of Ag-doped TiO_2 nanoparticles coated with carbon nanotubes by the sol−gel method and their antibacterial activities. *Journal of Sol-Gel Science and Technology*, *90*(3), 498–509. doi:10.1007/s10971-019-04973-w

Mohammed, Z. G., Ahmad, F., Ibrahim, S. A., & Zaidi, S. (2022). Application of nanotechnology in different aspects of the food industry. *Discover Food*, *2*(1), 12. doi:10.1007/s44187-022-00013-9

Mohan Gokhale, M., & Ravindra Somani, R. (2015). Fullerenes: Chemistry and its applications. *Mini-Reviews in Organic Chemistry*, *12*(4), 355–366. doi:10.2174/1570193X12666150930224428

Mohiuddin, A. K. (2019). Chemistry of secondary metabolites. *Annals of Clinical Toxicology*, *2*(1), 1014.

Mondal, A., Dubey, B. K., Arora, M., & Mumford, K. (2021). Porous media transport of iron nanoparticles for site remediation application: A review of lab scale column study, transport modelling and field-scale application. *Journal of Hazardous Materials*, *403*, 123443. doi:10.1016/j.jhazmat.2020.123443 PMID:32798796

Mondal, S., & Mishra, S. (2020). Nano-based thermal insulation of buildings. In *Emerging Nanotechnologies for Green Building* (pp. 1–28). Elsevier.

Morresion, G., & Hark, L. (1999). *Medical Nutrition and Disease*. BlackWell Science.

Mousavi, S. R., & Rezaei, M. (2011). Nanotechnology in agriculture and food production. *Journal of Applied Environmental and Biological Sciences*, *1*(10), 414–419.

Mpongwana, N. & Rathilal, S. (2022). A Review of the Techno-Economic Feasibility of Nanoparticle Application for Wastewater Treatment. *Water, 14*(10), 1550. doi:10.3390/w14101550

Mubarok, S., Nuraini, A., Hamdani, J. S., Suminar, E., Kusumiyati, K., Budiarto, R., Lestari, F. W., Rahmat, B. P. N., & Ezura, H. (2024). Antioxidative response of parthenocarpic tomato, iaa9-3 and iaa9-5, under heat stress condition. *Plant Physiology and Biochemistry*, *207*, 108333. doi:10.1016/j.plaphy.2024.108333 PMID:38181640

Muhammad, M., Basit, A., Majeed, M., Shah, A. A., Ullah, I., Mohamed, H. I., Khan, A., & Ghanaim, A. M. (2024). Bacterial pigments and their applications. In *Bacterial Secondary Metabolites* (pp. 277–298). Elsevier. doi:10.1016/B978-0-323-95251-4.00019-3

Muhammad, M., Waheed, A., Wahab, A., Majeed, M., Nazim, M., Liu, Y. H., Li, L., & Li, W. J. (2024). Soil salinity and drought tolerance: An evaluation of plant growth, productivity, microbial diversity, and amelioration strategies. In *Plant Stress* (Vol. 11). Elsevier B.V. doi:10.1016/j.stress.2023.100319

Mukharjee, D., Lokwani, S., Dave, A., Gonsalves, C., & Sarkar, C. (2021). Bhringraj: A pharmaceutical treasure trove. *Indian Journal of Natural Sciences*, *12*(68), 35231–35239.

Mukherjee, A., Maity, A., Pramanik, P., Shubha, K., Joshi, D. C., & Wani, S. H. (2019). Public perception about use of nanotechnology in agriculture. In *Advances in phytonanotechnology* (pp. 405–418). Academic Press. doi:10.1016/B978-0-12-815322-2.00019-5

Mukherjee, A., Majumdar, S., Servin, A. D., Pagano, L., Dhankher, O. P., & White, J. C. (2016). Carbon nanomaterials in agriculture: A critical review. *Frontiers in Plant Science*, *7*, 172. doi:10.3389/fpls.2016.00172 PMID:26941751

Mukhopadhyay, S. S. (2014). Nanotechnology in agriculture: Prospects and constraints. *Nanotechnology, Science and Applications*, 63–71. doi:10.2147/NSA.S39409 PMID:25187699

Mukhopadhyay, S. S., Gurav, R. V., & Sabarinathan, N. (2020). Nano amendments for improving soil fertility and crop productivity. In *Nanomaterials for Agriculture and Forestry Applications* (pp. 111–141). Elsevier.

Muktar, Y., Bikila, T., & Keffale, M. (2015). Application of Nanotechnology for Animal Health and Production Improvement: A Review. *World Applied Sciences Journal*, *33*(10), 1588–1596. doi:10.5829/idosi.wasj.2015.33.10.96253

Munir, N., Gulzar, W., Abideen, Z., Hasanuzzaman, M., El-Keblawy, A., & Zhao, F. (2023). Plant–Nanoparticle Interactions: Transcriptomic and Proteomic Insights. *Agronomy (Basel)*, *13*(8), 2112. doi:10.3390/agronomy13082112

Munns, R., & Tester, M. (2008). Mechanisms of salinity tolerance. *Annual Review of Plant Biology*, *59*(1), 651–681. doi:10.1146/annurev.arplant.59.032607.092911 PMID:18444910

Munyi, M., Wainaina, I., & Njage, E. (2020). Nanotechnology applications in crop production and food systems: A review. *Journal of Applied Science & Environmental Management*, *24*(4), 655–667.

Mura, S., Greppi, G., Irudayaraj, J., & Seetharaman, K. (2016). Nanosensors for food and agriculture. In *Nanoscience in Food and Agriculture 1* (pp. 259–284). Springer.

Mustafa, G., & Komatsu, S. (2016). Insights into the response of soybean mitochondrial proteins to various sizes of aluminum oxide nanoparticles under flooding stress. *Journal of Proteome Research*, *15*(12), 4464–4475. doi:10.1021/acs.jproteome.6b00572 PMID:27780359

Nadi, E., Aynehband, A., & Mojaddam, M. (2019). Effect of nano-iron fertilizer on yield and yield components of chickpea (Cicer arietinum L.) under water deficit stress. *Journal of Crop Science and Biotechnology*, *22*(3), 249–256. doi:10.1007/s12892-018-0152-0

Naikoo, M. I., Dar, M. I., Khan, F. A., Raghib, F., & Rajakaruna, N. (2019). Trophic transfer and bioaccumulation of lead along soil–plant–aphid–ladybird food chain. *Environmental Science and Pollution Research International*, *26*(23), 23460–23470. doi:10.1007/s11356-019-05624-x PMID:31201701

Naikoo, M. I., Kafeel, U., Naushin, F., & Khan, F. A. (2020). Halophytes in India and Their Role in Phytoremediation. In M.-N. Grigore (Ed.), *Handbook of Halophytes: From Molecules to Ecosystems towards Biosaline Agriculture* (pp. 1–21). Springer International Publishing. doi:10.1007/978-3-030-17854-3_115-1

Nair, R., & Chung, Y. T. (2014). Environmental transformation of silver nanoparticles: Effects on aquatic organisms. *Environmental Toxicology and Chemistry*, *33*(2), 462–467.

Nair, R., Varghese, S. H., Nair, B. G., Maekawa, T., Yoshida, Y., & Kumar, D. S. (2010). Nanoparticulate material delivery to plants. *Plant Science*, *179*(3), 154–163. doi:10.1016/j.plantsci.2010.04.012

Nakabayashi, R., Hashimoto, K., Toyooka, K., & Saito, K. (2019). Keeping the shape of plant tissue for visualizing metabolite features in segmentation and correlation analysis of imaging mass spectrometry in Asparagus officinalis. Metabolomics. *Metabolomics*, *15*(2), 24. doi:10.1007/s11306-019-1486-5 PMID:30830429

Nandhakumar, M. R., Muthukrishnan, R., Kumar, T. P., Sureshkumar, P., & Nandini, S. (2023). Nano Fertilizer in Crop Production: The Changing Scenario. *International Journal of Environment and Climate Change*, *13*(6), 158–170. doi:10.9734/ijecc/2023/v13i61811

Nandhakumar, P., Kim, G., Park, S., Kim, S., Kim, S., Park, J. K., Lee, N.-S., Yoon, Y. H., & Yang, H. (2020). Metal Nanozyme with Ester Hydrolysis Activity in the Presence of Ammonia-Borane and its Use in a Sensitive Immunosensor. *Angewandte Chemie International Edition*, *59*(50), 22419–22422. doi:10.1002/anie.202009737 PMID:32875647

Nankar, R., Prabhakar, P. K., & Doble, M. (2017). Hybrid drug combination: Combination of ferulic acid and metformin as anti-diabetic therapy. *Phytomedicine*, *37*, 10–13. doi:10.1016/j.phymed.2017.10.015 PMID:29126698

Narayanan, A., Sharma, P., & Moudgil, B. M. (2013). Applications of engineered particulate systems in agriculture and food industry. *Kona Powder and Particle Journal*, *30*(0), 221–235. doi:10.14356/kona.2013021

Naseer, B., Srivastava, G., Qadri, O. S., Faridi, S. A., Islam, R. V. I., & Younis, K. (2018). Importance and health hazard of nanoparticles used in food industry. *Nanotechnology Reviews*, *7*(6), 623–641. Doi.org/10.1515/intrev-2018-0076. doi:10.1515/ntrev-2018-0076

Naseri, R., & Davoodi, R. (2011). Commercialization of nanotechnology in developing countries. *3rd International Conference on Information and Financial Engineering*, 385-389.

Nasirian, R., & Nouri, S. (2019). Improving tomato growth and yield by using of nano-silica coating on the greenhouse cover. *Acta Scientiarum Polonorum. Hortorum Cultus*, *18*(1), 3–13.

Nasir, S., Hussein, M. Z., Zainal, Z., & Yusof, N. A. (2018). Carbon-based nanomaterials/allotropes: A glimpse of their synthesis, properties and some applications. *Materials (Basel)*, *11*(2), 295. doi:10.3390/ma11020295 PMID:29438327

Nasrollahzadeh, M., Sajadi, S. M., Sajjadi, M., & Issaabadi, Z. (2019). Applications of nanotechnology in daily life. *Interface Science and Technology, 28*, 113-143.

Nasseri, M., Golmohammadzadeh, S., Arouiee, H., Jaafari, M. R., & Neamati, H. (2016). Antifungal activity of Zataria multiflora essential oil-loaded solid lipid nanoparticles in-vitro condition. *Iranian Journal of Basic Medical Sciences.*, *19*(11), 1231. PMID:27917280

Nastasijevic, I., Mitrovic, R., & Jankovic, S. (2021). Biosensors for animal health and meat safety monitoring: Farm-to-slaughterhouse continuum. *IOP Conference Series. Earth and Environmental Science*, *854*(1), 012063. Advance online publication. doi:10.1088/1755-1315/854/1/012063

Navarro, S., Fenoll, J., Vela, N., Ruiz, E., & Navarro, G. (2009). Photocatalytic degradation of eight pesticides in leaching water by use of ZnO under natural sunlight. *Journal of Hazardous Materials*, *172*(2–3), 1303–1310. doi:10.1016/j.jhazmat.2009.07.137 PMID:19729242

Naveed, M. Muhammad Majeed, Khizra Jabeen, Nimra Hanif, Rida Naveed, Sania Saleem & Nida Khan (2024). Plant Disease Diagnosis with Artificial Intelligence (AI). Microbial Data Intelligence and Computational Techniques for Sustainable Computing (pp 217–234). Springer. doi:10.1007/978-981-99-9621-6_15

Naveed, Memon, Abdullah, Makhdoom, Azeem, Mehmood, Salahuddin, Rajpoot, & Majeed. (2024). Use Cases and Future Aspects of Intelligent Techniques in Microbial Data Analysis. In Microbial Data Intelligence and Computational Techniques for Sustainable Computing (pp 259–280). Springer. doi:10.1007/978-981-99-9621-6_17

Naveed, Zahid, Fatima, Saleem, Majeed, Abid, Javed, Wazir, & Qasim. (2024). Disease Diagnosis with Artificial Intelligence (AI). In Microbial Data Intelligence and Computational Techniques for Sustainable Computing (pp 235–258). Springer. doi:10.1007/978-981-99-9621-6_16

New, W. X., Ogbezode, J. E., & Gani, P. (2023). Nanoparticles in Soil Remediation: Challenges and Opportunities. *Industrial and Domestic Waste Management*, *3*(2). Advance online publication. doi:10.53623/idwm.v3i2.357

Nguyen, N. H., & Falagan-Lotsch, P. (2023). Mechanistic Insights into the Biological Effects of Engineered Nanomaterials: A Focus on Gold Nanoparticles. *International Journal of Molecular Sciences*, *24*(4), 4109. doi:10.3390/ijms24044109 PMID:36835521

Nguyen, N. T. T., Nguyen, L. M., Nguyen, T. T. T., Tran, U. P. N., Nguyen, D. T. C., & Van Tran, T. (2023). A critical review on the bio-mediated green synthesis and multiple applications of magnesium oxide nanoparticles. *Chemosphere*, *312*, 137301. doi:10.1016/j.chemosphere.2022.137301 PMID:36410506

Nguyen, T. M. H., Le, T. T. H., Vu, T. T., Nguyen, T. H., Nguyen, V. H., & Tran, L. D. (2021). A graphene-based electrochemical biosensor for rapid detection of Escherichia coli O157:H7 in fresh produce. *Sensors and Actuators. B, Chemical*, *329*, 129196. doi:10.1016/j.snb.2020.129196

Nhan, L. V., Ma, C., Rui, Y., Liu, S., Li, X., Xing, B., & Liu, L. (2015). Phytotoxic\ mechanism of nanoparticles: Destruction of chloroplasts and vascular bundles and alteration of nutrient absorption. *Scientific Reports*, *5*(1), 11618. doi:10.1038/srep11618 PMID:26108166

Nhiem, N. X., Van Kiem, P., Van Minh, C., Tai, B. H., Cuong, N. X., Thu, V. K., Anh, H. L. T., Jo, S. H., Jang, H. D., Kwon, Y. I., & Kim, Y. H. (2010). A new monoterpenoid glycoside from *Myrica esculenta* and the inhibition of angiotensin I-converting enzyme. *Chemical & Pharmaceutical Bulletin*, *58*(10), 1408–1410. doi:10.1248/cpb.58.1408 PMID:20930414

Nicolini, V., Gambuzzi, E., Malavasi, G., Menabue, L., Menziani, M. C., Lusvardi, G., Pedone, A., Benedetti, F., Luches, P., D'Addato, S., & Valeri, S. (2015). Evidence of Catalase Mimetic Activity in Ce3+/Ce4+ Doped Bioactive Glasses. *The Journal of Physical Chemistry B*, *119*(10), 4009–4019. doi:10.1021/jp511737b PMID:25710332

Nie, D., Li, J., Xie, Q., Ai, L., Zhu, C., Wu, Y., & Tan, W. (2023). Nanoparticles: A Potential and Effective Method to Control Insect-Borne Diseases. *Bioinorganic Chemistry and Applications*, *2023*, 1–13. Advance online publication. doi:10.1155/2023/5898160 PMID:37213220

Nile, S. H., Baskar, V., Selvaraj, D., Nile, A., Xiao, J., & Kai, G. (2020). Nanotechnologies in food science: Applications, recent trends, and future perspectives. *Nano-Micro Letters*, *12*(45), 1–34. doi:10.1007/s40820-020-0383-9 PMID:34138283

Nima, Z. A., Lahiani, M. H., Watanabe, F., Xu, Y., Khodakovskaya, M. V., & Biris, A. S. (2014). Plasmonically active nanorods for delivery of bio-active agents and high-sensitivity SERS detection in planta. *RSC Advances*, *4*(110), 64985–64993. doi:10.1039/C4RA10358K

Nkurunziza, D., Niyigena, C., & Mugenzi, I. (2019). The effect of nanofertilizers on growth and yield of soybean (Glycine max L. Merr). *Journal of Nanomaterials*.

Noman, M., Ahmed, T., Hussain, S., Niazi, M. B. K., Shahid, M., & Song, F. (2020). Biogenic copper nanoparticles synthesized by using a copper-resistant strain Shigella flexneri SNT22 reduced the translocation of cadmium from soil to wheat plants. *Journal of Hazardous Materials*, *398*, 123175. doi:10.1016/j.jhazmat.2020.123175 PMID:32768848

Nowack, B., & Mitrano, D. M. (2018). Metal and metal oxide nanoparticles in aquatic systems: An analysis of data and publications. *The Science of the Total Environment*, *628*, 748–768.

Nuruzzaman, M. D., Rahman, M. M., Liu, Y., & Naidu, R. (2016). Nanoencapsulation, nano-guard for pesticides: A new window for safe application. *Journal of Agricultural and Food Chemistry*, *64*(7), 1447–1483. doi:10.1021/acs.jafc.5b05214 PMID:26730488

Nuss, E. T., & Tanumihardjo, S. A. (2010). Maize: A paramount staple crop in the context of global nutrition. *Comprehensive Reviews in Food Science and Food Safety*, *9*(4), 417–436. doi:10.1111/j.1541-4337.2010.00117.x PMID:33467836

Oberdorster, G., Oberdorster, E., & Oberdorster, J. (2005). Nanotoxicity: An emerging discipline evolving from studies of ultrafine particle. *Environmental Health Perspectives*, *113*(7), 823–839. doi:10.1289/ehp.7339 PMID:16002369

Ocsoy, I., Paret, M. L., Ocsoy, M. A., Kunwar, S., Chen, T., You, M., & Tan, W. (2013). Nanotechnology in plant disease management: DNA-directed silver nanoparticles on graphene oxide as an antibacterial against Xanthomonas perforans. *ACS Nano*, *7*(10), 8972–8980. doi:10.1021/nn4034794 PMID:24016217

Olaniran, A. O., & Igbinosa, E. O. (2011). Chlorophenols and other related derivatives of environmental concern: Properties, distribution and microbial degradation processes. *Chemosphere*, *83*(10), 1297–1306. doi:10.1016/j.chemosphere.2011.04.009 PMID:21531434

Oleg, F. (2017). *Dmitry, Beilin*. Nanotechnology in Agriculture. doi:10.1201/9781315229287-7

Oliveira, T. M. B. F., Ribeiro, F. W. P., Sousa, C. P., Salazar-Banda, G. R., de Lima-Neto, P., Correia, A. N., & Morais, S. (2020). Current overview and perspectives on carbon-based (bio) sensors for carbamate pesticides electroanalysis. *Trends in Analytical Chemistry*, *124*(115779), 115779. doi:10.1016/j.trac.2019.115779

Omanović-Mikličanina, E., & Maksimović, M. (2016). Nanosensors applications in agriculture and food industry. *Bull Chem Technol Bosnia Herzegovina*, *47*, 59–70.

Onyeka, H., Passaretti, P., & Miri, T. (2022). The safety of nanomaterials in food production and packaging. *Current Research in Food Science*, *5*, 763–774. doi:10.1016/j.crfs.2022.04.005 PMID:35520272

Otchere, E., McKay, B. M., English, M. M., & Aryee, A. N. A. (2023). Current trends in nano-delivery systems for functional foods: A systematic review. *PeerJ*, *11*, e14980. doi:10.7717/peerj.14980 PMID:36949757

Oughton, D. H., & Salbu, B. (2012). Ecological risk assessment. In A. Pieczynska (Ed.), *Uranium: Characteristics, occurrence and human exposure* (pp. 295–310). Nova Science Publishers.

Ozcakir, G. (2023). Applications of nanomaterials in food industry: A review. *Materials Proceedings*, *5*, x. doi:10.3390/xxxxx

Ozdemir, M., & Kemerli, T. (2016). Innovative applications of micro and nanoencapsulation in food packaging. *Encapsulation and controlled release technologies in food systems*, 333-378.

Özer, E. A., Özcan, M., & Didin, M. (2014). Nanotechnology in food and agriculture industry. *Food processing: Strategies for quality assessment*, 477-497.

Ozimek, L., Pospiech, E., & Narine, S. S. (2010). Nanotechnology in food and meat processing. *Acta Scientiarum Polonorum. Technologia Alimentaria*, *9*(4), 401–412.

Paarakh, P. M. (2010). *Terminalia arjuna* (Roxb.) Wt. and Arn.: A review. *International Journal of Pharmacology*, *6*(5), 515–534. doi:10.3923/ijp.2010.515.534

Paidari, S., Tahergorabi, R., Anari, E. S., Nafchi, A. M., Zamindar, N., & Goli, M. (2021). Migration of various nanoparticles into food samples: A review. *Foods*, *10*(9), 214. doi:10.3390/foods10092114 PMID:34574224

Pal, G., Rai, P. & Pandey, A. (2019). Green synthesis of nanoparticles: A greener approach for a cleaner future. *Green Synthesis, Characterization and Applications of Nanoparticles*, 1–26. doi:10.1016/B978-0-08-102579-6.00001-0

Palchoudhury, S., Jungjohann, K. L., Weerasena, L., Arabshahi, A., Gharge, U., Albattah, A., Miller, J., Patel, K., & Holler, R. A. (2018). Enhanced legume root growth with presoaking in α-Fe2O3 nanoparticle fertilizer. *RSC Advances*, *8*(43), 24075–24083. doi:10.1039/C8RA04680H PMID:35539206

Palmqvist, N. M., Seisenbaeva, G. A., Svedlindh, P., & Kessler, V. G. (2017). Maghemite nanoparticles acts as nanozymes, improving growth and abiotic stress tolerance in Brassica napus. *Nanoscale Research Letters*, *12*(1), 631. doi:10.1186/s11671-017-2404-2 PMID:29260423

PandaP.ChakrobortyS.BabuS.KrishnaN. (2023). *The Use of Silver Nanoparticles in Environmental Remediation*. doi:10.20944/preprints202301.0330.v1

Pandey, P., Tripathi, A., Dwivedi, S., Lal, K., & Jhang, T. (2023). Deciphering the mechanisms, hormonal signaling, and potential applications of endophytic microbes to mediate stress tolerance in medicinal plants. In Frontiers in Plant Science (Vol. 14). Frontiers Media SA. doi:10.3389/fpls.2023.1250020

Pandey, A. C., Sanjay, S. S., & Yadav, R. S. (2018). Fe3O4 nanoparticles-based magnetic solid-phase extraction coupled with high-performance liquid chromatography for organophosphorus pesticides monitoring in agricultural and environmental water samples. *Environmental Science and Pollution Research International*, *27*(29), 36679–36691.

Pandey, G. (2018). Challenges and future prospects of agri-nanotechnology for sustainable agriculture in India. *Environmental Technology & Innovation*, *11*, 299–307. doi:10.1016/j.eti.2018.06.012

Pandey, G., Maheshwari, D. K., & Samreen, S. (2018). Nanotechnology in agro-food industry. In *Nanotechnology* (pp. 147–190). Springer. doi:10.1007/978-981-10-8780-8_6

Pandey, V. C., Abhilash, P. C., & Singh, N. (2019). Sustainable phytoremediation and practices in the management of soil pollution. In *Phytomanagement of Polluted Sites* (pp. 1–19). Elsevier. doi:10.1016/B978-0-12-813912-7.00001-6

Pandita, D., & Pandita, A. (2021). Secondary metabolites in medicinal and aromatic plants (MAPs): potent molecules in nature's arsenal to fight human diseases. *Medicinal and Aromatic Plants: Healthcare and Industrial Applications*, 41-84.

Pandohee, J., Basu, R., Dasgupta, S., Sundarrajan, P., Shaikh, N., Patel, N., & Noor, A. (2023). Applications of multi-omics approaches for food and nutritional security. In *Sustainable Agriculture in the Era of the OMICs Revolution* (pp. 103–118). Springer International Publishing. doi:10.1007/978-3-031-15568-0_5

Pan, M., Yin, Z., Liu, K., Du, X., Liu, H., & Wang, S. (2019). Carbon-based nanomaterials in sensors for food safety. *Nanomaterials (Basel, Switzerland)*, *9*(9), 1–23. doi:10.3390/nano9091330 PMID:31533228

Panpatte, D. G., Jhala, Y. K., Shelat, H. N., & Vyas, R. V. (2016). Nanoparticles – The next generation technology for sustainable agriculture. In D. P. Singh, H. B. Singh & R. Prabha (Eds.), Microbial inoculants in sustainable agricultural productivity, Functional applications, 2 (pp. 289–300). Springer.

Paret, M. L., Vallad, G. E., Averett, D. R., Jones, J. B., & Olson, S. M. (2013). Photocatalysis: Effect of light-activated nanoscale formulations of TiO2 on Xanthomonas perforans and control ofbacterial spot of tomato. *Phytopathology*, *103*(3), 228–236. doi:10.1094/PHYTO-08-12-0183-R PMID:23190116

Parikh, S. J., & James, B. R. (2012). Soil: The foundation of agriculture. *Natl Educ. Knowl.*, *3*, 2.

Parisi, C., Vigani, M., & Rodríguez-Cerezo, E. (2015). Agricultural nanotechnologies: What are the current possibilities? *Nano Today*, *10*(2), 124–127. doi:10.1016/j.nantod.2014.09.009

Park, H. J., Kim, S. H., Kim, H. J., & Choi, S. H. (2006). A new composition of nanosized silica-silver for control of various plant diseases. *The Plant Pathology Journal*, *22*(3), 295–302. doi:10.5423/PPJ.2006.22.3.295

Patel, M. M., Hall, A. J., Vinjé, J., & Parashar, U. D. (2009). Noroviruses: A Comprehensive Review. *Journal of Clinical Virology*, *44*(1), 1–8. doi:10.1016/j.jcv.2008.10.009 PMID:19084472

Patil, H. J., & Solanki, M. K. (2016). Microbial inoculant: modern era of fertilizers and pesticides. Microbial inoculants in sustainable agricultural productivity: vol. 1: research perspectives, 319-343.

Patil, S. & Chandrasekaran, R. (2020). Biogenic nanoparticles: a comprehensive perspective in synthesis, characterization, application and its challenges. *Journal of Genetic Engineering and Biotechnology, 18*(1), 1–23. doi:10.1186/s43141-020-00081-3

Patra, J. K., & Baek, K. H. (2017). Antibacterial activity and synergistic antibacterial potential of biosynthesized silver nanoparticles against foodborne pathogenic bacteria along with its anticandidal and antioxidant effects. *Frontiers in Microbiology*, *8*, 167. doi:10.3389/fmicb.2017.00167 PMID:28261161

Pattanshetti, A., Pradeep, N., Chaitra, V., & Uma, V. (2020). Synthesis of multi-walled carbon nanotubes (MWCNTs) from plastic waste & analysis of garlic coated gelatin/MWCNTs nanocomposite films as food packaging material. *SN Applied Sciences*, *2*(4), 730. doi:10.1007/s42452-020-2442-8

Paul, N., & Bhojane, P. (2021). Role of nanoparticles in agriculture and food industry. *Journal of Emerging Technologies and Innovative Research*, *8*(11), 221–234.

Pavlidou, S., & Papaspyrides, C. D. (2008). A review on polymer–layered silicate nanocomposites. *Progress in Polymer Science*, *33*(12), 1119–1198. doi:10.1016/j.progpolymsci.2008.07.008

Payal, A., Krishnamoorthy, S., Elumalai, A., Moses, J. A., & Anandharamakrishnan, C. (2021). A review on recent developments and applications of nanozymes in food safety and quality analysis. *Food Analytical Methods*, *14*(8), 1–23. doi:10.1007/s12161-021-01983-9

Payne, P. R., Finlay-Smits, S., Small, B., Cave, V., & Kean, J. (2023). What's that bug? Community participation in biosecurity in Mount Maunganui, New Zealand. *Biological Invasions*, *25*(2), 593–610. doi:10.1007/s10530-022-02937-w

Pereira, A. E. S., Sandoval-Herrera, I. E., Zavala-Betancourt, S. A., Oliveira, H. C., Ledezma-Pérez, A. S., Romero, J., & Fraceto, L. F. (2017). γ-Polyglutamic acid/chitosan nanoparticles for the plant growth regulator gibberellic acid: Characterization and evaluation of biological activity. *Carbohydrate Polymers*, *157*, 1862–1873. doi:10.1016/j.carbpol.2016.11.073 PMID:27987906

Periakaruppan, R., Vanathi, P., & Vidhya, D. (2023). Management of soil nutrient deficiency by nanometal oxides. *Nanometal Oxides in Horticulture and Agronomy*, 291–320. doi:10.1016/B978-0-323-91809-1.00003-2

Petkoska, A. T., Daniloski, D., D'Cunha, N. M., Naumovski, N., & Broach, A. T. (2021). Edible packaging: Sustainable solutions and novel trends in food packaging. *Food Research International*, *140*, 109981. doi:10.1016/j.foodres.2020.109981 PMID:33648216

Petryayeva, E., & Krull, U. J. (2011). Localized surface plasmon resonance: Nanostructures, bioassays and biosensing—A review. *Analytica Chimica Acta*, *706*(1), 8–24. doi:10.1016/j.aca.2011.08.020 PMID:21995909

Phenrat, T., Kim, H. J., Fagerlund, F., Illangasekare, T., Tilton, R. D., & Lowry, G. V. (2009). Particle size distribution, concentration, and magnetic attraction affect transport of polymer-modified Fe0 nanoparticles in sand columns. *Environmental Science & Technology*, *43*(13), 5079–5085. doi:10.1021/es900171v PMID:19673310

Phogat, N., Ali Khan, S., Shankar, S., & Ansary, A., & Uddin, I. (2016). Fate of inorganic nanoparticles in agriculture. *Advanced Materials Letters*, *7*(1), 3–12. doi:10.5185/amlett.2016.6048

Pimentel, D. (1995). Amounts of pesticides reaching target pests: Environmental impacts and ethics. *Journal of Agricultural & Environmental Ethics*, *8*(1), 17–29. doi:10.1007/BF02286399

Pinmai, K., Chunlaratthanabhorn, S., Ngamkitidechakul, C., Soonthornchareon, N., & Hahnvajanawong, C. (2008). Synergistic growth inhibitory effects of *Phyllanthus emblica* and *Terminalia bellerica* extracts with conventional cytotoxic agents: Doxorubicin and cisplatin against human hepatocellular carcinoma and lung cancer cells. *World Journal of Gastroenterology*, *14*(10), 1491–1497. doi:10.3748/wjg.14.1491 PMID:18330936

Pinto, V. E. F., & Patriarca, A. (2017). Alternaria Species and Their Associated Mycotoxins. *Methods in Molecular Biology (Clifton, N.J.)*, *1542*, 13–32. doi:10.1007/978-1-4939-6707-0_2 PMID:27924529

Pokropivny, V., Lohmus, R., Hussainova, I., Pokropivny, A., & Vlassov, S. (2007). *Introduction to nanomaterials and nanotechnology*. Tartu University Press.

Poursani, A. V., Gerami, M., & Nezhad, M. N. (2020). Zinc oxide nanoparticles improve heat tolerance in potato (Solanum tuberosum L.) plants. *Journal of Plant Physiology*, *255*, 153293.

Pradhan, N., Singh, S., Ojha, N., Shrivastava, A., Barla, A., Rai, V., & Bose, S. (2015). Facets of nanotechnology as seen in food processing, packaging, and preservation industry. *BioMed Research International*, *2015*, 2015. doi:10.1155/2015/365672 PMID:26613082

Pradhan, S., Patra, P., Das, S., Chandra, S., Mitra, S., Dey, K. K., Akbar, S., Palit, P., & Goswami, A. (2013). Photochemical modulation of biosafe manganese nanoparticles on Vigna radiata: A detailed molecular, biochemical, and biophysical study. *Environmental Science & Technology*, *47*(22), 13122–13131. doi:10.1021/es402659t PMID:24144189

Pramanik, P. K. D., Solanki, A., Debnath, A., Nayyar, A., El-Sappagh, S., & Kwak, K. S. (2020). Advancing Modern Healthcare with Nanotechnology, Nanobiosensors, and Internet of Nano Things: Taxonomies, Applications, Architecture, and Challenges. *IEEE Access : Practical Innovations, Open Solutions*, *8*, 65230–65266. doi:10.1109/ACCESS.2020.2984269

Prasad, R. (2014). Synthesis of silver nanoparticles in photosynthetic plants. *Journal of Nanoparticles*.

Prasad, M., Lambe, U. P., Brar, B., Shah, I., Manimegalai, J., Ranjan, K., & Prasad, G. (2018). Nanotherapeutics: An insight into healthcare and multi-dimensional applications in medical sector of the modern world. *Biomedicine and Pharmacotherapy*, *97*, 1521–1537. doi:10.1016/j.biopha.2017.11.026 PMID:29793315

Prasad, R., Bhattacharyya, A., & Nguyen, Q. D. (2017). Nanotechnology in sustainable agriculture: Recent developments, challenges, and perspectives. *Frontiers in Microbiology*, *8*, 1014. doi:10.3389/fmicb.2017.01014 PMID:28676790

Prasad, R., Kumar, V., & Prasad, K. S. (2014). Nanotechnology in sustainable agriculture: Present concerns and future aspects. *African Journal of Biotechnology*, *13*(6), 705–713. doi:10.5897/AJBX2013.13554

Prasad, R., Pandey, R., & Barman, I. (2016). Engineering tailored nanoparticles with microbes: Quo vadis? *Wiley Interdisciplinary Reviews. Nanomedicine and Nanobiotechnology*, *8*(2), 316–330. doi:10.1002/wnan.1363 PMID:26271947

Preethi, B., Sivasubramanian, V., & Muthuraman, P. (2021). Nano-sensors for measuring soil moisture: A review. *Nanotechnology for Environmental Engineering*, *6*(1), 1–14.

Pretty, J., & Bharucha, Z. P. (2014). Sustainable intensification in agricultural systems. *Annals of Botany*, *114*(8), 1571–1596. doi:10.1093/aob/mcu205 PMID:25351192

Pruksasri, S., Lanner, B., & Novalin, S. (2020). Nanofiltration as a potential process for the reduction of sugar in apple juices on an industrial scale. LWT- Food. *Sciences et Techniques (Paris)*, *133*, 110118.

Qian, T., Zhao, D., Zhang, F., & Zhang, W. (2019). Nanotechnology for soil remediation: Applications vs. implications. *ACS Sustainable Chemistry & Engineering*, *7*(5), 4395–4404.

Qiao, A., Kalevaru, V. N., Radnik, J., Duvel, A., Heitjans, P., Kumar, A. S. H., Prasad, P. S. S., Lingaiah, N., & Martin, A. (2014). Oxidative dehydrogenation of ethane to ethylene over V2O5/Al2O3 catalysts: Effect of source of alumina on the catalytic performance. *Industrial & Engineering Chemistry Research*, *53*(49), 18711–18721. doi:10.1021/ie5008344

Qiao, Y., Wu, J., Xu, Y., Fang, Z., Zheng, L., Cheng, W., Tsang, E. P., Fang, J., & Zhao, D. (2017). Remediation of cadmium in soil by biochar-supported iron phosphate nanoparticles. *Ecological Engineering*, *106*, 515–522. doi:10.1016/j.ecoleng.2017.06.023

Qi, M., Liu, Y., & Li, T. (2018). Chitosan nanoparticles enhance the antioxidant activity of ascorbic acid in wheat seedlings under heat stress. *Molecules (Basel, Switzerland)*, *23*(11), 2908. PMID:30413008

Quintero-Fabián, S., Ortuño-Sahagún, D., Vázquez-Carrera, M., & López-Roa, R. I. (2013). Alliin, a garlic (*Allium sativum*) compound, prevents LPS-induced inflammation in 3T3-L1 adipocytes. *Mediators of Inflammation*, *2013*, 381815. doi:10.1155/2013/381815 PMID:24453416

Quintero-Quiroz, C., Botero-Botero, L. R., & Zárate-Triviño, D. G. (2019). Nanotechnology: An alternative for the development of potent antioxidant systems in plants under abiotic stress conditions. *Journal of the Science of Food and Agriculture*, *99*(14), 6141–6155.

Qu, J. H., Wei, Q., & Sun, D. W. (2018). Carbon dots: Principles and their applications in food quality and safety detection. *Critical Reviews in Food Science and Nutrition*, *58*(14), 2466–2475. doi:10.1080/10408398.2018.1437712 PMID:29420064

Qureshi, A., Singh, D. K., & Dwivedi, S. (2018). Nano-fertilizers: A novel way for enhancing nutrient use efficiency and crop productivity. *International Journal of Current Microbiology and Applied Sciences*, *7*(2), 3325–3335. doi:10.20546/ijcmas.2018.702.398

Rademacher, W. (2015). Plant growth regulators: Backgrounds and uses in plant production. *Journal of Plant Growth Regulation*, *34*(4), 845–872. doi:10.1007/s00344-015-9541-6

Radoi, A., Targa, M., Prieto-Simon, B., & Marty, J. L. (2008). Enzyme-Linked Immunosorbent Assay (ELISA) based on superparamagnetic nanoparticles for aflatoxin M1 detection. *Talanta*, *77*(1), 138–143. doi:10.1016/j.talanta.2008.05.048 PMID:18804611

Rafique, M., Sadaf, I., Rafique, M. S., & Tahir, M. B. (2017). A review on green synthesis of silver nanoparticles and their applications. *Artificial Cells, Nanomedicine, and Biotechnology*, *45*(7), 1272–1291. doi:10.1080/21691401.2016.1241792 PMID:27825269

Ragaei, M., & Sabry, A. K. H. (2014). Nanotechnology for insect pest control. *International Journal of Science, Environment and Technology*, *3*(2), 528–545.

Raghib, F., Naikoo, M. I., Khan, F. A., Alyemeni, M. N., & Ahmad, P. (2020). Interaction of ZnO nanoparticle and AM fungi mitigates Pb toxicity in wheat by upregulating antioxidants and restricted uptake of Pb. *Journal of Biotechnology*, *323*, 254–263. doi:10.1016/j.jbiotec.2020.09.003 PMID:32905797

Rai, P. K., Lee, S. S., Zhang, M., Tsang, Y. F., & Kim, K. H. (2019). Heavy metals in food crops: Health risks, fate, mechanisms, and management. *Environment International*, *125*, 365–385. doi:10.1016/j.envint.2019.01.067 PMID:30743144

Rai, V., Acharya, S., & Dey, N. (2016). Implications of nanobiosensors in agriculture. *Journal of Biomaterials and Nanobiotechnology*, *3*(2), 315–324. doi:10.4236/jbnb.2012.322039

Rai, V., Das, G., & Saxena, S. (2019). Nanomaterials toward the detection of abiotic stress in plants. In P. K. Singh, A. Kumar, & M. W. Siddiqui (Eds.), *Controlled Release Fertilizers for Sustainable Agriculture* (pp. 213–235). Academic Press.

Rajiv, P., Rajeshwari, S., & Venckatesh, R. (2013). Bio-Fabrication of zinc oxide nanoparticles using leaf extract of Parthenium hysterophorus L. and its size-dependent antifungal activity against plant fungal pathogens. *Spectrochimica Acta. Part A: Molecular and Biomolecular Spectroscopy*, *112*, 384–387. doi:10.1016/j.saa.2013.04.072 PMID:23686093

Rajput, A., Sharma, R., & Bharti, R. (2021). Pharmacological activities and toxicities of alkaloids on human health. *Materials Today: Proceedings*, *48*(5), 1407–1415.

Rajput, P., Kumar, P., Priya, A., Kumari, S., Shiade, S. R. G., Rajput, V. D., Fathi, A., Pradhan, A., Sarfraz, R., Sushkova, S., Mandzhieva, S., Minkina, T., Soldatov, A., Wong, M. H., & Rensing, C. (2024). Nanomaterials and biochar mediated remediation of emerging contaminants. *The Science of the Total Environment*, *916*, 170064. doi:10.1016/j.scitotenv.2024.170064 PMID:38242481

Rajput, V. D., Kumari, A., Minkina, T., Barakhov, A., Singh, S., Mandzhieva, S. S., Sushkova, S., Ranjan, A., Rajput, P., & Garg, M. C. (2023). A practical evaluation on integrated role of biochar and nanomaterials in soil remediation processes. *Environmental Geochemistry and Health*, *45*(12), 9435–9449. doi:10.1007/s10653-022-01375-w PMID:36070110

Rajput, V. D., Minkina, T., Ahmed, B., Singh, V. K., Mandzhieva, S., Sushkova, S., Bauer, T., Verma, K. K., Shan, S., van Hullebusch, E. D., & Wang, B. (2022). Nano-biochar: A novel solution for sustainable agriculture and environmental remediation. *Environmental Research*, *210*, 112891. doi:10.1016/j.envres.2022.112891 PMID:35183514

Rajput, V. D., Singh, A., Minkina, T., Rawat, S., Mandzhieva, S., Sushkova, S., & Upadhyay, S. K. (2021). Nano-Enabled Products: Challenges and Opportunities for Sustainable Agriculture. *Plants*, *10*(12), 2727. doi:10.3390/plants10122727 PMID:34961197

Raliya, R., Tarafdar, J. C., Gulecha, K., Choudhary, K., Ram, R., Mal, P., & Saran, R. P. (2013). Scope of nanoscience and nanotechnology in agriculture. *Journal of Applied Biology and Biotechnology, 1*(3), 41-44.

Raliya, R., Biswas, P., & Tarafdar, J. C. (2015). TiO2 nanoparticle biosynthesis and its physiological effect on mung bean (Vigna radiata L.). *Biotechnology Reports (Amsterdam, Netherlands)*, *13*, 21–29. doi:10.1016/j.btre.2014.10.009 PMID:28626678

Raliya, R., Saharan, V., Dimkpa, C., & Biswas, P. (2018). Nanofertilizer for precision and sustainable agriculture: Current state and future perspectives. *Journal of Agricultural and Food Chemistry*, *66*(26), 6487–6503. doi:10.1021/acs.jafc.7b02178 PMID:28835103

Raliya, R., & Tarafdar, J. C. (2013). ZnO nanoparticle biosynthesis and its effect on phosphorous-mobilizing enzyme secretion and gum contents in clusterbean (*Cyamopsis tetragonoloba* L.). *Agricultural Research*, *2*(1), 48–57. doi:10.1007/s40003-012-0049-z

Raliya, R., Tarafdar, J. C., & Biswas, P. (2016). Enhancing the mobilization of native phosphorus in the mung bean rhizosphere using ZnO nanoparticles synthesized by soil fungi. *Journal of Agricultural and Food Chemistry*, *64*(16), 3111–3118. doi:10.1021/acs.jafc.5b05224 PMID:27054413

Ramaiyan, S., Vijayakumar, P., Balasubramanian, R. A. B., & Nallathambi, M. (2023). Influence of nano-zinc oxide and fortified rice residue compost on rice productivity, zinc biofortification, zinc use efficiency, soil quality, zinc fractions and profitability in different rice production systems. *Journal of Plant Nutrition*, *46*(17), 1–22. doi:10.1080/01904167.2023.2212722

Rameshaiah, G. N., Pallavi, J., & Shabnam, S. (2015). Nano fertilizers and nano sensors–an attempt for developing smart agriculture. *International Journal of Engineering Research and General Science*, *3*(1), 314–320.

Ramkumar, C., Vishwanatha, A., & Saini, R. (2019). Regulatory aspects of nanotechnology for food industry. Nanotechnology applications in dairy science. doi:10.1201/9780429425370-7

Rane, J., Deshpande, M., Jagadish, S. V. K., Nagarajan, M., & Gambhir, P. (2015). Improving water use efficiency by using nano-fertilizers. *Indian Journal of Plant Physiology / Official Publication of the Indian Society for Plant Physiology*, *20*(4), 357–362.

Rangari, V. D. (2009). Pharmacognosy & phytochemistry. Career publications.

Rani, A., Rani, K., Tokas, J., Anamika, Singh, A., Kumar, R., & Kumar, S. (2020). Nanomaterials for agriculture input use efficiency. *Resources use efficiency in agriculture*, 137-175.

Rani, A., Rani, K., Tokas, J., Anamika, Singh, A., Kumar, R., ... Kumar, S. (2020). Nanomaterials for agriculture input use efficiency. *Resources use efficiency in agriculture*, 137-175.

Rani, C., Chawla, S., Mangal, M., Mangal, A. K., Kajla, S., & Dhawan, A. K. (2012). *Nyctanthes arbor-tristis* Linn. (Night Jasmine): A sacred ornamental plant with immense medicinal potentials. *Indian Journal of Traditional Knowledge*, *11*(3), 427–435.

Rani, K., & Yadav, S. (2020). Nano-encapsulation in agriculture: Emerging opportunities and challenges. *Journal of Plant Biochemistry and Biotechnology*, *29*(4), 606–621.

Ranjan, S., Dasgupta, N., & Lichtfouse, E. (2019). *Nanoscience in food and agriculture*. Springer International Publishing.

Rao, A. V., Fleshner, N., & Agarwal, S. (1999). Serum and tissue lycopene and biomarkers of oxidation in prostate cancer patients: A case-control study. *Nutrition and Cancer*, *33*(2), 159–164. doi:10.1207/S15327914NC330207 PMID:10368811

Rao, A. V., & Rao, L. G. (2007). Carotenoids and human health. *Pharmacological Research*, *55*(3), 207–216. doi:10.1016/j.phrs.2007.01.012 PMID:17349800

Rastogi, R. P., & Mehrotra, B. N. (1995). *Compendium of Indian medicinal trees*. Central Drug Research Institute.

Rathinamoorthy, R., & Thilagavathi, G. (2014). Terminalia chebula-review on pharmacological and biochemical studies. *International Journal of Pharm Tech Research*, *6*, 97–116.

Rathoure, A. K. (2020). Nanomaterials for wastewater treatment and water purification. In *Nanomaterials for Sustainable Energy and Environmental Remediation* (pp. 1–23). Elsevier.

Rawat, H., Verma, Y., Ayesha, N. S., Negi, N., Pant, H. C., Mishra, A., & Gaurav, N. (2021). *Nyctanthes arbor-tristis*: A traditional herbal plant with miraculous potential in medicine. *International Journal of Botany Studies*, *6*, 427–440.

Raza, A. (2022). Metabolomics: A systems biology approach for enhancing heat stress tolerance in plants. *Plant Cell Reports*, *41*(3), 741–763. doi:10.1007/s00299-020-02635-8 PMID:33251564

Reema, S., Deepali, T., Gauri, S., & Pankaj, S. (2018). Nyctanthes arbor-Tristis: A Wonder Indian herbal drug needs healthcare attention. *Biomedical Journal of Scientific & Technical Research*, *5*(3), 1199.

Reginatto, C., Cecchin, I., Heineck, K. S., Reddy, K. R., & Thomé, A. (2020). Use of nanoscale Zero-Valent Iron for Remediation of Clayey Soil Contaminated with Hexava lent chromium: Batch and Column Tests. *International Journal of Environmental Research and Public Health*, *17*(3), 1001. doi:10.3390/ijerph17031001 PMID:32033384

Rehman, A., Aziz, A., Anwar, M. M., Majeed, M., Albanai, J. A., Almohamad, H., & Abdo, H. G. (2023). Quantifying the impacts of urbanization on urban green, evidences from Maga City, Lahore Pakistan. *Discover Sustainability*, *4*(1), 48. Advance online publication. doi:10.1007/s43621-023-00169-z

Reina, G., Gonzalez-Dominguez, J. M., Criado, A., Vazquez, E., Bianco, A., & Prato, M. (2017). Promises, facts and challenges for graphene in biomedical applications. *Chemical Society Reviews*, *46*(15), 4400–4416. doi:10.1039/C7CS00363C PMID:28722038

Remya, S., Mohan, C. O., Sreelakshmi, K. R., & Bindu, J. (2019). Packaging applications of nanomaterials. ITEC training programme on 'Protocols for the production of high value secondary products from industrial fish and shellfish processing from 25/11/2019 to 21/12/2019 at ICAR-CIFT, Kochi. 215-221pp.

Rezaei, A., Daeihamed, M., Capanoglu, E., Tomas, M., Akbari-Alavijeh, S., Shaddel, R., Khoshnoudi-Nia, S., Boostani, S., Rostamabadi, H., Falsafi, S. R., & Jafari, S. M. (2021). Possible health risks associated with nanostructures in food. In *Safety and regulatory issues of nanoencapsulated food ingredients*. Elsevier Inc. doi:10.1016/B978-0-12-815725-1.00002-1

Rico, C. M., Majumdar, S., Duarte-Gardea, M., Peralta-Videa, J. R., & Gardea-Torresdey, J. L. (2011). Interaction of nanoparticles with edible plants and their possible implications in the food chain. *Journal of Agricultural and Food Chemistry*, *59*(8), 3485–3498. doi:10.1021/jf104517j PMID:21405020

Rimpika, A., Anushi, Manasa, S., Sharma, S., Thakur, A., Shilpa, & Sood, A. (2023). An Overview of Precision Farming. *International Journal of Environment and Climate Change*, *13*(12), 441–456. doi:10.9734/ijecc/2023/v13i123701

Rinschen, M. M., Ivanisevic, J., Giera, M., & Siuzdak, G. (2019). Identification of bioactive metabolites using activity metabolomics. *Nature Reviews. Molecular Cell Biology*, *20*(6), 353–367. doi:10.1038/s41580-019-0108-4 PMID:30814649

Rizwan, M., Ali, S., Adrees, M., Ibrahim, M., & Rehman, M. Z. u., Farid, M., Qayyum, M. F., Nawaz, R., Hussain, A., Chatha, S. A. S., & Abbas, F. (2016). Mechanisms of silicon-mediated alleviation of drought and salt stress in plants: A review. *Environmental Science and Pollution Research International*, *23*(5), 4416–4431. PMID:26335528

Rizwan, M., Ali, S., Qayyum, M. F., Ok, Y. S., Adrees, M., Ibrahim, M., Zia-ur-Rehman, M., Farid, M., & Abbas, F. (2017). Effect of metal and metal oxide nanoparticles on growth and physiology of globally important food crops: A critical review. *Journal of Hazardous Materials*, *322*, 2–16. doi:10.1016/j.jhazmat.2016.05.061 PMID:27267650

Rizwan, M., Ali, S., Rehman, M. Z. U., Adrees, M., Arshad, M., Qayyum, M. F., Ali, L., Hussain, A., Chatha, S. A. S., & Imran, M. (2019). Alleviation of cadmium accumulation in maize (Zea mays L.) by foliar spray of zinc oxide nanoparticles and biochar to contaminated soil. *Environmental Pollution*, *248*, 358–367. doi:10.1016/j.envpol.2019.02.031 PMID:30818115

Rizwan, M., Ali, S., Rehman, M. Z., Riaz, M., Adrees, M., Hussain, A., Zahir, Z. A., & Rinklebe, J. (2021). Effects of nanoparticles on trace element uptake and toxicity in plants: A review. *Ecotoxicology and Environmental Safety*, *221*, 112437. doi:10.1016/j.ecoenv.2021.112437 PMID:34153540

Rizwan, M., Ali, S., Rizvi, H., Rinklebe, J., Tsang, D. C., Meers, E., Ok, Y. S., & Ishaque, W. (2016). Phytomanagement of heavy metals in contaminated soils using sunflower: A review. *Critical Reviews in Environmental Science and Technology*, *46*(18), 1498–1528. doi:10.1080/10643389.2016.1248199

Robinson, K. E., Worrall, E. A., & Mitter, N. (2014). Double stranded RNA expression and its topical application for non-transgenic resistance to plant viruses. *Journal of Plant Biochemistry and Biotechnology*, *23*(3), 231–237. doi:10.1007/s13562-014-0260-z

Rocca, L. M., Gentili, A., PerezFernandez, V., & Tomai, P. (2017). Veterinary Drugs Residues: a Review of the Latest Analytical Research on Sample Preparation and LC-MS Based Methods. Food Additives Contaminants A-Chemistry Anal. Control. Expo. Risk Assess., 34, 766–784.

Rodrigues, S. M., Trindade, T., Duarte, A. C., Pereira, E., Koopmans, G. F., & Römkens, P. F. (2016). A framework to measure the availability of engineered nanoparticles in soils: Trends in soil tests and analytical tools. *Trends in Analytical Chemistry*, *75*, 129–140. doi:10.1016/j.trac.2015.07.003

Rodriguez-Eugenio, N., McLaughlin, M., & Pennock, D. (2018). *Soil pollution: a hidden reality*. Food and Agriculture Organization of the United Nations.

Rodríguez, J., Martín, M. J., Ruiz, M. A., & Clares, B. (2016). Current encapsulation strategies for bioactive oils: From alimentary to pharmaceutical perspectives. *Food Research International*, *83*, 41–59. doi:10.1016/j.foodres.2016.01.032

Rodríguez-Seijo, A., Vega, F. A., & Arenas-Lago, D. (2020). Assessment of iron-based and calcium-phosphate nanomaterials for immobilisation of potentially toxic elements in soils from a shooting range berm. *Journal of Environmental Management*, *267*, 110640. doi:10.1016/j.jenvman.2020.110640 PMID:32421672

Rodziewicz, P., Swarcewicz, B., Chmielewska, K., Wojakowska, A., & Stobiecki, M. (2014). Influence of abiotic stresses on plant proteome and metabolome changes. *Acta Physiologiae Plantarum*, *36*(1), 1–19. doi:10.1007/s11738-013-1402-y

Rohilla, D., Chaudhary, S., & Umar, A. (2021). An Overview of Advanced Nanomaterials for Sensor Applications. *Engineering and Science*, *16*, 47–70. doi:10.30919/es8d552

Romeh, A. A. (2021). Synergistic effect of Ficus-zero valent iron supported on adsorbents and Plantago major for chlorpyrifos phytoremediation from water. *International Journal of Phytoremediation*, *23*(2), 151–161. doi:10.1080/15226514.2020.1803201 PMID:32772548

Romero, N., Castro, G. R., & Gagneten, A. M. (2021). Ecotoxicologic effects of silver nanoparticles on freshwater nontarget species. In *New Trends in Removal of Heavy Metals from Industrial Wastewater* (pp. 705–733). Elsevier. doi:10.1016/B978-0-12-822965-1.00028-3

Rosen, J. E., Yoffe, S., Meerasa, A., Verma, M., & Gu, F. X. (2011). Nanotechnology and diagnostic imaging: New advances in contrast agent technology. *Journal of Nanomedicine & Nanotechnology*, *2*(5), 115–126. doi:10.4172/2157-7439.1000115

Rungsung, W., Ratha, K. K., Dutta, S., Dixit, A. K., & Hazra, J. (2015). Secondary metabolites of plants in drugs discovery. *World Journal of Pharmaceutical Research*, *4*(7), 604–613.

Sabir, S., Arshad, M., & Chaudhari, S. K. (2014). Zinc oxide nanoparticles for revolutionizing agriculture: Synthesis and applications. *TheScientificWorldJournal*, *2014*, 2014. doi:10.1155/2014/925494 PMID:25436235

Sachan, D. S., Reddy, K. J., Sulochna, Saini, Y., Rai, A. K., Singh, O., & Laxman, T. (2023). Assessing grain yield and achieving enhanced quality in maize by next generation fertilizer: A review. *International Journal of Environment and Climate Change*, *13*(8), 626–637. doi:10.9734/ijecc/2023/v13i81991

Sadeghi, R., Rodriguez, R. J., Yao, Y., & Kokini, J. L. (2017). Advances in nanotechnology as they pertain to food and agriculture: Benefits and risks. *Annual Review of Food Science and Technology*, *8*(1), 467–492. doi:10.1146/annurev-food-041715-033338 PMID:28125343

Sagadevan, S., & Periasamy, M. (2014). Recent trends in nanobiosensors and their applications-a review. *Reviews on Advanced Materials Science*, *36*, 62–69.

Sagadevan, S., Periasamy, M., Ling, T. C., & Dhanalekshmi, K. I. (2019). Application of nanosensors in smart agriculture. In *Nanosensors for smart manufacturing* (pp. 247–268). Elsevier.

Saha, A., Sekharan, S., & Manna, U. (2020). Superabsorbent hydrogel (SAH) as a soil amendment for drought management: A review. *Soil & Tillage Research*, *204*, 104736. doi:10.1016/j.still.2020.104736

Saharan, V., Kumaraswamy, R. V., Choudhary, R. C., Kumari, S., Pal, A., Raliya, R., & Biswas, P. (2016). Cu-chitosan nanoparticle mediated sustainable approach to enhance seedling growth in maize by mobilizing reserved food. *Journal of Agricultural and Food Chemistry*, *64*(31), 6148–6155. doi:10.1021/acs.jafc.6b02239 PMID:27460439

Sahoo, J. P., Behera, L., Sharma, S. S., Praveena, J., Nayak, S. K., & Samal, K. C. (2020). Omics studies and systems biology perspective towards abiotic stress response in plants. *American Journal of Plant Sciences*, *11*(12), 2172–2194. doi:10.4236/ajps.2020.1112152

Sahoo, M., Panigrahi, C., Vishwakarma, S., & Kumar, J. (2022). A review on nanotechnology: Applications in food industry, future opportunities, challenges and potential risks. *Journal of Nanotechnology and Nanomaterials*, *3*(1), 28–33.

Sahu, A., Kumar, N., Singh, C. P., & Singh, M. (2023). Environmental DNA (eDNA): Powerful technique for biodiversity conservation. *Journal for Nature Conservation*, *71*, 126325. doi:10.1016/j.jnc.2022.126325

Saikanth, D. R. K., Singh, B. V., Rai, A. K., Kumar, U. S., Surender, Yadav, B., & Singh, O. (2023). Biochar implementation in rice paddies for addressing greenhouse gas emissions and nutrient loss: A review. *International Journal of Plant and Soil Science*, *35*(18), 610–623. doi:10.9734/ijpss/2023/v35i183326

Saklani, S. A., Chandra, S. U., Mishra, A. P., & Badoni, P. P. (2012). Nutritional evaluation, antimicrobial activity and phytochemical screening of wild edible fruit of *Myrica nagi* pulp. *International Journal of Pharmacy and Pharmaceutical Sciences*, *4*(3), 407–411.

Salah, E. A. M., Zaidan, T. A., & Al-Rawi, A. S. (2015). Assessment of heavy metals pollution in the sediments of Euphrates River, Iraq. *Journal of Water Resource and Protection*, *7*(13), 1009–1023.

Salamanca-Buentello, F.; Persad, D. L.; Court, E. B.; Martin, D. K.; Daar, A. S. & Singer, P. A. (2005). Nanotechnology and the developing world. PLos Medicine, 2(5), 383-386.

Saleh, H. M., & Hassan, A. I. (n.d.). *Environmental Impact and Remediation of Heavy Metals*. Academic Press.

Salgado, C. M., Palacio, L., Prádanos, P., Hernández, A., González-Huerta, C., & Pérez-Magari, S. (2015). Comparative study of red grape must nanofiltration: Laboratory and pilot plant scales. *Food and Bioproducts Processing*, *94*, 610–620. doi:10.1016/j.fbp.2014.08.007

Salimi, F., Moradi, M., Tajik, H., & Molaei, R. (2021). Optimization and characterization of eco-friendly antimicrobial nanocellulose sheet prepared using carbon dots of white mulberry (*Morus Alba* L.). *Journal of the Science of Food and Agriculture*, *101*(8), 3439–3447. doi:10.1002/jsfa.10974 PMID:33289129

Salinas, M., & Baudet, C. (2020). The Challenges of Agriculture: Feeding the World of Tomorrow, on a Transitioning and Endangered Earth. *OIDA International Journal of Sustainable Development*, *13*(5), 21–58.

Samanta, A., Deng, Z., & Liu, Y. (2016). Nanomaterials for microbial biosensors. In *NanoBiosensors* (pp. 135–166). Academic Press.

Sangeeta Dash, E., Yadav, A. L., Bora, R., & Kumar Pandey, A. (2023). *Research Trends in Agriculture Science* (Vol. I). www.bhumipublishing.com

Sangeetha, J., Thangadurai, D., Hospet, R., Harish, E. R., Purushotham, P., Mujeeb, M. A., & Prasad, R. (2017). Nanoagrotechnology for soil quality, crop performance and environmental management. *Nanotechnology: an agricultural paradigm*, 73-97.

Sanghera, G. S., Wani, S. H., Hussain, W., & Singh, N. B. (2011). Engineering cold stress tolerance in crop plants. *Current Genomics*, *12*(1), 30–43. doi:10.2174/138920211794520178 PMID:21886453

Santoso, D., Lefroy, R. D., & Blair, G. J. (1995). Sulfur and phosphorus dynamics in an acid soil/crop system. *Soil Research (Collingwood, Vic.)*, *33*(1), 113–124. doi:10.1071/SR9950113

Sari, G., & Piyal, B. (2019). Nanomaterials and their effects on health. *Istanbul Medical Journal, 20*(6), 497–501. doi:10.4274/imj.galenos.2018.71598

Saritha, G. N. G., Anju, T., & Kumar, A. (2022). Nanotechnology-Big impact: How nanotechnology is changing the future of agriculture? *Journal of Agriculture and Food Research, 10*, 100457. doi:10.1016/j.jafr.2022.100457

Sarkar, M. M., Mathur, P., Mitsui, T., & Roy, S. (2022). A review on functionalized silica nanoparticle amendment on plant growth and development under stress. *Plant Growth Regulation, 98*(3), 421–437. doi:10.1007/s10725-022-00891-0

Sarno, M., Casa, M., Cirillo, C., & Ciambelli, P. (2017). Complete removal of persistent pesticide using reduced graphene oxide-silver nanocomposite. *Chemical Engineering Transactions, 60*, 151–156.

Sarwar, N., Imran, M., Shaheen, M. R., Ishaque, W., Kamran, M. A., Matloob, A., Rehim, A., & Hussain, S. (2017). Phytoremediation strategies for soils contaminated with heavy metals: Modifications and future perspectives. *Chemosphere, 171*, 710–721. doi:10.1016/j.chemosphere.2016.12.116 PMID:28061428

Sastry, R. K., Rao, N. H., & Cahoon, R. (2017). Can nanotechnology provide the innovations for a second green revolution in Indian agriculture? In *Nanotechnology in agriculture: Prospects and constraints* (pp. 3–21). Springer.

Schofield, C. L., Haines, A. H., Field, R. A., & Russell, D. A. (2006). Silver and gold glyconanoparticles for colorimetric bioassays. *Langmuir, 22*(15), 6707–6711. doi:10.1021/la060288r PMID:16831017

Scrinis, G., & Lyons, K. (2007). The emerging nano-corporate paradigm: Nanotechnology and the transformation of nature, food and agri-food systems. *International Journal of Sociology of Agriculture and Food, 15*(2), 22–44.

Sedghi, M., Hadi, M., & Toluie, S. G. (2013). Effect of nano zinc oxide on the germination parameters of soybean seeds under drought stress. Annals of West University of Timisoara, ser. *Biology (Basel), 16*, 73–78.

Sekhon, B. S. (2014). Nanotechnology in agri-food production: An overview. *Nanotechnology, Science and Applications*, 31–53. doi:10.2147/NSA.S39406 PMID:24966671

Semwal, A., Kumar, A., & Kumar, N. (2023). A review on pathogenicity of *Aeromonas hydrophila* and their mitigation through medicinal herbs in aquaculture. *Heliyon, 9*(3), e14088. doi:10.1016/j.heliyon.2023.e14088 PMID:36938468

Sertova, N. M. (2015). Application of nanotechnology in detection of mycotoxins and in agricultural sector. *Journal of Central European Agriculture, 16*(2), 117–130. doi:10.5513/JCEA01/16.2.1597

Servin, A. D., & White, J. C. (2016). Nanotechnology in agriculture: Next steps for understanding engineered nanoparticle exposure and risk. *NanoImpact, 1*, 9–12. doi:10.1016/j.impact.2015.12.002

Shafi, A., Qadir, J., Sabir, S., Zain Khan, M., & Rahman, M. M. (2020). Nanoagriculture: a holistic approach for sustainable development of agriculture. Handbook of Nanomaterials and Nanocomposites for Energy and Environmental Applications, 1-16.

Shafique, S., Shafique, S., Ahmed, A., & Shabbir, R. (2020). A nano-biosensor for early detection and management of Tomato leaf curl New Delhi virus. *Scientia Horticulturae, 265*, 109211. doi:10.1016/j.scienta.2020.109211

Shahadat, M., Teng, T. T., Rafatullah, M., & Arshad, M. (2015). Titanium-based nanocomposite materials: A review of recent advances and perspectives. *Colloids and Surfaces. B, Biointerfaces, 126*, 121–137. doi:10.1016/j.colsurfb.2014.11.049 PMID:25543989

Shahi, U. P., Singh, O., Singh, V. K., Shivangi, P. K., Singh, R., & Singh, V. K. (2024). Nanotechnology in rice farming: Optimizing nutrient management with nanofertilizers. In Sustainable Agriculture: Nanotechnology and Biotechnology for Crop Production and Protection. Walter de Gruyter GmbH. doi:10.1515/9783111234694-003

Shahid, M., Shamshad, S., Rafiq, M., Khalid, S., Bibi, I., Niazi, N. K., Dumat, C., & Rashid, M. I. (2017). Chromium speciation, bioavailability, uptake, toxicity and detoxification in soil-plant system: A review. *Chemosphere*, *178*, 513–533. doi:10.1016/j.chemosphere.2017.03.074 PMID:28347915

Shah, M., Fawcell, D., Sharma, S., Tripathy, S. K., & Poinern, G. E. J. (2015). Green synthesis of metallic nanoparticles via biological entities. *Materials (Basel)*, *8*(11), 7278–7308. doi:10.3390/ma8115377 PMID:28793638

Shalaby, T. A., Bayoumi, Y., Alshaal, T., Elhawat, N., Sztrik, A., & El-Ramady, H. (2018). Selenium fortification induces growth, antioxidant activity, yield and nutritional quality of lettuce in salt-affected soil using foliar and soil applications. *Plant and Soil*, *423*(1), 319–332.

Shan, B., Cai, Y. Z., Brooks, J. D., & Corke, H. (2007). The in vitro antibacterial activity of dietary spice and medicinal herb extracts. *International Journal of Food Microbiology*, *117*(1), 112–119. doi:10.1016/j.ijfoodmicro.2007.03.003 PMID:17449125

Shan, J., Ji, R., Yu, Y., Xie, Z., & Yan, X. (2015). Biochar, activated carbon, and carbon nanotubes have different effects on fate of (14)C-catechol and microbial community in soil. *Scientific Reports*, *5*(1), 16000. doi:10.1038/srep16000 PMID:26515132

Shapira, P., & Youtie, J. (2015). The economic contributions of nanotechnology to green and sustainable growth. *Green processes for nanotechnology: From inorganic to bioinspired nanomaterials*, 409-434.

Sharara, M. S. (2021). Application of nanotechnology in food industry and some of its hazard effect on human body: An overview. *Alexandria Journal of Food Science and Technology*, *18*(2), 23–32. doi:10.21608/ajfs.2022.217792

Sharf-Eldin, A. A., Alwutayd, K. M., El-Yazied, A. A., El-Beltagi, H. S., Alharbi, B. M., Eisa, M. A., & Ibrahim, M. F. (2023). Response of maize seedlings to silicon dioxide nanoparticles (SiO2NPs) under drought stress. *Plants*, *12*(14), 2592. doi:10.3390/plants12142592 PMID:37514207

Sharifi, M., Faryabi, K., Talaei, A. J., Shekha, M. S., Ale-Ebrahim, M., Salihi, A., Nanakali, N. M. Q., Aziz, F. M., Rasti, B., Hasan, A., & Falahati, M. (2020). Antioxidant properties of gold nanozyme: A review. *Journal of Molecular Liquids*, *297*, 112004. doi:10.1016/j.molliq.2019.112004

Sharma, J. G., Kumar, N., Mittal, P., & Chakrabarti, R. (2022). Evaluation of UV-B protective properties of leaves and seeds of *Achyranthes aspera* in Asian catfish *Clarias batrachus* (Linn.). *Photochemical & Photobiological Sciences*, *21*(8), 1341–1356. doi:10.1007/s43630-022-00222-2 PMID:35403962

Sharma, J. G., Kumar, N., Singh, S. P., Singh, A., Harikrishna, V., & Chakrabarti, R. (2019). Evaluation of immunostimulatory properties of prickly chaff flower *Achyranthes aspera* in rohu *Labeo rohita* fry in pond conditions. *Aquaculture (Amsterdam, Netherlands)*, *505*, 183–189. doi:10.1016/j.aquaculture.2019.02.065

Sharma, R., Kumar, R., Kodwani, R., Kapoor, S., Khare, A., Bansal, R., Khurana, S., Singh, S., Thomas, J., Roy, B., Phartyal, R., Saluja, S., & Kumar, S. (2016). A review on mechanisms of anti-tumor activity of chalcones. *Anti-cancer Agents in Medicinal Chemistry*, *16*(2), 200–211. doi:10.2174/1871520615666150518093144 PMID:25980813

Shen, N., Wang, T., Gan, Q., Liu, S., Wang, L., & Jin, B. (2022). Plant flavonoids: Classification, distribution, biosynthesis, and antioxidant activity. *Food Chemistry*, *383*, 132531. doi:10.1016/j.foodchem.2022.132531 PMID:35413752

Shi, C., Zhu, N., Cao, Y., & Wu, P. (2017). Biosynthesis of gold nanoparticles assisted by the intracellular protein extract of Pycnoporus sanguineus and its catalysis in degradation of 4-nitroaniline. *Nanoscale Research Letters*, *12*(1), 1–9. PMID:28050875

Shi, J., Tian, X., Liu, F., Luo, Y., & Lan, S. (2016). An electrochemical sensor based on Pt nanoparticles and molecularly imprinted polymer for sensitive detection of methyl parathion. *Sensors and Actuators. B, Chemical, 227*, 583–590. doi:10.1016/j.snb.2015.12.092

Shipley, H. J., Engates, K. E., & Grover, V. A. (2010). Removal of Pb (II), Cd (II), Cu (II), and Zn (II) by hematite nanoparticles: Effect of sorbent concentration, pH, temperature, and exhaustion. *Environmental Science and Pollution Research International, 20*(3), 1727–1736. doi:10.1007/s11356-012-0984-z PMID:22645012

Shipley, H. J., Engates, K. E., & Guettner, A. M. (2011). Study of iron oxide nanoparticles in soil for remediation of arsenic. *Journal of Nanoparticle Research, 13*(6), 2387–2397. doi:10.1007/s11051-010-9999-x

Shirsat, S., & K, S. (2023). Iron oxide nanoparticles as iron micronutrient fertilizer—Opportunities and limitations. *Journal of Plant Nutrition and Soil Science*, jpln.202300203. Advance online publication. doi:10.1002/jpln.202300203

Shoala, T. (2023). The dual role of nanotechnology in the management of biotic and abiotic stresses in plants. In *Nanotechnology in Agriculture and Agroecosystems* (pp. 51–70). Elsevier. doi:10.1016/B978-0-323-99446-0.00011-8

Shrivastava, P., & Kumar, R. (2015). Soil salinity: A serious environmental issue and plant growth promoting bacteria as one of the tools for its alleviation. *Saudi Journal of Biological Sciences, 22*(2), 123–131. doi:10.1016/j.sjbs.2014.12.001 PMID:25737642

Shtenberg, G., Massad-Ivanir, N., & Segal, E. (2018). Optical and electrochemical sensing platforms based on nanoporous alumina: A review. *Nanomaterials (Basel, Switzerland), 8*(11), 918. PMID:30405041

Shubair, T., Eljamal, O., Khalil, A. M. E., & Matsunaga, N. (2018). Multilayer system of nanoscale zero valent iron and Nano-Fe/Cu particles for nitrate removal in porous media. *Separation and Purification Technology, 193*, 242–254. doi:10.1016/j.seppur.2017.10.069

Shukla, P. K., Misra, P., & Kole, C. (2019). *Nanotechnology for agriculture*. Elsevier.

Shukla, P., Chaurasia, P., Younis, K., Qadri, O. S., Faridi, S. A., & Srivastava, G. (2019). Nanotechnology in sustainable agriculture: Studies from seed priming to post-harvest management. *Nanotechnology for Environmental Engineering, 4*(1), 1–11. doi:10.1007/s41204-019-0058-2

Simonin, M., Cantarel, A. A., Crouzet, A., Gervaix, J., Martins, J. M., & Richaume, A. (2020). Negative effects of copper oxide nanoparticles on carbon and nitrogen cycle microbial activities in contrasting agricultural soils and in presence of plants. *Frontiers in Microbiology, 11*, 1769. PMID:30619181

Simonin, M., Richaume, A., Guyonnet, J. P., Dubost, A., Martins, J. M., & Pommier, T. (2018). Titanium dioxide nanoparticles strongly impact soil microbial function by affecting archaeal nitrifiers. *Scientific Reports, 8*(1), 1–12. PMID:29311619

Singh, A., Agrawal, S., Rajput, V. D., Ghazaryan, K., Movsesyan, H. S., Minkina, T., . . . Gupta, S. K. (2023). Seed Nanopriming: An Innovative Approach for Upregulating Seed Germination and Plant Growth Under Salinity Stress. In Nanopriming Approach to Sustainable Agriculture (pp. 290-313). IGI Global.

Singh, A., Agrawal, S., Rajput, V. D., Ghazaryan, K., Movsesyan, H. S., Minkina, T., . . . Gupta, S. K. (2023c). Seed Nanopriming. *Nanopriming Approach to Sustainable Agriculture*, 290–313. doi:10.4018/978-1-6684-7232-3.ch013

Singh, A., Margaryan, G., Harutyunyan, A., Movsesyan, H., Khachatryan, H., Rajput, V., . . . Ghazaryan, K. (2024b). Advancing Agricultural Resilience in Ararat Plain, Armenia: Utilizing Biogenic Nanoparticles and Biochar under Saline Environments to Optimize Food Security and Foster European Trade. *Egyptian Journal of Soil Science, 64*(2). doi:10.21608/ejss.2024.257000.1703

Singh, A., Rajput, V., Varshney, A., Sharma, R., Ghazaryan, K., Minkina, T., . . . El-Ramady, H. (2024b). Revolutionizing Crop Production: Nanoscale Wonders - Current Applications, Advances, and Future Frontiers. Egyptian Journal of Soil Science, 64(1). doi:10.21608/ejss.2023.246354.1684

Singh, A., Rawat, S., Rajput, V. D., Ghazaryan, K., Minkina, T., Al Tawaha, A. R. M., & Varshney, A. (2023g). Impact of Nanofertilizers for the Mitigation of Multiple Environmental Stresses. Nanotechnology in the Life Sciences, 431–454. doi:10.1007/978-3-031-41329-2_16

Singh, A., Rawat, S., Rajput, V. D., Ghazaryan, K., Minkina, T., Al Tawaha, A. R. M., . . . Sandhu, K. S. (2023). Nanoparticle-Based Priming-Enhanced Agricultural Practices: Promoting Sustainable Plant Growth and Development. In Nanopriming Approach to Sustainable Agriculture (pp. 142-157). IGI Global.

Singh, A., Rawat, S., Rajput, V. D., Ghazaryan, K., Minkina, T., Al Tawaha, A. R. M., . . . Sandhu, K. S. (2023b). Nanoparticle-Based Priming-Enhanced Agricultural Practices. *Nanopriming Approach to Sustainable Agriculture*, 142–157. doi:10.4018/978-1-6684-7232-3.ch006

Singh, A., Sharma, R., Rajput, V. D., Ghazaryan, K., Minkina, T., Al Tawaha, A. R. M., & Varshney, A. (2023f). Green Synthesis of Nanofertilizers and Their Application for Crop Production. *Nanotechnology in the Life Sciences*, 205–231. doi:10.1007/978-3-031-41329-2_8

Singh, A., Sharma, R., Rawat, S., Singh, A. K., Rajput, V. D., Fedorov, Y., . . . Chaplygin, V. (2022). Nanomaterial-plant interaction: Views on the pros and cons. Toxicity of Nanoparticles in Plants, 47–68. doi:10.1016/B978-0-323-90774-3.00015-5

Singh, B. K., Delgado-Baquerizo, M., Egidi, E., Guirado, E., Leach, J. E., Liu, H., & Trivedi, P. (2023). Climate change impacts on plant pathogens, food security and paths forward. In Nature Reviews Microbiology (Vol. 21, Issue 10, pp. 640–656). Nature Research. doi:10.1038/s41579-023-00900-7

Singh, O., Shahi, U. P., Shivangi, Singh, V. K., Singh, P. K., Singh, R., & Singh, A. (2024). Enhancing crop nutrition with zinc nanoparticles: A novel approach to cereal crop micronutrient management. In Sustainable Agriculture: Nanotechnology and Biotechnology for Crop Production and Protection. Walter de Gruyter GmbH. doi:10.1515/9783111234694-006

Singh, P., Kumari, K., Vishvakrma, V. K., Mehrotra, G. K., Chandra, R., Kumar, D., ... & Shahare, V. V. (2017). Metal NPs (Au, Ag, and Cu): Synthesis, stabilization, and their role in green chemistry and drug delivery. *Green Technologies and Environmental Sustainability*, 309-337.

Singh, S. P., & Batra, R. D. (2018). Nanotechnology in wastewater treatment: A review. In Novel applications in polymers and waste management (pp. 173–182). Apple Academic Press. Crossref

Singh, A. V., Katz, A., Maharjan, R. S., Gadicherla, A. K., Richter, M. H., Heyda, J., del Pino, P., Laux, P., & Luch, A. (2023). Coronavirus-mimicking nanoparticles (CorNPs) in artificial saliva droplets and nanoaerosols: Influence of shape and environmental factors on particokinetics/particle aerodynamics. *The Science of the Total Environment*, *860*, 160503. doi:10.1016/j.scitotenv.2022.160503 PMID:36442637

Singh, A., Hussain, I., Singh, N. B., & Singh, H. (2018). Uptake, translocation and impact of green synthesized nanoceria on growth and antioxidant enzymes activity of Solanum lycopersicum L. *Ecotoxicology and Environmental Safety*, *158*, 345–353. doi:10.1016/j.ecoenv.2018.04.039 PMID:31284122

Singh, A., Kumar, A., Jaswal, A., Singh, M., & Gaikwad, D. S. (2018). Nutrient use efficiency concept and interventions for improving nitrogen use efficiency. *Plant Archives*, *18*(1), 1015–1023.

Singh, A., Rajput, V. D., Pandey, D., Sharma, R., Ghazaryan, K., & Minkina, T. (2023c). Nano Zinc-Enabled Strategies in Crops for Combatting Zinc Malnutrition in Human Health. *Frontiers in Bioscience (Landmark Edition)*, *28*(8), 158. doi:10.31083/j.fbl2808158 PMID:37664935

Singh, A., Rajput, V. D., Sharma, R., Ghazaryan, K., & Minkina, T. (2023d). Salinity stress and nanoparticles: Insights into antioxidative enzymatic resistance, signaling, and defense mechanisms. *Environmental Research*, *235*, 116585. doi:10.1016/j.envres.2023.116585 PMID:37437867

Singh, A., Rajput, V. D., Tawaha, A. R. M., Al Zoubi, O. M., Habeeb, T., Rawat, S., Ghazaryan, K., & Minkina, T. (2023a). A Review on Crop Responses to Nanofertilizers for Mitigation of Multiple Environmental Stresses. *Ecological Engineering & Environmental Technology*, *24*(7), 280–296. doi:10.12912/27197050/169313

Singh, A., Rajput, V. D., Varshney, A., Ghazaryan, K., & Minkina, T. (2023e). Small Tech, Big Impact: Agri-nanotechnology Journey to Optimize Crop Protection and Production for Sustainable Agriculture. *Plant Stress*, *10*, 100253. doi:10.1016/j.stress.2023.100253

Singh, A., Rawat, S., Rajput, V. D., Ghazaryan, K., Minkina, T., Al Tawaha, A. R. M., & Varshney, A. (2023). Impact of Nanofertilizers for the Mitigation of Multiple Environmental Stresses. In *Nanofertilizers for Sustainable Agroecosystems: Recent Advances and Future Trends* (pp. 431–454). Springer Nature Switzerland.

Singh, A., Sengar, R. S., Rajput, V. D., Al-Ghzawi, A. L., Shahi, U. P., Ghazaryan, K., Minkina, T., Tawaha, A. R. M. A., Al Zoubi, O. M., & Habeeb, T. (2024a). Impact of Salinity Stress and Zinc Oxide Nanoparticles on Macro and Micronutrient Assimilation: Unraveling the Link between Environmental Factors and Nutrient Uptake. *Journal of Ecological Engineering*, *25*(2), 1–9. doi:10.12911/22998993/172947

Singh, A., & Singh, N. B. (2019). Application of nanotechnology in agriculture. In *Nanomaterials in biofuels research* (pp. 151–187). Springer.

Singh, H., & Gupta, R. (2016). Synthesis and characterization of carbon nanotubes for advanced research and development. In *Nanocarbon and Its Composites* (pp. 79–94). Elsevier.

Singh, H., Sharma, A., Bhardwaj, S. K., Arya, S. K., Bhardwaj, N., & Khatri, M. (2021). Recent advances in the applications of nano-agrochemicals for sustainable agricultural development. *Environmental Science. Processes & Impacts*, *23*(2), 213–239. doi:10.1039/D0EM00404A PMID:33447834

Singh, J., Dutta, T., Kim, K. H., Rawat, M., Samddar, P., & Kumar, P. (2018). 'Green' synthesis of metals and their oxide nanoparticles: Applications for environmental remediation. *Journal of Nanobiotechnology*, *16*(1), 1–24. doi:10.1186/s12951-018-0408-4 PMID:30373622

Singh, J., Kumar, S., Alok, A., Upadhyay, S. K., Rawat, M., Tsang, D. C. W., ... Bolan, N. (2018). The potential of green synthesized zinc oxide nanoparticles as a nutrient source for plant growth. *Journal of Cleaner Production*, *214*, 1061–1070. doi:10.1016/j.jclepro.2019.01.018

Singh, J., Singh, A. P., & Singh, A. P. (2021). *Nyctanthes arbor-tristis*: A comprehensive review. *World Journal of Current Medical and Pharmaceutical Research*, *3*(4), 74–78. doi:10.37022/wjcmpr.v3i4.181

Singh, N., Bhalla, M., de Jager, P., & Gilca, M. (2011). An overview on Ashwagandha: A Rasayana (rejuvenator) of Ayurveda. *African Journal of Traditional, Complementary, and Alternative Medicines*, *8*(5S), 208–213. doi:10.4314/ajtcam.v8i5S.9 PMID:22754076

Singh, O. (2023). Nanotechnology for sustainability and food security in agriculture: A nanopriming long story in short. In A. Singh, V. Rajput, K. Ghazaryan, S. Gupta, & T. Minkina (Eds.), *Nanopriming Approach to Sustainable Agriculture* (pp. 315–339). IGI Global. doi:10.4018/978-1-6684-7232-3.ch014

Singh, O., Shahi, U.P., Dhyani, B.P., Kumar, S., Vivek, S., & Sengar, R.S., Shivangi, and Singh, A. (2023). Effect of zinc oxide nanoparticles application on growth and yield of basmati rice (Oryza sativa) in alkaline soil. *Indian Journal of Agronomy*, *68*(3), 1–6.

Singh, O., Singh, S., Singh, V. K., & Singh, A. (2023). Biochar: An organic amendment for sustainable soil health. In *Encyclopedia of Green Materials* (pp. 265–271). Springer Nature Singapore Pte Ltd. doi:10.1007/978-981-16-4921-9_265-1

Singh, P., Kim, Y. J., Wang, C., Mathiyalagan, R., El-Agamy Farh, M., & Yang, D. C. (2015). Biogenic silver and gold nanoparticles synthesized using red ginseng root extract, and their applications. *Artificial Cells, Nanomedicine, and Biotechnology*, *44*, 811–816. doi:10.3109/21691401.2015.1008514 PMID:25706249

Singh, R. P., Handa, R., & Manchanda, G. (2021). Nanoparticles in sustainable agriculture: An emerging opportunity. *Journal of Controlled Release*, *329*, 1234–1248. doi:10.1016/j.jconrel.2020.10.051 PMID:33122001

Singh, R. P., Singh, A., Srivastava, S., & Kumar, R. (2018). Gold nanoparticle-based colorimetric sensor for detection of Fusarium graminearum in wheat. *Analytical Methods*, *10*(20), 2374–2380. doi:10.1039/C8AY00473A

Singh, R., Chauhan, R. S., Semwal, A., & Kumar, N. (2023). Efficacy of 17α-methyl testosterone on growth and survival of fry of swordtail, *Xiphophorus helleri. Journal of Krishi Vigyan*, *11*(2), 65–69. doi:10.5958/2349-4433.2023.00017.X

Singh, R., Dutt, S., Sharma, P., Sundramoorthy, A. K., Dubey, A., Singh, A., & Arya, S. (2023). Future of nanotechnology in food industry: Challenges in processing, packaging, and food safety. *Global Challenges (Hoboken, NJ)*, *7*(4), 1–28. doi:10.1002/gch2.202200209 PMID:37020624

Singh, S., Chakma, S., Alawa, B., Kalyanasundaram, M., & Diwan, V. (2023). Identification, characterization, and implications of microplastics in soil – A case study of Bhopal, central India. *Journal of Hazardous Materials Advances*, *9*, 100225. doi:10.1016/j.hazadv.2022.100225

Singh, S., Singh, M., Agrawal, V. V., & Kumar, A. (2010). An attempt to develop surface plasmon resonance based immunosensor for Karnal bunt (Tilletia indica) diagnosis based on the experience of nano-gold based lateral flow immuno-dipstick test. *Thin Solid Films*, *519*(3), 1156–1159. doi:10.1016/j.tsf.2010.08.061

Singh, S., Singh, O., & Shahi, U. P. (2023). Nanotechnology: A New Innovation in Biofortification and Food Safety. In A. Singh, V. Rajput, K. Ghazaryan, S. Gupta, & T. Minkina (Eds.), *Nanopriming Approach to Sustainable Agriculture* (pp. 340–351). IGI Global. doi:10.4018/978-1-6684-7232-3.ch015

Singh, T., Shukla, S., Kumar, P., Wahla, V., Bajpai, V. K., Rather, I. A., Song, Y., Lu, F., Li, H., Wang, H., Zhang, M., Liu, Y., & Kang, Z. (2018). Degradable Carbon Dots from Cigarette Smoking with Broad- Spectrum Antimicrobial Activities against Drug-Resistant Bacteria. *ACS Applied Bio Materials*, *1*(6), 1871–1879. doi:10.1021/acsabm.8b00421 PMID:34996288

Singh, V. K., Gill, A. A. S., Singh, O., Singh, S., & Shahi, U. P. (2024). Zeolite: A Natural Mineral for Sustainable Agriculture. In *Encyclopedia of Green Materials* (pp. 1–10). Springer Nature Singapore Pte Ltd. doi:10.1007/978-981-16-4921-9_263-1

Singh, Y. K., Singh, B. V., Katiyar, D., Saikanth, D. R. K., Kumar, K., Singh, O., Verma, S., & Kumar, P. (2023). Efficacy of nano fertilizers on yield, attributes and economics of wheat. *International Journal of Environment and Climate Change*, *13*(7), 291–297. doi:10.9734/ijecc/2023/v13i71879

Sinha, D., Maurya, A. K., Abdi, G., Majeed, M., Agarwal, R., Mukherjee, R., Ganguly, S., Aziz, R., Bhatia, M., Majgaonkar, A., Seal, S., Das, M., Banerjee, S., Chowdhury, S., Adeyemi, S. B., & Chen, J. T. (2023a). Integrated Genomic Selection for Accelerating Breeding Programs of Climate-Smart Cereals. In Genes (Vol. 14, Issue 7). Multidisciplinary Digital Publishing Institute (MDPI). doi:10.3390/genes14071484

Sinha, T., Bhagwatwar, P., Krishnamoorthy, C., & Chidambaram, R. (2019). Polymer based micro-and nanoencapsulation of agrochemicals. *Polymers for agri-food applications*, 5-28.

Sinha, K., Ghosh, J., & Sil, P. C. (2017). New pesticides: a cutting-edge view of contributions from nanotechnology for the development of sustainable agricultural pest control. In *New pesticides and soil sensors* (pp. 47–79). Academic Press. doi:10.1016/B978-0-12-804299-1.00003-5

Smith, J., Owen, E., Earis, J., & Woodcock, A. (2006). Effect of codeine on objective measurement of cough in chronic obstructive pulmonary disease. *The Journal of Allergy and Clinical Immunology*, *117*(4), 831–835. doi:10.1016/j.jaci.2005.09.055 PMID:16630941

Society of Nematologists. (1961). *Journal of Nematology*, *55*(1), 20230047. Advance online publication. doi:10.2478/jofnem-2023-0047

Sodano, V. (2018). Nano-food regulatory issues in the European union. *AIP Conference Proceedings*, 1–8. doi:10.1063/1.5047772

Soliman, A. S., Hassan, M., Abou-Elella, F., & Ahmed, A. H. H. (2020). Application of nanomaterials for precision agriculture and food quality assessment: A review. In *Nanotechnology Applications in Environmental Engineering* (pp. 240–263). IGI Global.

Soltani, A. M., & Pouypouy, H. (2019). Standardization and regulations of nanotechnology and recent government policies across the world on nanomaterials. In *Advances in phytonanotechnology* (pp. 419–446). Academic Press. doi:10.1016/B978-0-12-815322-2.00020-1

Songca, S. P. (2022). Applications of nanozymology in the detection and identification of viral, bacterial and fungal pathogens. *International Journal of Molecular Sciences*, *23*(9), 1–25. doi:10.3390/ijms23094638 PMID:35563029

Song, M. R., Cui, S. M., Gao, F., Liu, Y. R., Fan, C. L., Lei, T. Q., & Liu, D. C. (2012). Dispersible silica nanoparticles as carrier for enhanced bioactivity of chlorfenapyr. *Journal of Pesticide Science*, *37*(3), 258–260. doi:10.1584/jpestics.D12-027

Sood, P., & Shri, R. (2018). A review on ethnomedicinal, phytochemical and pharmacological aspects of *Myrica esculenta. Indian Journal of Pharmaceutical Sciences*, *80*(1), 2–13.

Sopi, R. B., & Khan, M. F. H. (2013). Bronchodilatory effect of ethanolic extract of the leaves of *Nyctanthes arbor-tristis. Pharmacognosy Research*, *5*(3), 169–172. doi:10.4103/0974-8490.112422 PMID:23901212

Soria, N. G. C., Montes, A., Bisson, M. A., Atilla-Gokcumen, G. E., & Aga, D. S. (2017). Mass spectrometry-based metabolomics to assess uptake of silver nanoparticles by Arabidopsis thaliana. *Environmental Science. Nano*, *4*(10), 1944–1953. doi:10.1039/C7EN00555E

Soto Hidalgo, K. T., Carrión-Huertas, P. J., Kinch, R. T., Betancourt, L. E., & Cabrera, C. R. (2020). Phytonanoremediation by Avicennia Germinans (black mangrove) and Nano Zero Valent Iron for Heavy Metal Uptake from Cienaga Las Cucharillas Wetland Soils. *Environmental Nanotechnology, Monitoring & Management*, *14*, 100363. doi:10.1016/j.enmm.2020.100363

Sparks, T. C., & Nauen, R. (2015). IRAC: Mode of action classification and insecticide resistance management. *Pesticide Biochemistry and Physiology*, *121*, 122–128. doi:10.1016/j.pestbp.2014.11.014 PMID:26047120

Späti, K., Huber, R., & Finger, R. (2021). Benefits of Increasing Information Accuracy in Variable Rate Technologies. *Ecological Economics*, *185*, 107047. Advance online publication. doi:10.1016/j.ecolecon.2021.107047

Sponton, O. E., Perez, A. A., Grabois, M., & Santiago, L. G. (2019). Nanosupplements: improvements in nutritional functionality of food products. Food Applications of Nanotechnology, 99-119. doi:10.1201/9780429297038-5

Srilatha, B. (2011). Nanotechnology in agriculture. *Journal of Nanomedicine & Nanotechnology*, *2*(7).

Srinivasan, C., & Saraswathi, R. (2010). Nano-agriculture—Carbon nanotubes enhance tomato seed germination and plant growth. *Current Science*, *99*(3), 274–275.

Stadler, T., Buteler, M., Valdez, S. R., & Gitto, J. G. (2018). Particulate nanoinsecticides: A new concept in insect pest management. *Insectic. Agricul. Toxicol.*, *83*, 72448. doi:10.5772/intechopen.72448

Stadler, T., Buteler, M., & Weaver, D. K. (2010). Novel use of nanostructured alumina as an insecticide. *Pest Management Science: Formerly. Pest Management Science*, *66*(6), 577–579. doi:10.1002/ps.1915

Stahl, W., Heinrich, U., Wiseman, S., Eichler, O., Sies, H., & Tronnier, H. (2001). Dietary tomato paste protects against ultraviolet light-induced erythema in humans. *The Journal of Nutrition*, *131*(5), 1449–1451. doi:10.1093/jn/131.5.1449 PMID:11340098

Steiner, A., Aguilar, G., Bomba, K., Bonilla, J. P., Campbell, A., Echeverria, R., & Zebiak, S. (2020). Actions to transform food systems under climate change. https://hdl.handle.net/10568/108489

Stephenson, G. R., Coats, J. R., & Yamamoto, H. (2001). Pesticide use and world food production: risks and benefits. In *Expert Committee on Weeds Comité d'experts en malherbologie. Proceedings of the 2000 National Meeting* (pp. 9-15). Citeseer.

Stogiannidis, E., & Laane, R. (2015). Source characterization of polycyclic aromatic hydrocarbons by using their molecular indices: an overview of possibilities. In D. M. Whitacre (Ed.), *Reviews of environmental contamination and toxicology* (Vol. 234, pp. 49–133). Springer. doi:10.1007/978-3-319-10638-0_2

Stolker, A. A. M., & Brinkman, U. A. T. (2005). Analytical Strategies for Residue Analysis of Veterinary Drugs and Growth-Promoting Agents in Food-Producing Animals-A Review. *Journal of Chromatography. A*, *1067*(1-2), 15–53. doi:10.1016/j.chroma.2005.02.037 PMID:15844509

Struik, P. C., & Kuyper, T. W. (2017). Sustainable intensification in agriculture: The richer shade of green. A review. *Agronomy for Sustainable Development*, *37*(5), 39. doi:10.1007/s13593-017-0445-7

Subramanian, K. S., Manikandan, A., Thirunavukkarasu, M., & Rahale, C. S. (2015). Nano-fertilizers for balanced crop nutrition. In *Nanotechnologies in food and agriculture* (pp. 69–80). Springer. doi:10.1007/978-3-319-14024-7_3

Su, H., Fang, Z., Tsang, P. E., Zheng, L., Cheng, W., Fang, J., & Zhao, D. (2016). Remediation of hexavalent chromium contaminated soil by biochar-supported zero-valent iron nanoparticles. *Journal of Hazardous Materials*, *318*, 533–540. doi:10.1016/j.jhazmat.2016.07.039 PMID:27469041

Suh, S., Meng, X., & Ko, S. (2016). Proof of concept study for different-sized chitosan nanoparticles as carbon dioxide (CO_2) indicators in food quality monitoring. *Talanta*, *1*(161), 265–270. doi:10.1016/j.talanta.2016.08.055 PMID:27769405

Sukul, U., Das, K., Chen, J. S., Sharma, R. K., Dey, G., Banerjee, P., Taharia, M., Lee, C. I., Maity, J. P., Lin, P. Y., & Chen, C. Y. (2023). Insight interactions of engineered nanoparticles with aquatic higher plants for phytoaccumulation, phytotoxicity, and phytoremediation applications: A review. *Aquatic Toxicology (Amsterdam, Netherlands)*, *264*, 106713. doi:10.1016/j.aquatox.2023.106713 PMID:37866164

Sun, C. Q. (2007). Size dependence of nanostructures: Impact of bond order deficiency. *Progress in Solid State Chemistry*, *35*(1), 1–159. doi:10.1016/j.progsolidstchem.2006.03.001

Sun, D., Hussain, H. I., Yi, Z., Rookes, J. E., Kong, L., & Cahill, D. M. (2016). Mesoporous silica nanoparticles enhance seedling growth and photosynthesis in wheat and lupin. *Chemosphere*, *152*, 81–91. doi:10.1016/j.chemosphere.2016.02.096 PMID:26963239

Sun, H., Zhou, Y., Ren, J., & Qu, X. (2018). Carbon Nanozymes: Enzymatic Properties, Catalytic Mechanism, and Applications. *Angewandte Chemie International Edition*, *57*(30), 9224–9237. doi:10.1002/anie.201712469 PMID:29504678

Sun, M., Zhao, C., Shang, H., Hao, Y., Han, L., Qian, K., White, J. C., Ma, C., & Xing, B. (2023). ZnO quantum dots outperform nanoscale and bulk particles for enhancing tomato (Solanum lycopersicum) growth and nutritional values. *The Science of the Total Environment*, *857*, 159330. doi:10.1016/j.scitotenv.2022.159330 PMID:36228785

Sun, X., Zhang, Y., Chen, G., & Gai, Z. (2017). Application of nanoparticles in enhanced oil recovery: A critical review of recent progress. *Energies*, *10*(3), 345. doi:10.3390/en10030345

Sun, Y. F., Liu, S. B., Meng, F. L., Liu, J. Y., Jin, Z., Kong, L. T., & Liu, J. H. (2012). Metal oxide nanostructures and their gas sensing properties: A review. *Sensors (Basel)*, *12*(3), 2610–2631. doi:10.3390/s120302610 PMID:22736968

Sun, Y., Wang, Y., Deng, X., Liang, H., & Guo, Y. (2020). Nanomaterials for the remediation of soils contaminated by heavy metals and metalloids: A review. *Reviews in Environmental Science and Biotechnology*, *19*(4), 769–792.

Suresh, V., Jaikumar, S., & Arunachalam, G. (2010). Anti diabetic activity of ethanol extract of stem bark of *Nyctanthes arbor-tristis* linn. *Research Journal of Pharmaceutical, Biological and Chemical Sciences*, *1*(4), 311–317.

Suriyaprabha, R., Karunakaran, G., Yuvakkumar, R., Rajendran, V., & Kannan, N. (2014). Silica nanoparticles for increased silica availability in maize (Zea mays L.) seeds under hydroponic conditions. *Current Nanoscience*, *8*(6), 902–908. doi:10.2174/157341312803989033

Surjushe, A., Vasani, R., & Saple, D. G. (2008). *Aloe vera*: A short review. *Indian Journal of Dermatology*, *54*(3), 163–166. doi:10.4103/0019-5154.44785 PMID:19882025

Suthar, T. R., Gavane, A. B., Shah, A. Y., & Devkatte, A. N. (2021). Nanotechnoogy, a revolutionary technique in the food industry: A systematic review. Sustainability, Agriculture. *Food and Environmental Research*, *9*(1), 127–141. doi:10.7770/safer-VONO-art2256

Su, Y., Wu, D., Chen, J., Chen, G., Hu, N., Wang, H., Wang, P., Han, H., Li, G., & Wu, Y. (2019). Ratiometric Surface Enhanced Raman Scattering Immunosorbent Assay of Allergenic Proteins via Covalent Organic Framework Composite Material Based Nanozyme Tag Triggered Raman Signal "Turn-On" and Amplification. *Analytical Chemistry*, *91*(18), 11687–11695. doi:10.1021/acs.analchem.9b02233 PMID:31418273

Swain, R., Sahoo, S., Behera, M., & Rout, G. R. (2023). Instigating prevalent abiotic stress resilience in crop by exogenous application of phytohormones and nutrient. In *Frontiers in Plant Science* (Vol. 14). Frontiers Media S.A., doi:10.3389/fpls.2023.1104874

Szakal, C., Roberts, S. M., Westerhoff, P., Bartholomaeus, A., Buck, N., Illuminato, I., Canady, R., & Rogers, M. (2014). Measurement of nanomaterials in foods: Integrative consideration of challenges and future prospects. *ACS Nano*, *8*(4), 3128–3135. doi:10.1021/nn501108g PMID:24673283

Szunerits, S., & Boukherroub, R. (2018). Graphene-based nanomaterials in innovative electrochemistry. *Current Opinion in Electrochemistry*, *10*, 24–30. doi:10.1016/j.coelec.2018.03.016

Tabasum, A., Alghuthaymi, M., Qazi, U. Y., Shahid, I., Abbas, Q., Javaid, R., Nadeem, N., & Zahid, M. (2020). UV-accelerated photocatalytic degradation of pesticide over magnetite and cobalt ferrite decorated graphene oxide composite. *Plants*, *10*(1), 6. doi:10.3390/plants10010006 PMID:33374688

Tadros, T., Izquierdo, P., Esquena, J., & Solans, C. (2004). Formation and stability of nano-emulsions. *Advances in Colloid and Interface Science*, *108*, 303–318. doi:10.1016/j.cis.2003.10.023 PMID:15072948

Takshak, S. (2018). Bioactive compounds in medicinal plants: A condensed review. *Journal of Pharmacy and Natural Medicines*, *1*(1), 1–35.

Talarska, P., Boruczkowski, M., & Żurawski, J. (2021). Current knowledge of silver and gold nanoparticles in laboratory research—Application, toxicity, cellular uptake. *Nanomaterials (Basel, Switzerland)*, *11*(9), 2454. doi:10.3390/nano11092454 PMID:34578770

Tamjidi, F., Shahedi, M., Varshosaz, J., & Nasirpour, A. (2013). Nanostructured lipid carriers (NLC): A potential delivery system for bioactive food molecules. *Innovative Food Science & Emerging Technologies*, *19*, 29–43. doi:10.1016/j.ifset.2013.03.002

Tao, X., Wang, X., Liu, B., & Liu, J. (2020). Conjugation of Antibodies and Aptamers on Nanozymes for Developing Biosensors. *Biosensors & Bioelectronics*, *168*, 112537. doi:10.1016/j.bios.2020.112537 PMID:32882473

Tarigholizadeh, S., Sushkova, S., Rajput, V. D., Ranjan, A., Arora, J., Dudnikova, T., Barbashev, A., Mandzhieva, S., Minkina, T., & Wong, M. H. (2023). Transfer and Degradation of PAHs in the Soil–Plant System: A Review. *Journal of Agricultural and Food Chemistry*, *72*(1), 46–64. doi:10.1021/acs.jafc.3c05589 PMID:38108272

Tariq, A., Mumtaz, F., Majeed, M., & Zeng, X. (2023a). Spatio-temporal assessment of land use land cover based on trajectories and cellular automata Markov modelling and its impact on land surface temperature of Lahore district Pakistan. *Environmental Monitoring and Assessment*, *195*(1). doi:10.1007/s10661-022-10738-w

Taslim, N. A., Nurkolis, F., Hardinsyah, H., Yusuf, V. M., Gunawan, W. B., Samtiya, M., Mayulu, N., Assa, Y. A., & Tallei, T. E. (2023). Foodomics approaches to facilitate the verification of the authenticity of foods: A possible strategy to screen, validate, and standardize food matrices. *Nutrición Clínica, Dietética Hospitalaria*, *43*(1), 73–80. doi:10.12873/431hardinsyah

Tassadduq, S. S., Akhtar, S., Waheed, M., Bangash, N., Nayab, D. E., Majeed, M., Abbasi, S., Muhammad, M., Alataway, A., Dewidar, A. Z., Elansary, H. O., & Yessoufou, K. (2022). Ecological Distribution Patterns of Wild Grasses and Abiotic Factors. *Sustainability (Basel)*, *14*(18), 11117. Advance online publication. doi:10.3390/su141811117

Tayel, A. A. (2016). Microbial chitosan as a biopreservative for fish sausages. *International Journal of Biological Macromolecules*, *93*, 41–46. doi:10.1016/j.ijbiomac.2016.08.061 PMID:27565293

Thakur, M., Smita, S., Gupta, P., & Singh, D. (2020). Impact of nanoparticles on the rhizosphere microbiome and its ecological significance. *Archives of Agronomy and Soil Science*, *66*(14), 1977–1994.

Thirunavukkarasu, M., Subramanian, K. S., Kannan, P., & Balaji, T. (2018). Response of nano-sulphur to the groundnut. *International Journal of Chemical Studies*, *6*(3), 2067–2072.

Thompson, P. B. (2015). Risks and ethics in the context of food nanotechnology and the delivery of bioactive ingredients. In Nanotechnology and functional foods: Effective delivery of bioactive ingredients. John Wiley and Sons, Ltd. doi:10.1002/9781118462157.ch19

Thornhill, S., Vargyas, E., Fitzgerald, T., & Chisholm, N. (2016). Household food security and biofuel feedstock production in rural Mozambique and Tanzania. *Food Security*, *8*(5), 953–971. doi:10.1007/s12571-016-0603-9

Tian, H., Liang, Y., Yang, D., & Sun, Y. (2020). Characteristics of PVP–stabilised NZVI and application to dechlorination of soil–sorbed TCE with ionic surfactant. *Chemosphere*, *239*, 124807. doi:10.1016/j.chemosphere.2019.124807 PMID:31520982

Tian, L., Zhang, Y., Wang, L., Geng, Q., Liu, D., Duan, L., Wang, Y., & Cui, J. (2020). Ratiometric Dual Signal-Enhancing-Based Electrochemical Biosensor for Ultrasensitive Kanamycin Detection. *ACS Applied Materials & Interfaces*, *12*(47), 52713–52720. doi:10.1021/acsami.0c15898 PMID:33170623

Tohge, T., & Fernie, R. (2017). An overview of compounds derived from the shikimate and phenylpropanoid pathways and their medicinal importance. *Mini-Reviews in Medicinal Chemistry*, *17*(12), 1013–1027. doi:10.2174/13895575166 66160624123425 PMID:27342231

Tombuloglu, H., Slimani, Y., Tombuloglu, G., Almessiere, M., & Baykal, A. (2019). Uptake and translocation of magnetite (Fe3O4) nanoparticles and its impact on photosynthetic genes in barley (Hordeum vulgare L.). *Chemosphere*, *226*, 110–122. doi:10.1016/j.chemosphere.2019.03.075 PMID:30925403

Torabian, S., Zahedi, M., & Khoshgoftar, A. H. (2017). Effects of foliar spray of nano-particles of FeSO4 on the growth and ion content of sunflower under saline condition. *Journal of Plant Nutrition*, *40*(5), 615–623. doi:10.1080/019041 67.2016.1240187

Tóth, G., Hermann, T., Da Silva, M. R., & Montanarella, L. (2016). Heavy metals in agricultural soils of the European Union with implications for food safety. *Environment International*, *88*, 299–309. doi:10.1016/j.envint.2015.12.017 PMID:26851498

Tripathi, D. K., Shweta, Singh, S., Singh, S., Pandey, R., Singh, V. P., Sharma, N. C., Prasad, S. M., Dubey, N. K., & Chauhan, D. K. (2017). An overview on manufactured nanoparticles in plants: Uptake, translocation, accumulation and phytotoxicity. *Plant Physiology and Biochemistry*, *110*, 2–12. doi:10.1016/j.plaphy.2016.07.030 PMID:27601425

Tripathi, D. K., Singh, S., Singh, S., Chauhan, D. K., & Dubey, N. K. (2015). Silicon nanoparticles more efficiently alleviate arsenate toxicity than silicon in maize cultivar and hybrid differing in arsenate tolerance. *Frontiers in Environmental Science*, *3*, 46.

Tundisi, L. L., Ataide, J. A., Costa, J. S. R., Coêlho, D. de F., Liszbinski, R. B., Lopes, A. M., Oliveira-Nascimento, L., de Jesus, M. B., Jozala, A. F., Ehrhardt, C., & Mazzola, P. G. (2023). Nanotechnology as a tool to overcome macromolecules delivery issues. *Colloids and Surfaces. B, Biointerfaces*, *222*, 113043. doi:10.1016/j.colsurfb.2022.113043 PMID:36455361

Twaij, B. M., & Hasan, M. N. (2022). Bioactive secondary metabolites from plant sources: Types, synthesis, and their therapeutic uses. *International Journal of Plant Biology*, *13*(1), 4–14. doi:10.3390/ijpb13010003

Tyagi, A., Ali, S., Ramakrishna, G., Singh, A., Park, S., Mahmoudi, H., & Bae, H. (2023). Revisiting the Role of Polyamines in Plant Growth and Abiotic Stress Resilience: Mechanisms, Crosstalk, and Future Perspectives. In Journal of Plant Growth Regulation (Vol. 42, Issue 8, pp. 5074–5098). Springer. doi:10.1007/s00344-022-10847-3

Tyagi, G., Jangir, D. K., Singh, P., & Mehrotra, R. (2010). DNA interaction studies of an anticancer plant alkaloid, vincristine, using fourier transform infrared spectroscopy. *DNA and Cell Biology*, *29*(11), 693–699. doi:10.1089/dna.2010.1035 PMID:20662555

Uddin, M. K. (2017). A review on the adsorption of heavy metals by clay minerals, with special focus on the past decade. *Chemical Engineering Journal*, *308*, 438–462. doi:10.1016/j.cej.2016.09.029

Ullah, I., Aslam, B., Shah, S. H. I. A., Tariq, A., Qin, S., Majeed, M., & Havenith, H. B. (2022). An Integrated Approach of Machine Learning, Remote Sensing, and GIS Data for the Landslide Susceptibility Mapping. *Land (Basel)*, *11*(8), 1265. Advance online publication. doi:10.3390/land11081265

United Nations. (2021). The Sustainable Development Goals Report 2021. United Nations. https://unstats.un.org/sdgs/report/2021/

Usha Rani, P., Madhusudhanamurthy, J., & Sreedhar, B. (2014). Dynamic adsorption of α-pinene and linalool on silica nanoparticles for enhanced antifeedant activity against agricultural pests. *Journal of Pest Science*, *87*(1), 191–200. doi:10.1007/s10340-013-0538-2

Usman, M., Farooq, M., Wakeel, A., Nawaz, A., Cheema, S. A., Rehman, H. U., Ashraf, I., & Sanaullah, M. (2020). Nanotechnology in agriculture: Current status, challenges and future opportunities. *The Science of the Total Environment*, *721*, 137778. doi:10.1016/j.scitotenv.2020.137778 PMID:32179352

Valdés, A., Álvarez-Rivera, G., Socas-Rodríguez, B., Herrero, M., Ibáñez, E., & Cifuentes, A. (2022). Foodomics: Analytical opportunities and challenges. *Analytical Chemistry*, *94*(1), 366–381. doi:10.1021/acs.analchem.1c04678 PMID:34813295

Van HaC.Van NguyenD.NguyenH. M.LeN. T.NguyenK. H.LeH. M.. (2020). Copper nanoparticle application enhances plant growth and grain yield in maize under drought stress conditions. bioRxiv [Preprint]. doi:10.1101/2020.02.24.963132

Vandana Ghormade, V. G., Deshpande, M. V., & Paknikar, K. M. (2011). Perspectives for nano-biotechnology enabled protection and nutrition of plants. Academic Press.

Vannini, C., Domingo, G., Onelli, E., De Mattia, F., Bruni, I., Marsoni, M., & Bracale, M. (2014). Phytotoxic and genotoxic effects of silver nanoparticles exposure on germinating wheat seedlings. *Journal of Plant Physiology*, *171*(13), 1142–1148. doi:10.1016/j.jplph.2014.05.002 PMID:24973586

Vanti, G. L., Nargund, V. B., N, B. K., Vanarchi, R., Kurjogi, M., Mulla, S. I., Tubaki, S., & Patil, R. R. (2018). Synthesis of *Gossypium hirsutum*-derived silver nanoparticles and their antibacterial efficacy against plant pathogens. *Applied Organometallic Chemistry*, *33*(1), e4630. doi:10.1002/aoc.4630

Vardumyan, H., Singh, A., Rajput, V., Minkina, T., El-Ramady, H., & Ghazaryan, K. (2024). Additive-Mediated Phytoextraction of Copper-Contaminated Soils Using Medicago lupulina L. *Egyptian Journal of Soil Science*, *64*(2), 599–618. doi:10.21608/ejss.2024.266169.1714

Varga, B., Priksz, D., Lampé, N., Bombicz, M., Kurucz, A., Szabó, A. M., Pósa, A., Szabó, R., Kemény-Beke, Á., Remenyik, J., Gesztelyi, R., & Juhász, B. (2017). Protective effect of *Prunus cerasus* (sour cherry) seed extract on the recovery of ischemia/reperfusion-induced retinal damage in zucker diabetic fatty rat. *Molecules (Basel, Switzerland)*, *22*(10), 1782. doi:10.3390/molecules22101782 PMID:29065463

Vasile, C. (2018). Polymeric nanocomposites and nanocoatings for food packaging: A review. Materials, 11, 1834. doi:10.3390/ma1110183

Veera kumar, P., Sangili, A., Saranya, K., Pandikumar, A., & Lin, K. C. (2021). Palladium and silver nanoparticles embedded on zinc oxide nanostars for photocatalytic degradation of pesticides and herbicides. *Chemical Engineering Journal, 410*, 128434. doi:10.1016/j.cej.2021.128434

Vega-Vásquez, P., Mosier, N. S., & Irudayaraj, J. (2020). Nanoscale drug delivery systems: From medicine to agriculture. *Frontiers in Bioengineering and Biotechnology*, *8*, 79. doi:10.3389/fbioe.2020.00079 PMID:32133353

Velten, S., Leventon, J., Jager, N., & Newig, J. (2015). What is sustainable agriculture? A systematic review. *Sustainability (Basel)*, *7*(6), 7833–7865. doi:10.3390/su7067833

Verma, K. K., Song, X. P., Joshi, A., Tian, D. D., Rajput, V. D., Singh, M., & Li, Y. R. (2022). Recent trends in nanofertilizers for sustainable agriculture under climate change for global food security. *Nanomaterials (Basel, Switzerland)*, *12*(1), 173. doi:10.3390/nano12010173 PMID:35010126

Verma, N., & Bhardwaj, A. (2015). Biosensor technology for pesticides—A review. *Applied Biochemistry and Biotechnology, 175*(6), 3093–3119. doi:10.1007/s12010-015-1489-2 PMID:25595494

Verpoorte, R., & Memelink, J. (2002). Engineering secondary metabolite production in plants. *Current Opinion in Biotechnology, 13*(2), 181–187. doi:10.1016/S0958-1669(02)00308-7 PMID:11950573

Vickers, N. J. (2017). Animal communication: When I'm caching you will answer food? *Current Biology, 27*(14), 713–715. doi:10.1016/j.cub.2017.05.064 PMID:28743020

Victoria, J., Tripathi, S., Prakash, V., Tiwari, K., Mahra, S., Sharma, A., Rana, S., Kandhol, N., Sahi, S., Tripathi, D. K., & Sharma, S. (2024). Encapsulated nanopesticides application in plant protection: Quo vadis? *Plant Physiology and Biochemistry, 206*, 108225. doi:10.1016/j.plaphy.2023.108225 PMID:38147708

Vidotti, M., Carvalhal, R. F., Mendes, R. K., Ferreira, D., & Kubota, L. T. (2011). Biosensors based on gold nanostructures. *Journal of the Brazilian Chemical Society, 22*(1), 3–20. doi:10.1590/S0103-50532011000100002

Vijayvargia, P., & Vijayvergia, R. (2014). A review on *Limonia acidissima* L.: Multipotential medicinal plant. *International Journal of Pharmaceutical Sciences and Research, 28*(1), 191–195.

Vijayvargia, P., & Vijayvergia, R. (2016). Assessment of phytochemicals and antioxidant activity of *Murraya koenigii* Linn. *International Journal of Pharmaceutical Sciences and Research, 7*(5), 2163.

Vinayaka, A. C., Basheer, S., & Thakur, M. S. (2009). Bioconjugation of CdTe quantum dot for the detection of 2, 4-dichlorophenoxyacetic acid by competitive fluoroimmunoassay based biosensor. *Biosensors & Bioelectronics, 24*(6), 1615–1620. doi:10.1016/j.bios.2008.08.042 PMID:18930650

Vinayaka, A. C., & Thakur, M. S. (2010). Focus on quantum dots as potential fluorescent probes for monitoring food toxicants and foodborne pathogens. *Analytical and Bioanalytical Chemistry, 397*(4), 1445–1455. doi:10.1007/s00216-010-3683-y PMID:20401471

Visentin, C., Trentin, A. W. da S., Braun, A. B., & Thomé, A. (2021). Nano scale zero valent iron production methods applied to contaminated sites remediation: An overview of production and environmental aspects. *Journal of Hazardous Materials, 410*, 124614. doi:10.1016/j.jhazmat.2020.124614 PMID:33246810

Viswanathan, S., & Radecki, J. (2008). Nanomaterials in electrochemical biosensors for food analysis-a review. *Polish Journal of Food and Nutrition Sciences, 58*(2).

Vurro, M., Miguel-Rojas, C., & Pérez-de-Luque, A. (2019). Safe nanotechnologies for increasing the effectiveness of environmentally friendly natural agrochemicals. *Pest Management Science, 75*(9), 2403–2412. doi:10.1002/ps.5348 PMID:30672106

Vurukonda, S. S. K. P., Vardharajula, S., Shrivastava, M., & SkZ, A. (2016). Enhancement of drought stress tolerance in crops by plant growth promoting rhizobacteria. *Microbiological Research, 184*, 13–24. doi:10.1016/j.micres.2015.12.003 PMID:26856449

Waghmode, M. S., Gunjal, A. B., Mulla, J. A., Patil, N. N., & Nawani, N. N. (2019). Studies on the titanium dioxide nanoparticles: Biosynthesis, applications and remediation. *SN Applied Sciences, 1*(4), 310. doi:10.1007/s42452-019-0337-3

Waheed, M., Arshad, F., Majeed, M., Fatima, S., Mukhtar, N., Aziz, R., Mangrio, W. M., Almohamad, H., Al Dughairi, A. A., Al-Mutiry, M., & Abdo, H. G. (2022). Community Structure and Distribution Pattern of Woody Vegetation in Response to Soil properties in Semi-Arid Lowland District Kasur Punjab, Pakistan. *Land (Basel), 11*(12), 2145. Advance online publication. doi:10.3390/land11122145

Waheed, M., Arshad, F., Majeed, M., Haq, S. M., Aziz, R., Bussmann, R. W., Ali, K., Subhan, F., Jones, D. A., & Zaitouny, A. (2023). *Potential distribution of a noxious weed (Solanum viarum Du-nal), current status, and future invasion risk based on MaxEnt modeling*. Geology, Ecology, and Landscapes. doi:10.1080/24749508.2023.2179752

Walk, A. M., Khan, N. A., Barnett, S. M., Raine, L. B., Kramer, A. F., Cohen, N. J., Moulton, C. J., Renzi-Hammond, L. M., Hammond, B. R., & Hillman, C. H. (2017). From neuro-pigments to neural efficiency: The relationship between retinal carotenoids and behavioral and neuroelectric indices of cognitive control in childhood. *International Journal of Psychophysiology*, *118*, 1–8. doi:10.1016/j.ijpsycho.2017.05.005 PMID:28528704

Walkey, C., Das, S., Seal, S., Erlichman, J., Heckman, K., Ghibelli, L., Traversa, E., McGinnis, J. F., & Self, W. T. (2015). Catalytic properties and biomedical applications of cerium oxide nanoparticles. *Environmental Science. Nano*, *2*(1), 33–53. doi:10.1039/C4EN00138A PMID:26207185

Wang, G., Peng, Z., & Li, Y. (2019). Synthesis, anticancer activity and molecular modeling studies of novel chalcone derivatives containing indole and naphthalene moieties as tubulin polymerization inhibitors. *Chemical & Pharmaceutical Bulletin*, *67*(7), 725–728. doi:10.1248/cpb.c19-00217 PMID:30982797

Wang, H., Kim, B., & Wunder, S. L. (2015). Nanoparticle-supported lipid bilayers as an in situ remediation strategy for hydrophobic organic contaminants in soils. *Environmental Science & Technology*, *49*(1), 529–536. doi:10.1021/es504832n PMID:25454259

Wang, H., Wan, K., & Shi, X. (2019). Recent Advances in Nanozyme Research. *Advanced Materials*, *31*(45), 1805368. doi:10.1002/adma.201805368 PMID:30589120

Wang, K., Lu, X., Lu, Y., Wang, J., Lu, Q., Cao, X., Yang, Y., & Yang, Z. (2022). Nanomaterials in Animal Husbandry: Research and Prospects. *Frontiers in Genetics*, *13*(June), 1–11. doi:10.3389/fgene.2022.915911 PMID:35846144

Wang, L., Li, X., Zhang, G., Dong, J., & Eastoe, J. (2007). Oil-in-water nanoemulsions for pesticide formulations. *Journal of Colloid and Interface Science*, *314*(1), 230–235. doi:10.1016/j.jcis.2007.04.079 PMID:17612555

Wang, L., Ning, C., Pan, T., & Cai, K. (2022). Role of silica nanoparticles in abiotic and biotic stress tolerance in plants: A review. *International Journal of Molecular Sciences*, *23*(4), 1947. doi:10.3390/ijms23041947 PMID:35216062

Wang, P., Lombi, E., Menzies, N. W., Zhao, F. J., & Kopittke, P. M. (2018). Engineered silver nanoparticles in terrestrial environments: A meta-analysis shows that the overall environmental risk is small. *Environmental Science. Nano*, *5*(11), 2531–2544. doi:10.1039/C8EN00486B

Wang, P., Lombi, E., Zhao, F. J., & Kopittke, P. M. (2016). Nanotechnology: A new opportunity in plant sciences. *Trends in Plant Science*, *21*(8), 699–712. doi:10.1016/j.tplants.2016.04.005 PMID:27130471

Wang, S. (2020). A Comprehensive Review of Wearable Applications and Material Construction. *Ozean Journal of Applied Sciences*, *10*(06), 364–408. doi:10.4236/ojapps.2020.106027

Wang, W., Qian, S., Gong, L., Ni, Z., & Ren, H. (2021). Effects of carbon nanoonions on the tribological performance of food grade calcium sulfonate complex grease. *Lubrication Science*, *33*(8), 460–470. doi:10.1002/ls.1567

Wang, X.-F., Zhou, Y., Xu, J.-J., & Chen, H.-Y. (2009). Signal-on Electrochemiluminescence Biosensors Based on CdS-Carbon Nanotube Nanocomposite for the Sensitive Detection of Choline and Acetylcholine. *Advanced Functional Materials*, *19*(9), 1444–1450. doi:10.1002/adfm.200801313

Wang, X., Guo, J., Feng, J., & Shi, Z. (2018). Nanozyme-based environmental sensing and remediation. *NanoImpact*, *10*, 21–30. doi:10.1016/j.impact.2018.01.003

Wang, X., Liu, X., Chen, J., Han, H., & Yuan, Z. (2014). Evaluation and mechanism of antifungal effects of carbon nanomaterials in controlling plant fungal pathogen. *Carbon*, *68*, 798–806. doi:10.1016/j.carbon.2013.11.072

Wang, Y., Cen, C., Chen, J., & Fu, L. (2020). MgO/carboxymethyl chitosan nanocomposite improves thermal stability, waterproof and antibacterial performance for food packaging. *Carbohydrate Polymers*, *236*, 116078. doi:10.1016/j.carbpol.2020.116078 PMID:32172891

Wang, Y., Cui, H., Sun, C., Zhao, X., & Cui, B. (2014). Construction and evaluation of controlled-release delivery system of Abamectin using porous silica nanoparticles as carriers. *Nanoscale Research Letters*, *9*(1), 1–6. doi:10.1186/1556-276X-9-655 PMID:26088998

Wang, Y., Li, C., Wang, Y., Zhang, Y., & Li, X. (2018). Compound pesticide controlled release system based on the mixture of poly (butylene succinate) and PLA. *Journal of Microencapsulation*, *35*(5), 494–503. doi:10.1080/02652048.2018.1538265 PMID:30395751

Wang, Y., O'Connor, D., Shen, Z., Lo, I. M. C., Tsang, D. C. W., Pehkonen, S., Pu, S., & Hou, D. (2019). Green synthesis of nanoparticles for the remediation of contaminated waters and soils: Constituents, synthesizing methods, and influencing factors. *Journal of Cleaner Production*, *226*, 540–549. doi:10.1016/j.jclepro.2019.04.128

Wang, Y., & Qu, J. (2009). Research progress of soil moisture sensor based on capacitance. *Sensors & Transducers*, *108*(9), 73–81.

Wang, Z., Wei, F., Liu, S. Y., Xu, Q., Huang, J. Y., Dong, X. Y., & Chen, H. (2010). Electrocatalytic oxidation of phytohormone salicylic acid at copper nanoparticles-modified gold electrode and its detection in oilseed rape infected with fungal pathogen Sclerotinia sclerotiorum. *Talanta*, *80*(3), 1277–1281. doi:10.1016/j.talanta.2009.09.023 PMID:20006087

Wan, Y., Li, J., Ren, H., Huang, J., & Yuan, H. (2014). Physiological investigation of gold nanorods toward watermelon. *Journal of Nanoscience and Nanotechnology*, *14*(8), 6089–6094. doi:10.1166/jnn.2014.8853 PMID:25936063

Wanyika, H. (2014). Sustained release of fungicide metalaxyl by mesoporous silica nanospheres. In Nanotechnology for Sustainable Development (pp. 321-329). Springer International Publishing.

Weigl, B., Domingo, G., LaBarre, P., & Gerlach, J. (2008). Towards non-and minimally instrumented, microfluidics-based diagnostic devices. *Lab on a Chip*, *8*(12), 1999–2014. doi:10.1039/b811314a PMID:19023463

Wei, H., & Wang, E. (2013). Nanomaterials with enzyme-like characteristics (nanozymes): Next-generation artificial enzymes. *Chemical Society Reviews*, *42*(14), 6060–6093. doi:10.1039/c3cs35486e PMID:23740388

Wei, J., Yang, L., Luo, M., Wang, Y., & Li, P. (2019). Nanozyme-assisted Technique for Dual Mode Detection of Organophosphorus Pesticide. *Ecotoxicology and Environmental Safety*, *179*, 17–23. doi:10.1016/j.ecoenv.2019.04.041 PMID:31022651

Wibowo, D., Zhao, C. X., Peters, B. C., & Middelberg, A. P. (2014). Sustained release of fipronil insecticide in vitro and in vivo from biocompatible silica nanocapsules. *Journal of Agricultural and Food Chemistry*, *62*(52), 12504–12511. doi:10.1021/jf504455x PMID:25479362

Winckler, C., & Grafe, A. (2001). Use of Veterinary Drugs in Intensive Animal Production. *Journal of Soils and Sediments*, *1*(2), 66–70. doi:10.1007/BF02987711

Worheide, M. A., Krumsiek, J., Kastenmuller, G., & Arnold, M. (2021). Multi-omics integration in biomedical research - A metabolomics-centric review. *Analytica Chimica Acta*, *2*(1141), 144–162. doi:10.1016/j.aca.2020.10.038 PMID:33248648

World Health Organization. (2013). *WHO Traditional Medicine Strategy: 2014-2023*. World Health Organization.

Wu, L., Zhou, S., Wang, G., Yun, Y., Liu, G., & Zhnag, W. (2021). Nanozyme appications: a glimpse of insight in food safety. Frontiers in Bioengineering and Biotechnology, 9, 1-15. doi:10.3389/fbioe.2021.727886

Wuana, R. A., & Okieimen, F. E. (2011). Heavy metals in contaminated soils: A review of sources, chemistry, risks and best available strategies for remediation. *International Scholarly Research Notices*.

Wu, H., Santana, I., Dansie, J., & Giraldo, J. P. (2017). In vivo delivery of nanoparticles into plant leaves. *Current Protocols in Chemical Biology*, *9*(4), 269–284. doi:10.1002/cpch.29 PMID:29241293

Wu, H., Shabala, L., Shabala, S., & Giraldo, J. P. (2017). Enzyme nanoparticles enhance plant growth by delivering nutrients and protecting against abiotic stresses. *ACS Nano*, *11*(10), 9650–9661. PMID:28872828

Wu, J., Wang, X., Wang, Q., Lou, Z., Li, S., Zhu, Y., Qin, L., & Wei, H. (2019). Nanomaterials with enzyme-like characteristics (nanozymes): Next-generation artificial enzymes (II). *Chemical Society Reviews*, *48*(4), 1004–1076. doi:10.1039/C8CS00457A PMID:30534770

Wu, J., Yang, Q., Li, Q., Li, H., & Li, F. (2021a). Two-Dimensional MnO2 Nanozyme-Mediated Homogeneous Electrochemical Detection of Organophosphate Pesticides without the Interference of H2O2 and Color. *Analytical Chemistry*, *93*(8), 4084–4091. doi:10.1021/acs.analchem.0c05257 PMID:33588528

Wu, L., Li, G., Xu, X., Zhu, L., Huang, R., & Chen, X. (2019b). Application of Nano-ELISA in Food Analysis: Recent Advances and Challenges. *Trends in Analytical Chemistry*, *113*, 140–156. doi:10.1016/j.trac.2019.02.002

Wu, L., Liu, M., & Liang, R. (2016). Preparation and properties of a double-coated slow-release NPK compound fertilizer with superabsorbent and water-retention. *Bioresource Technology*, *99*(3), 547–554. doi:10.1016/j.biortech.2006.12.027 PMID:17320380

Wu, L., Yan, H., Wang, J., Liu, G., & Xie, W. (2019c). Tyrosinase Incorporated with Au-Pt@SiO2 Nanospheres for Electrochemical Detection of Bisphenol A. *Journal of the Electrochemical Society*, *166*(8), B562–B568. doi:10.1149/2.0141908jes

Wu, Y., Qi, Y., Song, B., Zhang, M., Jiang, L., & Wang, Y. (2019). Nanoparticles inhibit the microbial activities of As (III)-oxidizing bacteria and the transformation of As. In soil. *Environmental Pollution. III*, *255*, 113134.

Xianyu, Y., Wang, Z., & Jiang, X. (2018). A plasmonic nanosensor for immunoassay via enzyme-triggered click chemistry. *ACS Nano*, *8*(12), 12741–12747. doi:10.1021/nn505857g PMID:25423357

Xie, Y., He, Y., Irwin, P. L., Jin, T., & Shi, X. (2011). Antibacterial activity and mechanism of action of zinc oxide nanoparticles against Campylobacter jejuni. *Applied and Environmental Microbiology*, *77*(7), 2325–2331. doi:10.1128/AEM.02149-10 PMID:21296935

Xing, Y., Yan, W., Zhang, Y., Bian, G., Zhang, Y., Yang, G., & Wang, Q. (2019). Effects of nano-clay composite coatings on the quality of strawberries under ambient temperature. *Coatings*, *9*(6), 360.

Xu, C., Peng, C., Sun, L., Zhang, S., Huang, H., Chen, Y., & Shi, J. (2021). Distinctive effects of TiO2 and CuO nanoparticles on soil microbes and their community structures in flooded paddy soil. *Soil Biology & Biochemistry*, *43*(8), 1406–1414. doi:10.1016/j.soilbio.2011.03.010

Xue, J., Luo, Z., Li, P., Ding, Y., Cui, Y., & Wu, Q. (2014). A residue-free green synergistic antifungal nanotechnology for pesticide thiram by ZnO nanoparticles. *Scientific Reports*, *4*(1), 5408. doi:10.1038/srep05408 PMID:25023938

Xue, W., Huang, D., Zeng, G., Wan, J., Cheng, M., Zhang, C., Hu, C., & Li, J. (2018). Performance and toxicity assessment of nanoscale zero valent iron particles in the remediation of contaminated soil: A review. *Chemosphere*, *210*, 1145–1156. doi:10.1016/j.chemosphere.2018.07.118 PMID:30208540

Xue, W., Huang, D., Zeng, G., Wan, J., Zhang, C., Xu, R., Cheng, M., & Deng, R. (2018). Nanoscale zero-valent iron coated with rhamnolipid as an effective stabilizer for immobilization of Cd and Pb in river sediments. *Journal of Hazardous Materials*, *341*, 381–389. doi:10.1016/j.jhazmat.2017.06.028 PMID:28806558

Xu, J., Li, M., Niu, W., Wu, J., & Yang, W. (2019). Preparation and application of phase change microcapsules in thermal energy storage and temperature control of greenhouses. *Journal of Cleaner Production*, *247*, 119111.

Xu, L., Cao, L. D., Li, F. M., Wang, X. J., & Huang, Q. L. (2014). Utilization of chitosan-lactide copolymer nanoparticles as controlled release pesticide carrier for pyraclostrobin against Colletotrichum gossypii Southw. *Journal of Dispersion Science and Technology*, *35*(4), 544–550. doi:10.1080/01932691.2013.800455

Xu, L., Xing, H., Huang, Y., Dai, X., Zhang, W., & Chen, Y. (2020). Biochar-based fertilizer: A green method for enhancing nitrogen retention and reducing nitrogen leaching in vegetable soils. *Chemosphere*, *255*, 126978.

Xu, L., Zhang, W., & Ding, Y. (2014). Review of nano sensors for monitoring the soil moisture. *Applied Mechanics and Materials*, *651*, 1517–1522. doi:10.4028/www.scientific.net/AMM.494-495.1517

Xu, T., Wang, L., Tao, Y., Ji, Y., Deng, F., & Wu, X. H. (2016). The function of naringin in inducing secretion of osteoprotegerin and inhibiting formation of osteoclasts. *Evidence-Based Complementary and Alternative Medicine*, *2016*, 8981650. doi:10.1155/2016/8981650 PMID:26884798

Xu, Z. P., Stevenson, G. S., Lu, C. Q., Lu, G. Q., Bartlett, P. F., & Gray, P. P. (2006). Stable suspension of layered double hydroxide nanoparticles in aqueous solution. *Journal of the American Chemical Society*, *128*(1), 36–37. doi:10.1021/ja056652a PMID:16390109

Yadav, A., Yadav, K., & Abd-Elsalam, K. A. (2023). Nanofertilizers: Types, Delivery and Advantages in Agricultural Sustainability. *Agrochemicals*, *2*(2), 296–336. doi:10.3390/agrochemicals2020019

Yadav, A., Yadav, K., Ahmad, R., & Abd-Elsalam, K. A. (2023). Emerging Frontiers in Nanotechnology for Precision Agriculture: Advancements, Hurdles and Prospects. *Agrochemicals*, *2*(2), 220–256. doi:10.3390/agrochemicals2020016

Yadav, S. K. (2010). Heavy metals toxicity in plants: An overview on the role of glutathione and phytochelatins in heavy metal stress tolerance of plants. *South African Journal of Botany*, *76*(2), 167–179. doi:10.1016/j.sajb.2009.10.007

Yang, F., Jiang, X., Zhong, X., Wei, S., & Yuan, R. (2018). Highly Sensitive Electrochemiluminescence Detection of Mucin1 Based on V2O5 Nanospheres as Peroxidase Mimetics to Catalyze H2O2 for Signal Amplification. *Sensors and Actuators. B, Chemical*, *265*, 126–133. doi:10.1016/j.snb.2018.03.031

Yang, K., Lin, D., & Xing, B. (2009). Interactions of humic acid with nanosized inorganic oxides. *Langmuir*, *25*(6), 3571–3576. doi:10.1021/la803701b PMID:19708146

Yang, M., Wu, X., He, C., Zhang, J., Hou, J., & Lin, D. (2023). nZVI-induced iron poisoning aggravated the toxicity of TCEP to earthworm in soil. *Environmental Pollution*, *317*, 120785. doi:10.1016/j.envpol.2022.120785 PMID:36460191

Yang, X., Liang, Q., Chen, Y., & Wang, B. (2019). Alteration of methanogenic archaeon by ethanol contribute to the enhancement of biogenic methane production of lignite. *Frontiers in Microbiology*, *10*, 2323. doi:10.3389/fmicb.2019.02323 PMID:31649654

Yan, L., Li, P., Zhao, X., Ji, R., & Zhao, L. (2020). Physiological and metabolic responses of maize (*Zea mays*) plants to Fe_3O_4 nanoparticles. *The Science of the Total Environment*, *718*, 137400. doi:10.1016/j.scitotenv.2020.137400 PMID:32105936

Yanthan, M., & Misra, A. K. (2013). Molecular approach to the classification of medicinally important actinorhizal genus Myrica. *Indian Journal of Biotechnology*, *12*, 133–136.

Yao, K. S., Li, S. J., Tzeng, K. C., Cheng, T. C., Chang, C. Y., Chiu, C. Y., Liao, C. Y., Hsu, J. J., & Lin, Z. P. (2009). Fluorescence silica nano probe as a biomarker for rapid detection of plant pathogens. *Advanced Materials Research*, *79*, 513–516. doi:10.4028/www.scientific.net/AMR.79-82.513

Yerasu, S. R., Reddy, B. R., Singh, D. P., & Singh, J. (2022). Omics studies for vegetable improvement. In *Omics in Horticultural Crops* (pp. 215–238). Academic Press., doi:10.1016/B978-0-323-89905-5.00018-5

Yin, C., Zgao, W., Liu, R., Liu, R., Wang, Z., Zhu, L., Chen, W., & Liu, S. (2017). TiO_2 particles in seafood and surimi products; attention should be paid to their exposure and uptake through food. *Chemosphere*, *188*, 541–547. doi:10.1016/j.chemosphere.2017.08.168 PMID:28910729

Ying, S., Guan, Z., Ofoegbu, P. C., Clubb, P., Rico, C., He, F., & Hong, J. (2022). Green synthesis of nanoparticles: Current developments and limitations. *Environmental Technology & Innovation*, *26*, 102336. doi:10.1016/j.eti.2022.102336

Yoon, K. Y., Byeon, J. H., Park, J. H., & Hwang, J. (2007). Susceptibility constants of Escherichia coli and Bacillus subtilis to silver and copper nanoparticles. *The Science of the Total Environment*, *373*(2-3), 572–575. doi:10.1016/j.scitotenv.2006.11.007 PMID:17173953

Younas, A., Yousaf, Z., Riaz, N., Rashid, M., Razzaq, Z., Tanveer, M., & Huang, S. (2020). Role of nanotechnology for enhanced rice production. *Nutrient dynamics for sustainable crop production*, 315-350.

Younis, S. A., Kim, K. H., Shaheen, S. M., Antoniadis, V., Tsang, Y. F., Rinklebe, J., Deep, A., & Brown, R. J. (2021). Advancements of nanotechnologies in crop promotion and soil fertility: Benefits, life cycle assessment, and legislation policies. *Renewable & Sustainable Energy Reviews*, *152*, 111686. doi:10.1016/j.rser.2021.111686

Younis, U., & Malik, A. H. (2021). Nanotechnology and plant growth-promoting rhizobacteria for sustainable agriculture. In *Advances in soil microbiology* (pp. 169–186). Springer.

You, Y., Zhang, L., Zhang, J., Zheng, J., Zhang, Y., & Chen, Y. (2017). Stimulatory effects of multi-walled carbon nanotubes on plant growth: Relationship with nanotube functionalization. *Plant Physiology and Biochemistry*, *110*, 97–106.

Yunlong, C., & Smit, B. (1994). Sustainability in agriculture: A general review. *Agriculture, Ecosystems & Environment*, *49*(3), 299–307. doi:10.1016/0167-8809(94)90059-0

Yu, V. N., Povet'eva, T. N., Suslov, N. I., Shults, E. E., Ziuz'kov, G. N., Aksinenko, S. G., Krapivin, A. V., & Kharina, T. G. (2015). Anxiolytic activity of diterpene alkaloid songorine pharmacology and toxicology. *Bulletin of Experimental Biology and Medicine*, *159*(5), 620–622. doi:10.1007/s10517-015-3029-z PMID:26468026

Yuvaraj, M., Sathya Priya, R., Jagathjothi, N., Saranya, M., Suganthi, N., Sharmila, R., Cyriac, J., Anitha, R., & Subramanian, K. S. (2023). Silicon nanoparticles (SiNPs): Challenges and perspectives for sustainable agriculture. *Physiological and Molecular Plant Pathology*, *128*, 102161. doi:10.1016/j.pmpp.2023.102161

Yu, X., Chattopadhyay, D., Galeska, I., Papadimitrakopoulos, F., & Rusling, J. F. (2003). Peroxidase activity of enzymes bound to the ends of single-wall carbon nanotube forest electrodes. *Electrochemistry Communications*, *5*(5), 408–411. doi:10.1016/S1388-2481(03)00076-6

Zaghloul, S. M., El-Shishtawy, R. M., Atia, A. A., Mansour, N. A., & El-Sherbiny, I. M. (2021). Graphene oxide-based biosensor for detection of Candidatus Liberibacter asiaticus bacterium causing citrus greening disease. *Talanta*, *223*, 121714. doi:10.1016/j.talanta.2020.121714

Zain, M., Ma, H., Ur Rahman, S., Nuruzzaman, M., Chaudhary, S., Azeem, I., Mehmood, F., Duan, A., & Sun, C. (2024). Nanotechnology in precision agriculture: Advancing towards sustainable crop production. In Plant Physiology and Biochemistry (Vol. 206). Elsevier Masson s.r.l. doi:10.1016/j.plaphy.2023.108244

Zain, M. E. (2011). Impact of Mycotoxins on Humans and Animals. *Journal of Saudi Chemical Society*, *15*(2), 129–144. doi:10.1016/j.jscs.2010.06.006

Zanet, V., Vidic, J., Auger, S., Vizzini, P., Lippe, G., Iacumin, L., Comi, G., & Manzano, M. (2019). Activity evaluation of pure and doped zinc oxide nanoparticles against bacterial pathogens and *Saccharomyces cerevisiae*. Journal of Applied Microbiology, 127(5), 1391–1402. doi:10.1111/jam.14407

Zeng, R., Wang, J., Cui, J., Hu, L., & Mu, K. (2010). Photocatalytic degradation of pesticide residues with Re3+-doped Nano-TiO2. *Journal of Rare Earths*, *28*, 353–356. doi:10.1016/S1002-0721(10)60329-8

Zhang, C., Wang, H., Hao, H., Li, Y., Jiang, L., & Zhou, J. (2015). Functionalized quantum dots for biosensing applications. In *Nano-Oncologicals* (pp. 73–88). Springer.

Zhang, F., He, M., Zhang, C., Lin, D., & Zhang, J. (2021). Combined toxic effects of dioxin-like PCB77 with Fe-based nanoparticles in earthworm Eisenia fetida. *The Science of the Total Environment*, *766*, 144347. doi:10.1016/j.scitotenv.2020.144347 PMID:33418254

Zhang, H., Han, W., De Smet, I., Talboys, P., Loya, R., Hassan, A., Rong, H., Jürgens, G., Paul Knox, J., & Wang, M. H. (2020). Rapid activation of heat stress response pathways by small molecules promotes plant thermotolerance. *Nature Chemical Biology*, *16*(2), 148–152. PMID:31974524

Zhang, J. L., Zhou, Z. P., Pei, Y., Xiang, Q. Q., Chang, X. X., Ling, J., & Chen, L. Q. (2018). Metabolic profiling of silver nanoparticle toxicity in Microcystis aeruginosa. *Environmental Science. Nano*, *5*(11), 2519–2530. doi:10.1039/C8EN00738A

Zhang, L., & Fang, M. (2010). Nanomaterials in pollution trace detection and environmental improvement. *Nano Today*, *5*(2), 128–142. doi:10.1016/j.nantod.2010.03.002

Zhang, N., Wang, M., & Wang, N. (2002). Precision agriculture: A worldwide overview. *Computers and Electronics in Agriculture*, *36*(2-3), 113–132. doi:10.1016/S0168-1699(02)00096-0

Zhang, Y. J., Nagao, T., Tanaka, T., Yang, C. R., Okabe, H., & Kouno, I. (2004). Antiproliferative activity of the main constituents from *Phyllanthus emblica. Biological & Pharmaceutical Bulletin*, *27*(2), 251–255. doi:10.1248/bpb.27.251 PMID:14758047

Zhang, Y. J., Tanaka, T., Iwamoto, Y., Yang, C. R., & Kouno, I. (2001). Novel sesquiterpenoids from the roots of *Phyllanthus Emblica. Journal of Natural Products*, *64*(7), 870–873. doi:10.1021/np010059z PMID:11473414

Zhang, Y., Yang, J., Zhong, L., & Liu, L. (2018). Effect of multi-wall carbon nanotubes on Cr(VI) reduction by citric acid: Implications for their use in soil remediation. *Environmental Science and Pollution Research International*, *25*(24), 23791–23798. doi:10.1007/s11356-018-2438-8 PMID:29876853

Zhang, Z., Tian, Y., Huang, P., & Wu, F. Y. (2020). Using Target-specific Aptamers to Enhance the Peroxidase-like Activity of Gold Nanoclusters for Colorimetric Detection of Tetracycline Antibiotics. *Talanta*, *208*, 120342. doi:10.1016/j.talanta.2019.120342 PMID:31816712

Zhao, L., Huang, Y., Zhou, H., Adeleye, A. S., Wang, H., Ortiz, C., & Keller, A. A. (2016). GC-TOF-MS based metabolomics and ICP-MS based metallomics of cucumber (Cucumis sativus) fruits reveal alteration of metabolites profile and biological pathway disruption induced by nano copper. *Environmental Science. Nano*, *3*(5), 1114–1123. doi:10.1039/C6EN00093B

Zhao, L., Lu, L., Wang, A., Zhang, H., Huang, M., Wu, H., Xing, B., Wang, Z., & Ji, R. (2020). Nano-biotechnology in agriculture: Use of nanomaterials to promote plant growth and stress tolerance. *Journal of Agricultural and Food Chemistry*, *68*(7), 1935–1947. doi:10.1021/acs.jafc.9b06615 PMID:32003987

Zhao, L., Wang, W., Fu, X., Liu, A., Cao, J., & Liu, J. (2022). Graphene oxide, a novel nanomaterial as soil water retention agent, dramatically enhances drought stress tolerance in soybean plants. *Frontiers in Plant Science*, *13*, 810905. doi:10.3389/fpls.2022.810905 PMID:35242153

Zhao, P., Cao, L., Ma, D., Zhou, Z., Huang, Q., & Pan, C. (2017). Synthesis of pyrimethanil-loaded mesoporous silica nanoparticles and its distribution and dissipation in cucumber plants. *Molecules (Basel, Switzerland)*, *22*(5), 817. doi:10.3390/molecules22050817 PMID:28509872

Zhao, Q., Xu, J., Chen, J., Zhang, L., & Zhang, W. (2020). Slow-release of potassium fertilizer encapsulated by starch-based superabsorbent polymer and its effect on growth of cotton. *Carbohydrate Polymers*, *241*, 116255.

Zhao, X., Liu, W., Cai, Z., Han, B., Qian, T., & Zhao, D. (2016). An overview of preparation and applications of stabilized zero-valent iron nanoparticles for soil and groundwater remediation. *Water Research*, *100*, 245–266. doi:10.1016/j.watres.2016.05.019 PMID:27206054

Zhou, J., Chan, L., & Zhou, S. (2012). Trigonelline: A plant alkaloid with therapeutic potential for diabetes and central nervous system disease. *Current Medicinal Chemistry*, *19*(21), 3523–3531. doi:10.2174/092986712801323171 PMID:22680628

Zhou, J., Liu, G., Guo, Z., Wang, M., Qi, C., Chen, G., Huang, X., Yan, S., & Xu, D. (2023). Stimuli-responsive pesticide carriers based on porous nanomaterials: A review. *Chemical Engineering Journal*, *455*, 140167. doi:10.1016/j.cej.2022.140167

Zhou, Y., Wei, Y., Ren, J., & Qu, X. (2020). A Chiral Covalent Organic Framework (COF) Nanozyme with Ultrahigh Enzymatic Activity. *Materials Horizons*, *7*(12), 3291–3297. doi:10.1039/D0MH01535K

Zhu, X., Mao, X., Wang, Z., Feng, C., Chen, G., & Li, G. (2017). Fabrication of Nanozyme@DNA Hydrogel and its Application in Biomedical Analysis. *Nano Research*, *10*(3), 959–970. doi:10.1007/s12274-016-1354-9

Zhu, Y., Liu, X., Hu, Y., Wang, R., Chen, M., Wu, J., Wang, Y., Kang, S., Sun, Y., & Zhu, M. (2019). Behavior, remediation effect and toxicity of nanomaterials in water environments. *Environmental Research*, *174*, 54–60. doi:10.1016/j.envres.2019.04.014 PMID:31029942

Zia-ur-Rehman, M., Anayatullah, S., Irfan, E., Hussain, S. M., Rizwan, M., Sohail, M. I., Jafir, M., Ahmad, T., Usman, M., & Alharby, H. F. (2023). Nanoparticles assisted regulation of oxidative stress and antioxidant enzyme system in plants under salt stress: A review. *Chemosphere*, *314*, 137649. doi:10.1016/j.chemosphere.2022.137649 PMID:36587917

Zuverza-Mena, N., Armendariz, R., Peralta-Videa, J. R., & Gardea-Torresdey, J. L. (2016). Effects of silver nanoparticles on radish sprouts: Root growth reduction and modifications in the nutritional value. *Frontiers in Plant Science*, *7*, 90. doi:10.3389/fpls.2016.00090 PMID:26909084

About the Contributors

Athanasios Alexiou has a BSc in Applied Mathematics from the National and Kapodistrian University of Athens, a Ph.D. in Bioinformatics Algorithms and Mathematical Modeling in Biology, and a two-year Postdoc Research Fellowship in Neuroinformatics Applications from the Ionian University. He currently collaborates with research teams from King Abdulaziz University (Saudi Arabia) and the University of Sharjah (United Arab Emirates). He also serves as a Faculty Member at the Department of Science and Engineering at the Novel Global Community Education Foundation (Educational platform registered in Australia) and a senior researcher at the AFNP Med Company in Austria. Dr. Alexiou's current research interest falls into Computational Biomedicine. It focuses on developing novel mathematical models and algorithms in neurodegenerative diseases, prognostic Bayesian tools for Alzheimer's disease, identifying molecular mechanisms of infectious diseases, especially the SARS-CoV-2 virus, and the ethical implications of scientific research. He authorizes approximately 200 peer-reviewed scientific papers/chapters/books in high-impact publications and 4 patents.

* * *

Crl Ohikhatemen Ahonsi is a seasoned Researcher with kin interest in Medicinal Aromatic and Pesticidal Plant Research, Formulation and Development of Herbal Medicine for the Management and Treatment of plants, human and livestock diseases, Bioremediation of polluted environments, Antimicrobial Resistance of Clinical and Environmental Isolates, Molecular Studies of Microorganisms and Biotechnology. He holds a B.Sc (Hons) in Microbiology from Ambrose Ali University Ekpoma, M.Sc in Environmental and Public Health Microbiology and a PhD in Environmental and public health Microbiology both from the University of Benin, Benin City Nigeria. He was a Visiting Researcher to the Department of Natural Resources and Environmental system North Carolina A&T State University Greensboro USA. Participated at the International Public Health Management Development Program Organized by Department of Community Medicine & School of Public Health Post Graduate Institute of Medical Education and Research, Chandigarh, a programme Sponsored by Indian Technical & Economic Cooperation Programme (Ministry of External Affairs, Govt. of India). Currently, he is a Chief Research Officer in Product Development/Quality Assurance Department, Nigeria Natural Medicine development Agency, Federal Ministry of Innovation Science and Technology Lagos Nigeria. He has published many peer-reviewed Research articles in his research area and he is a member of Nigeria society for Microbiology (NSM), Nigeria Society for Biotechnology (BSN), Federation of European Microbiology Society, American Society for Microbiology and Society for applied Microbiology.I am passionate about interdisciplinary collaboration especially in medicinal plant research and drug Development.

Knsley Enerijiofi holds a Ph D degree in Environmental and Public Health Microbiology from University of Benin. He is an Associate Professor in the Department of Biological Sciences and the Acting Dean, College of Basic and Applied Sciences, Glorious Vision University formerly Samuel Adegboyega University, Ogwa, Edo State, Nigeria. His research interests are pollution studies, public Health, Nanotechnology and Environmental Sustainability.

Alh Bakhsh Gulshan is Associate Professor at Department of Botany, Ghazi University-32200, Dera Ghazi Khan, Punjab, Pakistan.

Arr Hussain is a young researcher in the field of probiotics and pursuing his Ph.D. degree from ICCBS, University of Karachi. Mr. Hussain received his Master's degree in chemistry/biochemistry from the University of Peshawar, Pakistan. He also obtained C.T., B.Ed., and M.Ed. degrees in Science education. Recently he completed an M.Phil./MS degree in biochemistry from HEJ Research Institute of Chemistry, ICCBS, University of Karachi. Currently, he is pursuing his Ph.D. in bioorganic chemistry with a research focus on the probiogenomics analysis of probiotics. He is also a professional speaker and delivered numerous lectures on different topics. Attending seminars, and conferences, and participating in scientific events are the marking agents of his enthusiasm. He has strong research skills and is able to work in a team and independently. He also trained many internship students in his laboratory. Mr. Hussain is a man of rule, has a keen interest in science, have excellent research, writing, and presentation skills. He felt more than happy to help other students and guide them in their careers.

Neesh Kumar is currently working as an Assistant Professor at the Department of Aquaculture, College of Fisheries, Rani Lakshmi Bai Central Agricultural University (RLBCAU), Jhansi, Uttar Pradesh, India. He is the gold medalist in the master's degree program and he got a DST-Inspire fellowship for his doctoral degree program from the Department of Science and Technology (DST), Ministry of Science and Technology, Government of India. Dr. Kumar has published various articles in peer-reviewed international journals (Elsevier, Springer, Frontiers, Cell Press, Taylor & Francis Group publishers) and has also published a basic fisheries book. Dr. Kumar served as the editorial board member and reviewer of several reputed journals. He is a life member of various international and nationally reputed societies. Dr. Kumar also attended various conferences, workshops, seminars, training, etc. During international conferences, he received a 'Best Young Scientist Award' and 'Philippe Bodak Top HotDoc Award' for his oral presentation.

Nuwe Maduka is an academic staff in the Department of Microbiology, Faculty of Science, Federal University, Otuoke, Bayelsa State.

Muhammad Majeed, a recent Doctor of Philosophy in Botany graduate from the University of Gujrat, Punjab, Pakistan, has distinguished himself as a prominent figure in botanical research. His thesis, titled "Phytosociological and Ethnobotanical Enumeration of District Jhelum, Punjab, Pakistan," reflects his expertise and dedication to advancing the understanding of plant ecosystems and their cultural significance. Completed rigorous studies, passing 28 courses from Master's to Doctorate level. Demonstrated expertise in a wide array of botanical disciplines, including Floristics studies, Applied Ecology, Biodiversity & Conservation, Remote Sensing, and GIS. Possesses advanced skills in geospatial data science, climate change impact assessment, and species distribution modeling. Authored

28 original research and 1 review articles in esteemed international journals, with 8 articles as the first author, contributing significantly to the field. Established a notable impact with an overall publication impact factor of 102.5, Research Interest Score of 830.8, 667 citations, and an h-index of 15. Pioneered research on indigenous flora, invasive species assessment, and ecological status in semi-arid regions of Pakistan. Actively engaged in collaborative research with 40 universities in Pakistan and 25 universities worldwide, fostering cross-cultural and interdisciplinary exchange. Founded and currently leads the Pakistan Botanical Organization (PBO), facilitating collaboration among 150 researchers nationally and internationally. Dr. Majeed Played a pivotal role in organizing 47 research sessions at national and international levels, contributing to knowledge dissemination and networking. Led 05 major regional field surveys and ongoing research activities, demonstrating proficiency in project management and execution. Authored 08 book manuals and 08 book chapters, expanding the reach of his research to wider audiences. Serves as a reviewer for 15 articles in international and national journals, ensuring the quality and integrity of scholarly publications. Taught 13 courses and conducted 5 practical labs, sharing his expertise with students. Mentored and supervised 4 Doctorate, 8 M.Phil., and 6 MS research scholars, guiding them through novel research conceptualization and methodology design. Adapted to online teaching methodologies during the COVID-19 pandemic, ensuring continued support for research scholars through virtual platforms. Dr. Muhammad Majeed's commitment to botanical research, leadership in academic organizations, and dedication to mentoring the next generation of scholars mark him as a prominent figure in the field. His contributions significantly advance our understanding of plant ecosystems and their implications for conservation and cultural practices. Researchgate link: ORCID link: https://orcid.org/0000-0001-7663-2563.95ea7489-9189-43c9-8dbd-414b5003db74

Ttana Minkina is the head of Soil Science and Land Evaluation Department Scientific area: Soil Science, Biogeochemistry of Trace Elements, Environmental Soil Chemistry. Soil monitoring, assessment, modeling and remediation using physicochemical treatment methods. Scopus Author ID: 15063165400 Web of Science ResearcherID: AAC-3561-2020 https://orcid.org/0000-0003-3022-0883.5c78fbd0-0d8b-4e98-b62b-dc9ccf43b486

Wjeha Noor, a dynamic and result-oriented medical researcher, is driven by a passion for innovation and a commitment to making meaningful contributions to the field. She holds a strong academic background in biotechnology with distinction from the University of Central Punjab. Currently, she has been selected as a fellow by the prestigious American Society for Microbiology (ASM). In this book, Wajeeha Noor played a dedicated research role at the forefront of applying nanotechnology techniques in agriculture. Her academic journey exemplifies excellence, focusing on cutting-edge technologies and emerging trends in the healthcare & scientific sector. Notably, her final year project involved designing a nanoparticle-based drug targeting antibiotic-resistant bacterial genes for cancer treatment. Besides this, she has remarkably contributed to the design of a pharmacophore-based drug targeting the ETHE1 gene and a food review article that is in the process of publication. Throughout her tenure at the university, Wajeeha distinguished herself not only academically but also through her active involvement in various extracurricular activities. To broaden her skills and knowledge, she embarked on a series of courses, including cancer informatics, precision medicine, and immunology. Wajeeha's commitment to continuous learning is evident. She is an active member of various international organizations such as ASCP & ASM. Wajeeha is dedicated to fostering international collaborations and staying abreast of the latest advancements in biotechnology . Her exceptional multidisciplinary skill set encompasses laboratory diagnostics, in silico tools, and systematic analysis, making her a valuable asset in scientific research and drug discovery. Her proficiency in scientific writing and keen analytical skills have earned her recognition as both a collaborative manuscript reviewer at UCP and a top-rated freelance writer. She is passionate about strategic biomedical innovation and is determined to leverage her expertise to develop novel solutions to complex healthcare challenges. With a track record of successful achievement and a commitment to excellence, she is poised to make significant contributions to the field of science and biotechnology.

Mohamed Nouri Ph.D. in Environmental Sciences. He is, since 2017, Assistant Professor in Environmental Engineering and Microbiology at the Department of Biology, Faculty of Polydisciplinary, University Sultan Moulay Slimane, Beni Mellal. He was graduated in Food technology (2003) by the University Cadi Ayyad, Marrakesh and later on he got the Master of Science and Technology option: "Quality Management in food industries"; (2009) and Ph.D. (2014) degrees in the University Sultan Moulay Slimane in Beni Mellal. PhD in Environmental Sciences, he has been developing his research activities in the Risk Assessment of Contaminated Sites and Ecotoxicology. As a result, some papers were already published in peer-reviewed international journals, as well as, it allowed her to participate in different international meetings. Meanwhile, he has been collaborating in different projects related with environmental sciences and site-specific risk assessment of areas contaminated by mining activities.

Khojimatov Olim K. was born in 11.07.1968 in Samarkand into an intelligent family, Uzbek nationality, higher education, in 1990 he graduated from the Faculty of Biology and Soil Science of Tashkent State University. Work experience of Khojimatov O.K. 33 years. Developed at the Institute of Botany of the Russian Academy of Sciences in 1990-2022. Since January 2023 - Acad at the Botanical Institute. Head of the laboratory "Dorivor plant introduction" Tashkent Botanical Garden named after F. N. Rusanov. Doctor of Biological Sciences, Professor. Khojimatov O.K. is the author of more than 250 scientific and popular articles and theses. This scientific work was published at prestigious journals and international scientific conferences of the republic and several foreign countries (USA, Great Britain, Korea, Switzerland, Japan). Including 7 monographs, including the 2-volume Springer Publishing House (Switzerland), 2 textbooks and 1 patent, 1 copyright certificate. In addition, he participated in two foreign grant projects as a botanist. The Hirsch index based on Scopus is 26 (). The Hirsch index based on Web of Science is 24 ().

Ruth Osaro-Mathew is an academic staff in the Department of Microbiology, College of Natural Science, Michael Okpara University of Agriculture, Umudike Abia State, Nigeria.

Aajeet Singh is working as an Assistant Professor, Department of Zoology, Sri Venkateswara College, University of Delhi, Delhi, India. He received his bachelor's, and master's degree in Zoology from University of Delhi. He received his doctorate from the Department of Zoology, University of Delhi. He has published several research articles in peer-reviewed international journals such as Aquaculture (Elsevier), Fish and Shellfish Immunology (Elsevier), Turkish Journal of Fisheries and Aquatic Sciences, Journal of Applied Aquaculture (Taylor & Francis) and Heliyon (Cell Press) etc. He received a prestigious DST-INSPIRE scholarship during his graduation and post-graduation studies and DST-INSPIRE fellowship during his doctoral studies. He received the DST-AWSAR award for popular science writings on National Science Day, 28 February 2021. His research area is fisheries and aquaculture. He has presented his work in several international and national conferences/workshops/seminars. He has 73 citations, h-index is 4 and i10-index is 3 as per the Google Scholar citation report.

Okr Singh, born on September 1, 1995, into a farming family in Uttar Pradesh, India, has emerged as a dedicated researcher and scholar in the field of Soil Science and Agricultural Chemistry. He earned his Ph.D. from S.V.P. University of Agriculture and Technology, Meerut, following an M.Sc. in Agriculture from N.D. University of Agriculture and Technology. Growing up in a farming environment, Omkar possesses a profound understanding of the challenges faced by the farming communities in the eastern Indo-Gangetic plains. His research focus includes the impact of climate change on agriculture, particularly in the context of sustainable and resilient cropping systems. As a Senior Research Fellow at SVPUA&T, Meerut, he actively engages in field experiments, soil, and plant sample analyses, and data interpretation. Omkar has published extensively, contributing to reputable journals with articles on topics ranging from the effects of nanoparticles on rice growth to sustainable soil management practices for climate resilience. He has also been a part of various national and international conferences, workshops, and training programs.

Dapayan Sinha is an Assistant Professor in Botany, Government General Degree College, Mohanpur, Paschim Medinipur, West Bengal. He has more than 14 years of teaching experience in the field of Botany and Plant Science and is also actively involved in research activities. He holds a number of publications in Journals and Books of International repute.

Index

P

R

S

T

V

www.ingramcontent.com/pod-product-compliance
Lightning Source LLC
LaVergne TN
LVHW080114011026
849160LV00013B/1097